MINE VENTILATION

PROCEEDINGS OF THE TENTH US / NORTH AMERICAN MINE VENTILATION SYMPOSIUM, 16–19 MAY 2004, ANCHORAGE, ALASKA, USA

Mine Ventilation

Edited by

R. Ganguli
Associate Professor of Mining Engineering, University of Alaska Fairbanks

S. Bandopadhyay
Dean, School of Mineral Engineering and Professor of Mining Engineering, University of Alaska Fairbanks

A.A. BALKEMA PUBLISHERS LEIDEN / LONDON / NEW YORK / PHILADELPHIA / SINGAPORE

Sponsored by

School of Mineral Engineering, University of Alaska Fairbanks
Society for Mining, Metallurgy and Exploration, Inc.
Silent Seal (Fomo Products)
M&I Power Technology Inc.
Rocvent Inc.
Schauenburg Flexadux Corporation
Spendrup Fan Company
Shaft Drillers International

Published by: A.A. Balkema Publishers, a member of Taylor & Francis Group plc
www.balkema.nl and www.tandf.co.uk

ISBN 90 5809 633 5

Printed in The Netherlands

Mine Ventilation, Ganguli & Bandopadhyay (eds)
© 2004 Taylor & Francis Group, London, ISBN 90 5809 633 5

Table of Contents

Ventilation and control systems

Mine fires and explosions

Dust generation, transport and control

Strata gas and control

Face ventilation/case studies

Advances, modeling and software

Case studies

Mine Ventilation, Ganguli & Bandopadhyay (eds)
© 2004 Taylor & Francis Group, London, ISBN 90 5809 633 5

Foreword

Experience in the last four decades has shown that the development of larger and more powerful machines for increasing higher production rates calls for increased attention to mine ventilation planning. Changing conditions, such as higher production from fewer mines, low grade deposits, increasing depth of mines, extraction of thinner seams, and growing recognition of health and safety liability, have had a profound influence on mine ventilation planning and design of mines. While ventilation has a significant impact on the health and safety of the workers in the short term, it also has a significant impact over time relative to chronic respiratory diseases. This presents us with many challenges and ventilation professionals have made significant progress and have contributed significantly to the overall improvement in underground ventilation systems, and minimized the cost of implementation and operating these systems. Since the enactment of the 1969 Mine safety and Health Act, the number of injuries per 1000 miners has reduced significantly; major strides have been made in the control of respirable dust and the incidence of fire and exploration in mines. Regulatory control has been a major factor and will continue to be so in the human health and safety aspect of mining.

The changes that have occurred in the mining industry in the past four decades have been driven by safety and economic considerations. The primary forces for future changes, we believe, will be the major advancements in the computer technology with respect to reduction in size and increased computation speed, accompanied by development of efficient computer algorithm, new sensor technologies, advanced monitoring and control systems, newer communication technologies, societal demands for safe working place and quality of life. There will be continued demand for developing methods to improve the safety and health of the miners.

The US Mine Ventilation Symposia are intended primarily for the exchange of professional expertise, exchange of ideas, and information. The evolution of mining methods and systems, through the application of innovative thinking and techniques, are vital to the success of our mining industry. I am confident that this symposium will bring together people with concern for mine ventilation to contribute to the further enhancement of this critical aspect of mining.

We wish to extend our personal thanks to the members of both the SME Underground Mine Ventilation Committee and the Symposium Executive Organizing Committee for their valuable input and assistance in organizing the sessions, providing session chairmen, and in giving valuable services.

School of Mineral Engineering
University of Alaska Fairbanks

Rajive Ganguli and Sukumar Bandopadhyay
Editors

Mine Ventilation, Ganguli & Bandopadhyay (eds)
© 2004 Taylor & Francis Group, London, ISBN 90 5809 633 5

Organizing committee

Euler De Souza, Queens University, Canada
Richard Ray, Parsons Brinckerhoff Inc., USA
Raja Ramani, Penn State University, USA
Chris Pritchard, FMC Corporation, USA
Tom McNider, Jim Walter Resources, USA
Keith Wallace, Mine Ventilation Service, USA
Andrew Wala, University of Kentucky, USA
John Kelleher, Schauenburg Flexadux Corporation, USA
Jan Mutmansky, Penn State University, USA
Ian Duckworth, P.T. Freeport, Indonesia
Pramod Thakur, Consol Coal, USA
Kirk McDaniel, Earth Tech, USA
Perre Mousset-Jones, University of Nevada Reno, USA
Satya Harpalani, Southern Illinois University, USA
Robert Timko, NIOSH, USA

Ventilation systems design

Mine Ventilation, Ganguli & Bandopadhyay (eds)
© 2004 Taylor & Francis Group, London, ISBN 90 5809 633 5

Comparative evaluation of block cave ventilation systems

F. Calizaya
Associate Professor, University of Utah, Salt lake City, Utah, USA

K.R. Mutama
Senior Ventilation Engineer, Barrick Gold Strike Mines Inc., Elko Nevada, USA

ABSTRACT: Block cave mining is a concatenated process of several activities, all of which require the utilization of costly equipment, and the participation of personnel of various disciplines. The process usually results in high ore production rates, but it brings along challenges that may impact the work area negatively. One of such inconveniences is the generation of excessive amounts of dusty air, smoke, and waste heat which if not dealt with adequately, can impair the working environment and limit the efficiency of the mining equipment. This study presents a comparative evaluation of four ventilation systems for block cave mine operations. The systems are illustrated with real mine examples showing the critical design aspects, the basic requirements, and the limitations beyond which the system becomes inefficient.

1 INTRODUCTION

Block caving is a bulk mining method used to extract metallic ore bodies from below ground. Once the mine is developed and the caving started, the rest of the mining process is relatively simple. The broken ore fills the draw points and the muck is removed by loaders or scrapers. The larger the bucket size the larger the production rate. A mine can reach a production rate of 25,000 tpd with twelve loaders and five trucks. In practice however, several factors will undermine this ideal situation. Often, the drifts are damaged by blasting or blocked by roof failure and need to be repaired, the draw points are plugged by boulders and need to be dislodged, or finally, the loaders are too smoky and need to be maintained. All these factors affect the ventilation system either by increasing the mine resistance or by increasing the amount of contaminants generated in the mine.

The success of a mining method, in addition to the rock-ore properties, depends on the efficiency of its ventilation system. For a block cave mine, the ventilation system is designed based on the air contaminants generated, and the location of the mining activities. In mechanized block cave mines, most of the active workings are concentrated in a few places, of which the production and haulage drifts are the most critical ones. Air contaminants such as rock dust, smoke, and waste heat are generated in these workings. The vent system should be designed to deliver large volumes of fresh air to these workings, and adjust to varying conditions.

The ventilation infrastructure depends on two factors: (1) number and size of mine airways, and (2) capacity of the main fans. A block cave ventilation system usually requires larger volumes of air than those required by other non-block cave mines. The flow requirement usually ranges from 1.7 to 5.0 tons of air per ton of ore. For high ore tonnage mines, this results in large volumes of air. To handle such large quantities, the ventilation system must include dedicated drifts and shafts, and large capacity fans. In block cave mines, it is not unusual to find 6 m diameter-concrete lined shafts for ventilation. The systems are equipped with main fans to pressurize the air and flow control devices to distribute it to where it is needed. The primary fans can be installed as blowers or exhausters. However, due to practical reasons, mine operators prefer exhaust fans. An exhaust system can considerably reduce the need for ventilation doors. The number and size of these fans will depend on various factors including the ore production rate, and the manner in which the air is distributed to the workings.

2 BASIC REQUIREMENTS

A ventilation system is designed to supply the mine workings with sufficient quantity of air to remove the air contaminants from the mine or to dilute them to levels that can be tolerated by mine workers. Most of these levels are established by specialized agencies such as the US Mine Safety and Health Administration (MSHA) and the American Conference of Governmental Industrial Hygienists (ACGIH). Others are practical figures set by mine operators under which workers can perform their jobs safely with the minimum risk of contracting an occupational illness.

In block cave mines, air flow requirements are calculated based on the following rules:

- Critical Air Velocity, v = 0.76 m/s (150 fpm). This is the minimum air velocity that must be maintained in all active workings where limited vehicular traffic is allowed. These include most of the development headings, haulage drifts and underground shops. For a 4 × 4 m cross-sectional area drift, this yields an airflow requirement of 12.2 m^3/s (26,000 cfm.).
- Minimum Air Quantity for Diesel Fumes, q = 7.9 $m^3/s/100$ kW (125 cfm/bhp). This is the quantity of fresh air needed to dilute diesel products to their threshold limit values. For a 213 kW (285 hp) loader, this yields a minimum quantity of 16.8 m^3/s (36,000 cfm). If two loaders are used in the same drift, this quantity is doubled.
- Maximum Wet Bulb Temperature, tw = 29.4°C (85°F). This is a safe temperature index used in most deep mines. Air temperatures higher than this can only be tolerated for short periods of time. In practice, this temperature should not be allowed to exceed 32°C (90°F).

The air quantities calculated based on the above rules should be regarded as strictly fresh air quantity estimates for sea level conditions. In practice, these quantities are adjusted to account for local factors. The adjustments will always increase the final estimates.

3 AIRFLOW RATES FOR BLOCK CAVE MINES

Table 1 shows a summary of airflow rates circulated through various block cave mines. This table also shows the ore production rates. Based on these figures, one can determine that on the average 2.9 tons of air per ton of ore is required. This rate is about one fourth of that required by other non-block cave metal mines (Wallace, 2001).

In the DOZ mine, at the time of this writing, the goal was to increase the ore production rate to 40,000 tpd and also to increase the flow quantity by upgrading the fan system. At the San Manuel mine, due to the ore characteristics, diesel powered units were not required once the caving process was initiated, but during primary development these machines were employed (Mutama & Pelletier, 1999).

4 MAJOR MINING ACTIVITIES

A block cave mine usually includes a fixed number of workings that need to be ventilated. This number is determined by the ore production rate. These workings are constructed and equipped to last for a long time. From a ventilation point of view, these activities are divided into five groups.

4.1 *Development and undercut drifts*

These include all blind headings constructed for uses such as exploration, pre-production, and undercutting. The air quality in these headings is highly dependent on auxiliary ventilation; the longer the heading the greater the dependency. For headings shorter than 200 m, a single blower fan system can be sufficient. However, for long headings, a multiple-fan system is required. This system includes a blower fan near the face and exhaust fans in rigid ducts near the return. The flow requirement is determined based on the amount of contaminants generated at the face.

4.2 *Panel drifts*

Once the cave is initiated and the broken ore allowed to flow into the draw points, the next job is to haul the ore from the draw points to the nearest ore pass. This can be accomplished by means of loaders, scrapers, or finger raises. In mechanized mines, load-haul-dump units (LHDs) are used, particularly those equipped

Table 1. Summary of airflow rates for block cave mine operations.

Mine description	Mining equipment	Production rate (tons/day)	Ventilation system	Total quantity (m^3/s)
DOZ, Indonesia, 2002	Diesel	30,000	Exhaust	1130
Henderson, USA, 1981	Diesel	30,000	Push/pull	1038
Henderson, USA, 2000	Diesel	36,000	Push/pull	1750
San Manuel, USA, 1998	Grizzly	50,000	Booster Fans	800
El Teniente, Chile, 1992	Grizzly and slushers	99,000	Push/pull	1162
Palabora, SA, 2000	Diesel and electric	30,000	Push/pull	600

with diesel powered engines. These are preferred because of their flexibility to adjust to varying operating conditions. The main drawback however is that these machines require large volumes of air for ventilation. Within the drifts, the air is also contaminated by dust and smoke produced by two other activities: secondary blasting and drift repair work, which usually involves the utilization of diesel powered units. When the cave is initiated, the ore often breaks into a mixture of fines and large lumps of rock. These lumps can plug the draw points and represent a safety hazard to the mine workers. They should be fragmented, often by means of secondary blasting. The broken material is then dumped onto a grizzly where it is further reduced by means of rock breakers. During the process more contaminants are added to the air stream. All these factors should be considered in the estimation of flow requirements.

4.3 *Reduction level*

Most block cave mines are now equipped with underground crushers located at strategic points below the production level. The ore is transported from the draw points to the crushers by means of transfer raises, gathering conveyors, trucks or trains. The unloading process is often accompanied by the production of large amounts of dust. At the crusher, the ore is reduced to small fragments (<0.10 m in diameter) and dropped into large bins. The ore reduction is carried out under dry conditions, thus adding more dust to the air stream. Sometimes, exhaust fans equipped with wet scrubbers are used to reduce the dust emissions. When large capacity trucks are used to haul the ore, in addition to diesel products, waste heat becomes an important design factor.

4.4 *Haulage drifts*

This is the lowest level of a mine. On this level the ore is collected from chutes located at the bottom of each ore bin and transported to the mill. Large capacity conveyors and trains are used to remove the ore from the mine. In deep mines, the ore is extracted by means of gathering conveyors and skips run on rope guides. On this level, the ore flow is fairly smooth, limited only by the capacity of the crushers. The major air contaminants are dust particles from the conveyors and transfer chutes, and the waste heat from surrounding rock and machinery.

4.5 *Fixed stations*

These include all underground shops, office facilities, storage rooms, ore flow control stations, and powder magazines. The underground shops can be further divided into repair and maintenance stations, lube bays, compressor rooms, etc. In each of these facilities more air contaminants are generated. These, if not diluted or removed, can build up within the room and reach unsafe levels rapidly. In one mine, the air flow requirement for a repair shop was equal to 50 m^3/s. The powder magazine is another facility which requires unobstructed ventilation. The quantity of air for this facility will vary with the amount of explosives stored and the air temperature. However, it is not unusual to find a powder magazine equipped with a vent system of 15 m^3/s of flow capacity.

The above workings all represent sources of air contaminants, rock dust, mine gases, smoke and waste heat. For existing mines, the emission rates should be monitored carefully at these places and used in determining the air flow requirements.

5 MINE VENTILATION SYSTEMS

The ventilation systems used in block cave mines are directly related to the degree of mechanization used with the method. Each system includes a number of intakes, returns, main fans, and flow control devices. In general, a block cave mine can be ventilated by one of the following four systems: (1) central suction, (2) side suction, (3) ore pass exhaust and (4) conventional push–pull system. The characteristics of each system are described below.

5.1 *Central suction system*

This is based on the single air pass principle, i.e. the air, once contaminated, should be returned to surface directly. This is accomplished by means of dedicated vent drifts and shafts, and exhaust fans of surface installation. In addition to main intakes and returns, the system requires an exhaust manifold located between the ore production and reduction levels. The manifold, a collection of drifts, cross-cuts and raises, is used to collect the contaminated air from the workings and direct it to the main returns. All workings including the panel drifts, haulage loops, and shops are connected to this manifold by means of ramps, drifts and vent raises (Figure 1). On the production level, for each panel, the fresh air is drawn from the access drifts located at the fringes of the ore body, circulated through the draw points, and directed to a mid panel raise. Often, the raise is located near an ore pass such that the dust and heat from the dumping process is downcast to the ventilation manifold (Calizaya & Karmawan, 2001). With this arrangement, two loaders can operate in the same drift and always remain in fresh air. A similar arrangement is used on the reduction level, except that this time the return air is upcast to the ventilation level. For a balanced flow distribution, the system requires booster fans and regulators.

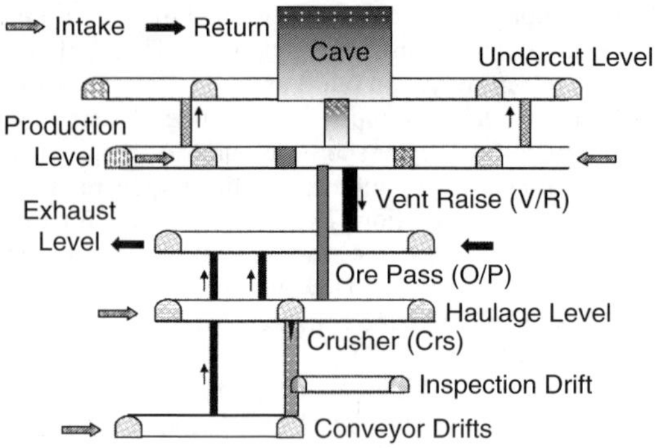

Figure 1. Ventilation schematic for central suction system.

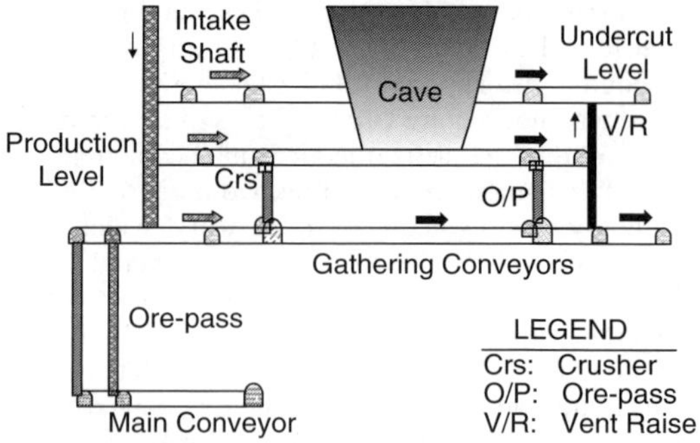

Figure 2. Ventilation schematic for side suction system.

Booster fans, often installed as blowers, are used to pressurize and direct the air to the active workings. To provide some flexibility, the fans are equipped with variable frequency drives. Regulators, located at the bottom of each raise, are used to reduce the flow rate in the less resistive airways and increase the availability of fresh air for other workings.

The main advantage of this system is that all workings are always in fresh air. Another advantage is that in case of panel failure draw points in damaged drifts can still be operated. One disadvantage is that it requires a large number of regulators to distribute the air. Regulators, just like doors and steel bulkheads, are subject to corrosion, and damage from blasting. To be functional, they should be installed in concrete bulkheads, equipped with automated control devices, and maintained regularly. Another disadvantage is the need for booster fans. Again, these should be properly sized and maintained regularly. Further, to reduce the onset of uncontrolled recirculation, the fans must be interlocked electrically with the power supplies to the surface fans and inby sections such that if for any reason the surface fans fail, the power to the boosters and inby sections is de-energized automatically.

5.2 *Side suction system*

This system requires a ventilation manifold located at the boundary of the undercut level. The manifold is connected to the production and reduction levels by means of vent raises and ramps. On the production level, there is one exhaust raise for every two panel drifts. Each raise is located adjacent to an ore pass at the downstream end of the drifts. The fresh air is supplied from main access drifts, circulated through the workings and exhausted to the vent raise. The ore pass is not used for ventilation (Figure 2). This method is used in mines where the ore body does not cave well and the broken material contains a significant portion of oversized lumps that require secondary blasting (Brunner & McPherson, 1988). Under these conditions, only one loader per drift is allowed. If two loaders were allowed, the operator of the second loader would be always working in return air. This situation does not improve much with an increased airflow rate, especially in small drifts with large amounts of settled dust, because an increased airflow rate will only cause the settled dust to enter the air stream. For this reason, the air velocity in these drifts should be restricted to a maximum of 2.5 m/s. The same arrangement of intakes and returns is used to ventilate the reduction and haulage levels.

The main advantage of this system is that it requires fewer regulators than the central suction system. The main disadvantage is that it only allows one loader per panel drift, reducing the mine output significantly. Under these conditions, the mine output can be increased only by increasing the capacity of the loaders or by adding new panel drifts. Another disadvantage is that, due to the location of vent raises, booster and jet fans cannot be used efficiently. The utilization of these fans will only promote flow recirculation. The problem is even worse when the ore passes are left open.

5.3 *Ore pass-exhaust system*

This system depends on open ore passes for exhaust. It is used in deep mines with large elevation differences between the production and haulage drifts. From the beginning, ore passes are constructed to accomplish two functions: (1) to serve as conduits to transfer the ore from one level to another, and (2) to direct the return air to the vent drifts, located beneath the production level. With this arrangement the number of ventilation raises can be reduced significantly. The fresh air is supplied from main access and service drifts, distributed to the workings, and directed through open ore passes down to the ventilation drifts (Figure 3). To facilitate the transfer of ore to centralized bins, adjacent ore passes are joined at some points below the production level to form a tree

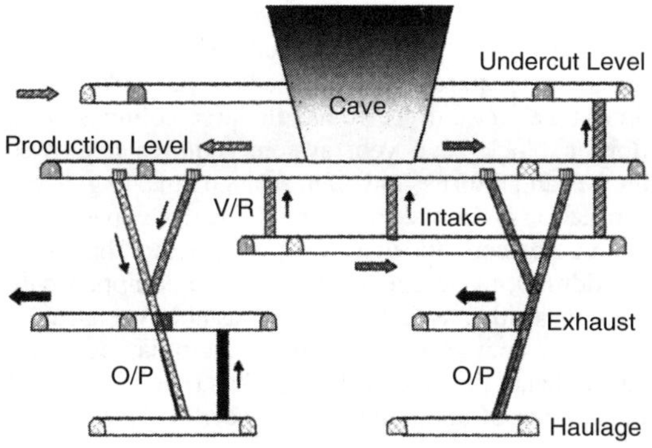

Figure 3. Ventilation schematic for ore-pass exhaust system.

of branch raises. On the ventilation level, these branches are intercepted by short drifts and cross-cuts to collect the return air and direct it to the main exhaust shafts (Rech et al., 2000).

This is what happens to the system operated on this principle: On the top of an ore pass, the air is subject to two descending forces, one created by the exhaust fans and another by the weight of the falling rock which tends to drag the air down. The system works as long as the ore level in the main ore pass is maintained just below the point where the branch raises meet. If too much ore is drawn, then the airflow direction in one of these raises will be reversed; the longer the ore pass the larger the backflow of dust particles. For long ore passes, this could be a serous problem due to pressure build up in the bin.

One of the main disadvantages of this system is the difficulty of controlling this back flow of dust and rock particles on the production level, especially on top of the inactive ore passes. The dust levels may be such as to impair the visibility of the mine workers. Another disadvantage is the excessive accumulation of coarse dust and rock fragments in the ventilation drifts. To attain acceptable ventilation efficiency, the return drifts should be cleaned regularly.

5.4 *Conventional push–pull system*

This system is used with the grizzly and slusher methods, where the major air contaminants are dust and smoke from secondary blasting. The system relies on booster fans to pressurize the air, and mine openings of inactive blocks for return ways. Once the mine is developed, booster fans are installed in the main access drifts, either as blowers or exhausters. The fans are sized to handle the total quantity of air for a section. The fresh air to the active workings is delivered through vent raises, circulated through fringe laterals and grizzly drifts, and directed down to the exhaust level. Grizzly drifts and transfer raises of inactive blocks are used to exhaust the return air. Doors, jet fans and temporary bulkheads are frequently used to direct the air through these openings (Figure 4). Air movers based on compressed air are also used on the production level. The quantity of air per drift is determined based on the minimum air velocity of 0.76 m/s. Typically, in a grizzly drift with a number of ore draw points the airflow requirement would vary from 2.35–4.7 m^3/s (5000–10,000 cfm). On the haulage level, where the main drifts are in line with the booster fans, airlock doors and regulators are used to control the air flow rate. Upon reaching the main return, the contaminated air is discharged to surface through the exhaust shafts. Sometimes, these shafts are equipped with surface fans (Mutama, 1999).

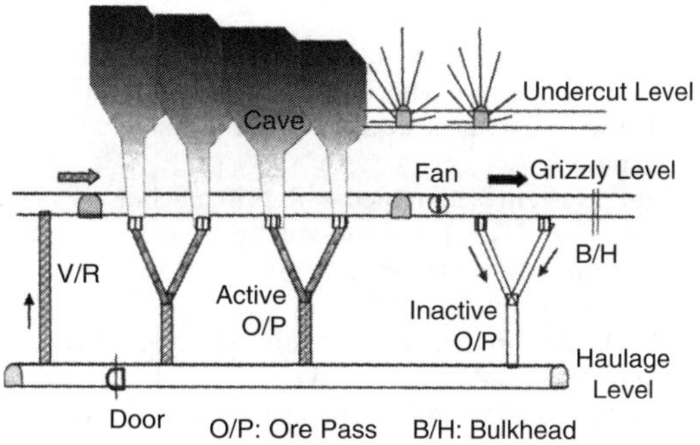

Figure 4. Ventilation schematic for push–pull system.

The main advantage of this system is that the losses of fresh air flow through mined out areas can be reduced considerably. This increases the volumetric efficiency of the ventilation system and decreased the power requirement. The main disadvantage is the onset of flow recirculation through active workings, especially when the system depends on temporary bulkheads, doors, and regulators for air flow distribution. These constructions are frequently damaged by concussion and other effects of secondary blasting.

6 COMPARATIVE ANALYSES

This analysis is restricted to the four systems presented in this paper. It highlights the similarities and differences of these systems, the basic requirements and the limitations beyond which the system becomes inefficient.

6.1 *Similarities and differences*

The systems presented have a number of similarities and differences. These are described below.

6.1.1 *Similarities*

- All require dedicated intakes and returns and depend on main surface and booster fans for ventilation. These components are sized to handle large volumes of air.
- Each system is equipped with primary fans of variable duty. The fan duty can be varied by either changing the motor speed or by adjusting the position of the inlet vanes.
- Each system depends on booster and auxiliary fans, bulkheads, doors, and regulators for flow distribution. The number of these varies with the size and the complexity of the mine.

6.1.2 *Differences*

- A suction system depends on surface exhaust fans for primary ventilation. A push–pull system depends on underground booster fans.
- In a suction system the loss of air through bulkheads, doors, and regulators is a major problem. In a push–pull system, the leakage flow is controlled to some degree by the booster fans, but flow recirculation is the major challenge.
- A suction system requires large diameter intakes and returns for ventilation. In a push–pull system, the size of an opening is defined by the mining equipment.
- A suction system depends on doors and regulators for flow distribution; a push–pull system depends on booster fans and bulkheads. Doors and regulators often get damaged or disabled. Booster fan are generally well protected.
- In an ore pass-exhaust system, dust build-up in return drifts and cross-cuts can be a problem. In a suction system the rock dust is controlled by local exhaust systems and dust collectors.
- A suction system is used in mechanized mines with large airflow requirements. A push–pull system is used in mines with grizzly or slusher drifts. The ratio of flow requirements for these two systems can be as high as 4 to 1.

6.2 *Selection criteria*

In the selection of a vent system, the following factors should be examined.

6.2.1 *Mine size*

This is defined by the ore production rate. This rate, in addition to rock and ore characteristics, depends on the type and size of mining equipment used. Currently, block cave mines are developed to achieve high production rates (from 30,000 to 50,000 tpd.) In most cases, diesel powered units are used to achieve these rates. The utilization of these units requires large volumes of air for ventilation. On the average, the flow requirement is about 2.9 tons of air per ton of ore extracted.

6.2.2 *Ventilation infrastructure*

When a gravity method is used (grizzly or slusher drifts), this requirement is reduced considerably (sometimes by a factor of 4). To handle large volumes of air flow, a block cave vent system requires dedicated intakes and returns, and main surface fans. In general, ventilation drifts and shafts should be developed to the maximum economic diameter (often greater than 6 m). In addition, these openings should be equipped with main fans either as blowers or exhausters. When trucks, trains, or conveyors are used for main haulage, surface exhaust fans are preferred. This practice reduces the number of doors in the access drifts considerably.

6.2.3 *Ventilation efficiency*

This is measured by the volumetric efficiency of the whole system, i.e., the ratio of two airflow rates: quantity delivered to the workings over the total quantity handled by the main fans. For an acceptable design, this ratio should be greater than 75%.

6.2.4 *Limitations*

In a suction system, the quantity of air circulated through the mine is fixed by the operating point of the main fan(s). This capacity can be reduced to adjust to varying conditions, but can seldom be increased. Reductions in flow rate can be achieved by lowering the fan speed or by changing the position of the inlet vanes. The system capacity can be increased only by decreasing the mine resistance (driving new airways), or increasing the fan speed (replacing fan motor.) In a push–pull system, the quantity varies with the number and size of booster fans. Since the system depends on several fans, an improved airflow distribution can be achieved. For instance, section fans can be stopped without affecting the air quality of other sections. However, this requires careful planning. Oversize booster fans can induce flow recirculation specially, when the fans are sized to develop high pressure differentials.

Table 2 summarizes the salient features of four block cave mine ventilation systems. It highlights the conditions under which a system can be selected, the basic requirements, and the potential problems that may arise when conditions are not met.

7 AIRFLOW CONTROL DEVICES

Once the primary ventilation system has been established the next step is to distribute the air to where it is needed in a manner that is economic and efficient. An *economic distribution* means to select the least resistive vent system. For instance, if the system requires regulators, use as few as possible, and if the production of air contaminants is sporadic, supply only the quantity

Table 2. Comparative analysis of block cave mine vent systems.

Vent system	Mine size	Required infrastructure	Air quality	Limitations
Central suction	Mechanized mines with high ore production rates	Dedicated airways, main and booster fans, large number of doors	Large flow rates, good air quality in all workings	Lack of flexibility, depends on booster fans, frequent maintenance
Side suction	Mechanized mines with moderate ore production rates	Dedicated airways, main and booster fans doors and regulators	Large airflow rates, quantity declines with number of loaders	Requires large number of panel drifts, frequent maintenance
Ore pass – exhaust	Mechanized mines with moderate ore production rates	Dedicated intakes and returns, active ore pass shared with ventilation	Large airflow rates, possibility of airflow Reversal	Dust clean-up from vent drifts, onset of airflow recirculation
Conventional Push–pull	Used with grizzlies, low to moderate ore production rates	Extensive preproduction work, large number of doors and small fans	Moderate flow rates, mine air can be dusty not used with LHD's	Limited mechanization, onset of airflow recirculation

of air needed for dilution. An *efficient distribution* means to utilize the available ventilation energy in the best possible manner. For instance, if a system requires bulkheads and doors, construct them as permanent structures, and if a system requires booster fans, install them in concrete bulkhead and avoid any possibility of uncontrolled recirculation. In practice an economic and efficient distribution of air requires the utilization of control devices such as booster fans, auxiliary fans, bulkheads, doors, regulators, and finally, motors with variable frequency drives. Block cave mine ventilation systems depend on these devices greatly.

Common problems with these devices and possible solutions are discussed below.

7.1 *Booster fans*

In a block caving mine ventilation system, there are two types of booster fans serving two different functions. The large underground booster fan installation facilitates the movement of large airflow rates to serve multiple districts with the needed quantities and pressures. In so doing these booster fans, strategically installed underground, assist the main fans in handling the mine pressure or resistance, especially when the mine has increased in depth without the addition of new surface boreholes or shafts. For example, following rigorous ventilation modeling for an existing mine may indicate a requirement to move 94 m^3/s (200,000 cfm) of air to a number of mining panels at a given pressure. The only way this will be possible is to install a booster fan system for that area. Most underground booster fans tend to be of the axi-vane type because they can be installed easily and require very little space compared to centrifugal fans. Main underground booster fans should not be selected to overcome more than 40% of the system or mine overall pressure. For example if the main fan pressure was 4 kPa (16 inches) the booster fans ideally should be sized to handle much less than 1.6 kPa (6 inches) pressure; otherwise re-evaluation of the whole ventilation system should be performed.

The second type of booster fan used underground is used specifically for airflow distribution and tends to be small in size and input power. The fans are used where there are multiple parallel airways and certain quantities have to be provided for each airway. A typical example is in grizzly drifts on the production level. In a typical block caving operation there can be twenty or more grizzly drifts connecting to panel drifts. In each grizzly drift an airflow quantity of about 4.7 m^3/s (10,000 cfm) is required. This means for 20 grizzly drifts 94 m^3/s (200,000 cfm) total airflow will be required. Exhaust raises furnished with small fans might be installed every four grizzly drifts to handle a total of up to 18.9 m^3/s (40,000.) Typically, these fans are low speed fans, say 1200 rpm, 19 to 30 kW and 762–1067 mm diameter, depending on the required duty. Figure 5 shows a schematic of panel and grizzly drifts. Ventilation routes, ore transfer points and small booster fans can be seen clearly. Since each airway will have a different resistance to airflow, not all the grizzly drifts will have the same exact airflow rate of 4.7 m^3/s (10,000.) However, because there is a small booster fan to manage airflow for a given number of airways, the system will be easier to handle. Sometimes through the use of local compressed air jet fans (air movers) the quantities can be balanced. These fans operate at very low pressure differentials.

Booster fans are used in most block cave mines and in many cases successfully. Flow recirculation and low fan efficiencies are the common problems. Often, these are the results of poor installation and maintenance. To overcome these, it is important to standardize the fan operating procedures. The selection procedure for small booster fans should be thorough, to avoid poor fan efficiencies. Since these fans, unlike other booster fans, operate at almost a fixed point in the low pressure range, the fan should be selected to operate at high efficiency by choosing low pressure, low speed

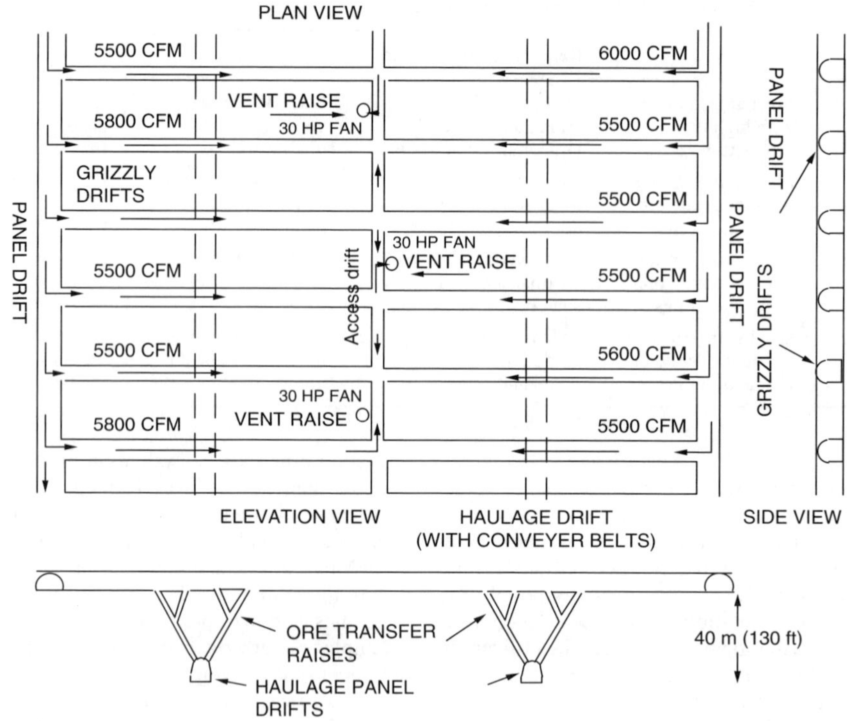

Figure 5. Schematic of panel and grizzly drifts showing ventilation routes ore transfer and vent fans (typical at San Manuel Mine).

fans with half the blades removed. The use of variable speed drives should be encouraged where possible. Poor installation and lack of a scheduled inspection often are the key drawbacks to required fan performance even if the fan is sized properly. Sometimes poor house keeping can result in poor fan performance for example, if debris is allowed to build up at the fan intake screen the fan will eventually be blocked and consequently operate in the stall zone. The result is that expected airflow quantity will not be delivered. Multiple small booster fan systems can collectively demand large amounts of electrical power and therefore careful ventilation planning, design and equipment selection should be exercised for an optimum system giving the best results.

The fans used in block caving for secondary distribution of air should be of a heavy-duty, silenced construction, with regards to fan housing, blades and electrical termination box. Repeated blasting of big rock hang-ups can result in serious and often costly fan damage, and create unsafe or unfavorable conditions for replacement or re-installation. Major incidents usually involve damage to fan rotor and blades. When the fan damage occurs, usually all the blades are lost or broken at the same time.

7.2 *Auxiliary fans*

Auxiliary fans are used to supply fresh air to dead ends. They are also used to ventilate isolated workings and shops located away from the main access drifts. Two system types are used in block cave mines namely, a blower system and an overlap system. Generally a blower system, equipped with flexible ducts, is used to ventilate headings of less than 200 m in length. For longer headings an overlap system is preferred. However, a correctly sized, single, large blower fan system with the correct diameter duct can ventilate headings 200 m or longer. The system employing both blower and exhaust fans requires a blower fan near the face and several exhaust fans in rigid ducts near the main return. The following rules should be observed in the installation of these:

- The blower fan capacity should be smaller than that of the exhaust fans.
- The overlapping length should be greater than 6 m (19.7 ft).
- The ducts should be manufactured of strong material to withstand pressure differentials larger than 2500 Pa (10 inches).
- The ducts should be joined by air-tight couplings.

One common problem with the utilization of auxiliary fans is the onset of flow recirculation. The efficiency of an auxiliary fan system depends on the manner in which the ducts are installed and the type of couplings used to join them. Rigid ducts and airtight couplings are recommended for permanent installations. Flexible ducts, including those reinforced with spiral wire, should not be used with exhaust systems. They always collapse under negative pressure.

Auxiliary fans should be sized properly for a block caving system, as in any mining method. The factors normally considered in selecting the right auxiliary fan are (1) sizes of the drifts driven, (2) number and sizes of diesel equipment to be used, (3) overall length of the heading, (4) fan set-up i.e. blower or exhaust system, and (5) minimum air velocities in the heading. All these factors will determine the airflow quantity and pressure required.

7.3 *Bulkheads*

These are permanent constructions built in underground openings away from active workings. They are used to separate main intakes from returns. They are also used to isolate active workings from abandoned areas and to direct the air where it is needed. To withstand heavy loads, static or dynamic, they should be constructed to meet specific design criteria. To reduce leakage, their edges should be keyed and anchored to the drift walls and their faces sealed, painted or covered with layers of concrete or shotcrete. For most applications, where pressure differentials are less than 2500 Pa (10 inches), 0.20 m (7.9 inches) thick bulkheads are sufficient. For higher pressure differentials such as those found around main fans, double brick or reinforced concrete bulkheads are preferred. Bulkheads are equipped with single or double doors. Often, these are major sources of leakage flow. For high pressure differentials, these should be installed in airtight structures, and equipped with actuators to open one door at a time. For safety reasons and to minimize the leakage flow, the number of these should be reduced to those absolutely necessary for emergency situations.

7.4 *Ventilation doors*

Similar to bulkheads, these structures are used to separate intakes from returns and enable the passage of personnel and equipment through the bulkheads. They should be hinged from steel frames built in concrete bulkheads, and equipped with mechanical devices to keep them closed. To reduce leakage, all door frames and joints should be covered with rubber gaskets and sealed. When the pressure differential across the door is above 500 Pa, it is recommended to install two or more airlock doors equipped with pneumatic or electrical actuators to open one door at a time. The space between the doors should be sufficient to enable one vehicle to pass through one door while the others are still closed. Sometimes, in addition to pneumatic doors, manual doors are added for the passage of personnel. The leakage through damaged doors is a serious problem. For health and safety reasons, these should be repaired immediately. In metal mines, the leakage flow through doors and bulkheads can be as high as 30%. The following measures should be applied in cutting down the leakage flow rate:

- Install concrete bumpers on both sides of a door to prevent it from being hit by a mining machine.
- Standardize all door parts; this enables one to replace the damaged parts immediately.
- Provide each door with good illumination.

7.5 *Regulators*

These constructions are built in a fashion similar to ventilation doors except that here the bulkheads include windows of adjustable cross-section instead of doors. They are installed in main returns or section exhaust ways where normally-circulated volumes of air are larger than required. To control the flow quantity distribution, they are equipped with dampers or sliding doors which are intentionally left open to pass a fixed volume of air. Often, regulators are made of steel frames and plates and installed in concrete bulkheads. To minimize air leakage, the following rules should be observed:

- All regulator components, including frames, sliding panels, and control devices should be made of durable materials, preferably steel, and installed in permanent bulkheads.
- They should be designed to operate at high pressure differential safely.
- They should be equipped with suitable mechanical or automatic control devices to operate under harsh conditions.

Regulators, in a manner similar to doors, are subject to external loads such as concussion from blasting, vibration, and corrosion. These factors can seriously limit the capacity of regulators. For efficient airflow distribution, regulators should be installed as permanent constructions and maintained regularly. Because regulators control the airflow distribution by increasing the resistance of the mine, their number should be reduced to the absolute minimum.

7.6 *Refrigeration plants and cooling systems*

In deep mines, in addition to rock dust, blasting fumes and diesel products, heat is a major air contaminant. The sources of heat include: auto-compression of the mine air, strata heat, heat removed from water and broken rock, and the heat released by the mining machinery. All these loads should be quantified to determine the ventilation requirements. However, there is a limit often referred to as the critical depth, beyond which ventilation is no longer the cheapest control measure for heat. Such heat loads should be controlled by refrigeration, which can include water chillers, surface cooling towers, and spot coolers. The correct choice depends on the mine depth, local power cost, and mining method. In hot mines, regardless of the cooling method chosen, the following rules should be observed (Mutama, 1999, Brake & Fulker, 2000):

- Select exhaust surface fans instead of blowers.
- Set up parking bays in the return airways.
- Replace diesel powered vehicles with electrical ones. These are more efficient and result in less waste heat and toxic fumes.
- Keep the intake air as dry as possible.

For ventilation design, ensure that the air velocity at the workings is in the 0.7–1.0 m/s (140–200 fpm) range and the wet bulb temperature is below 29.4°C (85°F).

At some point during the life of a deepening mine, a cooling system might be a consideration, because as the mine gets deeper the virgin rock temperature becomes higher.

There are many ways a primary cooling system can be set up. If mining is at a certain depth, less than about 914 m (3,000 ft) it might be advantageous to establish the primary chillers and cooling towers on the surface and also to cool the air in bulk air coolers or spray chambers located on surface. The mine inlet air temperature could be cooled just above the freezing point of water 0°C (32°F) to about 1.1°C (34°F). Some secondary cooling may be necessary in some hot headings to provide safe working temperatures. Underground chillers and spray chambers should be considered where this offers maximum advantage to the mining and other logistical reasons. The San Manuel mine cooling system employed three 1000 ton chillers on surface with the chilled water pumped underground in a closed system to a heat exchanger station at 880.9 m (2890 ft) level. From here secondary cooling took place when the chilled water fed underground spray chambers and cooling coils. Mining on the lowest levels would have been almost impossible without cooling. The rock temperature could be 49–60°C (120–140°F) or higher. Shaft deepening operations also needed dedicated cooling to the mine ventilation air. Even though the San Manuel cooling system worked very well for what it was intended, its efficiency was not very high. A redesign could have been required had mining continued.

Experience with cooling systems in other operating mines using differing mining methods can be incorporated for a particular block caving mine ventilation system. In South African mines cooling systems are very common, and here in the United States cooling systems have also been employed successfully over a number of years.

Designing and installing a cooling system is one thing but making sure it operates properly and is effective is another. Considerations have to be made for a reliable operating system and also one that can easily be maintained. Equipment such as chillers, heat exchangers types, and choice of refrigerant, pumps and distribution piping must be carefully selected.

8 CONCLUSIONS

Block cave mine operations require large volumes of air for ventilation. On the average, the flow requirement equals 2.9 tons of air per ton of ore. This quantity varies with the degree of mechanization, production rate, and mine depth. When diesel powered machines are employed, this quantity can be as high as 5 tons of air per ton of ore.

Block cave mines generate large amounts of rock dust, toxic fumes and waste heat. To mitigate these, ventilation systems are designed based on the single air pass principle; i.e. once the air is contaminated, it should be returned directly to surface. This requires large diameter airways.

Most block cave mines are equipped with surface fans for ventilation. Mines with heavy vehicular traffic utilize exhaust fans. This practice reduces the need for airlock doors considerably. Booster fans are used for flow distribution and to assist the surface fans, but not in place of them. Inadequate installations can results in uncontrolled recirculation. This is a major problem with a push–pull system.

Block cave mine ventilation systems depend on booster fans, doors, bulkheads and regulators for flow distribution. These devices should be installed and operated according to pre-established standards. In one mine, all ventilation doors had the same design (3 × 4 m), were installed in concrete bulkheads equipped with concrete bumpers, and were operated by the same type of pneumatic pull switches. This practice increased the ventilation efficiency at this mine considerably.

In a suction system, leakage through doors and bulkheads is a problem especially in areas where these devices are subject to shock waves from blasting and cave propagation. To reduce leakage these devices should be designed as permanent structures and installed to last for the life of the project.

REFERENCES

Brunner D.J. & McPherson M.J. 1988. Ventilation Planning Study of the El Teniente Block Caving Mine in Chile. *Proc. of the 4th International Mine Ventilation Congress*. Brisbane, Queensland, Australia, 603–610.

Brake R. & Fulker B. 2000. The Ventilation and Refrigeration Design for Australia's Deepest and Hottest Underground Operation – The Enterprise Mine. *Proc. of the MasMin 2000*. The Australasian Institute of Mining and Metallurgy, Brisbane, Australia, 611–621.

Calizaya F., Karmawan K. & Wallace K.G. 2001. PT Freeport DOZ Ventilation system. In S. Wasilewski (ed.), *Proc. of the 7th International Mine Ventilation Congress*. Research and Development Center for Electrical Engineering and Automation in Mining, Cracow, Poland, 950–953.

Mutama K.R. 1999. Performance Assessment of a Multi-Stage Cooling System at San Manuel Mine. *Proc. of the 8th US Mine Ventilation System*. University of Missouri, Rolla, Rolla Missouri, 477–484.

Mutama K.R. & Pelletier D.P. 1999. Axial Flow Fan Use at San Manuel Mine. *Proc. of the 8th US Mine Ventilation System*. University of Missouri, Rolla, Rolla Missouri, 365–373.

Rech W., Keskimaki K.M. & Steward D.S. 2000. An Update on Cave Development and Draw Control at the Henderson Mine. *Proc. of the MassMin 2000*. The Australasian Institute of Mining and Metallurgy, Brisbane Australia, 495–505.

Wallace K.G. 2001. General Operational Characteristics and Industry Practices of Mine Ventilation Systems. In S. Wasilewski (ed.), *Proc. of the 7th International Mine Ventilation Congress*. Research and Development Center for Electrical Engineering and Automation, Cracow, Poland, 229–234.

Mine Ventilation, Ganguli & Bandopadhyay (eds)
© 2004 Taylor & Francis Group, London, ISBN 90 5809 633 5

A study on event of unstable operation of an axial fan in a ventilation network

Jerzy Krawczyk
Polish Academy of Sciences Strata Mechanics Research Institute, Krakow, Poland

Stanislaw Wasilewski
Research Centre for Electrical Engineering and Automation in Mining EMAG Katowice, Poland

ABSTRACT: Unstable operation of a main fan in a ventilation system of an underground coal mine causes transient states of air and gas flow. Such disturbances can have a wide range and a considerable influence on operation of the ventilation system of a mine, and even cause life or health risk for miners. The paper presents the observation of air parameters during unstable operation of the fan-network system at Siemianowice hard coal mine. A reduced network of a mine in a process of closing was fitted with an axial fan. During recording of the fan characteristics, events of flow and pressure oscillations occurred. They not only affected the operation of the fan itself, but also extended over the whole network. A set of measuring devices was installed to investigate these phenomena. Disturbance of flow occurred when the increase of resistance of the inlet damper made the fan operate to the left of the peak of its performance curve. Pulsation had a periodic nature. For some periods the amplitude was stable, then, after a short transient, the flow became steady. The phenomena appeared in a random way. The whole event was recorded with a set of barometers placed in four points of the network and a logger collecting flow and pressure data in vicinity of the fan. An attempt was made to use compressible flow models for computer simulation of the phenomena.

1 INTRODUCTION

Transient states of an underground coal mine ventilation process are caused usually by disturbances of random nature, or by carrying out regulation activities. In certain circumstances, such transient states are characterized by rapid changes of air flow and pressure of large amplitude. Such changes may be of periodical character (*Bystroń 1961*), (*Trutwin 1968*) (e.g. during so-called pumping of a fan) or have traits of unstable flow, i.e. of such flow within which occur considerable changes in pressure and flow quantity in a coal mine ventilation system, resulted from minor disturbances of the flow conditions (*Kruszyński, Trutwin 1972*). Oscillations or unstable air flow may also happen during carrying out the flow regulation (*Litwiniszyn 1997*), (*Trutwin, Wasilewski, Sielski 1997*). The previous papers and dissertations (*Krawczyk 1994*), (*Trutwin 1968*), (*Trutwin 1972*), (*Wala 1972*) provide the numerical methods, and in particular computer simulation methods that may be applied for the analysis of the above-mentioned phenomena in such complex real systems as a fan station – a coal mine ventilation network. Until now a barrier for a broader implementation of those methods in the mining conditions was the lack of data that had to be determined by way of experiment in a coal mine.

After liquidation of the coal mine Barbara-Chorzów, within its territory a water pumping station was established, belonging to the area of the coal mine KWK Siemianowice. After replacement of WPR-200/1.8 fan with a lesser one WOK-1.5 with a saddle characteristic (Fig. 2), ventilation network of this coal mine (Fig. 1) has been used for recording of transient states of the air parameters caused by unstable performance of a fan-network system. At the time of recording of the new fan's characteristics, the tendencies to getting into oscillation were observed (*Trutwin, Wasilewski, Sielski 1997*). It turned out that the regulation range of a fan-network system allowed for shifting of the operating point to the left part of the characteristics. Such change of the operating point position was obtained as a result of a change of slope of the network characteristics, through changing the dampers' aperture level (Fig. 4). A number of experiments were performed in order to register transient states caused by unstable

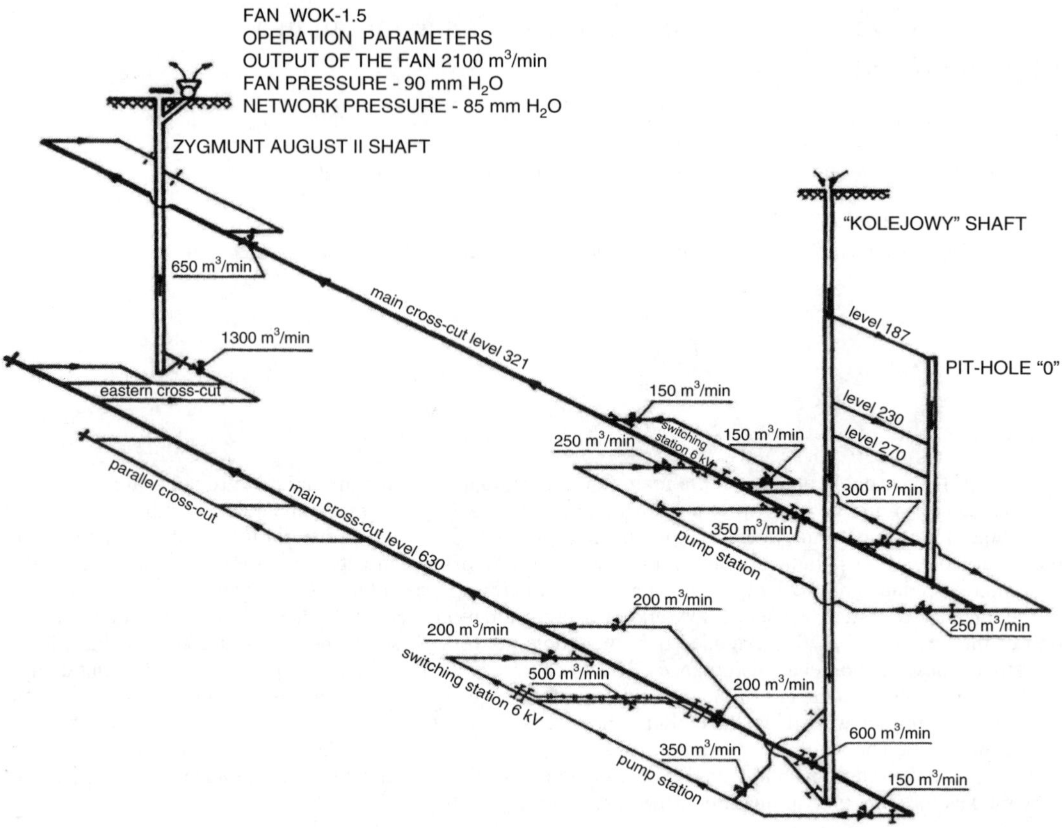

Figure 1. Spatial scheme of the water pumping station Barbara-Chorzów.

work of a fan-network system. However, the tests proved that transition of a system into oscillation state is of random character, and during a series of repeated experiments the full oscillation was not achieved, but merely a momentary "rocking" of the system.

Observations of transient states of this kind and their consequences in a real ventilation network of a coal mine were previously impossible, due to the lack of appropriate monitoring and measurement equipment. Specialised equipment allowing the changes of parameters to be stored directly in computer memory was therefore developed (*Wasilewski 1996*), enabling consequently such experiments at a whole coal mine scale to be performed. This refers in particular to a possibility of recording changes of the absolute pressure in a network, as well as changes in the air flow.

2 VENUE AND TIME OF EXPERIMENT

An experiment in a coal mine Siemianowice was performed in the area of a pump station Barbara-Chorzów. The region of the coal mine Siemianowice that presently serves as a water pumping station, established after liquidation of the coal mine production area, has two levels – of 321 m and 630 m (Fig. 1). The total length of ventilation paths comprise of workings of 6,150 m length and a downcast shaft "Kolejowy" and upcast shaft "Zygmunt August II" of the total length of both equal 1,310 m.

After completion of a coal mine liquidation process and simplification of its ventilation system, the main fan was replaced with a lesser one. Since Barbara-Chorzów coal mine region has been ventilated with a

Table 1. The current parameters of WOK-1.5 fans (the basic and reserve ones) installed at Zygmunt August II shaft.

Parameters	Unit	Fan No. 1	Fan No. 2
Fan total pressure	Pa	902.3	912.1
Flow quantity	m^3/min	2165.0	2200.0
Equivalent aperture	m^2	1.43	1.44
Efficiency	%	74.1	73.8
Power	kW	44.0	45.4
Rotational speed	l/min	735.0	735.0

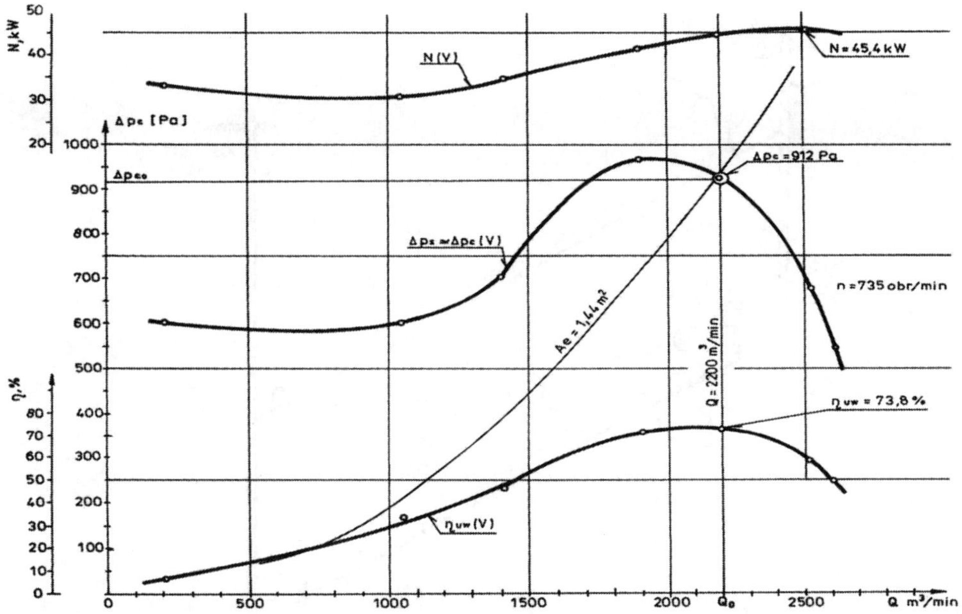

Figure 2. Characteristics of WOK-1.5-type fan No. 2 installed at Zygmunt August II shaft in coal mine Siemianowice.

fan of WOK-1.5-type installed at the Zygmunt August II shaft. The fan's parameters are given in Table 1, and its characteristics presented in Figure 2. In order to obtain the current operating point of a fan, the additional damping of flows at both 321 m and 630 m levels was necessary.

3 THE PURPOSE AND SCOPE OF EXPERIMENT

The purpose of experiments was obtaining and recording unstable (oscillatory) performance of a fan-network system at the time of the fan's operating point shifting within the characteristics. Changes in the network resistance by application of dampers were performed in the same way as during determination of the main fan's characteristics in a coal mine. Such state was possible thanks to the shape of axial fans' characteristics, of which the left side of characteristics shows the explicitly outlined shape of a saddle (Fig. 2). Changing a network resistance by means of dampers, a work on the left side of characteristics was effected (Fig. 4), enabling in this way the state of work to be unstable, when the network characteristics is crossing the fan's characteristics in two or three points. Parameters of a fan operating point switch at that time between those points, causing oscillation.

Testing proved that the transition of a fan-network system into oscillation of unstable kind, and also withdrawing from this state are both of entirely random nature, which is a confirmation of these (*Litwiniszyn, Trutwin*), that determination of circumstances of oscillation excitement is impossible. Repeated endeavours to recreate this phenomenon at other times were unsuccessful, and the full oscillations were not achieved, but merely a momentary "rocking" of the system. However, on the other hand such tests proved that oscillations had constant amplitude and frequency.

The essential part of experiments was the fact of simultaneous registering of the air parameters (absolute pressure, air velocity) in a ventilation channel and at several points of a main fan, as well as parameters of the air in underground workings network. In the station located at Zygmunt August II shaft, current recording of the following parameters were carried out:

- Hl – depression of fan,
- Ql – flow quantity of fan (dynamic pressure on Pitot-Prandtl tube),
- H2 – pressure in shaft.

For registration purpose, a multi-channel digital recorder of Campbell Scientific type was used (*Wasilewski 1996*), entering measurement signals into Data-logger memory at l-second intervals. Observation of propagation of disturbances generated by a fan in the coal mine ventilation network was executed by current registering of the following air parameters, in points set out in a diagram (Fig. 3):

- absolute pressure in a fan channel – Bl,
- the air velocity in a fan channel – V1,

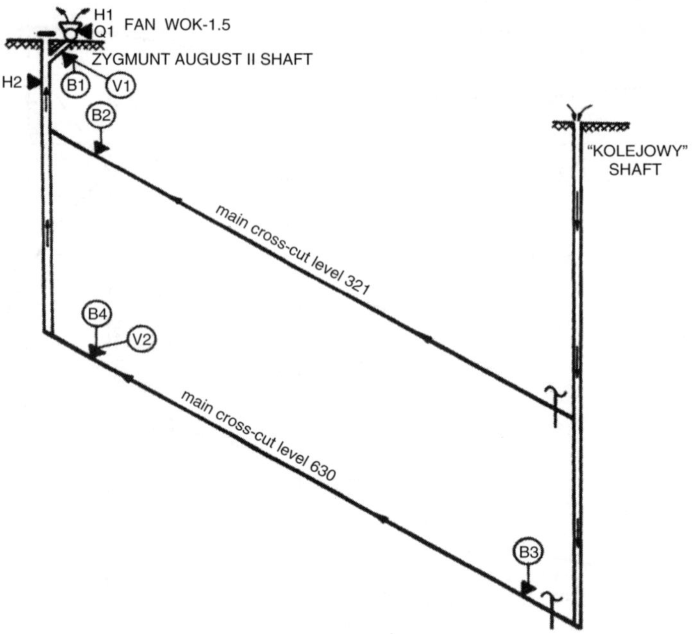

Figure 3. Distribution of measurement points in the course of experiment.

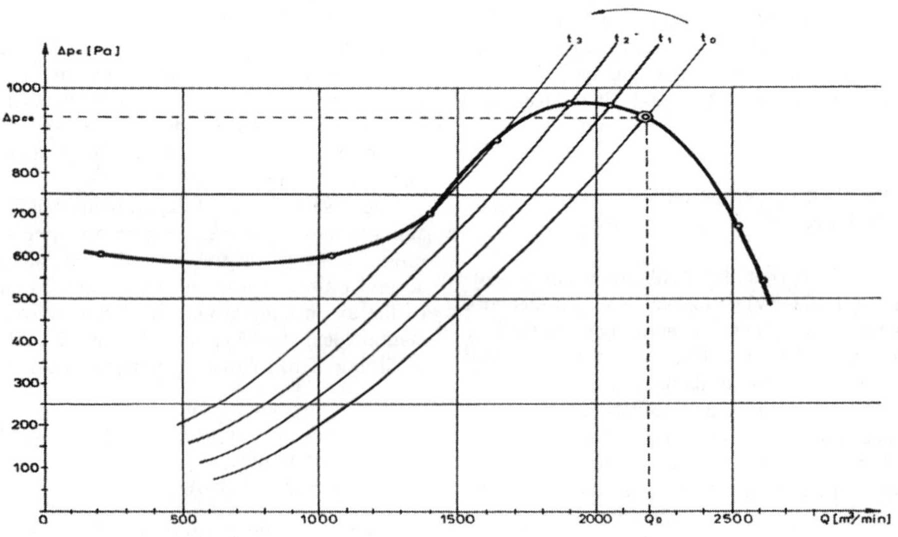

Figure 4. The network resistance's change in time, effected by means of dampers.

- absolute pressure in the main cross-cut at 321 m level (a few dozen metres from a shaft) – B2,
- absolute pressure in the main cross-cut at 630 m level (a few dozen metres from a shaft) – B4,
- the air velocity in the main cross-cut at 630 m level (a few dozen metres from a shaft) – V2,
- absolute pressure in the eastern cross-cut at 630 m level (a few dozen metres from the downcast shaft "Kolejowy") – B3.

Absolute pressure was registered by means of digital pressure sensors with recorders of μBAR-type

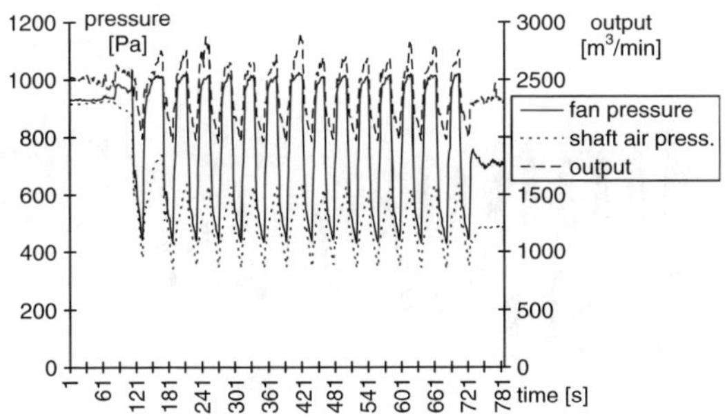

Figure 5. Parameters at the fans station.

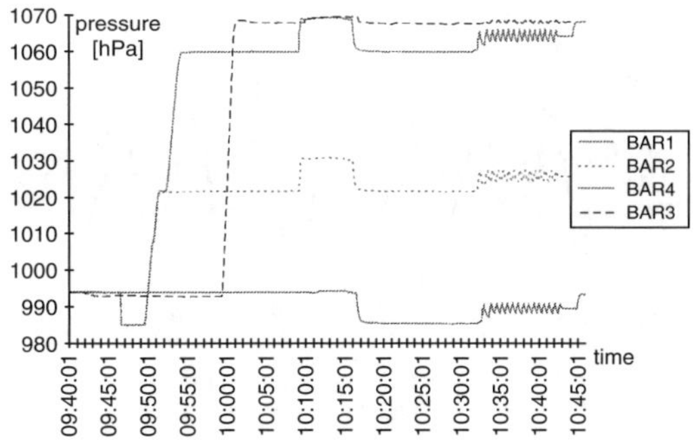

Figure 6. Registration of pressures in the Barbara coal mine.

(*Wasilewski 1996*), whereas the air velocity by means of digital anemometers with recorders of μAS-type. The parameters to be measured, were registered at 1-second intervals.

4 THE COURSE OF EXPERIMENT AND PRESENTATION OF RESULTS

At around 10^{05}, operation of the main fan No. 1 at Zygmunt August II shaft was switched off, and at around 10^{08} the fan No. 2 was switched on. After stabilisation of ventilation conditions in a network at around 10^{22}, a regulation of the operating point of the fan No. 2 was commenced. Shifting operating point by means of dampers to the left part of characteristics (Fig. 4), the unstable performance of the fan-network system was achieved. Such state was observed in a form of a fan's parameters' oscillation, i.e. oscillation of pressure and fan's output. Oscillations of pressure and fan's output lasted, at the fixed positions of dampers, for about 11 minutes, and then the new operating point of a fan-network system was automatically established in a point located about halfway on oscillations amplitude. The state of a fan oscillation was also distinctly audible at the fan station as pulsations (rumble), while the electric current meter of the engine oscillated at that time within the range of $I = <40.0 \text{ A} \div 63.5 \text{ A}>$. Establishing of the new operating point without pulsations occurred at about 10^{30} for the engine electric current value $I \sim 50.0$ A. This operating point did not shift even after the complete closing of dampers at about 10^{37}. Only after opening of dampers the new operating point was established, what happened at about 10^{40}.

As a result of the experiment being carried out, the unstable work of a fan-network system (*Trutwin, Wasilewski, Sielski 1997*) and apparent oscillation of a fan's parameters were obtained (Fig. 5). So occurred disturbances in the air pressure were propagated within the coal mine ventilation network (Figs. 6 and 7). It seems to be interesting that the amplitude (of about 300 Pa) and frequency of oscillation recorded in a fan channel, as well as in the neighbourhood of shaft at 320 m level, and also at 630 m level in Zygmunt August II shaft, were the same. This may prove the lack of damping of a pressure wave being propagated along the shaft. The vibration's period of this pressure wave was about 40 seconds. The effect of damping of disturbances in a ventilation network was visible not earlier than for pressure registration BAR3 at the furthermost point B3 at the bottom of the intake shaft "Kolejowy" at 630 m level, where the oscillation pressure amplitude did not exceed about 80 Pa, i.e. was four time smaller than in a fan channel.

4.1 *Examination of dynamics and interrelations between pressure signals*

For pressure signals being recorded in the course of experiment along the air flow path, i.e. from a fan channel through a shaft and subsequent points along the workings network B1, B2, B3 and B4, the particularly interesting course of a correlation function was obtained. Cross-correlation function is being applied broadly in automatics, among others for the time series and processes identification, as well as their interdependencies (*Wasilewski 1996*). One of the most important qualities of a cross-correlation function is detecting of a signal oscillation, whereas, what is particularly important, the course of a correlation function retains a period of the primary signal. The course (graph) of a correlation function for pressure signals (Fig. 8) shows the periodical nature of pressure changes and perfect accordance of the oscillation period for all registered signals within the entire scope of determined function. One may state that the nature of pressure changes (pulsations) is being transported against the current of the air flow with the ideal preservation of its period (about 40 seconds).

Simultaneously, with the absolute pressure registration, also registration of the air velocity was performed i.e. in a fan channel with Pitot-Prandtl tube

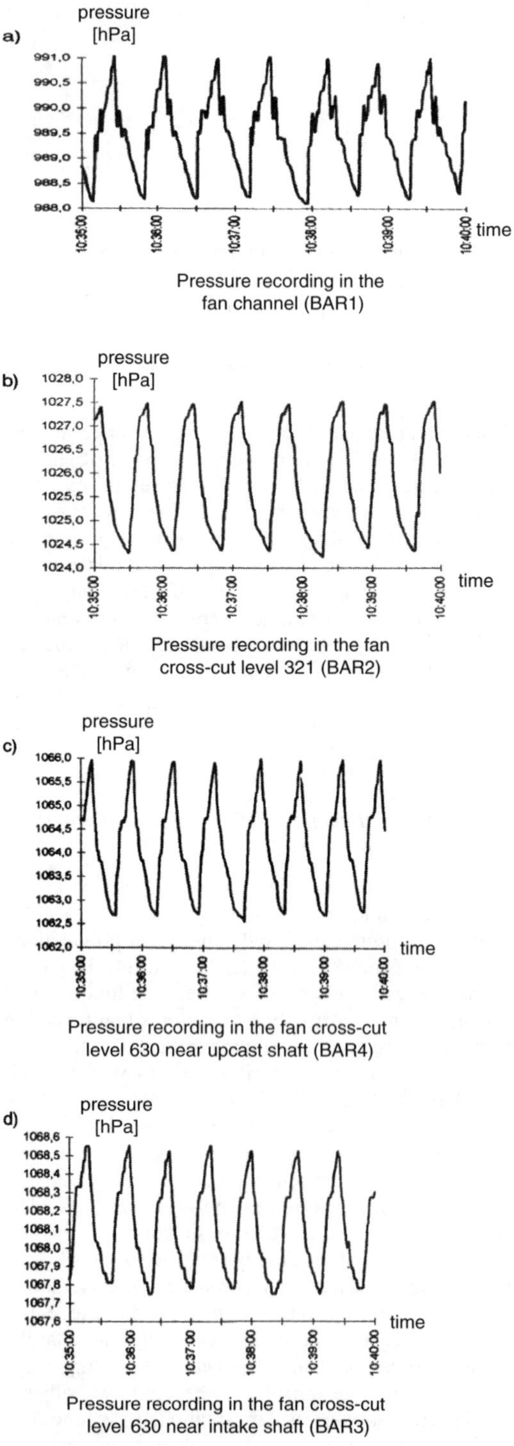

Figure 7. Pressure oscillations recorded with barometers BAR1 to BAR4 (located as shown in Fig. 3).

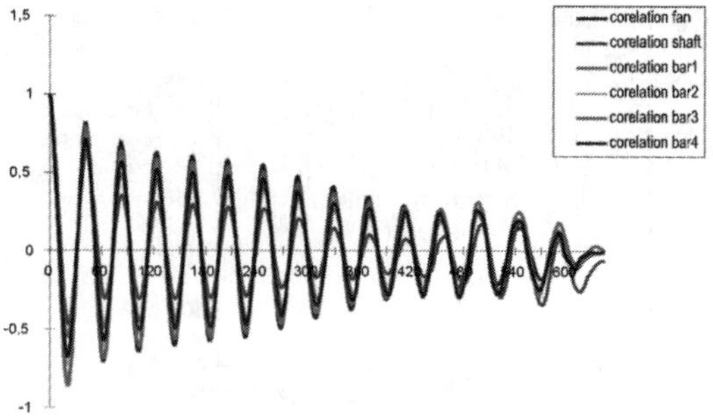

Figure 8. Correlation functions for pressure signals in Barbara coal mine.

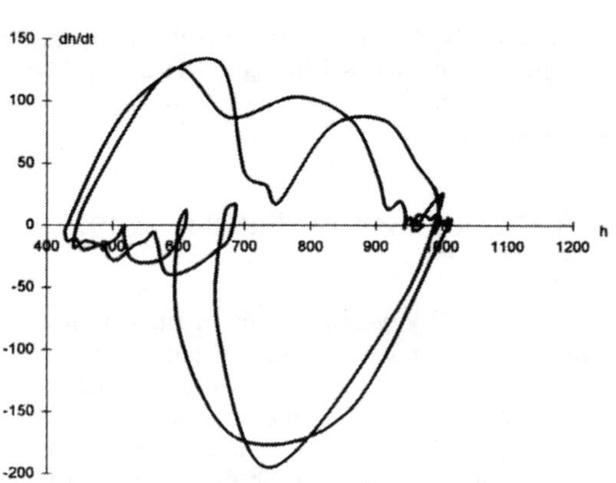

Figure 9. Phase portrait of fan's pressure $dh/dt = f(h)$.

and with two vane anemometers of μAS-type. One of them (V1) was installed in a fan channel, and the next one (V2) in a main cross-cut at 630 m level in the pressure measurement point (B4) – Figure 3. Registrations of air velocity, although are more disturbed, show also the moments of oscillations as regular signal pulsations. Confirmation of the simultaneous occurrence of pressure and air velocity pulsation shall be a course of correlation function, which maintains a period of pressure oscillation within the range of about 40 seconds.

One of the methods of testing dynamic systems in automatics is a method of phase plane. Closed curves at the phase plane are correlated with the periodic courses. Applying a method of phase plane for the analysis of a dynamic system fan-network, a phase trajectory of the tested system was determined (Fig. 9). In Figure 9 a phase portrait of a system for two cycles at the time of pumping of a fan is presented. So obtained closed curves $dh/dt = f(h)$ are correlated with the periodic courses, and are similar in quality aspects to those obtained in a work (*Trutwin 1968*).

One has to notice that, although vibrations are not of a fading nature, yet in certain moment they extinguish rapidly even without any external reason, and a system gets into a stable intermediate state, which becomes a stable state of a fan-network system. What is interesting, that obtaining of unstable state of work of a fan-network system is not automatic, and the endeavours to recreate such state were not always successful. That may confirm a random character of mechanism of achieving oscillation state of a fan within the network.

5 COMPUTER SIMULATIONS OF UNSTABLE WORK OF FAN-NETWORK SYSTEM

Theoretical analyses and models regarding a phenomenon of axial fan pumping of saddle characteristics are widely acknowledged. At the time of fan operation in the ascending part of its characteristics curve, a positive feedback between flow at the fan inlet and its pressure may occur, which in certain circumstances may cause the flow oscillation. Taking into consideration such properties of the air like its compressibility and inertia of motion in equations describing flow made possible the qualitative analysis of circumstances of initiation and proceeding of pumping in fans.

5.1 *Application of a model with lumped constants*

In 1957 W. Kazakiewicz described a compressor-ventilation channel system by means of a series of ordinary differential equations. This constitutes a model with lumped constants, where the inertia, compressibility and energy dissipation in a ventilation channel are distinctive, what is presented in a form of diagram in Figure 10.

For a ventilation channel of length L, cross-section F, resistance R and head loss W, we may write down as follows:

$$\begin{cases} \beta \dfrac{dQ_2}{dt} = h(Q_2) - p^* \\ C_{AC} \dfrac{dp^*}{dt} = [Q_2 - Q_1(p^*)] \end{cases}$$

where β: $= L/F$ means so-called acoustic mass, and C_{AC}: $= LF/c^2$ means acoustic capacity, while c is a local velocity of sound

$$p^* = R|Q_1|Q_1 \Rightarrow Q_1 = \operatorname{sgn}(p^*)\sqrt{\frac{|p^*|}{R}}$$

The analysis of local stability of the equation system led to formulating of two determinants of a fan's operating point stability, dependent on slopes of both fan and ventilation channel characteristics. The second determinant deciding on the possibility of oscillation encloses β and C_{AC} – so-called acoustic mass and capacity, i.e. the parameters representing inertia of motion and compressibility of the air in ventilation channel. It has been stated that pumping is an example of occurrence of the flow self-excited vibrations, and the analysis of equations with application of a phase plane method made possible to draw a trajectory of oscillation limit cycle. The character of excitation was estimated in accordance with Lapunow theory, with indication of a possibility of pumping occurrence not only in the event when a fan operates on the ascending part of characteristics, but also for its operating point positioned on the descending part of characteristics, in proximity of maximum, provided that certain additional conditions have been also fulfilled.

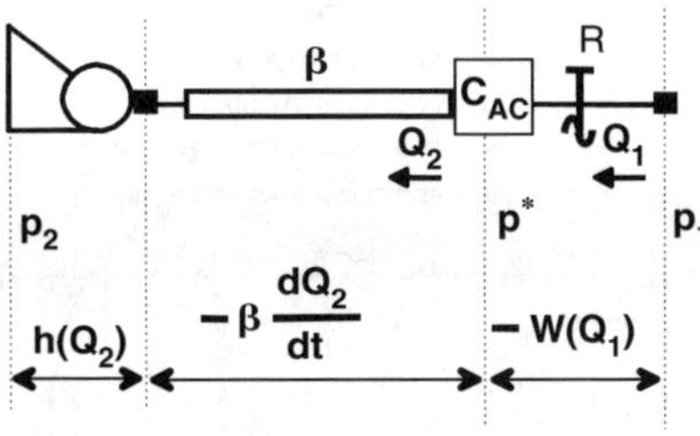

Figure 10. Diagram of a model with lumped constants.

During the first series of simulation the whole network was replaced with a system (Fig. 10): lumped resistance, capacity and acoustic mass (*Bystroń H. 1973, Krawczyk J. 1994*), with parameters selected in such way that resistance is equal to the total equivalent resistance of a network, acoustic mass corresponds with the ventilation shaft and acoustic capacity corresponds with a network's capacity. A system of ordinary differential equations was solved with Runge-Kutta method of variable integration step.

Computation started from a stable state with a damper opened, then at the moment t = 10 [s] closing of a damper was commenced, which lasted 40 [s]. The fan's operating point moved onto the ascending part of characteristics curve and the flow oscillations were observed (Fig. 11). Amplitude of such oscillations was greater than the one measured within a network, however a period of those oscillations (about 30 [s]) is of similar value like of the one measured during the experiment within a network.

5.2 *Application of a model of one-dimensional, transient flow of compressible liquid*

The broader display of a phenomenon one may obtain when using the more detailed description. A model of one-dimensional, transient flow of compressible liquid

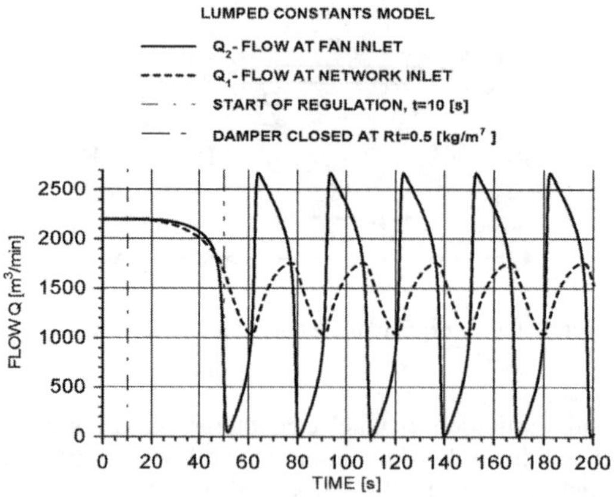

Figure 11. Excitement of flow oscillation in accordance with a model of lumped constants.

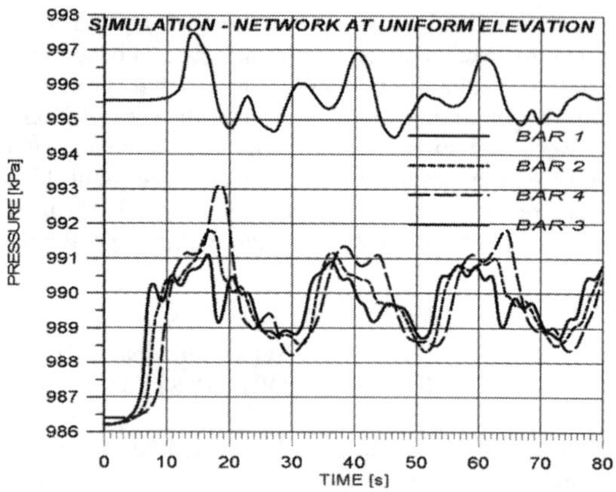

Figure 12. Occurrence of pressure oscillation in accordance with a model of one-dimensional compressible flow.

was here applied. In this description the equations of flow in the workings' cross-section plane were averaged. In this case, parameters like the pressure, density or flow velocity are represented by averages for cross-section. Such simplification enables the flow in complex networks to be modelled efficiently, even in circumstances of the present limitations of computers' performance.

In computer calculations the effort was made to recreate the course of experiment. Series of simulations for networks of various level of complexity were carried out. The simplest network comprised: downcast shaft, 630 m level with air stopping in the inlet proximity, and upcast shaft. Another variant included both levels considered as simple workings with air stoppings at their inlets. In a most complex case, in proximity of both levels' ran parallel airways, like in scheme 1. The network structure of the real object was even more complex. Longitudes and cross-sections areas of workings correlated with the real figures. Fan characteristics were approximated with a fourth-degree polynomial. Calculations were carried out for both networks with real elevations and with omission of the height differences – the results did not vary in qualitative aspect, whereas for networks with the same values of levelling spot heights the imaging of pressure wave propagation was not obscured by the auto-compression effect.

Simulation was commenced in normal circumstances of ventilation. Then a flow was choked by way of gradual increase of the fan's damper resistance until the moment when a fan worked on the ascending part of characteristics curve, in proximity of its peak. In so altered conditions the flow oscillations occurred, presented in Figure 12. Simulations results vary from data obtained in experiment, however, one can find certain similarities and which facilitates visualization of the phenomenon course. The following figure (Fig. 13) presents evolution of pressure distributions in the net mesh comprising a downcast shaft, 630 m level and upcast shaft together with a fan channel. Time is represented on one axis, on the other – distance from inlet s. Respectively, $s \in [0..660][m]$ corresponds with a downcast shaft, $s \in [660..760][m]$ – with a section of 630 m level up to the regulating air stopping, further: $s \in [760..1780]$, $s \in [1870..2100]$ $s \in [2100..2350]$ $[m]$ correspond with the consecutive workings at this level, whereas $s \in [2350..3040][m]$ – with upcast shaft and a fan channel.

Oscillations are accompanied by propagation of pressure and velocity waves that are reflected off such obstacles as crossings with levels 321 and 630 or the stoppings. Reflected waves are of the amplitude greater than for real objects. The oscillation period was influenced by the fan's characteristics and damper's aperture value.

Simulation does not provide an answer to the question, why oscillations extinguished.

6 SUMMARY

Observations presented above shall contribute to research on transient states of ventilation occurring at the time of switching on, switching off or regulation procedures carried out in regard to the main fans. Presented results in a quantitative aspect confirm the previous conclusions derived from theoretical works as well as from laboratory tests and modelling exercises. Those results encourage to perform the detailed analysis and further research leading to identification of mechanism of this phenomenon which, when left without control, may even cause damage of the system's main fan.

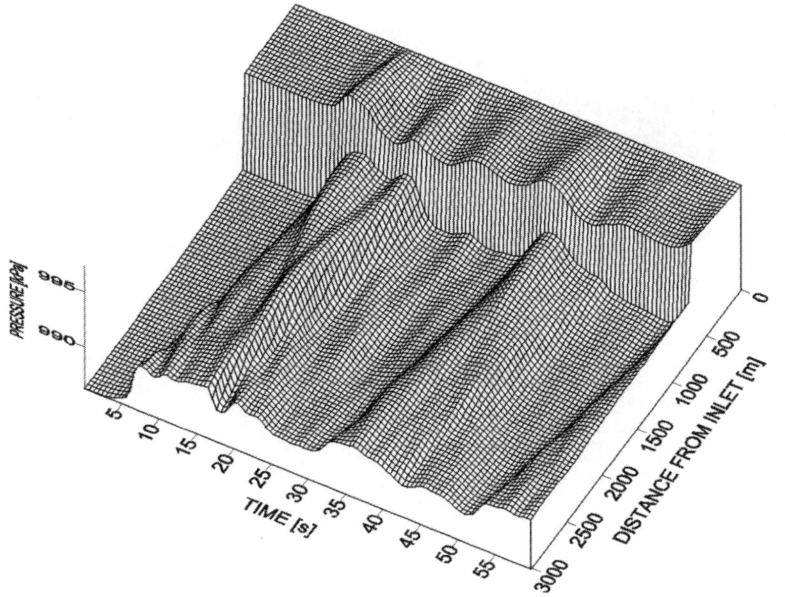

Figure 13. Propagation of pressure waves in accordance with a model of one-dimensional, compressible flow.

For the purpose of recreation of this phenomenon's course, a method of computer simulation was applied that proved to be an effective research tool. The values of pressures amplitudes and simulation results show that during unstable operation the fan's operating point moved – within the wide range – on the ascending part of characteristics curve, between the relative extreme values for $h_w = 570$ and $h_w = 950$ [Pa]. Pressure waves are being reflected off such obstacles as crossings and the air stoppings.

REFERENCES

Bystroń (1961) Quantitative representation of pumping of a main fan in coal mine. (In Polish) Mining Review [*Przegląd Górniczy*], No.11.

Bystroń (1973) Analytical Method for Investigating of the Stability of Operation of Main Fans in Mine Ventilation Networks Considered as System of Lumped Parameters. (In Polish) GIG Works. Communiqué No. 576, Katowice, pp. 54

Dziurzyński & Krawczyk (2001) Unsteady flow of gases in a mine ventilation network – a numerical simulation. Archives of Mining Sci., Vol. 46, Issue 2, pp. 119–137.

Kazakiewicz (1957) Self excited flow oscillations in compressors," (in Russian), DAN SSSR, v. 115, (4), pp. 677–680.

Krawczyk (1994) Examples of Application of Models with Lumped Constants for Simulation of Unsteady Air Flow in the System: Airway-Fan Archives of Mining Sciences Vol. 39, Issue 3, pp. 431–439.

Kruszyński & Trutwin (1972) Stability of the air flow in the coal mine ventilation networks. (In Polish) Books on Mining Sciences Issues [*Zeszyty Problemowe Górnictwa*] PAN Vol. 10, Book 2.

Litwiniszyn (1997) On stability of fans performance. (In Polish) Proceedings of the Conference of School for Underground Production, '97, Szczyrk.

Trutwin (1968) Modelling of transient states, in ventilation networks with application of analogue devices. (In Polish) Scientific Books of the Academy of Mining and Metallurgy, Books on Mining Sciences Issues [*Zeszyty Naukowe AGH*]. Vol. Mining, Book 19, Kraków.

Trutwin (1972) Numerical simulation of transient states in ventilation process and regulation of coal mines ventilation networks. (In Polish) Books on Mining Sciences Issues [*Zeszyty Problemowe Górnictwa*] PAN, Vol. 10, Book 2.

Trutwin, Wasilewski & Sielski (1997) Research on the unsteady operation of a fan-network in hard coal mine KWK Siemianowice. (In Polish) Proceedings of the Science & Technology Conference AEROLOGIA '97, Kraków. Mechanisation and Automation in Mining, No. 11/327, pp. 18–24.

Wala (1972) Examination of dynamic propensities of the air flow regulation systems in coal mine workings with application of modelling methodology. (In Polish) Books on Mining Sciences Issues [*Zeszyty Problemowe Górnictwa*] PAN Vol. 10, Book 2.

Wasilewski (1996) The new possibilities of measurement and recording of parameters in coal mine ventilation network. (In Polish) Proceedings of the Conference Excavation of hard coal bed at great depth and in difficult geothermal conditions. Świeradów Zdrój.

Mine Ventilation, Ganguli & Bandopadhyay (eds)
© 2004 Taylor & Francis Group, London, ISBN 90 5809 633 5

Airflow in leaking ducts with multiple fans

J. Partyka & O. Johnson
Laurentian University, Sudbury, Ontario, Canada

ABSTRACT: An analytical solution to predict flow conditions in a leaking out/in ventilation column is presented. The solution predicts the pressure and the airflow for fans delivering/removing a required amount of air to or from mine workings located at various depths or at high altitudes. The solution can be used to calculate pressure loss due to friction, the amount of air leaking out or in and electrical motor input power of all fans involved. It can also be used to examine the performance of existing ventilation systems and to classify the operating ducts as excellent, good, fair, or poor. The quality of a column is defined by a friction factor and an effective area of leaking spots. A leaky ventilation column requires more electrical power than a non-leaking one. The additional power increases almost proportionally with the length in systems less than 1,000D long. In a longer duct this power raises as dimensionless length to the power of two or three. New and well-installed ventilation systems have the effective leaking area of 5 to 10 mm^2/m^2 (representing an excellent or a very good duct). However, the leaking area for ducts in service may increase gradually with time to 20 to 60 mm^2/m^2 (making it a fair or a poor duct). A lack of proper maintenance can already increase the electrical power consumption by 10% on systems only 100D long. Auxiliary ventilation systems with L/D >1,000 should be served by two or more fans. Multiple fans should never be installed in one location, at one end of a column or in the middle of a column. They should be initially distributed along a column so that each fan draws the same amount of electrical power regardless of different pressure-volume duties. However, as duct qualities undergo changes with time equal power criterion will be destroyed.

1 INTRODUCTION

Ventilation ducts are formed from several sections connected in series. The joints and unexpected holes as well as damaged places are responsible for air leakage. When air leaks in or out through openings distributed evenly along a duct, air quantity changes continuously with distance, while static pressure drops differently than in an airtight duct. The amount of leaking air is proportional to square root of duct pressure, and the airflow inside a duct depends on the duct pressure as well. This coupled relationship between airflow and duct pressure is governed by a system of ordinary, non-linear, differential equations. In this paper, the mathematic solution is represented in dimensionless form that may show great value for ventilation engineers.

2 ANALYTICAL CONSIDERATIONS

First, we will develop the general relationships that apply for a leaking duct under positive gauge pressure in which a fan moves air along the x-axis. We consider a section of a column having length of Δx (Figure 1). We can write the following:

1. Pressure change due to friction (uniform duct)

$$\Delta P_f = f\left(\frac{\rho}{2}\right)\left(\frac{\Delta x}{D}\right)U^2 \quad (1)$$

where f is the friction factor (dimensionless); U represents mean duct velocity in axial direction (m/s);

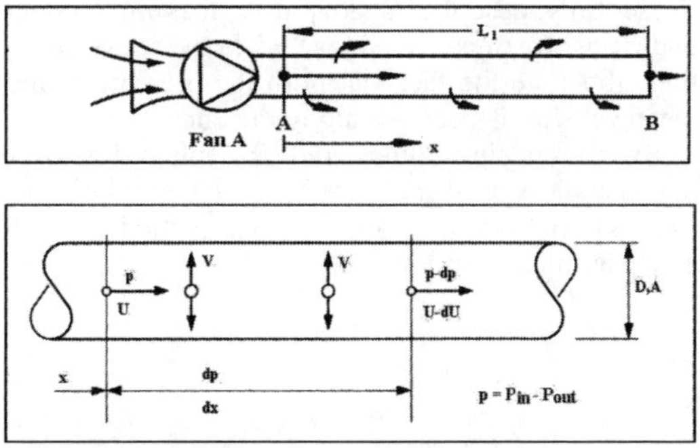

Figure 1. Leaking duct – blowing system.

ρ is air density (kg/m^3); and D is the duct diameter (m).

The mass flow of air along the duct changes due to leakage. The amount of air leaking through the pipe's outer area depends strongly on the pressure difference, $p(x) = P_{in}(x) - P_{out}(x)$; where P_{in} represents absolute barometric pressure inside a duct, and P_{out} is the barometric pressure outside the duct. The velocity of air leaking through a hole or a leaking opening, V, can be calculated using the following formula:

$$V = \sqrt{\frac{p}{\frac{\rho}{2}(1+K_o)}} \quad m/s \tag{2}$$

where K_o is the loss coefficient of a leaking spot.

The amount of air leaking through the duct at Δx-length is

$$\Delta m = (\pi D \Delta x) A_S \sqrt{2\rho p} \quad \frac{kg}{s} \tag{3}$$

where $A_S = A_R / \sqrt{1+K_o}$ is a specific leaking area in m^2/m^2 and A_R represents the relative leakage area (m^2 per m^2 of duct surface area). The mass balance for Δx-section gives the following relation:

$$\frac{\Delta U}{\Delta x} = \left(\frac{4A_S}{D}\right)\sqrt{\frac{2p}{\rho}}. \tag{4}$$

2.1 *Blowing system*

Equations 1 and 4 form two pressure-dependent differential equations;

$$\frac{dp}{dx} = -\left(\frac{f}{D}\right)\left(\frac{\rho U^2}{2}\right) \tag{5}$$

$$\frac{dU}{dx} = -\left(\frac{4A_s}{D}\right)\sqrt{\frac{2p}{\rho}}. \tag{6}$$

These fully describe airflow in a leaking column under positive pressure. Signs "$-$" in the above equations illustrate the fact that both the pressure p and mean velocity U decrease along the duct.

By introducing dimensionless length $\zeta = x/L_1$, dimensionless pressure $\psi_\zeta = 2\,p_\zeta/(\rho U_B^2)$, and dimensionless axial velocity $\phi_\zeta = U_\zeta/U_B$. Equations 5 and 6 can be written as follows:

$$d\psi = -f\,\phi^2\left(\frac{L}{D}\right)d\zeta \tag{7}$$

$$d\phi = -(4A_s)\left(\frac{L}{D}\right)\sqrt{\psi}\,d\zeta. \tag{8}$$

Equations 7 and 8 can be combined into a single differential ordinary equation

$$\phi^2 d\phi = \left(\frac{4A_s}{f}\right)\sqrt{\psi}\,d\psi \tag{9}$$

where L_1 is the duct length and U_B denotes mean axial velocity at a duct outlet.

In the case when the specific leaking area A_S is uniformly distributed along distance ζ, Equation 9 has the following solution (Maple V 1997):

$$\phi^3 = \left(\frac{8A_s}{f}\right)\psi^{3/2} + C_1 \tag{10}$$

where C_1 is the integration constant. At $\zeta = 1$; $\phi = \phi_B = 1$; $U = U_B$; $\psi = \psi_B = \psi_L$; therefore

$$C_1 = 1 - \left(\frac{8A_s}{f}\right)\psi_B^{3/2}.$$

Equation 10 can now be written as:

$$\psi = \left[\left(\frac{f}{8A_s}\right)(\phi^3 - 1) + \psi_B^{3/2}\right]^{2/3} \tag{11}$$

Equation 11 governs the relationship between the dimensionless pressure ψ and dimensionless velocity ϕ.

2.1.1 *Velocity distribution along a duct*

We can now develop equations which allow us to observe changes in velocity and pressure along distance x. For instance, substituting Equation 11 into Equation 8 we get

$$d\phi = -a^{1/3}\left[(\phi^3 - 1) + b^3\right]^{1/3} d\zeta \tag{12}$$

$$\frac{\theta\, d\theta}{1-\theta^3} = -a^{1/3} d\zeta \tag{13}$$

where $a = 8(fA_s^2)\left(\frac{L}{D}\right)^3$; $b = 2\left(\frac{A_s}{f}\right)^{1/3}\sqrt{\psi_L}$ and

$\theta = \left(1 + \frac{b^3 - 1}{\phi^3}\right)^{1/3}$ is the new dimensionless velocity.

Integrating both sides of Equation 13 gives (Maple V 1997)

$$\int_\theta^b \frac{\omega\, d\omega}{1-\omega^3} = -\frac{1}{6}ln\left(\frac{1+\theta+\theta^2}{(1-\theta)^2}\right) + \frac{1}{\sqrt{3}}arctan\left(\frac{2\theta+1}{\sqrt{3}}\right)$$

$$+\frac{1}{6}ln\left(\frac{1+b+b^2}{(1-b)^2}\right) - \frac{1}{\sqrt{3}}arctan\left(\frac{2b+1}{\sqrt{3}}\right) \tag{14}$$

$$-a^{1/3}\int_{\zeta}^{1} d\tau = a^{1/3}(\zeta - 1) \qquad (15)$$

$$\frac{1}{6}ln\left(\frac{\theta^2+\theta+1}{(1-\theta)^2}\right) - \frac{1}{\sqrt{3}}arctan\left(\frac{2\theta+1}{\sqrt{3}}\right) = a^{1/3}(1-\zeta)$$
$$+\frac{1}{6}ln\left(\frac{b^2+b+1}{(1-b)^2}\right) - \frac{1}{\sqrt{3}}arctan\left(\frac{2b+1}{\sqrt{3}}\right) \qquad (16)$$

where we denote that when $\zeta = 1$, $\phi = \phi_B = 1$, $\theta = \theta_B = b$.

Solving Equation 16 for $\theta = \theta_\zeta$ at assumed ζ and recalling that $\phi_\zeta = \left(\frac{b^3-1}{\theta_\zeta^3-1}\right)^{1/3} = \frac{U_\zeta}{U_L}$, the duct mean velocity at any cross-section can be determined as follows:

$$U_\zeta = U_B\left\{\frac{b^3-1}{Q_\zeta^3-1}\right\}^{1/3} \qquad (17)$$

At $x = 0$: $\zeta = 0$, $\phi_A = U_A/U_B$,

$$\theta_A = \left(1 + \frac{b^3-1}{\phi_A^3}\right)^{1/3}.$$

Therefore, at the duct inlet, dimensionless velocity θ_B is determined by the following equation:

$$\frac{1}{6}ln\left(\frac{\theta_A^2+\theta_A+1}{(1-\theta_A)^2}\right) - \frac{1}{\sqrt{3}}arctan\left(\frac{2\theta_A+1}{\sqrt{3}}\right)$$
$$= a^{1/3} + \frac{1}{6}ln\left(\frac{b^2+b+1}{(1-b)^2}\right) - \frac{1}{\sqrt{3}}arctan\left(\frac{2b+1}{\sqrt{3}}\right) \qquad (18)$$

while the duct inlet velocity is

$$U_A = U_B\left\{\frac{b^3-1}{\theta_A^3-1}\right\}^{1/3}. \qquad (19)$$

2.1.2 *Distribution of pressure along a duct*

Value U_ζ, calculated by means of Equation 17 can be substituted into Equation 11. In this way we derive the following formula:

$$p_\zeta = \left(\frac{\rho U_B^2}{2}\right)\left\{\left(\frac{f}{8A_s}\right)\left[\left(\frac{U_\zeta}{U_B}\right)^3 - 1\right] + \psi_B^{3/2}\right\}^{2/3} \qquad (20)$$

which describes distribution of static pressure along a duct.

2.1.3 *Application of analytical equations*

We now consider a forcing column when the gauge static pressure at the exit is zero, i.e., $p_B = 0$. The following sketch shows the flow conditions at a duct inlet and outlet.

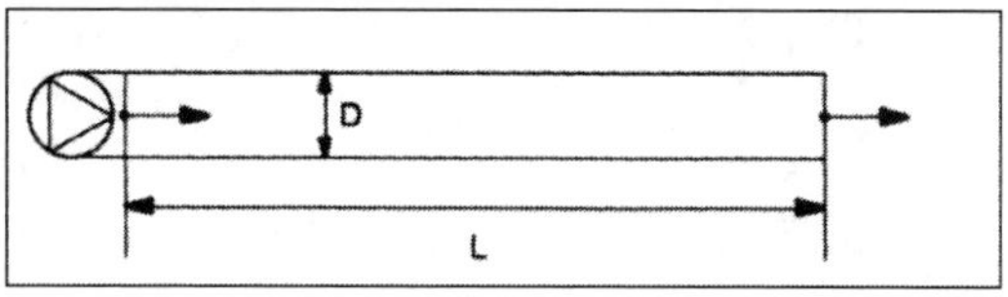

p_B; $\qquad$ $p_B = 0$;

U_A; $\qquad$ U_B;

$\phi_A = U_A/U_B \geq 1$; $\qquad$ $\phi_B = 1$;

$$\frac{1}{6}ln\left(\frac{\theta_A^2+\theta_A+1}{(1-\theta_A)^2}\right) - \frac{1}{\sqrt{3}}arctan\left(\frac{2\theta_A+1}{\sqrt{3}}\right)$$
$$= a^{1/3} - \frac{1}{\sqrt{3}}arctan\left(\frac{1}{\sqrt{3}}\right); \qquad \Rightarrow \theta_A < 1;$$

$$\psi_A = p_A/\left(\rho U_B^2/2\right) = \left[\left(\frac{f}{8A_s}\right)\left(\phi_A^3 - 1\right)\right]^{2/3} > 0;$$

$$\phi_A = \left[\frac{1}{1-\theta_A^3}\right]^{1/3}; \qquad \psi_B = 0;$$

$$a = 8\left(fA_s^2\right)\left(\frac{L}{D}\right)^3; \qquad \theta_B = b = 0; \qquad b = 0.$$

The practical range for f is from 0.01 (very smooth pipe) to 0.035 (relative roughness of 0.006), while the specific leakage area can vary from zero (leakproof pipe) to 60 mm^2/m^2 of outer surface of a duct (Le Roux 1986, MVS 1992, SIA 1998). Based on these values, an arbitrary classification has been introduced to describe a duct quality with respect to friction and leakage (Partyka & Invarsson 2001). The dimensionless velocity ϕ_A depends only on one parameter a, which represents a combination of the f, A_S, and L/D. The dependent variable ψ_A depends not only on parameter a, but also involves a second independent variable, i.e., the duct category or parameter f/A_S.

2.2 *Exhausting system*

Figure 2 illustrates an exhaust ventilation system with its section having a length of Δx. The following system of equations

$$\frac{dp}{dx} = \left(\frac{f}{D}\right)\left(\frac{\rho U^2}{2}\right) \qquad (21)$$

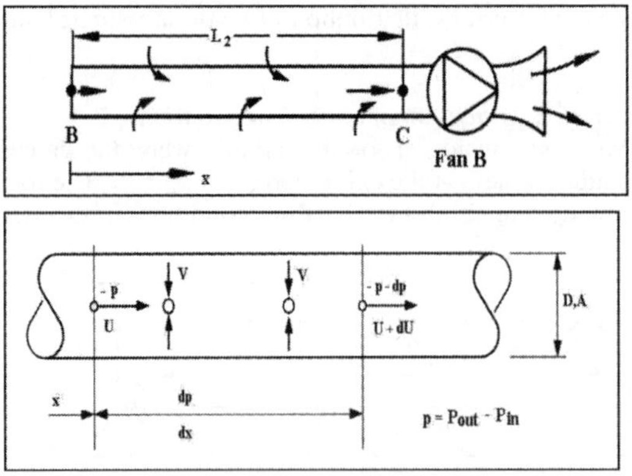

Figure 2. Leaking duct – exhausting system.

$$\frac{dU}{dx}=\left(\frac{4A_s}{D}\right)\sqrt{\frac{2p}{\rho}} \tag{22}$$

describes now the flow of air along a leaking column under negative gauge pressure.

Using dimensionless length $\zeta = x/L_2$, dimensionless pressure $\psi_\zeta = 2p_\zeta/(\rho U_B^2)$ and dimensionless axial velocity $\phi_\zeta = U_\zeta/U_B$, Equations 21 and 22 can be written as follows:

$$d\psi = f\,\phi^2\left(\frac{L_2}{D}\right)d\zeta \tag{23}$$

$$d\phi=\left(4A_s\right)\left(\frac{L_2}{D}\right)\sqrt{\psi}\,d\zeta \tag{24}$$

Again, these can be combined into a single differential ordinary equation

$$\phi^2\,d\phi=\left(\frac{4A_s}{f}\right)\sqrt{\psi}\,d\psi \tag{25}$$

where L_2 is the duct length and U_B denotes mean axial velocity at a duct inlet.

In the case when the specific leaking area A_S is uniformly distributed along distance ζ, Equation 25 has the following solution

$$\phi^3=\left(\frac{8A_s}{f}\right)\psi^{3/2}+C_2 \tag{26}$$

where C_2 is the integration constant. At $\zeta = 0$;

$\phi=\phi_B=1;\quad U=U_B;\quad \psi=\psi_B;$ and

$$C_2=1-\left(\frac{8A_s}{f}\right)\psi_B^{3/2}.$$

Equation 26 can now be written as

$$\psi=\left[\left(\frac{f}{8A_s}\right)\left(\phi^3-1\right)+\psi_B^{3/2}\right]^{2/3} \tag{27}$$

Equation 27 is identical to Equation 11.

2.2.1 *Velocity distribution along a duct*

We can now develop equations which allow us to observe changes in velocity and pressure along distance x. For instance, when substituting Equation 27 into Equation 24 it becomes

$$d\phi=a^{1/3}\left[\left(\phi^3-1\right)+b^3\right]^{1/3}d\zeta \tag{28}$$

$$\frac{\theta\,d\theta}{1-\theta^3}=a^{1/3}d\zeta \tag{29}$$

where $a=8\left(fA_s^2\right)\left(\frac{L}{D}\right)^3$; $b=2\left(\frac{A_s}{f}\right)^{1/3}\sqrt{\Psi_B}$ and

$\theta=\left(1+\frac{b^3-1}{\phi^3}\right)^{1/3}$ is the new dimensionless velocity.

Integrating both sides of Equation 29 gives (Maple V 1997)

$$\int_b^\theta\frac{\omega\,d\omega}{1-\omega^3}=\frac{1}{6}ln\left(\frac{1+\theta+\theta^2}{(1-\theta)^2}\right)-\frac{1}{\sqrt{3}}arctan\left(\frac{2\theta+1}{\sqrt{3}}\right)$$
$$-\frac{1}{6}ln\left(\frac{1+b+b^2}{(1-b)^2}\right)+\frac{1}{\sqrt{3}}arctan\left(\frac{2b+1}{\sqrt{3}}\right) \tag{30}$$

$$a^{1/3}\int_0^\zeta d\tau=a^{1/3}\zeta \tag{31}$$

$$\frac{1}{6}ln\left(\frac{\theta^2+\theta+1}{(1-\theta)^2}\right)-\frac{1}{\sqrt{3}}arctan\left(\frac{2\theta+1}{\sqrt{3}}\right)=a^{1/3}\zeta$$
$$+\frac{1}{6}ln\left(\frac{b^2+b+1}{(1-b)^2}\right)-\frac{1}{\sqrt{3}}arctan\left(\frac{2b+1}{\sqrt{3}}\right) \tag{32}$$

where we denote that when $\zeta = 0$, $\phi = \phi_B = 1$, $\theta = \theta_B = b$.

Solving Equation 32 for $\theta = \theta_\zeta$ at assumed ζ and recalling that $\phi_\zeta=\left(\frac{b^3-1}{\theta_\zeta^3-1}\right)^{1/3}=\frac{U_\zeta}{U_B}$, the duct mean velocity at any cross-section can be determined as follows:

$$U_\zeta = U_B \left\{ \frac{b^3 - 1}{\theta_\zeta^3 - 1} \right\}^{1/3} \tag{33}$$

At $x = 0$: $\zeta = 0$, $\phi_B = 1$, $\theta_B = b$. Therefore, at the duct exit, dimensionless velocity θ_C is determined by the following equation;

$$\frac{1}{6} ln\left(\frac{\theta_C^2 + \theta_C + 1}{(1-\theta_C)^2} \right) - \frac{1}{\sqrt{3}} arctan\left(\frac{2\theta_C + 1}{\sqrt{3}} \right)$$
$$= a^{1/3} + \frac{1}{6} ln\left(\frac{b^2 + b + 1}{(1-b)^2} \right) - \frac{1}{\sqrt{3}} arctan\left(\frac{2b+1}{\sqrt{3}} \right) \tag{34}$$

while the duct inlet velocity is

$$U_C = U_B \left\{ \frac{b^3 - 1}{\theta_C^3 - 1} \right\}^{1/3}. \tag{35}$$

2.2.2 *Distribution of pressure along a duct*

Value U_ζ, calculated by means of Equation 33 can be substituted into Equation 27. In this way, we derive the following formula

$$p_\zeta = \left(\frac{\rho U_B^2}{2} \right) \left\{ \left(\frac{f}{8A_s} \right) \left[\left(\frac{U_\zeta}{U_B} \right)^3 - 1 \right] + \psi_B^{3/2} \right\}^{2/3} \tag{36}$$

which describes the difference in static pressure between the outside and inside pressures along a duct.

2.2.3 *Application of analytical equations*

We consider now an exhaust column when the gauge static pressure at the inlet is zero, i.e., $p_B = 0$. The following sketch shows the flow conditions at a duct inlet and outlet.

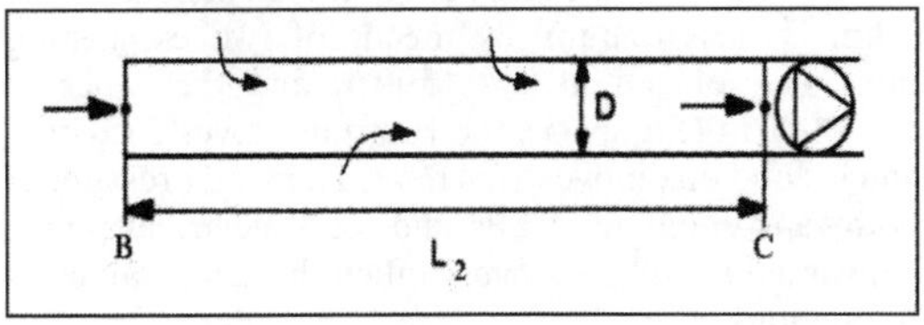

$p_B = 0$; $\quad$ p_C;

U_B; $\quad$ U_C;

$\phi_B = 1$; $\quad$ $\phi_C = U_C / U_B \geq 1$;

$\theta_B = 0$;

$$\frac{1}{6} ln\left(\frac{\theta_C^2 + \theta_C + 1}{(1-\theta_C)^2} \right) - \frac{1}{\sqrt{3}} arctan\left(\frac{2\theta_C + 1}{\sqrt{3}} \right)$$
$$= a^{1/3} - \frac{1}{\sqrt{3}} arctan\left(\frac{1}{\sqrt{3}} \right); \qquad \Rightarrow \theta_C < 1$$

$$\psi_B = 0 \qquad \psi_C = \left(\frac{2p_C}{\rho U_B^2} \right) = \left[\left(\frac{f}{8A_s} \right) \left(\phi_C^3 - 1 \right) \right]^{2/3} > 0$$

$$b = 0; \qquad \phi_C = \left(\frac{1}{1-\theta_C^3} \right)^{1/3}; \quad a = 8\left(fA_s^2 \right) \left(\frac{L_2}{D} \right)^3$$

3 ELECTRICAL POWER REQUIREMENT FOR LEAKING DUCTS

The motor input power of *Fan A* (Figure 1) used to deliver air quantity Q_B to workings along a uniformly leaking duct is equal to

$$kW_A = \frac{(p_A \cdot Q_A)_{Leaking\ duct}}{\eta_{Fan\ A} \cdot \eta_{Motor\ A}} = \frac{\Psi_A \left(\frac{\rho U_B^2}{2} \right) \phi_A \cdot Q_B}{\eta_{Total\ A}} \tag{37}$$

while for a non-leaking duct (pipe) it is

$$kW_{pipe} = \frac{f \cdot \left(\frac{L}{D} \right) \left(\frac{\rho U_B^2}{2} \right) \cdot Q_B}{\eta_{Total\ A}} \tag{38}$$

Fan motor input power is higher for a leaking duct than that for a pipe. If $\eta_{Total\ A}$ does not change, the analytical power ratio (*APR*) is:

$$APR_{blower} = \frac{(p_A \cdot Q_A)_{Leaking\ duct}}{(p_A \cdot Q_B)_{Pipe}} = \frac{\psi_A \phi_A}{f\left(\frac{L_1}{D} \right)} \tag{39}$$

The motor input power of *Fan B* (Figure 2) used to withdraw air quantity Q_B from workings along a uniformly leaking duct is equal to

$$kW_B = \frac{(p_C \cdot Q_C)_{Leaking\ duct}}{\eta_{Fan\ B} \cdot \eta_{Motor\ B}} = \frac{\psi_C \left(\frac{\rho U_B^2}{2} \right) \phi_C \cdot Q_B}{\eta_{Total\ B}} \tag{40}$$

and the power ratio is

$$APR_{exhaust} = \frac{(p_C \cdot Q_C)_{Leaking\ duct}}{(p_C \cdot Q_B)_{Pipe}} = \frac{\psi_C \, \phi_C}{f\left(\frac{L_2}{D}\right)} \tag{41}$$

When forcing and exhaust columns are identical, i.e. they have the same f, A_s and L/D ratio, then the blower and exhaust fans;

1. draw the same flows, i.e. $Q_A = Q_C$ (Equations 19 and 35);
2. have the same static pressures, i.e. $p_A = p_C$ (Equations 20 and 36);
3. consume the same power, i.e. $kW_A = kW_B$ (Equations 37 and 40).

4 MULTIPLE FANS

We now consider a 3000-m-long, one meter in diameter leaking column delivering 10 m^3/s of air by means of one or two fans. These fans will be located differently. The following options are calculated:

1. Blower at $x = 0$ m
2. Exhaust at $x = 3000$ m
3. Two blowers: one at $x = 0$ m; one at $x = 1500$ m
4. Two exhaust fans: one at $x = 1.5$ km; one at $x = 3$ km
5. One fan at each end
6. Two blowers: one at $x = 0$ m; one at $x = 2000$ m
7. Two blowers: one at $x = 0$ m; one at $x = 1.0$ km
8. Two exhaust fans: one at $x = 1.0$ km; one at $x = 3$ km
9. Two exhaust fans: one at $x = 2$ km; one at $x = 3$ km

Each option is evaluated based on power consumption for five different duct qualities. The duct quality is based on categorization discussed in Partyka & Invarsson 2001. These are characterized as:

(a) Excellent (class S): $f = 0.015$; $A_s = 5$ mm^2/m^2
(b) Very good (class A): $f = 0.020$; $A_s = 10$ mm^2/m^2
(c) Good (class B): $f = 0.025$; $A_s = 20$ mm^2/m^2
(d) Fair (class C): $f = 0.030$; $A_s = 40$ mm^2/m^2
(e) Poor (class D): $f = 0.035$; $A_s = 60$ mm^2/m^2

5 RESULTS

Final results are based on total electrical power required to deliver 10 m^3/s of air by blowers to a working place located ahead 3 km (Equation 37), or to exhaust that amount and move it along the 3-km long column by means of exhausting fans (Equation 40). Total power is based on total pressure and the air quantity passing through each fan placed in a system. The total pressure includes static pressure of each fan, the velocity pressure of the first fan, the additional velocity pressure added by second fan, and total minor losses. These minor pressure loses are due to an inlet or outlet and losses caused by a screen placed at the inlet of each fan. In all cased motor efficiency is 0.92 and fan efficiency is 0.75.

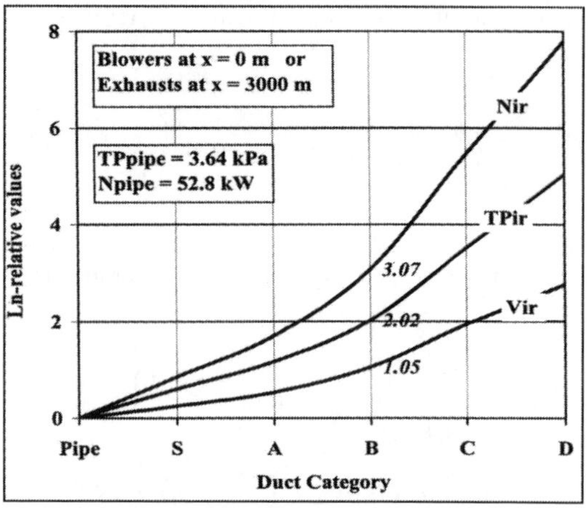

Figure 3. Ventilation systems with fans placed only on inlet or outlet.

Figure 3 shows that a blower installed at the inlet of a B-class duct handles 10 ≅ exp(1.05) = 28.6 m^3/s or 2.86 times more air than in a leakproof pipe.

Fan total pressure increases then exp(2.02) = 7.54 times or equals 3.64 ≅ 7.54 = 27.44 kPa. The total electrical power consumed is 52 ≅ exp(3.07) = 1,137 kW, 21.5 times higher. The blower and exhaust fan have the same pressure-volume operating point.

When a system has to deliver 10 m^3/s at the end of a 3-km B-class column by means of two blowers, one placed at inlet and other in the middle of the column, blower at $x = 0$ m moves 23.6 m^3/s whereas blower located at $x = 1500$ m moves 15.4 m^3/s. The total pressure of both blowers equals 28 kPa and total electrical power used equals 717 kW (Figure 4).

When a ventilation system has to remove 10 m^3/s of air from one place and deliver it to at the end of the 3-km B-class column by means of two exhausting fans, one placed at $x = 1500$ m and the other at $x = 3000$ m (Figure 5), the first fan moves 23.9 m^3/s and second one moves 15.4 m^3/s. The total pressure of both fans equals 22.9 kPa and total electrical power consumed is 709.5 kW, very much the same values as in previous case.

Figure 6 illustrates a system in which one blower and one exhaust fan are placed at the ends of the B-column. Each fan uses 41 kW, handling 10 m^3/s at total pressure of 2.81 kPa. The same fan located in a

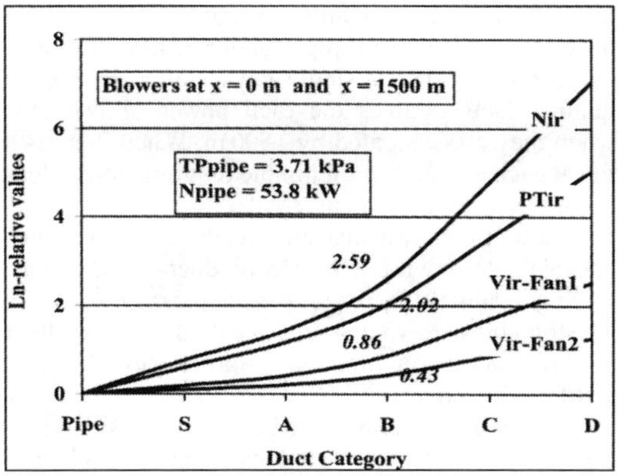

Figure 4. Column served by two blowers 1500 m apart.

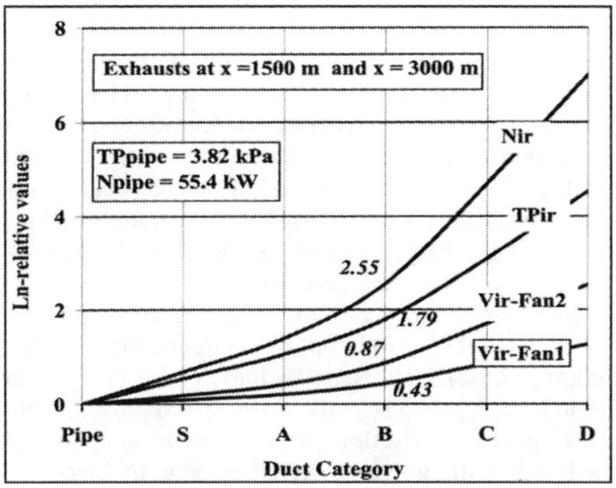

Figure 5. Column served by two exhausts 1500 m apart.

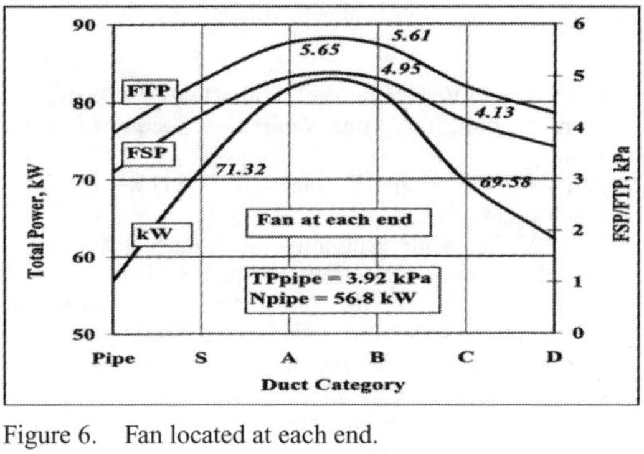

Figure 6. Fan located at each end.

pipe will generate total pressure of 1.96 kPa using 28.4 kW of electrical power.

Table 1 contains electrical powers of ventilation systems delivering/exhausting 10 m^3/s along 3-km, 1-m-diameter ventilation column. These values are

Table 1. Electrical power requirements.

Option ↓	Electrical power, kW				
Class →	Pipe	S	A	B	C
1	52.8	123.1	287.9	1.138	12.235
2	52.8	123.1	287.9	1.138	12.235
3	53.8	115.6	226.8	715.4	6.344
4	53.4	109.1	221.6	708.6	6.105
5	56.8	71.3	81.8	81.4	69.6
6	55.2	108.5	216.9	702.1	6.407
7	58.4	112.9	232.5	766.7	6.810
8	57.0	97.1	214.4	687.0	6.182
9	55.4	113.2	232.8	763.9	6.678

Table 2. Total pressure of all fans.

Option ↓	Total pressure, kPa				
Class →	Pipe	S	A	B	C
1	3.64	6.62	11.7	27.5	122.1
2	3.64	6.62	11.7	27.5	122.1
3	3.71	6.90	11.2	23.2	87.5
4	3.82	6.53	10.9	23.0	84.3
5	3.92	4.92	5.65	5.61	4.80
6	3.81	6.14	10.5	21.8	83.5
7	4.03	6.63	11.2	24.2	94.9
8	3.93	6.00	10.4	21.5	81.0
9	3.83	6.65	11.2	24.3	94.1

based on total fan pressure and the quantity of each fan, located in the column (options 1 to 9) for five duct quality (classes). Class D (poor) is omitted in this table. Actually, classes B, C, and D should not be considered for this type of duty because the demand for power is extremely high and impractical.

Table 1 illustrates that the S-type duct requires approximately twice, A-duct four times, B-duct more than ten times, and C-duct more than a hundred times as much power as a pipe. Furthermore, the power consumption increases when a single fan is placed at the end of the column. When there is a fan at each end of a column the power required to remove or deliver 10 m^3/s increases by less than a half.

Table 2 illustrates that the S-type duct requires approximately twice, A-duct three times, B-duct six times, and C-duct twenty times as much pressure as a pipe. However, the total pressure can rise more than thirty times when a single fan is placed at the end of the column. When there is a fan at each end of a column the power required to remove or deliver 10 m^3/s increases by less than a half.

Table 3 shows fan duties, pressures and volumes, serving S and A quality ducts. For instance, a single fan placed in A-duct handles 17 m^3/s at total pressure

Table 3. Duty of fans – Class S/A.

Option	First fan Quantity (m^3/s)	First fan Total pressure (kPa)	Second fan Quantity (m^3/s)	Second fan Total pressure kPa
1	12.8/17.0	6.6/11.7	–	–
2	12.8/17.0	6.6/11.7	–	–
3	12.1/15.1	3.8/6.2	11.0/12.3	3.1/4.5
4	11.0/12.3	3.0/4.3	12.1/15.1	3.5/6.6
5	10.0/10.0	2.5/2.9	10.0/10.0	2.4/2.8
6	12.1/15.3	4.5/7.9	10.5/11.2	1.6/2.7
7	12.1/15.3	2.7/5.2	11.5/13.6	3.6/5.5
8	10.5/11.2	2.1/2.8	10.5/15.3	3.9/7.6
9	11.5/13.6	4.1/6.2	12.1/15.3	2.6/5.0

of 11.7 kPa. When two fans are considered, each fan in options other than 5 has a different operating point. In option 8 (A-class duct) the first fan moves $11.2\,m^3/s$ at pressure of 2.8 kPa, while the second moves $15.3\,m^3/s$ at 7.6 kPa.

6 CONCLUSION

An exact analytical solution has been developed that allows prediction of the operating condition of one, two or more fans installed in a uniformly leaking column of constant diameter. The fans can operate as blowers, exhausts or as a combination of both. Two implicit formulae describe flow and pressure along a leaking duct, and these can easily be solved in spreadsheet form, e.g., by means of the Microsoft Excel Solver package.

The duty of fans delivering a required quantity of air to underground workings or exhausting that amount of air and exiting it some distance ahead by means of a leaking column of a specified length and diameter depends strongly on the quality of the ventilation column. The friction factor and the specific leakage area values must be known before any calculations can be performed. Another possibility is to designate a duct quality by its class, as suggested in this paper. The duct quality may range from perfect to poor, which covers realistic values for the friction factor (0.01 to 0.035) and the leaking area (0.01 to $60\,mm^2/m^2$).

To ventilate a working place, by removing or delivering $10\,m^3/s$ of air by means of one-metre in diameter and 3-km long (L/D = 3,000), an excellent duct requires approximately twice, very good duct – four times, good duct – more than ten times, and fair duct more than a hundred times as much power as that needed for a leakproof pipe (about 55 kW). Furthermore, the power of a single fan installed in B or D quality ducts doubles the total power of two fans when they are separated by 1500 m. When there is a fan at each end of the column the required power does not increase by more than a half.

The study shows that an excellent duct requires approximately twice, a very good duct – three times, and a good duct – six times as much pressure as a pipe. The total pressure of two fans installed in an excellent or a very good column already approaches 6 kPa and 11 kPa, respectively.

Ventilation systems with L/D > 1,000 should be served by two or more fans located so that each will draw equal amount of power. They should never be installed in one location, at one end of a column or in the middle of a column. A single fan in ducts with L/D = 1,000 moves a flow increased by a factor of 1.1 (excellent duct) to 2.2 (poor duct) and uses increased power by a factor of 1.2 (excellent duct) to 4 (poor duct). A very long duct, e.g. these having L/D = 5,000, may need at inlet, anywhere from 1.8 (excellent duct) to 7 (good duct) times more air and from 2.8 (excellent duct) to 70 (good duct) times higher electrical power than a perfect duct delivering the same quantity of air to workings. Poor or fair ducts should never be considered for distances L/D >400.

In general, when multiple fans are installed in a leaking ventilation column, each fan operates at a different point of its pressure-volume curve; a fan with higher total pressure will move less than one with lower pressure. Multiple fans can be located so that each fan will draw the same amount of electrical power. However, when in use the equal power criterion will not be sustained when duct quality changes.

REFERENCES

Le Roux, W.L., 1986, Mine Ventilation Notes for Beginners, Third Edition, The Mine Ventilation Society of South Africa.

Maple, V., 1997, "Maple V Release 4: The Power Edition," Waterloo Maple Inc.

MVS, 1992, The Mine Ventilation Practitioner's Data Book, The Mine Ventilation Society of South Africa.

Partyka, J., Invarsson B., 2001, Analytical Design of Ventilation Systems for Underground Mining Development, Proceedings of the 7th International Ventilation Congress, Krakow, Poland.

SIA, 1998, "Baulüftung im Untertagbau," Schweizer Norm SN, Bauwesen 531 196, SIA 196 (1983), Schweizerischer Ingenieur-und Architekten-Verein, Zürich.

Mine Ventilation, Ganguli & Bandopadhyay (eds)
© 2004 Taylor & Francis Group, London, ISBN 90 5809 633 5

Assessing the performance and efficiency of booster and main fans in a large ventilation network

K.R. Mutama
Senior Ventilation and Mechanical Engineer, Barrick Goldstrike Mines Inc., Underground Division, Elko, Nevada, USA

ABSTRACT: This paper describes the operation of main and booster fans at Barrick's Meikle mine which comprises three large mining areas with an interacting ventilation system. The performance and efficiency of booster and main fans in this large mine ventilation system have been analyzed. It is found that while a fan can satisfy performance in terms of pressure and quantity it does not always operate in the high efficient envelope. The ventilation system in a large network with many interacting fans become very dynamic and therefore performance rather than efficiency become the main criteria for selecting a fan for a particular duty. Using some automated fans whose blade pitch can be adjusted automatically and variable speed drive (VFD) axial vane fans is very beneficial to the system because it provides more airflow to certain areas when demand arises and reduces quantity in areas where activities are less. Main fans show stable operation for a given duty despite the addition of secondary or booster fans in mining districts. Interesting observations are made when operating a combination of booster fans while others are turned off in other parts of the network. The unique ventilation system at the Meikle underground complex provides a good in depth understanding of the behavior of axial fans in large networks covering different areas in multiple levels.

1 INTRODUCTION

Booster fans play a very important role in large underground multi-level mines especially when the ventilation system is complex. The primary ventilation system involves the main fans, which may be located on surface or underground. The purpose of the main fans is to provide the primary total intake fresh air from surface that the mine needs for adequate ventilation to be achieved for the underground operations. In most cases the mine is spread out and with many mining horizons or districts. To move large volumes of air flow to these areas requires dedicated booster fans feeding them. Without underground booster fans most of the ventilation air would short-circuit to surface via less resistant airways. Moreover more controls in the form of doors and regulators would be required to achieve the required airflow balance. Even then it still will be very difficult to achieve the right airflow quantities in certain areas of the mine. Booster fans help over come the mine resistance in those zones to move the desired airflow.

The size of booster fans may vary from say twenty-four inches in diameter (610 mm) to as large as eighty-four inches (2,134 mm) in diameter or more. The power required to drive these fans can be as little as 5 hp (3.73 kW) to as large as 1,000 hp (746 kW) or more. The differential pressure that these booster fans have to overcome may be less than 0.5 inches (124.6 Pa) to as much as 8.0 inches water gauge (1,993 Pa). The volume moved can be as little as 10,000 cfm (4.7 m^3/s) to 500,000 cfm (235 m^3/s) or more. Selecting the correct booster fan for the required duty requires setting up a ventilation network model for the mine with the specified quantities in those areas. The model is able to predict the required pressure that the booster fan will be operating. However, this is only part of the task because in a practical situation there are many variables and constraints in an operating mine ventilation system.

2 BOOSTER FAN GOVERNING FACTORS

There are a number of factors to be considered when it comes to booster fan operating parameters. These are fan diameter, rotational speed, pressure, volume and operating horsepower. Conventional terms that describe fan performance can be defined in the following manner. The static pressure (P_s) needed to move airflow through the system is proportional to the square of the quantity delivered. The static pressure is negative if it is below atmospheric pressure, and positive if it is above atmospheric pressure. The static pressure is expressed in inches w.g. or Pascal pressure. The airflow traveling at a velocity V (m/s or ft/min) will create a dynamic or kinetic energy pressure called velocity pressure (P_v) commonly expressed in inches w.g. or Pascal. The velocity pressure P_v is positive and is always exerted in the direction of airflow.

The energy transferred to the airflow by the fan wheel results in the increase of both the static and velocity pressures. The total pressure is the sum of both the static and velocity pressures.

$$P_s \propto Q^2 \quad (1)$$

$$P_v = (V/4005)^2 \text{ (inches w.g.)} \quad (2)$$

$$P_t = P_s + P_v \quad (3)$$

Equation (3) is a form of the Bernoulli equation and can be expressed in various forms including the one below for inlet conditions to the fan wheel at point 1 and exit at point 2. H being the change in total pressure, H_d is fan head and H_{loss} is the losses across the system. The symbols p, ρ, g, Z are the pressure, density, gravity constant and elevation respectively.

$$H = (p/\rho g + V^2/2g + Z)_2 - (p/\rho g + V^2/2g + Z)_1 = H_d - H_{loss} \quad (4)$$

It is the difference in fan total pressure across the fan wheel that is of most importance in fan installation systems and in the case of a booster fan installed in a bulkhead, differential pressure across the bulkhead is used to assess fan performance. The power delivered to the airflow is equal to the specific weight times the discharge quantity times the net head change.

$$Pw = \rho g Q H \quad (5)$$

This is called the air horsepower. The power required to drive the fan wheel is called the brake horsepower (bhp).

$$Bhp = \omega T \quad (6)$$

where ω is the shaft angular velocity and T is the shaft torque. If losses did not exist the bhp and the air power Pw would be equal. Pw is less than the brake horsepower and the efficiency of the fan or system installation can be defined as

$$\eta = P_w/bhp = \rho g Q H/\omega T \quad (7)$$

The efficiency is composed of various parts at the particular installation and can be described as the volumetric, hydrodynamic and mechanical efficiency. The hydrodynamic efficiency is composed of shock losses at inlet and exit to the fan intake, friction losses in the blade passages and circulation loss at the exit side of the blades. Mechanical losses are due to friction in the bearings and other contact points. The overall efficiency is the product of all the other efficiencies. The efficiency of booster fan systems can vary from as little as 20% to as high as 80% for well-installed systems.

In the fan performance literature it is often common to see simplified equations for quick determination of important parameters. Conversion factors normally apply. Fan brake horsepower can be also be expressed as,

$$Bhp = CFM \times \text{total pressure}/(6356 \times \text{efficiency}) \quad (8)$$

For a three phase motor,

$$bhp = (Vo \times I \times \text{motor efficiency} \times Pf \times 1.73)/746 \quad (9)$$

The fan motor input power in kilowatts for a three-phase motor is expressed as follows,

$$\text{Power (kW)} = (Vo \times I \times Pf \times 1.73)/(1000) \quad (10)$$

The symbol Vo is the average phase voltage, I is the average current and Pf is the power factor normally ranging from 0.7 to 0.8 in most systems.

In addition fan laws governing quantity Q, speed RPM, pressure P and horsepower HP can be written as follows,

$$\text{Volume } Q_2 = Q_1 (RPM_2/RPM_1) \quad (11)$$

In this case air density is constant.

$$\text{Pressure, } P_2 = P_1 (RPM_2/RPM_1)^2 \quad (12)$$

Pressure varies directly with density ratio

$$P_2 = P_1 (\rho_2/\rho_1) \quad (13)$$

The horsepower;

$$HP_2 = HP_1 (RPM_2/RPM_1)^3 \quad (14)$$

The above equations are used in the practical assessment of all fan systems including booster fans. Details complicated turbomachine equations describing fans exist in the literature and for this paper the simplified expressions are sufficient.

Other points of interest in this paper can also be mentioned. In a given system the efficiency does not change with speed however as can be seen in equations (11) to (14) quantity, total pressure and power input depend on the speed. For a given pressure, the noise level is proportional to the tip speed of the fan wheel and to the air velocity leaving the wheel. Regardless of the blade type fan noise is roughly proportional to the pressure developed.

3 DESCRIPTION OF THE MEIKLE MINE VENTILATION SYSTEM

The underground complex of the Meikle mine covers three areas, which are large enough to be stand alone mines by themselves. These are the Meikle main zone, the Griffin area and the Rodeo mining complex. Within Meikle is another mining horizon called South Meikle with many levels of its own that uses drift and fill mining method. The other areas use longhole stoping. The ventilation system for these areas is interconnected and is very complex. There are a total of four shafts. One intake shaft and one exhaust shaft at the Meikle main complex and a similar set up at the Rodeo area. Rodeo and Meikle are 1.6 kilometers apart (1 mile) and are connected by two long roadways on two different levels. In between Meikle and Rodeo is the Griffin area (see Figure 1). The total air moved is 846 m^3/s (1,800,000 cfm). The Meikle shaft downcasts about 470 m^3/s exhausting 404.2 m^3/s (860,000 cfm) using 2 × 1750 hp centrifugal fans located on surface. The Rodeo intake shaft downcasts 376 m^3/s (800,000 cfm) and the exhaust shaft exhausts about 441.8 m^3/s (940,000 cfm) using 2 × 1500 hp axial flow fans located on surface. On each intake shaft there is a bulk air cooler where approximately 75% of the downcast air is cooled to 45°F (7.2°C) and four 250 hp axi-vane fans with variable speed drives push the air through each cooler at Rodeo and Meikle respectively.

The lower half of Rodeo area is ventilated using 188 m^3/s (400,000 cfm) via the lower connecting haul way connecting Meikle and Rodeo. There are 2 × 500 hp fans at the Meikle side feeding Lower Rodeo and Lower Griffin located underground. The Griffin area has no dedicated intake or exhaust borehole of its own. The upper part of Griffin mining zone receives 112.8 m^3/s from the Rodeo intake shaft using 2 × 200 hp booster fans via the upper haul way connecting Meikle and Rodeo. The lower portion of the Griffin levels is ventilated by 65.8 m^3/s (140,000 cfm) from the Meikle intake shaft. All the airflow from Griffin is exhausted at the Meikle shaft. Thus the mine ventilation system in these areas is very large and complicated to manage.

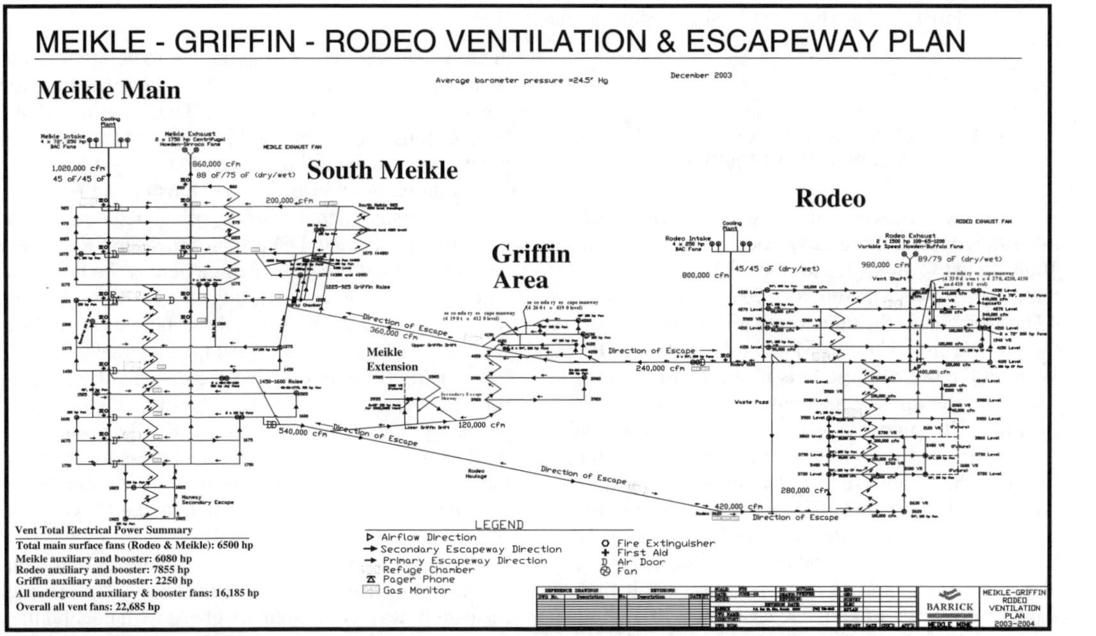

Figure 1. Meikle–Griffin – Rodeo ventilation and escapeway plan.

Figure 2. 54″-150 hp fan installed in a shotcrete bulkhead.

Figure 3. Installation of a booster fan on plate above a vent raise.

4 ASSESSING BOOSTER FAN PERFORMANCE AND EFFICIENCY

All the booster fans used in the Meikle mine are of the axial flow type and are of various sizes ranging from 1066 to 2134 mm (42 to 84 inches) in diameter with horsepower of 100 to 500 hp. There are over 48 booster fans in the mine ventilation system. Some of these fans run at 1200 RPM and others at 1800 RPM. Without the use of these fans it will be impossible to distribute the airflow adequately in various parts of the mine. The objective is to ensure that there is adequate airflow at each level of the mine. The manner, in which these fans are installed and used, has a lot to do with their measured efficiency. The majority of the booster fans are installed in shotcrete bulkheads as shown in Figure 2 and others are installed directly above vent raises and set on a plate with a hole matching the diameter of the fan as shown in Figure 3. Some of the booster fans play a dual role of moving airflow from a vent raise from the level below and the airflow ducted directly from the fan to ventilate headings in the area (see Figure 4). This tends to eliminate the need for an auxiliary fan. The location of the booster fans can be seen in Figure 1 and their relative position in the overall mine ventilation system.

Figure 4. Two 48″-150 hp vent raise booster fans (in parallel) serving as auxiliary fans, steel ducting to be attached to the fans for heading ventilation.

The large number of booster fans operating in the system demanded that a survey be carried out, of their operating efficiency. The primary measurements were simply the total pressure change across the bulkhead where the fans were operating, the airflow quantity, the voltage and current that these fans were pulling. By using equation (7) and (10) the efficiency of these fans could be computed. The exercise involved taking power consumption over many days that the fans were operating. The delivery volume of the fans was measured using a vane anemometer. The pressure drop was measured using a special barometer that could be set to determine pressure change across the bulkhead or plate where the fan was installed. Some of the fans have electronic pressure differential devices that monitor pressure change on a continuous basis. The pressure measured was compared to pressure data obtained from the mine network model. In most cases this was very close.

Table 1. Booster fan operating performance and efficiency.

Description	Power (hp)	Diameter (inches)	Speed (RPM)	Pressure Drop (W.G.)	Airflow (Kcfm)	Voltage (V)	Current (A)	Power Consumption (kW)	Efficiency (%)
Meikle Area									
925/1075 Vent Raise	100	60	1200	0.5	119.3	480	92	50.86	13.79
4480	150	54	1800	3.76	98.4	480	93	51.42	84.59
4450	150	48	1800	4.0	76.1	480	100	55.29	64.73
4450	100	48	1800	4.0	76.1	480	110	60.82	58.85
4540 Sou. Meikle	100	42	1800	5.33	34.8	480	62	34.28	63.61
4525	150	54	1800	5.33	57.6	480	108	59.71	60.44
1175/1225 Vent Raise	15	32	1800	0.5	34.9	480	8	4.42	46.38
1175/1300 Vent Raise	125	45	1800	0.66	71.1	480	70	38.70	14.25
1225/1450 Vent Raise	150	54	1800	2.19	106.6	480	110	60.82	45.13
1300/1375 Vent Raise	100	42	1800	0.14	57.9	480	100	55.29	1.72
1450/1600 Vent Raise	500	84	1200	2.76	280	4160	32	153.33	59.25
1450/1600 Vent Raise	500	84	1200	2.76	280	4160	26	124.58	72.92
S.HAULAGE	150	48	1800	5	78.8	480	105	58.05	79.79
1600/1525 Vent Raise	100	42	1800	0.29	57.6	480	100	55.29	3.55
East Zone	100	42	1800	1.16	55.4	480	90	49.76	15.18
East Zone	100	42	1800	1.16	55.4	480	75	41.47	18.22
1600/1750 W. Raise	125	60	1200	1.36	69.9	480	100	55.29	20.21
Griffin									
3920–3980 Vent raise	150	54	1800	0.18	96.6	480	70	38.70	5.28
DD8 Airdoors	200	54	1800	0.45	137.4	480	210	116.10	6.26
DD8 Airdoors	125	54	1800	0.45	137.4	480	135	74.64	9.74
4050–4120 Vent raise	100	42	1800	0.5	79	480	81	44.78	10.37
4050–4190 Vent raise	125	49	1800	2.02	76.7	480	81	44.78	40.67
NFW	150	48	1800	0.23	57.73	480	95	52.52	2.97
SFW	150	48	1800	2.32	76.33	480	79	43.68	47.66
Lower Airdoor	200	54	1800	4.24	100.1	480	213	117.76	42.37
Lower Airdoor	200	54	1800	4.24	100.1	480	176	97.31	51.28
Rodeo									
4330–5920 Vent Raise	125	54	1800	0.5	58.2	480	98	54.18	6.31
4330–2130 Vent Raise	200	78	1200	3.5	219.6	480	219	121.08	74.62
4330–2130 Vent Raise	200	78	1200	3.5	219.6	480	216	119.42	75.66
4210–5920 Vent Raise	150	54	1800	1.6	155	480	122	67.45	43.22
4210–5290 Vent Raise	150	54	1800	1.22	77.6	480	117	64.69	17.21
4270–2130 Vent Raise	200	72	1800	2.06	180.4	480	132	72.98	59.86
4270–1940 Vent Raise	200	72	1800	2.06	180.4	480	158	87.35	50.01
4100–5920 Vent Raise	125	54	1800	0.36	99.4	480	98	54.18	7.76
4150–1940 Vent raise	150	60	1200	0.44	134.2	480	76	42.02	16.52
4100–1940 Vent raise	150	60	1200	0.62	125.9	480	105	58.05	15.81
3620–5480 Vent Raise	100	42	1800	0.67	56.6	480	78	43.12	10.34
3620–5480 Vent Raise	150	54	1800	2.5	94.7	480	122	67.45	41.26
3790–2630 Vent raise	125	60	1200	0.68	133.8	480	38	21.01	50.91
3790 NFW Vent raise	125	54	1200	0.69	97.1	480	84	46.44	16.96
3720–2630 Vent raise	100	60	1200	0.75	133.2	480	76	42.02	27.95
3620–5480 Vent raise	125	54	1200	0.55	91.2	480	84	46.44	23.1
3620–2630 Vent raise	125	54	1200	1.0	98.7	480	90	49.76	23.2

5 DISCUSSION OF RESULTS OF FAN PERFORMANCE AND EFFICIENCY

The results of booster fan performance and efficiency at the Meikle mine are shown in Table 1. The fans are installed in vent raise bulkheads or directly above the raise sitting on a plate. The pressure change across the fan bulkhead needed to move the measured quantities tend to vary between 0.5 to 5 inches w.g. depending on the location of the fan in the ventilation circuit. The booster fans delivered the expected volume but at very low efficiencies. A total of forty three fans are presented in Table 1 and out of the total only 14 fans can be seen to be operating above 50% efficiency

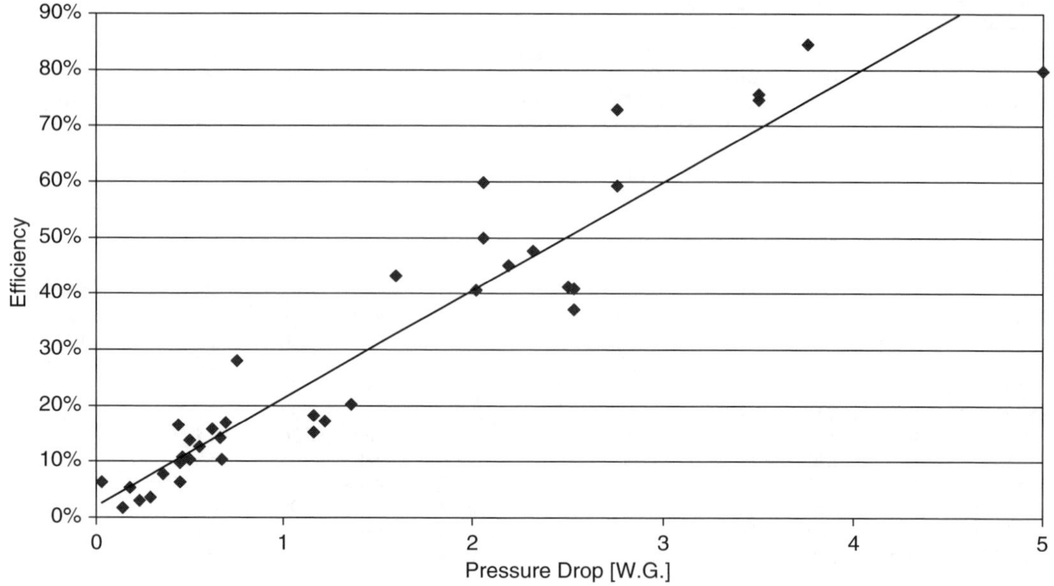

Figure 5. Booster fan pressure drop vs. fan efficiency.

with twenty nine operating below 50%. Twenty fans in the circuit were operating above 40% efficiency. Sixteen fans were running below 20% at the time the survey was carried out. Figure 5 presents fan efficiency against pressure developed to move the required quantities. It can be seen clearly that there is a direct correlation that can be made between fan efficiency and pressure.

- Fan operating in the very low-pressure range tend to have poor efficiencies.
- Fans operating in the high-pressure range operate at higher efficiencies.
- Efficiency as defined by equation (7) is improved by the fan requiring less input electric energy and at the same time developing high pressure or large volume flow or both.

This is to be expected because on every fan curve from any supplier the low-pressure range is the poor efficiency envelope. However the fans in the Meikle mine are capable of operating at better efficiencies than presented. A number of reasons can be given to explain the low efficiencies.

- The mine ventilation system has many transient phases. The main fans operate at different flow volume from time to time during the day. The Meikle main zone surface centrifugal fans are reduced in flow capacity when heavy equipment is lowered or hoisted in the exhaust shaft.
- The Rodeo surface 2 × 1500 hp axial flow fans have variable speed drives (VFD) and are sometimes slowed down when mucking is occurring at the bottom of the intake shaft.
- Because of the presence of a large fleet of diesel equipment operating everywhere in the mine, somewhere in the circuit an air door is being opened or left open for a long time and that tends to affect the operation of booster fans in the circuit.
- Poor efficiency can largely be attributed to poor installation problems of the booster fans. This is because fans are located close to production areas where there is frequent blasting in the headings or stopes. The blasting damages bulkheads in which the fans are located. After sometime there are holes in the bulkhead with the result that the fan is not able to develop expected operating pressure and also re-circulates some airflow.
- Further more because the fans in most cases are installed without an inlet diffuser and an exit bell or evasee, shock losses contribute significantly to poor efficiencies.
- Some of the fans are installed above an air door in the bulkhead. If the door starts to leak badly the efficiency and performance of the fans are also affected. A typical example is the Griffin air door 2 × 200 hp fans that are mounted above an air door. These fans are monitored on a frequent basis and it is obvious when they start showing poor efficiency and performance; mainly because of the status of the door below them.
- The Griffin 4120 level DD8 air door two booster fans are operating at efficiencies of between 6 and 10%

Table 2. Surface main fan efficiency data.

Description	Speed (RPM)	Pressure Drop (W.G.)	Airflow (kcfm)	Voltage (V)	Current (A)	Power Consumption (kW)	Efficiency (%)
Meikle Fan # 1 (Exhaust)	890	14.2	430	4160	212	1015.81	74.4
Meikle Fan # 2 (Exhaust)	890	14.2	430	4160	219	1049.35	74.4
BAC Fan #1 (Intake) VFD	1080	5.0	167	480	202	111.55	87.7
BAC Fan # 2 (Intake) VFD	1080	5.0	167	480	202	111.55	87.7
BAC Fan # 3 (Intake) VFD	1080	4.8	171	480	188	103.82	86.8
BAC Fan # 4 (Intake) VFD	1080	4.6	180	480	188	103.82	91.4
Rodeo Fan # 1 VFD	1001	9.6	470	2900	142	640.00	82.6
Rodeo Fan # 2 VFD	1001	9.6	470	2900	144	640.00	82.6
BAC Fan #1 (Intake) VFD	960	1.7	148	480	134	73.99	39.8
BAC Fan # 2 (Intake) VFD	960	1.7	146	480	131	72.34	40.2
BAC Fan # 3 (Intake) VFDs	960	1.8	147	480	133	73.44	42.2
BAC Fan # 4 (Intake) VFD	960	2.0	144	480	133	73.44	45.9

because they are used as jet fans or induction fans and they move a total of $131.6\,m^3/s$ (280,000 cfm) without a bulkhead to develop pressure against. The original door was removed to allow for large production diesel equipment. It was determined however that these fans could still move a significant air flow volume if they operated as jet fans because of the pull from the main surface fans. Less than 125 Pa (0.5 inches w.g.) was needed to move the expected quantity.

A number of remedies can be applied to the Meikle mine booster fan system. Some of the fans operating in the circuit were installed during the initial phases of mine development when the ventilation circuit was less complex. These fans are no longer at the same pressure and quantity as intended. Even though they still supply the expected volume flow they have become inefficient over time. In some cases fan blades are heavily coated with mud and dust to the detriment of efficiency. It is also possible that some fan blades can be worn out at the leading and trailing edges resulting in the fan losing volumetric and hydrodynamic efficiency. Fans with excessive dirty blades need to be stopped and cleaned regularly to improve their performance and efficiency. Those with worn out blades need to be replaced with new ones.

Looking at the fan performance curves of the fans presented in Table 1 it can be seen that most of these fans operate at speeds of 1200 and 1800 RPM and they are full bladed fans. It is possible in some cases to do a number of things to these fans, especially those operating at less than 250 Pa (1 inch) pressure or less.

- Switch to a fan of the same diameter but with a lower horsepower motor or use a different fan that will handle the same airflow with a different size motor. The fan can have both low horsepower and speed but with a favorable operating efficiency envelope.
- Most fan manufacturers or suppliers can provide a similar size fan with half the number of blades on the hub to operate in the low-pressure range of 1 kPa (four inches) or less with a lower horsepower motor whilst providing better efficiency.

The problem with low pressure fans such as half bladed ones is that they are good in a stable ventilation circuit but when the situation changes temporarily these fans are likely to stall when the pressure surges in the system such as at the Meikle mine. Some of the fans are installed on the same vent raise system on different levels of the mine. Allowance for pressure handling is necessary since the fans compete to pull airflow from the same vent raise system. It is common at the Meikle mine to have a vent raise that passes through many levels on each footwall drift with a fan to either pull air flow on that level or exhaust it into the raise. Sometimes there are vent raises on both ends of the level where booster fans are operating.

In Table 2 the operating efficiency of the main surface fans are presented. Looking at the efficiency data for the main fans it is obvious that they operate in higher efficiency range than underground booster fans. For the Meikle main zone centrifugal surface 1750 hp fans the efficiency is higher than 70%. The Rodeo main surface variable speed 1500 hp fans are operating at efficiency higher than 80% at that speed. The Rodeo variable speed fans are at the moment being operated at 1000 RPM. The maximum possible speed is 1200 RPM. The Rodeo surface 250 hp fans that are used to overcome the pressure drop across the bulk air cooler show low operating efficiency. These fans are operated at 80% of their maximum speed of 1200 RPM. Comparing surface fans and underground booster fans is difficult because surface fans were installed to satisfy all aerodynamic requirements to make them efficient. The smaller numerous booster fans in bulkheads are not always installed to satisfy aerodynamic efficiency but to satisfy performance for that given time. The large underground booster fans in the diameters of 72, 78 and 84 inches also show reasonable operating efficiency higher than 50%. The two large 84 inch 500 hp booster fans installed at the 1450 level to supply airflow to half of Rodeo zone and Lower Griffin were installed with large amount of space to give them better performance. These fans operate side by side in parallel and move about 263 m^3/s (560,000 cfm). They operate at 59% and 73% efficiency respectively.

6 RESPONSE OF VENTILATION SYSTEM TO DYNAMIC CHANGES

The Meikle ventilation system is a very dynamic one and every so often something is changing in the circuit. This means that the booster fans are also responding to the system changes. When major booster fans that supply a certain district are turned off, the airflow in some areas will reverse and go to the other areas. When Meikle main zone surface fans are running at a reduced capacity more airflow tend to be handled by the main fans at the Rodeo side. This also tends to affect the operation of some of the underground booster fans. The air quantities in certain districts is affected by the status of the surface fans and other activities in the system such as the status of major air doors and whether or not all fans that should be operating in the system are doing so. The operation of controllable pitch fans is very advantageous because the blade angle can be adjusted while the fan is running. This means that when volume requirements change momentarily the fan blade angle can be varied to achieve the desired quantities. Efficiency can be maintained over a large blade angle positions. The overall electrical energy savings can be very high. In the underground complex at Meikle four controllable pitch fans are in use and are very beneficial when it comes to supplying airflow where it is most needed. However because these fans are not being operated at high pressure the efficiency tends to very poor at very high volumes.

7 CONCLUDING REMARKS

A booster and main fan system for a large complex ventilation system has been described in terms of efficiency and performance. It is found that the ventilation system containing numerous number of booster fans can be very dynamic in terms of the pressure and volumes that fans experience from time to time whenever changes occur temporarily in the system. The efficiency of most the booster fans underground is found to be in the low range with only a few booster fans running at efficiencies higher that 60%. When the pressure that the booster fan has to overcome is low (less than 124 Pa or 0.5 inches) the efficiency is very low $<40\%$ in most cases. The higher the pressure and volume the fan handles the better the efficiency. Installation problems contribute significantly to poor efficiency and it is not always possible to follow strict and sound installation guidelines because of limited space that might be available to install the fan. The large main surface fans are found to have better efficiencies when compared to underground booster fans. This can be attributed to better installation set up on surface and the fact that these fans operate in the high volume and high pressure range where the efficiency is much higher.

REFERENCES

Anon, Meikle Mine Ventilation Reports 2000–2003
Anon, Fan Performance Curves from Various source

Mine Ventilation, Ganguli & Bandopadhyay (eds)
© 2004 Taylor & Francis Group, London, ISBN 90 5809 633 5

A comprehensive approach to improve ventilation for operating underground mines*

Q.X. Yun, Y.F. Chen & C.W. Liu
Xian University of Architecture and Technology, P. R. China

M.J. Lian, Z.F. Zhang & X.B. Wang
Hanxing administration bureau of Metallurgical mines, P. R. China

ABSTRACT: Based on a three-year project in Xishemen mine, China, this paper describes a comprehensive approach to improve ventilation for operating underground mines. Firstly, a detailed investigation is necessary to reveal all the problems in ventilation. Then, some modification is needed to reform the ventilation system, particularly for the modification of opening network. Thirdly, ventilation facilities such fans and walls should be supplemented to increase ventilation capacity. Finally, special software is useful to carry out the complex calculation for ventilation network.

1 INTRODUCTION

Ventilation is vital for the safety and environmental condition in underground mines. By means of ventilation the quality of air can be improved and the harmful contents can be eliminated in underground working faces. Particularly, it is essential for the safety in underground coalmine because underground gas has to be attenuated and exhausted by ventilation. Therefore, ventilation has been a permanent subject for mining engineering.

However, the ventilation capacity usually decreases with the progress of underground mining, which results from the extension of openings, air leakage or plug of ventilation system as well as the wear and tear of ventilators. Therefore, it is necessary to improve the quality of ventilation during the later stage of operating mine.

This paper provides a comprehensive approach to enhance the ventilation, which is based on the experience of a three-year project in Xishimen underground mine, China.

Xishimen underground mine is one of the largest underground mine in China. It has a mining history of more than 30 years. In general speaking, its extension of ore body is almost from north to south. There are two main cage shafts at the middle of the mine as the main entrance for fresh air. In addition, a skip shaft nearby works as main path for ore transportation. At the same time a set of ventilation inclined shafts are allocated at different sections to exhaust contaminated air, which forms three sections for mining and ventilation.

The three-year project focuses its attention to south section, where an exhausted inclined shaft is allocated at the middle of the section as the main exit for contaminated air. In the section fresh air comes both from two main cage shafts at the north of the section and an additional ventilation inclined shaft at the south of the section, which constitutes a boundary ventilation system. Usually, there are three working levels at least in the section. The mining method is sublevel caving method.

2 INVESTIGATION OF PROBLEMS

In order to reveal all issues in ventilation for the mine, it is essential to implement a full-scale investigation to the whole ventilation network, including the measurement of air pressure, air volume, ventilation resistance, ventilation leakage ratio between effective ventilation volume and total air input.

2.1 *Measurement of ventilation volume*

There are more than 60 sites for the measurement of ventilation air volume. They are allocated at the main

* Supported by Natural Science Foundation of Shaanxi, China.

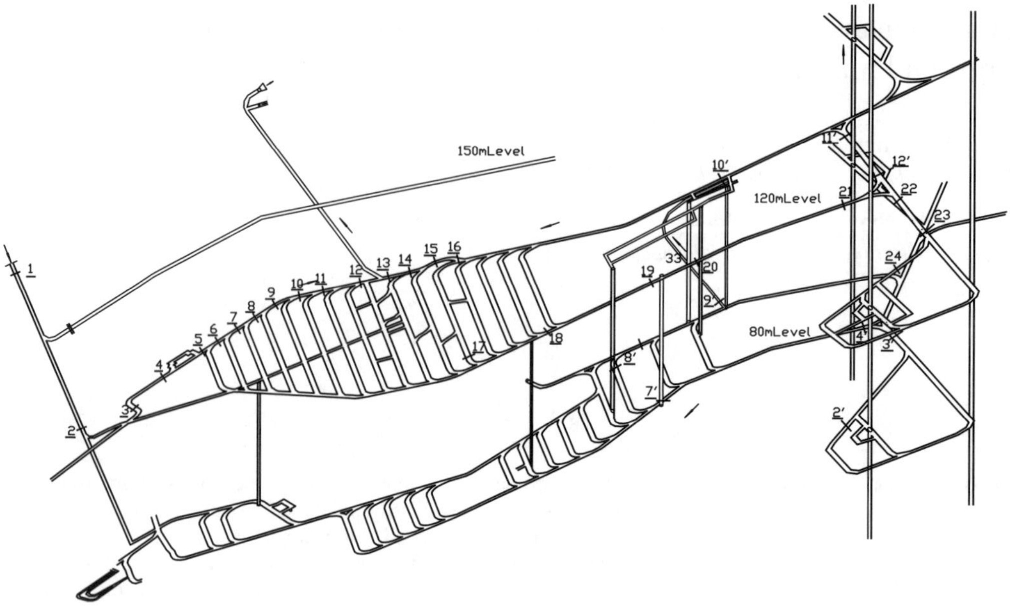

Figure 1. The location of sites for measurement.

entrances, main exits, different entrances and exits for different levels as well as different working faces for development and mining operations. Figure 1 shows the location of sites for measurement.

The leakage ratio and the efficiency of air usage for the mine can be calculated as:

$$L_o = \frac{Q_o}{Q_m}, \quad \eta = \frac{Q_e}{Q_m}, \quad L_o = \frac{Q_o}{Q_m}$$

where: L_o – leakage ratio of air volume outside the mine, %; Q_o – leakage of air volume outside the surface, m^3/s; Q_m – total air volume supplied by ventilator, m^3/s; η – effective ratio of air volume for the whole mine; Q_e – effective air volume, it can be regards as the total fresh air entering to all levels, m^3/s; L_i – leakage ratio of air volume inside the mine, %; Q_i – leakage of air volume in underground, m^3/s.

According to the measurement and calculation, leakage ratio of air volume outside the mine L_o is 17.2%, leakage ratio of air volume inside the mine L_i is 72.8%, effective ratio of air volume for the whole mine η is 10.2%. The air velocity in working openings is only 0.1–0.18 m/s. Therefore, in Xishimen mine the leakage of air volume is serious and the air condition in working face is poor. The reason to cause the poor ventilation condition results mainly from a lot of small mining operations by illegal invaders, which destroy many openings and ruin the ventilation system.

Table 1. Measurement result for air resistance.

Items	Pressure, mm H_2O	% in the total resistance
Resistance in entering part	3.67	3.98
Resistance in working part	10.2	11.26
Resistance in exiting part	78.41	84.76
Resistance in the whole mine	92.5	

2.2 *Measurement of air resistance*

In order to measure air resistance four paths are allocated in the mine, including the paths from the entrance for fresh air, to the exit for contaminated air as well as the paths that go through different kinds of working faces. Table 1 illustrates the measurement result for air resistance.

It can be seen from table 1 that the distribution of air resistance is very unacceptable: air resistance in working part is only 11.26%, which is far less than 30% as required; meanwhile the resistance in exiting part is 84.76% which is much higher than the normal percentage 40%. In addition, air resistance in entering part is 3.98%, which implies a lot of air coming from unknown void caves or fissures.

2.3 *Measurement of ventilator*

In order to investigate ventilator, several investigations are carried out for ventilation fan, electrical motor and air tunnel. Table 2 shows its result.

Table 2. Measurement result for ventilator.

Statistic pressure of the fan	92.5 mm H_2O
Dynamic pressure of the fan	2.8 mm H_2O
Total pressure of the mine	95.3 mm H_2O
Total air volume of the mine	126.3 m^3/s
Statistic power of the fan	114.5 KW
Statistic efficiency of the fan	38.9%

Table 3. Example of 3-point crossover.

Crossover point	1 2 3
Parent 1:	1 1 0 0 1 0 0 0 1 1 1 0 1 1 1 0
Parent 2:	1 0 1 1 0 0 1 0 0 0 1 1 1 0 0 1
Offspring 1:	1 1 0 0 0 0 1 0 1 1 1 0 1 0 0 1
Offspring 2:	1 0 1 1 1 0 0 0 0 0 1 1 1 1 1 0

It is seen from table 2 that working situation of ventilation fan is unexpected; its efficiency only reaches 38.9%. Moreover, some cracks are discovered in leafs of the fan because of an accidence in history. It means that the fan is too old to work in the future.

In general speaking, ventilation situation of the mine is serious: leakage of air volume is high, distribution of air resistance is unexpected, and working situation of ventilator is out of order. Therefore, it is necessary to reform the ventilation network and facility by additional technical means.

3 OPTIMIZATION OF VENTILATION SYSTEM

The original ventilation system might be perfect from the point of planning in mine during the early stage, however some modification to the original system is necessary because of the impact from mining activities. Here, genetic algorithm is used to optimize ventilation network. The basic principle of the new method in this paper is to select the optimum ventilation system from a variety of ventilation openings such as shaft, inclined shaft, drift or adit with the help of genetic operations (Haup & Haup, 1998).

3.1 *Establish ventilation alternatives by genetic operations*

Genetic algorithm is a new kind of optimization technique by self-adaptive search. Based on a set of feasible solutions, it can approximate to the optimal solution one iteration after another by means of genetic operations. Here, genetic algorithm is employed to generate ventilation alternatives.

(1) Coding. In genetic algorithms, binary strings are used to represent problem. In this paper a set of binary strings are used to represent different ventilation alternatives. The length of string is equal to the total number of openings proposed. Each digit in string represents a given opening. If the digit is equal to 1, it means that the opening will exist. If it is 0, the opening will be deleted. For example, string [1100101011] represents a ventilation alternative which is selected from 10 potential openings, the alternative is constituted by openings #1, #2, #5, #7, #9 and #10.

(2) Generate initial population. In genetic algorithm a group of individuals constitute a population. To generate the initial population, random generation is the common method to be used.

(3) Reproduction. Reproduction is the main operation to carry out Darwinian Principle – the survival of the fittest. *Here*, fitness-proportion selection is used for reproduction, which ensure the quality of next generation is better than the previous one, i.e.

$$p(i) = f_i / \sum_{i=1}^{m} f_i$$

where: $p(i)$ = probability to be selected of individual i; f_i = fitness of individual i; m = number of individuals in a population.

(4) Crossover. Crossover is a main tool to generate new individuals in genetic algorithms. Since the length of strings in this paper is long, 3-point crossover is used to improve the variety of individuals (Michalewicz, 1996). Table 3 illustrates the principle of 3-point crossover.

(5) Mutation. Mutation is another way to generate new individuals in genetic algorithms. By means of mutation, a digit in a selected string will be changed from 1 to 0 or from 0 to 1.

In general, the binary strings, which represent different ventilation alternatives, will represent a lot of combination of openings by means of previous genetic operations.

3.2 *Verify the ventilation alternatives by graphic theory*

From the points of mining engineering, all the feasible ventilation alternatives should meet the following requirements:

(1) All openings in a ventilation alternative are connected with each other without separate one. In terms of graphic theory it means that the graph should be a connective one.

(2) There should be two accesses to reach surface at least for any ventilation alternative. On other words, there should be one entrance and one exit at least for any alternative.

(3) Except of entrance and exit, there should not be any terminal end for opening in an alternative for

better ventilation. In terms of graphic theory it means that the degree of node should be larger than 1.

To verify the ventilation alternatives generated by genetic algorithm, following techniques are employed.

(1) Matrix representation. Usually, incidence matrix is the common one to represent network as follows:

$$A = \begin{bmatrix} a_{11} & \cdots & a_{1j} & \cdots & a_{1m} \\ \cdots & & \cdots & & \cdots \\ a_{i1} & \cdots & a_{ij} & \cdots & a_{im} \\ \cdots & & \cdots & & \cdots \\ a_{n1} & \cdots & a_{nj} & \cdots & a_{nm} \end{bmatrix}$$

where: m = number of arcs; n = number of nodes;

$$a_{ij} = \begin{cases} +1 & \text{if arc j is incident to node i} \\ -1 & \text{if arc j is incident from node i} \\ 0 & \text{otherwise} \end{cases}$$

In order to reduce the memory of computer, incidence matrix A can be substituted by two index matrices D and E as follows:

$$D = \begin{bmatrix} d_{11} & \cdots & d_{1j} & \cdots & d_{1m} \\ d_{21} & \cdots & d_{2j} & \cdots & d_{2m} \end{bmatrix}$$

$$E = \begin{bmatrix} e_{11} & e_{12} & \cdots & e_{1P} \\ \cdots & \cdots & \cdots & \cdots \\ e_{i1} & e_{i2} & \cdots & e_{iP} \\ \cdots & \cdots & \cdots & \cdots \\ e_{n1} & e_{n2} & \cdots & e_{nP} \end{bmatrix}$$

where: d_{1j} = source node of arc j; d_{2j} = sink node of arc j; e_{ij} = arc j which is incident with node i; p = the maximum degree of nodes.

If matrix A is substituted by matrices D and E, the larger the m is, the less the memory will be.

(2) The depth-first search. The depth-first search is used to check all the nodes in a graph if they are connected. To implement the search, table L and stack S are established. At the beginning, source node of network is the first one to be accessed. After source node is pushed into table L, all its conjoint nodes will be pushed into stock S. Then, a conjoint node V is pulled out from stock S and moved it to table L. All conjoint nodes with V will be found out in sequence and stock S only saves the new conjoint nodes which do not exist in table L and stock S. Repeat the procedure until stock S is empty. When stock S is empty, it means that the network is a connected one.

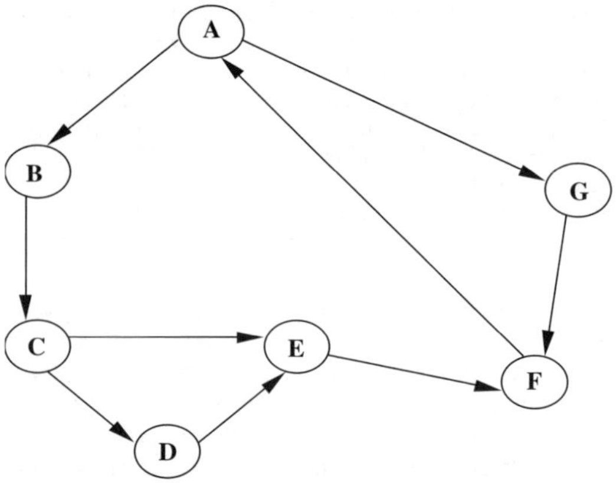

Figure 2. Depth-first search.

Figure 2 illustrates the depth-first search. Suppose that node A is the first one to push in table L, then table L and stock S become as follows:

L = [A]
S = [F, B, G]

After G is pulled out from stock S and pushed into table L, the list of L and S turn out to be:

L = [A, G]
S = [F, B]

Since the conjoint node F for G has already in stock S, B is the next nodes to be pulled out from stock S and moved into table L. The conjoint node of B is node C. Since node C is not in the list of L or S, it will be pushed into stock S. The new lists of L and S are as follows:

L = [A, G, B]
S = [F, C]

The 4-th iteration will change the lists of L and S as:

L = [A, G, B, C]
S = [F, D, E]

The 5-th iteration will result in:

L = [A, G, B, C, E]
S = [F, D]

The 6-th iteration will make the lists as:

L = [A, G, B, C,E, D]
S = [F]

Finally the list of stock S becomes empty after node F is moved to table L, i.e.:

L = [A, G, B, C, E, D, F]

Therefore, the network is proved to be a connected one (Yun & Huang, 2000).

(3) Degree of node. This technique is used to check if there is terminal end for openings in a ventilation system. According to the definition of graphic theory, degree of a node is the total number of arcs incident from and into the node. Obviously, if the degree of node is larger than 1, there is no terminal end of opening in the node; otherwise, there is a terminal end.

For example in Figure 2, degree of node A is equal to 3 while degree of node B is 2 and so on.

(4) Module. In ventilation system there are some openings, which should exist no matter what alternative it is, such as entrance or exit or the drifts to reach ore-body. In order to simplify this verification, module is used. Module here means a given string where some digits are fixed to 1, which imply that these openings must exist. For example, if module T is [#1##1###], where # could be 1 or 0, and a string of ventilation alternative N is [10011101], it means that #2 opening is absent according to the requirement of module T. Therefore, #2 opening should be added and alternative N will be modified as [110111101].

3.3 *Fitness*

In genetic algorithms, fitness is a measurement to evaluate the quality of individuals. In this paper annual cost C is used as the fitness to evaluate ventilation alternatives, i.e.

$$C = D + P \cdot \frac{(1+i)^n \cdot i}{(1+i)^n - 1}$$

where: D = annual operation cost: P = construction investment; i = annual interest rate; n = life of mine.

(1) Construction investment. Construction investment includes opening construction cost and equipment purchase. In order to calculate opening construction cost, openings are divided into three categories: vertical, inclined and horizontal openings with different specific cost. Then, the total construction cost is the sum of the product of opening length and its specific cost.

(2) Operation cost. Operation cost here includes transportation cost and ventilation cost. Transportation cost is the product of transportation length and specific cost. The latter is also divided into three categories: vertical, inclined and horizontal transportation.

For ventilation there is special software to deal with in this paper. Ventilation cost V can be calculated by air volume and air pressure as:

$$V = e \cdot H \cdot Q \cdot \zeta \cdot T$$

where: e = specific cost of electricity; H = total air pressure; Q = total air volume; ζ = efficiency of ventilation machine; T = working time.

Air volume and air pressure are calculated by iterations as follows:

$$\sum_{i=1}^{n-1} \sum_{j=1}^{m} a_{kj} a_{ij} c_j p_i = q_k - \sum_{j=1}^{m} a_{kj} c_j (H_{fj} + H_{ej}) \quad (k=1,2,\cdots, n-1)$$

where: a_{ij} = element of row i and column j in incidence matrix; p_i = total air pressure of node i; q_k = total air volume of node k; H_{fj} = mechanical air pressure of arc j; H_{ej} = difference between static pressure of arc j; R_j = resistance coefficient of ventilation for arc j; Q_j = air volume of arc j; $c_j = 1/(R_j|Q_j|)$ ($j = 1, 2, \ldots, m$).

3.4 *Termination*

Genetic algorithm is an algorithm by iteration; there should be some criterion to terminate the iteration.

In this paper two criterions are set:

(1) Once the iteration times reach given maximum times, usually it is 200, the algorithm will stop.
(2) When the difference between the minimum fitness and average fitness is less than 3–5%, the algorithm will stop.

According to the result of genetic algorithm, some ventilation raises with diameter $4 \times 4\,m^2$ is added as the pass for contaminated air. An additional inclined shaft under 80 m level is also constructed as a new path for the entering of fresh air.

4 SELECTION AND ALLOCATION OF VENTILATION FACILITIES

4.1 *Selection of ventilation fan*

As time rolls on, it is also necessary to supplement some ventilation facilities in order to meet the new requirements for better ventilation. Main ventilation fan is the focus among all facilities. In Xishimen mine, there are 7 alternatives of ventilation fan can be selected, their parameters and characteristics are shown in table 4.

Finally, alternative 7 is selected as the final decision, in which two main fans with axial-flow are established at 80 m level in underground as exhausted ventilators. The advantage of underground fan is:

(1) Main fan can be established in the section where air leakage is less than others.
(2) Main fan is close to working area so that the efficiency of air usage is higher.
(3) Many openings can be used simultaneously for air entrance and air exit so that the air resistance is less.

Table 4. Alternatives of ventilation fan.

	1	2	3	4	5	6	7
Alternative	2K56 No.30, 600 r/min	2K56 No.24, 750 r/min	62A14-11No.24, 750 r/min	62A14-11No.24, 1000 r/min	DK62-10-No.30, 590 r/min	K45-6-No.20 × 2,980 r/min	K45-6-No.19 × 2,980 r/min
Required air volume, m^3/s	135	135	135	135	135	135	135
Required air pressure, mmH_2O	200	200	200	200	200	185	185
Working air volume, m^3/s	140	133	142	135	150	75 × 2	80
Working air pressure, mmH_2O	200	200	200	200	200	184	182
Efficiency, %	83	73	76.5	75	81	77	76
Power of motor, Kw	400	475	460	460	160 × 2	250 × 2	200 × 2
Equipment cost, million Yuan	1.086	0.806	0.719	0.357	0.882	0.42	0.362
Construction cost, million Yuan	0.04	0.04	0.03		0.086	0.03	0.03
Total cost	1.126	0.846	0.749	0.357	0.968	0.45	0.392
Electric cost, million Yuan/year	0.825	0.917	0.877	0.884	0.829	0.949	0.819
Year of capital return	4.03	4.62	3.18	1.21	3.45	2.59	1.03

(4) Distribution of air pressure is smoother, part of pressure is positive while the other is negative.
(5) Alternative 7 is also the best one from the point of economics.

4.2 *Assembly of ventilation fans*

As to the assembly of underground fans, there are two alternatives: one is established at the total exit for contamination air with serious connection of fans, the other is established at two exits of ventilation raises for contamination air with parallel connection of fans in distance.

The latter is selected since it has following advantages:

(1) Main fan is easy to fixed in the opening since the diameter of the fan is similar to the size of opening.
(2) There is better ventilation condition since leakage part is located behind the fan while solid part is ahead the fan.
(3) It is easier to adjust ventilation by means of turn on/off one of the fan.

4.3 *Ventilation structures*

Because of the serious air leakage in mine, it is necessary to construct a set of ventilation structures to improve ventilation in underground.

Ventilation wall is used to separate the waste openings in order to reduce air leakage. At the entrance of all waste openings concrete wall is set to plug opening. More than 12 walls are set in Xishimen mine.

Dredging is another measure to improve ventilation. For the plugged openings due to rock fall, dredging is used to restore the original size of openings. Some openings are even enlarged in section in order to reduce air resistance. There are more than 130 m^3 rock is removed in Xishimen mine.

Ventilation gate is used to separate the openings for ventilation while it is also used for transportation. Four gates are set in 80 level and 40 level separately.

5 DEVELOPMENT OF VENTILATION SOFTWARE

Since it is difficult and time-consuming for the calculation in ventilation network, special software is developed to meet the frequent modification of ventilation in Xishimen mine. Its function includes drawing of ventilation network, calculation of ventilation pressure, volume and resistance. Figure 3 shows the structure of software for ventilation.

The characteristics of software are:

(1) Complete functions. The functions of software include data treatment, iteration for network solution, adjustment of air distribution, calculation of air volume and resistance as well as information search, delete/add and help function. In addition,

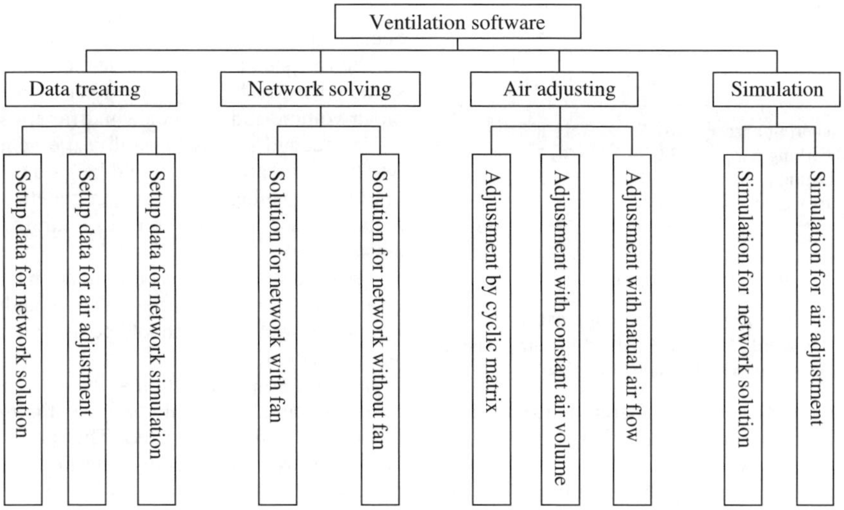

Figure 3. The structure of software for ventilation.

there are different algorithms for the calculation of ventilation resistance and air volume as well as air adjustment. User can chose any one as needed.
(2) Visualization. By means of interface, any complex ventilation network can be vividly represented on screen and directly perceived through senses.
(3) Combined technology. Based on the principles of object-oriented design, software employs many computer techniques such as database, applied mathematics and computer drawing.
(4) Friendly interface. Software is developed in accordance with the traditional procedure in ventilation management, therefore, it is easy to learn and to use.

The most important characteristic in software is the combination of data and figure together. All calculation can be carried out by means of triggering the figure of network on screen, including node or arc. Its operations include:

(1) Whole ventilation network is represented on screen. User can add or delete any opening for ventilation network directly on screen. By means of function 'zoom', one can check any part of network in detail.
(2) At the end of each calculation, ventilation network on screen represents the direction of airflow by different color on openings.
(3) After trigger of any opening on screen, one can obtain all its parameters and information.
(4) Ventilation fan can be added or deleted by user and displayed on screen, related new calculation can be carried out and its impact is illustrated on screen.

Table 5. Comparison in air volume and velocity.

Main fan	Rate of useful air volume, %	Air velocity in drilling drift, m/s	Air velocity in drawing drift, m/s
62A14-11-No.24 (Surface)	10.2	0.15–0.18	0.08
K45-6-No.19 (underground)	60.9–61.9	0.54	0.93

6 IMPROVEMENT OF VENTILATION REFORM

After the ventilation reform in Xishimen mine, including the modification of ventilation system and replacement of main fans as described in previous parts, its ventilation condition is improved distinctly.

6.1 *Improvement on air volume*

Table 5 shows the comparison in air volume and velocity by ventilation reform. It can be seen from the table that the improvements of reform are as follows:

(1) Leakage rate of air volume is decreased from 89.8% to 38.1–39.1%, it meets the requirement of safety regulation in China.
(2) Air velocity is increase almost 6 times in working area, it improve the air quality in underground.
(3) The decision is correct to move main fan from surface to underground.

Table 6. Measurement result on air resistance.

Items	Pressure, mm H_2O	% in the total resistance
Resistance in entering part	10.7	11.8
Resistance in working part	33.3	36.7
Resistance in exiting part	46.8	51.5
Resistance in the whole mine	90.8	

6.2 *Improvement on air resistance*

Table 6 illustrates the measurement result on air resistance after ventilation reform. In comparison with Table 2, it is seen the improvement as:

(1) Resistance in working part increases from original 11.3% to 36.7%, it means that a lot of fresh air goes through working area.
(2) Resistance in entering part increase from original 4% to 11.8%, it means that fresh air mostly comes from cage shaft and inclined shaft instead of void caves with dust.
(3) Resistance in exiting part decreases from original 84.7% to 51.5%, it means that contamination air escapes more fluently.

In general, ventilation reform in Xishimen mine obtains significant improvement in ventilation.

7 CONCLUSION

Ventilation is important to the underground mine due to its serious impact. As time goes on, ventilation in an operating mine should be reformed to improve its network and facilities. This paper presents a comprehensive approach to carry out such reform.

The first step is to investigate the whole ventilation system attentively, including a set of measurements on air volume and resistance in different sites.

The second step is to modify the ventilation network. Special attention should be paid to the leakage or plug in openings. In this paper genetic algorithm is recommended as a new tool to optimize ventilation network.

The third step is to replace ventilation fan if it is necessary. Even it is not some ventilation facilities such as ventilation wall or gate should be set to improve ventilation situation.

In addition, a ventilation software is useful in the reform since the calculations in ventilation are time-consuming and complicated. This paper introduces new software with friendly interface and manifold functions.

The application in Xishimen mine proves that the previous recommendation is reasonable and useful.

REFERENCES

Haupt, R.L. & Haupt, S.E. 1998. Practical genetic algorithms. New York: John Wiley & Sons, Inc.

Michalewicz, Z. 1996. Genetic algorithms + data structure = evolution programs. New York: Springer-Verlag, Inc.

Yun, Q.X. & Huang, K.M. 2000. Determination of ventilation system for mines by genetic algorithms. Proceedings of the 6-th international symposium on environment issues and waste management in energy and mineral production, Canada.

Mine Ventilation, Ganguli & Bandopadhyay (eds)
© 2004 Taylor & Francis Group, London, ISBN 90 5809 633 5

Ventilation systems for underground trona mines

C. Pritchard
FMC-Westvaco

B. Glackin
OCI Wyoming, L.P.

L. Patton
General Chemical (Soda Ash) Partners

L. Refsdal
Solvay Chemicals, Inc.

R.T. Leigh
OCI Wyoming, L.P.

ABSTRACT: This paper presents an overview of the mine ventilation systems utilized by each of the four underground mines comprising the Wyoming trona basin located near Green River, Wyoming. Each mine is owned and operated by separate companies: FMC-Westvaco, General Chemical (Soda Ash) Partners, OCI Wyoming and Solvay Chemicals, Inc. Total mine airflows, number of shafts and fans, fan quantities and operating pressures, number of operating production units, methane emission rates and other mine-wide ventilation data will be reviewed for each operation. In addition, face ventilation methods for both continuous miner and longwall sections utilized at the various operations will be discussed.

1 INTRODUCTION

1.1 *Wyoming trona*

Wyoming is often referred to as the "Soda Ash Capital of the World" because of the enormous world-class trona reserves found in the Green River Basin. Trona is a sodium sesquicarbonate ($Na_2CO_3 \cdot NaHCO_3 \cdot H_2O$) mineral. Trona was precipitated in a lacustrine environment during the restrictive stages of Lake Gosiute of the Wilkins Peaks Member of the Eocene Green River Formation, approximately 50 million years ago as shown in Figure 1. The bedded trona is composed of 70% sodium carbonate and is referred to as natural soda ash. The single major industrial product obtained from refining trona is soda ash. Four Wyoming trona companies produce over 90% of the United States' soda ash output. This equates to over 30% of the world's soda ash production.

The total resource of Wyoming trona deposits is estimated at 127 billion short tons, encompassing

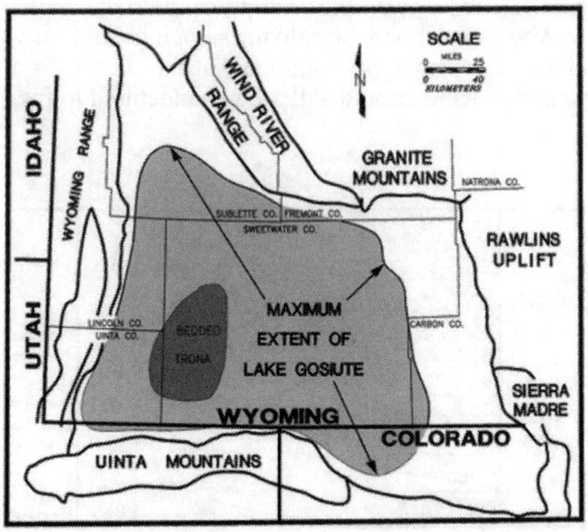

Figure 1. Lake Gosiute.

approximately 3,720 km^2 (1,300 mi^2). Of this, 40 billion tons of economical trona reserves are mineable by mechanized extraction including drum type miner and boring machines, and hydraulically supported longwall shearing.

1.2 *Geologic setting*

Twenty-five trona-halite beds, exceeding 0.9 m (3 ft) in thickness and 259 km^2 (100 mi^2) in areal extent, have been recognized along with 17 minor beds. Identified in ascending order, these thicker trona beds are numbered 1 through 25, from the oldest (stratigraphically lowest) to the youngest (stratigraphically highest). Four companies currently produce soda ash from trona beds. Specifically, FMC – Westvaco, General Chemical (Soda Ash) Partners, and Solvay Chemicals, Inc. all mine Bed 17. OCI Wyoming, L.P. operates in shallower beds 24 and 25. See Figure 2 for a cross-section through the trona beds in the Green River Basin.

1.3 *General ventilation requirements*

Trona mines are metal/nonmetal mines, classified gassy (Class III) by Mine Safety and Health Administration (MSHA) under 30 CFR Part 57 of the mining regulations due to their non-combustible composition and their liberating a potentially explosive concentration of methane. Ventilation systems are based on typical room and pillar mining but have some significant differences; such as, the belt conveyor galleries are not isolated and are utilized as intake airways. Intake and return airways are to be separated. Permissible mining and haulage equipment is required inby and in face areas. Comprehensive pre-shift and on-shift examinations for methane and ventilation airflow quantities are also required. The "150 ft" outby rule for permissible longwall face equipment does not apply.

Wood, metal and Styrofoam stoppings and polyurethane sealants are commonly used and perform well in moving ground. All are manufactured to high

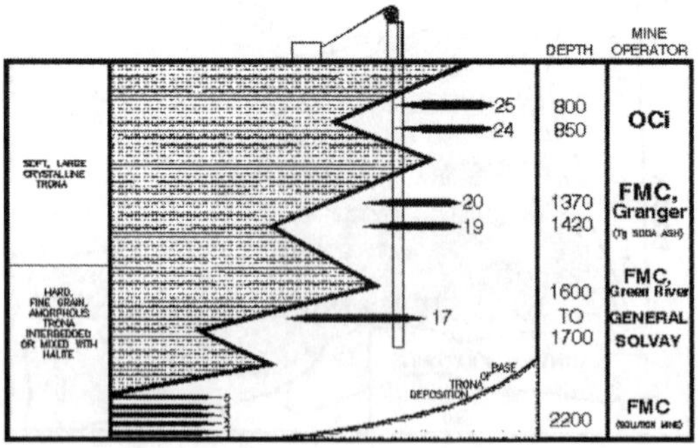

Figure 2. Green River Basin cross-section.

flame retardant standards. Booster fans are allowed with restrictions but district recirculation is not.

Methane sources are from shales located above and below the trona beds. MSHA mandates spot inspections with increasing methane liberation from every 15, 10 and 5 days to assure adequate ventilation of faces and bleeders.

2 FMC – WESTVACO

2.1 *General description of mine ventilation system*

FMC began excavating trona in 1947 and has mined over 136 Mt (150 Mst) since. Present annual mining capacity is 4.1 Mt (4.5 Mst) with approximately 3.2 Mt (3.5 Mst) derived from longwall mining with the rest from borer miner development. The longwall face is 230 m (750 ft) wide and panels approach 2,700 m (9,000 ft) in length. Development haulage is electric shuttle cars. Support equipment is primarily rubber-tired diesel.

Total mine airflow averages 670 m^3/s (1.4 Mcfm) with methane emissions of 93,500 m^3/d (3.3 Mft^3/d). At full longwall operation methane emissions can exceed 142,000 m^3/d (5 Mft^3/d). Diluted return air methane is approximately 0.2%.

Main fan data is included in Table 1 and mine shaft data in Table 2. Fans are operated in blowing mode, as

Table 1. FMC – Main fan data.

Location	5 Shaft	7 Shaft	8 Shaft
Manufacturer	Jeffrey	Jeffrey	Jeffrey
Type	2-stage	2-stage	2-stage
Model	8HU-117	8HU-117	8HU-117
Size m (in)	3.0 (117)	3.0 (117)	3.0 (117)
RPM	710	710	710
Motor size kW (BHP)	1120 (1500)	1120 (1500)	1120 (1500)
SP* Kpa (in-wg)	2.1 (8.5)	1.2 (4.9)	2.1 (8.4)

* Static pressure at the operating point.

Table 2. FMC – Shaft data.

	Size			Airflow	
Location	m	ft	Type	m^3/s	kcfm
1 Shaft	3.7	12	Plugged		
2 Shaft	4.3	14	Return/Hoist	35	75
3 Shaft	5.5	18	Return	71	150
4 Shaft	7.3	24	Return/Hoist	222	450
5 Shaft	5.5	18	Intake	217	460
6 Shaft	6.7	22	Return	316	670
7 Shaft	6.7	22	Intake	191	405
8 Shaft	6.7	22	Intake	175	370
9 Shaft	5.5	18	Exhaust (2004)		

regulations requiring exhaust ventilation were not in force when the mine operation commenced in the late 40's. Direct-fired natural gas burners heat intake air. Mined out areas comprising approximately 50% of the total mine have been sealed and contribute minimal methane. Mine development entries are mined utilizing borer continuous miners that open areas of 9 to 12 m^2 (100 to 130 ft^2).

2.2 *Section ventilation*

Face ventilation in mains and longwall development can be seen in Figures 3 and 4. Main and sub-main development is from 4 to 8 entries depending on airflow demand.

Maximum main entry airflow is 47 m^3/s (100 kcfm) per entry or a rule of thumb of one intake entry for each mining panel. Face ventilation is by blowing fans and entry development crosscuts are from 30 to 60 m (100 to 200 ft) centers on advance (see Table 3).

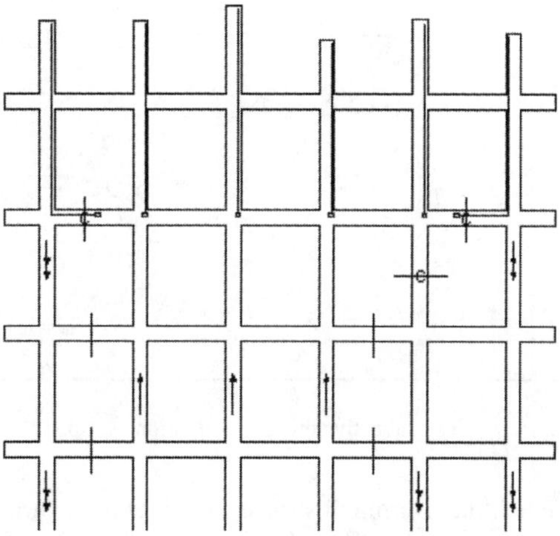

Figure 3. Typical mains development.

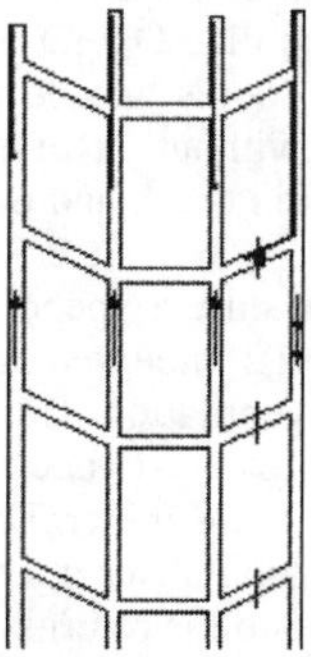

Figure 4. Typical longwall development ventilation.

Longwall face ventilation (Figure 5) is unique in that the belt entry is in a middle entry, with a "Zero" room located in the longwall block for maintenance access. Airflow quantities of 57 to 95 m^3/s (120 to 200 kcfm) are used to ventilate the face and gob areas.

2.3 *Challenges*

Ventilation challenges at FMC are effectively ventilating faces with long development crosscuts, assuring less than 1.0% methane at bleeder mixing points and designing ventilation systems to keep development and longwall return air separate. Moderate floor heave problems require diligence maintaining stopping integrity, which is minimized by utilizing Styrofoam structures.

3 OCI WYOMING

3.1 *General description of mine ventilation system*

OCI began mining trona at the Big Island Mine in 1962 and has mined 112 Mt since. Present annual mining

Table 3. FMC – Auxiliary fan data.

		Motor	
Manufacturer	Motor	kW	HP
Joy	21¼-14-3450	11	15
SMJ	AFSP Series 2	22	30
Spendrup	AMF 160-12-11	11	15
Spendrup	M60X2-30-3600	22	30
Spendrup	AMF 160-14-11-22	22	30

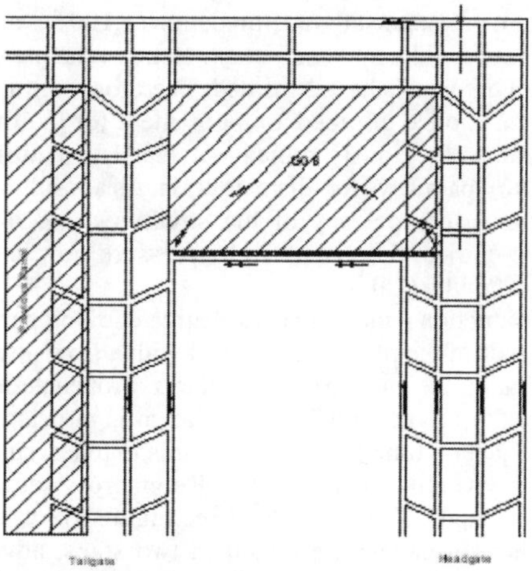

Figure 5. Typical longwall retreat.

Table 4. OCI – Main fan data.

Location	West	East
Manufacturer	Jeffrey	Jeffrey
Type	8HU-117	8HU-117
Size m (in)	3.0 (117)	3.0 (117)
RPM	894	691
Motor size kW (BHP)	224 (300)	224 (300)
SP* Kpa (in-wg)	1.4 (5.6)	1.4 (5.6)

* Static pressure at the operating point.

Table 5. OCI – Shaft data.

	Size			Airflow	
Location	m	ft	Type	m^3/s	kcfm
1 Shaft	6.1	20	Intake	132	280
2 Shaft	6.1	20	Return/Hoist	264	560
3 Shaft	6.1	20	Intake/Hoist	132	280

capacity is 3.5 Mt (3.9 Mst) with all production derived from continuous drum miner panels. The panels average 290 m (950 ft) wide and range from 609 to 2,450 m (2,000 to 8,000 ft) long.

Total mine airflow averages 264 m^3/s (560 kcfm) with methane emissions of 9,910 m^3/d (0.4 Mft^3/d) at approximately 0.05% at the return shaft. All air is exhausted from the mine thru the No. 2 Shaft. Two Jeffrey fans operate in parallel. See Tables 4 and 5.

3.2 *Section ventilation*

Since 1998, Joy continuous drum miners provide all tonnage mined. Due to the significantly lower overburden pressures at <335 m depth (1,100 ft) mined entries are wider and competent. None of the previous mined out panels have subsided. Therefore ventilation structures once installed are permanent for the life of the mine. Previously mined out panels are utilized as return pathways for all ventilation. As a result of the long-term competency of the structures and of the return entries, the overall mine pressure drop is less than 0.25 kPa (1.0 in).

Production panels are a 30-degree chevron pattern. One half of the panel is advanced with a flow through ventilation design. The barrier pillar is "broken through" every 5th to 7th room and ventilation is directed into the adjacent mined out panel. In this manner, supplementary ventilation is avoided. Room cross-sectional area averages 27 m^2 (290 ft^2) (See Figure 6).

Chevron panels are mined in two steps, advance and retreat. During the advance step, the right side of the panel is mined concurrent with the roadway and

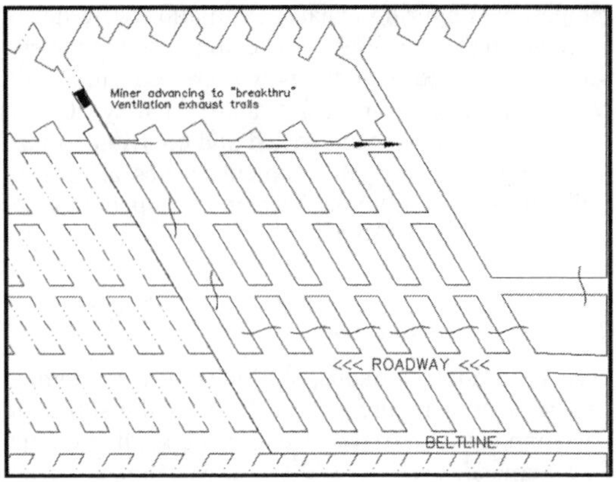

Figure 6. Flow through ventilation on advance.

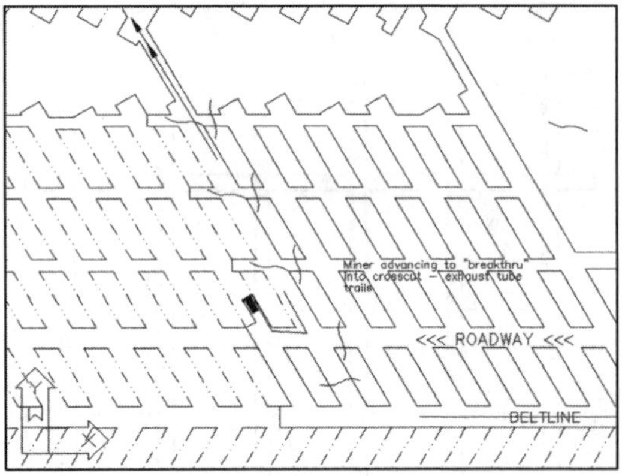

Figure 7. Typical active mining ventilation set-up.

the beltline. During this stage, (see Figure 7) curtain lines are maintained in the pillar set adjacent to the roadway and in the crosscuts up to the last open crosscut. This line confines the return air to a section of the mined out rooms adjacent to the roadway. This area is relieved via a breakthrough into the adjacent panel through the barrier pillar. Operational experience has determined that 7 rooms can be mined on advance with curtain lines without adverse impacts upon air entering the last open break and exiting the last open crosscut.

During a mining shift, a Spendrup permissible fan (Model 159-40-FPHE) mounted on top of each miner maintains the face ventilation. This fan provides up to 8 m^3/s (17 kcfm) of exhaust ventilation from the face immediately ahead of the miner. The fan exhausts into flexible tubing, which discharges through the last open crosscut and into the returns.

During the mining of a room, efforts are made to "break thru" into crosscuts mined forward from the

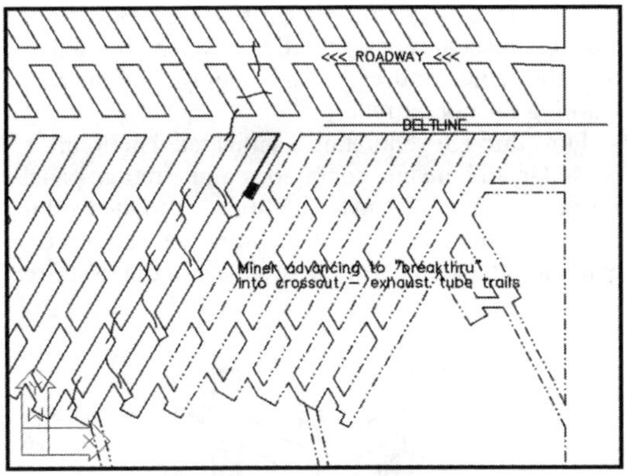

Figure 8. Typical retreating ventilation.

previous rooms. This facilitates ventilation and eliminates the need to install curtain lines into a mined out room. The crosscuts themselves are ventilated by means of a curtain line in the returns to direct ventilation across the face.

On retreat, (see Figure 8) the left side of the panel is mined out as utilities and the beltline are retreated. Once again, curtain lines are retreated as well in the previous rooms crosscuts. Curtains are also hung into the advancing crosscuts so that airflow across the faces is maintained.

Main development has not been required in recent years and therefore, prepared methodologies are not discussed.

3.3 *Challenges*

OCI continues to expand the lateral extent of the mine. Ventilation challenges include maintaining an overall pressure drop of less than 1.5 kPa (6.0 in wg.). This preserves the integrity of the curtain wall in the exhaust shaft. Traditionally, maintaining low mine resistances and improving section efficiencies has been the main avenue to achieve this goal. With the increasing mine extent, the resistances actually drop since more entries are available for air to migrate back to the shaft, however, the fresh air entries will begin to exert greater resistances as more air must be delivered to sections farther away from the intakes.

4 GENERAL CHEMICAL (SODA ASH) PARTNERS

4.1 *General description of mine ventilation system*

General Chemical (Soda Ash) Partners began mining trona in 1966, 488 m (1,600 ft) below the surface in Bed 17 and has mined 109 Mt (120 Mst) since.

Table 6. GCSAP – Main fan data.

Location	1 Shaft	4 Shaft	5 Shaft
Manufacturer	Joy	Joy	Joy
Type	Axivane	Axivane	Axivane
Model	H84-50D	H84-50D	H84-50D
Size m (in)	2.1 (84)	2.1 (84)	2.1 (84)
RPM	1175	1175	1175
Motor size kW (BHP)	313 (420)	313 (420)	313 (420)
SP* Kpa (in-wg)	1.97 (7.9)	1.69 (6.8)	1.84 (7.4)

* Static pressure at the operating point.

Table 7. GCSAP – Shaft data.

	Size			Airflow	
Location	m	ft	Type	m^3/s	kcfm
1 Shaft	3.7	12	Exhaust	125	265
2 Shaft	6.1	20	Intake	123	260
3 Shaft	6.1	20	Intake	163	345
4 Shaft	4.0	13	Exhaust	170	360
5 Shaft	4.7	15	Exhaust	326	690
6 Shaft	7.9	26	Intake	335	710

The mine works cover 11.3 km by 6.4 km (7 mi by 4 mi) or a total of 64.8 km^2 (25 mi^2). The room and pillar method used employs continuous miners (both bore and drum type). Present annual mining capacity is 4.3 Mt (4.7 Mst) with approximately 2.8 Mt (3.1 Mst) derived from bore miner production and the remainder derived from continuous miner production.

Total mine airflow averages 621 m^3/s (1,315 kcfm) with methane emissions ranging from 0 to 70,800 m^3/d (0 to 2,500 kft^3/d) with less than 0.1% recorded at the return shafts.

There are four main exhaust fans on the surface ventilating the mine works. All of the fans are identical, Joy direct-drive Axivane fans, details shown in Table 6.

There are six airshafts connecting the mine to the surface. Intake air is provided through 2, 3 and 6 Shafts and mine air is exhausted through 1, 4 and 5 Shafts. The 4 Shaft and 5 Shaft both have two fans set-up in parallel. Only one fan operates at 4 Shaft at a time. The two fans on 5 Shaft operate in parallel. Table 7 shows the shaft data.

Fresh air is heated either by steam coils or direct natural gas-fired heaters. Heating of mine air is required eight months of the year.

4.2 *Section ventilation*

Continuous miners develop the initial eight crosscuts for the bore miners as well as perform all mains and sub-main development work. Mains are developed with seven entries and sub-mains are typically developed with five entries, on 24.4 m (80 ft) by 30.5 m (100 ft) centers. Room sizes are approximately 5.5 m (18 ft) wide and 2.7 m (9 ft) high. Bore miners develop and pillar three-entry panels. The retreated bore miner panels are 204 m (670 ft) wide and 1,611 m (5,450 ft) in length. Room sizes for bore miner panels are ovular in cross-section with an axis of 4.3 m (14 ft) and a height of 2.4 m (8 ft).

Mine air efficiency is 40% of the total fresh air; therefore, a balance is achieved with the 264 m^3/s (560 kcfm) between the eight production machines. Airflow quantities of 16.5 m^3/s (35 kcfm) to 30.7 m^3/s (65 kcfm) are recorded for ventilating each panel based on mining activity. Foamed permanent stoppings, regulators, curtains, overcasts and airlock doors are the main device controls used. There are three types of face ventilation systems currently used for the various mining configurations: exhausting, blowing and flow through.

The exhausting system, as illustrated in Figures 9 and 10, is used in advancing continuous miner panels. One or two fans will be set just out-by the last open crosscut in the panel return airway. Rigid tubing is extended from the fan inlet to each mining face. The fan draws air, gases and dust from the face through the tubing and discharges it through the fan into the return airway.

The blowing system, as illustrated in Figure 11, is used in advancing and retreating bore miner panels. Fans are set just out-by the last open crosscut in fresh air and flexible tubing is extended from the discharge side of the fans to each face. The fans blow fresh air into the face sweeping the dust and gases from the face and discharging it into the last open crosscut and into the panel return(s) if advancing and in-by the working face on retreat.

Flow through ventilation, as illustrated in Figure 12, is used in all bore and continuous miner retreat panels. Fresh air enters the panel and is directed to the working face using line curtain in the continuous miner retreat panel and fans and flexible tubing in the bore retreat

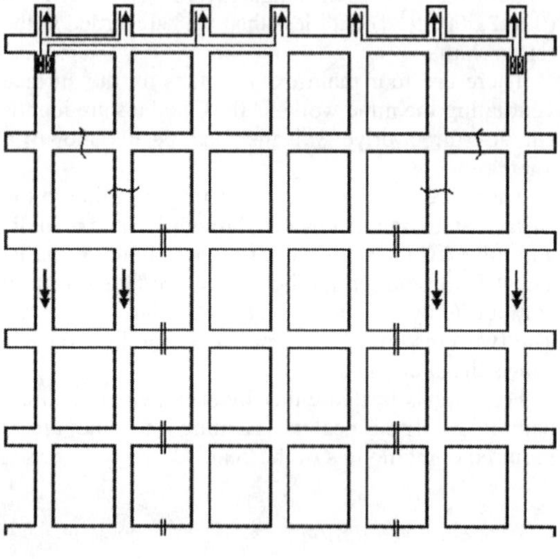

Figure 9. Typical 7-entry advancing continuous miner panel with a double return and four exhausting fans.

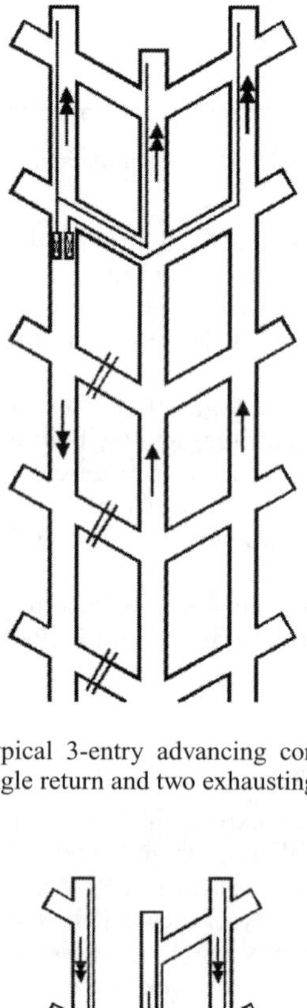

Figure 10. Typical 3-entry advancing continuous miner panel with a single return and two exhausting fans.

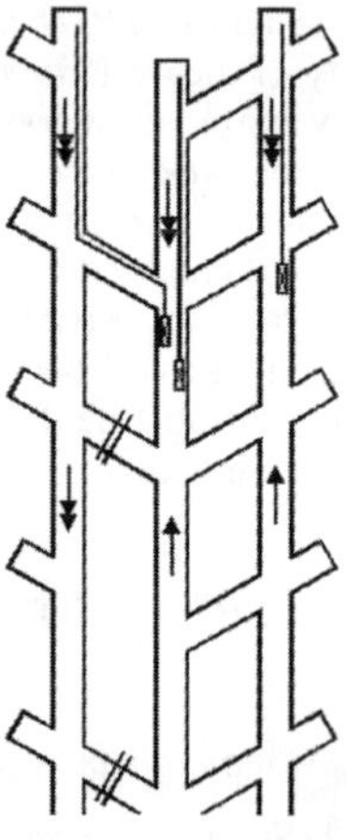

Figure 11. Typical 3-entry advancing bore miner panel with a single return and three blowing fans.

panels. The air passes over the working face and into the mined out area in-by to the end of the panel and is discharged directly to a bleeder at the back of the panel or to a return airway in the panel where the stopping line was left intact. The return air is brought back adjacent to the mining area and discharged through a regulator at the head of the panel directly to a main return.

Once air has entered the return airway, it is coursed directly to one of the return airshafts where it is discharged directly to the outside. It is not used to ventilate any other portion of the mine.

Table 8 lists the types of auxiliary fans used for both exhausting and blowing ventilation for the various mining configurations.

4.3 *Challenges*

The greatest ventilation challenges facing the General Chemical operation are shaft placement and maintaining the main returns for the next 40 years.

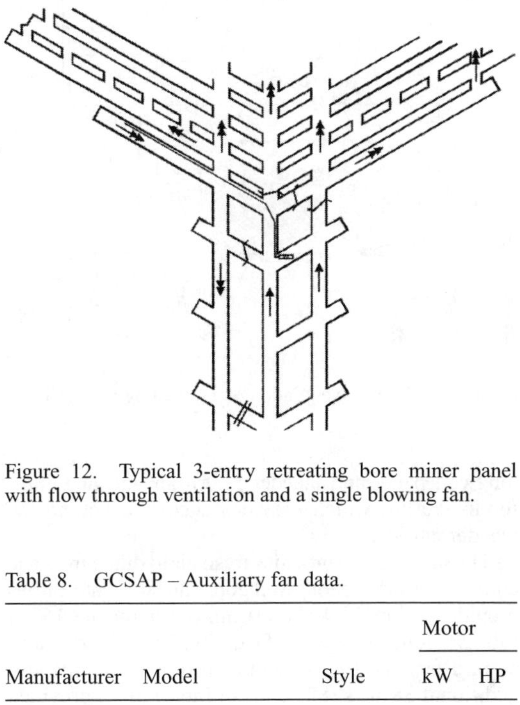

Figure 12. Typical 3-entry retreating bore miner panel with flow through ventilation and a single blowing fan.

Table 8. GCSAP – Auxiliary fan data.

			Motor	
Manufacturer	Model	Style	kW	HP
Joy 2 stage	23¼-17½-3450	Exhaust	42.8	57.5
Joy 2 Stage	28¼-17½-3450	Exhaust	37.3	50
Joy	28¼-17½-3450	Blowing	37.3	50
Joy	23¼-17½-3450	Blowing	18.7	25
SMJ	AFSP	Exhaust	37.3	50
SMJ	AFSP	Blowing	37.3	50
Spendrup	AMF-160-14-16	Blowing	22.4	30

5 SOLVAY CHEMICALS, INC.

5.1 *General description of mine ventilation system*

Solvay began mining trona (Bed 17) in 1981 and has mined 49 Mt (54 Mst) to date. Present annual mining capacity is 4.1 Mt (4.5 Mst) from one longwall, eight bore-type continuous miners and recovery of solutions from tailings disposal. The areal extent of the mine is 6.4 by 6.4 km (4 by 4 mi) or approximately 24.6 km^2 (9.5 mi^2) of mined area of which 100% is ventilated with an exhausting system.

Total mine airflow averages 448 m^3/s (950 kcfm) with methane emissions ranging from 42,480 to 70,800 m^3/d (1.5 to 2.5 Mft^3/d). Methane concentrations range from 0.11% to 0.18% at the return shaft bottom. The methane released from the caving behind the longwall accounts for over one-half of the total methane.

The Solvay trona mine is ventilated with one, axial-vane exhaust fan located at the surface of the return shaft. A second fan in parallel is available as a spare. To heat the air in the two intake shafts, centrifugal and axial-vane fans blow air through fin-tube heat exchangers. Table 9 shows the main fan data.

The mine level is accessed from the surface through three, circular, concrete-lined shafts. The shafts vary in depth from 495 to 504 m (1,625 to 1,653 ft). Additional mine shaft data is shown in Table 10.

Table 9. SCI – Main fan data.

Location	1 Shaft	2 Shaft	3 Shaft
Manufacturer	Buffalo forge	SMJ (4 fans)	SMJ (2 fans)
Type	Centrifugal	Axial vane	Axial vane
Model	2175L-39 DWDI	AFS-2.0	AFS-3.2
Size m (in)	2.78 (107)	2 (79)	3.2 (126)
RPM	400	1180	880
Motor size kW (BHP)	447 (600)	93 (125)	1490 (2000)
SP* Kpa (in-wg)	0.37 (1.5)	0.37 (1.5)	1.74 (7.0)

* Static pressure at the operating point.

Table 10. SCI – Shaft data.

	Size			Airflow	
Location	m	ft	Type	m^3/s	kcfm
1 Shaft	7.9	26	Intake	236	500
2 Shaft	5.5	18	Intake	212	450
3 Shaft	7.9	26	Exhaust	448	950

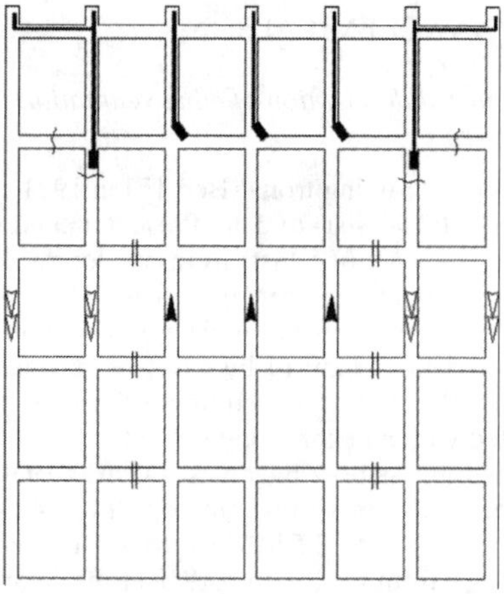

Figure 13. Typical main development with blowing face ventilation and dual split returns.

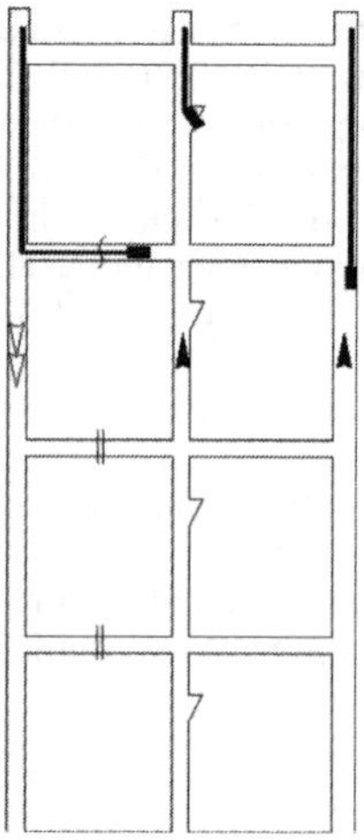

Figure 14. Typical longwall head gate development with blowing face ventilation.

5.2 *Section ventilation*

Main developments are typically seven entries with 30.5 m (100 ft) entry centers and 38.1 m (125 ft) cross cut centers. The center three entries are intake with a dual split return. Sub-main developments are typically five entries with 27.4 m (90 ft) entry centers and 38.1 m (125 ft) cross cut centers with the exception of panel stubs in which the crosscut centers are reduced to 19 m (62.5 ft). The center three entries are intake with a single split return. Typical airflow quantities are 23.6 m^3/s (50 kcfm) intake and 11.8 m^3/s (25 kcfm) in each return. Mains and sub-mains are developed with bore miners which cut oval openings 4.5 m (14.8 ft) wide by 2.7 m (9 ft) high. Stoppings are constructed of Styrofoam blocks sealed with expandable foam. Face ventilation is accomplished with 2-stage, 22-kW (30 hp) fans blowing through 41 cm (16 in) diameter "blow bag" ducting. Figure 13 shows a typical seven-entry main development.

Longwall head gate developments are three entries with 44.2 m (145 ft) entry centers and 53.3 m (175 ft) cross cut centers. The head gate is advanced 1,737 m (5,700 ft) before developing a four-entry bleeder, which is connected to the previous head gate. The head gates are developed with bore miners which cut an oval opening 4.5 m (14.8 ft) wide by 2.7 m (9 ft) high. Typical airflow quantity is 16.5 m^3/s (35 kcfm). Stoppings are constructed of metal panels and sealed with expandable foam. Face ventilation is accomplished with 2-stage, 22 kW (30 HP) fans blowing through 41 cm (16 in) diameter "blow bag" ducting. Figure 14 shows a typical longwall head gate development.

Solvay began operating a longwall in April 2003. The first panel is currently being mined. A five-entry tailgate was developed to provide four entries for longwall panel and bleeder return air and one entry for intake air, which provides access to behind the bleeder entries.

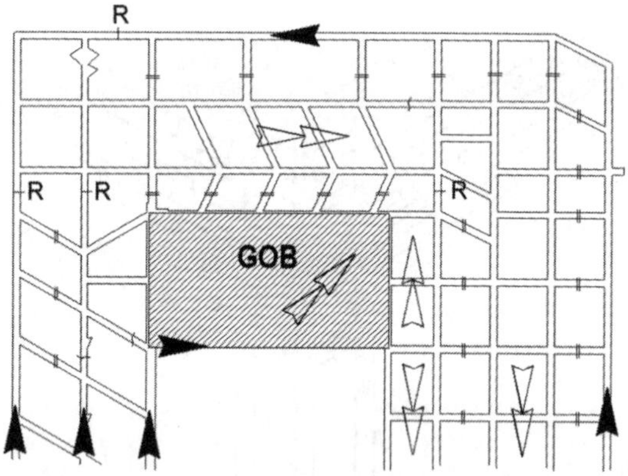

Figure 15. Typical ventilation of a retreating longwall.

The intake entry provides fresh air to dilute methane emissions from the longwall gob. The head gate entries provide 28.3 m^3/s (60 kcfm) intake air for the 152 m (500 ft) long longwall face. Typical air quantity at mid-face is 19 to 24 m^3/s (40 to 50 kcfm). An additional 38 m^3/s (80 kcfm) of intake air is provided to the longwall bleeder entries to dilute methane concentrations exiting from the longwall gob area to below 2% in the bleeder entries and 1% in the main returns. A total airflow of 66 m^3/s (140 kcfm) is required to dilute total methane emissions from the longwall to below 1%. Figure 15 shows ventilation arrangement for the longwall.

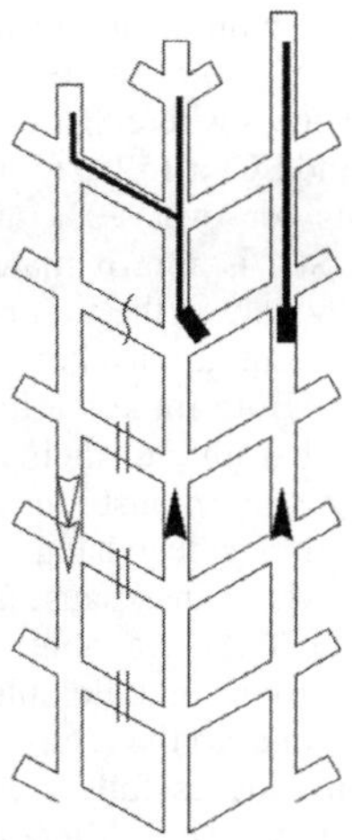

Figure 16. Typical room-and-pillar development with blowing face ventilation.

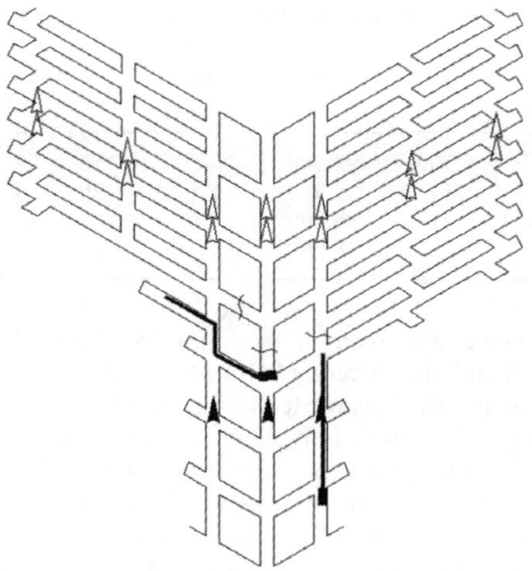

Figure 17. Typical room-and-pillar retreat with blowing face ventilation.

A total of six bore miner panels are ventilated at any one time, five of which are active and one spare. Room-and-pillar panel developments are three entries with 19 m (62.5 ft) entry centers and 22.9 m (75 ft) cross cut centers. The panels are advanced approximately 1,524 m (5,000 ft) in length at which time a connection is made to a bleeder system. The panels are developed with bore miners which cut an oval opening 4.4 m (14.3 ft) wide by 2.6 m (8.5 ft) high. Typical airflow quantity is 16.5 m^3/s (35 kcfm). Stoppings are constructed of metal panels and sealed with expandable foam. Face ventilation is accomplished with 2-stage, 22 kW (30 hp) fans blowing through 41 cm (16 in) diameter "blow bag" ducting. Figure 16 shows a typical room-and-pillar development.

Room-and-pillar panel retreat is accomplished by mining 75 m (245 ft) rooms with 11.4 m (37.5 ft) centers on both sides of the three-entry development. With the use of the bleeder system, all three-development entries are opened for intake air during retreat mining. Return air is regulated at the end of the panel. Typically, two cross cuts are mined between each retreat room. Airflow quantity is 16.5 m^3/s (35 kcfm). The same face ventilation fans are used in retreat as in advance of the panel. Figure 17 shows a typical room-and-pillar retreat panel.

Table 11 shows a breakdown of the total air used in the mine.

5.3 *Challenges*

Maintaining the mine's ventilation system to minimize both leakage and static fan pressure as well as provide adequate airflows to the operating sections of the mine to minimize heat and dust levels at the working faces.

Table 11. SCI – Mine air distribution.

	Airflow	
Mine area	m^3/s	kcfm
Bore miner panels	101	215
Longwall panel	66	140
Tailings disposal	109	230
Shops	47	100
Idle panels	28	60
Leakage	97	205
Total	448	950

6 SUMMARY

Four underground mines near Green River, Wyoming mine trona, but with few actual similarities. The mineral is the same, but utilized mining methods and ventilation systems are different.

The similarities are that permissible mining equipment, similar to that used in coal mining, are utilized to extract trona: bore miners, drum miners and longwall. Methane is liberated during extraction from the surrounding oil shales.

A common challenge among mines is to maintain ventilation systems for long periods of time, as each operation possesses many years of mining reserves. Shaft location and system maintenance is critical to ventilation efficiency.

It is at this point that differences become more obvious. As seen in Table 12, there is a great range in areal extent of the mines, which is related to the time

Table 12. Mine summary data.

Mine	Airflow m^3/s	Methane m^3/d	Tonnage Mt/a	Extent km^2	Shafts
FMC	670	93,500	4.1	118	8
OCI	264	9,910	3.5	39	3
GCSAP	621	35,000	4.3	65	6
SCI	448	56,000	4.1	24	3

since commencement of operation. FMC started after WWII and most recently SCI in 1981. The number of ventilation shafts is related to the mine's areal extent. Methane liberation is highest in the two operations utilizing a longwall, less where room and pillar retreat is used and lowest at OCI due to mining of a different trona bed and reduced ground pressures. There is also a great difference in room dimensions (OCI), which can be attributed to the specific bed being mined, ground conditions and individual mining equipment.

FMC and SCI both utilize longwalls, but FMC has a fourth entry "zero" room inside the longwall block for maintenance access where SCI does not. FMC has a 230 m (750 ft) and SCI a 152 m (500 ft) face. FMC's longwall accounts for over 90% of mine methane emissions where SCI is approximately one-half, but mining is presently only in the first panel.

Auxiliary face ventilation choices are related to mining machinery. Bore miners utilize blowing fans and flexible blow bag due to their filling the entire drift. Drum miners use exhaust fans and rigid duct as space is available alongside to hang tubing. Again, the exception is OCI, where an exhaust fan is installed on the drum miner due to the available 3.0 m (10 ft) mining height of the room and flexible blow bag that extends from the miner to the return airway.

Four operations successfully mine trona in the Green River Basin with very different machinery and ventilation practices. As with underground coal mines, if you have seen one, you haven't seen them all.

Mine Ventilation, Ganguli & Bandopadhyay (eds)
© 2004 Taylor & Francis Group, London, ISBN 90 5809 633 5

Measurement of frictional pressure differentials during a ventilation survey

Brian S. Prosser
Bechtel SAIC Company

Ian M. Loomis
Mine Ventilation Services, Inc.

ABSTRACT: During the course of a ventilation survey, both airflow quantity and frictional pressure losses are measured and quantified. The measurement of airflow has been extensively studied as the vast majority of ventilation standards/regulations are tied to airflow quantity or velocity. However, during the conduct of a ventilation survey, measurement of airflow only represents half of the necessary parameters required to directly calculate the airway resistance. The measurement of frictional pressure loss is an often misunderstood and misapplied part of the ventilation survey. This paper compares the two basic methods of frictional pressure drop measurements; the barometer and the gauge and tube. Personal experiences with each method will be detailed along with the authors' opinions regarding the applicability and conditions favoring each method.

1 INTRODUCTION

The quantification of differential pressures and airflows is essential for the development of an accurate ventilation model. An emphasis is generally placed on the measurement of airflow as it is required for the fulfillment of most governmental ventilation regulations. Various techniques and instruments have been developed to provide accurate airflow measurements (averaged spot measurements, anemometer traverses, ultrasonic measurements, vane anemometers, hotwire anemometers, tracer gases, velometers, vortex shedding anemometers, and velocity pressure measurements using a pitot tube). These methods have been studied extensively in symposium papers and doctorial theses. However, since the measurement of frictional pressure drops is generally required only for the quantification of airway resistance and is not generally required in the day-to-day operation of a mine, very little is actually studied about this method. Almost all mining references contain a section on pressure measurement (typically detailing fan measurements), however, there are few texts that actually detail the practicalities involved with conducting a pressure survey.

2 GENERALITIES

In general, two practical methods are used to determine pressure differentials in a mine. These are the barometer (altimeter) survey and the gauge and tube survey. Each survey technique has its advantages, is accurate under certain circumstances, and each can be incorporated into an overall ventilation survey. However, each method has certain disadvantages that should also be considered.

The use of barometers (altimeters) in ventilation surveys has long been established and various techniques have been developed to incorporate their readings into the structure of a ventilation survey. However, with improvements in micromanometers and for reasons discussed in the following section, the usefulness of the barometer method is seen as being limited to measurement of shafts, across inaccessible areas, or areas of high pressure differential. The use of micromanometers has improved the accuracy and convenience of differential pressure measurement across bulkheads and ventilation structures as well as increased the accuracy of measurements in areas of smaller frictional pressure losses. Although a barometric survey can be accomplished by a single person taking measurements at specific locations and a gauge and tube survey requires a team of at least two people, the increased accuracy of the manometer survey is a significant benefit. If a ventilation survey is for the purposes of establishing an accurate ventilation model of a mine, then that survey should be conducted in a manner so as to provide the most accurate and useful information possible. The degree of accuracy of the data needs to be determined prior to the selection of the survey method.

3 BAROMETER SURVEY

There are two basic approaches to using a barometer during a ventilation survey. These are the roving, and the leapfrogging methods. The roving method can be accomplished by a single person as long as a recording surface barometer is used. This method involves continuously or incrementally recording a stationary "surface" barometer along with the discrete roving barometric pressure at points in the mine. The barometric pressure at each transit point can then be corrected for changes in the background pressure over the course of the shift by using the surface barometer reading. This method assumes that the barometric pressure in the mine will fluctuate at exactly the same time as the fluctuation is noted on surface. This assumption and inherent weakness can be mitigated by using the "leapfrog" technique. This method involves two measurement teams recording the barometric pressure simultaneously at two points underground. A surface barometer is not needed as the measurements are taken simultaneously. This method is inherently more accurate than the roving method but requires constant communication between the measurement teams and two instruments that are very similar in accuracy and precision.

For each method, the barometer should be placed in approximately the same spatial location within the drift for each measurement, either the center of the drift, near the rib, etc. This will allow the turbulence conditions to be relatively constant for each measurement. Individual measurements taken in areas of high turbulence should be avoided as the measurement could be improperly influenced. Regardless of which method is used there is a substantial amount of data reduction involved in the determination of the pressure differential between two points. Differential pressures across bulkheads, doors, and fans should still be measured with a manometer. These measurements are required to balance the differential pressures measured along the airways using Kirchhoff's second law (the sum of all pressures around a circuit must equate to zero) to ensure accuracy. Depending upon the method used for the reduction of the barometric pressure data, psychrometric data is required to be measured at each measurement location along with the barometric pressure. The accuracy of the psychrometric data should be held to as high a standard as possible as the errors in the psychrometric measurements and barometric pressures will have a compounding effect. In addition to those values measured during the ventilation survey, the results will be entirely dependant upon elevation data obtained separately through the mine survey department. In many cases the elevations could have been measured years prior to the ventilation survey, or the survey markers moved to more convenient locations.

4 THEORETICAL REDUCTION OF BAROMETRIC SURVEY DATA

Three different methods were used to determine the frictional pressure differential in this section. For this example the frictional pressure differential in a shaft is determined. The roving method is used and the following data presented in Table 1 was measured.

4.1 *Method 1: Direct application of the steady flow energy equation*

Subsurface Ventilation and Environmental Engineering (McPherson 1993, Section 6.3) provides the following procedure for reduction of the data taken during a barometric pressure survey. Equation 1 (Steady Flow Energy Equation) is used to evaluate the work done against friction as the air travels between two stations.

$$F_{12} = \frac{u_1^2 - u_2^2}{2} + (Z_1 - Z_2)g - R(T_2 - T_1)\frac{\ln(P_2 / P_1)}{\ln(T_2 / T_1)} \quad (1)$$

where F – Work done against friction (J/kg), P – Barometric pressure (kPa), T – Absolute temperature (Kelvin), Z – Elevation of barometer location (m), u – Air velocity at the barometer location (m/s), R – Mean gas constant (J/kg K), g – Gravitational acceleration (9.81 m/s^2).

This work term is then converted into a frictional pressure drop using Equation 2.

$$p_{12} = \rho_a F_{12} \quad (2)$$

Table 1. Barometer survey data reduction.

Location	Surface barometer (kPa)	Roving barometer (kPa)	Dry bulb temp. (°C)	Relative humidity (%)	Air velocity (m/s)[1]	Elevation (m)[2]
Top of shaft	98.782	103.750	15.6	13.0	2.03	2652.0
Bottom of shaft	98.800	104.610	17.2	14.2	1.52	2573.0

[1] Davis Anemometer calibrated in feet, converted to SI (m/s) in spreadsheet.
[2] Elevation given by site surveyors and measured independently of ventilation survey.

where p_{12} – Frictional pressure drop (Pa), ρ_a – Average density of air between two stations (kg/m^3).

If the barometric pressures at two stations are not read simultaneously it is necessary to apply a correction to one of the values to incorporate any changes in the surface atmospheric pressure. By assuming that a series of polytrophic processes link the control barometer to the roving unit underground, then the following correction is applied as described by Equation 3.

$$P_1^{'} = \Delta P_c \frac{P_1}{P_c}, \tag{3}$$

where P_1 – Updated value for barometric pressure at station 1, P_1 – Raw data for barometric pressure at station 1, ΔP_c – Change in surface atmospheric pressure, P_c – Surface atmospheric pressure taken at the same time as station 1 reading.

4.2 *Method 2: Approach recommended by the Mine Ventilation Society of South Africa (MVSSA)*

Environmental Engineering in South African Mines (Burrows J. et al 1989, Chapter 6) uses the following approach outlined in Equations 4 and 5 to determine the pressure loss:

$$p_{12} = -(P_2 - P_1) - g\int wdZ \tag{4}$$

where the term $g\int wdZ$ is the theoretical increase in pressure. The difficulty with this approach is in evaluating the integral term for the change in air density as a function of depth. A series of stations could be established between the points, however this can result in an excessive amount of data reduction. Hence, an assumption is made that the density varies linearly with elevation, as given in Equation 5:

$$\int wdZ = \frac{1}{2}(\rho_1 + \rho_2)(Z_1 - Z_2) \tag{5}$$

The error associated with this assumption is particularly severe when the elevation change is large. For elevation changes of less than 300 m the equations are considered adequate. Another assumption with this method is that the airflow present at each station is representative of the airway in between. This is valid in certain cases, but is obviously not correct when complex networks are involved.

4.3 *Method 3: Exact density solution – Per Hall (1981)*

Mine Ventilation Engineering (Hall C.J. 1981, Chapter 8) presents an exact solution for barometric pressure data that uses a density analysis (similar to the MVSSA method presented above). In this case a frictionless pressure (P_{2calc}) is determined from Equation 6.

$$P_{2calc} = P_2\left(\frac{2P_1 + Dg\,\rho_1}{2P_2 - Dg\,\rho_2}\right) \tag{6}$$

where D – Depth below datum (m)

The pressure drop due to friction, shock and increases in kinetic energy is given by the following equation:

$$p_{12} = P_{2calc} - P_2 \tag{7}$$

4.4 *Comparison of data reduction methods*

As can be seen from the calculations below (Table 2, and Table 3, and Table 4), all three methods provide very similar answers. The method recommended by Dr. McPherson results in a value of 127 Pa, compared to 125 and 126 Pa for the MVSSA and Hall method respectively. However, this difference is very small, and provided that the difference in elevation is not too great, each of these methods is acceptable for reducing survey data. Based upon the results of a sensitivity analysis, the factor that results in the greatest error for the calculated differential pressure is the station elevation. With this example a 1% error in elevation difference between stations results in a 5% change in calculated differential pressure. However, as the elevation difference between stations lessens, the calculation becomes more sensitive. A 1% error in temperature resulted in a negligible error in calculated differential pressure.

4.5 *Instrumentation*

There are four basic types of barometers used in mining. They can be classified into precision electronic barometer (Setra), reduced accuracy electronic barometer (Air Instruments), precision aneroid barometer (Wallace & Tiernan), and reduced accuracy aneroid barometer (Taylor). The manufacturers listed above are referenced based upon the authors experience and are not intended as an inclusive list.

In general, the most accurate instrumentation should be used for differential pressure calculations. The span encompassed by the resolution of reduced accuracy barometers, in some cases, can be greater than the value of the frictional pressure drop attempted to be measured. The aneroid barometers have the advantage of being permissible and allowed into the return airways in coal mines, however they can be difficult to read/interpolate and take longer to stabilize than the electronic barometers. The electronic barometers

Table 2. Complete barometric data reduction after McPherson (1993).

Input data sheet for barometric survey – Wet bulb/dry bulb input

Station number	Time	Traverse barometer P (mbar)	Temperatures td (C)	tw (C)	Elevation Z (m)	Velocity u (m/s)	Control barometer (mbar)
1	13:42	1037.50	15.60	13.00	2652.0	2.03	987.8
2	14:05	1046.10	17.20	14.20	2573.0	1.52	988.0

Calculated data

RH calc (%)	Actual density (kg/m^3)	Gas constant (J/kgC)	Diff pressure (Pa)	Corrected pressure (kPa)	Work done 1–2 (J/kg)	Pressure drop 1–2 (Pa)	Press Dr. british (m. in. w.g.)
74.7	1.246	288.431		103.769	–	–	–
72.2	1.249	288.517	0.018	104.610	101.60	127	508.75

Sat vap esw (Pa)	Moist sat Xs (kg/kg dry)	Latent Lw (J/kg)	Sigma S (J/kg/dry)	Moist X (kg/kg dry)	Vap press E (Pa)	Enthalpy h (J/kg dry)	Dry vap esd (Pa)
1497.280	0.009108	2471482	35575	0.008035	1323.106	36012.4	1771.767
1618.941	0.009777	2468619	38408	0.008536	1416.258	38915.1	1961.783

Table 3. Barometer data reduction after MVSSA (1989).

Location	Input: Corrected BP (kPa)	Density (kg/m^3)	Output: Int. wdZ (kg/m^2)	Pressure loss (Pa)
Test 1	103.769	1.246		
Test 2	104.610	1.249	−98.5327	125

Table 4. Barometer data reduction after Hall (1981).

Location	Input: Corrected BP (kPa)	Density (kg/m^3)	Output: P_{ecalc} (kPa)	Pressure loss (Pa)
Test 1	103.769	1.246		
Test 2	104.610	1.249	104.7358	126

have a finer resolution, foolproof digital readout, and rapid stabilization but are tied to a battery, and are somewhat fragile to the harsh atmospheric conditions experienced in mines. But under the controlled conditions of a ventilation survey and limited contact with the mining atmosphere the precision electronic barometers should be used in order to obtain the most accurate results. Setra Instruments offer high-resolution digital barometers that have been found to have excellent accuracy and sensitivity. Although the Setra barometer has a significant weight, the increased accuracy of the measurement more than offsets the discomfort of carrying the instrument during the survey. In addition to the barometers, the instrument elevations must be surveyed accurately and the survey stations well marked. The psychrometric properties of the air must also be accurately quantified. A digital relative humidity/dry bulb temperature instrument can greatly speed up the measurement of psychrometric properties.

5 GAUGE AND TUBE METHOD

Frictional pressure drops through airways can accurately be determined using the gauge-and-tube technique. The gauge and tube (or trailing hose) method allows the direct measurement of frictional pressure differentials using a digital manometer (or magnehelic gauge) connected to a length of tubing, the ends of which are connected to the total pressure ports of pitot static tubes. Lengths of 6 mm (¼ inch) nylon tube of up to 300 m(1,000 ft) (limited by practicality) can be used. The nylon (strong, semi rigid, relatively inflexible) tube is strung along a drift from one marked station to another marked station. Both ends of the tube are connected to pitot tubes which are positioned facing into the airstream. A manometer is placed in-line as shown in Figure 1. In this way the difference in total pressure between the two stations can be measured. When the measurement is finished (the manometer has stabilized around an average

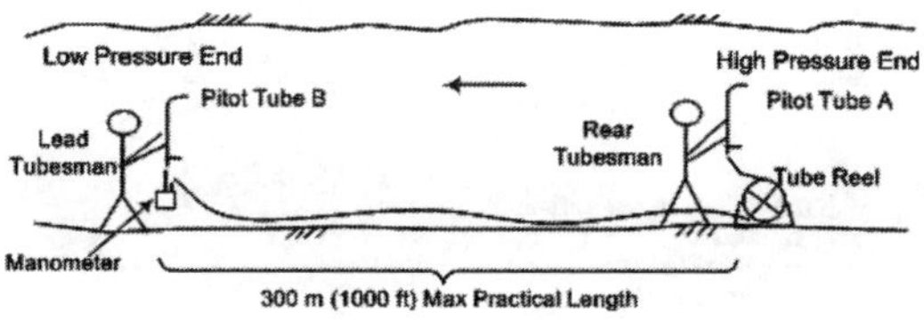

Figure 1. Gauge and tube technique.

value) the person at the rear of the tube is signaled (verbal, cap lamp or tug signal) and the tube is then advanced along the drift to the next measurement location, the person at the rear of the tube stops at the marked measurement station. The survey must be planned in advance to determine the most appropriate tube length, and approximate measurement station locations. Quality control dictates that the tube must be pressure tested prior to each shift to ensure against leaks. In general the rear tube measurement location should be up wind of the manometer, or positive side of the manometer. In this way the tube can be judged to be functioning properly by examining the response of the manometer. If the pressure displayed on the manometer shows no fluctuation from the previous measurement a kink is indicated, if it displays little to no pressure differential then a leak in the tube may be indicated. The pressure should build slowly from zero to the actual pressure differential and then oscillate around an average value. Although the manometer will rarely settle steadily on a value because of turbulence, it will stabilize around an average value.

This measurement technique is independent of minor changes in elevation (e.g. approximately 300 m), psychrometric parameters, or independent air velocity measurements. No additional equipment, other than the pitot tubes, nylon connection tube, and manometer need be used. The simplicity of the measurement allows for rapid reduction of data, field accuracy/verification checks, and is unhindered by the need for additional parameters to be measured by the survey team. However, this method requires at least two people for the measurement, and can be exceedingly difficult and cumbersome to use if improperly planned. Each airway must be measured from junction to junction, which can cause logistical difficulties if tube length is not adequately matched. This method can and has been used to accurately measure frictional pressure drops down to 1 Pa (0.004 in. w.g.).

Measurements taken in shafts or long ramp systems with significant changes in elevation may be subject to a correction factor. This correction factor is necessary as the air in the measurement tube is stationary, and not affected by friction, which will result in a slightly higher pressure inside of the tube than exists in the airway (McPherson 1993, Section 6.3.1). This correction can be approximated by Equation 8 (Hinsley 1962).

$$p_{12} = \Delta P \times \frac{P_m}{P_L} \quad \text{(Pa)} \tag{8}$$

where p_{12} – Frictional pressure drop from point 1 to point 2, P_m – Mean barometric pressure in the shaft, P_L – Barometric pressure at measurement location (at either point 1 or point 2), ΔP – Raw differential pressure measurement.

Shafts of up to 300 m(1,000 ft) in depth can be directly measured without the need for correction, however, the application of the correction factor should be used for shaft measurements over 300 m(1,000 ft) in depth. The correction for elevation differences is insignificant for measurements taken along normal working levels in mines. For long ramp systems with significant changes in elevations a correction could be applicable if the entire ramp were to be measured with a single tube measurement, however, this is not generally practical. In practice ramps measurements are broken into many smaller segments and generally measured with tube lengths of less than 150 m(500 ft) to avoid excessive drag on the tube as it wraps around the various bends encountered in a ramp.

5.1 *Instrumentation*

The equipment used to measure the differential pressure must be accurate, incorporate a fine resolution, and be durable. One of the best instruments for measuring the differential pressure along airways has been found to be the Zephyr pressure gauge, it incorporates a "damping" feature in the instrument, is self zeroing, fine resolution, high accuracy, shock proof, and incorporates a water resistant case. But it is not classified as permissible so it requires permission as a low-voltage diagnostic device to be used in coal mine return airways. Examples of lower accuracy manometers include both the Taylor and PDM 304 manometers. The Taylor manometer generally does not have the accuracy or resolution to measure the often low pressures encountered in along an airway, however it is an excellent device to measure differential pressures across bulkheads, regulators, and doors. The PDM 304 has the same problems as the Taylor with respect to accuracy and resolution, but it is one of the only permissible electronic manometers available. The 60 Pa (¼ inch) Magnehelic gauge is the instrument originally used for this technique. It is a very sensitive mechanical pressure measurement device that can be used in any environment. However, it is quite sensitive to horizontal alignment which makes field measurements with it extremely troublesome as a few degrees off level results in a widely varying measurement. An example of each instrument is shown in Figure 2. With the availability of suitable mechanical and electronic instruments, the inclined water or oil filled manometers are not considered practical for the

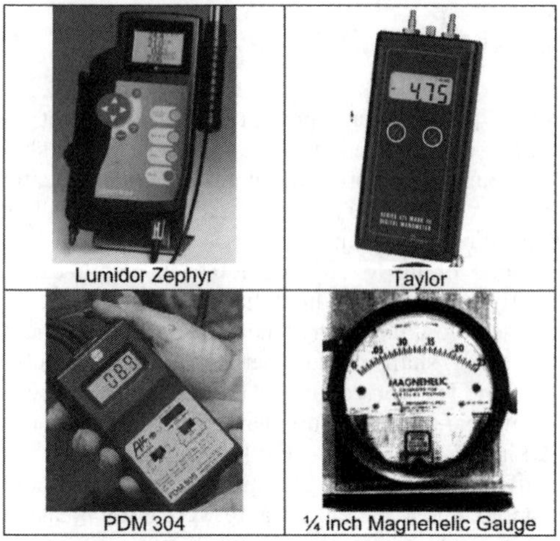

Figure 2. General types of manometers.

performance of a modern ventilation survey. With the continual upgrade and development of electronic manometers the process of measuring pressure drops is greatly streamlined.

6 EXAMPLES OF PARALLEL MEASUREMENTS AND CASE STUDY MEASUREMENTS (EXAMPLES)

During a recent ventilation survey the frictional pressure drop was measured in a set of airways. The pressure differential was measured with both the gauge and tube, and barometer method in order to provide a set of checks. Figure 3 (not to scale) shows a schematic with the gauge and tube measurements taken during the ventilation survey and the measurement points for the barometer survey (points A through D). An effort was made to measure everything as accurately as possible within the time allotted for the measurements. The gauge and tube measurements took approximately 1-½ shifts to complete, and the barometer measurements took approximately half a shift to complete. If the barometer measurements were accurate enough a substantial time savings could be realized in the field.

Get ride of the crosscut references. They can reveal the mine and otherwise don't add to the illustration. The application of the gauge and tube method during this survey involved extensive measurements in ramps, along main haul routes, and through high velocity drifts. The direction of airflow along the path of this section of the survey was constant, although the airflow through the measured drifts increased and decreased at the various junctions. The gauge and tube measurements were broken at each junction where a significant amount of air entered or exited the

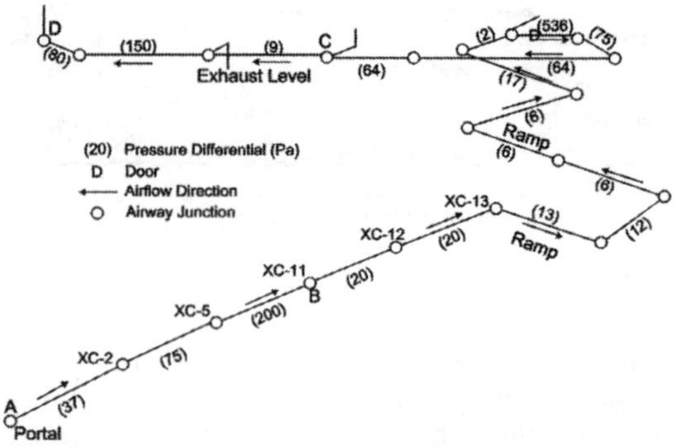

Figure 3. Pressure survey measurements.

measurement route. The field measurements are summarized in Table 8.

The data obtained during the barometric portion of the ventilation survey are shown in Table 5. Measurements of barometric pressure, dry bulb temperature, relative humidity, and air velocity were taken at each station shown on Figure 3. A control barometer was set up at the engineering office outside of the mine to measure changes in barometric pressure over time at a fixed location. The elevations of each barometric measurement station were provided by the mine survey department.

Tables 6 and 7 show the data reduction for these measurements.

The data was reduced using McPherson's method of a direct application of the steady flow energy equation.

Table 8 shows a comparison between the two measurement techniques for this portion of the survey. As can be seen from the comparison one of the measurements (portal A to end of exhaust level D) compares very closely between the two measurement techniques. However, the rest of the measurements compare very poorly. In the case of measurement stations C and D the barometer reduction came out in the negative indicating that the air would be flowing opposite the direction of presentation (from D to C) which is incorrect as personnel in the drift verified the direction of airflow from C to D. The substantial differences between the two measurement techniques could have been resolved by taking additional or redundant barometric measurements. However, these differences were discovered during data reduction and would have necessitated re-measuring this portion of the mine. There are many variables with regards to the barometric survey that could have introduced error into the measurements. The most likely cause of the extreme error would be with regards to the elevations measured by the mine surveyors. This is usually the weak link in the barometric survey as airway geometry changes with time due to ground movements, sloughage, milling and mucking of floor/sill, and roof falls.

Table 5. Barometric survey measurement data.

Barometer survey measurement input data

Location		Time	Traverse barometer (kPa)	Dry bulb temperature (C)	Relative humidity (%)	Velocity (m/s)	Elevation (m)	Control barometer (kPa)
From:	A	12:04	72.350	13.0	77.7	0	2946.0	72.391
To:	B	11:55	70.755	10.9	90.8	6.6	3112.8	72.419
From:	B	11:55	70.755	10.9	90.8	6.6	3112.8	72.419
To:	C	11:13	68.630	13.6	92.3	1.1	3197.7	72.437
From:	C	11:13	68.630	13.6	92.3	1.1	3197.7	72.437
To:	D	11:37	68.518	13.1	97.0	10.1	3239.6	72.437
From:	A	12:04	72.350	13.0	77.7	0	2946.0	72.391
To:	D	11:37	68.518	13.1	97.0	10.1	3239.6	72.437

Table 6. Barometer survey data reduction calculations.

Barometer survey data reduction calculations

Location		Absolute temp (K)	Partial press. e (Pa)	Moisture content X (kg/kg dry)	Gas constant (J/kg C)	Mean gas constant (J/kg C)	ΔSur press. (kPa)	Cor. pres (kPa)	Frict. work (J/kg)
From:	A	286.150	1163.3869	0.0101652	288.796	–	–	–	–
To:	B	284.050	1183.6568	0.0105824	288.867	288.831	0.028	72.378	209.259
From:	B	284.050	1183.6568	0.0105824	288.867	–	–	–	–
To:	C	286.750	1437.1741	0.0133038	289.331	289.099	0.018	70.773	1724.939
From:	C	286.750	1437.1741	0.0133038	289.331	–	–	–	–
To:	D	286.250	1461.8898	0.0135602	289.374	289.352	0.000	68.630	–325.785
From:	A	286.150	1163.3869	0.0101652	288.796	–	–	–	–
To:	D	286.250	1461.8898	0.0135602	289.374	289.085	0.046	72.396	1624.014

Table 7. Barometer survey results.

Barometer survey output

Location		Density (kg/m^3)	Specific volume (m^3/kg)	Frictional pressure drop (Pa)	Frictional pressure drop (milli-in w.g.)
From:	A	0.875	1.142	181.8	730
To:	B	0.862	1.160		
From:	B	0.862	1.160	1457.2	5850
To:	C	0.827	1.209		
From:	C	0.827	1.209	–269.5	–1082
To:	D	0.827	1.209		
From:	A	0.875	1.142	1382.6	5551
To:	D	0.827	1.209		

Table 8. Comparison of survey results.

Barometer/gauge and tube comparison

Location	Frictional pressure drop (barometer method) (Pa)	Frictional pressure drop (gauge and tube method) (Pa)
From: A To: B	182	312
From: B To: C	1457	841
From: C To: D	–270	239
From: A To: D	1383	1392

Although the gauge and tube measurements in this example are not shown in a loop so closure can not be determined, the magnitude of gauge and tube measurements seems appropriate. During the survey the manometer stabilized around a (fairly) steady value which tends to lend confidence to the measurement. The gauge and tube measurements were later used with other pressure measurements to show that these readings were accurate.

Table 9. Brief comparison of methods.

Barometric method		Gauge and tube method	
Advantage	Deficiency	Advantage	Deficiency
Requires minimal personnel	Relies on more than a single pressure measurement (velocity, elevation, psychrometric)	Instantaneous pressure measurement	Travel must be accessible between the two measurement stations
Establishes density at each point	Data reduction is not done in the field, erroneous measurements are not easily identified during the survey	Closure determined in the field	Requires fit personnel for measurements in order to carry tube between stations through degraded airways, over roof falls, and exposure to temperature extremes
Rapid measurement	Lengthy data reduction	Higher accuracy	Requires experience
Travel between measurement stations not required (gob measurements, abandoned areas, black box models)		Relies on only one measurement (pressure)	Requires accurate digital manometer for best results Measurements limited to around 6 km/day in large high seam coal mine.

7 CONCLUSIONS

Based upon the performance of many pressure surveys in both metal mines and coal mines, is the author's belief that the gauge and tube technique is more preferable than the barometric pressure technique. Each technique is applicable for certain aspects of a pressure survey, but for general measurements the gauge and tube method provides for more rapid evaluation and more accurate results. Although a "basic" barometric pressure survey can be accomplished faster than a gauge and tube survey, once the necessary back-up or redundant barometric measurements are taken into consideration and the time involved in reducing the data, the measurement time difference is not so great. The gauge and tube method would be much more preferable than the barometric method for discrete measurements such as those required for friction factor determination and across ventilation structures such as regulators and bulkheads. A comparison between the barometric, and gauge and tube methods is presented in Table 9.

When evaluating gob areas, abandoned areas, substantial lengths of shafts (greater than 600 m), black box models, or areas of a mine not safe to travel through, a barometric survey can be preferable over a gauge and tube survey. While the authors have taken direct pressure measurements, using the gauge and tube method, in shafts exceeding 600 m in length, factors such as general tube handling, coupling of multiple tube lengths and the time element tend to favor the barometric method.

Each survey method has a purpose and a preferred application. Prior to conducting a ventilation survey the pressure portion must be planned and the various methods chosen according to their applicability. There is no single "one size fits all" methodology, but rather a combination of methods should be incorporated into the overall survey plan.

ACKNOWLEDGEMENTS

I would like to acknowledge the contribution of Dr. Ian Duckworth in providing assistance with the description of barometer survey data reduction techniques.

REFERENCES

Hinsley, F.B. 1962. "The assessment of Energy and Pressure Losses Due to Airflow in Shafts, Airways, and Mine Circuits", *Mining Engineering*. Vol: 121, pp. 761–777.

McPherson, M.J. 1993. *Subsurface Ventilation and Environmental Engineering.* New York, New York: Chapman & Hall.

Burrows, J. et al 1989. *Environmental Engineering in South African Mines*. The Mine Ventilation Society of South Africa.

Hall, C.J. 1981. *Mine Ventilation Engineering*. Society of Mining Engineers of The American Institute of Mining, Metallurgical and Petroleum Engineers, Inc. New York, New York.

Mine Ventilation, Ganguli & Bandopadhyay (eds)
© 2004 Taylor & Francis Group, London, ISBN 90 5809 633 5

Ventilation system of Pyhäsalmi mine, Finland – a case study

M.S. Pulkkinen
M.Sc., INMET Mining, Pyhäsalmi Mine Oy, Pyhäsalmi, Finland

A.L. Martikainen
M.Sc., Helsinki University of Technology, Espoo, Finland

ABSTRACT: This paper discusses ventilation at the Pyhäsalmi mine, concentrating on the New Mine development stage and experiences during first operating years. The New Mine was built under the existing mine, between 1050 and 1445 meters underground. The mine depth set increased demands on the mine ventilation system. A major factor is the higher temperature of the rock at the increased depth therefore demanding cooling of air during the hottest summer days. Air volume blown to the mine is automatically controlled through pressure settings. All secondary fans are remote-controlled and several CO, SO_2 and NO_X detectors monitor the quality of air. Helsinki University of Technology performed a ventilation survey. The aim was to evaluate the efficiency of the existing system, develop a model with computer simulation software in order to improve long-term ventilation planning, to create a tool to evaluate costs, and as a result to ensure an efficient ventilation system.

1 INTRODUCTION

1.1 *Description of the mine*

The Pyhäsalmi mine is located in Central Finland. The Pyhäsalmi ore body was discovered in 1958 and production began in 1962, first in open pit and since 1976 only from underground operations. In the late 1990s, known ore reserves were only 5 million tons and the expected lifetime for the mine was until the end of year 2000. During 1996 exploration drilling proved that mineralisation continues at least to 1300 meters below the surface. Intensive diamond drilling discovered the new ore body, up to 20 million tons of ore, with the bottom located on the 1410 level.

The New Mine project commenced in April 1998 and the New Mine was built under the existing mine between 1050 and 1445 meters beneath the surface. The project was executed during 1998 and 2001 and production from the New Mine started in July 2001. Production level in 2003 is 1.3 Mt of zinc-copper ore, average grades, 1.1% Cu and 2.7% Zn.

The New Mine is the deepest operating metal mine in Europe and the mine depth set new and increased demands on the mine ventilation system. The bottom of the mine is down on the 1445 level, maintenance facilities are on level 1410 and ore production will take place between levels 1050 and 1425. A new factor was the higher temperature of the rock at the increased depth, between 20° and 22°C.

2 EXISTING VENTILATION

2.1 *Ventilation design principles*

The major parameters to consider, while the ventilation system was designed, were comfort and good working conditions furthermore the safety of personnel and security of clean air delivery to the worksites. Other criteria included as simple a layout as possible and efficient flow-through ventilation. An additional criterion was easy and open controllability of the ventilation through computers and mine radio without specialized ventilation personnel.

In the capacity evaluations of the fan the amount of diesel powered equipment and their installed power was the main factor.

The main fan is made by ABB, equipped with a 450 kW electric motor. Design capacity is 120 m^3/s and measured is 140 m^3/s. The fan and airflow volume is pressure controlled in automatic mode, but also manual setting is always possible. The fan is located on the surface in the middle of the industrial yard in a special noise dampened building. In the same building there

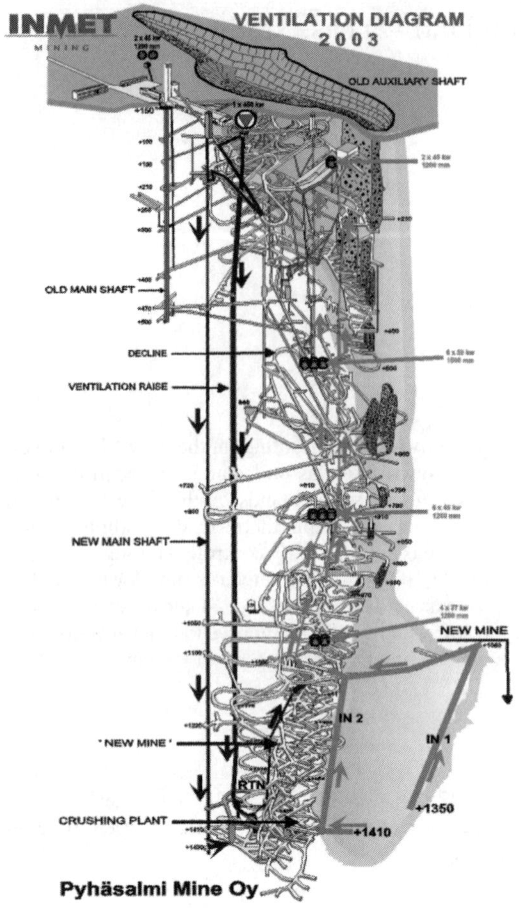

Figure 1. Mine layout and ventilation principles.

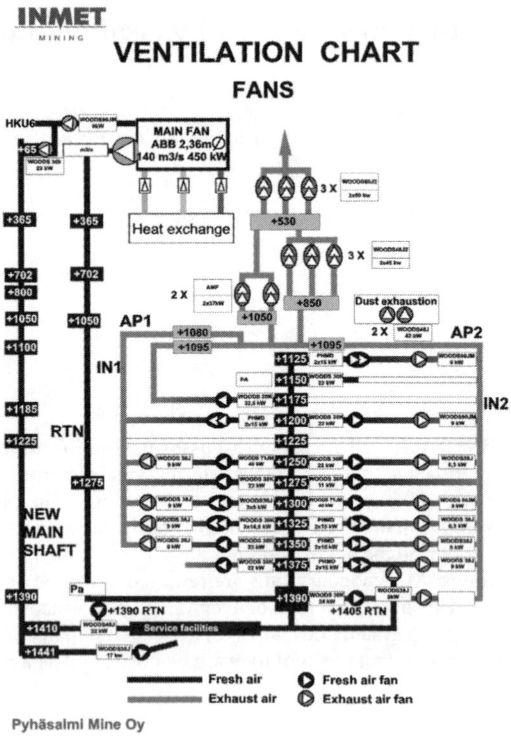

Figure 2. Ventilation chart.

are air heating/cooling installations and ventilation control computers.

Air is conducted into the mine via a new 1390 meters deep ventilation raise, diameter 3.1 m. With the full ventilation capacity the speed of fresh air is about 18.5 m/s. Fresh air is distributed to worksites through another raise, which is located in the middle of the ore body. From this raise, fresh air is subdivided to worksites by means of remote controlled fans and ventilation duct network. Exhaust air from the mining area is collected into two exhaust raises, situated at either end of the mining area. Vitiated air is discharged to surface mainly through a single exhaust raise. A small portion is used for the ventilation of the ramp to surface.

Temperatures in Middle Finland vary during the year from −30°C in winter to +30°C in summer. Cold fresh air is heated during the wintertime by using warm water from mill and mine as an energy source. Warm water from the mine is pumped to heat exchangers and the collected heat is fed in to thermo blocks, by which the fresh intake air is heated. Use of warm water as energy source saves 500 t/a fossil fuel and reduces emission by 1500 t/a. During hot summer days air is cooled by using cold mine water collected from the upper parts of the mine.

2.2 *Control system*

The main fan air volume to the mine is automatically controlled through pressure settings. All secondary fans are remote-controlled and they can be started/switched off through a few computers. Air quality control is implemented by using several CO, SO_2 and NO_X detectors at locations, which are critical for ventilation and safety reasons.

All information collected by gas monitors as well as the set-up of the main fan and other fans are available on the computers, which are connected into the general data-handling network.

2.3 *Blasting ventilation*

Blasting takes place every day after the evening shift. Before the blast, the mine is totally evacuated of all personnel due to the possibility of sulfide dust explosions. All fans are shut down automatically and, after the blast, local fans at each level where blasting took place as well as the main fan and exhaust fans, will be run with full capacity until the exhaust air of the mine has achieved the pre-set values for all noxious gases. After that, ventilation will be set to the energy saving level, which means that only 20–30% of the ventilation capacity is in use during the rest of the night. Over the weekend ventilation will be in the same energy saving mode.

2.4 *Safety*

Fresh air is directed straight from the fresh air raise to the underground canteen and to the main rescue chamber situated on the main level. The same applies for the underground maintenance facilities. On each operating level rescue chambers are in connection with the fresh air raise.

As the main shaft is the only access in and out of the mine in the case of a possible fire, a special attention has been put on the security of the main shaft. On each level access to the shaft there are indicators for the doors and pressure measuring gauges. In the shaft there are three sets of smoke, carbon monoxide and sulfur dioxide detectors. In the case of a fire, before using the shaft, one must check that all doors to the shaft are closed, the shaft is positively pressured, and monitored gases and smoke values are at acceptable levels.

2.5 *Experiences*

During the first years of operation in the New Mine it has been found that the design principles of the ventilation system are correct and the ventilation of the mine has performed as expected. Air quality at the work sites has stayed good, and the amount of noxious gases has been below all accepted exposure limits.

During the hottest summer days the cooling capacity has not been efficient enough and the temperature in the service facilities of mine on level 1410 has been close to 30°C, which is unpleasant for the personnel.

In the inclined ramp, which is used for material transporting, sometimes fog appears due to the high humidity of the air and the decreased temperature. This is one issue to resolve in the future.

Crushing plant at 1390 level creates at times unacceptable amount dust, even when the moistening and dust collection systems are in use. Dust from crushing plant will spread into decline and further in the mine. That requires also consideration in the future.

2.6 *Future plans*

Due to the design of the ventilation system, the energy consumption in ventilation is considerable and reduction of consumed power of the vent fans is one target, especially on the exhaust side of the ventilation system. One prospect to study is the possibility of replacing the more resistant part of the existing exhaust raise by a new raise of larger cross section.

One opportunity to control local fans is offered by tagging systems, which are under evaluation. All mobile equipment can be fitted with it's own tag and, when passing the reader situated on each level it informs the ventilation control the model of equipment and amount of air it requires when operating. Ventilation control will start fans and when the equipment leaves the level it will switch ventilation off after a preset extra ventilation time.

3 STUDY

3.1 *General*

Helsinki University of Technology performed a ventilation survey in the mine during the year 2003. The aim of the study was to evaluate the efficiency of the existing ventilation system, develop a ventilation model with computer simulation software to improve long-term ventilation planning, create a tool to evaluate ventilation costs, and as a result to ensure an efficient ventilation system.

The ventilation survey started with collecting the data of the current ventilation system. It was renewed during the design and building of the New Mine. The data was only a couple of years old and the planning was performed to enable current mining stage, the New Mine, until the year 2015. The problematic part was the information concerning the old mine, which was fractured, hard to interpret. Because of that a series of mine audits was required in order to verify some of the older data.

3.2 *Measurements*

The project started with air velocity and cross-sectional area measurements. Air velocity measurements with a thermoanemometer were performed several times in order to check the effects of different outside surface temperatures on the airflows, and the distribution of air with different fan airflow and pressure settings.

Then humidity, temperature, pressure, and particle measurements as well as a flue gas analysis were performed. Humidity and temperature measurements with a hygro- and thermometer were done due to many reasons that were for example the depth of the mine, which causes high temperature problems, fog in the decline, and exploring the effects of humidity and temperature differences on overall ventilation. Pressure measurements were made with a barometer to compare the planning of the New Mine with the current reality, to check for pressure losses, and to examine the pressure conditions of the mine. The motivation for the particle measurements was to check the distribution of particles in different parts of the mine, especially in the working areas and nearby the crushing station, and to ensure that the contents were under the limits given by The Finnish Ministry of Social Affairs and Health (Sosiaali- ja terveysministeriö, 2002). An aerosol meter, which has a large particle size range of 0.5 μm–10 μm, was used to unravel the total amount of breathable particles, PM_{10}, in the air. Flue gas analysis was made mainly for comparison purposes, since the mine has an independent monitoring system. In the flue gas analysis the values of O_2, CO, CO_2, NO, and NO_X were measured.

4 MODELING

4.1 *General*

In the modeling Ventsim ventilation simulation software Version 3.32 was used. Since the new part of the mine was modeled with Surpac, these files were used, and transferred into Ventsim. The model was built by combining the usable Surpac parts, and the information received from the old mine maps and coordinates.

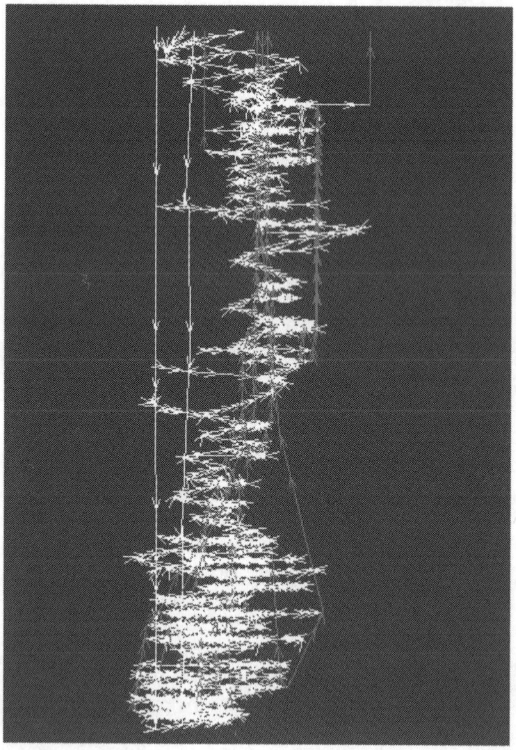

Figure 3. Ventilation model.

Fan curves were found from every fan type used in the mine. Also airway information concerning the New Mine was very thorough and accurate. The ventilation control program used in the mine gave reference values, like pressure and airflow values, that could be used in modeling.

4.2 *Encountered difficulties*

The problems encountered in modeling were the huge amount of airways, the complicated structure of the mine, and the missing parts of the old material. There were some areas that one could not check because they were totally closed and some estimation was required about the characteristics of those areas. Still another difficulties appeared in modeling the old backfilled areas, because air leakages were discovered. The amount of leakage air, the routes it chose, places where it merged with other airflows, the resistances, and the shock losses were difficult to predict and to model. During testing of the model it was noticed that also the stepless invert control of some of the fans was complicated to model.

5 RESULTS

5.1 *Measurement results and comments*

Humidity measurements showed that after entering the mine the air absorbs some water on the way. The humidity values rise a little already in the fresh-air shaft probably because of the moist rock surface. Even higher rise is indicated when airflow reaches the tunnel network. The use of water in drilling and in dust prevention increases the moisture of the air. Also, the leakages in upper parts of the decline through joints and broken rock areas increase humidity. The New Mine area is much dryer, having a good rock quality with no leakages.

The flue gas analysis confirmed the knowledge about the flue gases by the continuous detectors installed in the mine. The values were similar and well under all limits. The results from the particle measurements were more interesting. There were increased values noticed in a couple of working areas. The scaling site with mechanized scaling equipment was one of them. Another site with high values was one drift face. The limit value of 10 mg/m^3 was, however, only exceeded in the scaling site, and the high values observed near scaling and loading work was considered temporary. In both cases operators were working in isolated cabins equipped with air-conditioning. Areas around the crushing plant actually gave lower values than expected. The reason for this was most likely the recently improved moistening system. The values in the decline were also somewhat higher than the values of the horizontal tunnels, especially in foggy areas. The decline is used as one of the exhaust air channels and speed of airflow varies along it. The particle graph is shown in Figure 4.

The program used in particle analysis is TrakPro 3.3, which was acquired with the measurement device. The first high and narrow peak is measured in the scaling site, the other high peak, reaching above the value of 6 mg/m^3, is from the mucking site. Other higher values are almost entirely from the decline.

Air velocity measurements gave values that proved the ventilation system of the New Mine working well. Since these measurements were the only ones performed more than once, quite reliable results are considered to be achieved. In the other measurement results temporary disturbances were noticed, and after reporting, these were ignored in analyzing functioning of the ventilation system as an entity.

5.2 *Results of the modeling*

In modeling the biggest difficulty was that the adjustment of the main fan affected almost all airways by changing the air pressures, velocities and thus also the airflows. In some airways even the direction of the airflow changed with the fan changes. This caused difficulties in both modeling and interpretation of the model. Commenting things like placing of ventilation walls seemed purposeless because they proved to be useful in one situation and a hindrance in the other because of the restrictions of modeling. All this also made changing of the network and then checking the results of the changes quite unstable. Some assumptions, like the velocity changes in the decline were made and these proved to be correlating well with the measurement results, but the thorough knowledge of the network that was aimed for wasn't achieved. To enable more complete understanding thorough re-research of the results and further testing of the model are currently being worked.

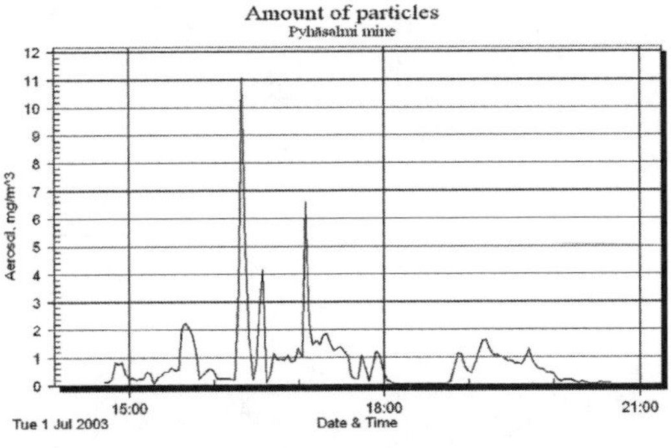

Figure 4. Particle graph.

When the model was deemed sufficient for comparison purposes two different situations of two months, March and April 2003, were examined and compared. The comparison showed that opening of new stopes changes the air circulation mostly on the New Mine area, where all these changes occur, as expected. In levels with fans and ducts the air circulation was affected less than in levels with no active working areas.

5.3 *Ventilation requirements*

The ventilation requirements based on the usage of diesel equipment, was performed in the planning stage of the New Mine and in this study it was only updated. Still the updated version of the model showed that the amount of needed fresh air is well below the current maximum fresh air capacity. This capacity meets current Finnish mining standards. There is some re-circulation of exhaust air in the lowest levels. The re-circulation is, however, quite well controlled, and occurs only near working sites with minor diesel emissions.

5.4 *The energy survey*

The energy survey as well as the cost analysis based on it has not yet been completed by October 2003. The installation of energy measurement devices has been delayed. As the energy measurement starts to work as planned, the comparison between the results of the model and the measurement results will be completed. Also a comparison between the costs of the current exhaust system and a less complicated system presented earlier in the text in Chapter 2.6, will be done in order to check how much additional expenses the higher resistance of the current system causes. Then the savings will be compared with the price of a new, shorter route, and possibly with fewer fans, and report on the matter given accordingly.

6 CONCLUSIONS

As the project still continues all these issues are further investigated. The model, even if still imperfect, seems to provide a considerable help in understanding the overall situation, and hopefully later also the details. Gathering the results of the study, completing the computer model, and building up improvement suggestions is still a work in progress.

Changing of the airflows and their directions due to fan control cannot be avoided with the current ventilation system of Pyhäsalmi mine. The complexity of the system and frequent changes in the mine structure imply that modeling could be helpful if performed in the future based on the model. This way the effect of changes could be investigated in advance and the achieved results then compared with the updated model.

All in all the ventilation system of the mine is working well. A lot of thought has been put to safety issues. The air quality is good, and air movement control has been updated frequently.

ACKNOWLEDGEMENTS

The authors would like to thank Inmet Mining for the possibility to study the mine and to write this paper as well as Helsinki University of Technology for providing the measurement devices and the opportunity to the work.

REFERENCE

Sosiaali- ja terveysministeriö 2002. *HTP – arvot 2002*. Tampere: Kirjapaino Öhrling. ISBN 952-00-1107-2, ISSN 1456-2561

Mine Ventilation, Ganguli & Bandopadhyay (eds)
© 2004 Taylor & Francis Group, London, ISBN 90 5809 633 5

A ventilation study of the Detroit Salt Mine

E.Z. Manos
The Detroit Salt Company LLC, Detroit, Michigan, USA

A.M. Wala
University of Kentucky, Mining Engineering Dept. Lexington, Kentucky, USA

ABSTRACT: The Detroit Salt Mine is an underground rock salt mine located in urban area within the city of Detroit, Michigan. Development of the mine started with the sinking of a shaft in 1906, and continues today as a modern industrial mineral mine producing over a million tons of crushed and screened rock salt per year. On September 27, 2003, a comprehensive ventilation investigation was performed at this historic salt mine. The purpose of the study was to determine the existing operating parameters of the mine's current ventilation system, including main fan characteristics, pressure and air quantity, shaft resistance and brattice wall leakage. This paper will discuss the results of this study in regards to historical ventilation; future mine planning and ventilation upgrades concerning underground mining's new health and safety regulations.

1 INTRODUCTION

The Detroit Salt Mine is a historical underground rock salt mine within the confines of a heavily populated and industrial area of Detroit, Michigan. From the mines beginning in 1906 to the present day, the salt mines ventilation system has continued to evolve as equipment technology, mining methods and health and safety regulations have changed and expanded.

The Detroit Salt Mine is classified non-gassy and has employed different versions of the room and pillar mining method throughout its history. From 1910 to the 1940's the mines equipment was electric, including locomotive haulage, electric drill jumbo, under-cutter and shovels for loading thus requiring a limited and simple ventilation system. With the advent of diesel equipment in 1950's, the mine upgraded their ventilation system with a new fan, brattice walls and auxiliary ventilation fans. Remarkably, these upgrades allowed the mine to progress the room and pillar workings to a distance of over 3.5 miles from the intake and return shafts. In 1983, active salt extraction ceased for economic reasons and the mine went under care and maintenance until the year 1997. The mine was rehabil itated and put back into production in 1998 (Head & Manos 2000).

From 1998 to the present the mine has employed the room and pillar mining method with a combination of electrical and diesel equipment to limit the ventilation requirements needed to comply with the Health and Safety standards set by law. The recently promulgated regulation imposed on the underground mining industry has been the Diesel Exhaust Particulate Matter Standard. In July 2002, Mine Safety and Health Administration (MSHA) has set DPM exposure limits of 400 microgram per cubic meter ($\mu g/m^3$). A final exposure limit of 160 micrograms per cubic meter will become effective as of January 20, 2006, (MSHA 2001).

Currently, the Detroit Salt Mine meets the intern DPM limit of 400 $\mu g/m^3$, with existing mining and ventilation equipment. Due to the final DPM limit of 160 $\mu g/m^3$ imminent, the ventilation investigation commenced on September 27, 2003. The major goals of this investigation were: (1) perform a pressure-quantity survey to evaluate and analyze the existing ventilation system and develop numerical model of the system; (2) determine pros and cons to main fan, restrictive areas, shaft resistance, brattice wall leakage and check in lesser part, the significance of Natural Ventilation Pressure (NPV).

2 DETROIT SALT MINE VENTILATION SYSTEM

The Detroit Salt Mine ventilation system consist of two shafts, #1 and #2, that are 1140 ft. (348 m) in depth,

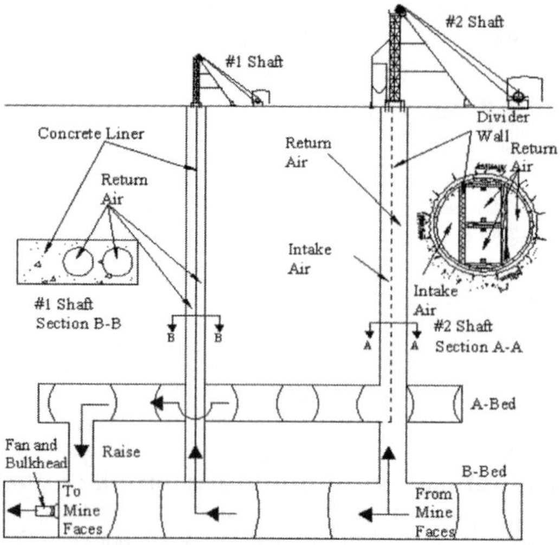

Figure 1. Diagram of the ventilation system at the Detroit Salt Mine (Manos 2001).

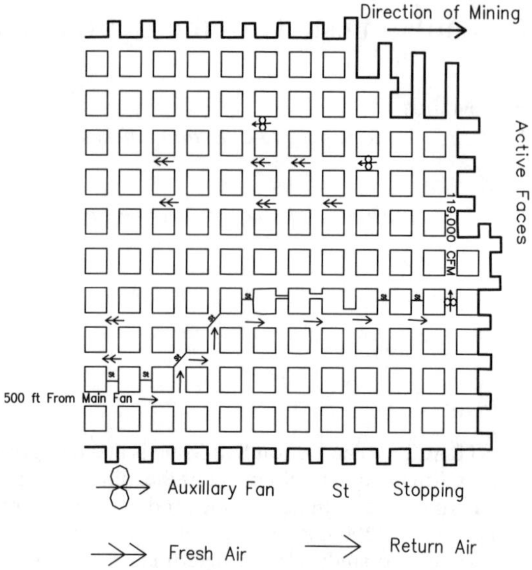

Figure 2. Typical face production ventilation system.

to the B-Bed mining level. There is however another salt bed called A-Bed, this bed is above B-Bed and provides an air course for fresh air into the B-Bed mining level through the raise.

The #1 Shaft was sunk in 1906–1909 and was originally a 9 ft. by 16 ft. (2.7 m × 4.9 m), three compartment shaft, that was reconfigured in 1928 to two, 42 in. (107 cm) diameter tubes separated and backfilled with cement its entire length to the B-Bed mining level.

The #2 Shaft was sunk during the years 1922–1926 and is a 16 ft (4.9 m) diameter concrete lined shaft that consists of 10 in. by 10 in. (25.4 cm × 25.4 cm) timbered sets on 5 ft (1.52 m) centers with wood guides for the skip to travel on. The #2 Shaft has many uses; it is used for skipping the salt, man trips and all the service requirements for the mine. The #2 Shaft connects to A-Bed with the fresh air plenum and extends to the B-Bed mining level.

A diagram of the ventilation system in use at the mine is shown in Figure 1. Intake air travels down a half moon section of #2 Shaft with an effective area of 34 ft^2 (3.2 m^2) into an upper salt bed (A-Bed), where it travels through multiple entries sized 40 ft by 20 ft (12.2 m × 6.1 m), and down a 10 ft by 10 ft (3.1 m × 3.1 m) raise to be drawn by the main mine fan located on B-Bed mine level, close to the base of the raise. The intake air then courses through the B-Bed workings to sweep the active mine faces; see Figure 2, and returns out naturally through the B-Bed workings and exhausts up #1 Shaft and the remaining compartments of #2 Shaft. Figure 2 shows the typical components of the mining face ventilation system at the Detroit Mine.

The room and pillar mining method is used at the Detroit Mine. Pillars are 70 ft (21 m) square and the rooms are 40 ft (12 m) wide by 22 ft (6.7 m) high. Due to the size of the mined openings, leakage between intake and return air has been a concern at the mine from the startup in 1998. The mine, however, has made its stopping wall construction a priority to eliminate as much leakage as possible. These stoppings consist of a plastic grid material (Tensar Geo Grid) with a non-woven felt glued on one side of the grid. The 2 in. × 2 in. (5.1 cm × 5.1 cm) grid comes in 25 ft (7.6 m) wide by 100 ft (30.5 m) rolls and placed in a typical room opening 23 ft (7 m) high × 40 ft (12.2 m) wide. The grid is then sprayed with a cement-type mine sealant up to ¼ in. (6.3 mm) thick on both sides. These walls are flexible enough to handle the concussion of the air blasts during blasting, and the minimal roof convergence that occurs at the Detroit Mine. These permanent stoppings however are not inexpensive and are labor intensive; but the low maintenance; non-flammable material and low air leakage more than offsets their initial construction costs. As an alternative construction method to the plastic grid and cement sprayed walls, a yellow brattice material (Venta Tex by ABC Industries Inc.) placed within a 2 in. by 2 in. (5.1 cm × 5.1 cm) metal tube frame system is also employed in the mine. When sprayed with an expanding sealant foam around the edges, also provides a nearly leak proof wall and is more economical and easier to install but doesn't have the durability of the sprayed walls.

The Detroit Mine uses a Jeffrey 8H-60 Aerodyne axial vane fan (serial number 9009), operated at 1600 rpm with a 200 HP (149.2 KW) motor to ventilate the mine workings. This fan was installed in 1949 and to date, continues to be used as the main mine fan. To distribute the air in the face area four auxiliary, 10 ft (3.05 m) diameter, 50 HP (37.3 KW), Jeffrey Aerodyne Jr. fans are being used. These fans running at 345 rpm produce 138,000 cfm (65.1 m^3/s) in a free discharge position (Dalzell 1969).

3 VENTILATION SURVEY

On September 27, 2003, a one-day ventilation survey, under normal operating conditions, was conducted at the Detroit Salt Mine.

The ventilation system at the Detroit Mine is a combination a forcing and exhausting system. The main fan located at the B-Bed has both, an inlet and discharge airways. This arrangement for the main fan is called a booster system (Hartman 1995).

The ventilation survey consists of pressure and flow measurements to determine the airways resistance needed to develop a numerical model of the existing ventilation system. This numerical model will be used to perform a computer simulation study for long-term environmental and ventilation planning.

3.1 *Pressure survey*

The barometric method with base was used to determine the pressure gradients and pressure losses in the system. This method involved setting up one altimeter at the base (surface or underground), where atmospheric pressure changes were measured and recorded relative to time. Another altimeter, used as a roving instrument was taken underground to measure the absolute pressures in previously pre-selected locations throughout the ventilation system. At each of these locations the time, air velocity and dry and wet bulb temperatures were also measured and recorded.

3.2 *Air quantity survey*

Prior to taking the measurements, the location of each station for air quantity measurements was selected and marked on the mine map. A vane anemometer (Davis Instrument), mounted to a telescoping survey range pole was used to determine the velocity of airflow at various locations throughout the mine. From knowing the area or size of the opening being measured, a volume of air passing that specific location can be determined. Due to the large size of the mine openings, traversing must be done with the outmost care, and diligence to measure accurately. Many locations were traversed during the ventilation study. These intake air locations included the base of the intake air plenum into A-Bed, and at the bottom of the intake ventilation raise between A, and B salt beds. Many locations on the main fan's, outby the last open cross cut before the fresh air enters the working faces and at the top of the #1, and #2 Shaft. These locations were chosen in order to provide the best possible data to allow for an accurate accountability of the mines air leakage between intake and return air, an inherent difficulty encountered when using a booster system such as the Detroit Mine.

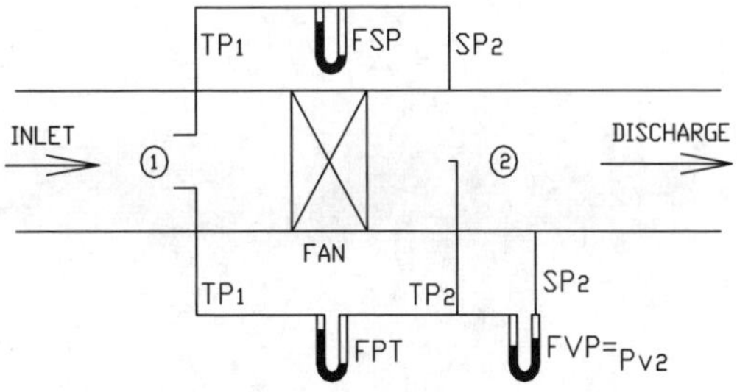

$$FPT = TP_2 - TP_1$$
$$FST = SP_2 - TP_1$$

Figure 3. Main fan booster system (McPherson 1993).

3.3 *Main fan test*

The single axial fan, Jeffrey Aerodyne Model 8HU-60, 1600 rpm, ventilates the Detroit Salt Mine. The ventilation system at the Detroit Mine is a combination a forcing and exhausting system. The main fan located in the B-Bed has both, an inlet and discharge airways.

As depicted from Figure 3, to determine the total, static, and velocity pressure for the Jeffery's 8HU-60 Aerodyne fan a digital manometer (Dwyer Model, Series 475 Mark III), and a 36 in. (1 m) long Pitot tube were used. Fan total pressure (FTP) is the algebraic difference between the total pressure at the fan discharge (TP_2) and the total pressure at the fan inlet (TP_1) or $FTP = TP_2 - TP_1$. Fan static pressure (FSP), is the difference between static pressure on the discharge (SP_2), and the total pressure at the fan inlet (TP_1) or $FSP = SP_2 - TP_1$. Fan velocity pressure (FVP) is the difference between the total pressure (TP_2), and the static pressure (SP_2) both on the discharge.

3.4 *Natural ventilation pressure*

An attempt to directly measure the Natural Ventilation Pressure (NPV) and its effects on the ventilation system were made, however, the time limitation didn't allow for completion of these measurements. NPV has been found to play a significant role in other northern

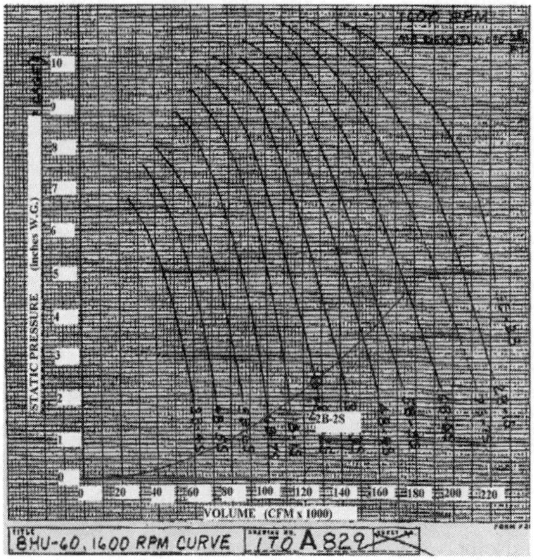

Figure 4. Main fan operating point and mine characteristic.

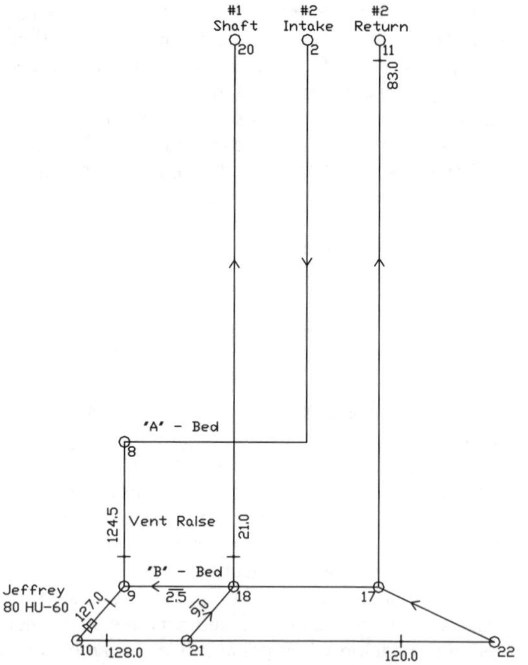

Figure 5. Location and measured flow distribution in the Detroit Salt Mine (kcfm).

climate mines (Wala et al., 2002) and plays a certain role in the Detroit Mine's ventilation system as well.

4 DISCUSSION OF THE SURVEY RESULTS

4.1 *Fan data*

Using the digital manometer and Pitot tube, total static and velocity pressures were measured for the Jeffrey's Model 8HU-60 Aerodyne axial vane fan running at 1600 rpm at the 2B-2S-blade setting. According to these measurements taken on the day of the survey, the fan was operating at 127,000 cfm (60 m^3/s) with a static pressure at 2.3 in. of water gage (572 N/m^2). The fan operating point, as an intersection between the fan characteristics for the 2B-2S setting, and the mine characteristics, is shown in Figure 4. The fan total pressure was measured at 7.1 in. of water gage (1,700 N/m^2). As can be seen from the Figure 4, the main fan is operating very inefficiently. An estimated efficiency is around 40 percent.

4.2 *Flow and pressure losses*

From the airflow survey it was determined that the mine fan is producing 127,000 cfm (60 m^3/s) and the fresh air intake entries are delivering 119,000 cfm (56.2 m^3/s) to the last open cross cut on the mining section approximately 5,500 feet in length as shown on Figure 5.

The low leakage measurement reflects the mines commitment to constructing and maintaining a competent, stopping wall system, to provide fresh air to its single production section. Figure 5 shows the locations and measured flows (kcfm) in the ventilation system as on the day of survey.

Concerning the pressure losses along the Detroit Mine, using absolute pressures with altimeters, there is a pressure loss of 2.0 in. water gage (500 N/m^2) from the intake plenum down #2 Shaft through A-Bed and down the ventilation raise to the fan and a pressure loss of 0.5 in. water gage (125 N/m^2) for the mine workings and the return air shafts.

5 DEVELOPMENT OF THE NUMERICAL MODEL

In essence, the completed mine ventilation survey provided the input data for the computer analysis.

This data includes the information about network structure, airway resistance, and fan performance. To establish the true representation of the mine's ventilation system as it existed at the time of the survey, the mine ventilation simulation software VnetPC, was used to develop the numerical model.

During the development process this model was analyzed using computer simulation method to correlate the measured and computed airflows and pressure distribution. Figures 6 and 7 show the simulated

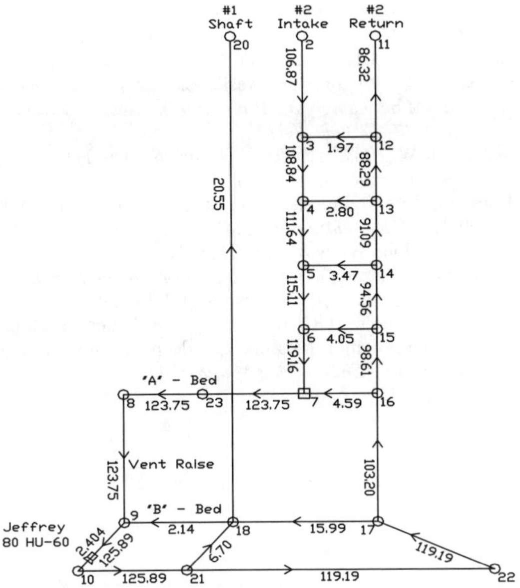

Figure 6. Simulated airflow (kcfm) distribution in the ventilation system of the Detroit Salt Mine.

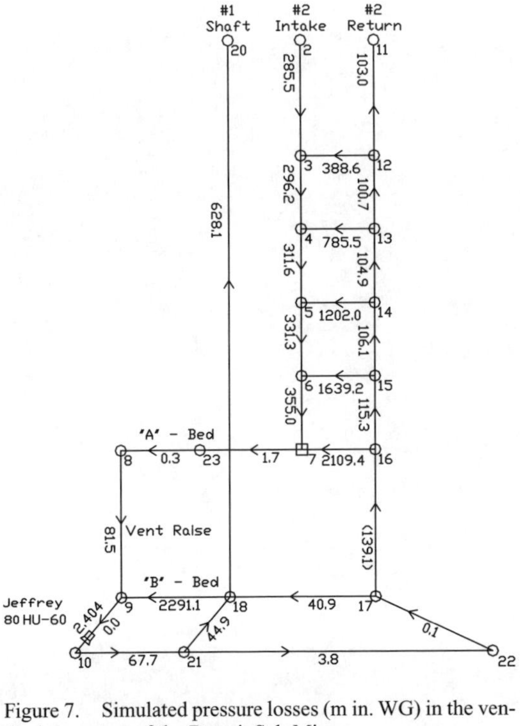

Figure 7. Simulated pressure losses (m in. WG) in the ventilation system of the Detroit Salt Mine.

pressure losses and flow distribution in the Detroit Salt Mine, respectively.

By comparing the simulation results with measured data it can be seen that both of them are very close. This means that numerical model could be used in the future for system analysis.

6 RECOMMENDATIONS

The major objective of the ventilation survey was to find the strengths and any shortcomings of the Detroit Mines ventilation system. A list of current recommendations and control strategies to compensate for the shortcomings has been compiled to address these issues in relation to future mine planning and the final DPM limit of 160 $\mu g/m^3$.

- Replace the 50 year old, inefficient main mine fan with a properly sized new fan that will almost double the quantity of air delivered for ventilation while maintaining the integrity of the mines stopping wall system.
- Upgrade the stoppings at all high-pressure locations, including those walls closest to the main fan and the divider wall down #2 Shaft.
- Structurally modify the transition from the #2 Shaft intake plenum to the A-Bed to reduce the pressure loss. The high-pressure loss at this location is caused by a narrow, sharp turn and high air velocities of 3750 fpm (18.6 m/s).
- In addition to the ventilation improvements, using alternative fuels such as Bio-Diesel and installing particulate filters on the larger pieces of diesel machinery may be needed.

7 CONCLUSIONS

All of the major objectives and goals of this ventilation study at the Detroit Salt Mine were met excluding, Natural Ventilation Pressure. To be able to determine the NVP effect on the ventilation system at the Detroit Salt Mine, the measurement will be taken using direct method (Wala et al., 2002) for extreme conditions (cold winter and hot summer days).

The pressure and quantity survey were performed and a numerical model of the Detroit Salt Mines ventilation system was developed. Any modifications and improvements can now be analyzed using computer simulation before implementation.

The strength of the current ventilation system was found to be, the low leakage though-out the entire mine. The low leakage measurements from this study are invariably due to the competent stopping system installed in the mine and the low pressures generated by the booster arrangement of the main mine fan.

ACKNOWLEDGEMENTS

The authors would like to express their gratitude to the management of Detroit Salt Company for their permission to publish this paper and to Viorel Mosincat (DSC Mine Engineer) for his depictions.

REFERENCES

Head, H.J. & Manos, E.Z. 2000. Detroit Salt Mine – Past and Future. In Rob M. Geertman (ed.), *Salt 2000 – 8th World Salt Symposium: Proc. intern. symp., The Hague, The Netherlands, 7–11 May 2000.* Amsterdam: Elsevier Science B.V.

MSHA, 2001. Diesel particulate matter exposure of underground metal and nonmetal miners. *Federal Register, Vol. 66, No. 13, Pages 5706–5910. 19 January 2001. US Mine Safety and Health Administration.*

Manos, E.Z. 2001. Detroit Salt Mine Ventilation Practices 1906–2001. In Stanslaw Wasilewski (ed.), *7th International Mine Congress; Proc. Intern. Symp., Cracow, Poland, 17–22 June 2001.*

Dalzell, R.W. 1969. *Absolute Pressure Survey*, Feb. 9–14, 1969, International Salt Co. Detroit, Mich.

Hartman, H.L., Mutmansky, J.M., Ramani, R.V. & Wang, Y.J. 1995. *Mine Ventilation and Air Conditioning,* Third Edition, John Wiley & Sons, New York.

McPherson, M.J. 1993. *Subsurface Ventilation and Environmental Engineering,* Chapman & Hill, London.

Wala, A.M., Stoltz, J.R., Thompson, E. & Pattee, S. 2002. Natural ventilation pressure in a deep salt mine-a case study, *Mining Engineering March 2002.*

Diesel in underground mines

Mine Ventilation, Ganguli & Bandopadhyay (eds)
© 2004 Taylor & Francis Group, London, ISBN 90 5809 633 5

Testing and reduction strategies for diesel emissions at PT Freeport Indonesia

Ian J. Duckworth, Ketut Karmawan & John Barber
PT Freeport Indonesia, West Papua, Indonesia

ABSTRACT: PT Freeport Indonesia (PTFI) operates a mining complex located in the highlands of West Papua, Indonesia. The underground workings comprise of a block cave mine, and a network of conveyors and orepasses to transport surface and underground ore to the concentrator. Presently underground operations are heavily reliant on mobile diesel equipment. PTFI is required to apply Indonesian mining regulations for occupational safety and health, which presently do not address diesel particulate matter (DPM). In an effort to ensure the continued well being of the workforce, PTFI have made the decision to move towards compliance with the US Mine Safety and Health Administration's regulations for DPM. This paper discusses diesel control strategies that are presently being evaluated, and in some cases already implemented at PTFI's underground operations. The work encompasses baseline monitoring, including vehicle tailpipe gas tests and ambient DPM and indicator gas surveys. The paper discusses various practical methods for achieving significant reduction in DPM concentrations. At PTFI these include fuel changes, improved fleet management and engine specifications, continued focus on vehicle maintenance, worker training, improved personal protection, rapidly addressing shortcomings in the primary and auxiliary ventilation systems, and a study on the suitability of electric production loaders. The ongoing diesel control work at PTFI is broad and long term. The foundation for this work has been the development of a comprehensive database of area and occupational DPM exposures.

1 INTRODUCTION

PT Freeport Indonesia's (PTFI's) project site is located in the Sudirman Mountain range of Papua (formerly Irian Jaya). Papua is the Eastern-most province of Indonesia, which occupies the Western half of the island of New Guinea. PTFI's first underground mine in Indonesia was the GBT Block Cave, which started production in 1980. By 1986 this mine was producing about 15,000 tonnes/day (tpd). In 1987 the underground operation was expanded to incorporate the DOZ Open-Stope Mine, with a total underground tonnage of 21,000 tpd achieved in 1988. In 1996 the IOZ Block Cave Mine started production with an initial tonnage of 10,000 tpd. Final production from the IOZ Mine exceeded 25,000 tpd. Development of the DOZ Block Cave Mine started in 1996, with initial production in 2000. The layout of the block cave deposits, and the associated mine infrastructure are shown in Figure 1.

Presently PTFI are mining from the Grasberg Open Pit and the DOZ Block Cave Mine, which total about 230,000 tpd of ore. At the time of writing (October 2003) approximately 45,000 tpd is being produced from the DOZ Mine.

The tonnages underground are expected to keep increasing as new underground mines are brought online to replace tonnage from the open pit. A final underground tonnage of about 150,000 tpd is planned once the Grasberg Open Pit is exhausted (projected closure date is 2014). With such a commitment to underground mining, PTFI have made the decision to investigate the requirements for full compliance with the US Mine Safety and Health Administration's (MSHA) diesel particulate matter (DPM) initiatives. The first step in this process has been the development of a monitoring program to determine baseline workplace and occupational exposures. This has taken the form of periodic DPM surveys throughout the DOZ Mine, with the samples sent off-site for analysis per MSHA guidelines. During the last 12-month period three complete DPM surveys have been conducted in the underground mine with about 240 samples collected. These surveys were conducted during August 2002, April 2003 and August 2003. This publication summarizes the measurement strategies and

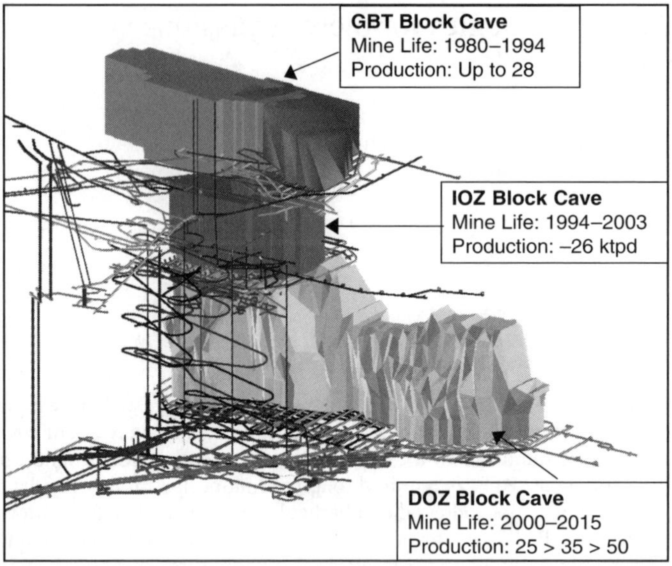

Figure 1. Developed block cave deposits at PTFI.

discusses the DPM reduction initiatives that are being examined.

2 DPM INFORMATION

The exhaust associated with diesel engines consists of various gases and diesel particulates. The gases include nitrogen, carbon monoxide, carbon dioxide, oxides of nitrogen, and sulfur dioxide. The particulates consist of soot, unburned fuel and aldehydes. Diesel particulates are generally of an extremely small size (less than one micron [μm]), which causes them to be easily inhaled and retained in the body.

The US mining industry is phasing in stringent regulations relating to diesel particulates in underground mines. In many cases these regulations represent the benchmark for other countries. Exposure limits for DPM have been adopted by MSHA for metal/nonmetal underground mines. This rule introduced an interim threshold limit value (TLV) for the ambient concentration of diesel particulates at 400 $\mu g/m^3$ (mass of DPM per volume of air). The value for the final TLV is still undergoing examination by MSHA.

These limits are as measured by the NIOSH 5040 method, with the DPM defined by total carbon (TC) content. In addition to setting these limits, MSHA have also developed an engine approval program, and mandates that fuel be low sulfur (less than 500 ppm [<0.05%] sulfur by weight), and that all fuel additives be approved by the US Environmental Protection Agency.

3 BASELINE MEASUREMENTS BASELINE INFORMATION

When analyzing data obtained from DPM surveys it is important to first understand the various contributing factors. At PTFI these are Table 1.

Three mine-wide DPM surveys have been conducted by PTFI to quantify worker exposure levels. The surveys have included personal sampling (for occupational exposure) and area sampling.

3.1 *Occupational exposure*

Personal sampling for DPM exposure at PTFI consists of fitting different workers with DPM pumps and filter assemblies at the start of the shift, recording cumulative exposure throughout the shift, then collection of the pumps at the end of the shift. Observations and measurements are taken throughout the shift to record such factors as equipment details, ventilation, work practices, and any unusual events. Through careful selection of individuals according to job description and work location, a baseline exposure level has been determined for the different jobs conducted underground. This has allowed high-risk jobs to be identified and steps to be taken to reduce worker exposure. Figure 2 shows relative DPM exposures based on job description, date that the measurements were taken, and the number of samples that are associated with each occupational group. The results do fluctuate, as is expected in a dynamic mine environment. However, it is possible to formulate a ranking for the risk associated with the

Table 1. Typical factors contributing to DPM concentration.

Maintenance	A comprehensive maintenance program is in place that incorporates an aggressive preventative maintenance schedule.
Engine types	The DOZ Mine has a large diesel fleet to support the development, construction and production activities. The majority of the loaders in the mine are Elphinstone R1600 (201 kW) and R1700 (231 kW) units that have either Tier 1 or 2 rated engines. The production trucks are Elphinstone AD-55 units that all have Tier 2 rated engines (485 kW). There are also many service vehicles in the mine including light trucks, graders, scissor lifts, tractors, drills, buses and concrete/shotcrete trucks, many of which do not have Tier-rated engines.
Fuel and additives	Presently the underground operations use approximately 2–3% of the total diesel fuel consumed on the site. The daily site consumption is about 850,000 liters of diesel fuel used for the mining operations, transport fleet, and generators. The bulk of this fuel is shipped to Amamapare, Papua (lowlands port) from a refinery in Singapore, and is pumped up to the mine via a 100 km long pipeline. The fuel has an average sulfur content of about 0.2%. Additives are not included with the underground fuel. The present fuel specification requires a minimum cetane index of 47 with a cloud point of no higher than 0°C.
After treatment	Presently there are no particulate filters or scrubbers used on the underground equipment.
Ventilation	PTFI ventilation systems have historically been designed based on an airflow allowance of 7.9 $m^3 s/100$ kW of diesel equipment. Control of airflow in the block cave mine is affected by the many parallel ventilation splits, the requirement for zero obstruction along haulage routes, and the need for the flexibility of multiple working areas.
Work practices & training	PTFI have a comprehensive equipment operator training program. The present training does not specifically address DPM mitigation issues.
Fleet management	The identification, removal, and repair/refit of gross polluters is conducted based on feedback from tail pipe testing, visual observation and preventative maintenance.
Enclosed cabs	The production trucks are all equipped with enclosed cabs. The loaders do not have enclosed cabs.
Respirators	Most production loader operators wear respirators. There are however other underground occupations and locations where respirators are not worn. The use of respirators is strongly advised for all underground workers.

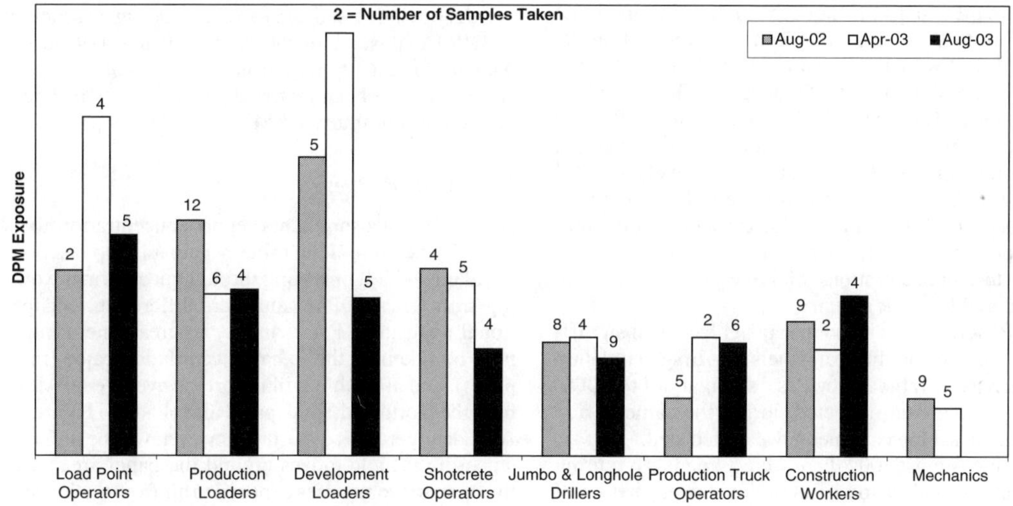

Figure 2. Occupational exposure to DPM.

different jobs, as given in Table 2. The highest occupational exposure is ranked 1.

Presently PTFI are using the MSHA interim TLV as the target (400 μg/m^3) for compliance purposes. Once MSHA release final ruling on the DPM TLV, PTFI will amend as required. The data indicate a general reduction in DPM occupational exposure underground over the past year. This is the result of increased airflow quantities through the production areas, and an ongoing effort to remove gross-polluters from the

Table 2. Ranking of occupational exposure to DPM.

Job description	Rank
Load point operators	1
Production loader operators	2
Construction workers	3
Development loader operators	4
Production truck operators	5
Shotcrete operators	6
Jumbo and longhole drillers	7
Mechanics	8

Table 3. Ventilation details for the DOZ block cave mine.

	Date of Measurements		
Parameter	8-02	4-03	8-03
Airflow (m^3/s)			
Total all panels (ext level)	269	349	500
Truck level	371	385	333
Other areas (shops, belts)	446	399	347
Total mine	*1,086*	*1,133*	*1,180*
Number of active panels	14	17	20
Average dry tonnes	28,500	40,400	41,600
Tonnes air/tonnes ore	3.1	2.3	2.3
Airflow/panel (m^3/s)	19.2	20.5	25.0

underground fleet. The loader and load point (LP) operators are the groups that have the highest occupational exposure to DPM. Loading points are located on the Truck Level, which allow ore dumped into the panel grizzlies above to be fed into the trucks. The LP workers are located above the trucks and close to the flowing ore, which places them in an area with less ventilation, and potentially exposes them to higher levels of DPM and dust. The construction workers fall into the next category. The drillers, production truck operators, shotcrete operators and mechanics are all exposed to relatively low concentrations of DPM and as such are considered low risk groups.

The increase in airflow per panel has been significant even as the number of panels requiring ventilation has increased. Table 3 provides a summary of the DOZ Mine airflow data collected during the same period that the diesel measurements were recorded.

During August 2002 there were 14 panels requiring ventilation, with a total Extraction Level airflow of 269 m^3/s. This corresponds to an average airflow per panel of 19 m^3/s. By August 2003 the number of open panels was 20 (+43%), the average daily production was 41,600 tpd (+46%), and the total airflow on the Extraction Level increased to 500 m^3/s (+86%). The airflow increase within the panels was possible through improved control and a general reduction of airflow supplied to other mine areas. During the last year the total mine exhaust increased by 94 m^3/s (+9%), which is primarily the result of the upgrade of a main mine exhaust fan.

Figure 3 shows a plan view of the DOZ Extraction Level as per October 2003. Panels #13–24 are in production, with a further eight panels open for development and construction activities (with the exception of Panel #9, Panels #6–27 have all been broken through). The arrows denote the direction of airflow movement. It can be seen that each panel is broken into two ventilation segments with fresh air being delivered from the North and South fringe drifts, and exhausted through the central panel exhaust raises. Typically there are two production loaders per panel, with one operating in the North ventilation split and one in the South. The layout therefore allows for single-pass ventilation across the equipment. Fresh air is delivered to the North by a dedicated intake drift supplying the West side, and a booster fan installation on the East (not shown). The shaded section to the West denotes the undeveloped DOZ reserve. The other outlines indicate reserves that may be added to the DOZ production footprint.

The reduction of DPM exposure associated with loader operators is more obvious when the sample data are examined independently.

Figure 4 plots relative DPM concentration against panel number for the three surveys. Different numbers of samples were taken during each survey depending on the particular focus and objectives at the time, however an obvious trend can be seen showing a reduction in DPM exposure. In the case of Panel #19 during August 2002 a ventilation issue was identified. This issue has since been rectified, and a reduction should be seen during future DPM surveys in this panel.

3.2 *Area DPM levels*

Area DPM measurements report concentration for a specific location in the mine, where a pump has been installed and left operating throughout a normal working shift. These DPM values are different to occupational exposures. For example, an area panel sample may be taken at the exhaust ventilation raise (mid-panel), and as such it will report the average DPM for both the North and South panel air splits (see Figure 3). A loader operator's exposure, however, will be different since the vehicle moves around the panel area, with the operator sometimes exposed to air from the loader's exhaust and other times is breathing relatively fresh air.

Figure 5 is a chart showing the average measured DPM levels for different areas of the DOZ Mine on a relative basis. The production panels and Truck Level have the highest DPM concentrations, however as with the occupational samples the average level has been reduced over the last year. The area concentrations in the development and construction panels, and in the

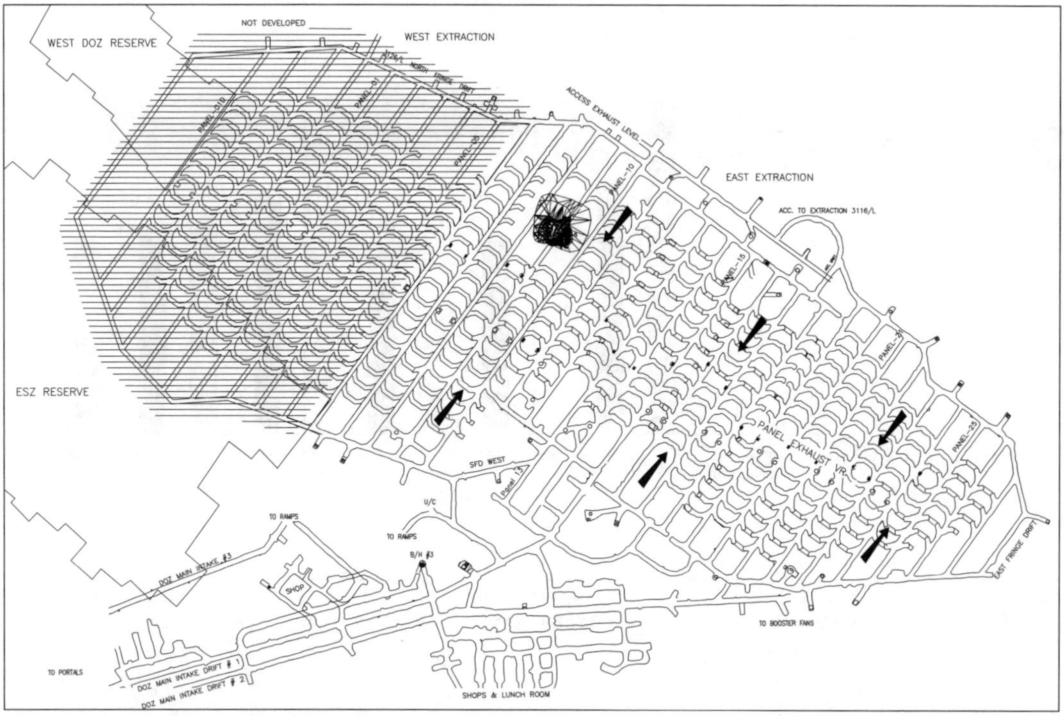

Figure 3. Plan of the DOZ extraction level.

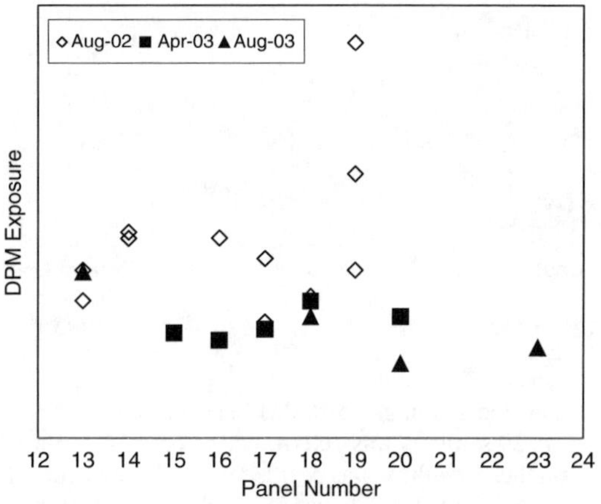

Figure 4. Loader operator exposure by panel number and date of measurement.

Undercut drifts are significantly lower. Figure 6 plots relative DPM concentration for a typical loop through the DOZ Mine. DPM is a cumulative pollutant since the particles typically remain entrained within the airstream. As such the DPM concentration increases gradually from the portal to the end of the intake drifts. The ambient DPM levels then rise rapidly when the air is used to ventilate the Undercut and Extraction Levels. Mixing of air on the main exhaust level from different sources results in a drop in average exposure at the end of this level.

4 OTHER STUDIES

Throughout the last year certain other studies have been conducted using data obtained from the DPM surveys. These include:

- In August 2003 testing was conducted on a new auxiliary ventilation system to provide air to the LP operators on the Truck Level. An LP was fitted with a small forcing fan that picked air up from a selected location below the loading platform and blew the air onto the operator. DPM testing showed that there was a slight reduction in occupational exposure when the fan was running, however the difference was less than 5%. Due to the many variables involved the test was considered to be inconclusive, and additional data will be collected during the next DPM survey.
- In April 2003 a cassette study was undertaken. Standard 37 mm filter cassettes used for PTFI's dust sampling program cost about 15% of the specialized SKC DPM cassettes presently used for the DPM surveys. A decision was made to conduct parallel

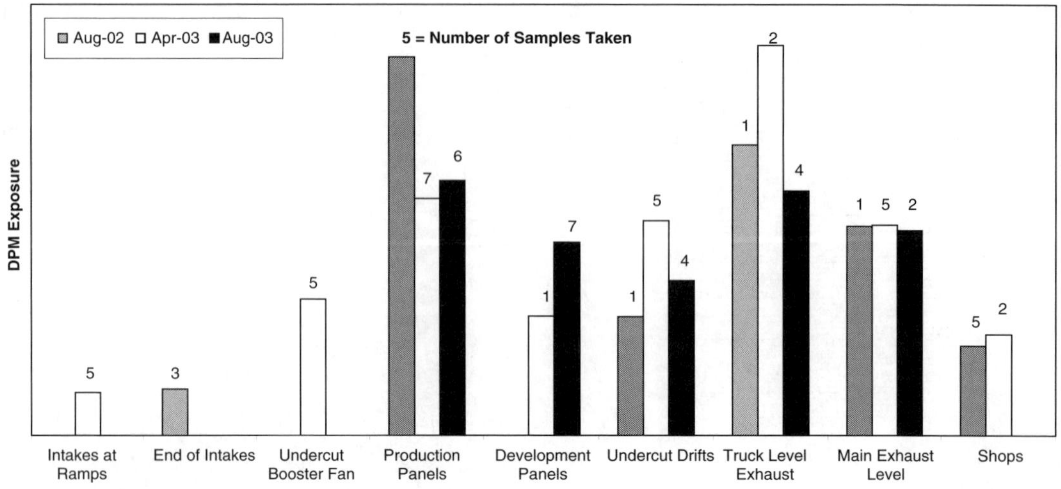

Figure 5. Area DPM measurements.

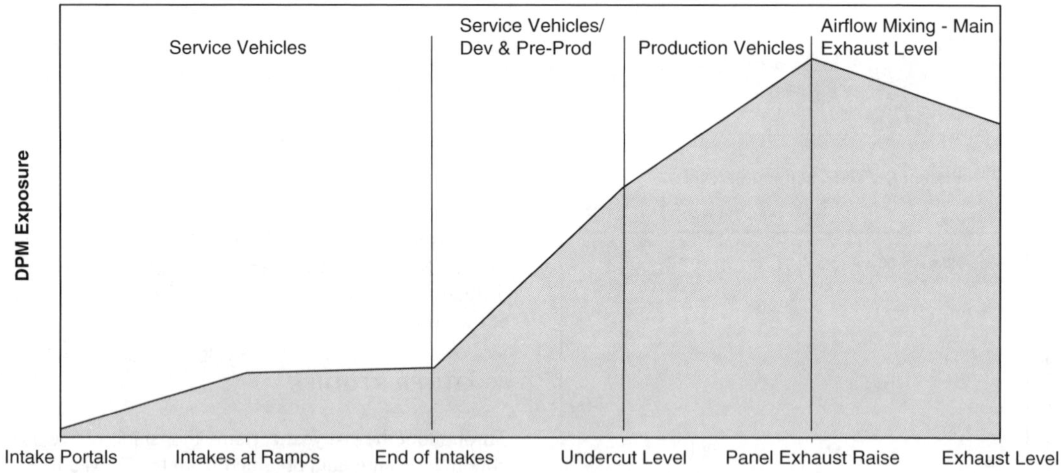

Figure 6. Area exposures through the DOZ loop.

sampling to determine the difference in DPM collection. The SKC DPM cassette is specifically designed for sampling in conditions where it is necessary to differentiate DPM from other respirable dust. The SKC DPM units consist of a plastic cassette within which are loaded a precision-jeweled impactor, impaction substrate, a heat-treated quartz filter for sample collection, and a second quartz filter to be used as a blank. The precision-jeweled impactor screens out respirable particles greater than one micron. Particles less than one micron are collected on the filter. A total of 20 parallel samples were collected and the results compared. In general it was found that the standard cassettes resulted in elevated readings for DPM. The average error for the 20 samples was determined to be +16%. Based on these results it was decided that there was not sufficient confidence to warrant a change of cassette.

- Another study was undertaken during April 2003 to monitor average Carbon monoxide (CO) levels in the same locations that DPM samples were being collected in an attempt to trend CO against DPM (potentially allowing CO to be used as a rough indicator for DPM). This test failed to provide an acceptable correlation between CO and DPM. There are many factors that impact the generation of DPM in exhaust, some of which appear not to effect gas emissions to the same degree.

5 REDUCTION INITIATIVES

PTFI have established a committee devoted to the long-term reduction of diesel particulates in the underground operations. This committee consists of representatives from the Underground Ventilation and Industrial Hygiene Group, Underground Engineering Department, Maintenance, Management and Freeport Corporate Engineering.

5.1 *Baseline monitoring and analysis*

PTFI have made a commitment to continue DPM monitoring on a 3–6 month basis or as needed to determine the impact of ventilation changes, fuel tests, and other studies. The number of samples taken per test will be in the order of 80. Presently all samples are sent to a laboratory in the US (Mine Ventilation Services, Inc.) that offers NIOSH 5040 analysis. This does result in a significant delay while the cassettes are shipped and analyzed, and there are additional costs. During 2004 it is planned that the underground industrial hygiene facilities will be expanded to incorporate a DPM laboratory.

5.2 *Engine types*

Presently PTFI are purchasing new equipment with Tier-rated clean burning engines. Furthermore, an initiative is underway to start replacing engines in the support fleet with Tier-rated units. This would be done on a gradual basis, with perhaps 10 engines replaced each year. In some cases this change is not straightforward and requires modifications to the vehicle chassis to fit the new engine.

5.3 *Diesel fuel*

Diesel fuel quality has been identified as a key component for improvement. However, logistically this will be challenging. Due to the extremely remote location of the mine, all fuel received must be pumped 100 km to the mines through dedicated pipelines. An assessment of the transportation process suggests that it will be difficult and expensive to separate out a different fuel for the Underground. However, through working with the fuel vendor it is anticipated that an affordable solution will be found such that an improved fuel can be provided to the job site.

5.4 *Work practices*

PTFI's operator training program will be expanded to emphasize the importance of limiting such practices as excessive engine idling and lugging, and to stress the recognition and removal of poorly functioning equipment from the active fleet.

5.5 *Ventilation*

At the end of 2004 the DOZ Mine ventilation system is scheduled to undergo a significant increase in primary fan capacity. This system upgrade will (1) increase the quantity of airflow through the Extraction and Truck Levels such that the optimum design airflows are achieved, and (2) continue to meet future demands as the mine expands to the West. The planned upgrade incorporates a new 5.5 × 6.5 m exhaust adit connecting the DOZ Mine exhaust level to the surface (approximately 2 km long). A primary underground fan installation will exhaust approximately 350 m^3/s through this new drift. The total airflow in the DOZ Mine will not increase by this full amount due to the existing primary fans being forced to operate slightly higher on the fan curves, however the total capacity increase will be considerable.

5.6 *LP areas*

Since LP operators are exposed to the highest levels of dust and DPM, three initiatives are underway to improve conditions. Two of these will help reduce exposure for the immediate future, with the third offering a long-term solution. For the near future an LP auxiliary ventilation system has been installed to provide fresh air to the loading pocket, and visors with powered filtration systems have been purchased and will be provided to the LP operators. For the long-term automated chute control by the truck drivers is being tested. From design work conducted so far this system will consist of CCTV linked to monitors in the trucks, and ultrasonic detection to ensure that the truck is correctly positioned under the chute. The system will also be tied to laser ore level detectors to ensure that the orepass is not fully emptied (this can result in higher dust concentrations and chute damage when ore is dumped from the top).

5.7 *Electric production loaders*

PTFI are considering the purchase of an electric production loader for testing in the East panels of the DOZ Mine. Such units have been successfully used in other caving mines including the North Parkes and Kiruna mines. Electric loaders have the advantage of zero engine pollution, and can rapidly provide 100% load without having to rev-up the engine (potential improvement in productivity). The heat produced by the equipment will also be considerably less per ton of rock moved compared to a diesel unit due to the electric engine being much more efficient. Disadvantages include safety and maintenance aspects of the trailing cable, and reduced flexibility. Ventilation will still be required to control dust and heat, however it is anticipated that the air quantity can be reduced by integrating dust suppression into the haulage design (may consist of sprays at the draw points and grizzly, and wetting of the panel drifts).

6 CONCLUSIONS

Significant progress has been made during the first year of DPM sampling and analyses at PTFI. Three detailed DPM surveys have been successfully conducted with accurate baseline data collected for all the main underground work places and occupational groups. These measurements have enabled risk levels to be categorized for the different occupations.

PTFI are committed to a program of continued reduction in worker exposure to DPM and have a series of initiatives in place that are aimed at improving conditions and achieving MSHA compliance.

ACKNOWLEDGEMENTS

The authors would like to thank PTFI Management for allowing the information contained within this text to be published. The support of Mine Ventilation Services, Inc. with data collection and analysis is also gratefully acknowledged.

REFERENCE

MSHA, *Code of Federal Regulations 30*, Part 75.

Mine Ventilation, Ganguli & Bandopadhyay (eds)
© 2004 Taylor & Francis Group, London, ISBN 90 5809 633 5

Conducting a comprehensive baseline study for Diesel Particulate Matter

J.D. Stinnette & K.G. Wallace, Jr.
Mine Ventilation Services, Inc., Fresno, California, USA

ABSTRACT: The Mine Safety and Health Administration (MSHA) now mandate that metal/non-metal mine operators comply with the newly enforced Diesel Particulate Matter (DPM) interim exposure limit of 400 $\mu g/m^3$ (30 CFR § 57.5060(a)). As a result, many operators are faced with difficult and sometimes costly decisions in how to reduce DPM exposure. Given the wide range of available controls, such as increased ventilation, improved equipment engine maintenance, diesel exhaust filters, and fuel additives, it is vital that a comprehensive DPM baseline study be performed in order to identify the most effective DPM reduction strategy for each individual mine. This paper describes the approach Mine Ventilation Services, Inc. (MVS) uses to perform such a study. A baseline study should be based upon a sampling program, which includes measurements of the present DPM exposure, diesel equipment utilization, equipment maintenance, ventilation rates, and other components that can affect the level of DPM exposure. The results of this sampling, when properly assessed, will assist operators in targeting the areas of greatest need/concern when developing a comprehensive compliance strategy for the reduction of DPM (e.g. increased ventilation in select areas with a combination of equipment exhaust filters and select engine replacement). The paper also describes the sampling strategy necessary to continue monitoring DPM exposure after control measures have been implemented, both for verifying the performance of any implemented controls and ensuring future compliance. The MSHA regulations and standards included in this paper are subject to change and revision, however, every effort was made to ensure that any such references were correct at the time of writing.

1 INTRODUCTION

As Diesel Particulate Matter (DPM) and Diesel emissions have become a major area of concern for many mine operators and workers, the options for the treatment and reduction of such emissions have increased. A myriad of control strategies and products are available to consumers, all of varying scopes, efficiencies, and expense. Noting the high cost of some of the available options, the question often arises of what controls to apply where, and how a combination of options can provide the most efficient use of the allocated resources. The basis for any complete DPM compliance strategy should be a comprehensive baseline study, including ambient air sampling, laboratory analysis, analysis of laboratory results and sampling data, and development of a realistic plan for DPM reduction. A thorough program of this nature can identify areas of concern within the system and provide the operator with the ability to target specific concerns with the control technology best suited for the application. Using the information gathered from the sampling and analysis, an initial compliance strategy may be developed. This "targeted" approach allows the operator to concentrate their efforts on the areas of greatest need, and addresses their concerns in the most efficient manner possible. An important part of this process is the realization that there is no one answer for the reduction of DPM exposure.

Once the compliance strategy has been implemented, continued sampling can measure the effects of any DPM reduction controls, and will indicate if control measures are functioning as intended. Additionally, this post-study sampling complies with the 30 CFR § 57.5071 requirements for sampling "as often as necessary" to ensure compliance and monitor DPM exposure levels throughout the mine

2 SAMPLE COLLECTION

Adequate sampling should form the basis for any comprehensive baseline study and compliance strategy. The conditions and exposure levels vary greatly from mine to mine, and between separate locations within the same mine. As a result, it becomes highly important to evaluate the DPM levels at each individual mine, and in as many areas and for as many processes as possible to ensure that no potential DPM sources are overlooked.

Ambient air samples replicate the process used for compliance determination by MSHA (30 CFR § 57.5061), and therefore, give results that are easier to correlate than do tailpipe emissions tests or other methods of DPM sampling. This method involves the use of a specialized, wearable pump designed to deliver a constant flow (volume rate) with an accurate timing device. The pump is connected to a "sample train" which includes an MSHA approved Jeweled-Impactor cassette (with quartz-fiber filter), cyclone, and holder assembly.

The purpose of the on-site sampling is to provide a "snap-shot" of the DPM levels throughout the mine under representative conditions. In order to accomplish this, it is necessary to record all of the pertinent parameters which potentially have an effect on either DPM emissions or the sampling results. Such critical factors include the pump calibration and timing process, the direction and quantity of the ventilation, the condition of local ventilation structures/controls, the type and number(s) of equipment in the sample area (as well as the engine type and manufacturer), visible clues such as smoke or haze, and other notes stemming from direct observation of the sampling process. The importance of accurately reporting such information cannot be overstated. Laboratory values of Total Carbon are utterly useless without them.

Consider a sample returned to a mine operator taken on a LHD operator which indicates a DPM exposure of 900 $\mu g/m^3$, more than twice the limit set by MSHA. At first glance, it would seem as though this is a clear example of an area which needs immediate attention. However, consider how the reaction of the mine operator might differ if he was also aware that during the testing, the auxiliary ventilation tubing was damaged, and the access drift in which the testing occurred received only 10% of its normal airflow. What would be done if notes indicated that the equipment was smoking heavily during the test, and had since been serviced by the maintenance shop?

Information gathered during on-site sampling is an integral part of the sampling process, and should be given equal weight in the development of a reduction strategy as the laboratory-defined carbon levels of the samples themselves. In light of this, MVS has developed and refined a sampling protocol intended to adequately record as much pertinent information as possible while sampling. This method has been used by MVS to sample DPM levels in mines and underground facilities both in the U.S. and abroad.

3 LABORATORY ANALYSIS

The method used by MSHA for the measurement of DPM exposure for the purpose of compliance determination is known as the NIOSH 5040 method. This method involves the determination of Organic Carbon (OC) and Elemental Carbon (EC) as a surrogate for DPM by a thermal optical analyzer. First, DPM is collected on a specially manufactured quartz fiber filter using standard air sampling equipment. A portion of the filter with a known area is then removed and placed in a quartz "oven". The Organic and Elemental Carbon in the sample are measured using a two-stage thermal-optical analysis technique. In the first stage, Organic and Carbonate Carbon are evolved in a helium atmosphere, where the temperature is gradually stepped up to 850°C. The evolved Carbon is oxidized catalytically to Carbon Dioxide (CO_2), and eventually reduced to Methane (CH_4) in a nickel-firebrick methanator. A flame ionization detector quantifies the CH_4. Stage Two involves the introduction of an Oxygen-Helium mixture into the atmosphere of the oven, and another gradual temperature increase, this time to 900°C. As this occurs, pyrolytic Carbon is oxidized and filter transmittance increases. The initial increase in laser transmittance is associated with pyrolytic Carbon. Once the transmittance of the filter reaches its initial value, all Carbon volatized after that is considered Elemental Carbon. A specially designed and application-specific software aids in the quantification of the EC.

MVS adheres to strict quality assurance procedures to ensure the integrity of the analysis at all stages. This process includes a procedure for logging the samples in and out of our refrigerated storage container, a record of all analyses, a system for monitoring the accuracy of the sampling apparatus, and the archiving of all samples for a minimum of six months following the analysis. Before any analysis is conducted, the accuracy of the thermal-optical analyzer is tested with a series of calibration standards. This is done by depositing known volumes of calibration solution with known carbon content onto pre-cleaned quartz filters. The filters are then analyzed in the instrument, and the measured carbon content of each sample correlated with the calculated theoretical content. So called "instrument blanks" are run at a rate of approximately one blank for every five samples. This is when a pre-cleaned filter is analyzed to purposefully obtain a "zero-point" for carbon levels and verify the instrument results. Additionally, duplicate samples are analyzed at the rate of approximately one in every ten samples and evaluated for consistency. The goal of MVS's quality assurance program is to ensure that the highest possible degree of confidence in the analysis results is maintained.

The results from the laboratory analysis are given in units of micrograms (μg) of Total Carbon (TC) present on the filter. TC is defined as the sum of OC and EC. The results from the thermal-optical analyzer are presented in units of micrograms of OC, EC, and TC per square centimeter of filter ($\mu g/cm^2$). In cases of carbonate interference, the carbonate fraction of

OC is removed by integration with the aid of specialized software. The TC present on the filter is calculated by multiplying the TC given from the analysis ($\mu g/cm^2$) by the filter deposit area (cm^2), resulting in TC on the filter (μg).

The original design of the Jeweled-Impactor cassettes (JIC) specified by MSHA for compliance testing was such that the filter deposit area varied significantly from cassette to cassette. However, a recent modification to the cassette design has resulting in a much more uniform deposit area. It is important to note that while the laboratory results are reported in units of μg of TC present on the filter, the compliance determination will be made based on the exposure of the filter in units of $\mu g/m^3$. This requires additional calculations beyond the scope of most laboratories.

4 ANALYSIS OF LABORATORY RESULTS

Once the results of the laboratory analysis are obtained, the exposure levels associated with each sample must be calculated. This requires the correlation of the pump calibration data and the corresponding TC level from the laboratory analysis. The exposure level of the sample is given by the following equation:

$$E = \frac{TC(Q_p)t}{1000}$$

where: E = DPM exposure ($\mu g/m^3$)
TC = total carbon (μg)
Q_p = pump flow-rate (l/min)
t = pump run-time (min)

Once all of the exposure rates have been calculated, a database of sampling information can be compiled, consisting of pertinent information recorded for each individual sample. This information includes the pump calibration information and the sampling notes, as well as the final exposure level. By grouping all of this information together, it is possible to evaluate the exposure data in a meaningful context. In order to aid this process, MVS has developed specialized software for the analysis of DPM exposure levels for the specific applications found in mining. MVS begins by sorting the data into a series of groups, based on various parameters associated with exposure levels. The exposure levels are analyzed in terms of ventilation rates and velocities, operators, activity type, equipment type, and location in order to spot trends, and identify specific problems within the system both on a macro and a micro scale.

For example, careful analysis of the mine-wide exposure trends may identify a recurring problem within an isolated mining district, or a single piece of equipment that continually exceeds the exposure limit, regardless of location or activity. Although both result in repeatedly high levels of DPM, the two situations described above are quite different in nature, and require two equally disparate solution strategies.

Such information about the nature of the problem is critical in identifying the action or combination of actions that result in the most effective application of resources, and the greatest reduction in DPM. Only after the entire database has been thoroughly evaluated in this manner should the process of developing a compliance strategy begin.

5 COMPLIANCE STRATEGY DEVELOPMENT

The development of a comprehensive DPM compliance strategy for a mine or underground facility should be based upon a thorough knowledge of the existing conditions of the underground. Once the nature of the problem is known, it is possible to go about devising a solution or solutions to correct any undesirable situations. During this process, it is important to consider that there is no panacea for the reduction of DPM, and that no one solution is suitable for all underground facilities or applications.

Although the operations and conditions at each underground mine or facilities are unique, several potential solutions will exist for the reduction of DPM. Based on the information provided by the analysis of the laboratory results, it will be possible to craft strategies incorporating the DPM reduction controls and techniques most likely to provide the greatest benefit, resulting in a "custom-fit" plan for emissions reduction.

This will likely incorporate multiple DPM controls applied in various scopes, from mine-wide application, to the modification or replacement of individual pieces of equipment. The types and quantities of the controls applied will have a direct effect on the results achieved by their application. However, it is important to note that in such a dynamic system, specific results are nearly impossible to predict. Each control applied, will not only effect the propagation of DPM in the underground, but may also affect the usefulness of other measures employed. Consider that the high sulfur content of some off-road Diesel fuel can completely preclude the use of most particulate filters and exhaust after treatment devices. Catalyzed Diesel particulate filters (DPF's) have high efficiencies and favorable regeneration characteristics, but can also generate increased NO_x emissions, which may place an added burden on the ventilation system. Although the individual problems identified in the analysis stage must each be addressed, it is important not to lose sight of the effects of any proposed changes on the whole system.

Once the compliance strategy is complete, the final product can be utilized to comply with the requirements of § 57.5062 of the 30 CFR, provision for a written "...Diesel particulate matter control plan..." if such a plan is deemed necessary.

6 MONITORING AND VERIFICATION

Once the compliance strategy has been completed and instituted, the question arises of how to monitor the effectiveness of the various changes initiated, and verify that the desired results are being achieved. Section 57.5071 of the 30 CFR also requires that air sampling occur "as often as necessary" to ensure compliance with the rule. This is done in a similar manner as the initial air sampling, with an added emphasis on any known problem areas, or hazards identified during the initial sampling process. Occasionally, when the desired results are not achieved by the initially prescribed solution, the process of evaluation, recommendation, and verification should be repeated until the pre-determined goals of the study are met.

7 SUMMARY

This paper describes a comprehensive approach to performing a baseline DPM study at an existing underground mine or facility, and using the data collected during that study into a dynamic compliance strategy for the mitigation of Diesel particulate emissions. By using this approach, including all of the relevant sampling parameters and complete analysis, it is possible to develop a plan to match specific control devices and techniques to specific applications, in order to achieve the greatest possible efficiency and to maximize the effects of any controls instituted as part of the compliance strategy. Such a study would satisfy the regulatory requirements of § 57.5071 of the 30 CFR, while assisting the mine operator in meeting the current 400 $\mu g/m^3$ DPM interim limit for metal/non-metal operations (30 CFR § 57.5060).

REFERENCES

Mine Safety and Health Administration, Federal Register 30 CFR Part 57, Final Rule – Diesel Particulate Matter Underground Metal and Non-Metal Mines, Vol. 66, No. 13, January 19, 2001

NIOSH Method ID 5040, "Elemental Carbon" (Diesel Particulate), revised January 1998

Standard Operating Procedure (SOP) For the Analysis of Organic and Elemental Carbon (OC/EC) on Quartz Fiber Filters, Sunset Laboratory Inc., March 8, 1999

Thermal/Optical Analyzer, Sunset Laboratory Inc., 1998

Mine Ventilation, Ganguli & Bandopadhyay (eds)
© 2004 M.J. Schultz, ISBN 90 5809 633 5

Using biodiesel fuels to reduce DPM concentrations; DPM results using various blends of biodiesel fuel mixtures in a stone mine

M.J. Schultz, K.G. Fields, D.J. Atchison, E.J. Gerbec, R.A. Haney & D.M. Tomko
Mine Safety and Health Administration, Pittsburgh Safety and Health Technology Center, Pittsburgh, Pennsylvania, USA

G. Love & G. Kuhnhein
Carmeuse North America, Inc., Maysville & Black River Mines, Maysville & Carntown, Kentucky, USA

ABSTRACT: This paper discusses a collaborative effort among Mine Safety and Health Administration (MSHA) Industry, and the Kentucky Department of Energy to test diesel particulate matter (DPM) emissions and exposures when using various blends of biodiesel fuels in two underground stone mines. The study was initiated by the industry partner, with MSHA and National Institute for Occupational Safety and Health (NIOSH) providing support for study design, data collection and sample and data analysis. Project funding was provided by Carmeuse North America, Inc., and the Kentucky Department of Energy through the Kentucky Clean Fuels Coalition.

The study evaluated mine DPM emissions when using various biodiesel to diesel fuel mixtures. Two types of biodiesel were evaluated, a recycled vegetable base biodiesel fuel mixture and a soy based biodiesel fuel mixture. Baseline conditions were established using 100% low sulfur No. 2 diesel.

1 INTRODUCTION

Diesel particulate surveys were conducted at the underground Maysville Limestone Mine, Maysville, Kentucky and Black River Limestone Mine, Butler, Kentucky. The purpose of the study was to evaluate the effect of utilizing biodiesel fuel mixtures on diesel particulate emissions and personal exposures in underground nonmetal mines. Management of Carmeuse North America, Inc., had obtained outside funding to evaluate the effectiveness of biodiesel fuels. The Kentucky Clean Fuels Coalition coordinated the funding, to conduct the surveys, which was provided from two outside sources. This funding was provided by The Kentucky Division of Energy and by Griffin Industries, Inc.

Two types of biodiesel fuels were studied. They were a recycled vegetable oil (RVO) and a virgin soy oil (VSO). Various concentrations of each fuel were tested. The biodiesel fuel mixtures and baseline study were tested during normal underground mining operations. The studies were jointly conducted by MSHA personnel and Carmeuse North America, Inc., personnel. MSHA personnel involved in these studies were from the Pittsburgh Safety and Health Technology Center, Dust Division and Metal and Nonmetal Mine Safety and Health, Southeastern District.

Table 1 shows the dates and the biodiesel fuel mixtures studied at the Maysville Limestone Mine.

Table 2 shows the dates and the biodiesel fuel mixtures studied at the Black River Limestone Mine.

Table 1. Maysville Mine biodiesel fuel mixtures.

Date	Biodiesel fuel
DEC 10–12, 2002	20–80% biodiesel mixture (RVO)
JAN 7–9, 2003	50–50% biodiesel mixture (RVO)
FEB 4–6, 2003	No. 2 low sulfur diesel fuel (baseline)
APR 1–3, 2003	50–50% biodiesel mixture (VSO)

Table 2. Black River Mine biodiesel fuel mixtures.

Date	Biodiesel fuel
MAR 18–19, 2003	No. 2 low sulfur diesel fuel (baseline)
APR 8–9, 2003	35–65% biodiesel mixture (RVO)
APR 29–30, 2003	35–65% biodiesel mixture (VSO)

Each study period lasted for two weeks. During the study period, all the underground diesel equipment was fueled using the appropriate blend of biodiesel fuel. The first week of each study was used to purge the equipment and fuel distribution system from the previous fuels used. Sampling was conducted during the second week of the survey.

2 BACKGROUND

2.1 *Maysville Limestone Mine*

The Maysville Mine, located in Mason County, Kentucky, is an underground limestone mine owned and operated by Carmeuse North America, Inc. The Camp Nelson Limestone formation is mined. The mine operates two 10-hour production shifts per day to produce approximately 3.5 million tons of limestone annually. The active mining area is approximately 1,000 feet deep. Mined entries are approximately 45 to 50 feet wide with the final mining height ranging from 50 to 60 feet. The limestone deposit is mined using a regular room-and-pillar, heading-and-bench mining method. The headings are approximately 20 to 24 feet high and the bench ranges in height from 30 to 40 feet. This process results in a mine layout consisting of an upper level, which eventually was shot down to the lower level creating room heights over 60 feet.

A conventional mining system, where the limestone is drilled and blasted, was used to advance heading or mine the benches. This process consisted of drilling the face or the floor and then loading the drilled holes with ammonium nitrate and fuel oil (ANFO). The blasting sequence was initiated at the end of each shift. A two-hour idle period followed blasting to allow for the gasses and other contaminants to be removed by the ventilation system. The broken stone was then loaded at the faces by front-end-loaders into 40-ton haulage trucks. The trucks transported the material to a crusher and a belt feeder. The conveyor system carried the stone from the crusher area and out of the mine via an 18° slope. On the surface, the stone was further crushed and screened. The diesel equipment used to mine limestone included: front-end-loaders, haul trucks, scalers, roof bolters, face drills, a grader, a dozer, a water truck, a service truck, an explosives truck, fork lifts, and tractors.

Primary airflow was induced into the mine using ventilation fans located underground at the base of two vertical shafts. The intakes were the elevator shaft and the 1 West shaft, located off of the M-roadway in 1 West drift between panel 4 North and 4 South. Each fan installation consisted of a set of fans that worked in parallel with each other, installed side-by-side.

The ventilation system induced an average mine airflow of approximately 878,000 cubic feet per minute (cfm) for the months of December, January, and February. An average mine airflow of approximately 892,000 cfm was calculated during the baseline survey and 748,000 cfm for the tests conducted during the month of April. The change in airflow was the result of a change in natural ventilating pressures (NVP) occurring between the baseline survey and the VSO 50–50% survey. Surface air temperature changed from 40° Fahrenheit during the baseline survey to 85° Fahrenheit during the VSO 50–50% survey. During the baseline survey, NVP assisted with the mine mechanical ventilation, but during the VSO 50–50%, NVP worked against the mine mechanical ventilation.

Air entered the mine at the elevator shaft and the 1 West intake shaft. Air was then coursed to the working areas by air walls. Air walls were constructed of conveyor belting material approximately 10-feet long and anchored to the mine roof. Recycled fines, or waste rock, that had been brought back into the mine were then placed under the belting to complete the air wall. Freestanding auxiliary fans, which had no ductwork or tubing assisted ventilating the working panels. Intake air was coursed throughout the mine to two exhaust areas: an exhaust shaft and the belt slope. An average airflow of 629,000 cfm was measured exhausting out of the main exhaust shaft and an average of 249,000 cfm was measured exhausting out the slope for the months of December, January, and February. During the baseline survey, an average airflow of 640,000 cfm was measured exhausting out of the main exhaust shaft and an average of 250,000 cfm was measured out of the slope. During the VSO 50–50% survey, an average airflow of 530,000 cfm was measured exhausting out of the main exhaust shaft and an average of 220,000 cfm was measured exhausting out of the slope. The exhaust shaft was located between 1 North and 2 North panels and the slope was located near the elevator shaft.

2.2 *Black River Limestone Mine*

The Black River Mine, located in Pendleton County, Kentucky, is an underground limestone mine, also owned and operated by Carmeuse North America, Inc. The Camp Nelson Limestone formation is mined. The mine operates two 10-hour production shifts, 4 days per week to produce approximately 2.5 million tons of limestone annually. The limestone deposit is mined using a regular room-and-pillar, heading-and-bench mining method. The headings are approximately 24 feet high and the bench ranges from 30 to 40 feet in height. Mined entries are 30 to 40 feet wide. This process results in a mine layout consisting of an upper level which eventually is shot down to a lower level.

A conventional mining system where the limestone is drilled and blasted was used to mine the limestone deposit. This process consisted of drilling the face or the floor and then loading the drilled holes with ANFO.

The blasting sequence was initiated at the end of each shift. A two-hour and fifteen-minute idle period followed blasting to allow for the gasses and other contaminants to be removed by the ventilation system. The broken stone was then loaded at the faces by front-end-loaders into haulage trucks. The trucks transported the material to a jaw crusher which fed onto the mine belt. The belt conveyor system carried the stone from the crusher area out of the mine via the exhaust air slope. On the surface, the stone was further crushed and screened. The diesel equipment used to mine limestone included: front-end-loaders, haul trucks, scalers, face drills, a roof bolter, a grader, a grease rig, a water truck, a service truck, an explosives truck, and tractors.

Primary airflow was induced into the mine at two intake locations. Air was then coursed to the working areas by air walls and auxiliary fans. The freestanding auxiliary fans, which had no ductwork or tubing assisted ventilating the working panels. Intake air was coursed throughout the mine to two exhaust areas: an exhaust shaft and the belt slope. Airflow varied from 147,000 cfm to 179,000 cfm exhausting out of the main exhaust shaft during the study and varied from 67,000 cfm to 109,000 cfm exhausting out of the slope.

2.3 *Biodiesel*

Biodiesel is a methyl ester product produced by combining methanol oil or feedstock, then adding a catalyst. Glycerin is spun off during the refining process with the remaining product being termed a monoalkyl ester known as biodiesel. Biodiesel can be made from a variety of feedstocks including soybeans, rapeseed, canola, and palm oil as well as from recycled vegetable oils. The RVO, more commonly referred to as "yellow grease", was supplied by Griffin Industries, Inc. The VSO was supplied by Peter Cremer North America.

3 SAMPLING LOCATIONS

3.1 *Maysville Limestone Mine*

Six area and five personal samples were collected during each day of the study. Area samples were collected at two main intake locations: one at the bottom of the elevator shaft and one at the outlet end of the dual intake fans located at the 1 West shaft. Return samples were also taken: two side-by-side samples at the bottom of the return shaft and another two side-by-side samples approximately 400 feet up the slope entry.

Five personal samples were collected on each shift and included drillers, roof bolters, scalers, loaders, truck drivers, and a powderman. Smoking was permitted underground in the mine. There were not enough nonsmokers working production, therefore smokers and nonsmokers were selected for sampling.

3.2 *Black River Limestone Mine*

Seven area samples and five personal samples were collected during each day of the study. Area samples were collected at the two main intake locations: near the North intake airshaft and the South intake airshaft. Return side-by-side area samples were also taken at the bottom of the return airshaft and the exhaust slope. The North jaw crusher dump point was also sampled. Five personal samples were collected on each shift. Although individual workers were selected based on availability, the same occupations were sampled for each phase of the study. Occupations sampled were a loader operator, two truck drivers, a scaler, and a roof bolter. Smoking was permitted underground in the mine. Since individual workers varied at the occupations during different phases of the test, their smoking and nonsmoking designations changed.

4 ANALYTICAL

4.1 *Sampling confidence range*

The study sampling was designed to determine whether concentration reductions were significant at the 95% confidence level. Prior to the test, it was assumed that the baseline exhaust concentration would be approximately 400 $\mu g/m^3$ with a standard deviation of 20% and the reduction from biodiesel fuel would be approximately 20%. A "t-test" was used to determine whether reductions were significant. The critical "t" values for a 95% confidence limit range from 2 to 3. Using the following "t-test" equation:

$$t = \frac{x_i - x_o}{s / (n)^{\frac{1}{2}}}$$

where:

x_i = initial concentration
x_o = final concentration
s = standard deviation
n = sample size

Solving for the sample size "n" gives:

$$n = \frac{t^2 \times s^2}{(x_i - x_o)^2} = \frac{2.5^2 \times (80)^2}{(400 - 320)^2} \approx 6 \textit{ samples}$$

To allow for variability and mine operational delays, a sample size of 6 was selected. This resulted in the collection of 2 samples per day for 3 days at each of the mine exhaust air locations (shafts and slope). The critical "t" value at 95% for a two-sided test with 5 degrees of freedom (6 – 1 samples) is 2.571. This value was used in the analysis of the data to confirm significance.

4.2 *Sampling procedure*

Individual area and personal samples were collected with SKC, Inc., diesel particulate sampling cassettes. This cassette includes a submicron impactor and a quartz fiber filter. All sampling units used 10-millimeter nylon preseparator cyclones. Samples were collected using SKC pumps and MSA Elf's calibrated and operated at 1.7 liters per minute (Lpm) of airflow.

The airborne carbon samples were analyzed by MSHA, Pittsburgh Laboratory and NIOSH, Pittsburgh Research Laboratory according to NIOSH Method 5040. Elemental carbon (EC), organic carbon (OC), and total carbon (TC) values were determined from the samples collected. This method uses a thermal/optical carbon analyzer to determine the OC and EC matter per square centimeter of filter surface. Separation of different types of OC is accomplished through temperature ramping over time and controlled atmospheric conditions. Carbonaceous minerals are separated at a temperature of 750°C (fourth OC peak). The carbonaceous mineral content, evolved at the 750°C peak, was subtracted from the OC portion of the analysis using the software capability of the analytical program. OC and EC were added together to obtain the TC. A field blank correction was also applied to the carbon measurements. If the field blank correction resulted in a negative carbon measurement, the carbon measurement was defaulted to zero. Concentrations of carbon were calculated from the following formulas:

$$carbon\ conc.\,(\mu g/m^3) = \frac{C(\mu g/m^3)\times A(cm^2)\times 1{,}000\left(\frac{L}{m^3}\right)}{1.7Lpm\times Time(min.)}$$

and

$$TC = EC + OC \ or \ TC = EC\times 1.3$$

where:
C = The corrected OC or EC, concentration measured in the thermal/optical carbon analyzer.
A = The surface area of the filter media used. The surface area of the filters is 8.04 cm^2.

All area sample concentrations were based on actual sampling time resulting in time weighted averages (TWAs). For MSHA enforcement activities, normal MSHA Metal and Nonmetal protocol is to base all personal samples as shift weighted averages (SWAs). SWA calculations use 480 minutes as the sampled time regardless of the time sampled. The personal samples are reported as SWAs.

5 RESULTS AND DISCUSSION

Table 3 contains the TWA summary of the average area DPM sampling for the Maysville Mine survey.

The weighted average DPM concentrations were obtained by multiplying each location's air quantity by its respective DPM concentrations, and then dividing the sum of these products by the combined total air quantity. Concentrations and percent reductions based on TC = EC + OC or TC = EC × 1.3 were similar.

The total exhaust airflows were similar for the baseline, RVO 20–80%, and RVO 50–50% surveys; therefore, a comparison of concentrations can be used to assess the impact of these blends of biodiesel fuels. The VSO 50–50% survey was conducted under different weather conditions with warm spring weather outside. The baseline survey was conducted during the mid-winter weather conditions. The weather conditions significantly affected the amount of air ventilating the mine. Total return air quantities measured during the baseline survey averaged 892,000 cfm. This average air quantity was reduced to 748,000 cfm during the VSO 50–50% survey. The air quantity ventilating the mine has a direct affect on the DPM concentrations in the mine. Because the airflows are different in these two surveys, the comparison of effectiveness must be made using the total mass of

Table 3. Maysville Mine average area samples.

	Baseline		RVO 20–80%		RVO 50–50%		VSO* 50–50%	
	TC = EC × 1.3 (μg/m³)	TC = EC + OC (μg/m³)	TC = EC × 1.3 (μg/m³)	TC = EC + OC (μg/m³)	TC = EC × 1.3 (μg/m³)	TC = EC + OC (μg/m³)	TC = EC × 1.3 (μg/m³)	TC = EC + OC (μg/m³)
Weighted return	352	321	235	225	109	121	178	175
Percent reduction	–	–	33%	30%	69%	62%	49%	45%

* results adjusted for changes in airflow and intake concentrations.

particulate emitted. This value is obtained by multiplying the concentration by the airflow.

Table 4 contains the TWA summary of the average area DPM sampling for the Black River Mine survey.

The airflows during the baseline and RVO 35–65% surveys were similar; therefore, a direct evaluation of DPM concentrations can be made. A direct comparison of DPM concentrations could not be used to assess the VSO 35–65% survey, because during this survey there was a decrease in airflow due to the changes in the natural ventilation of the mine. For this comparison the mass of diesel particulate (airflow times concentration) from the baseline was compared to the mass diesel particulate from the VSO 35–65% test.

Figure 1 shows a graph of the biodiesel fuel type versus percent DPM reduction for the Maysville Mine. The percent DPM reductions are shown for the RVO 20–80%, RVO 50–50%, and VSO 50–50% surveys for TC = EC × 1.3 and TC = EC + OC. The biodiesel fuels were shown to be effective in the decrease of DPM exposures with the greatest DPM decrease utilizing RVO 50–50%.

Figure 2 shows a graph of the biodiesel fuel type versus percent DPM reduction for the Black River Mine. The percent DPM reductions are shown for the RVO 35–65% and VSO 35–65% surveys for TC = EC × 1.3 and TC = EC + OC. The biodiesel fuels were shown to be effective in the decrease of DPM exposures with the greatest DPM decrease utilizing RVO 35–65%.

Table 5 shows the SWA summary of the average personal DPM concentrations inside and outside of the cabs for the Maysville Mine using TC = EC × 1.3. This TC formula was used for all personal sample calculations because it removes all carbonaceous interferences.

Table 6 shows the SWA summary of the average personal DPM concentrations inside and outside of the cab for the Black River Mine using TC = EC × 1.3. This TC formula was used for all personal sample calculations because it removes all carbonaceous interferences.

Personal samples had many variables affecting their exposure to DPM concentrations; therefore, they are not as useful in determining the effectiveness of the biodiesel fuels as are the area samples.

Nitrogen dioxide (NO_2) diffusion tubes were also collected with all of the samples taken. The highest

Table 4. Black River Mine average area samples.

	Baseline		RVO 35–65%		VSO* 35–65%	
	TC = EC × 1.3 (μg/m³)	TC = EC + OC (μg/m³)	TC = EC × 1.3 (μg/m³)	TC = EC + OC (μg/m³)	TC = EC × 1.3 (μg/m³)	TC = EC + OC (μg/m³)
Weighted return	693	626	480	439	582	526
Percent reduction	–	–	31%	30%	16%	16%

* concentrations adjusted due to decrease in airflow.

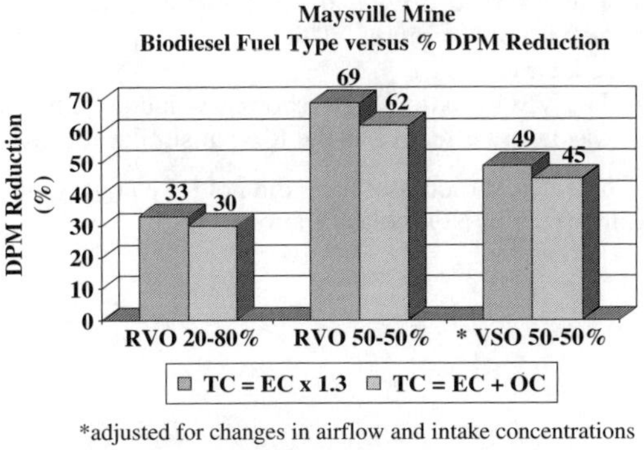

Figure 1. Graph of Maysville Mine biodiesel fuel type versus percent DPM reduction.

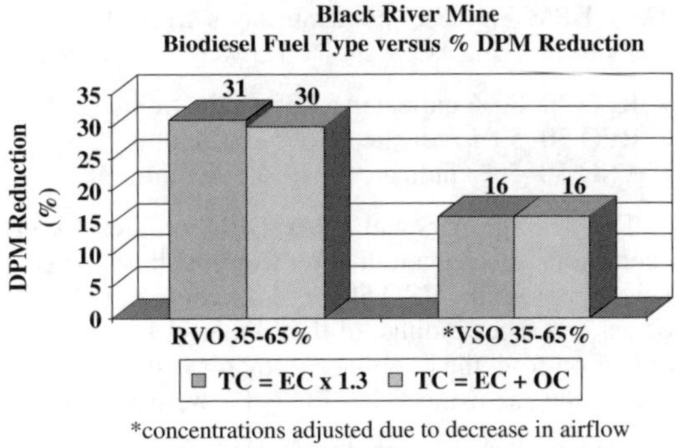

Figure 2. Graph of Black River Mine biodiesel fuel type versus percent DPM reduction.

Table 5. Maysville Mine average personal DPM concentrations inside and outside of cab; TC = EC × 1.3 (μg/m³).

	Baseline	RVO 20–80%	RVO 50–50%	VSO 50–50%
Average workers inside of cab	220	219	89	212
Percent reduction	–	0%	60%	4%
Average workers outside of cab	300	208	216	313
Percent reduction	–	31%	28%	4%

Table 6. Black River Mine average personal DPM concentrations inside and outside of cab; TC = EC × 1.3 (μg/m³).

	Baseline	RVO 35–65%	VSO 35–65%
Average workers inside of cab	1,611	731	967
Percent reduction	–	55%	40%
Average workers outside of cab	1,369	771	1,013
Percent reduction	–	44%	26%

concentrations of NO_2 recorded were 1.5 parts per million (ppm) on the high scaler and the loader operator. The roof bolter, the downhole driller, and the loader operator had concentrations of 1.0 ppm. These numbers did not significantly change from survey to survey. There was no indication that NO_2 concentrations increased during the RVO 20–80%, RVO 50–50%, and VSO 50–50% surveys from the baseline survey.

6 FINDINGS

6.1 *Maysville Limestone Mine*

The Maysville Mine reductions in the weighted exhaust TWA DPM biodiesel concentrations from the baseline survey using TC = EC × 1.3 were as follows:

- RVO 20–80% indicated a 33% reduction
- RVO 50–50% indicated a 69% reduction
- VSO 50–50% indicated a 49% reduction

The average personal SWA DPM biodiesel concentrations showed a reduction from the baseline survey, except for the VSO 50–50% survey when outside of the cab. The findings of the biodiesel fuel percent reduction from the baseline survey for inside and outside of the cab using TC = EC ×1.3 were as follows:

- RVO 20–80% indicated a 0% inside cab
- RVO 20–80% indicated a 31% outside cab
- RVO 50–50% indicated a 60% inside cab
- RVO 50–50% indicated a 28% outside cab
- VSO 50–50% indicated a 4% inside cab
- VSO 50–50% indicated a 4% outside cab

6.2 *Black River Limestone Mine*

The Black River Mine reductions in the weighted exhaust TWA DPM biodiesel concentrations from the baseline survey using TC = EC ×1.3 were as follows:

- RVO 35–65% indicated a 31% reduction
- VSO 35–65% indicated a 16% reduction

A direct comparison of DPM concentrations could not be used to assess the VSO 35–65% survey, because there was a decrease in airflow due to the changes in the natural ventilation of the mine; therefore, the mass of the diesel particulate from the baseline survey was compared to the mass of the diesel particulate from the VSO 35–65% survey.

The average personal SWA DPM biodiesel concentrations showed a reduction from the baseline survey. The findings of the biodiesel fuel percent reduction from the baseline survey for inside and outside of the cab using TC = EC × 1.3 were as follows:

- RVO 35–65% indicated a 55% inside cab
- RVO 35–65% indicated a 44% outside cab
- VSO 35–65% indicated a 40% inside cab
- VSO 35–65% indicated a 26% outside cab

7 CONCLUSIONS

The use of biodiesel fuel mixtures reduces DPM concentrations in underground nonmetal mines:

- Two entire underground limestone mines had been switched over to various blends of biodiesel fuel mixtures. Subsequent DPM sampling has shown reductions in the weighted exhaust DPM concentrations at both mines from all biodiesel mixtures tested.
- The RVO 50–50% had the greatest reduction in DPM concentration. This mixture had a 69% reduction in TC.
- The VSO mixtures also showed reductions, but it was not as effective as the RVO at similar blends.

The use of biodiesel fuels did not have any significant effect on NO_2 concentrations.

Mine Ventilation, Ganguli & Bandopadhyay (eds)
© 2004 B. Rubeli, ISBN 90 5809 633 5

Evaluation of the contribution of light-duty vehicles to the underground atmosphere diesel emissions burden

B. Rubeli, M. Gangal & K. Butler
Natural Resources Canada, CANMET Mining and Mineral Sciences Laboratories, Ottawa, Ontario, Canada

W. Aldred
Falconbridge Limited, Kidd Creek Mining Division, Timmins, Ontario, Canada

ABSTRACT: A series of in-mine field studies were performed to determine the relative contributions of light and heavy-duty vehicles to the underground diesel emissions burden in a metal mine. Based on these tests, a model of the Kidd Creek mine diesel fleet was developed. Light-duty vehicles were responsible for 47% of the underground diesel particulate matter (DPM) burden of the model fleet, while heavy-duty (HD) vehicles were responsible for 53%.

1 INTRODUCTION

The aim of this work is to characterize diesel particulate matter (DPM) exhaust emissions from light-duty (LD) and heavy-duty (HD) vehicles in a metal mine and to estimate the relative contributions of both types of vehicles to the overall underground DPM burden.

The field study was divided into two parts. Part one focused on measurement of DPM from seven heavy-duty vehicles including five production load haul dump (LHD) vehicles and two haulage trucks. Part two focused on measurement of DPM emissions from eight light-duty vehicles including three pickup trucks, two tractors and three utility vehicles.

2 BACKGROUND

The multi-stakeholder, Diesel Emissions Evaluation Program (DEEP), is presently involved in two major field studies aimed at evaluating high-efficiency diesel filtration technology from the standpoint of their impact on the underground environment and their implementation in underground mines. These studies are primarily focusing on heavy-duty vehicles, although two light duty vehicles were tested as part of the study at INCO Stobie mine. The constantly increasing number of light duty vehicles in underground mines and limited knowledge on their contribution to overall diesel emissions motivated this study.

Data showing the relative contribution of light-duty vehicles to DPM concentrations in a test mine can be used to determine whether further research and work aimed at light-duty engine emissions are needed. The hypothesis is that enhanced efforts in emission control and maintenance aimed at the light-duty fleet may benefit the underground environment.

Traditionally, the size of the engine has been used to differentiate between light, and heavy-duty vehicles. In this study, vehicles that are not used in regular production cycles will be categorized as light-duty units regardless of horsepower. Higher horsepower units involved in ore, waste or fill handling will be considered heavy-duty vehicles. Because of the large size of the engines involved and the intensive nature of the work performed, underground diesel emissions are thought to originate mostly from heavy-duty production equipment. While heavy-duty vehicles are no doubt major contributors, new data suggests that light-duty vehicles could be responsible for a significant and possibly increasing portion of the airborne diesel emissions burden in the mining environment.

3 KIDD CREEK MINE DIESEL FLEET

The Kidd Creek Mine is a copper/zinc operation located in Timmins, Ontario. The ore body has been mined through three separate shafts known as the No. 1, No. 2 and No. 3 mines. There is ramp access from the surface to all areas of the mine. Kidd Creek mined 2.2 million tones of ore in 2002.

The Kidd Creek diesel equipment fleet comprises approximately 156 vehicles used underground (Table 1).

Table 1. Breakdown of Kidd Creek U/G diesel fleet.

Vehicle type	Number
Load haul dump (LHD)	20
Haulage trucks	2
Pickup trucks U/G	42
Tractors	18
Utility vehicles	74
Total U/G	156

Table 2. Vehicles selected for the U/G field study.

Vehicle type	Number
Load haul dump (LHD)	5
Haulage trucks	2
Pickup trucks U/G	3
Tractors	2
Utility vehicles	3
Total U/G	15

The fleet was classified into two main groups; light-duty (LD) and heavy-duty (HD) vehicles. The HD group was further classified into two types: load haul dump (LHD) vehicles and haulage trucks. The LD group was further classified into three types: pickup trucks, tractors and utility vehicles.

It was technically feasible to test ten percent of the Kidd Creek Mine fleet, or 15–16 vehicles. The final selections were based on observation of the fleet, communications with mine staff and technical requirements. The following vehicles were selected (Table 2).

Part One: Heavy Duty: The HD vehicle selections were made in the Phase I report (Grenier 2000). The vehicles tested were different from the original group selected because of availability problems, however the test vehicles are believed to be more representative of fleet operations.

Table 3 shows the HD vehicle duty vehicle testing matrix.

Part Two: Light Duty: The LD vehicle selections were made after observation of the fleet and discussion with operators, supervisors and management. Table 4 shows the light-duty vehicle test matrix.

Table 3. Heavy duty vehicles tested during the field study (part 1).

Number	Vehicle ID	Task	Engine	Power (HP)
33635	SCOOPTRAM, TORO 501D	Stope Mucking LHD	Detroit Diesel Series 60, 11.1 L	285
33638	SCOOPTRAM, TORO 501D	Stope Mucking LHD	Detroit Diesel Series 60, 11.1 L	285
33638	SCOOPTRAM, TORO 501D (repeat)	Stope Mucking LHD	Detroit Diesel Series 60, 11.1 L	285
33616	SCOOPTRAM, TORO 400D	Stope Mucking LHD	Detroit Diesel Series 50, 8.5 L	250
33626	SCOOPTRAM, WAGNER ST3.5	Backfill Mobile Filling No. 1 Mine	Detroit Diesel 4-71, 4.65 L	180
33661	TRUCK, EJC430	Haulage Truck	Detroit Diesel Series 60, 11.1 L	300
33661	TRUCK, EJC430 (repeat)	Haulage Truck	Detroit Diesel Series 60, 11.1 L	300

Table 4. Light duty vehicles tested during the field trials (part 2).

Number	Vehicle ID	Task	Engine	Power (HP)
33348	TRUCK, CHEV 2500 [2000]	4600 Level Mobile Shop	General Motors L65 6.5L TD	195
33325	TRUCK, FORD SUPER DUTY [2002]	4600 Level Mobile Shop	Navistar 99F 7.3L TD	150
33336	TRUCK, DODGE RAM 2500 [2001]	Mine "D" Construction	Cummins ISB 5.9L TD	230
33966	TRACTOR, KUBOTA 5030	Ore & Waste Mobile Equipment No. 3 Mine	Kubota S20802-DI-A 2.8L	54
33952	TRACTOR, KUBOTA 5030	Construction Upper Mine No. 1	Kubota S20802-DI-A 2.8L	54
33973	SHOTCRETE, SPRAYER DRIFTEC	Lateral Development	Cummins QSB5.9-240 5.9L	240
33872	SCISSORLIFT, GETMAN A64	Supply & Sanitation	Deutz F5L413FRW 8.0L	110
33899	BOOM TRUCK, GETMAN A64	Construction Upper Mine No. 1	Deutz F6L912W 5.6L	82

These vehicles were selected to be representative of the Kidd Mine fleet. The only key vehicle missing from the trial was a jumbo drill rig. It was not possible to schedule a test for that type of vehicle because of production commitments. It may be possible to simulate drill rig operation from available test data as its duty cycle is similar to the Driftec shotcrete sprayer and it uses the same engine as the Getman boom truck.

4 TEST METHODOLOGY

This study focused on diesel particulate matter (DPM) emissions. No attempt was made to look at other gaseous emissions. In order to compare DPM emissions from LD and HD vehicles, it was necessary to monitor three major parameters; engine exhaust flow, DPM mass and duty cycle.

4.1 *DPM production measurement*

The ultimate goal of this study is to determine the amount of DPM produced by selected diesel powered vehicles over a specified sampling period and ultimately over a full-shift period. The following general formula can be used to illustrate:

$$M_{DPM} = C_{DPM} \times V_{EXH} \quad (1)$$

where M_{DPM} = Mass of DPM (in milligrams) produced by the vehicle during the sampling period; C_{DPM} = Concentration of DPM in exhaust as measured by the DPM sampling apparatus (milligrams per cubic meter at standard conditions) during the sampling period; V_{EXH} = Total dry exhaust volume produced during the sampling period (cubic meters).

A vehicle that is performing significant work for the mine will be emitting a large volume of exhaust gas. In addition, because of the higher volumes produced, such vehicles have the potential to contaminate the mine workings more quickly.

4.2 *Vehicle duty cycle evaluation*

Evaluation of each vehicle's duty cycle will ensure that it is classified correctly as light or heavy-duty. It will also provide insight into the utilization for each vehicle and vehicle type.

All the HD, electronically controlled Detroit Diesel engines provide a percent engine load parameter on the ECU data stream. This data was recorded during each test. In addition, the shotcreter and pick-up truck were also equipped with electronically controlled engines.

Duty cycle data for mechanically controlled engines was determined based on engine rpm, exhaust flow and temperature. Charts were prepared for each vehicle showing the duty cycle distribution.

4.3 *Testing sequence*

Each day the vehicle to be tested was brought to either the 4600 or 2800 level service shop. CANMET-MMSL personnel installed the DPM sampling system, data loggers and ambient DPM and dust monitors. Each operator wore a personal sampling pump and train to measure personal exposure.

The vehicle would then proceed to its normal working area. CANMET-MMSL personnel would ride along with the operator to periodically change sample filters and make notes pertaining to the duty cycle.

The test vehicle would perform its normal duties during the shift. The cycle would be interrupted briefly (1–2 minutes) to change the DPM sample filter as required. After a sufficient number of duty cycles were recorded, the vehicle would return to the shop area where the sampling system was removed in preparation for the next shift.

5 SAMPLING SYSTEM CONSTRUCTION

Preliminary DPM sampling system construction is explained in the Phase I report (Grenier 2000). After the release of that report, it was decided to modify the system for proportional flow. A control loop was integrated into the DPM sampling system to provide isokinetic sampling flow.

This is achieved by measuring the exhaust flow and temperature. As exhaust mass flow increases, the DPM sampling pump flow is increased proportionally. DPM mass collected on the sample filter is therefore weighted proportionally to the engine operating modes. The system has a fast response and calibrates well against laboratory instruments (Figure 1 and Figure 2).

The sampling system uses a series of interchangeable, averaging pitot tubes that allow exhaust flow feedback control for a the complete range of vehicles tested at the Kidd Creek Mine. Each pitot tube was

Figure 1. DPM sampling system installed on Wagner ST-3.5 LHD.

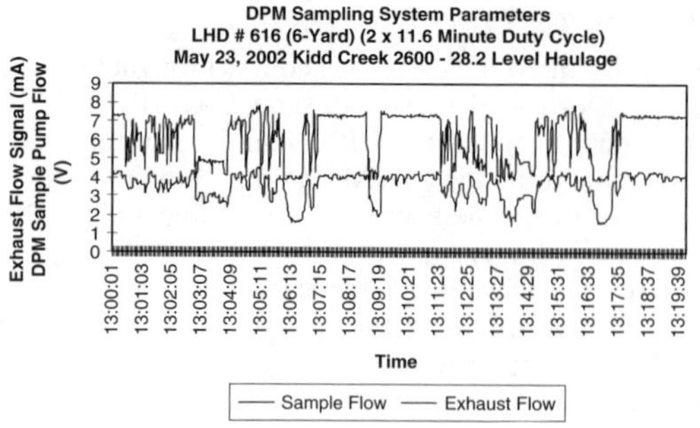

Figure 2. DPM sampling system response.

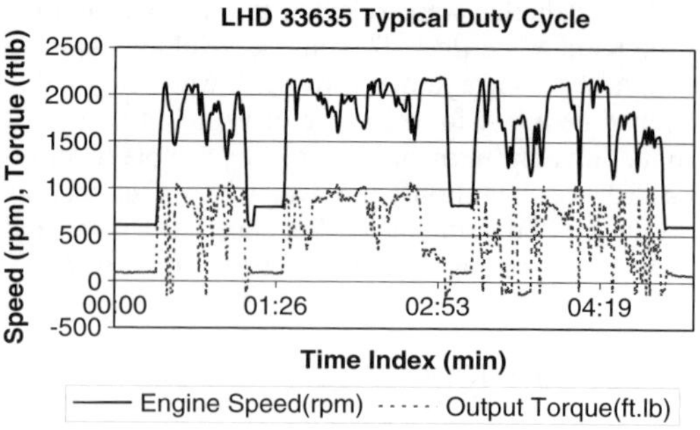

Figure 3. LHD 33635 typical duty cycle.

Table 5. LHD 33635 DPM emissions summary.

Vehicle	Duty cycle description	DPM (mg/m^3)	DPM mass (g)	Run time (min)	DPM mass rate (g/hr)
LHD 33635	4600 fuel bay to 4800	12.22	1.86	10	11.13
	Mucking on 4800	17.26	6.69	23	17.45
	Mucking on 4800	16.34	9.75	24	24.37
	Mucking on 4800	25.15	7.72	16	28.96
	Mucking on 4800	15.19	6.16	17	21.73
	Mucking on 4800	24.21	7.90	15	31.59
	Mucking on 4800	15.33	5.74	20	17.23
	Up ramp 4800 to 4600S	N/A	N/A	N/A	N/A

individually calibrated against laboratory instruments at the CANMET-MMSL Diesel Emissions test facility before use underground.

6 HEAVY-DUTY VEHICLE TESTING

Several different types of heavy-duty vehicles were tested for the study, however, only two will be discussed here. Data from the others is included in the summary table (Table 5).

6.1 *Load haul dump (LHD) vehicles*

6.1.1 *LHD # 33635 Toro 501D*

This LHD is used for normal production on 4800 Level. The duty cycle is high torque and approximately five minutes in duration (Figure 3).

The field exhaust DPM concentrations are consistent with laboratory data obtained from a similar engine. The DPM mass emission rate while mucking can be three times higher than when traveling to the work area because of transients (Table 5).

6.2 *Haulage trucks*

6.2.1 *Truck # 33661*

When descending the ramp empty, the haulage truck engine speed is high, while the developed torque is low (Figure 4). Since the ramp is not wide enough for two-way traffic, the truck must stop repeatedly to allow smaller vehicles time to reverse and clear the way. This is reflected by short idle periods in the chart. Occasionally, if convenient, the truck will reverse and yield.

The engine operating near the peak torque point characterizes the up ramp cycle (Figure 5). The idle sections are where personnel were exchanging DPM filters.

The truck operation modes are best defined as descending the ramp empty, idling, and hauling up the ramp loaded. The lowest DPM production rate occurs at the idle mode (Table 6).

It was initially thought that the descending ramp mode would also have a low DPM emission rate due to the fuel being shut off under compression braking. The field trials have shown that, contrary to this theory, the engine is still developing power as the operator attempts to maintain a vehicle speed consistent with production goals.

This places the engine in a high speed/low torque mode where combustion temperatures are low leading to poor in-cylinder DPM control and higher emissions. In addition, the turbocharger boost pressures at that operating mode are not elevated enough for efficient combustion.

Once the truck has been loaded and is ascending the ramp, the engine can achieve steady state operating conditions, and relative low DPM emissions.

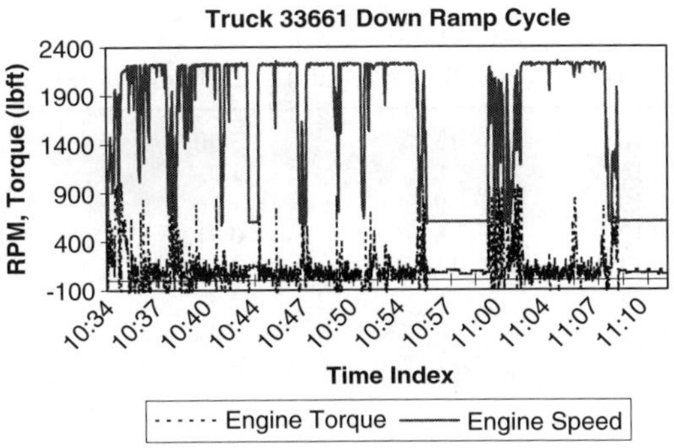

Figure 4. Haulage truck 33661 down ramp cycle.

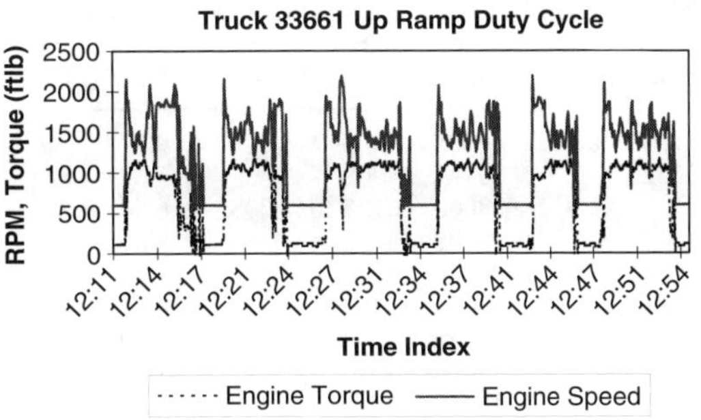

Figure 5. Truck 33661 up ramp cycle.

Table 6. Truck 33661 DPM emissions summary.

Vehicle	Duty cycle description	DPM (mg/m^3)	DPM mass (g)	Run time (min)	DPM mass rate (g/hr)
Truck 33661	4600S to 6400	28.13	14.68	24	36.71
	6400 to 6900	25.10	5.74	12	28.70
	Idle on 6900	12.77	0.36	5	4.33
	6900 to 7400	25.11	9.38	19	29.63
	7400 to 7200	18.51	3.79	8	28.43
	7200 to 7000	10.34	1.43	6	14.26
	7000 to 6900	11.30	2.46	8	18.46
	6900 to 6600	11.18	1.77	6	17.70
	6600 to 6400	3.68	0.66	7	5.62
	6400 to 6200	2.59	0.47	8	3.49
	6200 to 5800	7.89	1.68	8	12.60
	5800 to 5600	5.91	1.08	8	8.08
	5600 to 5400	5.57	0.73	5	8.70
	5400 to 5200	4.61	0.89	7	7.64
	5200 to 5100	10.09	1.93	9	12.85
	5100 to 4600S	9.37	3.40	12	17.01

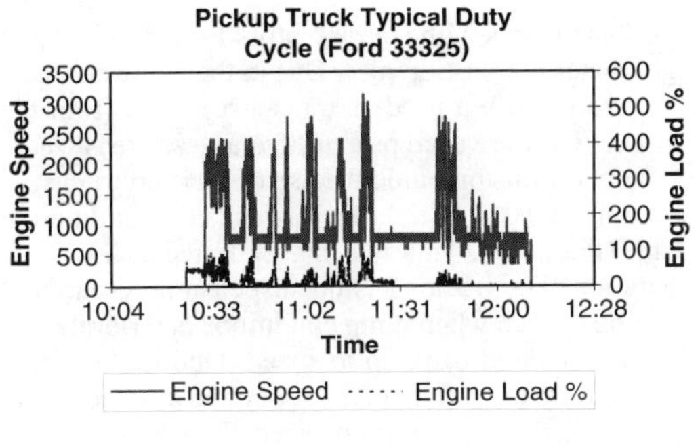

Figure 6. Pick-up truck typical duty cycle (Ford 33325).

7 LIGHT-DUTY VEHICLE TESTING

Several different types of light-duty vehicles were tested for the study, however, only two will be discussed here. Data from the others is included in the summary table (Table 6).

7.1 *Pick-up trucks*

All three pick-up trucks had a similar duty cycle, serving mainly as personnel and light material transport through the mine. Kidd Mine currently uses pick-up trucks supplied by three major manufacturers; Chevrolet, Ford and Dodge.

All the pick-ups have a large percentage of idle time due to the nature of their work. In addition, the average engine load is 10 to 15% (Figure 6). There are no areas where sustained high power is required. Even when ascending the ramp, the pick-ups have so much extra power that it is possible to quickly exceed recommended maximum ramp travel speeds.

The pick-up trucks are inefficient because the mine must still provide ventilation for the engine rated power according to regulations even though it is almost impossible to apply that power towards useful work.

Table 7. Ford pick-up 33325 DPM emissions summary.

Vehicle	Duty cycle description	DPM (mg/m^3)	DPM mass (g)	Run time (min)	DPM mass rate (g/hr)
Ford pickup 33325	4600S to 36-3	9.19	0.76	13	3.50
	36-3 to 2800S	3.67	0.16	8	1.22
	2800S to 28-7 to 2800S	6.54	0.19	9	1.26
	2800S to 36-3	3.62	0.12	8	0.91
	36-3 to 2800S	8.49	0.43	11	2.33
	2800S to 36-3	8.31	0.29	6	2.90
	Idle on 36-3	4.20	0.10	16	0.37
	36-3 to 4600	3.30	0.17	11	0.93
	4600 Level run	9.38	0.36	17	1.26

The trucks do have the advantage of high speed transport through the mine and relatively low DPM emission rate due to their lightly loaded duty cycles (Table 7).

The pick-up trucks have four common duty cycle modes. These modes are: ascending the ramp, descending the ramp, idling, and running on a level.

The idle mode has the lowest DPM mass emissions, ascending the ramp has the highest followed by descending and running on a level. While descending the ramp, the pickup trucks exhibited similar behavior as the HD haulage trucks. It was anticipated that ramp descent would yield the lowest DPM emission rate because of fuel shut off during compression braking. Instead, because of the limited number of transmission ratios available, compression braking in first gear results in too slow a descent.

Shifting to a higher gear increases the vehicle speed to a dangerous level. The operator is then forced to alternately brake and accelerate down the ramp to achieve an acceptable rate of descent, consistent with local ramp conditions. Use of the accelerator pedal injects fuel into the engine, which cannot be burned effectively due to the low in-cylinder temperatures and lack of turbo boost.

The Ford pick-up has a similar modal DPM emissions breakdown as the Chevrolet. The Ford truck was de-rated to approximately 150 HP with an aftermarket device that modifies the demand signal sent to the electronic fuel injection system. The device is effective in reducing the DPM mass emission rates. The Ford had the lowest DPM emission rates (per mode) of any vehicle tested.

Based on the available data, it is possible to rank the pick-up trucks in order of best DPM emissions as Ford first, Dodge second and Chevrolet third. However, it should be noted that only one vehicle of each type was tested. Further testing of more pickups should be done over a longer period of time to develop a more sound opinion. In addition, the effect of the de-ration device on the Ford pick-up should be further examined, perhaps looking at gaseous emissions also.

Finally, the Kidd Mine has begun purchasing new Chevrolet pickups equipped with the 6.6L Duramax 6600 engine. This engine is rated at 300 HP, over 35% more power than the L65 engine it replaces. This vehicle should be tested also.

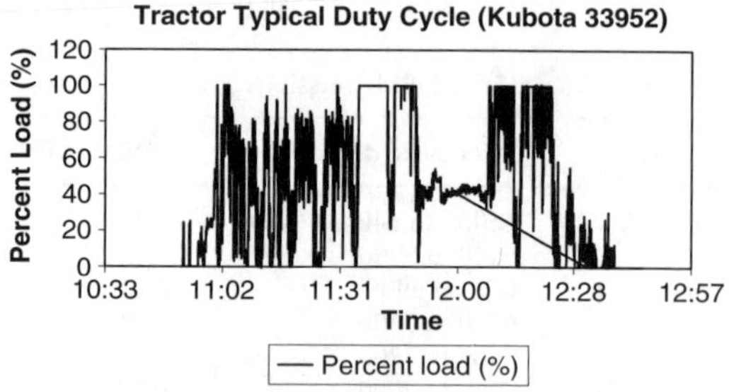

Figure 7. Tractor typical duty cycle (Kubota 33952).

7.2 *Tractors*

The Kidd Creek Mine uses tractors for personnel and light material transport also. Due to their slower speed, they are not often used as transport to and from the surface. They are used primarily to access areas where drift conditions preclude the use of pick-up trucks.

From 11:10 to 11:33, the tractor is descending the ramp (Figure 7). This is a highly transient mode of operation. The tractor has no suspension system and must be slowed when ramp conditions deteriorate and then accelerated back up to speed. Due to the effective gearing, the tractor can be operated at steady state conditions during the ramp ascent. The Kubota tractors typically have a higher power utilization averaging 45–50%.

The Kubota tractor is powered by an older, mechanically controlled engine. It is not turbocharged and uses old style pencil-type injectors. As a result the DPM mass emission rates at high engine loads are greater than for either the 8-yard LHD or the 30 ton haulage truck.

Table 8. Kubota tractor # 33952 DPM emissions summary.

Vehicle	Duty cycle description	DPM (mg/m^3)	DPM mass (g)	Run time (min)	DPM mass rate (g/hr)
Kubota tractor 33952	5100 to 5400	34.59	0.89	8	6.68
	5400 to 5800	31.73	0.85	7	7.26
	5800 to 6400	30.64	0.93	8	6.97
	6400 to 5800	105.56	5.40	8	40.53
	5800 to 5400	64.90	2.90	7	24.82
	Idle on 5400	9.74	0.29	15	1.14
	5400 to 5100	68.48	2.73	7	23.38
	4600 level run	36.39	2.21	13	10.21

Table 9. Sample fleet DPM emissions.

Vehicle	Total mass (g)	DPM rate (g/hr)
LHD 33635 8-yd	45.8	22.0
LHD 33616 6-yd	43.2	15.6
LHD 33626 3.5-yd	37.8	23.6
LHD 33638 8-yd	43.1	22.1
Haulage Truck 33661	50.4	19.9
Truck 33661 repeat	43.5	12.0
Chevrolet truck 33348	4.0	3.3
Ford truck 33325	2.6	1.6
Dodge truck 33336	4.3	3.1
Kubota Tractor 33966	24.5	22.7
Kubota Tractor 33952	16.2	13.3
Shotcreter 33973	12.4	6.3
Scissorlift 33872	10.8	6.0
Boom Truck 33899	19.5	12.9

8 SAMPLE FLEET DPM EMISSIONS

Table 9 summarizes the DPM emissions for all the vehicles tested at the Kidd Creek Mine. Total DPM mass emissions represent the total mass of DPM in grams emitted by the vehicle into the mine environment during the testing period. The next column is the average DPM emission rate (in grams per hour) for each vehicle, weighted against the time spent at each distinct operating mode.

8.1 *General discussion*

Several key points are identified on Table 9. Generally, the HD LHD production vehicles have the highest DPM mass emission rates, followed by the production haul trucks, tractors, utility vehicles, with pick-up trucks having the lowest.

The older technology two-stroke engine on the Wagner ST3.5 (33626) is the likely reason that it has the highest emission rate of the LHD's. The two 8-yard LHD's have similar average DPM emission rates for similar duty cycles, which supports the repeatability of the test method.

Haulage truck 33661 (repeat) spent almost 40% of the test time idling. This is reflected in the lower DPM emission rate from Truck 33661. Interestingly, during those 84 minutes of idling, 3.25 grams of DPM (or 7.5% of the total) were released into the mine environment. Simply shutting the engine down could have stopped this pollution. Although 3.25 grams may seem small, it is more than the total amount released by the Ford pick-up during the entire trial period.

The Kubota tractors demonstrate the wide emissions variability seen with in-use mechanically controlled engines. Fuel system problems are suspected with tractor 33966 and it should be pulled from service for repair. The DPM emission rate from tractor 33966 exceeds all of the HD production vehicles except the two-stroke Wagner ST3.5. This is important because at approximately 50% power utilization, the mine is only getting 27 HP of work from the Kubota tractor but is absorbing the same DPM "burden" as an 8 yard LHD.

The utility vehicles have reasonable DPM emission rates compared to the HD production vehicles. They are usually purpose built for mining and as such the engine and transmission ratios are selected for optimum engine operation. The DPM emission rates of the Deutz 912W engine in the boom truck 33899 and the Deutz 413W engine in the scissorlift 33872 highlight the difference between these series of engines. The Ford pick-up had the lowest DPM emission rate of all the vehicles tested. This is likely due to the deration of the engine from factory specifications. The Dodge pick-up was next, due to the use of a base industrial diesel design with a modern electronically controlled radial fuel injection pump. While in last place among the pick-ups, the Chevrolet truck emissions were still significantly better than the next worst vehicle, the Getman scissorlift.

8.2 *Ambient operator exposures*

During the field study, ambient air samplers were mounted in each vehicle cab and worn by the vehicle operator. The following table highlights results of measurements of personal exposure of operators to DPM (Table 10).

Table 10. Operator exposures.

Operator's Vehicle	EC (mg/m^3)	OC (mg/m^3)	TC (mg/m^3)	EC/OC
LHD 33635 8-yd	0.094	0.098	0.191	0.96
LHD 33616 6-yd	0.137	0.282	0.419	0.49
LHD 33626 3.5-yd	0.229	0.173	0.401	1.32
LHD 33638 8-yd	0.088	0.122	0.210	0.72
Truck 33661	0.184	0.161	0.345	1.14
Truck 33661 repeat	0.180	0.159	0.339	1.13
Chev. truck 33348	0.206	0.170	0.376	1.21
Ford truck 33325	0.089	0.061	0.150	1.47
Dodge truck 33336	0.129	0.064	0.193	2.01
Kubota Tractor 33966	0.287	0.262	0.549	1.09
Kubota Tractor 33952	0.247	0.144	0.391	1.72
Shotcreter 33973	0.087	0.053	0.140	1.63
Scissorlift 33872	0.100	0.057	0.157	1.75
Boom Truck 33899	0.112	0.074	0.186	1.52

Table 11. Tailpipe EC/OC split for sample fleet.

Vehicle	Tailpipe EC/OC
LHD 33635 8-yd	4.47
LHD 33616 6-yd	6.04
LHD 33626 3.5-yd	0.74
LHD 33638 8-yd	3.09
Truck 33661	1.74
Truck 33661 repeat	2.56
Chevrolet truck 33348	2.63
Ford truck 33325	1.40
Dodge truck 33336	N/A
Kubota Tractor 33966	0.35
Kubota Tractor 33952	N/A
Driftec Shotcreter 33973	1.26
Getman Scissorlift 33872	0.50
Getman Boom Truck 33899	1.68

Due to the importance of obtaining real duty cycles, no attempt was made to isolate the test vehicles from other sources of DPM. In particular, DPM exposures of operators during cycles with large portions of ramp travel time should be very carefully considered.

In general, the ambient exposures are a good indicator of tailpipe emissions. The table correctly predicts the Kubota tractor operator is being exposed to high DPM concentrations and the Ford truck operator is exposed to lower levels.

8.3 *Tailpipe elemental carbon/organic carbon*

The following table (Table 11) shows the tailpipe EC/OC split for each of the vehicles in the test fleet. EC/OC split can provide information about the composition of DPM emitted by the engine. Electronically controlled engines operating at high load usually have a higher EC/OC split. Mechanical engines at low loads generally have lower EC/OC.

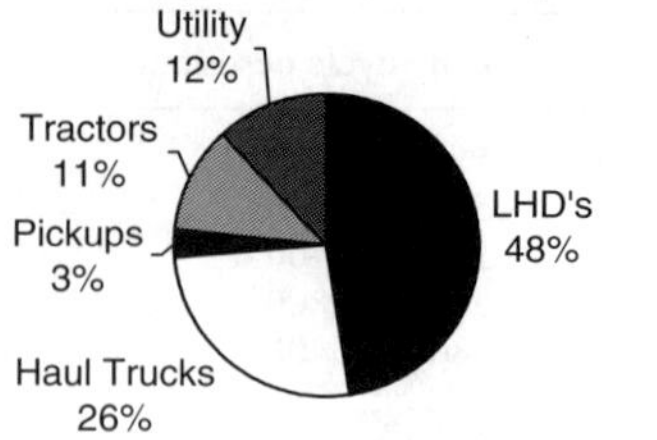

Figure 8. DPM burden distribution for 14-vehicle sample fleet.

HD LHD's have the highest EC/OC split, which is a well-known characteristic of modern electronically controlled diesel engines. Mechanical engines have a more even split, in some cases dropping below unity. LHD 33626 has an EC/OC split also lower than one, which is in accordance with observations previously made on the characteristic of two-stroke engines.

8.4 *Sample fleet DPM burden*

Figure 8 shows the division of total DPM burden among the sample test fleet. The breakdown is based on the weighted average DPM emission rate for each vehicle during the sampling period only.

HD vehicles including LHD's and haulage trucks were responsible for 74% of the total DPM burden emitted by the sample fleet. LD vehicles were responsible for 26%.

In order to adequately determine the relative contributions of LD and HD vehicles to the underground DPM burden, the sample fleet data must be extrapolated to the entire Kidd Creek Mine fleet.

9 KIDD CREEK MINE FLEET DPM EMISSIONS

To estimate the underground DPM burden at Kidd Creek Mine, the sample fleet test data was extrapolated to represent the entire Kidd Creek Mine diesel fleet. Not every type of vehicle underground could be tested so the following assumptions were made to simplify the comparison.

1) Total U/G fleet of 156 vehicles (from Kidd records, see Table 1).
2) The total fleet is composed only of vehicles from the 14-unit sample fleet.
3) The total fleet operates over the same duty cycles as the sample fleet.
4) The total fleet operates the number of hours outlined in Table 12 (hourly data provided by Kidd Maintenance).

Table 12. Model fleet comparison and DPM burden.

Vehicle Type	Number	DPM rate (g/hr)	Utilization (hr/yr)	DPM Burden (kg/yr)
8 yd LHD	10	22.05	2583	569.55
6 yd LHD	3	15.62	1880	88.07
3.5 yd LHD	7	23.63	1583	261.85
30t Trucks	2	15.95	1600	51.04
Chevrolet pickups	33	3.32	698	76.53
Ford pickups	1	1.56	698	1.09
Dodge pickups	8	3.07	698	17.13
Tractors	18	18.00	942	305.21
Shotcreters	25	6.35	651	103.33
Scissorlifts	24	6.03	686	99.34
Boom trucks	25	12.88	779	250.82

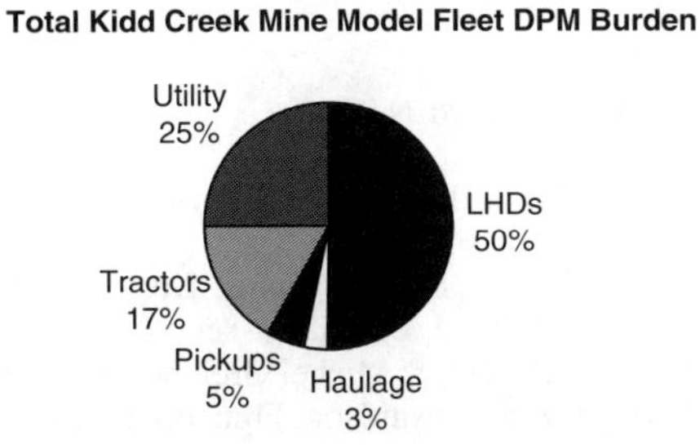

Figure 9. Total Kidd Creek Mine model fleet DPM emissions burden.

Thus, our model fleet is composed of 20 LHDs, ten 8-yard, three 6-yard, and seven 3.5-yard. Both haulage trucks are 30t EJC430s. The 42 vehicle pick-up truck fleet is 79% Chevrolet, 19% Dodge and 2% Ford. All 18 tractors are Kubotas (average of the two tested). The utility vehicle fleet is composed of 25 shotcrete sprayers, 24 scissor lifts and 25 boom trucks.

The fourth assumption about the utility vehicle breakdown is probably the least valid. The wide diversity of utility vehicles underground made selection difficult, because each vehicle is almost unique.

For the model fleet, HD vehicles (LHDs, haulage trucks) are believed to be responsible for approximately 53% of the total underground DPM burden. LD vehicle (tractors, pick-ups utility) are responsible for about 47% of the burden (Figure 9).

10 CONCLUSIONS

10.1 *General conclusions*

The Kidd Creek Mine model fleet is composed of 14% HD and 86% LD vehicles. Despite this fact, Table 12 shows the underground DPM burden is fairly evenly split between the HD and LD vehicles. HD vehicles and LHD's in particular remain the major contributors to DPM underground. However, LD vehicles are still producing 47% of the underground DPM burden and must be a part of any DPM control strategy.

The DPM mass emission data contained in this report could form the basis for a more comprehensive model of the Kidd Creek diesel fleet. A more refined model could allow strategic planning for DPM emissions and prevent local areas of high concentration by monitoring vehicle distribution in the mine. Emissions data should be used to develop a DPM reduction program including implementation of control technologies.

10.2 *Engine operating modes*

Steady state engine operation promotes low DPM emission rates. Emission control systems on HD vehicles such as turbo charging and charge air-cooling perform best over steady state duty cycles. Traffic management and roadway maintenance can help vehicles to achieve this on ramp sections. Somewhat surprisingly, ramp descent in particular can be a high emissions transient cycle, depending on local ramp conditions.

Unnecessary engine idling can be a significant source of DPM emissions, so vehicle engines should be shut off wherever possible.

10.3 *Engine type, technology and maintenance*

Even when electronically controlled, older technology two-stroke engines are not able to achieve the same DPM emissions performance as four-stroke engines.

Different engine models from the same manufacturer can have different DPM emission rates. Detroit Diesel Series 50 engines were found to be cleaner than Series 60, while Deutz 413W series were cleaner than 912W.

Engine maintenance has a large effect on DPM emissions. Of two identical tractors, the unit with an obvious fuel system defect had a 41% higher DPM emission rate than the unit in good condition. The operator of this particular vehicle was exposed to the highest ambient DPM concentrations observed during the test.

Engine de-rating, as demonstrated on the Ford pick-up, can have a beneficial effect on DPM mass emissions.

11 RECOMMENDATIONS

The Kidd Creek Mine should implement a DPM mitigation strategy, which includes the following components:

Start by applying DPM control technologies to as many 8-yard LHD's as possible. These vehicles are responsible for the largest single portion of the mine DPM burden. Retrofitting all the Toro 501D 8-yard LHD's would allow the maintenance department to maintain one common system and stock of spare parts for one type of vehicle.

Next, DPM control devices should be retrofitted to the Kubota tractor fleet, as they represent the second largest contributor to concentrations of DPM from a single type of vehicle.

After successful implementation of the above two systems, DPM controls should be considered for the utility vehicle fleet, starting with the Deutz 912W series engines. Integration of DPM controls for the utility fleet will be a challenge due to its diversity, but still important as they are a major contributor.

Pick-up trucks and haulage trucks are lesser contributors and can be scheduled for retrofit at a later date.

This paper will not provide any recommendations for DPM emission control devices. The reader is referred to the Diesel Emissions Evaluation Program's (DEEP) two major studies on these devices at www. deep.org.

The Kidd Creek Mine fleet DPM model should be further refined to more closely match the vehicles actually used underground. In addition, the DPM model could be combined with a mine ventilation model to provide the ventilation department with predictions of local ambient DPM concentrations throughout the mine.

The pick-up truck fleet has the highest turnaround of vehicles in the mine. Further emissions testing should be performed on the pick-up truck fleet including non-derated Ford trucks and newer GM Duramax 6600 engines to provide information for a vehicle purchasing strategy.

The Kubota tractor 33966 should be pulled from service and have the fuel system defect corrected, as it is exposing the operator to high DPM concentrations.

ACKNOWLEDGEMENTS

The authors would like to thank the following individuals for their assistance:

Falconbridge – Kidd Mining Division: Alain Landry, Fern Monette, Dennis Sasseville

NRCan – CANMET: Michel Grenier, Vince Feres, David Young, Eric Edwardson, Francois Tang

REFERENCES

Grenier, M., Gangal, M., Butler, K., Edwardson, E. & Feres, V. 2000, *Evaluation of the contribution of light-duty vehicles to the underground diesel emissions burden – Phase I report.* Diesel Emissions Evaluation Program, Canada.

Vergunst, J. & Paquette, C. 1996. *Ontario underground diesel equipment and procedure survey.* Ontario Ministry of Labour, Ontario, Canada.

Mine Ventilation, Ganguli & Bandopadhyay (eds)
© 2004 M. Gangal, ISBN 90 5809 633 5

DPM reduction by thermal regeneration systems

M. Gangal, B. Rubeli & D. Young
Natural Resources Canada, CANMET-MMSL, Ottawa, Ontario, Canada

P. Tumati
ArvinMeritor, Columbus, Indiana, USA

N. Birkby
ArvinMeritor, Warton, Preston, UK

ABSTRACT: World wide concerns over adverse health effects of diesel exhaust will require substantial reduction in exhaust pollutants, which can be achieved by proper application of clean engine technology, clean fuel, and exhaust gas control systems. Diesel equipment is extensively used in underground mines for increased productivity. Many underground mines are investigating application of various diesel particulate filters to reduce diesel particulate matter (DPM). This paper discusses the use of an active thermal regeneration system to ensure successful regeneration for diesel engines, with low exhaust gas temperatures in particular. A series of tests were conducted under controlled laboratory conditions to examine the filtration efficiency, regeneration effectiveness and emissions performance. The system successfully regenerated the particulate filter during operation in about seven minutes, reduced DPM by over 84% and 99% respectively based on particulate mass and number. During the short regeneration period the filter assembly's exterior temperature reached 322°C, and the exhaust carbon monoxide (CO) emission rate peaked to a value of 880 g/h before returning to its normal value of 8 g/h.

1 INTRODUCTION

Diesel machines have been employed in underground mining operations for the last fifty years. The performance advantages of diesel machines in terms of durability, flexibility and productivity are well understood. Carbon monoxide (CO) and sulfur oxides (SO_x) emissions have now become much less of a concern relative to other constituents. The main concern of diesel exhaust is diesel particulate matter (DPM) followed by concentration of nitrogen oxides (NO_x). DPM is emitted in the environment due to incomplete combustion of fuel in the combustion chamber, and NO_x is emitted due to extreme heat during the combustion process.

DPM is visible and has always been perceived as objectionable, and possibly noxious by miners. However, recent widespread concern over its health effects has prompted several control measures worldwide. Diesel particulate filters (DPF) are very efficient in removing diesel particulate matter and can remove up to 99% of the nanoparticles (Mayer et al. 2002).

Earlier field studies to evaluate DPF technologies in Canadian underground mines were conducted in the early nineteen eighties and have been summarized elsewhere (Dainty et al. 1986).

In 1996, American Conference of Governmental Industrial Hygienists (ACGIH) included DPM on their list of intended changes with a Threshold Limit Value (TLV) at 0.15 mg/m^3. Thereafter, a Diesel Emissions Evaluation Program (DEEP 1997) was formed by the collaboration of industry, labor, government and researchers, with a goal of reducing underground miner's exposure to diesel emissions. Two of the DEEP studies included extensive field trials to evaluate diesel particulate filter technologies in underground mines (McGinn 2003, Stachulak 2003).

Trapping particulate matter at a high efficiency is not a problem. However, maintaining an acceptable pressure drop over a long period of time is often difficult without frequent regeneration of the filter. In many underground mining applications, due to low exhaust gas temperatures, some type of DPF systems fail to regenerate as needed. Hence, there is a need to enhance or assist the regeneration of the diesel particulate filters by active means. One such method is the application of a fuel burner DPF. With this technique, a diesel fuel burner (thermal regenerator) is used to increase the temperature of the exhaust gas stream to a temperature above the auto ignition temperature of soot. A filter inlet temperature of 600°C is generally sufficient to initiate combustion of the trapped soot particulates. ArvinMeritor thermal regeneration system was tested in a controlled laboratory environment at CANMET-MMSL and the system performance is presented in this paper.

During these tests, the exhaust gas backpressure was continuously monitored to check whether the engine backpressure was within the engine manufacturer's recommended limits.

2 THERMAL REGENERATION SYSTEM

The thermal regeneration system (TRS) used for this study is shown in Figure 1. It consists of a flame holder plus DPF assembly, an electronic control unit (ECU), and an accessory control unit (ACU) box. The DPF assembly had a Corning Duratrap RC200 cordierite diesel particulate filter. The principle of this system involves drawing diesel fuel from the vehicle tank and burning it to increase the temperature of the exhaust gas to above 600°C (auto ignition temperature of the particulate matter), and enable complete combustion of the particulate matter trapped in the DPF.

For diesel engine applications, the system is installed in such a way that all the exhaust gas passes through the DPF assembly and the particulate matter is trapped in the filter. The ECU unit continuously monitors the backpressure and temperatures at various locations in the filter assembly. The ECU can also interface with the electronic engine control system to monitor additional parameters (engine speed and load, manifold pressure etc.) if required. The ECU also initiates the filter regeneration depending on the chosen criteria (back pressure, time limit etc.).

Figure 1. Thermal regeneration system (TRS).

Once the ECU triggers the regeneration process, the diesel fuel is drawn by a fuel pump (located in the ACU box) and pumped to a mixing chamber. Metered quantities of diesel fuel is atomized with high-pressure air and injected into the combustion chamber. The combustion air is supplied by an air-blower, which is also located in the ACU box. The fuel-air mixture is then ignited by means of electrodes. The fuel flow rate is regulated by the ECU so as to limit the combustion temperature in the filter assembly system well within safe limits. The fuel air ratio is always maintained to prevent soot formation and other products of incomplete combustion. The combustion in the flame holder assembly is allowed to complete before mixing with the diesel exhaust gases. The regeneration process typically takes 6 to 8 minutes and the fuel consumption is typically low. The development of the system is described elsewhere (Zelenka et al. 2002).

3 EXPERIMENTAL

The thermal regeneration system evaluation was performed at the CANMET-MMSL diesel test facility in Ottawa, Canada. A series of tests were conducted using an engine dynamometer. All engine parameters, regulated emissions and particulates concentrations were measured using standard equipment and methods (ISO 1996). This ISO standard specifies the measurement and evaluation methods for gaseous and particulate exhaust emissions from reciprocating internal combustion engines under steady state conditions on a test bed. Various combinations of engine load and speed reflect different engine applications. In this work, steady state and free acceleration transient test cycles were used. The steady state testing included four cycles of the 8-mode test cycle commonly used in off-road applications. These four mode cycle points are as follows:

Mode A – 1400 rpm at 100% load
Mode B – 1400 rpm at 50% load
Mode C – 2000 rpm at 50% load
Mode D – 2000 rpm at 100% load

In addition, a test was conducted at a fifth point (mode E), which is a repeat of mode A. The testing at mode E was conducted to check the repeatability of test data.

The engine used for this testing was a rebuilt engine described as:

Manufacturer:	Liebherr
Model:	D914T
Displaced volume:	6.11 L
Rated power:	105 kW @2000 rpm
Peak torque speed:	1400 rpm
Supercharging:	Turbocharger without cooling

The diesel fuel used for this study is a low sulfur fuel with following properties:

Sulfur	40 ppm
Cetane number	53.0
Density @15C	835.5 kg/m^3
Flash point	92°C
Ash	<0.001% wt

Prior to testing, the engine was conditioned on the dynamometer. The engine intake air restriction and exhaust backpressure at the rated power were set to the maximum values allowed by the engine manufacturer. In this case, intake air restriction was set at 3.75 kPa (15 in. H_2O), and the exhaust backpressure was set at 7.5 kPa (30 in. H_2O) while the engine was running at rated power. In order to determine the efficiency of the regeneration system, tests at four mode points were conducted with and without the thermal regeneration system (filter system). The testing without the filter is referred to as the baseline test. Testing with the filter was done both with a clean filter (after filter regeneration), and also with a charged filter (after loading the filter with particulates). The clean or regenerated filter was essentially free of any soot, while the charged filter was loaded with soot prior to any testing.

Gaseous emissions for carbon monoxide (CO) and carbon dioxide (CO_2) were measured using a non-dispersive infrared detector, while concentrations of nitric oxide (NO) and nitrogen oxides (NO_x) were measured using and a chemiluminescence detector. The concentration of nitrogen dioxide (NO_2) was calculated by subtracting the simultaneously measured value of NO from the value of NO_x. The concentrations of oxygen (O_2) and hydrocarbon (HC) were measured using paramagnetic and heated flame ionization detectors (HFID) respectively.

The particulate mass sampling was performed using a heated sampling line and partial flow dilution system. The particulate sample was collected on a pair of 90 mm filters and were conditioned and weighted in an environmentally controlled weighing chamber.

The exhaust gas particle size distribution measurements were conducted using a scanning mobility particle sizer (SMPS). The sampling included a partial dilution and a thermo-denuder system to condition the samples prior to measurements. The thermo-denuder system was set at a temperature of 300°C to remove moisture and volatile particles from the sample in order to measure dry particles.

An extensive amount of pioneering work on DPF evaluation and particulate size distribution has been conducted under a European VERT program. VERT is a European joint project team that focuses on reducing diesel exhaust emissions at tunnel sites. A summary of important VERT studies are published elsewhere (Mayer et al. 2002, Czerwinski & Mayer 2001).

4 RESULTS

Basic engine parameters measured at the test conditions without the DPF are listed in Table 1. Initially the engine was run at rated power to condition the engine with required settings (e.g. intake air restriction and exhaust backpressure). Then the engine was operated at mode point A and data were recorded. Similarly engine data was collected by running the engine at modes B, C, and D, and then back to the mode point A. This last mode point is denoted here by mode E. It is expected that the basic engine parameters measured at mode A should be essentially the same as mode E. The measured data (shown in Table 1) at modes A and E are within 1%, indicating identical operating conditions.

During testing, engine parameters, exhaust gaseous emissions concentrations, particulate mass and size distribution were monitored. The measured concentrations of undiluted exhaust gases were converted to gram per kilowatt-hour (g/kWh) for data analysis. During Normal operation, no change in exhaust gas concentrations is expected, as the thermal regeneration system does not contain any chemicals or catalysts to alter gas composition.

Figure 2 shows CO emission rate at four modes with and without the use of the filter. In general, there is a slight increase in CO emission rate. The CO increase

Table 1. Measured engine parameters.

Mode point	Speed (rpm)	Torque (nm)	Power (kW)	BSFC* (g/kWh)
A	1399	600	88.0	211.5
B	1400	298	43.6	213.2
C	2001	254	53.3	249.2
D	2001	515	108.0	226.1
E	1399	598	87.7	209.4

* Brake specific fuel consumption.

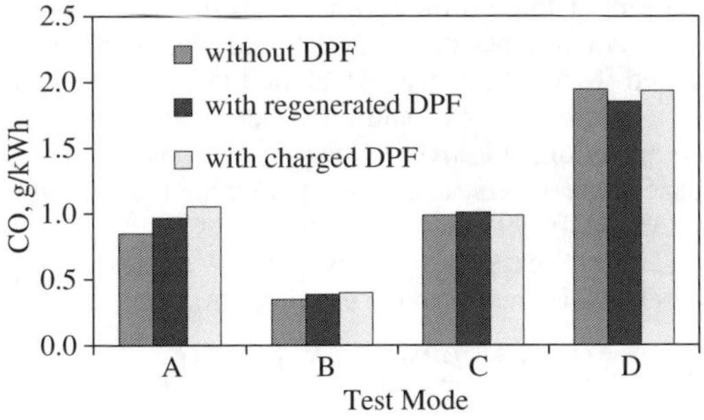

Figure 2. CO emission rate at steady state mode points with and without the DPF system.

was the highest at mode A, 14% for the regenerated filter and at 24% for the charged filter.

Figure 3 shows NO emission rates for all four modes with and without the use of the TRS system. In general, there is only a small change in NO values at all mode points. As mentioned earlier no change in NO emissions are expected.

Figure 4 shows NO_2 emission rate at four mode points with and without the use of the TRS system. At each mode point NO_2 reduction is significant. The NO_2 reduction with regenerated DPF varies from 16% at mode point A to 41% at mode point B, while the reduction with charged DPF varies from 83% at mode point D to 92% at mode point A. The reductions could be attributed to the oxidation of the soot by NO_2. The charged DPF typically contained about 60 gram of soot, and the exhaust gas temperature was high enough to react with the trapped soot.

Figure 5 shows DPM mass emission rate at four modes points with and without the use of the TRS system. As expected, DPM reduction at every mode point is very significant, varying from 67% at mode point B to 87% at mode point D for the regenerated filter. Similarly, DPM reduction with charged filter at mode point A is at 96%, and around 90% at all other mode points. The particulate reduction based on particulate number is described later.

4.1 *Exhaust quality index (EQI)*

A clean air mine environment is most desirable from a health and safety point of view. However, such an environment is often difficult to achieve in a production mine. There is obviously a need to balance the costs and benefits when diesel equipment is introduced into a mine. A CANMET contract study (French & Mildon 1978, 1984 & 1990) reviewed the world literature relevant to potential health effects of diesel exhaust, and concluded that DPM is the pollutant of greatest concern followed by NO_2. The same study developed an EQI expression as a tool to estimate the relative toxicity of diesel exhaust emissions. This expression is heavily weighted for DPM, as it appears at three different places in the EQI expression. For this reason an EQI based testing protocol called 'MAPTEST' was developed to evaluate emissions performance of add-on systems including, fuel composition, additives, catalytic converters, diesel particulate filters, and other treatment technologies (CANMET 1997).

The EQI expression is given below (gases in ppm and DPM in mg/m^3 in the exhaust gas).

$$EQI = \frac{CO}{50} + \frac{NO}{25} + \frac{DPM}{2} + 1.5\left(\frac{SO_2}{3} + \frac{DPM}{2}\right) + 1.2\left(\frac{NO_2}{3} + \frac{DPM}{2}\right) \quad (1)$$

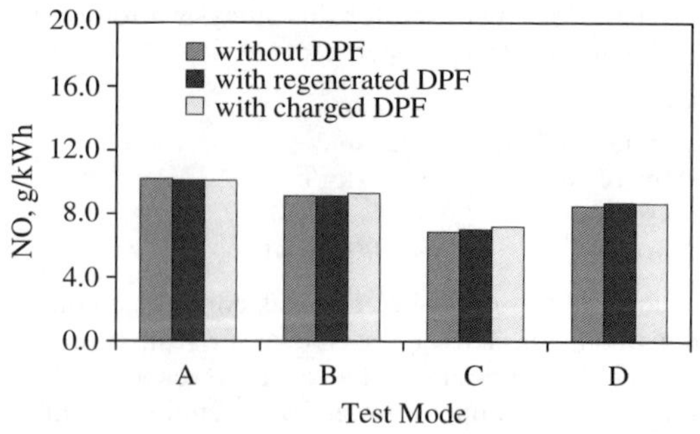

Figure 3. NO emission rate at steady state mode points with and without the DPF system.

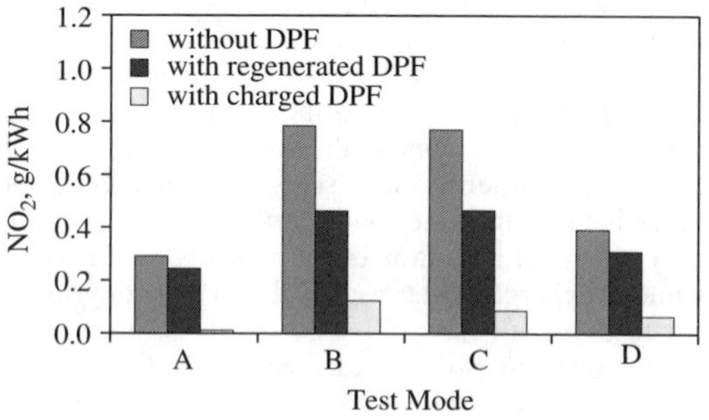

Figure 4. NO_2 emission rate at steady state mode points with and without the DPF system.

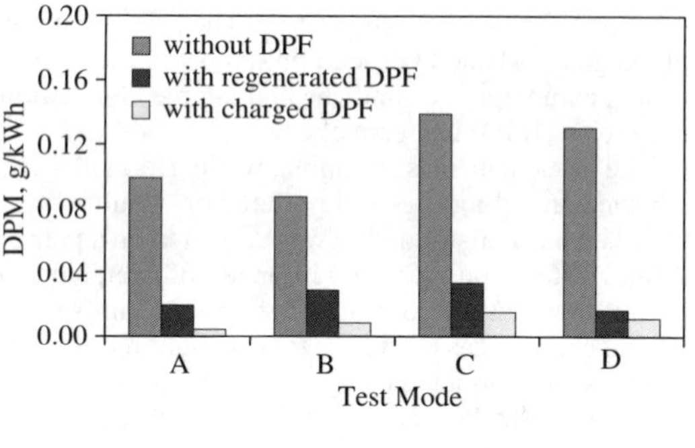

Figure 5. DPM emission rate at steady state mode points with and without the DPF system.

This expression is called Air Quality Index (AQI) when gases and DPM are measured in the ambient mine air. A maximum AQI value of three is recommended for exposure of miners to diesel pollutants.

The EQI expression is used as the exhaust toxicity criterion in the Canadian Standards Association (CSA) standards in order to define the ventilation dilution for engines destined for gassy underground coal mines

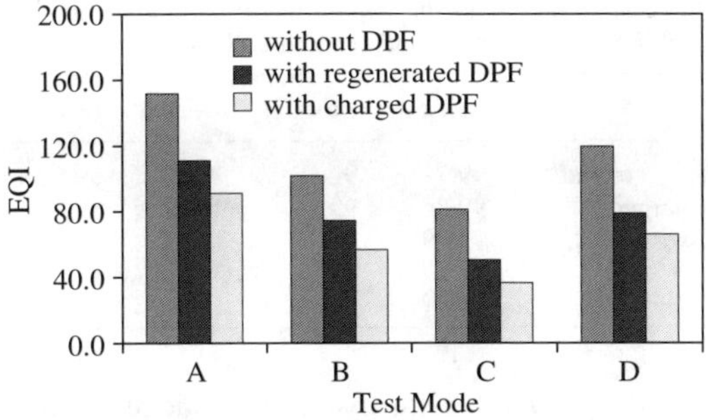

Figure 6. Exhaust quality index at four mode test points with and without the DPF system.

(CSA 1988) and for non-gassy underground mines (CSA 1990).

Figure 6 shows calculated values of Exhaust Quality Index (EQI) at four modes points with and without the use of the TRS system. In each case EQI reduction is significant. The EQI reduction varies from 27% at mode point A to 38% at mode point C with the regenerated filter, while the reduction with charged filter is much higher from 40% at mode A to 55% at mode point C. This reduction in EQI is mostly due to significant reduction in DPM, as the EQI expression is heavily weighted for DPM as shown in equation 1. Further, a reduction in EQI is equivalent to an increase in air quality. Therefore, a lower value of EQI is desired in areas with diesel activities.

4.2 *Particulate size distribution*

The size distribution data for all steady state mode points was collected for baseline, regenerated and charged DPF. Figure 7 shows the size distribution data at the rated speed and full load mode point with and without the regenerated filter. The size distribution without DPF is a normal curve with peak at 80–90 nm with a concentration of about 20 million particulates per cubic centimeter. The middle figure shows excellent filter efficiency by more than two orders of magnitude. The lower figure shows particulate penetration through the filter. The particulate penetration is calculated by dividing the particulate count with filter by the baseline count data.

Similarly, Figure 8 shows the histogram of particle number concentration in diameter range from 20 to 300 nm. The data points are taken from tests at rated speed and full load with and without the regenerated DPF. The number concentration without DPF varied from 22 million counts (range 20–30 nm) to 200 million counts (80–190 nm) per cubic centimeter. Similarly, the number concentration with regenerated DPF varied from 28 thousand (20–30 nm) to 383 thousand counts (80–190) per cubic centimeter. These

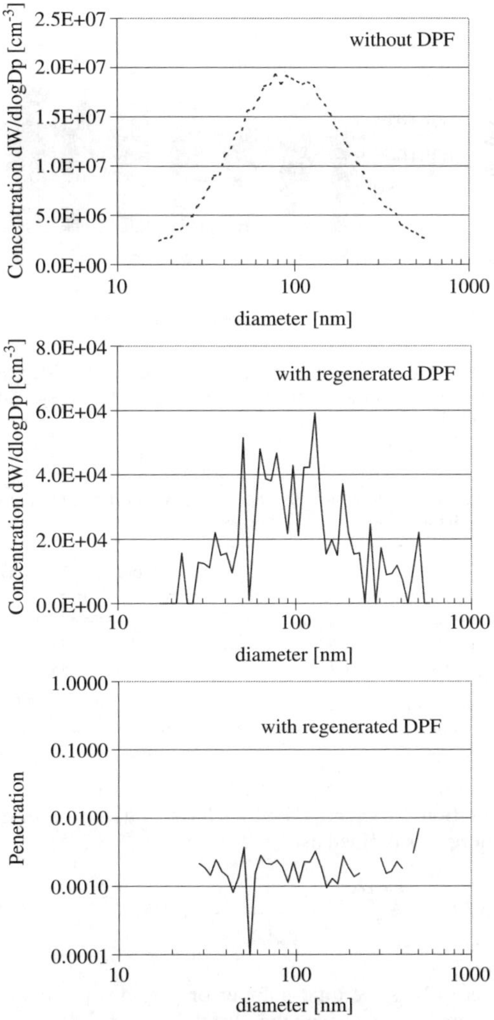

Figure 7. Particulates size distribution with and without the regenerated DPF system at rated speed and full load point (Mode D).

integrated values of particulate counts indicate better filtration efficiency at the low particle size ranges.

4.3 *Particulate filter efficiency*

The DPM trapping efficiency of the DPF system is calculated in two ways. The first one is Particulate Mass Filtration Efficiency (PMFE) based on particulate mass, and is defined as:

$$PMFE, \% = \frac{PM_{baseline} - PM_{filter}}{PM_{baseline}} \, x \; 100 \qquad (2)$$

where PM is particulate mass emission rate (g/kWh).

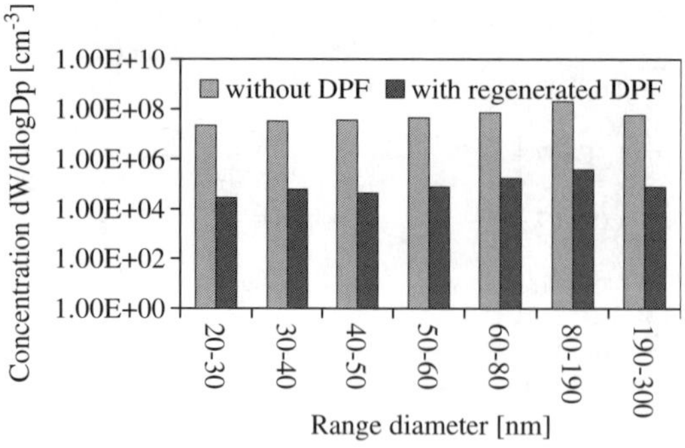

Figure 8. Integrated number of particles in different size spectra with and without the regenerated DPF system (Mode D).

Table 2. Filter efficiency according to particulate mass (PMFE) for regenerated and charged DPF system.

DPF	Mode A	Mode B	Mode C	Mode D
Regenerated	80.3	66.7	75.9	87.1
Charged	95.5	90.0	88.9	91.1
Average	87.9	78.3	82.4	89.1
Overall average	84.4			

Similarly, the second method Particulate Number Filtration Efficiency (PNFE) based on the particle numbers is defined as:

$$PNFE, \% = \frac{PN_{baseline} - PN_{filter}}{PN_{baseline}} \; x \; 100 \qquad (3)$$

where PN is the total number of particles per cubic centimeter in the size range from 20 nm to 200 nm.

The PMFE filter efficiency at all mode points is given in Table 2.

The efficiency of the DPF system after it was cleaned by regeneration of the filter varied from 66.7% to 80.3% with an average value of 77.5%. Similarly, the efficiency of the charged DPF system after it was loaded with about 60 grams of soot varied from 88.9% to 95.5% with an average efficiency of 91.4%. As expected, the charged filter efficiency at all mode points is higher than the regenerated filter efficiency, as the later was already loaded with soot and had a higher backpressure. The average efficiency of the system to trap particulates was measured at 84.4%.

The particulate number filter efficiency at all mode points for regenerated and charged filter is given in Table 3.

Table 3 shows that the efficiency of the DPF system after it was cleaned by regeneration varied from

Table 3. Filter efficiency according to particulate number (PNFE) for regenerated and charged DPF system.

DPF	Mode A	Mode B	Mode C	Mode D
Regenerated	99.71	97.60	99.86	99.82
Charged	99.86	99.04	99.91	99.90
Average	99.79	98.32	99/89	99.86
Overall average	99.46			

97.60% to 99.86% with an average value of 99.25%. Similarly, the efficiency of the charged DPF system after it was loaded with about 60 gram of soot was measured above 99% at all points. The overall efficiency of the system according to particle number (PNFE) was measured at 99.46%.

The number based filter efficiency at 99.5% is much higher than the mass based efficiency at 84.4%. Therefore, the thermal regeneration system is very effective in trapping nanoparticles.

4.4 *System regeneration*

A test was conducted to determine the effect of filter regeneration on exhaust gas concentrations. For this purpose the filter was loaded with approximately 60 grams of soot by running the engine at rated speed and full load. The regeneration ran for about 430 seconds while the engine was running steady state at 1400 rpm and 50% load. During the regeneration period, exhaust gas emissions concentration, flow rate, backpressure, flame temperature and filter inlet temperature were monitored. The engine exhaust gas flow rate prior to starting the burner (regeneration) was about 313 kg/h. During regeneration period the burner system added about 75 kg/h of fresh air to the exhaust gas stream. This additional air diluted the exhaust gas concentration of CO and NO_x during regeneration period.

Figure 9 shows the gas mass flow rates for CO and NO_x during the first 600 seconds. The CO peaked a high value of 880 g/h within 140 seconds of starting the regeneration, and then returned to its initial value of 8 g/h after the end of regeneration period. During the regeneration period of 432 seconds the average value of CO was 339 g/h (900 ppm). It should be noted that the value of NO_x did not significantly change during regeneration.

Similarly, Figure 10 shows the flame temperature, inlet DPF gas temperature and backpressure data during the regeneration period as measured by the thermal regeneration system's Electronic Control Unit (ECU).

The flame temperature, measured at 1070°C, raised the temperature of inlet DPF gas from 300°C to 650°C. This temperature is sufficient to ignite trapped soot in the filter. The regeneration reduced the exhaust backpressure from an initial value of 2.5 kPa (10 in. H_2O)

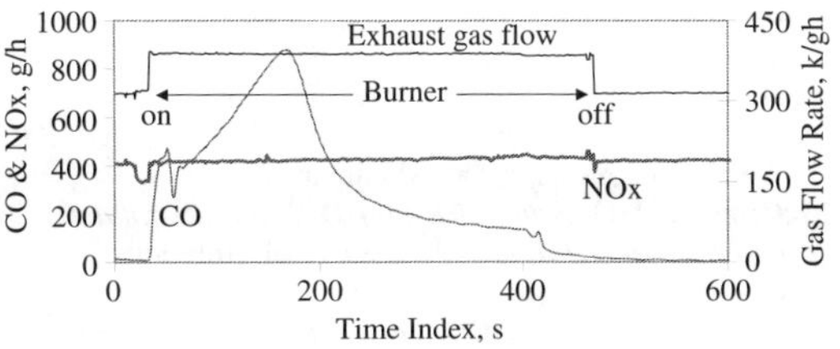

Figure 9. CO, NO_x and exhaust gas flow during regeneration at 1400 rpm and 50% load point.

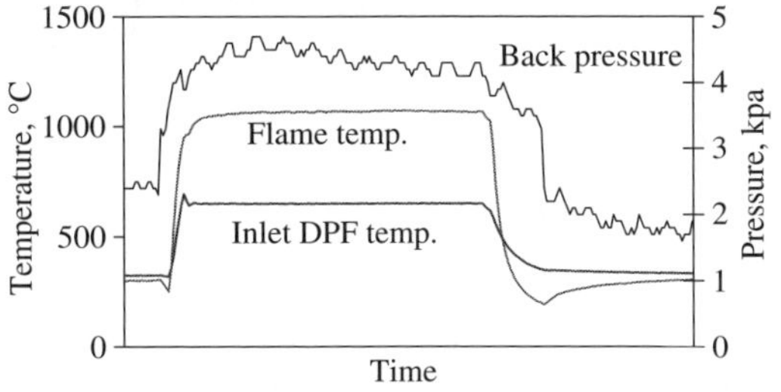

Figure 10. Temperature and backpressure data during regeneration at 1400 rpm and 50% load mode point.

to 1.7 kPa (7 in. H_2O). During regeneration the burner added 75 kg/h of air to the exhaust gas stream, and the backpressure at this increased exhaust gas flow was measured at 4.5 kPa (18 in. H_2O). As mentioned earlier, during engine conditioning, prior to beginning any test, the exhaust gas backpressure was set to 7.5 kPa (30 in. H_2O) at the rated speed and full load (mode point D), which is equivalent to an exhaust gas pressure of 1.6 kPa (6.3 in. H_2O) at mode B (engine speed of 1400 at 50% load). This suggests that after the regeneration, the exhaust gas pressure was reduced to 1.7 kPa (7 in. H_2O after regeneration compared to 6.3 in. H_2O under clean conditions before the start of the test), confirming that the filter was fully regenerated. The external maximum temperature of the filter assembly was measured at 322°C during regeneration. This regeneration consumed about 0.5 litre of fuel.

5 CONCLUSIONS

The thermal regeneration system application substantially reduced the Exhaust Quality Index at all test mode points. The reduction varied from 27% to 55%, which is equivalent to an increase in mine air quality or decrease in minimum ventilation requirement.

There was a slight increase in exhaust CO emission at some modes, and a decrease in NO_2 at all test modes. This change in exhaust gas emissions could be due to combustion of trapped soot in the filter.

Diesel particulate matter was significantly reduced at all mode points, with an overall reduction by 84.4% on mass basis, and 99.5% on number basis, demonstrating the significance of filter technology in removing nanoparticles.

The thermal regeneration system did not provide secondary emissions (during operation/loading), and is not affected by higher fuel sulfur.

The system was suitable in regenerating the filter during normal engine operation. However, CO was significantly increased for a short period of time during regeneration.

6 RECOMMENDATIONS

Since the laboratory testing of the thermal regeneration system was found suitable, the next step is to try the system in a production mine. This will provide long term evaluation of the system in harsh mining environment. The in-field evaluation should include emissions monitoring, and development of a trouble shooting and maintenance program.

The mine regulations should promote the application of proven emission control technologies with some incentive programs.

ACKNOWLEDGEMENTS

The authors express their gratitude for the assistance and cooperation of the following persons:

- Vince Feres, NRCan, CANMET-MMSL, Ottawa
- Francois Jaussi, Liebherr Machines Bulle S.A, Switzerland
- A. Mayer, Technik Thermische Maschinen, Switzerland
- J. Czerwinski, Abgasprüfstelle Fachhochschule Biel, Switzerland

REFERENCES

CANMET. 1997. *Manufacturer's protocol for exhaust systems testing (MAPTEST)*, Natural Resources Canada, Document No. MMSL 97-064 (CR).

CSA. 1988. *Flameproof non-rail-bounded diesel-powered machines for use in gassy underground coal mines*, CAN/CSA-M424.1-88, published by the Canadian Standards Association, Toronto, Ontario, Canada.

CSA. 1990. *Non-rail bounded diesel-powered machines for use in non-gassy underground mines*, CAN/CSA-M424.2-M90, published by the Canadian Standards Association, Toronto, Ontario, Canada.

Czerwinski, J. & Mayer, A. 2001. *Diesel particulate filter technology for construction machines – Swiss activities & experiences*. Proceedings of the 7th International Mine Ventilation Congress. Krakow, Poland: 921–928.

Dainty, E.D. et al. 1986. *A summary of underground mine investigations of ceramic diesel particulate filters and catalytic purifiers*. Heavy-duty diesel emission control: A review of technology. The Canadian Institute of Mining and Metallurgy, Special volume 36:54–77.

DEEP. 1997. *Diesel emissions evaluation program*. www.deep.org.

French, I.W. & Mildon, C.A. 1978, 1984 & 1990. *Health implications of exposure of underground workers to diesel exhaust emissions*, CANMET Contract Report Nos. 23SQ.23440-9-9142, 14SQ.23440-1-9003, 23SQ.23440-2-9062, o9SQ.23440-7-9042, Energy, Mines & Resources Canada, Ottawa.

ISO. 1996. ISO 8178-1. *Reciprocating internal combustion engines – Exhaust emission measurement*. Part 1: Test-bed measurement of gaseous and particulate exhaust emissions.

Mayer, 2002. *VERT partculate trap verification*. SAE paper number 2002-01-0435. SAE 2002 World Congress, Detroit, Michigan Mayer, 2002. *VERT partculate trap verification*. SAE paper number 2002-01-0435, SAE 2002 World Congress, Detroit, Michigan.

McGinn, S. 2003. *Brunswick mine DPF project-final results and recommendations*. Mining Diesel Emissions Conference, Toronto, Ontario, Canada.

Stachulak, J. 2003. *INCO/DEEP diesel particulate trap project*. Mining Diesel Emissions Conference, Toronto, Ontario, Canada.

Zelanka, P. et al. 2002. *Development of a full-flow burner DPF system for heavy duty diesel engines*. SAE paper number 2002-01-2187. SAE Powertrain & Fluid Systems, Conference & Exposition, San Diego, California.

Mine Ventilation, Ganguli & Bandopadhyay (eds)
© 2004 Taylor & Francis Group, London, ISBN 90 5809 633 5

CFD analysis of diesel emissions in a railroad tunnel

R.E. Ray, K. Kang & M. Gilbey
Parsons Brinckerhoff Quade & Douglas, Inc., New York, USA

ABSTRACT: Computational fluid dynamics (CFD) modeling can be used to predict three-dimensional distribution of diesel emissions gases in underground mines and tunnels. This paper presents the results of transient CFD simulations performed for a passenger railroad tunnel utilizing diesel locomotives. Both natural and mechanical ventilation systems were evaluated for a stopped passenger train operating in "hotel mode" (diesel engine running to supply lights and air conditioning/heating to passenger cars). The mass flow rate and temperature of diesel engine exhaust products were modeled at the locomotive exhaust port location. Concentrations of nitrogen dioxide (NO_2), the critical diesel exhaust product for railroad tunnels, were evaluated at the air intake port locations of the passenger cars. In addition, the simulations enabled NO_2 levels at the passenger car intake ports and the sampling locations for the NO_2 gas detection system used to control the ventilation system for the tunnel to be compared.

1 INTRODUCTION

In railroad tunnels utilizing diesel locomotives, ventilation is required to dilute contaminants contained in the diesel engine exhaust and to remove heat generated by the locomotives. Natural, train-generated "piston effect," or mechanical ventilation systems can be employed. For shorter tunnels or tunnels provided with closely spaced shafts, "piston effect" ventilation will typically produce sufficient airflow to remove heat and adequately dilute diesel exhaust products. As the train moves through the tunnel, air in the annular spaces between the train and tunnel walls flows in the same direction as the train. Ambient air is drawn in through the portal and shafts located behind the train and discharged through shafts and the portal located in front of the train.

For normal train operations, one-dimensional fluid flow models such as the Subway Environment Simulation (SES) program can be used to predict longitudinal airflow through railroad tunnels and shafts, as well as "average" diesel exhaust gas concentrations at junctions in the tunnel network model. The SES program is a time-dependent, one-dimensional ventilation model with subprograms for simulating airflow, heat transfer, environmental conditions and train movement in tunnels.

While one-dimensional fluid flow models can effectively simulate environmental conditions for normal train operations, they are not always best suited to analyze natural ventilation systems where no piston effect is available to generate longitudinal air movement. In railroad tunnels, this occurs when diesel trains stop and continue to operate in "hotel mode." In "hotel mode," the diesel engine still provides power for on-board auxiliaries such as air conditioning and lighting for the passenger cars.

Diesel engine exhaust gases are discharged from the top of the locomotive at high temperature and tend to stratify at the crown of the tunnel. When the train is stopped and operating in "hotel mode," no "piston effect" longitudinal airflow is generated to dilute and mix the engine exhaust products. For this situation, the use of three-dimensional fluid flow models can provide more useful quantitative design information regarding environmental conditions throughout the entire tunnel cross section and the length of tunnel segment of interest than one-dimensional models.

Computational Fluid Dynamics (CFD), a numerical technique for modeling fluid flow, heat transfer, and mass transport phenomena in three dimensions, has been used extensively to analyze ventilation problems for transit, rail, and vehicular tunnels. This paper describes CFD simulation analysis performed for the ventilation system required for diesel locomotive "hotel mode" operation for a rail overbuild created by the construction of a series of commercial high-rise residential buildings over existing passenger railroad tracks.

2 CFD METHODOLOGY

The steps required to conduct a CFD modeling exercise include creation of the geometric model, grid or mesh generation, establishment of boundary conditions, running the simulation to solve the problem, and post-processing of results. The tunnel ventilation analysis presented in this paper was performed using FLUENT, a general purpose CFD program, capable of simulating fluid flow and heat and mass transfer in two and three dimensions. FLUENT solves the conservation equations for mass, momentum, energy, and chemical species. These equations are described in the FLUENT User's Guide (1998) and by Ray & Zigh (1999).

FLUENT uses a control volume based finite difference method to solve the governing equations, which are discretized on a curvilinear grid to enable computation in complex geometries. The "computational domain," representing the process being analyzed is divided into a number of non-overlapping control volumes with a grid point placed at the center of each control volume. The value of an internal grid point is assumed to prevail over the control volume around it. Similarly, the value of a boundary grid point is assumed to prevail over the face of the associated control volume.

CFD simulations involve two types of variables: scalar and vector. Chemical species concentration is an example of a scalar variable, which has a value at a point, but no direction. Temperature is also a scalar variable. A vector variable, such as air velocity, has both a magnitude and a direction. The CFD simulation solves for the value of velocity, temperature, chemical species concentration, and any other variable of interest at every grid point in the computational domain. To obtain these values, the relevant differential equations are transformed into algebraic equations by integration over each control volume. The algebraic equations are subsequently solved for the dependent variables at the various grid points.

The simulation of diesel exhaust gas distributions utilize the basic conservation equations, the k-epsilon turbulence model and the mass transport model for emissions. The mass continuity and momentum equations are resolved with the SIMPLE algorithm. The standard k-epsilon two-equation turbulence model is used with buoyancy terms included in both the equations of kinetic energy and turbulence dissipation.

3 RAIL OVERBUILD DESCRIPTION

For this case study, CFD simulations were performed for a rail overbuild created by the construction of a series of commercial high-rise residential buildings over a pair of existing passenger railroad tracks in New York City. The initial phase of the project consisted

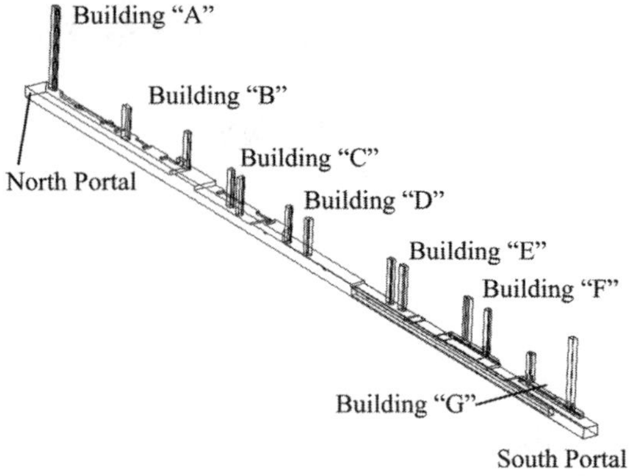

Figure 1. Overbuild geometry.

of the erection of three high-rise buildings (referenced from north to south as "C," "D" and "E"), each extending approximately one city block in the direction parallel to the north-south centerline of the tracks for a total length of approximately 256.03 m (840 feet). Subsequently, three more buildings have been constructed – "B" and "A" to the north, and "F" to the south. The addition of the seventh Building, "G," will increase the overall length of the overbuild to approximately 563.88 m (1,850 ft). Future construction planned by the developer may eventually extend the length of the overbuild to 975.4 m (3,200 ft).

A three-dimensional representation of the overbuild encompassing Buildings "A" through "G" is shown in Figure 1. The height of the enclosure varies from a minimum of 5.56 m (18.25 ft) above the top-of-rail beneath Building "E" to a maximum of 8.53 m (28 ft) beneath Building "B." Full-height crash walls running parallel to the tracks form the east and west boundaries of the overbuild and provide a constant width of 9.75 m (32 ft) for the southern half of the enclosure (beneath Buildings "G," "F," "E" and "D"). The east crash wall is dropped beneath Buildings "C," "B" and "A," allowing the overbuild width to increase to 15.24 m (50 ft). At that location, a pre-existing stone wall forms the east boundary.

4 OVERBUILD VENTILATION SYSTEM

Due to the uncertainty regarding the sequence of building construction and the total number of buildings to eventually be erected, the original ventilation analysis performed for the rail overbuild focused on modular ventilation concepts that could be employed throughout the progressive development of the project. Close spacing of ventilation shafts was possible since the developer allowed for shaft construction through the low-rise section of each building.

Based on steady state CFD simulation analysis documented by Ray & Zigh (1999), a natural ventilation system was provided for tunnel fire emergencies for the initial three-building enclosure, with two ventilation shafts constructed from the roof of the overbuild to the top of the low-rise portion of each building. Located directly over the railroad tracks in the approximate center of the overbuild, the natural ventilation shafts allow hot smoke and fire gases to exhaust naturally from the overbuild during a fire emergency. The shafts total approximately 23.22 m^2 (250 ft^2) in cross-sectional area per building and are spaced from 7.01 to 16.15 m (23 to 53 ft) apart within each structure. The distance between the northernmost shaft in each of the first three buildings and the southernmost shaft in the adjacent buildings varies from 51.82 to 71.63 m (170 to 235 ft).

For small fires that do not generate sufficient buoyancy to ventilate naturally, and for dilution of diesel exhaust gases produced by stopped trains operating in "hotel mode," each shaft in the original three buildings was provided with a mechanical exhaust ventilation system with a capacity of 14.16 m^3/s (30,000 cfm). The exhaust system consists of one 7.08 m^3/s (15,000 cfm) centrifugal fan connected to top of each shaft and a second 7.08 m^3/s (15,000 cfm) centrifugal fan connected to the top of an extraction duct which extends from the top of the shaft to a location inside the overbuild between adjacent shafts. A damper located over the top of each shaft is closed when the fans operate to enable the shaft fan to exhaust air from the overbuild through the shaft. To simplify operation of the mechanical ventilation system, all shaft and duct fans are operated in all buildings for fire emergencies and high NO_2 alarms.

The shaft fan/extraction duct fan system for small fires and diesel emissions was extended to Building "B" and the north shaft of Building "F." However, beginning with the single Building "A" shaft and the south shaft at Building "F"(both located to the east of the track area instead of directly over the tracks), vertically-mounted jet fans were installed in the shafts. The jet fans, which are axial fans with open inlets and outlets, were needed to generate sufficient exhaust flow from the offset shafts during fire emergencies. They also eliminated the need for shaft dampers and the centrifugal fan systems. Rather than operating against pressure in a ducted system, jet fans induce airflow through a tunnel or shaft by applying thrust to the airstream with a high discharge velocity (typically 30.48 m/s [6,000 fpm]). Like the centrifugal fans, all overbuild jet fans are also operated for fires and high NO_2.

5 BUILDING "G" CFD ANALYSIS

Due to space limitations, two closely spaced ventilation shafts located directly over the railroad tracks

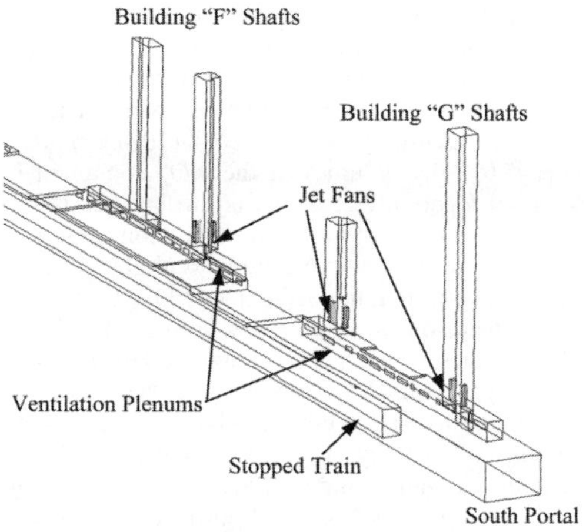

Figure 2. Building "F" and "G" shafts.

and situated near the longitudinal (north-south) center of the buildings (as were constructed for Buildings "C" through "E") could not be provided for Building "G." The Building "G" shafts are located at the extreme north and south ends of the building, approximately 39.62 m (130 ft) apart.

As shown in Figure 2, the north shaft is located to the east of the east crash wall while the south shaft overlaps the east crash wall and the track area. A common plenum located outside of the track area adjoins the upper portion of the east crash wall. The plenum is approximately 2.77 m (9.1 ft) wide by 2.77 m (9.1 ft) high, connects the bottom of the shafts and extends for the entire length of Building "G." A series of rectangular openings are provided on the track side of the plenum to allow smoke (for emergency ventilation) or diesel exhaust gases to be captured by the jet fan ventilation system located in the shafts. The plenum beneath Building "G" and the similar plenum beneath Building "F" were added to enhance the performance of the ventilation shafts since both shafts in each building are located completely or partially to the side of the track area, rather than directly over the tracks.

Transient CFD analysis was performed to evaluate the effectiveness of the offset shafts and plenum in capturing diesel exhaust products from stopped trains operating in "hotel mode." The transient (time dependent) simulation begins after the train has come to rest. Wintertime conditions are assumed, with an ambient air temperature of −10 degrees Centigrade (14 degrees Fahrenheit).

5.1 *Diesel emissions assumptions*

The most significant contaminants produced from diesel rail operations in overbuilds are the oxides of nitrogen (NO_x). As described by Bendelius (1982), previous emission studies have demonstrated that if concentrations of NO_x are maintained within acceptable levels, all other concentrations of diesel exhaust contaminants will also be maintained within acceptable levels. According to Hobbs et al. (1977), approximately 90% (by volume) of the NO_x contained in raw diesel locomotive exhaust is nitric oxide (NO). However, as a result of molecular oxidation caused by high temperatures and exposure to sunlight, some of the NO converts to nitrogen dioxide (NO_2). In ambient air, the conversion rate is quite slow compared to the air exchange rates experienced in tunnels. For the overbuild analysis, total NO_x emitted by the diesel locomotive engine exhaust was conservatively assumed to be comprised of 25% (by weight) NO_2 and 75% NO.

The diesel engine in diesel locomotives drives an electric generator that supplies power to the electric traction motor. The locomotive used on the railroad line discussed in this paper has eight throttle, or notch, settings that control engine speed. Emission levels vary depending on the notch setting. Based on recommendations from the locomotive manufacturer, a notch setting of 2 (approximately 499 kW [669 hp]) was assumed for "hotel mode" operation, with a corresponding NO_x emissions rate of 7,000 grams/hour.

5.2 *Air quality criteria*

Air quality requirements in the overbuild have been based on the recommended Threshold Limit Value – Short-Term Exposure Limit (TLV-STEL) and the Threshold Limit Value-Time-Weighted Average (TLV-TWA) established by the American Conference of Governmental Industrial Hygienists (ACGIH 1996). The STEL refers to a 15-minute time-weighted average concentration. The TLV-TWA values represent a "time-weighted average concentration for a conventional 8-hour workday and a 40-hour workweek, to which nearly all workers may be exposed, day after day, without adverse effect." The TLV-TWA levels have been incorporated in the Occupational Safety and Health Administration (OSHA) regulations limiting employee exposure to hazardous substances in the workplace.

The TLV-STEL and TLV-TWA for NO_2 concentrations, by volume, are 5 parts per million (ppm) and 3 ppm, respectively. No short-term exposure levels have been established for NO, while the 8-hour Time-Weighted Average concentration is limited to 25 ppm. The ACGIH recommends excursion limits for TWAs that do not have STELs of three times the TLV-TWA for no more than 30 minutes during a workday. The excursion limit for NO would be 75 ppm. Only NO_2 concentrations were evaluated for this study since the OSHA limits for NO are five times higher than for NO_2 and recommended short-term exposure levels are 25 times higher.

For diesel locomotive "hotel mode" operation, passengers would remain inside the train cars while the diesel engine continues to operate. The critical location for evaluating NO_2 concentrations would be at any of the four outside air intake ports of each car. An air intake port is located near each end of the car, on both sides, approximately 3.66 to 3.96 m (12 to 13 ft) above the top-of-rail. The overbuild atmosphere at these ports will be drawn into the cars, so the NO_2 concentrations at that location cannot be allowed to remain at 5 ppm for more than 15 minutes.

5.3 *CFD model description*

5.3.1 *Geometry*

A three-dimensional grid was developed to represent the overbuild volume enclosed by Buildings "G" through "A." Shafts previously constructed over Buildings "F," "E," "D," "C," "B" and "A" were incorporated in the model at their "as-built" locations. A 4.06-m (13.33-ft)-long by 3.15-m (10.33-ft) wide by 17.53-m (57.5-ft)-high shaft was modeled near the north end of Building "G" and a second shaft, with the identical cross sectional area as the north shaft and a height of 44.44 m (145.8 ft), was modeled near the south end. As described previously, a 2.77-m (9.1-ft)-wide by 2.77-m (9.1-ft)-high plenum connected to the shaft bottoms is modeled to the east of the northeast corner of the overbuild cross section and extends the length of the building.

The height of the overbuild beneath Building "G" and the street to the south of Building "G" is approximately 6.25 m (20.5 ft). The overbuild drops to an average height of approximately 5.64 m (18.5 ft) beneath the street to the north of Building "G" and increases to a height of approximately 7.01 m (23 feet) at the southern end of Building "F."

Portal openings to the atmosphere were treated as "outlet vents" in the model which allow natural ventilation airflow to enter or exit the overbuild. The top of shafts equipped with dampers (Buildings "B" through the north shaft in Building "F") were treated as velocity boundaries with an exhaust flow rate from the centrifugal fans connected to the top of the shafts of 7.08 m^3/s (15,000 cfm) per shaft. The inlet to each extraction duct (one per shaft for Buildings "B" through north shaft in Building "F") was also modeled as a velocity boundary with an exhaust flow rate of 7.08 m^3/s (15,000 cfm). Partial blockage of the natural ventilation shafts created by the vertical riser portion of the extraction ducts was also accounted for in the CFD grid.

5.3.2 *Jet fans*

For the south shaft in Building "F" and both shafts in Building "G" two 630-mm (24.8-in.), 29.83-kw (40-hp) vertically-mounted jet fans were located near the

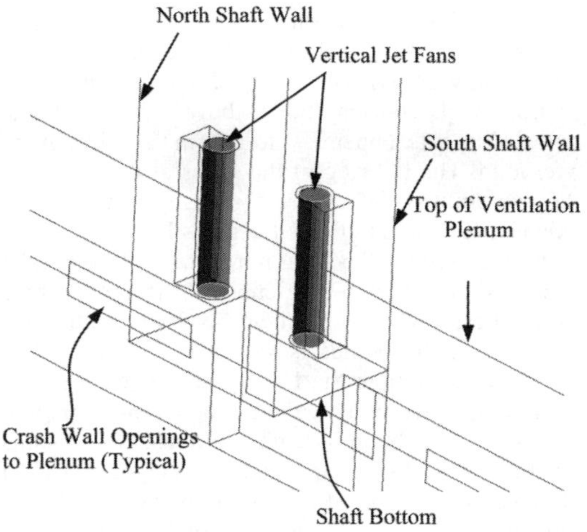

Figure 3. Shaft jet fan arrangement.

bottom of the north and south shaft walls, respectively, of each shaft (see Figure 3). The total length of the jet fans was modeled at 3289 mm (129.5 in.) to account for sound attenuators on each end.

Jet fans are selected based on the amount of thrust imparted to the airstream. To convert the momentum of the high velocity jet to the bulk shaft airflow, a sufficient length of shaft is required downstream of the fans. Maevski & Fay (2002) have shown that it is desirable to provide at least ten shaft/tunnel diameters downstream of the jet fans to achieve adequate momentum transfer. For the overbuild project, this required placing the fans as close to the shaft bottoms as possible. However, since the shaft bottoms represent a high turbulence zone with airflow turning 90 degrees from the plenum into the shafts, the jet fans were placed approximately one shaft diameter above the base of the shafts to improve fan inlet conditions. This resulted in a reduction in the downstream shaft length.

The thrust delivered by the fans is seldom wholly converted into useful thrust in a tunnel or shaft because of the proximity of the fans to obstructions to airflow. The ratio of useable thrust to theoretical thrust is termed the jet fan installation efficiency, often abbreviated by the Greek symbol beta (β). Studies by Saika et al. (2000) of the relationship between jet fan beta and installation characteristics indicate that there is a marked reduction in jet fan efficiency when jet fans are located in corners. For this project, a jet fan beta value of 0.55 was used since the fans are located in corners, are in close proximity to one another, and in some cases have other ventilation ducts located in the shaft. The extraction duct is located in the southeast corner of the entire height of the Building "F" south shaft, and a kitchen flue exhaust duct is located in the northeast corner of the north Building "G" shaft.

The jet fans in Buildings "F" and "G" are operated for both fire emergencies and control of NO_2.Emergency ventilation analysis previously performed for Building "F" indicated that shafts located to the side of the track area must exhaust a minimum of 47.20 m^3/s (100,000 cfm) of air each to maintain acceptable smoke conditions in the overbuild during a fire emergency. Thrust levels required to produce an exhaust airflow of approximately 47.20 m^3/s (100,000 cfm) per shaft were calculated. The thrust rating for the pair of jet fans (in pounds of force) in each shaft was determined by multiplying the plenum and shaft pressure drop associated with a 47.20 m^3/s (100,000 cfm) exhaust flow rate, expressed in units of force per unit area, by the shaft cross sectional area and dividing by the product of the jet fan efficiency and a de-rating factor for operation of the fans at high temperature. Using a pressure drop of 32 Pa (0.13 in. w.g.) through the plenum openings and shaft, a jet fan efficiency of 0.55, and de-rating the thrust for operation at 150 degrees C (302 degrees F) by utilizing a correction factor of 0.69, a total thrust requirement per shaft of 1,059 N (238 lb force) was calculated. Two jet fans, each with a thrust rating of 534 N (120 lb force) were selected. The jet fans are modeled as a momentum source for the CFD analysis.

5.3.3 *Train and diesel exhaust modeling*

To represent a "worst case" scenario for "hotel mode" operation, a 18.29-m (60-ft)-long by 3.2-m (10.5-ft)-wide by 4.27-m (14-ft)-high diesel locomotive and seven 25.9-m (85-ft)-long by 3.2-m (10.5-ft)-wide by 4.27-m (14-ft)-high passenger cars were modeled on the west (southbound) set of tracks. The west set of tracks is located furthest from the Building "G" and Building "F" shafts. The locomotive was situated so that the exhaust port on the top of the locomotive was located approximately midway between the Building "G" shafts. Only one train operating in "hotel mode" at a time was considered since the headway between trains using the overbuild is in excess of an hour.

A 508-mm (20-inch) by 254-mm (10-inch) rectangular exhaust port was modeled on the top center of the locomotive, approximately 11.80 m (38.7 feet) from the front end. Using the locomotive engine exhaust data for the notch setting 2, five gas species were modeled from a total engine exhaust mixture of 2.13 m^3/s (4,514 cfm) on a mass fraction basis for the notch 2 throttle setting: CO (0.0025), NO (0.00138), NO_2 (0.00046), hydrocarbons (0.00008) and air (remainder of mixture). Temperature of the exhaust was assumed to be 438 degrees C (820 degrees F).

Previous analysis using the SES program to evaluate normal train operations predicted average background NO_2 concentrations for the overbuild of less than 1 ppm. The transient CFD simulations for "hotel mode" operation were based on an ambient NO_2 concentration of 1 ppm.

5.4 *Transient simulation*

To effectively examine the conditions inside the overbuild after the diesel locomotive has stopped and is operating in "hotel mode," the transient simulation was performed with a constant exhaust rate from the diesel engine, but with a time-dependent set-up of the mechanical ventilation system. As designed, the mechanical ventilation system comprised of centrifugal shaft and duct fans at Buildings "B" through "F," and jet fans at Buildings "A," "F" and "G" is not energized until the NO_2 monitoring system for the overbuild detects an NO_2 concentration of 3 ppm or higher.

The overbuild NO_2 monitoring system draws air samples from 18 different locations in the overbuild. Due to potential interferences with future electrification of the railroad line, none of the sampling lines can be located over the track area. Instead, the open ends of each of the sampling lines are attached to the east and west crash walls, approximately 4.27 m (14 ft) above the top-of rail. Generally, one sampling line is terminated on each crash wall approximately midway between shafts in adjacent buildings. In buildings where the shafts are located at opposite ends of the building ("B" and "G"), sampling lines also terminate between shafts within the building.

The height of the sampling line inlet was selected to provide a safety factor above the 3.66 to 3.96-m (12 to 13-ft) zone where the passenger car fresh air intakes are located since the high temperature diesel exhaust gases tend to stratify at the ceiling of the overbuild and may drop down to lower elevations over time if not adequately ventilated. For the transient CFD simulation, a customized subroutine was written to interact with the CFD simulation solutions. For each 5 to 10-second time step in the transient simulation, NO_2 levels near the approximate location of the end of each sampling line were monitored. If any sampling line detected NO_2 concentrations greater than or equal to 3 ppm, the subroutine closed all shaft dampers and started all centrifugal shaft and duct fans, and the jet fans in the open shafts in Buildings "A," "F" and "G."

5.4.1 *Natural ventilation results*

The "hotel mode" transient simulation was started with the diesel locomotive stopped beneath Building "G," discharging engine exhaust from the stack located on top of the locomotive in a natural ventilation mode with no overbuild fans operating. The NO_2 sampling station located midway between the Building "G" shafts on the east crash wall detected NO_2 concentrations of 3 ppm at Time = 210 seconds, initiating operation of the shaft and duct exhaust fans in Buildings "B" through "F" and the jet fans in Buildings "A," "F" and "G."

Figure 4 depicts a cross section (looking north) through the overbuild and ventilation plenum at the approximate location of the locomotive exhaust port at Time = 210 seconds showing NO_2 distribution. At that location, the diesel exhaust gases are stratified horizontally across the overbuild and plenum cross section, with concentrations above 9 ppm (darkest color) filling the upper 0.91 to 1.22 m (3 to 4 ft) of the overbuild. The layering of the gases at the crown of the overbuild is due to the high temperature of the exhaust, 438 degrees C (820 degrees F).

A cross section (looking north) taken through the overbuild and ventilation plenum at the approximate location experiencing the highest NO_2 concentrations at Time = 210 seconds is shown in Figure 5. This section cuts through the first passenger car situated behind the locomotive. Figure 6, an orthographic view of the overbuild beneath Building "G," contains a plan view of NO_2 distribution at a height of 4.27 m (14 ft) above the top-of-rail that illustrates the longitudinal location of the high NO_2 concentration pockets with respect to the train.

An examination of Figure 6 indicates that at Time = 210 seconds, NO_2 concentrations at the height

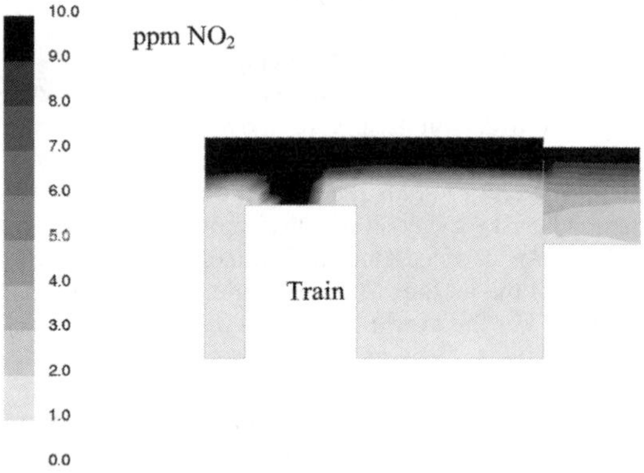

Figure 4. Time = 210 sec.: Cross section at exhaust port.

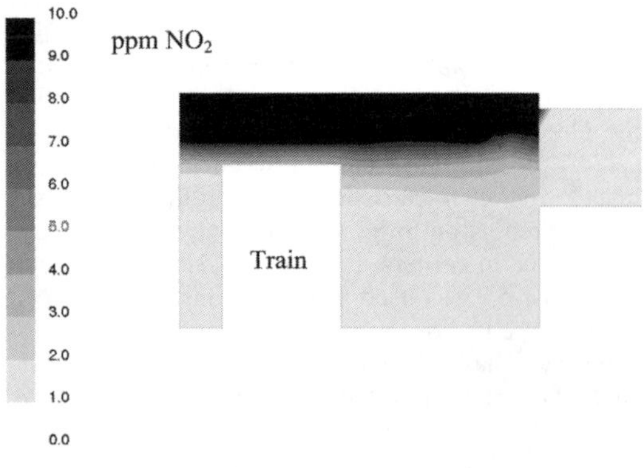

Figure 5. Time = 210 sec.: Cross section at passenger car.

of the gas sampling lines and top of the train cars, 4.27 m (14 ft), are in the range of 2 to 4 ppm against a portion of the passenger car immediately behind the locomotive. Small pockets of NO_2 against the east crash wall and inside the plenum near the Building "G" north shaft are in the range of 4 to 5 ppm. Figure 5 shows that at the time the fans start, NO_2 concentrations against the train are in the 0 to 2 ppm range at the height of the fresh air intakes, 3.66 to 3.96 m (12 to 13 ft).

Figure 7 is a longitudinal section (looking east) taken approximately 0.61 m (2 ft) east of the train and Figure 8 is a longitudinal section (looking east) taken through the Building "F" and "G" shafts and plenum. Both figures display NO_2 distribution. Figure 7 demonstrates that high concentrations of NO_2 (greater than 9 ppm) are concentrated at ceiling level, above the top of the train. The NO_2 layer extends from the south Building "F" shaft to the south portal, indicating that some of the gas is exhausting out of the portal. The portal is actually intaking ambient air below the gas layer. The heaviest concentration of NO_2 is located near the north Building "G" shaft, just north of the locomotive exhaust port.

The NO_2 distribution patterns shown in Figure 8 demonstrate that prior to starting the fan system, the south shaft in Building "G" is exhausting while the Building "F" south shaft and north shaft in Building "G" are intaking. At Time = 210 seconds, the south shaft in Building "G" is exhausting approximately 33.04 m^3/sec (70,000 cfm) of air, while the south Building "G" shaft and north Building "F" shaft are intaking less than 4.72 m^3/sec (10,000 cfm) each. Net airflow through the portal into the overbuild is approximately 7.08 m^3/sec (15,000 cfm).

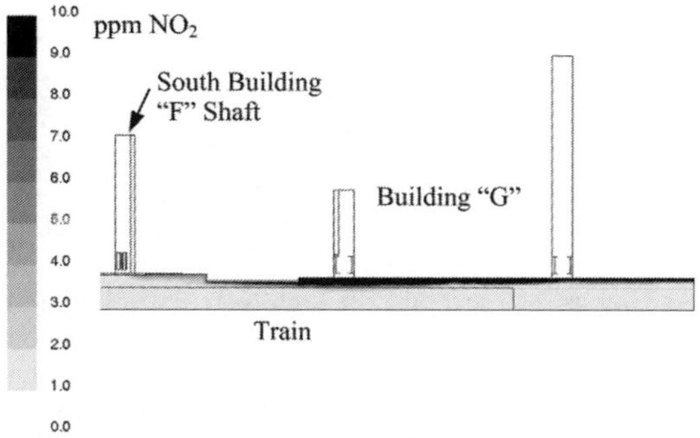

Figure 7. Time = 210 sec.: Longitudinal section east of train.

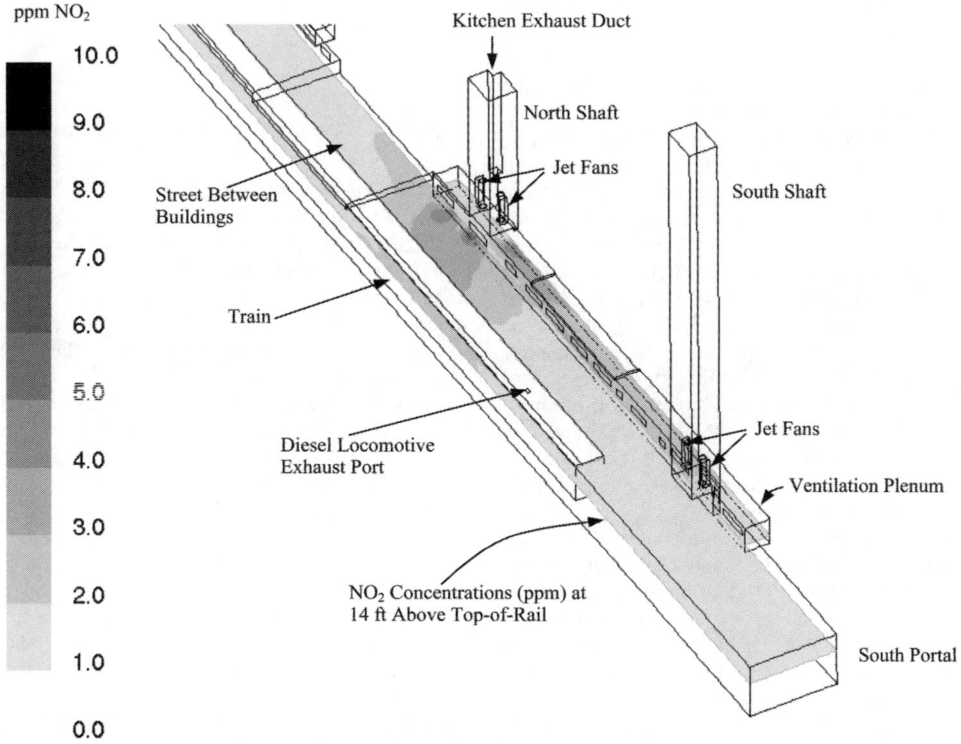

Figure 6. Time = 210 sec.: Orthographic view of overbuild beneath building "G".

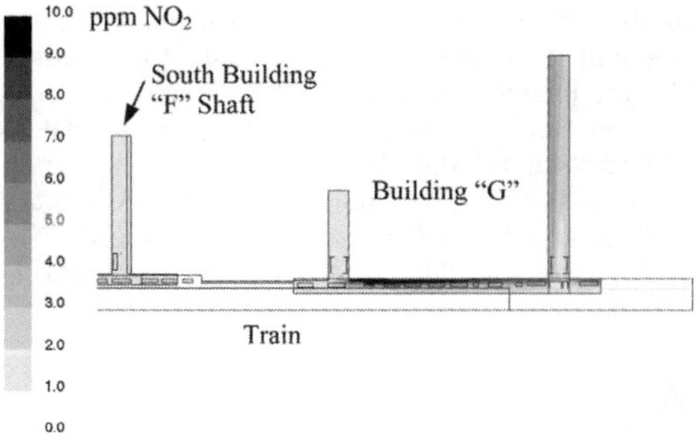

Figure 8. Time = 210 sec.: Longitudinal section at shafts.

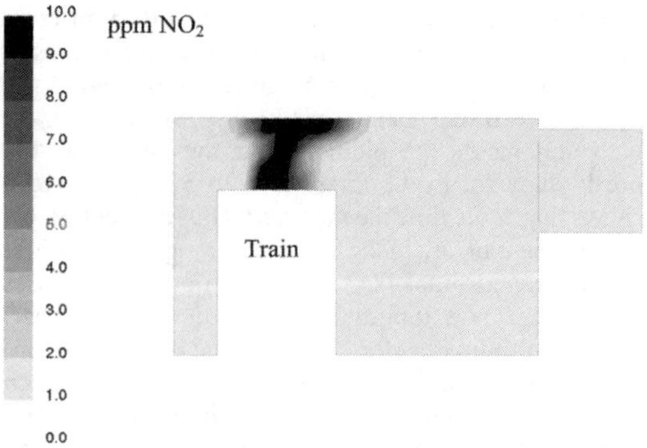

Figure 9. Time = 730 sec.: Cross section at exhaust port.

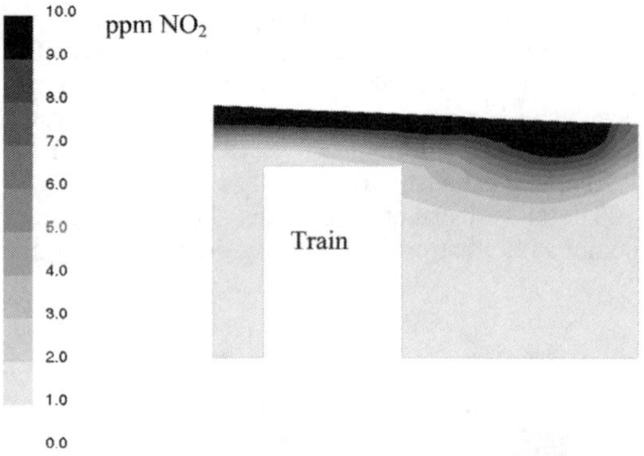

Figure 10. Time = 730 sec.: Cross section at passenger car.

5.4.2 *Mechanical ventilation results*

At Time = 210 seconds, the high NO_2 alarm initiated by the NO_2 detection system closes the dampers on top of the Building "B" through the north Building "F" shafts, and starts the shaft, duct and jet fans for Buildings "A" through "G." The mechanical system was allowed to run until Time = 730 seconds before the simulation was stopped.

Figure 9 depicts NO_2 distribution for the cross section (looking north) through the overbuild and the ventilation plenum at the approximate location of the locomotive exhaust port that is shown in Figure 4, but at Time = 730 seconds. Compared to Figure 4, NO_2 layers are restricted to a relatively narrow zone located directly above the center of the locomotive. The intake airflow drawn through the South Portal by the operation of the mechanical ventilation system eliminates the gas layer that previously stretched from the west to east crash walls and through the plenum.

Figure 10 displays NO_2 distribution for a cross section (looking north) taken through the south end of the street situated immediately to the north of Building "G." Similar to Figure 5, this represents the approximate location of the highest NO_2 concentrations against the train. The ceiling slopes downward from west to east beneath the street, from approximately 5.87 m (19.25 ft) at the west crash wall to 5.43 m (17.83 ft) at the east crash wall. NO_2 concentrations are in the 0 to 2 ppm range at the height of the passenger car fresh air intakes.

Figure 11, the same orthographic view of the overbuild beneath Building "G" shown in Figure 6, also contains a plan view of NO_2 distribution at a height of 4.27 m (14 ft) above the top-of-rail that illustrates the longitudinal location of the high NO_2 concentration pockets with respect to the train. An examination of Figure 11 indicates that at Time = 730 seconds, the diesel exhaust gases have been completely cleared below the 4.27-m (14-ft) level from the segment of the train adjacent to the Building "G" shafts and is limited to a small segment of the low height area under the street between Buildings "G" and "F."

Figure 12 depicts NO_2 distribution for the same longitudinal section (looking east) taken approximately 0.61 m (2 ft) east of the train in Figure 7. Figure 13 displays NO_2 distribution for the same longitudinal section (looking east) taken through the Building "F" and "G" shafts and plenum that is shown in Figure 8. Figure 12 shows that NO_2 is no longer exhausting out of the South Portal and has been virtually cleared to a location approximately midway between the two Building "G" shafts. Figure 13 indicates that the south shaft in Building "G" is also clear of NO_2, with the heaviest concentrations in the north Building "G" shaft. Some NO_2 can be seen in the south Building "F" shaft exhaust.

Analysis of the exhaust airflow data for the Building "G" shafts indicates flow rates of approximately 97.22 m^3/s (206,000 cfm) from the south Building "G" shaft, 75.51 m^3/s (160,000 cfm) from the north Building "G" shaft and 53.33 m^3/s (113,000 cfm) from the south Building "F" shaft. Approximately 207.66 m^3/sec (440,000 cfm) of supply air is drawn in

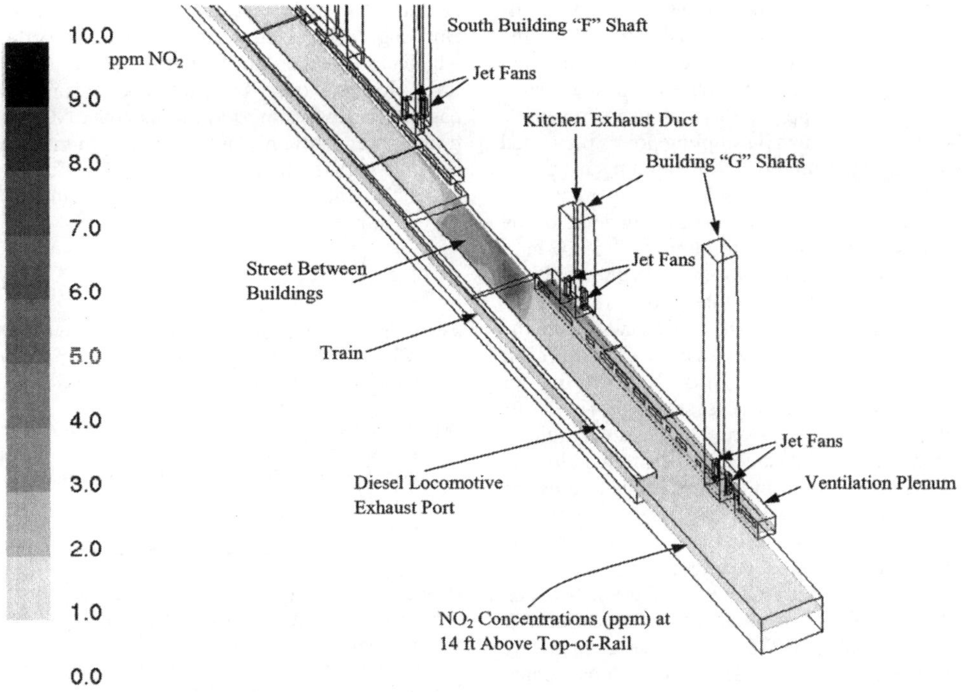

Figure 11. Time = 730 sec.: Orthographic view of overbuild beneath building "F" and "G".

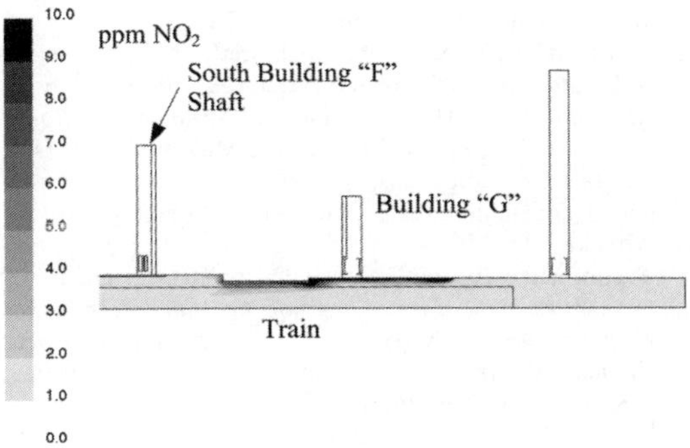

Figure 12. Time = 730 sec.: Longitudinal section east of train.

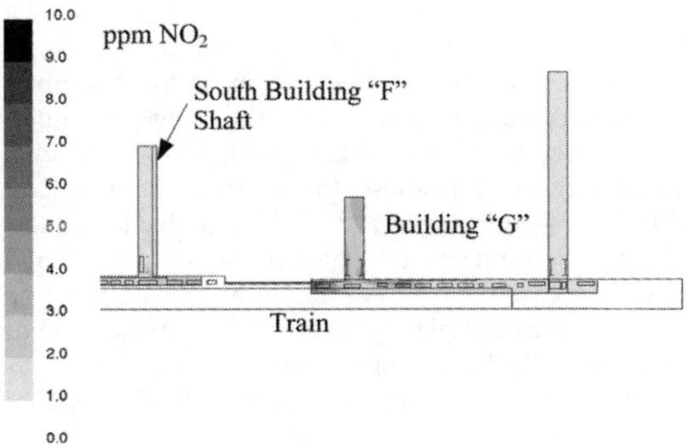

Figure 13. Time = 730 sec.: Longitudinal section at shafts.

through the South Portal by the operation of the overbuild mechanical ventilation system.

The difference in the exhaust flow rates between the shafts can be attributed to several factors:

1 proximity of the south Building "G" shaft to a supply air source (South Portal),
2 presence of the train obstruction adjacent to the north Building "G" and south Building "F" shafts,
3 obstructions in the form of extraction ducts and kitchen exhaust ducts in the south Building "F" and north Building "G" shafts, respectively,
4 and, a greater amount of straight duct downstream of the jet fans in the south Building "F" shaft due to the greater height of the shaft.

6 CONCLUSIONS

The transient CFD simulation for the rail overbuild extended to Building "F" demonstrated both the effectiveness of the NO_2 sampling system and the jet fans placed in the south Building "F" shaft and the two Building "G" shafts. The simulation showed that there was very little variation in NO_2 concentrations against the train versus at the crash wall where the ends of the gas sampling lines are located. Initiating operation of the mechanical ventilation system when NO_2 levels reached 3 ppm at the 14-ft elevation above top-of-rail at the crash wall sampling points prevented excessive NO_2 concentrations from building up against the train and being drawn into the passenger cars.

The predicted exhaust airflow generated by the jet fans in the south Building "F," 53.33 m^3/s (113,000 cfm), correlated closely with the required 47.20 m^3/s (100,000 cfm) flow rate used to size the fans. Predicted jet fan-generated exhaust flow rates from the two Building "G" shafts, 97.22 m^3/s (206,000 cfm) for the south shaft and 75.51 m^3/s (160,000 cfm) for the north shaft were well in excess of the 47.20 m^3/s (100,000 cfm) design flow rate. These results indicate that factors such as resistance in the immediate segment of the tunnel adjacent to the shaft caused by the train location, proximity to intake air sources (South Portal), obstructions in the shafts themselves, and the amount of straight shaft available downstream of the jet fans greatly impact airflow distribution between the shafts. Overall, it appears that the choice of the jet fan efficiency value of 0.55 may have been overly conservative, even with the duct obstructions located inside the shaft. However, it is believed that a conservative approach is appropriate when providing ventilation equipment for life safety systems that help control smoke from train fires and diesel exhaust gases in rail tunnels utilized by the public.

Finally, the continued operation of the overbuild mechanical ventilation system in the transient simulation cleared the diesel exhaust gas layers that formed in the vicinity of the diesel locomotive. The increased dilution of NO_2 began at the shaft with the highest exhaust flow rate (south shaft in Building "G") and worked its way north toward Building "F." Operation of the jet fans completely eliminated the exhausting of NO_2 out of the South Portal. The sudden drop in overbuild height in the street immediately to the north of Building "G," followed by an increase in height beneath Building "F," created a low spot in the overbuild in which gas layers dropped slightly below the top of the train. However, even if one of the intake air ports of a passenger car was located in this zone, NO_2 concentrations were still only in the 0 to 2 ppm range, which is below both the 8-hour exposure limit of 3 ppm NO_2 and the 15-minute exposure limit of 5 ppm NO_2.

REFERENCES

American Conference of Governmental Industrial Hygienists (ACIGH), 1996. *Threshold Limit Values for Chemical Substances and Physical Agent/Biological Exposure Indices*. Cncinnati, OH: ACIGH.

Bendelius, A., 1982. Tunnel Ventilation, In John Bickel & T.R. Kuesel (ed.), *Tunnel Engineering Handbook,* 546–549, 1982. New York: Van Nostrand Reinhold Company.

Fluent, Inc., 1998. FLUENT (Computational Fluid Dynamics Software), Release 5. Lebanon, NH: Fluent Incorporated.

Hobbs, J., Walter, R., Hard, T. & Devoe, D., 1977. Train Generated Air Contaminants in the Train Crew's Working Environment. US Department of Transportation, Report No. FRA/ORD-77/08, 16–23. Cambridge, MA.

Maevski, I. & Fay, J., 2002. Portal Jet Fans Performance for Tunnel Smoke Extraction. In *Proceedings of the Fourth International Conference on Tunnel Fires, 125–139, Basel. Switzerland, 2–4 December, 2002.* Kempston, UK: Independent Technical Conferences LTD.

Ray, R.E. & Zigh, A., 1999. CFD Applications in Tunnel Ventilation Analysis. In Jerry Tien (ed.), *Proceedings of the 8th U.S. Mine Ventilation Symposium, Rolla, Missouri, 11–17 June 1999.* Rolla, Missouri: University of Missouri-Rolla Press.

Saika, T., Nakajima, K., Setoyama, S., Haraguchi, S., Miyake, M. & Kanoh, T., 2000. Jet Fan Thrust Performance Evaluation Installed in Road Tunnels. In A. Caserta (ed.), *Proceedings of the 10th International Symposium on Aerodynamics and Ventilation of Vehicle Tunnels, Boston, MA, 1–3 November 2000.* Trowbridge, UK: The Cromwell Press.

Mine Ventilation, Ganguli & Bandopadhyay (eds)
© 2004 K. Fields, ISBN 90 5809 633 5

Study to assess the performance of ceramic diesel particulate filters for reducing diesel emissions

K. Fields, W. Pomroy, G. Saseen & R. Haney
Mine Safety and Health Administration, Pittsburgh Safety and Health Technology Center, Pittsburgh, Pennsylvania, USA

M. Good
Greens Creek, Kennecott Minerals, Juneau, Alaska

ABSTRACT: The Mine Safety and Health Administration and Greens Creek, Kennecott Minerals participated in a collaborative study to verify the efficiency of catalyzed ceramic diesel particulate filters for reducing diesel emissions. The purpose of the study was to determine the reduction in emissions and personal exposure that can be achieved when ceramic filters are used.

The study was conducted over a two week period. Three shifts were sampled with ceramic after-filters installed; and three shifts were sampled without the after-filters installed. Personal samples were collected to assess worker exposures. Area samples were collected to assess engine emissions. Both gaseous and diesel particulate measurements were taken.

Results indicated that ceramic diesel particulate filters may have a significant impact on personal exposures. Other factors such as intake air, stope ventilation parameters, and isolated atmospheres in vehicle cabs also impact diesel particulate matter exposures.

1 INTRODUCTION

This study was conducted in cooperation with the Greens Creek Mining Company and the Mine Safety and Health Administration (MSHA). The study was conducted on January 21–29, 2003.

The purpose of this study was to verify the effectiveness of catalyzed ceramic diesel particulate filters for reducing diesel emissions. The goal was the identification of site-specific, practical mine-worthy filter technology. Practical mine-worthy filter technology means feasible, effective, and durable filters which will enable the mine to comply with the diesel particulate matter (DPM) concentration limits specified in 30 C.F.R. §57.5060. These filters should consistently reduce DPM emissions by no less than 80% in actual conditions of use. These filters should provide this reduction without causing equipment damage or failure nor otherwise create safety hazards or health hazards such as unhealthful or impermissible levels of any air contaminant.

This series of tests was designed to determine the reduction in emissions and personal exposure that can be achieved when ceramic filters are installed on diesel equipment operating in a production stope. Relative engine gaseous and diesel particulate matter emissions were also determined for this equipment under a specific load condition.

2 BACKGROUND

Greens Creek, Kennecott Minerals Company (the "Company") operates an underground metal mine (the "mine") on Admiralty Island, south of Juneau, Alaska. The mining project is a Joint Venture between Kennecott Minerals and Hecla Mining Company. The ore body was discovered in 1975. Exploration drilling began in 1978, initial mine development in 1987, and full production in 1989. The mine was closed in 1993 due to low metal prices, and has been reopened since 1996 after completion of mine development work.

The Company utilizes a conventional stoping to develop a multilevel mine. Ore is shot in a heading and then loaded with a front-end–loader and transported to the surface by large diesel ore trucks equipped with Detroit Diesel, Series 60 (12.7 L) DDEC engines. The polymetallic (silver, zinc, gold, and lead) ore is transported to a surface mill and concentrator, which in turn produces three separate concentrates. These concentrates are shipped to various smelters throughout the world on a regular basis.

Since March 3, 2000, the mine has been testing on-board regenerating, 15 × 15 Engelhard, platinum catalyzed, diesel particulate filters on large horsepower production equipment. The focus of in-mine testing, to date, has been to assure that the filters could be properly installed and did not pose operational problems such as high engine back pressures, causing engine damage, or decreased performance. The longest running filter has been used for approximately 5,000 hours, since January 2001. The mine has purchased a total of 11 filters and installed 7 filters, five of which were in service prior to or at the time of the study.

3 SAMPLING AND ANALYSIS PROCEDURE

3.1 *Diesel particulate sampling*

In order to assess the performance of the diesel particulate filters, tests were conducted on a production unit when the filters were in use and on the same or similar production unit when the filters were not in use. Where possible, the same production equipment was used for both test configurations. DPM samples were collected using the SKC DPM sampling cassette with impactor and a cyclone. Samples were collected on three stope-loading cycles for each configuration of ceramic filter. Approximately 21 DPM samples were collected on each shift to assess ceramic filter performance. These samples included:

- 16 Area samples at 4 locations for emissions:
  - 6 samples at Intake 1 with filters,
  - 6 samples at Return 2 with filters,
  - 2 samples each at Intake 2 and Return 1,
- 2 Area samples (inside and outside loader cab), and
- 3 Personal samples (loader operator and 2 truck drivers).

Area samples for emissions were taken only in the test areas. Samples were collected up-wind and down-wind of the entrance to the stope. The minimum sample time was determined by the production cycle. Samples were collected during production and continue to be collected for 1/2 to 1 hour after the production in the stope was completed. Personal samples were collected for the full shift.

In addition to the DPM samples, the following data will be collected:

- Airflow in the stope,
- Nitrogen Dioxide measurements at the 4 area sample locations,
- Engine and emission data,
- Filter data,
- Time and motion data, and
- Engine tail pipe emissions with and without ceramic filters including:
- Gaseous emissions and Bosch numbers on the two trucks and loader (minimum).

In addition to collecting samples to assess performance of the catalyzed ceramic filters, DPM samples were collected to evaluate the performance of an environmental cab on loader LR 46. One area sample was collected inside and one area sample was collected outside the cab.

Both area and personal samples were collected with SKC, Inc., diesel particulate sampling cassettes. This cassette includes a submicron impactor and tandem quartz fiber filters. All samplers used a 10-millimeter nylon pre-separator cyclone. Samples were collected with SKC and MSA pumps pre-calibrated at 1.7 liters per minute (Lpm) and post-checks were made on all pumps used. At the time of the survey, all pumps were functioning properly.

Diesel particulate samples were analyzed at the MSHA Pittsburgh lab using NIOSH Method 5040. Elemental and organic carbon masses were determined from the samples collected. This process uses a thermal/optical carbon analyzer to determine the elemental and organic carbon matter per square centimeter of filter surface. Total carbon was estimated by multiplying the elemental carbon by a factor of 1.3 and determined by adding the organic carbon mass to the elemental carbon mass. Concentrations of total carbon were calculated from the following formulas:

Total Carbon Concentration ($\mu g/m^3$) =

$$\frac{EC(\mu g/cm^2) \text{ x } A(cm^2) \text{ x } 1{,}000 \text{ L/m}^3 \text{ x } 1.3}{1.7 \text{ Lpm x time (min)}}$$

and

$$\frac{[EC(\mu g/cm^2) + OC(\mu g/cm^2)] \text{ x } A(cm^2) \text{ x } 1{,}000 \text{ L/ m}^3}{1.7 \text{ Lpm x time (min)}}$$

MSHA's compliance limits are based on an 8-hour equivalent exposure concentration and are referred to as shift weighted averages (SWA). The SWA is calculated using 480 minutes as the sampling time.

3.2 *Evaluation of engine and ceramic filter*

The engine, in each of the test vehicles, was tested using a repeatable engine load test to determine the raw gaseous emissions before and after the catalyzed ceramic filter. This test was done prior to and repeated after the on-section test with the ceramic filters installed. The test was conducted using MSHA's ECOM multi-gas analyzer. Carbon Monoxide (CO), Carbon Dioxide (CO_2), Nitric Oxide (NO), and Nitrogen Dioxide (NO_2) were measured. The exhaust gas was sampled in the exhaust system upstream of the DPM filter location.

In addition to measuring the raw gaseous emissions, a smoke test was conducted before and after the on-section test with the ceramic filters installed. A sample of raw exhaust was collected before and after the ceramic filter on the ECOM's filter paper and the filter paper coloration was compared to a Bosch smoke number. This test was conducted under the engine operating condition at torque stall. Torque stall condition represents a loaded engine condition that would produce higher amounts of DPM. A 1.62 L sample of exhaust gas was passed through a quartz fiber filter. The degree of black/gray/white on the filter paper was compared to a Bosch smoke scale.

3.3 *Data analysis*

Because of the low emissions for the DDEC Series 60 engines and the short production cycle in a stope, this test was not able to resolve the reduction in engine emissions with a high degree of certainty. This study did determine:

- The results for the tail pipe test for each piece of equipment, with and without ceramic filters installed,
- the change in full shift personal exposure with and without ceramic filters installed,
- the level of NO_2 in the stope with and without the ceramic filters installed,
- the difference between measurements inside and outside the loader cab, with and without the ceramic filters,
- the potential impact of directing airflow into the stope through the use of an auxiliary fan and tubing system, and
- diesel emissions within the test areas with and without ceramic filters installed.

DPM sample results, airflow, and engine emission data were tabulated. Emissions within the test area were determined by taking the difference between the up-wind and down-wind concentration and multiplying by the airflow.

$$\text{Emissions} = (C2 - C1) \times Q$$

This gives the total emissions. The with- and without-filter scenarios were used to calculate reduction in emissions. The personal samples, emission data and airflow can be used to evaluate the current control strategy and to model future control strategies.

4 STUDY RESULTS AND DISCUSSION

4.1 *Results of tail pipe tests*

During typical mining operations, the mine utilizes one front-end-loader (FEL) and two or three haul trucks to remove ore from a stope. During the study, all of the equipment utilized was equipped with Detroit Diesel, Series 60 (12.7 L) DDEC engines. This engine is approved by MSHA (7E-B049-0). It is rated at 475 horsepower at 2100 RPM. The approved ventilation rate is 28,000 cfm and the Particulate Index (PI) is 8,500 cfm. The engine on the FEL had been de-rated to 300 horsepower. The corresponding ventilation rate and PI would be approximately 20,000 cfm and 5,000 cfm, respectively. All of the haul trucks and the loader utilized in the stopes where sampling occurred were equipped with working Engelhard DPM filters during the first three days of the study. The ceramic filters were heavily catalyzed and used with passive regeneration. As part of the study, tail pipe sampling was conducted on the FEL and four trucks with- and without-ceramic filters installed.

The Engelhard DPM filters installed on the machines were based on a Corning cordierite filtration media that is wash coated with a platinum catalyst. The platinum catalyst is a second generation coating which means less platinum deposited on the cordierite. The platinum based catalyst provides the means for passive regeneration at a specific exhaust gas temperature. Passive regeneration occurs when the exhaust gas temperature duration is sufficient during the duty cycle of the machine to burn off the deposited DPM on the filter. This is successfully accomplished at an exhaust gas backpressure that does not increase above the recommended value specified by the engine manufacturer. During the study, all indications were that the filters were properly regenerating during the normal production duty cycle of the machines.

During visual inspection of the ceramic filters by the Company and subsequently by MSHA personnel, it was noted that some of the 15×15 Engelhard filters had rotated in the canisters and were cracked. The filter that prior to the study was reported to have operated over 5,000 hours on a haul truck had completely failed. This filter was replaced. Two similar filters installed on haul trucks, which had operated for approximately 2,500 hours, also rotated and displayed cracks along the outer perimeter of the ceramic media. Figure 1 shows a typical installation of an Engelhard DPM filter installed on a haulage truck. From information presented by the Mine and visual inspection of the 5 ceramic filters that were currently being used, the rotation appeared to begin after approximately 600 hours of use. Cracks occur between

Figure 1. Installation of an Engelhard DPM filter on a Haulage truck.

600 hours and approximately 2,500 hours of use. Cracking of the media continues until the filter completely fails. MSHA discussed the problem with the Company and the filter manufacturer. Although the filters used in the study displayed some damage, they effectively filtered the DPM.

The gaseous exhaust emissions were measured with MSHA's ECOM AC portable gas analyzer on the four pieces of equipment used during the study. These measurements were made at four engine operating points. Based on the comparison of data, all of the engines tested were performing in the approved range. For each piece of equipment tested there was a general decrease in CO downstream of the ceramic filter and there was a general increase in the NO_2 downstream of the ceramic filter.

As part of the tail pipe emission tests, a smoke test was conducted upstream and downstream of the ceramic filter. The purpose of this test was to determine how well the filters were working by visually evaluating a smoke dot on a sample filter paper. Due to time restraints in the mine, the smoke tests were performed using a quartz fiber filter instead of a standard filter paper normally used for determining a Bosch Number. Therefore, the Bosch Numbers assigned to the quartz filter paper are only being used as a relative number for the purpose of this report. As discussed below, the DPM collected on the quartz fiber filter was used for a quantitative analysis.

Bosch Numbers range from 0 (no observed color) to 9 (black). Testing of the ceramic filters on loader LR-46 and haul truck HT-21 gave Bosch numbers reducing from 8 to 1. This range of Bosch numbers between unfiltered and filtered exhaust indicates the ceramic filters were working properly. HT-23 presented a less effective result with a reduction in Bosch

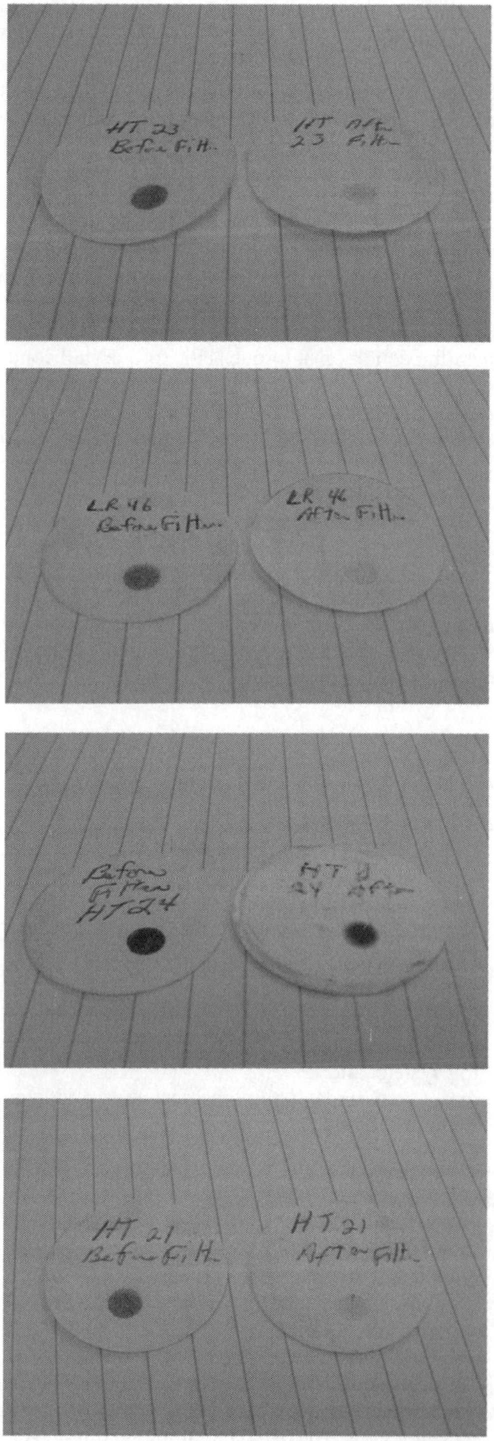

Figure 2. Smoke test comparison between damaged and undamaged DPM filters.

number from 8 to 3. This Bosch number range indicates that this filter was less effective than the LR-46 or HT-21 filters but much better than the HT-24 filter.

The smoke test of the DPM filter on haul truck HT-24 indicated a smoke number reduction from 9 to 7. This is the filter that had rotated in the housing. The Bosch numbers confirmed that it was not functioning as efficiently as the other filters. The ceramic filter on this truck was replaced with an undamaged filter prior to in-mine testing. An example of the color comparison between the damaged and undamaged filters is shown in Figure 2.

In addition to the visual method to assign a Bosch number, tests were conducted upstream and downstream of the ceramic filters using quartz fiber filters for the smoke test medium. This allows the DPM sample to be analyzed and quantified using the NIOSH 5040 analytical method. These results show a 84.8 to 99.0 percent reduction in total carbon reduction and a 91.7 to 99.6 percent reduction in elemental carbon. Even though there was some leakage due to filter rotation on several filters, the elemental carbon filter efficiency remained above 90 percent.

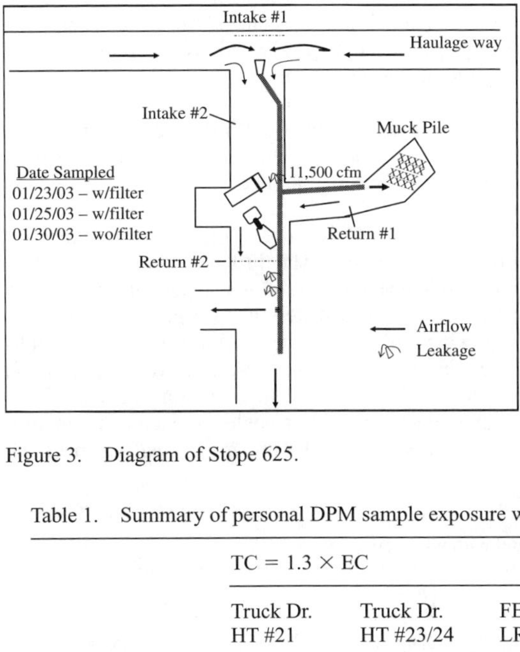

Figure 3. Diagram of Stope 625.

4.2 *Change in full shift personal exposure*

A diagram of a typical stope sampled during the study is presented in Figure 3. Area sample package locations, appropriate airflow direction and other pertinent information are also shown in this figure. The following paragraphs summarize the sampling results. Raw data is available upon request.

A summary of the personal diesel particulate samples collected with and with out the ceramic filters are given in Table 1. During all sampling, the equipment operators remained inside the cabs while loading in a stope. When ceramic after-filters were utilized on the production equipment, the highest personal TC exposure (EC $\times$ 1.3) was 139 $\mu g/m^3$. When filters were not utilized on the production equipment, personal TC exposures ranged from 240 to 423 $\mu g/m^3$. Without the after-filters, exposures would meet the interim exposure limit of 400 $\mu g/m^3$. With the after-filters the TC exposures would have met the existing final DPM limit.

Figure 4 shows a comparison of personal exposure for the two truck drivers and FEL operator, with and without the filters installed. The percent reduction in the personal samples between with and without using the filters is shown in Figure 5 in terms of $TC_{(EC*1.3)}$ and $TC_{(EC+OC)}$. For $TC_{(EC*1.3)}$ the percent reduction ranged from 57 to 70 percent. For $TC_{(EC+OC)}$ the percent reduction ranged from 41 to 62 percent. Note that the Driver of HT-38 is not shown due to lack of data with a filter installed.

The reduction in personal exposure is not indicative of after-filter performance because other factors such as variation in production, ventilation, up-wind equipment use and cabs, influence personal exposure.

Table 1. Summary of personal DPM sample exposure with and without filters.

	TC = 1.3 × EC			TC = OC + EC		
	Truck Dr. HT #21	Truck Dr. HT #23/24	FEL Oper. LR #46	Truck Dr. HT #21	Truck Dr. HT #23/24	FEL Oper. LR #46
Without filters						
1/28/03	280	324	185	267	340	173
1/29/03	423	405	240	417	410	223
1/30/03	223	283	240	275	294	229
Average	309	337	222	320	348	208
With filters						
1/24/03	138	106	37	159	273	57
1/25/03	132	93	103	145	134	103
Average	133	100	70	152	204	80
Percent reduction	57	70	68	53	41	62

4.3 *Level of NO_2 in the stope*

Personal gaseous exposures were measured using long term diffusion stain tubes. Due to logistic problems, these measurements could not be made on the first day of the study. A summary of personal gaseous CO and NO_2 measurements taken on equipment operators with and with out the ceramic filters are given in Table 2. In general the data indicates that the CO levels decreased by approximately one-half when the catalyzed filters were being used. While there appears to be an increase in NO_2, when catalyzed filters are being used, it is unclear whether this increase is due to data variability, changes in ventilation rate or use of the catalyzed filters. The approved ventilation rate for a single engine is 28,000 cfm. The airflow in only one of the production stopes exceeded this level.

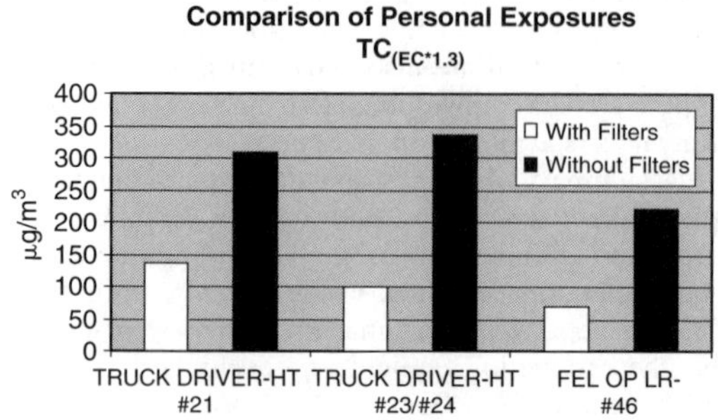

Figure 4. Comparison of personal exposures with and without after-filters installed.

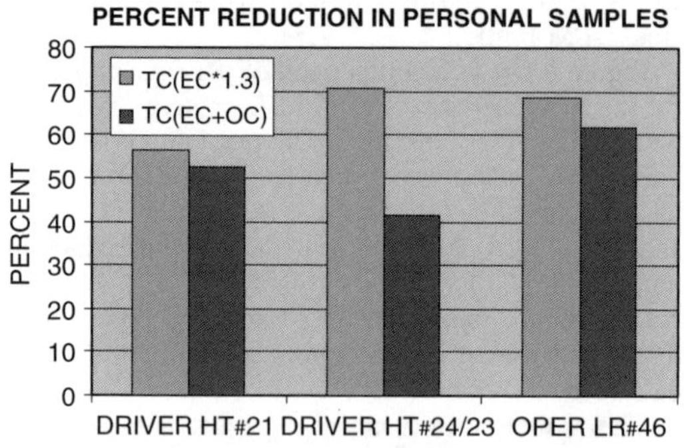

Figure 5. Percent reduction in personal samples with after-filters installed.

4.4 *Difference between measurements inside and outside the loader cab*

The results of the area samples collected inside and outside the cabs, with and without the filters are given in Table 3. Without the filters, the average DPM concentration outside the cab was 1351 $\mu g/m^3$ and the concentration inside the cab was 271 $\mu g/m^3$. With the filters, the average DPM concentration outside the cab was 193 $\mu g/m^3$ and the concentration inside the cab was 49 $\mu g/m^3$. This data indicates that under current conditions ventilation alone would not be adequate to maintain ambient diesel particulate concentrations below the interim standard. Either cabs or after-filters are needed to meet the interim standard. Both cabs and filters are needed to meet the existing final DPM limit.

The use of filters reduced the ambient concentration by 85 percent. The FEL cab filtration system reduced DPM concentrations inside the cab by 75 percent when after-filters were used and by 80 percent when after-filters were not in use. This data indicates that under current conditions ventilation alone would not be adequate to maintain ambient diesel particulate concentrations below the interim standard. Either cabs or after-filters are needed to meet the

Table 2. Summary of personal gaseous exposure with and without filters.

	CO			NO_2		
	Truck Dr. HT #21	Truck Dr. HT #23/24	FEL Oper. LR #46	Truck Dr. HT #21	Truck Dr. HT #23/24	FEL Oper. LR #46
Without filters						
1/28/03	6	6	13	0	0	0
1/29/03	11	9	19	0	0	1
1/30/03	0	6	15	0	0	3
Average	6	7	16	0	0	1.3
With filters						
1/24/03	–	–	–	1	1	4
1/25/03	8	6	6	1	1	3
Average	8	6	6	1	1	3.5

interim standard and both cabs and filters are needed to meet the existing final DPM limit.

4.5 *Potential impact of ventilating stopes with an auxiliary fan and tubing system*

In the drawing supplied by the Company for the original test plan, stopes were ventilated by a free standing fan setting on the floor at the mouth of the stope, blowing into the stope without ventilation tubing. During the study, this ventilation practice was not observed. All stopes were ventilated by fans using rigid and bag tubing to direct the airflow into the stopes. However, fans in two of the stopes were installed inside the mouth of the stopes rather than in the main intake. This practice causes recirculation of air and decreases the amount of fresh air available to ventilate a stope.

Airflow to the stopes from the auxiliary fan and tubing systems ranged from 11,500 to 39,000 cfm. Airflow in the intake to the stope ranged from 23,000 cfm to 58,500 cfm. Typically, the FEL loaded one truck at a time in the stope during the loading cycle. For the 300 horsepower FEL and a 475 horsepower truck, (725 horsepower), 100 cfm/hp would have required 72,500 cfm. The sum of the gaseous ventilation rates would be 45,600 cfm and the sum of the rates times the Particulate Indices would be 13,000 cfm.

The airflow supplied in the stopes was close to the Particulate Index, indicating a diesel particulate concentration increase, above the intake levels of approximately 1000 $\mu g/m^3$. DPM concentrations outside the cabs, without filters were consistent with this value. The low airflow relative to the gaseous airflow values indicated that elevated concentrations of gaseous emissions can occur either with or without the catalyzed ceramic filters.

4.6 *Diesel emissions within the test areas*

Due to time and production constraints, only one stope per day could be directly monitored for emissions. The average of the area sample packages collected at locations designated as Intake 1 and Return 2 were utilized to determine the diesel emissions generated in the stope during production activity.

Diesel emissions within the test area were determined by using the DPM concentrations entering the stope (Intake 1) and leaving the stope (Return 2). The results of the area samples and airflow measurements collected to assess emissions within the test areas (with and without ceramic filters installed) are given in Table 4. The average increase in concentration from stope intake to return when filters were used was 8 $\mu g/m^3$. The average increase in concentration from stope intake to return when filters were not used was 301 $\mu g/m^3$. The average diesel emission in the stopes when ceramic filters were utilized was 0.82 gm/hr and average diesel emission in the stopes when ceramic filters were not utilized was 22.46 gm/hr. Even though three trucks were operated during the tests with no filters, no adjustment in the number of vehicles operated was made to compare emissions because the time of the trucks inside the stope, with or without filters was similar. The average reduction in emissions, based on the DPM concentration, was 97 percent. The average reduction in emissions, based on engine emissions, was 96 percent.

Table 4 also shows that the average intake to the stope was reduced when ceramic filters were used. While this could have occurred and would not have

Table 3. Summary of front end loader cab performance based on inside and outside of cab DPM measurements.

	TC = 1.3 × EC			TC = OC + EC		
Date sampled	DPM conc. outside cab, $\mu g/m^3$	DPM conc. inside cab, $\mu g/m^3$	Percent reduction	DPM conc. outside cab, $\mu g/m^3$	DPM conc. inside cab, $\mu g/m^3$	Percent reduction
Without filters						
1/28/03	1079	363	66	926	334	64
1/29/03	1345	240	82	1158	211	82
1/30/03	1630	211	87	1438	217	85
Averages:	1351	271	80	1174	254	78
With filters						
1/24/03	203	46	77	229	62	73
1/25/03	182	53	71	187	62	67
Averages:	193	49	75	208	62	70
Percent reduction with vs. without filters						
	85	82		82	76	

Table 4. Summary of emissions data from area samples based on TC = 1.3 × EC.

Date	Stope	Stope intake concentration* μg/m³	Stope return concentration* μg/m³	Stope airflow cfm	Stope DPM emissions gm/hr
Without filters – includes 3 trucks and 1 FEL					
01/28/2003	030	307	816	39000	33.78
01/29/2003	490	123	408	23000	11.13
01/30/2003	625	536	647	34000	6.41**
Average		322	623		22.46
With filters – includes 2 trucks and 1 FEL					
01/23/2003	625	206	221	34000	0.86
01/24/2003	446	156	164	58500	0.78
01/25/2003	625	134	135	34000	0.03**
Average		165	173		0.82
Percent reduction					
		49	72		96.3

* Average concentration.
** Not included in average because FEL went beyond Return 2 Sample location.

been unexpected, the study was not designed to demonstrate this situation. The increased intake levels, without the filters, could have been attributed to the deeper location of operations in the mine and the increased amount of upwind equipment.

5 FINDINGS AND CONCLUSIONS

- The results of the raw exhaust gas measurements conducted during the study indicated that the engines were operating properly.
- The ceramic filters installed on the machines used in this study did not adversely effect the machine operation. Even with some apparent visual cracking from the rotation of the filter media, the ceramic filters effectively removed DPM. The filters passively regenerated during machine operation.
- The Bosch smoke test provides an indication of filter deterioration; however the colorization method does not quantify the results.
- Personal DPM exposure was reduced by 57 to 70 percent when after-filters were used.
- CO levels decreased by up to one-half when the catalyzed filters were being used. There appears to be an increase in NO_2 when catalyzed filters are being used, however, it was unclear whether this increase was due to data variability, changes in ventilation rate or the use of the catalyzed filters.
- The use of cabs reduced DPM concentrations by 75 percent when after-filters were used and by 80 percent when after-filters were not in use.
- Ventilation airflow was provided to the stopes through fans with rigid and bag tubing. Airflow was the same or greater than the Particulate Index, but typically lower than the gaseous ventilation rate.
- The use of ceramic after-filters reduced average engine emissions by 96 percent.
- The reduction in personal exposure was not indicative of only after-filter performance because other factors such as ventilation, up wind equipment use and cabs also influence personal exposure.

Mine Ventilation, Ganguli & Bandopadhyay (eds)
© 2004 Taylor & Francis Group, London, ISBN 90 5809 633 5

Measuring exposure to diesel engine exhaust emissions in UK coal and non-coal mines

J.R. Leeming
Health and Safety Executive, Sheffield, UK

D.W. Dabill
Health and Safety Laboratory, Sheffield, UK

ABSTRACT: At present the UK does not have an occupational exposure limit for Diesel Engine Exhaust Emissions (DEEEs) but under the Control of Substances Hazardous to Health (COSHH) Regulations, DEEEs are regarded as a substance hazardous to health and as such exposure should be either prevented or at least adequately controlled. In the confined spaces of coal and non-coal mines, where large diesel powered vehicles and machinery are in use, the potential for exposure to DEEEs is high. Drivers and workers positioned close to or down ventilation of a diesel engine will inevitably be exposed.

This paper describes work done for the UK Mines Inspectorate by the Health and Safety Laboratory to measure the exposure of mineworkers to DEEEs in UK coal and non-coal mines. Diesel particulate samples were collected in a number of mines and analysed for organic carbon (OC) and elemental carbon (EC), the latter providing a good surrogate measure of exposure to DEEEs. The paper presents EC exposure data from coal mines determined using the Personal Diesel Aerosol Sampler developed by the United States Bureau of Mines and the University of Minnesota. It compares EC exposure data obtained in coal and non-coal mines with surface industry EC exposure data and it describes a simple on-site method for determining exposure to DEEEs by measuring the 'blackness' of the diesel particulate samples and relating this to EC.

1 INTRODUCTION

There is evidence to suggest that DEEEs are carcinogenic and animal experiments using whole and filtered diesel exhaust have demonstrated that it is the fine and ultrafine diesel particulate matter (DPM) associated with DEEEs that is the causative agent. In 1989 the International Agency for Research on Cancer (IARC) declared 'diesel engine exhaust is probably carcinogenic to humans'. Epidemiological data suggests that the increased cancer risk from long term exposure to DEEEs is small but nevertheless it exists and everything possible should be done to keep exposure to DEEEs to a minimum.

At present the UK does not have an occupational exposure limit for DEEEs. The Health and Safety Executive (HSE) feels there is insufficient evidence for DEEEs to be classed as a carcinogen but it does believe it to be a substance hazardous to health and, under the control of substances hazardous to health (COSHH) regulations (2002), exposure should preferably be eliminated or at least reduced to as low as practicably possible.

Most coal and non-coal mines use diesel powered vehicles and equipment underground and the potential for exposure to DEEEs in the confines of a mine is high. Drivers and workers close to or down ventilation of the engine will inevitably be exposed but to what degree and what can be done to ensure their exposure is kept as low as practicably possible?

To answer these questions the UK mines inspectorate have supported a number of projects with the Health and Safety Laboratory, the main objectives of which were:

- To determine the exposure of coal and non-coal mine workers to DEEEs and to see how their exposure compares with surface industry workers.
- To look at ways of controlling exposure at source i.e. alternative fuels and exhaust filters.
- To try and develop a simple on-site method of determining exposure to DEEEs.

This paper summarises some of the work that has been carried out.

2 MEASURING EXPOSURE

DEEEs are a complex mixture of gases, vapors and particulates and identifying the optimum component(s) to provide a representative measure of exposure is problematic and a number of substances have been measured in the past. However, one method, which is thought by many to be the best, is to collect a respirable airborne diesel particulate sample onto a quartz filter and analyse it for organic carbon (OC) and elemental carbon (EC). OC is a measure of the volatile organic species adsorbed onto the carbon particles in the exhaust emissions without discriminating between the many substances that may be present. EC is a measure of the carbon (soot) particles themselves and it is the EC fraction that provides a measure of exposure. In most workplace environments the only source of fine airborne elemental carbon is from DEEEs and so the surrogate measure of EC is very specific to DEEEs. There are exceptions, coal mines being an obvious example.

The Health and Safety Laboratory (HSL) have developed an analytical method for measuring OC and EC (Dabill et al., 1996). Diesel particulate samples are collected onto quartz fibre filters using a respirable cyclone sampler and the analytical method uses coulometry to determine the carbon mass. The method is very specific for carbon and, with the exception of carbonates, which can easily be removed, other dusts do not interfere with the analysis. HSL has entered into a number of international round robins analysing OC and EC in diesel particulate samples and has always emerged as one of the top performing laboratories.

2.1 *Surface industry survey*

In the mid 1990's HSE carried out a technical development survey (TDS) of exposure to DEEEs in surface industry workplaces (Groves & Cain, 2000). In Table 1 is summarised the median EC exposure concentration for each work group.

The fork-lift truck drivers were found to be the highest exposed group with personal and background exposure levels of 75 and 92 $\mu g/m^3$, the mean exposure for all the groups was 33 and 32 $\mu g/m^3$ respectively. The TDS report also stated that for the purposes of control a personal exposure to elemental carbon in excess of 200 $\mu g/m^3$ would indicate high exposure to DEEEs.

From this information and from data collected in both coal and non-coal mines the view is that time weighted average (TWA) exposure levels to EC in the range 0 to 100 $\mu g/m^3$ EC should be considered low exposure; 100 to 200 $\mu g/m^3$ EC should be considered medium exposure and greater than 200 $\mu g/m^3$ should constitute high exposure. To provide more meaningful guidance the UK mines inspectorate would like all mines to have TWA EC exposure levels below 150 $\mu g/m^3$ with an ultimate target of below 100 $\mu g/m^3$.

To make this practically possible one objective of this work was to develop a simple and inexpensive on-site method which would allow mines to measure airborne particulate EC exposures themselves without recourse to laboratory methods, and this is described in section 7.

Table 1. TDS surface industry work group median EC exposure concentration data.

Work group	EC exp. conc. ($\mu g/m^3$) Personal	Background
Ambulance Depot	33	28
Ro-Ro ferry	49	26
Railway repair	18	17
Bus garage/repair	31	35
Vehicle testing	7	7
Fork-lift trucks	75	92
Toll booths, tunnels	17	17
Mean all work groups	33	32

3 RESEARCH

3.1 *Measuring exposure to DEEEs in coal mines*

Because the analytical method cannot discern between diesel particulate EC and coal dust EC a simple respirable cyclone sampler was not suitable for use in coal mines and either a different sampler or analytical method, that could distinguish between the two was required. It was important that the new method was well correlated with diesel particulate EC so direct comparisons with non-coal mine and surface industry exposure data could be made.

Initial work on a diesel test bed measured a number of gases and vapors in the diesel exhaust to see if any were correlated with the diesel particulate EC and could be used as a surrogate measure of EC exposure. None of the gases/vapors studied i.e. carbon monoxide (CO), carbon dioxide (CO_2), nitrogen oxides (NOx), unburnt hydrocarbons or acetaldehyde were found to be correlated with diesel particulate EC.

However, side-by-side sampling in a salt mine showed there to be a loose correlation between CO_2 and diesel particulate EC and these data are presented in Figure 1. These data show that monitoring CO_2 offers some measure of air quality in the salt mine where higher CO_2 levels correlate with higher diesel particulate EC concentrations and are indicative of higher DEEEs. It was hoped monitoring CO_2 in coal

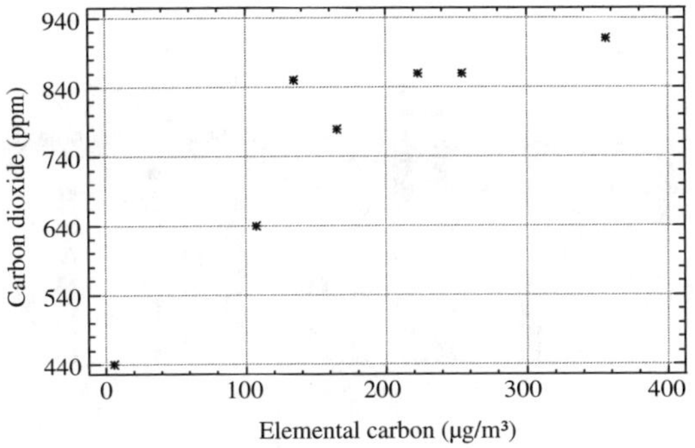

Figure 1. Concurrent sampling of CO_2 and diesel particulate EC in a salt mine.

mines would also offer some measure of exposure to DEEEs. However, initial trials in a coal mine showed there were large sources of CO_2 other than from diesel exhaust. It was clear this approach was not practical in coal mines.

3.2 *Personal Diesel Aerosol Sampler (PDAS)*

The PDAS was developed by the US Bureau of Mines and the University of Minnesota specifically for measuring diesel engine exhaust particulate in the presence of coal dust (McCartney & Cantrell, 1992). Essentially the sampler is a cyclone in series with a 0.8 μm impactor stage which removes the larger coal dust particles (98% coal dust >0.8 μm) but allows the finer diesel particulate (98% diesel particulate <0.8 μm) through to be collected onto a quartz filter for analysis. NIOSH kindly loaned ten PDAS impactors, plus the detailed information on how to make the impactor substrates to HSL and these were evaluated and samples collected in three UK coal mines.

More recently a commercial version of the PDAS sampler is now available. It is marketed by SKC Ltd. as a 'diesel particulate matter cassette with precision–jeweled impactor' Cat. No. 225–315. The cassette is designed for single use and after sampling it is re-sealed and sent for analysis. A limited experiment was carried out in a coal mine to compare the SKC and PDAS samplers and the results are reported in section 4.4.

3.3 *Preliminary experiments to measure the efficiency of the PDAS sampler at eliminating coal dust*

Parallel sampling with the PDAS and a standard respirable cyclone sampler was carried out in a nondiesel coal mine (only coal dust present) and in a diesel particulate sampling chamber (no coal dust present) to determine the efficiency of the PDAS at both eliminating coal dust and collecting diesel particulate. The results show that the mean mass of coal dust EC collected on the PDAS filter was approximately 6% of the coal dust EC mass collected with the respirable cyclone sampler and in the chamber the particulate EC losses with the PDAS were of the order of 8% of the sample collected with the respirable sample. To some degree these gains and losses will cancel each other out and the final error will be small. Therefore, it was decided to accept the PDAS EC result as a measure of diesel particulate EC with no attempt to compensate for DPM losses and coal dust gains.

4 MEASUREMENT OF DEEEs IN COAL MINES

4.1 *Outline*

Sampling was carried out in three coal mines. PDAS samplers were positioned in pairs on diesel powered free steer vehicles (FSVs) and in the main intake and return to the district where the FSVs were operating. The samplers on the vehicles were close to the driver and could be classed as personal samples but others were general body samples although nearly all the samples collected in the return roadways were close to where men were stationed and again would very much represent personal samples for them. In parallel with the PDAS samplers in the intake and return roadways was positioned a cyclone sampler to measure respirable dust levels.

Two of the coal mines monitored, A and B, used standard class A1 diesel fuel but coal mine C was using biodiesel fuel, a fatty acid methyl ester derived from rape seed oil. Research done by HSE/HSL (Dabill, 1999) has shown that biodiesel has much to offer in reducing emissions from diesel engines, especially particulates. When compared to standard diesel over the ISO eight mode test for off road vehicles, biodiesel produced 58% less particulate and a 30% biodiesel/class A1 blend produced 42% less. Similarly the coulometrically determined EC showed reductions of 76% and 48% for biodiesel and the blend respectively when compared to class A1 diesel fuel. Biodiesel also produced 32% less CO and 8.6% less NOx than class A1 diesel and there was no evidence that biodiesel produced any more aldehydes.

4.2 *Sampling details*

4.2.1 *Coal mine A*

At the time of sampling most of the activity in the mine was directed towards driving long headings for a new face. The tailgate heading was about 2760 metres in length and the coal gate heading about 1480 meters in length. Both headings were on forced ventilation and the ventilation volume and velocity were about

5 m^3/sec and 0.23 m/s respectively. Ventilation volumes in the main intake and return were about 20 m^3/sec and 18 m^3/sec respectively.

Sampling was over two days and on each day two FSVs were monitored. These were all engaged in ferrying equipment and supplies down the headings. Traveling inbye with the load forward and exhaust behind the driver, the ventilation would carry the exhaust fume away and the driver's exposure would be low. However, traveling outbye the FSV would be traveling faster than the ventilation and so running into the exhaust fume. In this situation the driver's exposure would be high.

4.2.2 *Coal mine B*

Sampling was in the NW district where there was one face producing coal and one face in salvage. The ventilation volumes on the face lines were 33 and 32 m^3/sec respectively. Ventilation volumes in the main intake and main return were 50 and 40 m^3/sec respectively, considerably higher than coal mine A.

Sampling was over two days and essentially it was the same as for coal mine A except that only one FSV was monitored on day two. The mine has two production areas and so monitors were positioned at the top of a drift close to the upcast shaft to monitor diesel fume from both districts.

4.2.3 *Coal mine C*

The mine had only recently started to use FSVs and as they were concerned about exposure to DEEEs they were using biodiesel fuel as a way to control workers exposure to DEEEs.

Two sets of samples were collected in the mine. One set before the introduction of FSVs which were used to determine the breakthrough of coal dust on the PDAS sampler and are reported above in section 4.1. The second set were collected after the introduction of FSVs and the mine had been using biodiesel fuel for about 4 months. The mine reported that they had had no operational problems with biodiesel and no complaints from workers about the smell or eye or throat irritation, which has been a reported problem with biodiesel in the past.

It was planned to monitor coal mine C again at a later date but unfortunately the mine suffered serious flooding and has subsequently been closed.

Samplers were fastened to two FSVs which were collecting supplies from the materials store at the loco junction and carrying them either to a new development or into D2s tailgate where coal was being cut. Samplers were also placed in the FSV garage, which was in the main return, and in the main intake. Ventilation volumes were ~18 m^3/sec in D2s tailgate and 23 m^3/sec in the intake and return roadways.

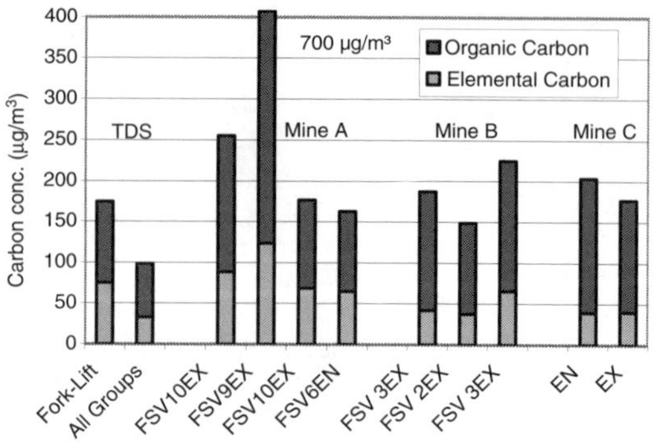

Figure 2. FSV OC and EC exposure data for three coal mines.

4.3 *OC and EC analysis*

All the particulate samples collected in the coal mines were analyzed for OC and EC and the following results obtained.

4.3.1 *FSVs*

Figure 2 compares the FSV results from the three mines with selected data from the TDS, i.e. the fork-lift truck drivers, the highest exposed group and the mean of all the work groups monitored. With the exception of FSV9 from Mine A which broke down in the roadway and had to be rescued, the total carbon results (OC + EC) for the other FSVs are similar to or slightly higher than the fork lift truck drivers result.

In terms of EC, the results from Mine A are higher than those from Mine B and this is almost certainly due to the fact that the ventilation rate in Mine A was much lower than in Mine B and consequently there was less dilution of the DEEEs. Nevertheless, the EC exposures are in all cases well below the 150 μg/m^3 target limit and most are below 100 μg/m^3. This is similar to the TDS fork-lift truck driver's result and in terms of exposure to DEEEs places FSV drivers towards the high end of the surface industry exposure range.

At first glance the EC data for mine C in Figure 2 does not appear to be any better than Mine B and biodiesel fuel does not appear to provide any benefit with regard to reduced particulate emissions. However, ventilation volumes in Mine C are only half those in Mine B. If ventilation volumes in both mines were the same then particulate levels in Mine C would have been lower than Mine B. It is more reasonable to compare Mine C with Mine A in which case the EC exposures are much lower and the advantages of biodiesel fuel are more apparent.

4.3.2 *Intake and return roadways*

Figure 3 presents the intake and return roadway data for all three mines. The intake EC results are all below

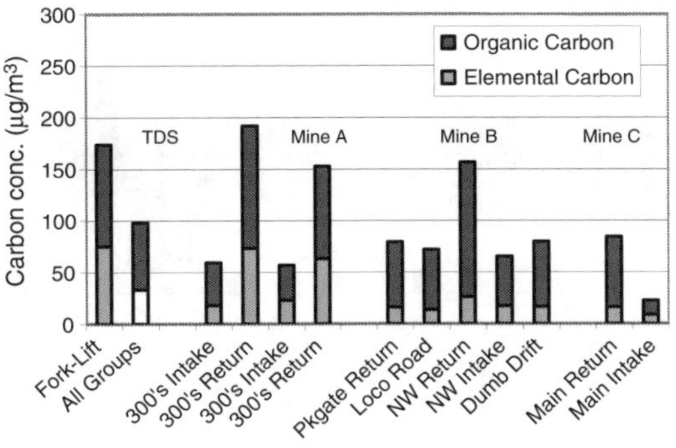

Figure 3. Intake and return roadway OC and EC exposure data for the three coal mines.

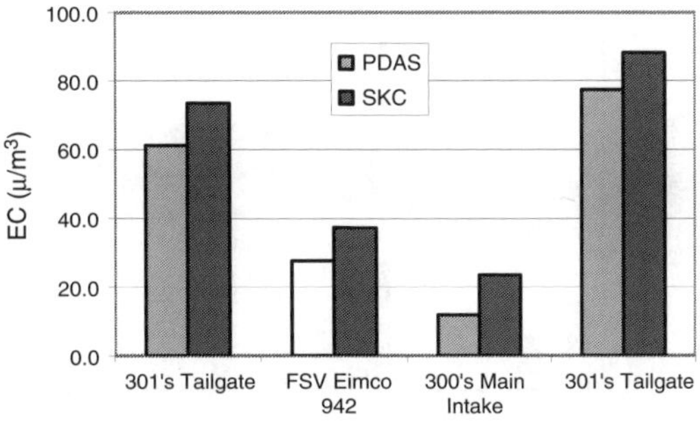

Figure 4. Comparing EC results obtained with the SKC and PDAS samplers.

25 µg/m³ and essentially this is a measure of the coal dust breakthrough with the PDAS sampler.

The EC return results for Mine A are typically three times those for mines B and C. However, they are all well below 100 µg/m³ target figure as outlined in section 2.1.

Overall the EC exposure results from the coal mines suggest that coal mines do not have much of a problem with regard to exposure to DEEEs and exposures are very much in line with the high end of the TDS surface industry data. The vigorous ventilation necessary in coal mines to reduce the risk of explosion also rapidly dilutes the diesel exhaust and prevents the build up of DEEEs. In all cases EC exposures were below the 150 µg/m³ target limit and most were below 100 µg/m³.

4.4 *SKC DPM cassette*

SKC DPM cassettes were operated side-by-side with PDAS samplers at four locations in a coal mine. One set were mounted on a Eimco 942 FSV as it transported a chock down the tailgate and onto the face line, one set was positioned in the main intake and two sets were positioned for different periods of time in the tailgate return. Sampling was from between 3 and 5.5 hours. The EC results are presented in Figure 4.

In all cases the SKC sampler result was higher than the PDAS sampler result and by about the same amount in each case i.e. 11.2 ± 0.9 µg/m³. It is thought this may be due to the SKC sampler having a slightly higher cut point than the PDAS sampler and consequently collecting slightly more of the fine airborne coal dust but more experiments need to be carried out to confirm this. Nevertheless, the SKC DPM cassette performed well and its availability has made it possible for coal mines to measure exposure to DEEEs, something that has not been possible in the past.

5 MEASUREMENT OF DEEEs IN NON-COAL MINES

Diesel particulate samples were collected in a salt mine and a silica sand mine. In each case mining was by the bord and pillar method. Standard respirable cyclone samplers operating at 2.2 l/min to provide a 4 µm cut were used. This type of sampler eliminates most of the other background dust whilst collecting all the very fine respirable diesel particulate. The samples were analysed for OC and EC and the following results obtained.

5.1 *Salt mine*

At the time of sampling a large CAT 990 front loader was running rock to a crusher. Diesel particulate samples were collected on the CAT 990 both inside and outside the filtered air cab; in both the intake and return air to the production area; in the roof and directly below at breathing level; in the workshops; on a scaling machine operating in a blind heading and close to the scaling machine. Ventilation volumes in the production area were of the order of 100 m³/sec which equates to a flow velocity of approximately 0.7 m/sec.

The results are presented in Figure 5. Three of the TWA EC results, all collected in and around the production area, are greater than 200 µg/m³ which puts them into the high exposure range. The highest result of 494 µg/m³ EC was measured close to the roof about 100 metres down ventilation of the CAT 990 activity. A corresponding sample taken directly below at breathing zone height was 50% lower at 245 µg/m³ EC. This vertical concentration gradient is due to the hot diesel exhaust from the CAT 990 layering at roof height. Eventually the exhaust gases cool and start to mix with the general body air and a homogeneous mix is established. This is a concern for electricians and scalers who work at roof height for a good deal of the time and who may unknowingly be more exposed to DEEEs than other workers in the mine.

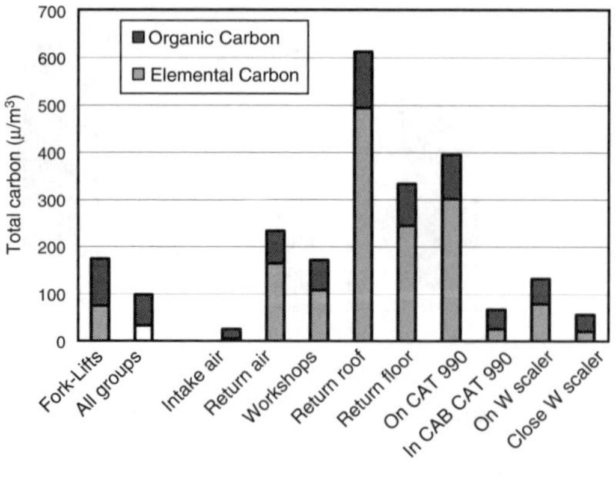

Figure 5. OC and EC data for samples collected in a salt mine.

The benefits of a filtered cab are demonstrated well with these data. Outside the cab on the CAT 990 the TWA EC concentration was 301 μg/m^3 whilst inside the filtered cab the exposure was 26 μg/m^3 EC, a 91% reduction.

5.2 *Silica sand mine*

At the time of sampling there were six diesel vehicles operating in the mine; two 30 ton Caterpillar dump trucks that transported the mineral from the production area to a stock pile at the surface; a Caterpillar 938G front end loader; a cherry picker elevator; an Atlas Copco boomer driller and a Massey Ferguson tractor. The dump trucks were virtually new with only 200 hours of use and the front end loader was a new vehicle. These all had filtered cabs. The remaining vehicles were several years old.

With the two dump trucks traveling in and out of the mine along the intake air route there were very few areas in the mine where there was no diesel particulate. Samples were collected both inside and outside the cabs of a dump truck and the loader; in the intake and return air well away from the production area and in the production area itself. One sample was collected in a blind heading where the loader was loading blasted rock into the dump trucks.

The OC and EC results are presented in Figure 6. All the EC results from the production area are well in excess of 200 μg/m^3 putting them into the high exposure range. These high results are purely down to the ventilation, which was inadequate for the number of diesel vehicles operating in the production area. The highest result was in a blind heading with a TWA EC concentration of 858 μg/m^3. This area was adjacent to the main ventilation flow but ventilation in the heading was extremely poor. Fortunately, only the loader and dump truck drivers were operating in this area and they were inside filtered cabs. The diesel particulate

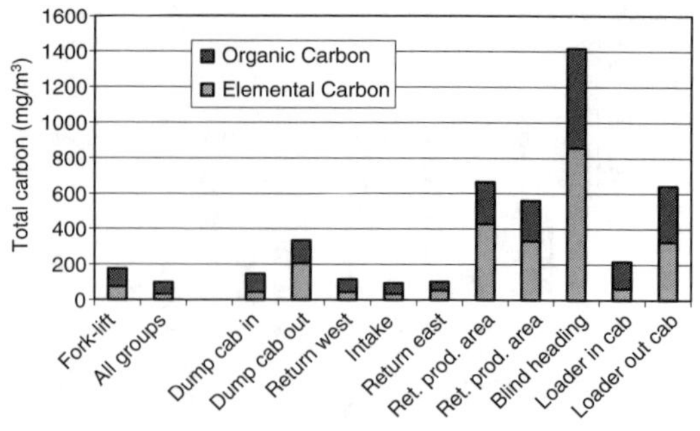

Figure 6. OC and EC data for samples collected in a silica sand mine.

EC exposures for the loader and dump truck drivers were respectively 81% and 79% lower than EC exposures measured outside the cab.

In general, exposure to DEEEs in non-coal mines was found to be higher than in coal mines. This is partly due to the larger number of diesel vehicles operating in non-coal mines but more likely it is due to the lower ventilation velocity. In the production areas for the two non-coal mines monitored, EC exposures were well in excess of 200 μg/m^3. This places them into the high exposure range. Subsequently both mines have taken measures to reduce exposures. The salt mine has moved away from diesel power and now uses an electric mining machine to cut the mineral directly onto a conveyor belt. The silica mine has opened up a new development with much improved ventilation.

6 SIMPLE ON-SITE METHOD FOR DETERMINING EXPOSURE TO DEEEs IN NON-COAL MINES USING A BOSCH SMOKEMETER

To make it possible for non-coal mines to monitor exposure to DEEEs themselves without recourse to laboratory methods a simple and inexpensive onsite method has been developed and is outlined below.

Within certain limits there exists a relationship between the mass of elemental carbon collected on a diesel particulate filter and the 'blackness' of the filter. Eventually, a point is reached where adding more diesel particulate to the filter cannot make it any 'blacker' and the relationship breaks down. But as long as the relationship is known and the sample 'blackness' lies within this limit it is possible to estimate how much EC is present on a diesel particulate filter simply by measuring its 'blackness'. This forms the basis of a simple on-site method for determining exposure to DEEEs.

The Bosch Smokemeter is a simple and inexpensive device designed for measuring diesel exhaust

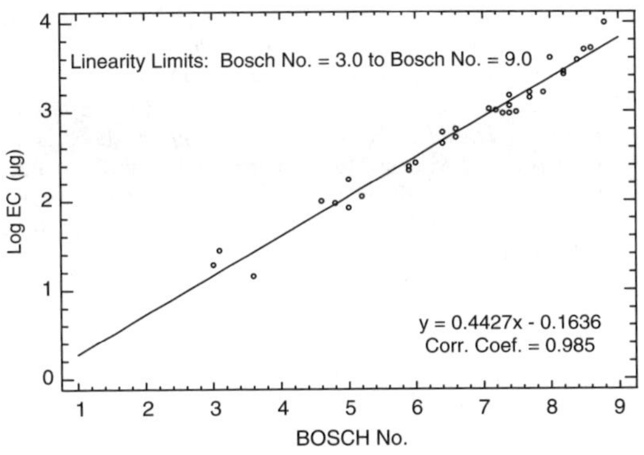

Figure 7. Relationship between Log EC and Bosch Number for gypsum mine data.

tailpipe particulate emissions. It comprises two parts, a sampling pump which is inserted into the exhaust pipe and which draws a defined amount of diesel exhaust through a paper filter; and an evaluator unit which measures the 'blackness' of the filter and provides a Bosch Number from 0.0 (white) to 9.9 (intense black). The more diesel particulate in the exhaust fume, the blacker the filter and the higher the Bosch Number.

A few UK mines already own a Bosch Smokemeter and use it to monitor at intervals the diesel exhaust particulate emissions from engines in the mine under some standard condition of operation. Over time a baseline norm for each engine is determined such that any deviations from the norm can be quickly identified and the engine attended to. The UK mines inspectorate are very keen to promote this approach and are trying to encourage all UK mines to purchase a Bosch Smokemeter and to regularly monitor tailpipe particulate emissions.

However, the smokemeter evaluator unit can also be used to determine personal or general body diesel particulate EC exposures. To do this a number of diesel particulate samples are collected in the mine using a respirable cyclone sampler covering a range of exposures. These are scanned for 'blackness' using the Bosch evaluator unit and their Bosch Number determined before being analysed for OC and EC. From these data a relationship between EC and the Bosch Number can be determined. An example is shown in Figure 7 using diesel particulate sample data collected in six different gypsum mines. The relationship may well be specific for each mine dependant on the colour of the mineral being mined.

From the equation of the line a simple look up table relating EC mass to Bosch Number can be produced making it easy to determine EC exposure by simply measuring the 'blackness' and knowing the sample volume. Monitoring is typically for two hours. Any longer than this and, in most situations, the filter becomes overloaded. However, if the mine activity over the sampling period is typical of the whole shift, then the determined TWA exposure will be representative of the whole shift.

This simple method allows mines to regularly monitor exposures to DEEEs and so ensure that all the controls in place to reduce exposure to DEEEs are working and that miners exposures are being kept to a level that is as low as practicably possible.

7 CONCLUSIONS

From the results presented here the following conclusions can be drawn:

- The diesel particulate EC exposure of the FSV drivers monitored in UK coal mines were in all cases below 150 µg/m^3 the recommended maximum target limit and most were below 100 µg/m^3. When compared to the surface industry TDS data this puts them towards the top end and similar to those of the fork-lift truck drivers, the highest exposed work group.
- Exposure in non-coal mines was higher than coal mines and in the production areas many EC results exceeded 200 µg/m^3 putting them in the high exposure range; some exceeded 300 µg/m^3 and higher. The main reason for the higher exposures is the reduced ventilation velocity when compared to coal mines. In coal mines the vigorous ventilation necessary to reduce explosion risk quickly dilutes the exhaust fume and carries it away. In non-coal mines the exhaust fume lingers longer in the area and the concentration builds up. Mines must ensure the maximum ventilation possible where FSVs are operating if exposure to DEEEs is to be kept to a minimum.
- Filtered cabs on modern vehicles are excellent at reducing exposure to DEEEs. Reductions of 80 to 90% were measured in this study.
- Biodiesel has much to offer in reducing exposure to DEEEs. Experiments on a diesel test bed comparing biodiesel and standard class A1 diesel showed diesel particulate EC reductions of 76% and 48% for a 30% biodiesel/class A1 blend. Other emissions i.e. carbon monoxide and nitrogen oxides were also reduced.
- A simple and inexpensive on-site method for measuring diesel particulate EC exposure in non-coal mines has been developed. The method allows these mines to monitor their own DEEEs exposures without recourse to laboratory methods. Some mines in the UK are already using the method and are making full use of the information to ensure exposures to DEEEs are being kept to a minimum.

REFERENCES

Control of Substances Hazardous to Health Regulations 2002. SI 2002/2677. ISBN 0 11 042919 2.

Dabill, D.W., Groves, J.A. & Hughes, H.J. 1996. *Development of a method for measuring organic and elemental carbon in diesel particulate*, HSE Internal Report, UK.

Dabill, D.W. 1999. *Engine Exhaust emissions from Biodiesel.* HSL Internal Report OMS/99/16.

Groves, J.A. & Cain, J.R. 2000. *A survey of exposure to diesel engine exhaust emissions in the workplace.* Ann Occup. Hyg., Vol. 44, No 6, pp. 435–447.

McCartney, T.C. & Cantrell, B.K. 1992. *A Cost Effective Personal Diesel Exhaust Aerosol Sampler.* Diesels in Underground Mines: Measurement and Control of Particulate Emissions. Proceedings: Bureau of Mines Information and Technology Transfer Seminar, Minneapolis, MN, September 29–30.

Ventilation and control systems

Mine Ventilation, Ganguli & Bandopadhyay (eds)
© 2004 Taylor & Francis Group, London, ISBN 90 5809 633 5

Development of a real time airflow monitoring and control system

A.D.S. Gillies & H.W. Wu
The University of Queensland, Brisbane, Australia

N. Tuffs
Ventilation & Gas Engineer, Capcoal, Australia

T. Sartor
Ventilation Engineer, BHP Billiton Cannington, Australia

ABSTRACT: Computer software has been developed to link real time information generated by underground mine ventilation airflow monitoring sensors into a network simulation program to undertake network simulations and allow interpretation of key system data and operational changes. Results were used in the development of a computerized monitoring and simulation system to provide immediate or real time data on air behaviour within each branch within an underground mine ventilation network through linking of sensors to the ventilation network simulation software. The outcome of the project is an online system which can report changes in the mine ventilation system, allow causes of changes to be isolated and rectified, improve balancing of available air throughout the mine, allow improved approaches to regulator setting and dispense with much of the labour used for underground ventilation measurement. The main work activities involved in the research program have involved examination and modelling of regulators used in two Australian mining (coal and metalliferous) environments, software modification and considerable mine site testing and optimising activities. The mathematical modelling of airflow through a variety of mine regulators was an important part of the project and this aspect is discussed. There is some discussion on approaches to control of flow through ventilation systems with increased information.

1 INTRODUCTION

The fundamental requirement of ventilation is to provide controlled air distribution within the underground mining operations to satisfy statutory and safety requirements with respect to ventilation quantity and quality. Ventilation design then facilitates this distribution while using modelling and predictive tools. The analysis of an operational ventilation network during both normal and abnormal conditions can be very complicated both on a theoretical and analytical level, requiring computer based solutions. Monitoring systems provide the ability to observe many aspects of the underground environment simultaneously. Robust, suitable and intrinsically safe instruments are available for measurement of, for instance, gas concentrations, air velocity and air pressure. These are often tied to extensive mine monitoring and communication systems.

One approach to establishing air quantity through a ventilation branch is through measurement of differential pressure across an opening or regulator. Mathematical relationships are available to relate (with some qualification) pressure drop and quantity through a regulator orifice placed symmetrically in a round flow conduit. However these can, at best, only be used to approximate mine regulator behaviour due to the irregularity of mine regulators in shape, symmetry and their positioning in mine airways, the construction of the mine regulator opening and uncontrolled air leakage through the regulator or adjacent bulkhead.

The study briefly describes efforts to characterize or mathematically model regulators. It then describes how this information is used in the development of a computerized monitoring and simulation system to provide immediate or real time information on each branch within an underground mine ventilation network through linking of sensors to the ventilation network simulation software.

The outcome of the study is an online system which can report changes in the mine ventilation system, allow causes of changes to be isolated and rectified and improve balancing of available air throughout the mine. It is envisaged that in time the real time model will be an integral part of an online mine wide planning, monitoring and control software platform that will be updated continuously along with the mine plan. The main steps involved in examination and modelling of regulators, software modification and considerable mine site testing and optimising activities are described.

2 THEORY OF REGULATORS

A regulator can be described as a large thin plate with an orifice installed in a fluid conduit. When a difference in pressure exists between the two sides fluid flows in the pattern shown in Figure 1. On the low pressure side the fluid issues as a converging jet in line with the centre of the orifice. The jet converges to its smallest area at a distance of about half the orifice diameter (Le Roux, 1990). This area is called the "vena contracta" (A_C at Fig. 1). The ratio between vena contracta and orifice area is the "coefficient of contraction", C_c (A_c/A_r at Fig. 1).

McElroy (1935) found that the C_c value is a relation between the ratio of the orifice and airway cross sectional area, N (A_r/A at Fig. 1). Bernoulli's equation can be applied to both sides of the orifice as shown in Figure 1 in order to calculate the velocity and hence the airflow quantity.

A correction must be made for the contraction of the jet at the vena contracta. The velocity equated based on Bernoulli's equations is the velocity at the vena contracta. Therefore, the velocity at the orifice can be obtained with the following equation:

$$V_2 = C_c\sqrt{\frac{2\Delta P_s}{\rho}}\frac{1}{\sqrt{1-N^2}} \tag{1}$$

Since airflow quantity through regulator $Q = V_2A_r$, it follows that:

$$Q = C_c\sqrt{\frac{2\Delta P_s}{\rho}}\frac{1}{\sqrt{1-N^2}}A_r \tag{2}$$

where A_r is orifice opening area in m^2.

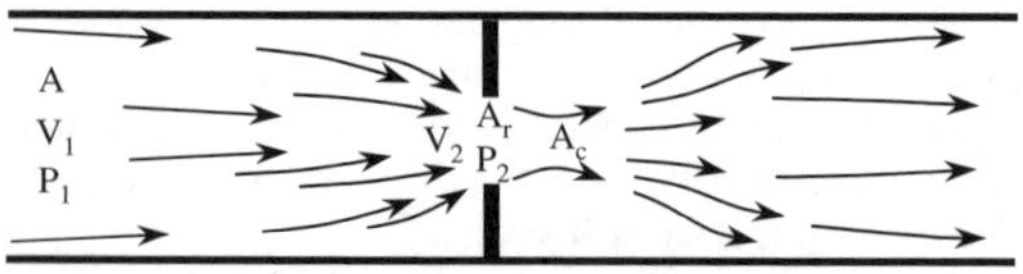

Figure 1. Airflow pattern through an orifice (after Burrows et al, 1989).

3 FIELD TESTS OF REGULATORS

Field tests on several types of regulators were conducted at various underground metalliferous and coal mines. Initially verification of air behaviour in flow through regulators was investigated. Parameters measured were airflow quantity and pressure drop across regulator. From pressure drop measurements, airflow quantity through regulators can be calculated with Equation 2. Results of this calculation can be compared with measured values and the reasons for significant differences investigated.

3.1 *Drop board regulator tests*

The regulator tested at the University of Queensland Experimental Mine (UQEM) is the drop board type as shown in Figure 2. Details of the tests and equations derived to account for leakage during measurements were described and discussed by Wu et al (2003) and Gillies et al (2002).

The relationship between the regulator opening area (A_r) and total resistance (R_t) can be derived as shown in Figure 3. Based on this, pressure and airflow quantity relationships can be calculated from mine regulator impedance characteristic curves. These can

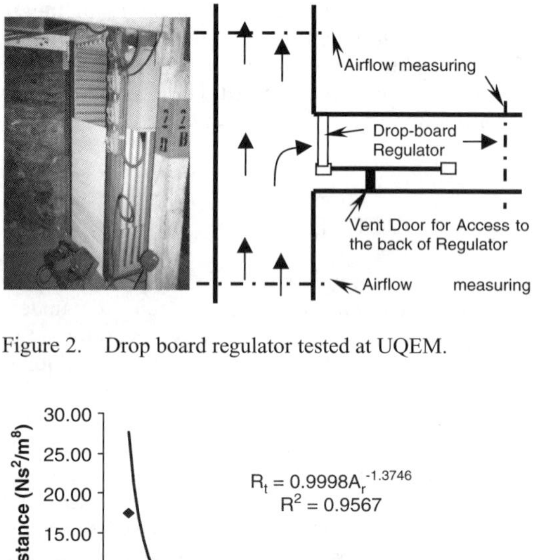

Figure 2. Drop board regulator tested at UQEM.

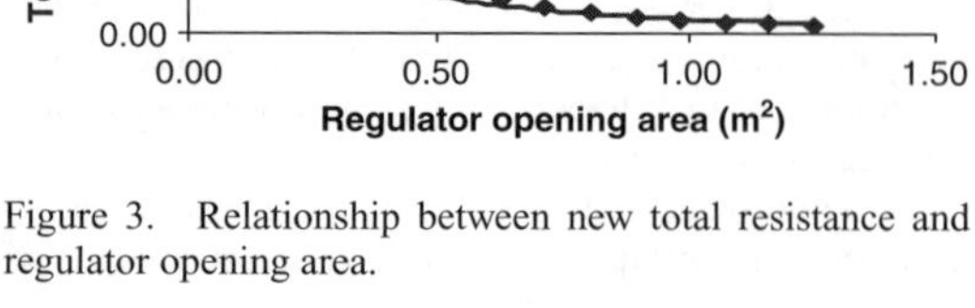

Figure 3. Relationship between new total resistance and regulator opening area.

be drawn for different mine configurations as shown in Figure 4.

3.2 *C-section regulator tests*

Similar tests were conducted on Drop Board style C-section regulators. Figure 5 shows a photographic view and the engineering drawing of a C-section regulator used by an Australian mine. The regulators were installed in either half or full sizes depending on the magnitude of the airflow regulation requirements and the locations.

A half size ventilation regulator may consist of up to a total of 16 C-section galvanised steel boards, which are secured with humpback split pins to the frame structure. The full sized version is two of these regulator frames placed side by side. The frame structure is secured in place using rock bolts to the concrete floor base. Packing is used to seal the bulkhead around the regulator frame structure. Dimensions of each C-section board are 1.65 m in width and 0.2 m in height. The maximum opening area of a half size C-section regulator is 5.28 m^2.

Several test series were also undertaken to verify the relationship between the equivalent total regulator resistance, R_t, and equivalent opening areas, A_r at various locations at the same mine. Figure 6 shows the calculated relationships between R_t and A_r for the C-section regulators. It can be seen that the relationships are similar to the relationship established from the drop board regulator tests.

In Figure 6, a theoretical relationship between the equivalent resistance and the opening area of a regulator similar to the approximation equation proposed by Le Roux (1990) was also included with assumptions made to account for the general mining conditions, for example, $C_c \cong 0.64$ and $\sqrt{1 - N^2} \cong 1$. Therefore a value of 1.1 was suggested to replace these terms and R_t can be calculated from the following equation at standard air density of 1.2 kg/m^3.

$$R_t = 1.452 A_r^{-2.0} \tag{3}$$

In Figure 6, it can be seen that the equivalent total resistance calculated from the tests was lower than the theoretical regulator resistance. As mentioned before, the equivalent total regulator resistance also takes into account resistances of the leakage paths which parallel air paths through the regulator opening. It is expected that the values of R_t will be lower than the regulator resistance itself.

3.3 *Louvre regulator tests*

Tests were carried out on louvre type regulators. This type of regulator is popular in Australian coal mines and an example is shown in Figure 7. Generally a double vehicle door design has louvres placed within panels of both doors. The louvre blades can be adjusted to various angles to control airflow. In this test both left and right doors could be set at nine positions. Tests were undertaken both by holding one side fixed and varying the other and by varying both sides. Relationships between resistance and the equivalent opening areas are shown in Figure 8.

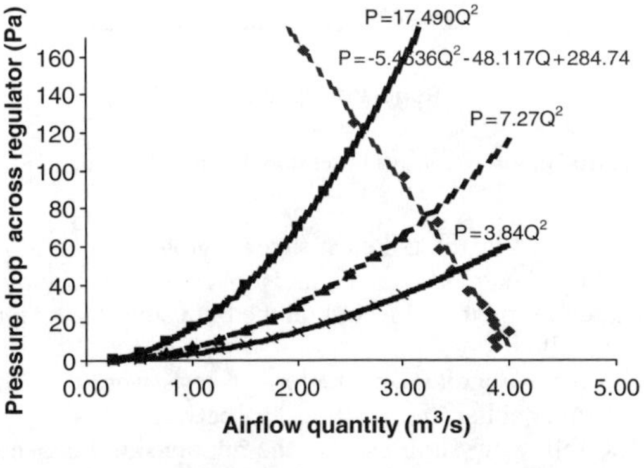

Figure 4. Drop board regulator characteristic curves.

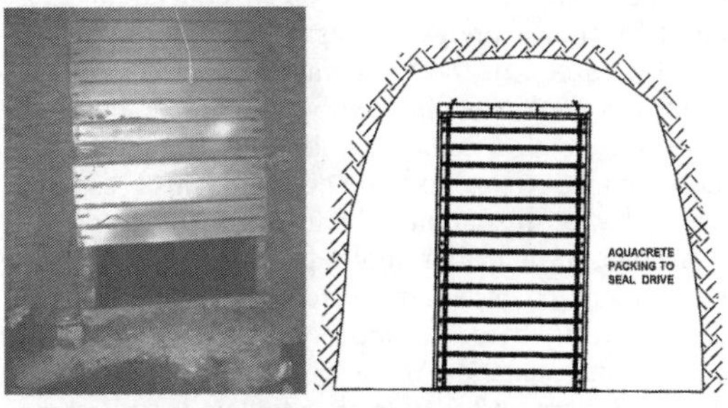

Figure 5. Photographic views and engineering drawing of a half size C-section regulator.

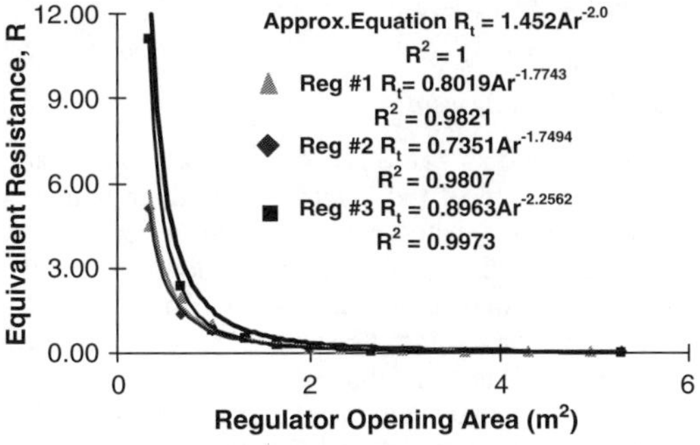

Figure 6. Relationship between R_t and A_r for C-section regulators.

Figure 7. An example of a louvre regulator.

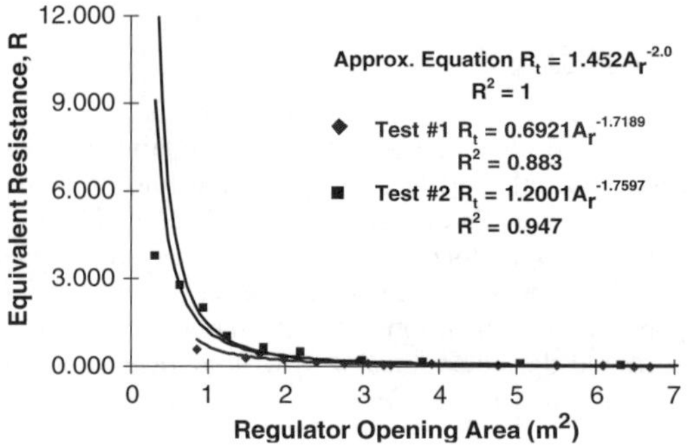

Figure 8. Relationship between R_t and A_r for louvre regulator.

It would appear that both aspects follow a similar relationship to that observed for the drop board or C-section regulators tested. There is no doubt that the resistance values increased as the equivalent opening areas decreased. However, it is suspected that due to the nature of fluid flow through louvre blade settings, the relationship between the resistance and equivalent opening areas is more complex than drop board flow behaviour. A literature review on louvre regulators indicated that limited research had been conducted on louvre flow behaviour.

3.4 *Roller door regulator tests*

Use of a roller door as a ventilation regulator was examined. Adjustment of the height of this door is undertaken via an automated control system. There is very little information on the use of roller doors as ventilation regulators and on the automation of such a system.

A roller door was installed for trial purpose as shown in Figure 9. Basically the roller door works like a Venetian blind, that is the lifting belt (made from sling material) is attached at the bottom corners of the door leaf and runs up inside the side guides to the drive assembly at the top. As the drive winds the belt, it lifts the bottom beam and each successive horizontal aluminium beam stacks on the top causing the vinyl coated fabric to billow out on the both sides as the air is expelled.

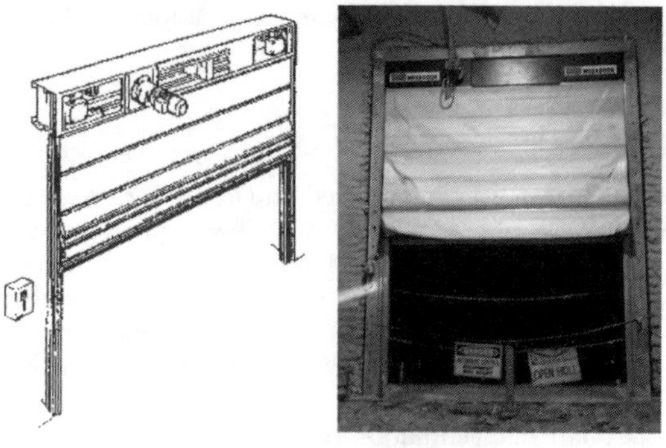

Figure 9. Schematic and photographic views of a roller door regulator.

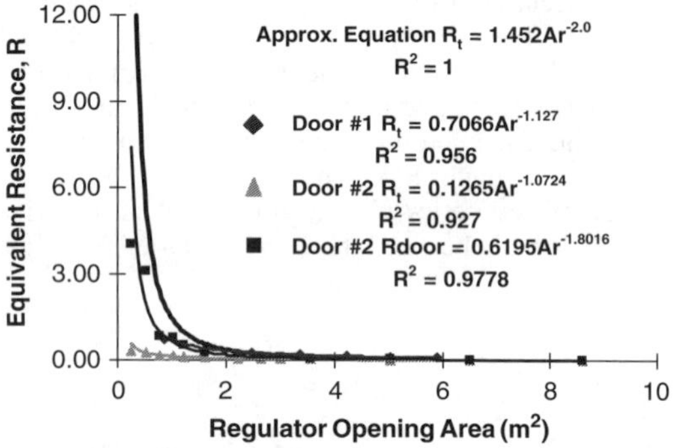

Figure 10. R_t vs. A_r for roller door regulator tested.

Several regulator characteristic tests were carried out to establish the relationship between R_t and A_r. The following figure shows the relationship between R_t and A_r for roller door tested. It should be noted that the air quantity and pressure was very low during the tests. Therefore, the calculation of equivalent resistance from measurements was difficult.

Also as one of the tests was conducted when the roller door was installed without a proper bulkhead built around it, a temporary brattice was erected to stop the airflows around the door. For this reason the leakage was substantial which means the equivalent resistance of the roller door regulator structure tested was much lower. As measurement was taken with the door fully closed it was possible to work out the equivalent resistance of the leakage paths around the door structure and then back calculate the resistance of the roller door only, R_{door}, without the leakage paths as shown in the Figure 10. It can be seen that it

Table 1. Summary of equations for various regulators tested.

Type of regulator	Regulator equations*
Drop boards	$R_t = 0.9998A_r^{-1.3746}$
C-sections	$R_t = 0.7685A_r^{-1.7619}$
Louvre	$R_t = 1.2009A_r^{-1.7597}$
Roller door	$R_t = 0.6195A_r^{-1.8016}$

*Equations derived from multiple test results.

gives a much better comparison with the theoretical relationship between R_t and A_r.

It is proposed that further tests should be undertaken under higher air pressure and quantity conditions with a proper bulkhead built around the roller door regulator. In summary, the behaviours of various types of regulators examined can be described by the equations shown in Table 1.

4 REAL TIME MINE VENTILATION SYSTEM

The aim of this mine ventilation research was to develop a computerised monitoring system to provide immediate or real time simulated information on each branch within an underground ventilation network. The system measures airflow or air pressure changes in selected ventilation branches and simulate flows through all other branches. This new approach to ventilation provides improved understanding of airflows through all mine sections. The popular Australian ventilation simulation modelling program "Ventsim" has been used as a simulation engine within the system. This software has been altered by the program author, Mr Craig Stewart, to accept real time information generated by underground mine ventilation monitoring sensors, undertake network simulations and interpret key system data and operational changes. Once the simulation program has updated readings it can remodel the whole mine system, report the flows in all branches and compare individual branch readings with expected values.

Initially, the UQEM was used to test the integration of a telemetry system into the "Ventsim" network analysis environment. The mine airflow monitoring system consisted of one El-Equip "Flosonic" and two vortex shedding Sieger BA5 air velocity sensors. The Flosonic air velocity sensor is an ultrasonic anemometer measuring the average air velocity value across a drift (Casten et al, 1995 & McDaniel et al, 1999). The aim of this testing was to use the system to monitor changing ventilation conditions, to establish airflow characteristics within the UQEM and to observe the resimulated network results. Details of the integration of the UQEM real time ventilation monitoring system including "Ventsim" modification have been described by Gillies et al (2000).

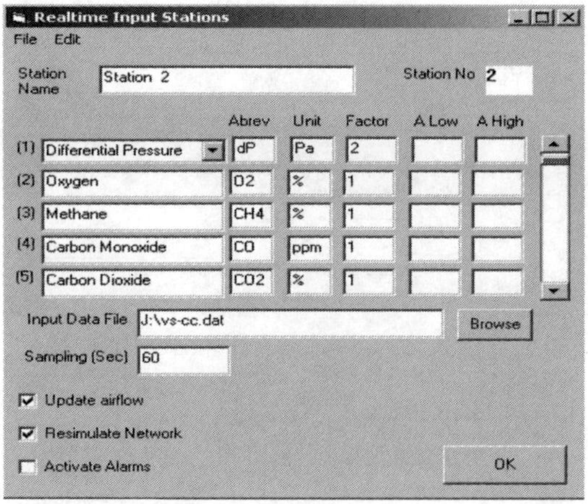

Figure 11. Example of information required for real-time input stations.

Table 2. Specific data format required by real time "Ventsim".

			Sensors				
Station	Date	Time	S1	S2	S3	S4	S5
1	13/06/2003	14:58:28	287	000	000	000	000
2	13/06/2003	14:58:28	438	000	000	000	000

4.1 *Real time "Ventsim" modifications*

Real time monitoring and simulation functions have been built into a recent version of the original "Ventsim" network simulation program. To run network simulation with real time data input, real time sensors installed in the ventilation network need to be initially numbered. Details of the real time sensors were input and edited in the Remote Station database as shown in Figure 11.

The Remote Station database consists of a list of remote locations and their detailed information such as type of sensors (up to five sensors for each station); unit of real time data, scale factor, alarm levels (low and high) for each sensors; location of real time input data file; sampling interval. Users have the options to select whether they want the real time "Ventsim" to update airflow values, re-simulate network and activate alarm or not. Up to 50 different remote stations may be stored within the current version of real time "Ventsim".

The real time data from underground remote stations must be made available to "Ventsim" through a text file with a specified format that can be recognised by "Ventsim" and stored in the directory as specified in the Input Data File path. A specific tab delineated text file format is required by real time "Ventsim" as in Table 2.

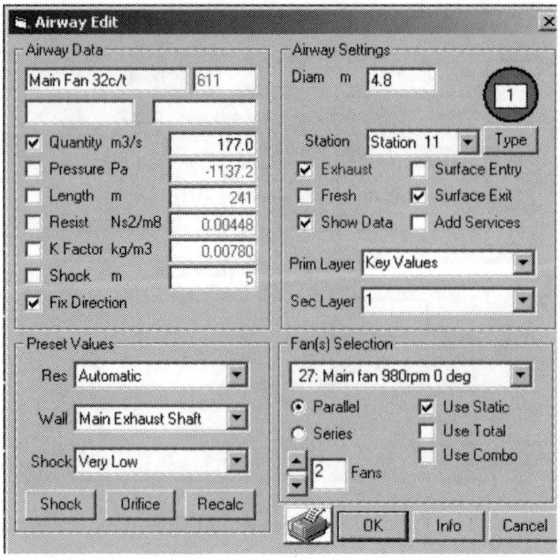

Figure 12. Airway Edit windows with remote station selection.

Table 3. Details of remote monitoring stations information at Central Colliery.

Stations	Locations	Type of data
1	31-32ct A Heading	Differential pressure
2	47-48ct A Heading	Differential pressure
3	311TG Homotropal regulator	Differential pressure
4	311 MG regulator	Differential pressure
5	311 MG 2ct B-C Heading	Differential pressure
6	312 MG 2ct B-C Heading	Differential pressure
7	64-65ct E Heading	Differential pressure
8	Surface Station	Barometric pressure
9	Borehole Fan 311	Air Quantity
10	Borehole Fan 311	Differential pressure
11	Main Fans	Air Quantity
12	Main Fans	Differential pressure

Once the details of each remote station are stored in the real time "Ventsim", a remote station number needs to be selected in the Airway Edit window as shown in Figure 12 for each airway containing real time sensors. If a remote station in a particular airway is selected all sensor data (for instance air velocity, quantity, differential pressure and various gas levels) will be available to "Ventsim" for display or for network simulation purposes. Once all airways with remote stations have been edited, the network can be simulated with real time data.

4.2 *Trial at Capcoal Central Colliery*

Capcoal's Central Colliery located in the Bowen Basin was the first modern underground longwall operation in Queensland. Central Colliery has an annual production of 2.5 million tonnes per year (mtpy). Longwall operations at the colliery have spanned over 16 years.

It was determined that the ventilation system at Central needed to be closely monitored with the installations of louvre type regulators and real time sensors at strategic locations throughout the mine as delivering required air quantities to the operating faces was becoming difficult due to the high mine resistance. Various airflow sensors were investigated initially for the project. Certain problems had been identified with air velocity sensors using vane anemometers, thermal mass cooling, and vortex shedding techniques. All of these sensor types measure spot air velocity instead of average air velocity. They are found to be difficult to re-calibrate to variable airflow requirements in operating mines and the output signal has a tendency to "wander away" immediately following re-calibration (McDaniel et al, 1999). The unreliability and lack of repeatability of these instruments will hamper the continued development and use of these types of sensors in mining industry. It was determined to install differential pressure sensors across regulators to measure the pressure drop induced by the shock loss.

A total of eight underground remote monitoring stations using intrinsically safe differential pressure sensors were installed for the trial at Central Colliery. An additional four remote stations with monitoring data for surface fans were also included in the system. Details of these monitoring stations and their status are shown in Table 3. In Capcoal Central Colliery, real time underground monitoring data are displayed on their mine control and monitoring system – Macroview. This information is displayed in a purpose made page as shown in the Figure 13.

An example display of the Capcoal Central Colliery real time "Ventsim" model with "live" data is as shown in Figure 14. Due to the site operational constraints, currently, only five real time sensors have been included in the Capcoal Central Colliery's real time "Ventsim" model. However, during the trial it was found that the system has the ability to update the mine ventilation network model and keep this mine planning tool current. Mine ventilation models are normally static simulation models that are accurate when calibrated after a mine ventilation survey. Even with care in frequent updating, models will tend to lose accuracy. The real time approach allows the model to be seen as a dynamic entity that can be tested for its accuracy at any time without the effort of undertaking a full ventilation survey.

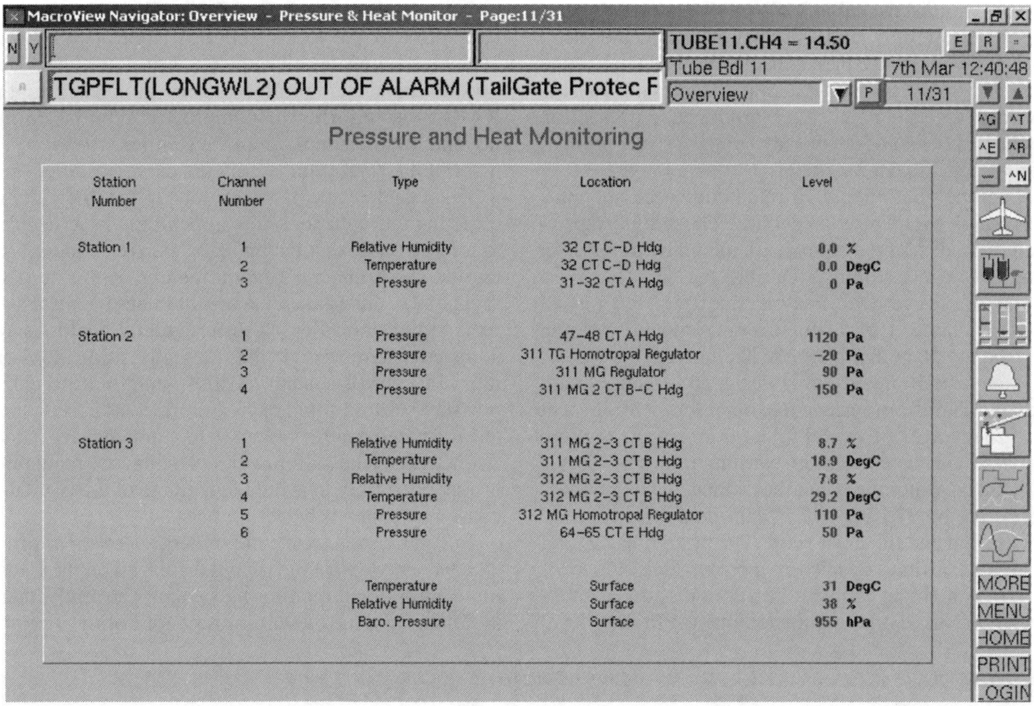

Figure 13. Pressure monitoring display at Central Colliery Macroview system.

In a typical coal mine operation, any ventilation change must be authorized before the change is made. Alternative options are evaluated through computer network simulation or manual calculations in the planning phases. Once the "best achievable" alternative is determined, authorization is gained and necessary adjustments to some of the system regulators made. Underground ventilation measurements can be conducted to verify the effects of the change.

A real time ventilation monitoring system can reduce or eliminate the need for numerous underground measurements necessary to verify the effects of ventilation system changes. The real time sensors installed in strategic locations will pick up airflow changes and subsequently make predictions of quantities in all other airways. However, after detection of changes, the real time "Ventsim" models should be modified to form an updated system reflecting changes that have occurred and placed in a state for future planning exercises. The project was initiated to achieve the following benefits from improved understanding of the mine ventilation system:

- Trouble shooting that would detect any abnormal situation in the underground ventilation system directly indicating where the trouble is located (i.e. indication of airway numbers) and what type of problem is involved (e.g. insufficient air quantity, air short circuiting and major leakage paths).
- Monitoring the air quantity and air quality in the return airway for determining the effectiveness of the ventilation design.
- Detecting regulator failure or incorrect setting from a sudden increase in air velocity or air quantity values and a decrease in differential pressure.
- Detecting stopping or seal failure from changes in airflow and the resultant potential for gas migration or spontaneous combustion.
- Detecting any blockage in the airway, which will indicate the potential of roof, back or rib failure or appropriate utilisation of equipment in the airway.
- Monitoring goaf leakage due to sudden change of air velocity or air quantity and increase of gas contamination levels through incorporated real time gas monitoring system.
- Providing an overall assessment of the fan performance in order to provide recommendations to alter the fan's ventilation capacity to its optimal level, and consequently saving of costs.

4.3 *BHP Billiton Cannington Trial*

The Cannington silver mine is located in north-west Queensland, 200 km south east of Mount Isa. The deposit was discovered in 1990 and the mine was commissioned in 1997. Full production was achieved in early 1999 and since then capacity has been expanded from 1.5 mtpy to over 2.2 mtpy. Cannington is the world's largest single silver producer, representing about six percent of the world's primary silver production (Anon, 2003).

Cannington is an underground mine accessed via a 5.2 m-high by 5.5 m-wide decline. The main hangingwall orebodies of the deposit are mined by transverse, longhole open sloping. During the first year of production, a vertical hoisting shaft with a finished internal diameter of 5.6 m was constructed from the surface to 650 m depth for ore haulage.

The ventilation system is designed around a main exhaust shaft from surface to a main return air level at the 325 m level (mLv). Return air from lower operating levels is exhausted through various return air raises (RARs) or sometime ore passes connected to a collection drive on 325 mLv. The main intake airways are the decline and the hoist shaft. The system has a total combined airflow capacity of approximately 450 m^3/s.

To maintain an efficient ventilation system, it was decided to have one unimpeded open split on a level towards the bottom of the mine and install a series of regulators in the levels above to control airflow in each levels. Levels below the open split level can have booster fans incorporated within bulkheads to the RARs to assist airflow and improve conditions during hot and humid summer months (Gillies, 1999).

Initially, C-section regulators were installed and tested at each levels. However, due to the difficulty in changing the regulator setting (opening), it was decided to replace the C-section regulator with roller-door type regulators. Adjustment of the height of this type of regulator is undertaken via an automated control system. At the start of every shift regulator positions are adjusted as required via the integrated mine monitoring and control system. Airflow and/or differential pressure sensors could be incorporated as part of such an automated control system to monitor changes made. At the end of the shift prior to blasting, the regulators are set open fully to avoid the damage of the regulator caused by shock waves from blasts.

In Cannington, real time underground monitoring data have been transferred and displayed on the Citect mine control and monitoring system. Currently, there are nine underground remote monitoring stations

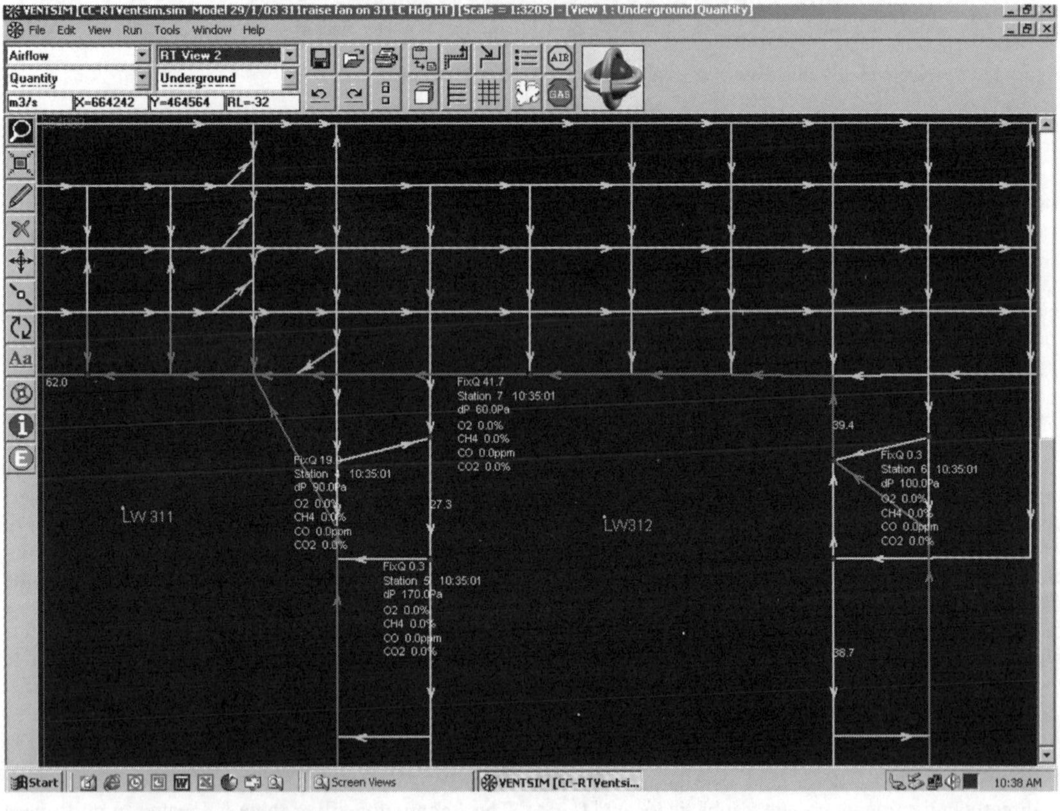

Figure 14. An example of the real time "Ventism" simulation display at Capcoal Central Colliery.

Table 4. Details of remote stations information at Cannington.

Stations	Locations	Type of data
1	Sa 50 RAR 575 mLv	Quantity/Pressure
2	Sa 50 RAR 550 mLv	Quantity/Pressure
3	Sa 50 RAR 520 mLv	Quantity/Pressure
4	Sa 50 RAR 500 mLv	Quantity/Pressure
5	Sa 50 RAR 475 mLv	Quantity/Pressure
6	Sa 50 RAR 450 mLv	Quantity/Pressure
7	Sa 50 RAR 425 mLv	Quantity/Pressure
8	Sa 50 RAR 375 mLv	Quantity/Pressure
9	Sa 50 RAR 350 mLv	Quantity/Pressure
10	Sa 50 Main fans	Quantity
11	Qb 80 Surface fans	Quantity

planned in the real time ventilation system. These stations are located at various levels accessing the Sa50 RAR and with both differential pressure sensors and ultrasonic airflow sensors installed. Sa50 and Qb80 surface fans are also monitored and included in the system. Details of these monitoring stations and their status are as shown in Table 4. Figure 15 shows an example of the computer screen displayed by Citect.

Figure 16 shows a snap shot display of the Cannington real time "Ventsim" model with "live" data input. Due to some specific site operational constraints, not all the regulators and sensors have been installed and calibrated to date. However, during the initial trial with only five stations online, it was found that the real time ventilation system at Cannington has the ability to detect changes in the ventilation circuit and re-simulate the network under the new conditions. Further trials will be required to evaluate the real time ventilation system behaviour more closely.

One major point of interest in such a system is the setting of the position of each remote controlled regulator at the beginning of each working shift. As the mine is not over-ventilated, it is important to maximise the availability of ventilation air to the active levels to effectively dilute the contaminants.

The setting of remote regulator positions is an iterative process as once one regulator is positioned such that the required quantities of airflow are delivered to working areas on a particular level; the airflows on other levels are disturbed. Adjusting the regulator

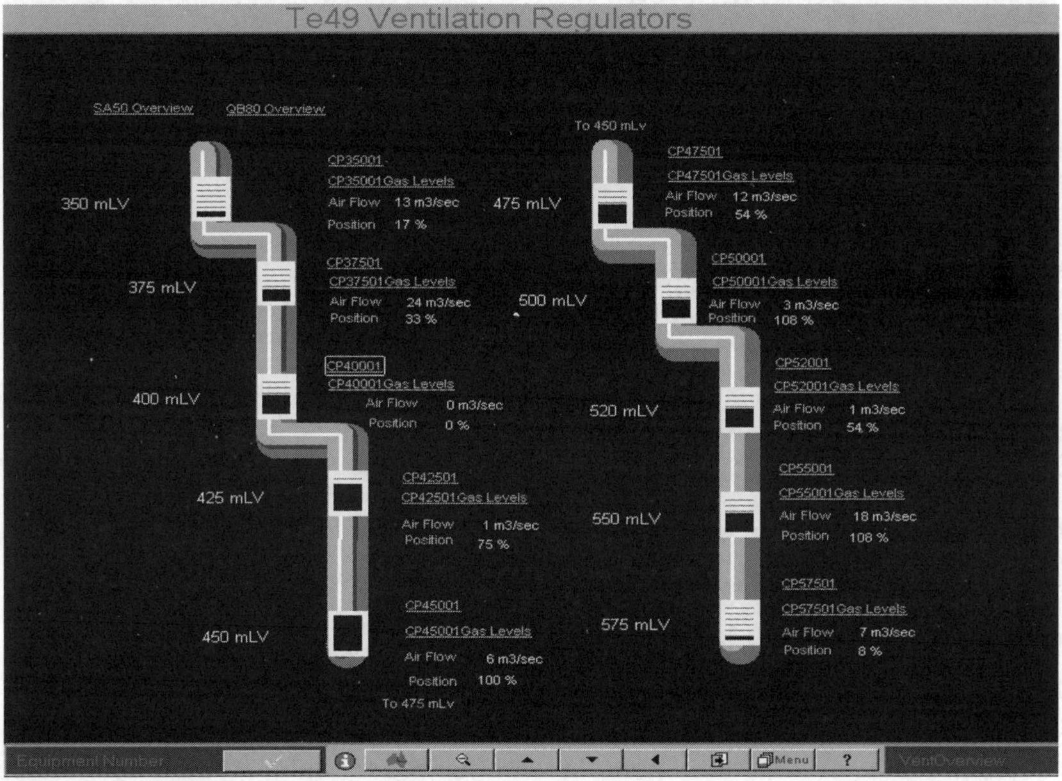

Figure 15. An example of the Citect mine monitoring system display at Cannington.

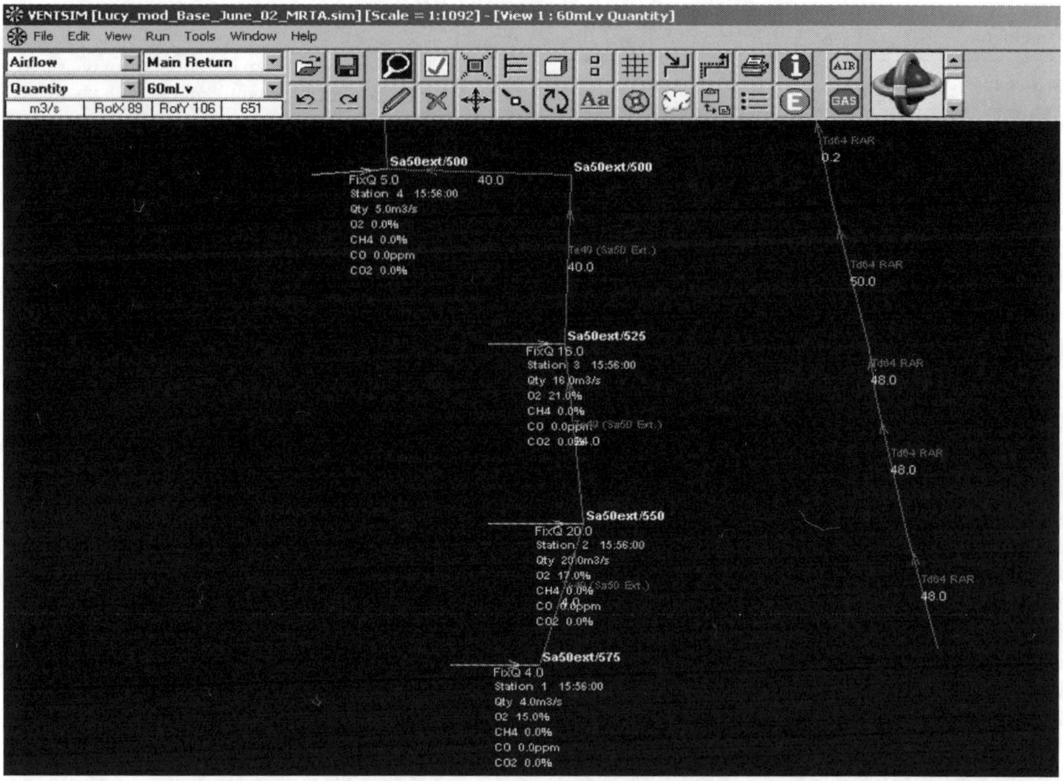

Figure 16. Snap shot of the real time "Ventsim" model at Cannington Sa50 RAR lower levels.

position on the other levels to deliver required airflows on these levels would consequently disturb the airflow on the first adjusted level and hence the cycle of obtaining correct airflow on one level whilst disturbing airflow on other levels continues. It is therefore necessary to investigate the best approach to this iterative process such that close to required quantities of airflow (particularly ensuring that the airflow quantities are not below the work place standards) are efficiently delivered to the working levels. It is important that the process is efficient with respect to the time it takes (it should be quick enough to perform before the underground mining work is underway each shift) and the ease in which it can be done i.e. it should be able to be learnt easily so that all personnel involved in the process can conduct it accurately from the early stages the process is put into place.

Ball, 2003 suggested that procedures for the ventilation and shift engineers to follow on a shift-to-shift basis should include

- periodic updating of the "Ventsim" model,
- calculations of shift-to-shift airflow requirements,
- real time linking to "Ventsim" and
- end of shift blast time remote regulator positioning to minimize re-entry times and risk of blast damage to the regulators.

Another major point of interest is the delay time or transient period between the instant of a change and when the system detects the change during the trials. The transient period in the UQEM is short and therefore is not of great significance in interpreting the network system. However, in large-scale mines like Cannington, the period can be up to 10 minutes. What this means is that reliance cannot be placed completely on "real time" airflow readings being instantaneously correct as reported for all branches within a mine ventilation simulated network. There is nothing that can be done to eliminate this characteristic as it is representative of the nature of airflow within underground mines. A change which leads to a hazardous condition may go unreported for the interval of this transient period. Of course changes in mine ventilation systems measured manually are rarely immediately picked up but this limitation of an automatically reporting real time system should be recognised.

5 CONCLUSIONS

Efforts to characterize or mathematically model a number of operating mine regulators have been described. Underground measurements have indicated that theoretical calculations to predict airflow quantity through practical mine regulators based on measured pressure drop are inadequate. The theoretical approaches are limited as they are based on prediction of fluid flow through a circular orifice in the middle of a plate whereas most mine regulators have a rectangular non-symmetrically positioned orifice. Also, most importantly, there is air leakage through the regulator bulkhead frame and gaps that increase actual quantity compared to that predicted.

The way to overcome this difference is to quantify the resistance of the leakage path based on regulator opening area and then to recalculate the total resistance of the regulators. The relationship between leakage path resistance and regulator opening area varies, but the resistance should increase along with an increase in opening area as observed in the field measurements.

The aim of the study was to gain greater understanding of a computerized monitoring system to provide immediate or real time simulated information in each branch of an underground ventilation network. The system measures airflow in selected ventilation branches and simulates flows through all other branches. Investigations were undertaken in two Queensland mining (coal and metalliferous) environments as to whether the Real Time Airflow Monitoring system can detect changes within the mine ventilation system, examine accuracy of the system and identify constraints that will limit performance of the system. As a result of trials, it was demonstrated that the system was able to detect changes occurring within the mine ventilation system and was also able to predict the changes accurately.

Maximising the benefit of the remote regulators has the capacity to improve mine ventilation system such that all working levels can have the best possible ventilation on demand. Consequently the risk of worker heat stress and fume inhalation will be minimized. Any improvement in ventilation in a mine has great potential to follow with improvements in production. Therefore the maximization of the beneficial use of the remote regulators in operations is highly likely to follow with increases in the mine's profitability.

Limitations caused by transient period delays have been examined. Updating of simulation models from use of real time data has also been discussed. It is envisaged in the future that the ventilation model would be an integral part of a real time mine wide planning, monitoring and control software platform from which the model would be updated in real time continually.

ACKNOWLEDGEMENT

The support of the University of Queensland, Mr Craig Stewart and a number of operations within the Australian mining industry in funding this study are acknowledged. Permission to publish this paper has been granted by BHP Billiton Cannington mine and Anglo Coal Central Colliery.

REFERENCES

Anon. 2003. http://cannington.bhpbilliton.com/index.asp.

Ball, L., 2003. *Remote Regulator Operation at Cannington Mine*, Bachelor of Engineering (Mining) Thesis, The University of Queensland, Brisbane, Australia.

Burrows, J., Hemp, R., Holding, W. & Stroh, R.M., 1989. *Environmental Engineering in South African Mines*, 2nd edition, pp 66–70 Mine Vent. Soc. South Africa, Johannesburg.

Casten, T., Mousset-Jones, P. & Calizaya, F., 1995. Ultrasonic Anemometry in Underground Excavations, *Proceedings of 7th US Mine Vent Sym*, Lexington, Kentucky, 1995, pp 429–434. Ed A.M. Wala. SME, Littleton, Colorado, USA.

Gillies, A.D.S., 1999. *Review of BHP Cannington mine ventilation and long term strategies*, report prepared for BHP Cannington Mine, December 1999.

Gillies, A.D.S., Mayes, T.I., Wu, H.W., Kizil, M. & Wang, N., 2000. The Development of a Real Time Airflow Monitoring and Control System, *Proceedings of 1st Mine Envirn and Vent Sym*, Dhanbad, India, 2000, pp 255–264. Ed D.C. Panigrahi. A.A. Balkema, Rotterdam.

Gillies, A.D.S., Wu, H.W., Mayes, T.I. & Halim, A., 2002. Measurement of Airflow Through Regulators and Real Time Integrated Monitoring. In E. DeSouza (ed.), *Proceedings of North American/9th US Mine Vent Sym*, Proc., Kingston, 8–12 June 2002. Rotterdam: Balkema.

Kingery, D.S., 1960. Introduction to Mine Ventilating Principles and Practices, *US Bureau of Mines Bul 589*, pp 89 (US Bureau of Mines, Washington DC).

Le Roux, W.L., 1990. *Mine Ventilation Notes for Beginners*, 3rd edition, pp 31–34 Mine Vent. Soc. South Africa, Johannesburg.

McDaniel, K., Duckworth, I.J. & Prosser, B.S., 1999. Evaluation of Different Airflow Sensors at the WIPP Facility, *Proceedings of 8th US Mine Vent Sym*, Rolla, Missouri, 1999, pp 519–525. Ed J.C. Tien. SME, Littleton, Colorado, USA.

McElroy, G.E., 1935. Engineering Factors in Ventilating Metal Mines, *US Bureau of Mines Bull 385*, pp 55–68 (US Bureau of Mines, Washington DC).

Wu, H.W., Gillies, A.D.S. & Mayes, T.I. The Measurement of Airflow Through Regulators, *Proceedings Fourth Australasian Coal Operators Conference*, Ed. N. Aziz and B. Kininmonth, Aus. Inst. Min. Metall., Melbourne, pp 212–224 February 2003.

Mine Ventilation, Ganguli & Bandopadhyay (eds)
© 2004 Taylor & Francis Group, London, ISBN 90 5809 633 5

Monitoring and control of underground ventilation systems using VUMA-network®

F.H. von Glehn
Bluhm Burton Engineering, Sandton, South Africa

I. Ox
CSIR Miningtek, Auckland Park, South Africa

ABSTRACT: Ventilation conditions in mines must be monitored to ensure safe and healthy conditions. However, the multiplicity of ventilation branches and equipment make it impractical to automate measurement of parameters all over a mine. It is more practical to monitor a few critical measurements that are fully reliable and extrapolate to 'predict' remaining conditions using ventilation network software. In the past, predictions made by VUMA-network have been verified against manual measurements and these have given confidence in predicted conditions in the remainder of the circuit. The paper describes work being undertaken to link the simulator with monitoring systems to enable accurate predictions of conditions in areas not covered by instrumentation. Information from a monitoring system will be linked to an existing network file. The measured and predicted data will be compared with predictions and verified predictions will be used to examine conditions in areas without instrumentation. Potentially unsafe situations will be highlighted and operators alerted to rectify the situation.

1 INTRODUCTION

In any underground mining operation, ventilation conditions must be monitored to ensure that safe and healthy conditions prevail and that effective operation and management is maintained. Mine-wide monitoring systems are used to a greater or lesser extent for this purpose on many mines. However, the multiplicity of ventilation branches and equipment, as well as the numerous ventilation parameters of interest, make it impractical to locate all measurement transducers all over a mine. Those systems that attempt to do so, generally become too large and complex and are often unreliable, lack credibility and are ultimately ineffective. Furthermore, the actual transducers for ventilation parameters are notoriously difficult to maintain [e.g. air flow and wet-bulb temperature sensors]. The installation of many of these transducers all over a mine network is very expensive in first cost and can only be effective with high maintenance efforts. Older mines with a limited life expectancy would not be able to justify the expense of an extensive communication and environmental monitoring system. In most mines these systems are generally doomed to unreliability. Even with a large budget, there will inevitably be areas in a mine that lack instrumentation.

It is more practical to have a few critical measurement sites that are fully reliable and, from these critical measurements, extrapolate to 'predict' conditions all over a mine network using ventilation network software. Thus in both the design phase and the operation of a mine, underground ventilation systems can be accurately simulated to ensure that they are efficient, safe and cost-effective. VUMA-network® simulation software has been and is extensively used to satisfy this role. In the past, predictions made by VUMA-network have been verified against manual measurements and these have given confidence in the predicted conditions in the remainder of the circuit. Work is now being undertaken to link the simulator with mine monitoring systems to enable accurate predictions of conditions in areas not covered by instrumentation. As part of the South African FutureMine collaborative research programme, work has commenced to integrate VUMA-network with automated monitoring systems. Information received from a mine monitoring system will be passed through trending/filtering routines to remove noise and the data will then be linked to an existing VUMA-network file, which accurately represents the current mine ventilation circuit. The measured data will be compared with predictions made by VUMA-network and any differences at measurement locations will be reported. The verified predictions will be used to examine conditions in areas of the mine that have no instrumentation. Potentially unsafe situations will be highlighted and operators can be alerted to carry out pre-determined tasks to rectify the situation. This system diagnostic tool will be tested towards the end of 2003 in conjunction with a deep hot gold mine monitoring system. Obviously, there is future potential for control functions to be automatically carried out. Future work will also examine the possibility of identifying locations/sources of differences and extracting reasons for the differences from a knowledge base.

2 VUMA-NETWORK

Increasingly stringent standards are being set with regard to occupational health and safety. In the mining industry, maximum allowable pollutant levels are being reduced, medical compensation levels have been increased, medical surveillance requirements are increasing, and standards regarding escape and rescue equipment are more stringent. Furthermore, production rates have increased, mining depths are greater, mines are more mechanised and the network of excavations more extensive. On the other hand, the majority of commodity prices are low and the mining of coal, gold, platinum, gemstones and base metals has become an extremely competitive business worldwide. One of the disciplines that affects the well-being and profitability of the modern mining industry is that of mine underground ventilation control. It has become essential for ventilation practitioners to plan and design ventilation and cooling systems to a high degree of accuracy. These systems need to take modern mining methods into account as well as the costly infrastructures needed for future ventilation and cooling requirements. To enable the practical and economical design of mine ventilation and cooling systems, ventilation practitioners need effective and modern tools.

Given that the planning and design of underground ventilation control systems for future mines will require extensive optimisation, it will be necessary to evaluate a number of scenarios, not only for the ventilation systems, but also for the overall design of mines. Due to the integrated and extended nature of high production mines, seemingly small changes in design could have significant global effects on the underground ventilation control system.

Such a process would require numerous predictions for network conditions, system design and performance. It would be virtually impossible to make such predictions without simulation tools that are accurate and flexible enough to evaluate rapidly all possible options and scenarios.

The VUMA-network simulation program is specifically designed to assist underground ventilation engineers to plan, design and operate mine ventilation systems. VUMA-network is an interactive 'MS-Windows' program that allows for the simultaneous steady-state simulation of airflow, air thermodynamics and gas and dust emissions in an underground mine. The program caters for a wide variety of mining methods. A fundamental feature of VUMA-network is user-friendly interfaces that allow simulation networks to be constructed and viewed graphically and what-if studies to be rapidly performed to determine optimal designs and system requirements.

Some other aspects of VUMA-network are:

- Network consists of branches, starting and ending with nodes, and depicting network components such as shafts, tunnels, etc.
- Input data for branches is used to calculate the air pressure drop and air thermodynamic and contaminant level changes in a specific component of a network.
- Input data for nodes consists of the X, Y and Z co-ordinates, Barometric pressure [BP], Virgin Rock Temperature [VRT], and air temperatures. The BP, VRT and wet-bulb and dry-bulb temperatures only need to be set for the start-node as these parameters are calculated for other nodes throughout the network.
- Simulation networks are constructed in a two-dimensional [2-D] graphical editor on a level-by-level basis. Different levels are then interconnected, typically by shafts, declines or production zones.
- Network is viewed in 2-D format in either geometric, strike, or section view.
- Input data for each branch is entered in a specific input screen for that branch-type before a solution is obtained for airflow, contaminants and air thermodynamic properties.
- If only an airflow solution is required, only information relating to the geometry and air resistance characteristics of the branches needs to be entered.
- Iterative network solution algorithms are used to solve for airflow, and updated heat flow models are used to calculate the thermodynamic changes in each branch.
- VUMA-network contains an extensive help function to assist with the development of a simulation model.
- Three-dimensional [3-D] graphical viewer is used to view the network and colour coded results.
- In addition to the 2-D and 3-D graphics output display, results may be viewed in a tabular format.

The solver within VUMA-network is the result of theoretical and empirical models that have been developed and verified by engineers and scientists at both BBE and CSIR Miningtek over a number of

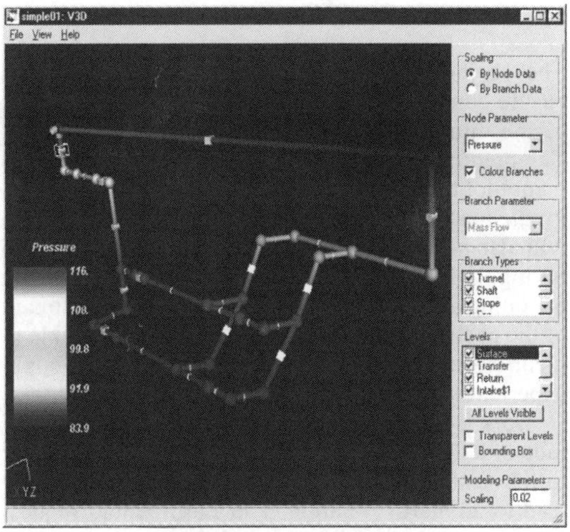

Figure 1. Network in 3-D viewer.

years. Further, details on the program are given by Marx *et al*, 2000.

3 LINK TO MINE MEASUREMENT SYSTEM

A project to examine the possibility of monitoring parameters at only a few key locations and then predicting conditions in the rest of the mine was undertaken as part of the FutureMine collaborative research programme. A survey of mine monitoring systems and the relevant ventilation instrumentation included in these systems was carried out on sponsoring mines. Of the 14 mines surveyed, all had systems in place that included automatic fire detection instruments and most of the mines monitored underground water flow, water temperature and dam levels. Six mines continuously monitored airflows and temperatures at key locations, while four mines were in the process of implementing such systems. Two mines monitored the status of ventilation doors, while five mines were in the process of installing such systems. Where large refrigeration plants were installed, remote access to the plant PLC systems is possible via the mine-wide monitoring/communication system.

The outcome of this project has been a number of software modules that are capable of:

- Communicating between the mine monitoring system and VUMA-network.
- Real-time solving and calibration of mine models.
- Anticipating potential problems by displaying warnings when changes in measured parameters exceed specified limits.

By linking the modules to VUMA-network, users can use existing network information and there is no need to switch between operations of two different programs. It is obviously important that the VUMA-network file accurately represents the current mine layout and that the user has verified the accuracy of input parameters such as airway friction factors, and linear heat loads. The system increases the coverage provided by existing instruments through extrapolation of measured values in VUMA-network to provide an expanded real-time view of the mine.

The modules include: Linker to a mine control and data acquisition [SCADA] system, Diagnostic faulty-instrument checker, Comparison-solver [which include calls to VUMA solver] and Reporter.

The Linker interacts with the mine SCADA system through a common open interface [OPC]. The Linker matches measurement locations with nodes within a VUMA-network data file and passes measured information to the Comparison solver.

The Linker screens the input data for 'noise' and also carries out some basic smoothing of data to minimise false alarms as a result of sudden but short-lived extreme changes in measured conditions.

The Diagnostic faulty-instrument checker scans the measured data for validity and consistency. If the measured reading is not valid or consistent with previous readings or with neighbouring measurements, a report is produced and the program is returned to the control of the Linker. The user will be requested to check the relevant instrument[s]. Checks carried out include whether the instrument is displaying a fault code, whether the measured data are within the range of the instrument and whether the measured value is within acceptable range for current conditions.

The Comparison solver calls the VUMA-solver to establish a reference prediction based on user input. The predicted values at the first measurement node are compared with the actual measurements and the difference displayed. For subsequent steps the parameters for that node are defined to be the same as the measured parameters so that downstream branches in the computer network will have the same starting point for calculation as the actual ventilation system. The VUMA solver is repeatedly called until the parameters at each measured node have been compared with the predicted values.

When the final measurement point has been compared or calibrated, all predicted values at measurement nodes will correspond exactly with measurements – any differences in intervening nodes/branches will be due to input parameters not corresponding with actual values or due to changes in the actual ventilation system which have not been taken into account.

Control is then handed back to the Linker to await the next series of measurements.

Figure 2 shows typical program logic.

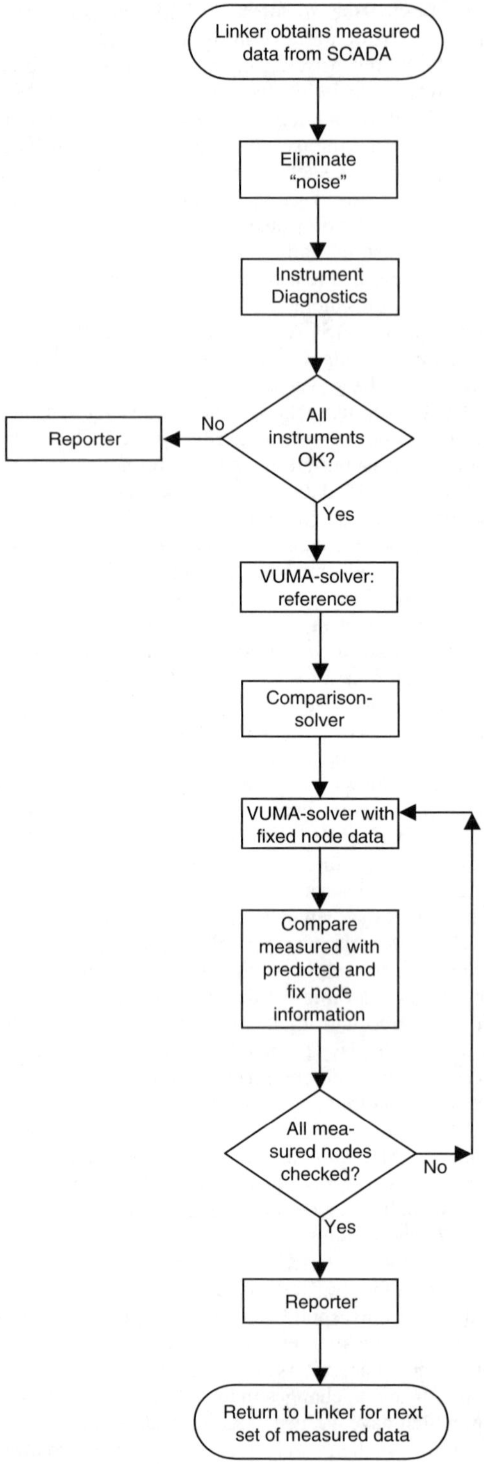

Figure 2. Program logic.

4 TRANSIENT ISSUES

The objective of the Comparison-solver will be to ensure confidence in predicted conditions in the rest of the mine where measurements are not available. The predictions will be used as an ongoing management tool to monitor conditions throughout the mine. Predictions based on accurate upstream measurements will also allow abnormal and emergency situations to be highlighted. For example, if an air cooler is not operating and the conditions immediately downstream of the cooler location are known, it will be possible to identify downstream areas of the mine, which could experience potentially dangerous conditions.

Work is currently underway to define what magnitude of temperature variation [from one measurement to the next or from an accepted mean value] should lead to an alarm situation. The alarm would request an operator to check/verify conditions at the measuring station. In addition the operator would check predicted downstream conditions to determine if alarms should be raised in other areas.

Although the thermal solution in VUMA-network is based on steady-state algorithms, minor temperature variations at a measurement location will not significantly affect the accuracy of predicted temperatures at a downstream location. In Figure 4, note how changes in inlet temperature conditions at '18 level mixing point' are not reflected in the conditions in the return airway. Smoothing/trending routines will be included in the Linker to remove minor variations and noise in the measured data.

Large variations in parameters at a measurement location will result in a variation downstream, but although the correct trend will be seen, the predicted value may differ from the actual value because of thermal inertia and other effects in the path from the upstream measuring station. Because of this effect an acceptance or uncertainty band around a predicted condition will have to be defined before an alarm situation is raised.

A companion program to VUMA-network, VUMA-transient® [Bluhm *et al*, 2002] has been developed to investigate the effect of dynamic temperature and flow variations in a network. This program will be used to examine the potential error that is obtained by using the steady state model of VUMA-network.

5 COMPARISON OF MEASURED AND PREDICTED PARAMETERS

As part of the FutureMine programme, management at Beatrix Gold Mine No. 4 Shaft agreed to provide data to allow the ventilation system to be set up in VUMA-network. The mine is ventilated using a

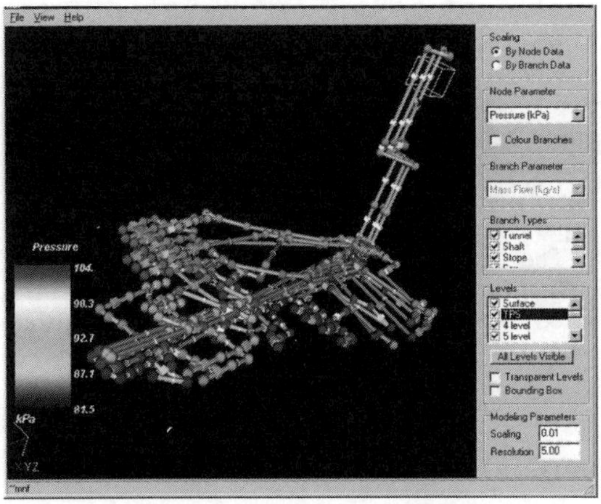

Figure 3. Network layout.

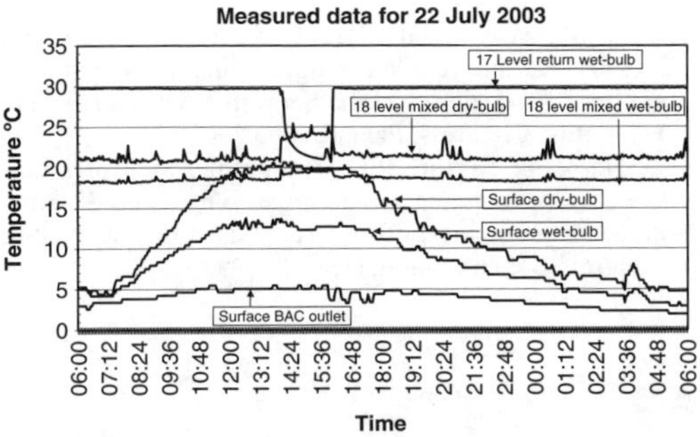

Figure 4. Typical measured data.

downcast airflow rate of 600 kg/s, which is cooled to a nominal mixed temperature of 7°C on surface, [Rose & Burton, 1992]. The fresh air is sent underground in two downcast shafts. The fresh air is supplemented by 300 kg/s of reconditioned, recirculated air underground to give a total of 900 kg/s to the workings. The recirculated air is cooled to a temperature of 18°C in three spray chambers on 17 Level [at a depth of 1855 m]. Inputs to the model were adjusted so that average predicted conditions corresponded with average conditions underground. Figure 3 shows part of the network.

A SCADA system is installed at the mine and temperature measurements were obtained at a number of strategic locations. For testing purposes the data were stored in data files in a format, which can be read by the Linker and then passed on to the Comparison-solver, as if the data were being obtained 'live' via the OPC interface. Data for a typical day are shown in Figure 4.

Return air from the mine arrives on 17 Level, is cooled and reintroduced into the ventilation circuit and mixes with fresh air from surface on 18 Level.

Table 1. Comparison of typical measured and predicted values.

Location	Time	Measured wet bulb °C	Predicted wet bulb °C	Measured dry bulb °C	Predicted dry bulb °C
18 Level mixing	06h00	18.38	18.70	21.33	21.58
17 Level return	06h00	29.98	30.57	32.86	32.68
18 Level mixing	13h00	18.56	18.72	21.14	21.57
17 Level return	13h00	29.93	30.56	32.75	32.69
18 Level mixing	17h00	18.92	18.73	21.58	21.61
17 Level return	17h00	29.95	30.42	32.78	32.68
18 Level mixing	20h29	19.46	18.71	23.70 [transient effect]	22.09
17 Level return	20h29	29.76	30.57	32.77	32.68

The blast occurs at about 14h00 each day and the recirculation system is automatically switched off for about 2 hours at this time. The drop in the measured 17 Level return air temperature in Figure 4 is due to the fact that the measurement location is downstream of the closed doors and the transducer does not experience any flow.

A comparison between measured and predicted temperatures at some measurement locations within the network is shown in Table 1.

The correspondence between the measured and predicted temperatures gives confidence in predicted values in the rest of the network. Further comparisons will be done at intermediate locations.

6 FUTURE WORK

Future work will try to identify the cause of any differences between measured and predicted parameters and will suggest ways of correcting the difference by altering network input data. For example, if the measured temperature downstream of a bulk air cooler is significantly higher than the predicted [or expected] value, the Comparison-solver would conclude that the air cooler is switched off and would display a message requesting the operator to check the status of the cooler. In the meantime, the program would switch off the cooler for subsequent iterations.

Other future work would be to identify 'events' that cause changes in the network. To implement this aspect of the program, a set of expert-system-type rules will have to be defined. The expert-system rules will be applied to each branch in an attempt to isolate and correct the cause of any difference. For example, if the predicted flow rate in a part of the network is too high, a regulator could be inserted to increase the overall network resistance. The operator would be requested to investigate a possible blockage or other restriction.

There is also future potential for control functions based on the analyses to be automatically carried out.

7 CONCLUSION

A set of software modules has been developed to manage measurement data obtained from a mine SCADA system via a Linker. The Linker passes the data to a Comparison-solver which assigns the measured values to nodes representing measurement locations and then calls VUMA-network to predict conditions throughout the rest of the network. Initial analysis has indicated that it is possible to confidently predict conditions in entire mine ventilation networks when parameters are measured at only a few key locations, thus minimising the costs of monitoring systems.

This system diagnostic tool will be tested in a real-time mode towards the end of 2003 in conjunction with a deep hot gold mine monitoring system.

ACKNOWLEDGEMENTS

The FutureMine Collaborative Research Programme was set up by South African gold-mining groups [Harmony/ARM, Anglogold, Gold Fields], CSIR-Miningtek and the National Research Foundation to ensure the long-term sustainability of gold mining in South Africa. The programme aims to improve safety and health in mines, increase the economic benefit, and minimise the environmental impact of gold mining. Permission from FutureMine to publish this paper is acknowledged.

The project team acknowledges the assistance of staff at Beatrix No. 4 Shaft [Oryx] in supplying measured data and setting up the trials of the software modules.

REFERENCES

Bluhm, S.J., Marx, W.M., von Glehn, F.H. & Biffi, M. VUMA mine ventilation software. Journal of the Mine Ventilation Society of South Africa, vol 54, July/September 2001.

Marx, W.M., von Glehn, F.H., Biffi, M. & Bluhm, S.J. VUMA – a mine ventilation and cooling network simulation tool. Proceedings of the Seventh International Mine Ventilation Congress, Poland, June 2001.

Rose, H.J.M. & Burton, R.C. The optimisation of a controlled recirculation system at Oryx gold mine. Proceedings of the Fifth International Mine Ventilation Congress, Johannesburg, October 1992.

Mine Ventilation, Ganguli & Bandopadhyay (eds)
© 2004 Taylor & Francis Group, London, ISBN 90 5809 633 5

Precision gas monitoring for methane drainage

A.E. Ketler & L.E. Sargent
Rel-Tek Corporation, Monroeville, PA USA

ABSTRACT: Draining methane through boreholes ahead of coal mining offers many benefits including environmental, safety and economic. However, in order to sell the gas to the utility, there must be continuous, precise analysis to assure high quality methane, with very little air intrusion. Sensors must read out 0–100% methane and 0–10% oxygen to a tight 1% tolerance – with long term accuracy – unattended. A high precision gas monitoring system was developed by Rel-Tek and installed at a remote borehole site – US Steel's Oak Grove Mine near Adger, Alabama. The paper describes this unique monitoring system and presents performance results showing the accuracy and stability requirements were met.

1 INTRODUCTION

Methane extraction through boreholes in advance of underground coal mining not only eliminates much of the hazardous underground gas emissions, but also provides an important profit center for sale of natural gas. The proviso for gas sales, of course, is that the natural gas be very nearly pure methane, and not contaminated with significant amounts of air.

Monitoring of the methane concentration in the gas extraction stream requires highly accurate sensors, which must be stable over time. Because access to bore holes is usually over hill and dale via poor or non-existent roads, frequent visits for calibrating and maintaining laboratory-quality instruments on site is out of the question. Besides, the high level of maintenance of such accurate and stable instruments can be an astronomical cost.

Figure 1. Bore hole methane drainage installation site near Adger, Alabama.

2 SOLUTION

Install Rel-Tek's MagiKal-II gas monitoring system, which assures continued high accuracy by providing frequent, high accuracy sensor calibrations. Calibrating the sensors every morning using cal-gas that is certified to NBS within 1% accuracy is a bit like setting your watch to the Naval Observatory every day. Before the instruments can think of drifting – bam – they're calibrated again. Drift of normal industrial instruments can be as high as a few percent per month, depending on changes in temperature, humidity, pressure and myriad other uncontrolled variables. However, by frequently calibrating using high accuracy cal gases, and by controlling sensor temperature – a major contributor to sensor errors – the sensors can be made to be very accurate. Although daily calibrations proved to be ample, for even higher precision the calibration schedule could be set for twice daily, or more often if needed. By calibrating frequently, the instruments never have an opportunity to drift significantly before – bam – they are again re-calibrated. The automatic calibration process usually takes about ten minutes. The trick for obtaining high precision is to calibrate automatically and frequently, using high accuracy cal-gas.

The automatic calibration procedure is invisible to any connected monitoring equipment. During the 10-minute automatic calibration process, the sensors

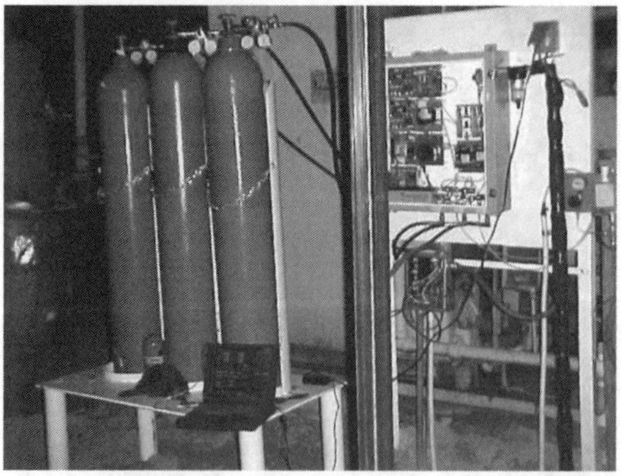

Figure 2. Three high-pressure tanks of certified calibration gas: one ZERO, one oxygen SPAN and one methane SPAN. The control panel is in the background inside shed.

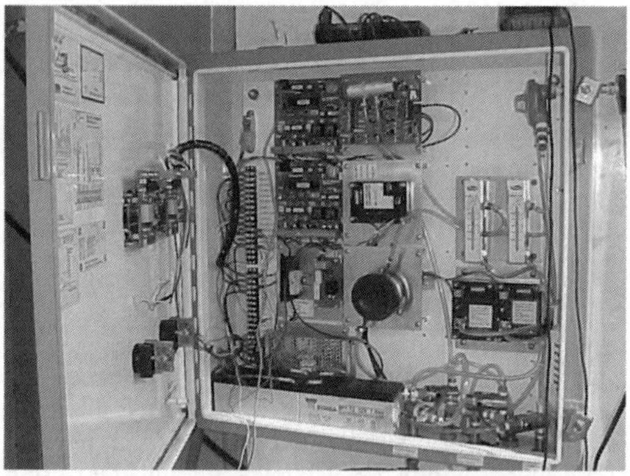

Figure 3. The 24 × 24 × 7″ control panel contains the two gas sensors, two MagiKal-II controllers and other electrical and pneumatic components.

are not actually monitoring the client's atmosphere or samples. Instead, the signal levels immediately before the calibration commences are continuously transmitted throughout the procedure, thus masking the temporary excursions due to calibration. The on-board computer controls the timing and orchestrates the calibration processes. The resulting, high-precision cal-parameters are then applied immediately after the calibration is completed. Because the calibration process is fully automated, there is no need for hands-on attention, and there is no need for anyone to waste time and effort to visit the site.

The photos show a MagiKal-II installation at US Steel Mining's Oak Grove Mine near Adger, Alabama. This methane drainage facility is in a hilly, hard to reach outpost, and maintenance visits are difficult and infrequent. The objective of the project was to monitor 0–100% methane and 0–5% oxygen to within 1% accuracy over long periods of time, unattended.

The 24 × 24 × 7″ MagiKal-II sensor and calibration pod shown in Figure 3 is monitored through the client's existing PLC. The high accuracy objective was met.

3 REMOTE ACCESS

Continuous monitoring over a LAN link provided remote supervision of the gas monitoring functions, blower and gas flaring alarms and calibration events, as well as any abnormalities such as power outages, lightning, vandalism, depletion of cal-gas supplies, or the failure of an instrument to calibrate properly.

Randall Townley of US Steel Mining handled installation of the gas monitoring/calibration system, with the assistance of Joe McDonald of Industrial-Mining Supply, Inc. of Hueytown, AL, a distributor of Rel-Tek products. Three tanks of calibration gas were purchased by IMSI from AirGas/Carbonic of Atlanta, GA, who certified the 1% accuracy. Ehren Graber, a Rel-Tek engineer, tested and commissioned the system in May 2003. The alarm thresholds were set up by USSM within their PLC to suit their particular methane gas sale requirements.

Oxygen sensor used was the OxyBoss*100 with a factory set range of 0–10% O_2. It employs an electrochemical-sensing element. The methane sensor was the GasBoss*1100 having a factory set range of 0–100% CH_4, using NDIR technology. The normal NEMA-4x sensor housings not used, enabling easier access to the sensor components, while enabling easier packaging inside the NEMA-12 steel panel.

Figure 4 shows a functional block diagram of the panel components. For visibility, only one sensor is shown, although both employed identical electrical and pneumatic connections. Only two cal gas tanks are shown, although there were three ... one 5% oxygen/nitrogen mixtures, one 99.7% methane/nitrogen, and one 99% nitrogen used as a shared zero gas.

A small volume of sample gas (about 1 liter/minute) was continuously extracted from the 8″ dia methane delivery pipe having 5–30 psig positive pressure. A heated 1/4″ dia tube helped maintain the tubing at mine temperature and minimize condensation, and a heat tape under the insulation avoids freezing. The box was electrically heated with three 100 W heaters to maintain the interior above the dew point. A dust/moisture filter and pressure regulator conditioned the incoming sample for introduction to the sensors. Flow controls and monitoring provided visibility and adjustment capability.

The raw (uncalibrated) 4–20 ma signal from a sensor was inputted to the MagiKal master control card.

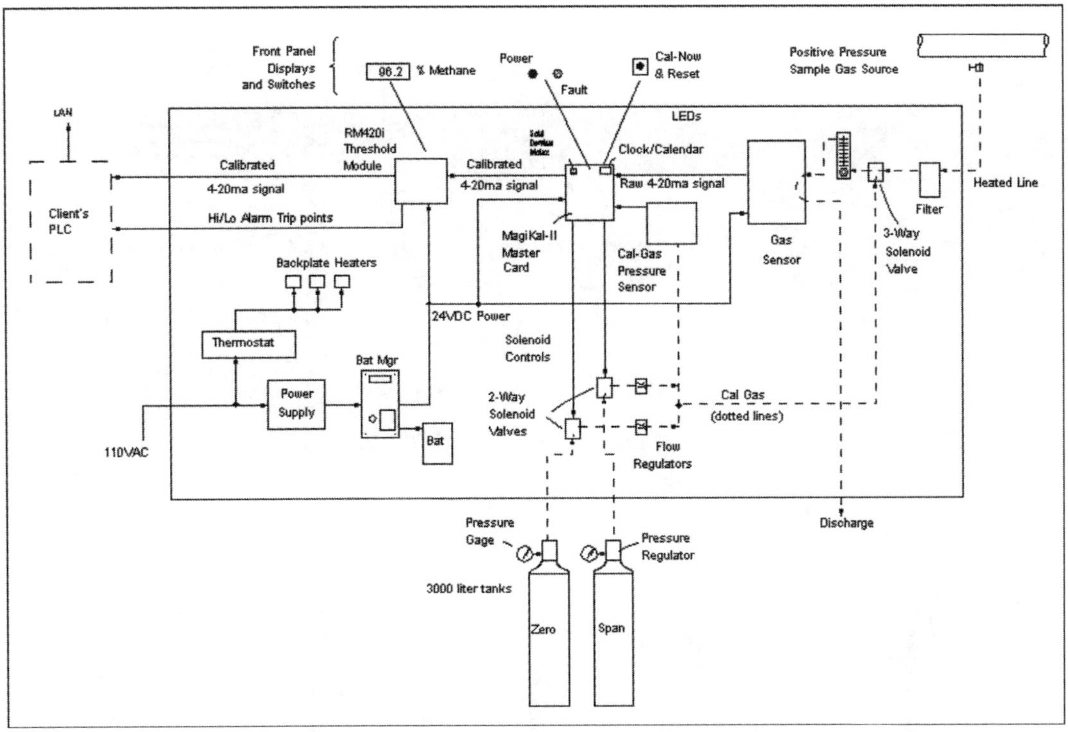

Figure 4. Functional block diagram of MagiKal components.

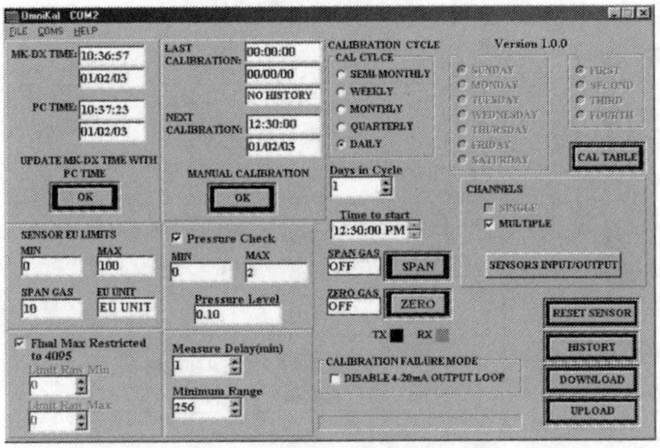

Figure 5. MagiKal II setup screen used to set parameters.

A calibrated 4–20 ma signal generated within the computer electronics was produced as an output for precise monitoring and alarm determination. Two RM420i modules provided front panel displays of the calibrated sensor signals before delivering the signal to USSM's external PLC monitoring equipment. The MagiKal master card contains a clock/calendar function, which signals the calibration procedure and sequences the span/zero cal gas solenoids on and off as required. A Cal-Now button on the front panel enables the operator to initiate additional calibrations at anytime needed.

The setup of the calibration parameters is made through a special software program running on the client's portable PC under Windows 98+. Figure 5 shows a typical screen format used to prompt all necessary information – sensor range, cal-gas concentrations, minimum acceptable output resolution, calibration schedule, sensor grouping and the like. The setup information is then downloaded into the master control card using an RS232 serial cable. Once downloaded, the PC is disconnected and would not be needed except to change the setup parameters. The PC can also upload the previous calibration information … date/time and the before/after calibration signal levels.

Figure 6 shows the single sensor and 16 sensor options for the MagiKal-II equipment. The master card can handle one sensor. Expanding to multiple sensors, however, requires the addition of as many as four slave cards, each capable of handling four sensors. For this two-sensor application it was decided to use two master cards, rather than one master and one slave card, as this provides independence and flexibility for treating each sensor independently, rather than as a group.

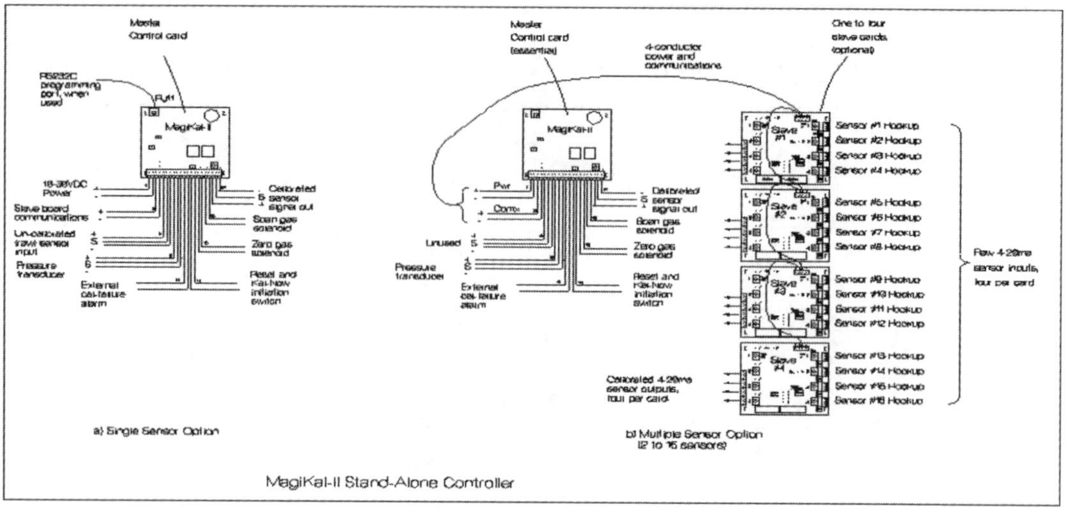

Figure 6. MagiKal II options.

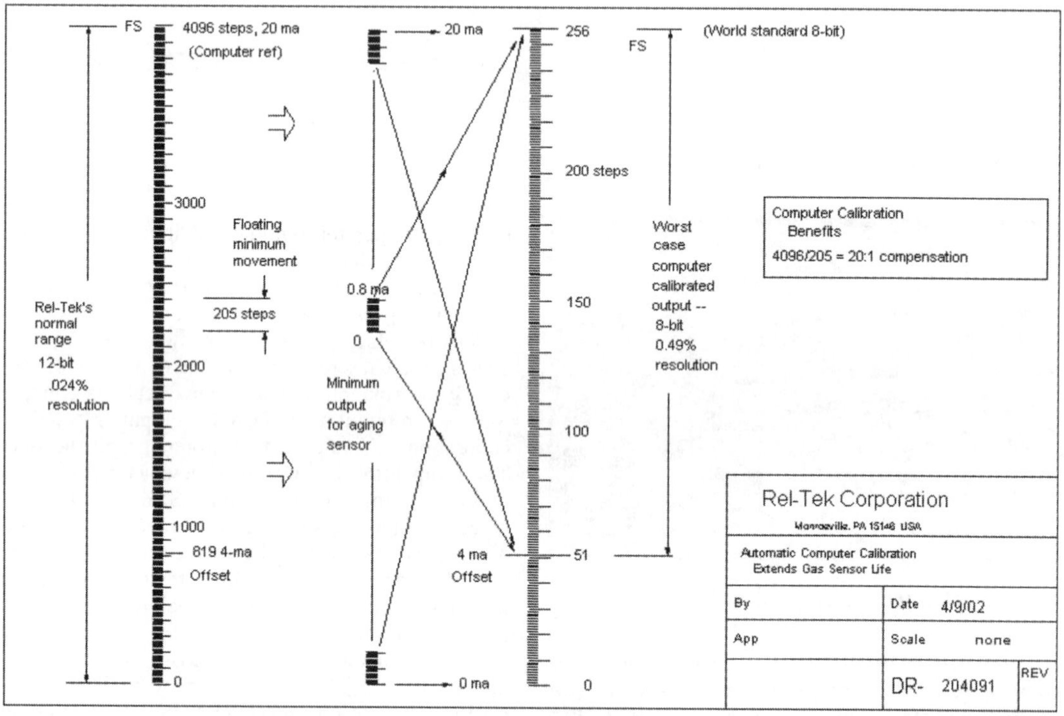

Figure 7. Illustrates the phenomenon that permits aging sensors to continue to operage.

Cal-gas pressure is monitored continuously during the calibration procedure to assure there is adequate gas to complete the procedure. If cal-gas pressure drops below an acceptable threshold, the calibration event is automatically canceled, the old cal values are re-applied, and an alarm is activated.

The unusually high 12-bit analog resolution used within the MagiKal computer, provides 4096 digitizing steps between zero and full scale. This ±.0244% resolution enables the system to accept a sensor signal as small as 0.8 ma (zero to FS range), and blow this up to a calibrated 4–20 ma level at normally acceptable 8-bit resolution. Figure 7 illustrates this. This phenomenon permits an aging sensor to continue to produce acceptable performance well after it is no longer able to generate a full 4–20 ma signal range ... thus extending the sensor life well beyond the expected retirement date.

4 CONCLUSION

By calibrating early and often (borrowing an old Chicago voting adage), a sensor can be made as accurate as the calibration gas used. The normal sensor drifting can be corrected by increasing the calibration frequency to a point where acceptable accuracy is reached. Manual calibration at this frequency may not be an acceptable option, depending on the cost of labor, the travel distances, etc. Automatic calibration is more dependable than a person, particularly with regard to scheduling, cal-gas flow timing, 12-bit parameter adjustment, cal-gas supply accuracy and pressure accuracy, precision of flow rates, and on, and on.

The **Cal-Now** push button calibration feature enables supervised calibrations, anytime, to verify the proper functioning of the automatic calibration system, and to enable double-checking the sensor performance after alarm events occur, after sensor maintenance or replacement, or just anytime one wants to verify a gas situation where precision is paramount.

Calibration gas is inexpensive. The 6000-liter tanks costing perhaps $500 a piece, which is normally purchased, will provide thousands of sensor calibrations. The annual cost for daily calibrating the two sensors is under $1,000. Compare this with the cost of manual calibration.

MagiKal-II is a stand-alone version of Rel-Tek's flagship MagiKal system (patents pending) that runs on a PC, providing logging, graphics and a host of other sophisticated automatic calibration features not available on this smaller version. The client had the PLC and LAN systems already in place, however, so the precision calibration function was performed using the smaller microprocessor version at low cost and with very adequate results.

Mine Ventilation, Ganguli & Bandopadhyay (eds)
© 2004 Taylor & Francis Group, London, ISBN 90 5809 633 5

Continual monitoring the natural ventilating pressure in a deep coal mine

F. Kopáček
VŠB – Technical University of Ostrava, Faculty of Mining and Geology, Ostrava, Czech Republic

R. Ceniga
OKD – Mine Darkov, Manager of Ventilation, Karviná, Czech Republic

M. Tabašek
Posgraduate doctoral student of the VŠB – Technical University of Ostrava, Faculty of Mechanical Engineering, Czech Republic

ABSTRACT: From November 2002, the continual monitoring of natural ventilating pressure has been performed in the 900 m deep of Darkov Mine, Ostrava–Karviná Coalfield, Czech Republic. The natural ventilating pressure is measured on the basis of temperatures measuring. The mine air thermometers are placed in downcast and upcast shafts while their measured levels are transmitted to the control computer. Also, the values of barometric and fan static pressures are simultaneously transmitted to the same computer algorithm. Time behaviors of temperatures, barometric and natural ventilating pressure are determined continually every time.

1 INTRODUCTION

The hard coal reserves of the Czech Republic are located in three basins: Czech part of Upper Silesian Basin, Czech part of Lower Silesian Basin and Central Bohemian Coal Basin. Within the Czech part of Upper Silesian Basin is Ostrava–Karviná coalfield (OKD company), where are the most significant Czech hard coal deposits, see Figure 1 and Figure 2.

Deep coal mines in OKD company region are from 900 m to 1100 m (2953 feet to 3609 feet) depth. Temperatures of the rock within the strata are about 32°C (90°F). Deep shafts can be found in geographical area where daily and seasonal ranges of surface air increase the mechanism of natural ventilating effect. Here the average seasonal temperature of "warm month", i.e. from May to October, is 15,2°C (59,4°F) and the contrary remaining "cold month" is 2,8°C (37°F), all within the last 13 year-period.

Under these conditions the natural ventilating pressure (NVP) is very important consequence of mining thermal energy. Especially, it becomes significant in case, where ventilation has to be run in combination of main fan (MF) and natural ventilation. It is also important that safety regulations in the Czech Republic for deep coal mines prescribe continual exhaust ventilation with two MF units (one is running and the other is spare one).

The authors have known many ways how to analyse NVP, however continual monitoring of running MF and NVP values for economic ventilation controls are necessary. From two or more reasons the authors have chosen a set of transmitting thermometer being placed in downcast and upcast shafts, where only dry temperatures are measured.

Dry air temperature sensors are very reliable. However dry air and water vapour modifications influence only a little difference to the numerical results obtained for a dry mine. On the other hand the authors have also known that water vapour measuring and its transmitting are very unreliable.

2 OSTRAVA–KARVINÁ REGION

OKD basin amounts about 20% of the total area of the Upper Silesian Coal Basin and is situated on the Czech–Poland frontier. It takes an area over 1600 km^2 and has 255 coal seams with a total net of 150 m thickness of coal. The hard coal of OKD exceeds 95% of the Czech Republic coal output.

The Mine Darkov is the part of the Ostrava–Karviná coalfield. Its history began in the year 1852 and a new plant started the mining in the year 1982. The Darkov mine area is 26 km^2 and the store of coal exploit makes about 100 million tons.

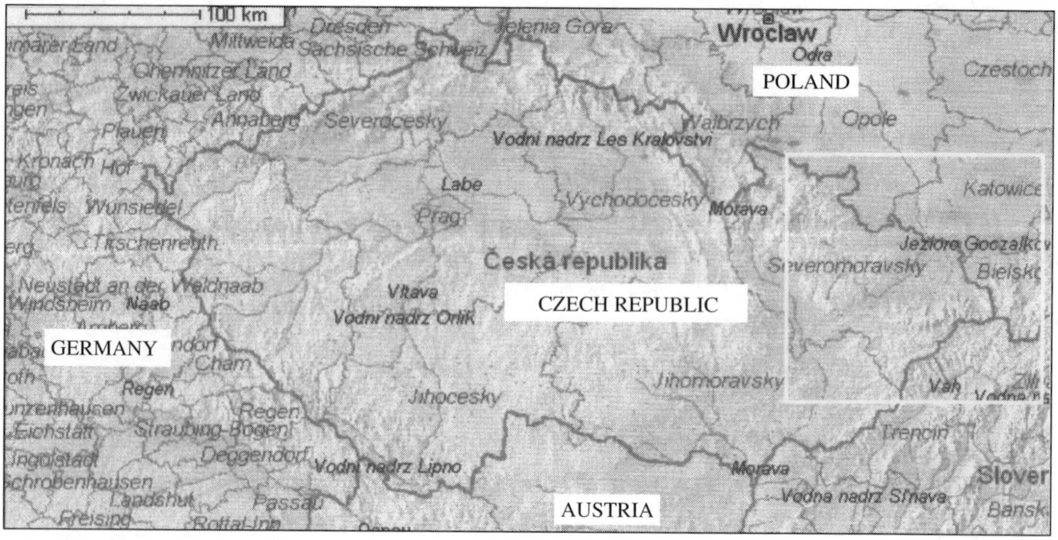

Figure 1. Czech Republic area.

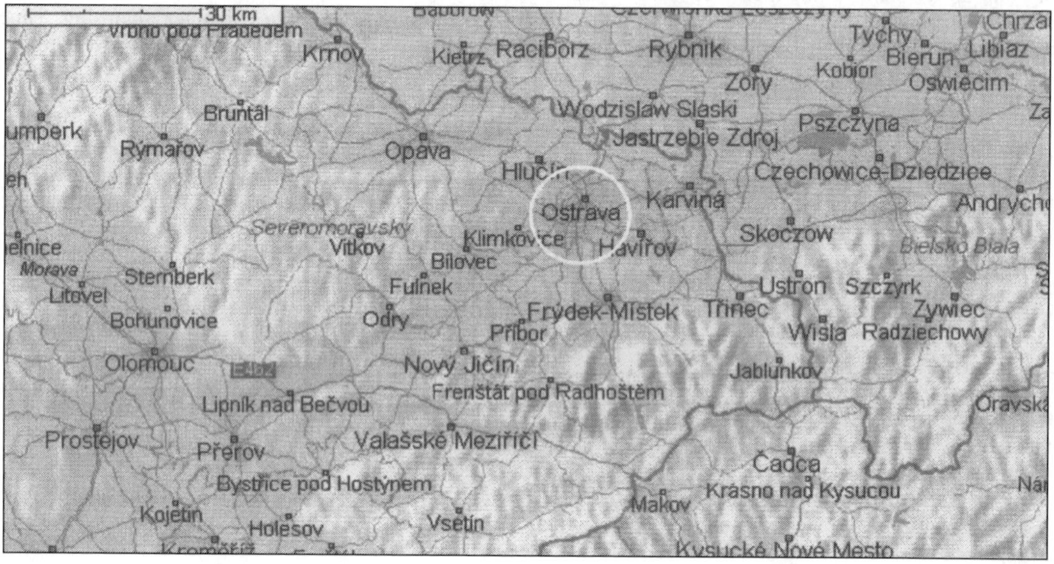

Figure 2. Ostrava–Karviná region.

The paper "Continual Monitoring the Natural Ventilating Pressure in a Deep Coal Mine" has been written after the performing of the total main ventilating reconstruction of the Mine Darkov, including Main Fan, axial type ARA – 2 – 3550 (ZVVZ – Milevsko, Czech Republic).

Therefore also the VŠB – Technical University of Mining and Metallurgy of Ostrava, is located in this part, in the middle of heavy engineering and mining region, where coal, coke, iron, steel and heavy chemical products are the key for the processing industry.

The following pictures in Figure 3 and Figure 4 present a view of the Darkov Mine area and a campus of the VŠB – Technical University of Mining and Metallurgy in Ostrava, respectively.

Figure 3. View of the Darkov Mine area, Czech Republic.

Figure 4. Campus of the VŠB – Technical University of Mining and Metallurgy in Ostrava.

3 SYSTEM DESIGN ANALYZE

Author's aim was to workout the computer algorithm of continual monitoring NVP under local mine conditions. The determination of NVP has been used in hydrostatic method, in the so-called mean density method (Mc Pherson 1993),

$$NVP = gH(\rho_{md} - \rho_{mu}) \qquad \text{[Pa]} \qquad (01)$$

where ρ_{md} = average density in the downcast shaft (DC); ρ_{mu} = average density in the upcast shaft (UC); H = vertical depth of shafts.

This method was chosen on the basis of skip or cage operating experience in the DC and UC shafts. However, the balance U – plastic tube method seems to be very questionable and unreliable. On the other hand the wet in UC shaft is often linked to water droplets that are not suitable for wet temperature monitoring. Therefore the simplest formulation of enhancement pressure directly to horizons and bottom of shafts from measured dry temperatures was chosen.

It means that air specific density increases progressively, but not linearly, toward the bottom of a vertical column (Weeks 1926), (Hartman et al. 1997).

Consider a column of dry air having a unit cross-sectional area S and a height a, and assume an incremental height da. Let the pressure on the base of this increment be dP. Then for forces can be written:

$$SdP = g\rho Sda = g\left(\frac{P}{RT}\right)Sda \quad \text{[N]} \tag{02}$$

Therefore, rearranging for integration:

$$\int_{P1}^{P2}\frac{dP}{P} = \frac{g}{0{,}5R(T_1+T_2)}\int_0^a da \quad \text{[-]} \tag{03}$$

After integrating for dry air, gas constant R = 287,1 $Jkg^{-1}K^{-1}$ and acceleration due to gravity g = 9,807 ms^{-2} is the following evident (see Figure 5):

$$P_2 = P_1 \exp\left[\frac{0{,}03416a_1}{0{,}5(273{,}1+t_1+273{,}1+t_2)}\right] \quad \text{[Pa]} \tag{04}$$

The average densities (01) are calculated by means of simplified scheme of Darkov Mine Ventilation, see Figure 5:

$$\rho_{md} = \frac{\sum_{i=1}^{i=5}\rho_i a_i}{H} \quad [\text{kgm}^{-3}] \tag{05}$$

$$\rho_{mu} = \frac{\sum_{j=6}^{j=13}\rho_j a_j}{H} \quad [\text{kgm}^{-3}] \tag{06}$$

where ρ_1 to ρ_5 = middle densities in the sections of DC; a_1 to a_5 = vertical lengths in the sections of DC; ρ_6 to ρ_{13} = middle densities in the sections of UC; a_6 to a_{13} = vertical lengths in the sections of UC.

Further, $H = \sum_{i=1}^{i=5} a_i = \sum_{j=6}^{j=13} = 886\,m$, is a vertical depth of both shafts under Mine Darkov conditions.

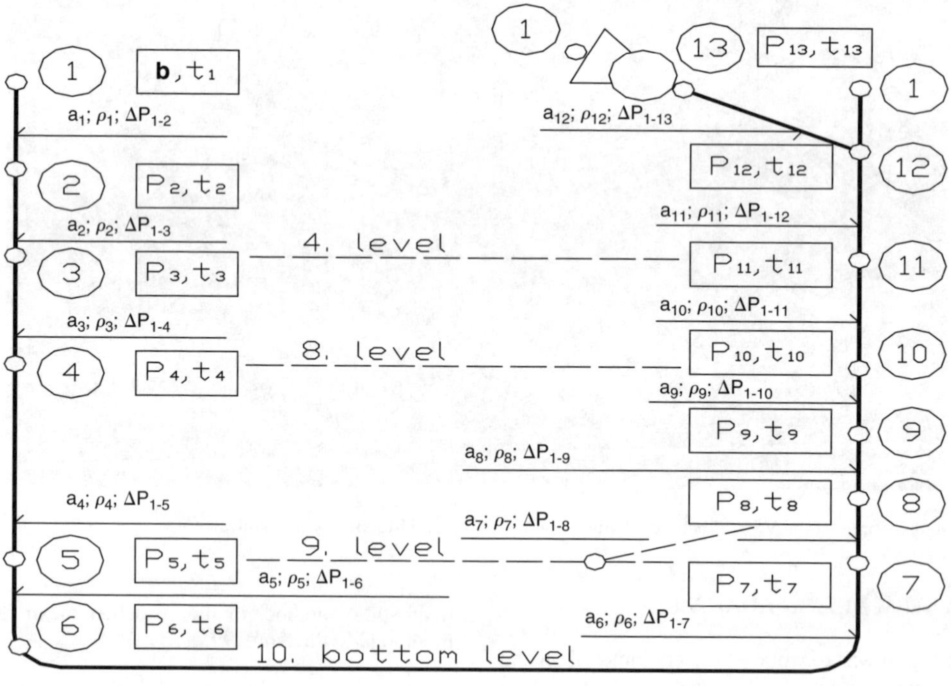

Figure 5. Darkov Mine Ventilation system.

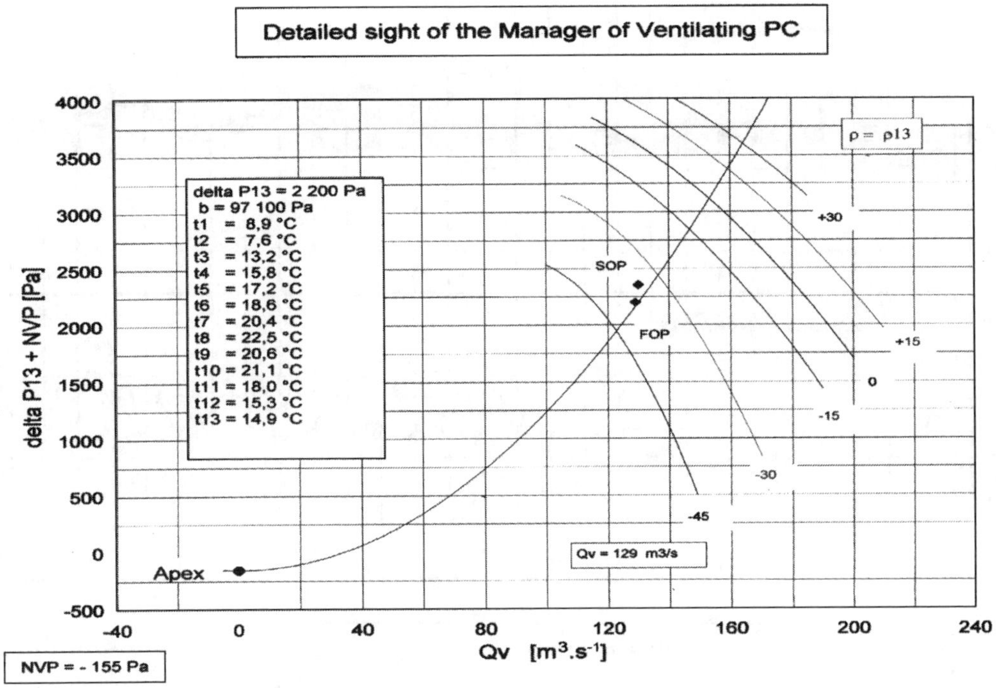

Figure 6. Control screen MF and NVP ventilation.

Figure 7. Control PC of MF and NVP in Manager Ventilation office.

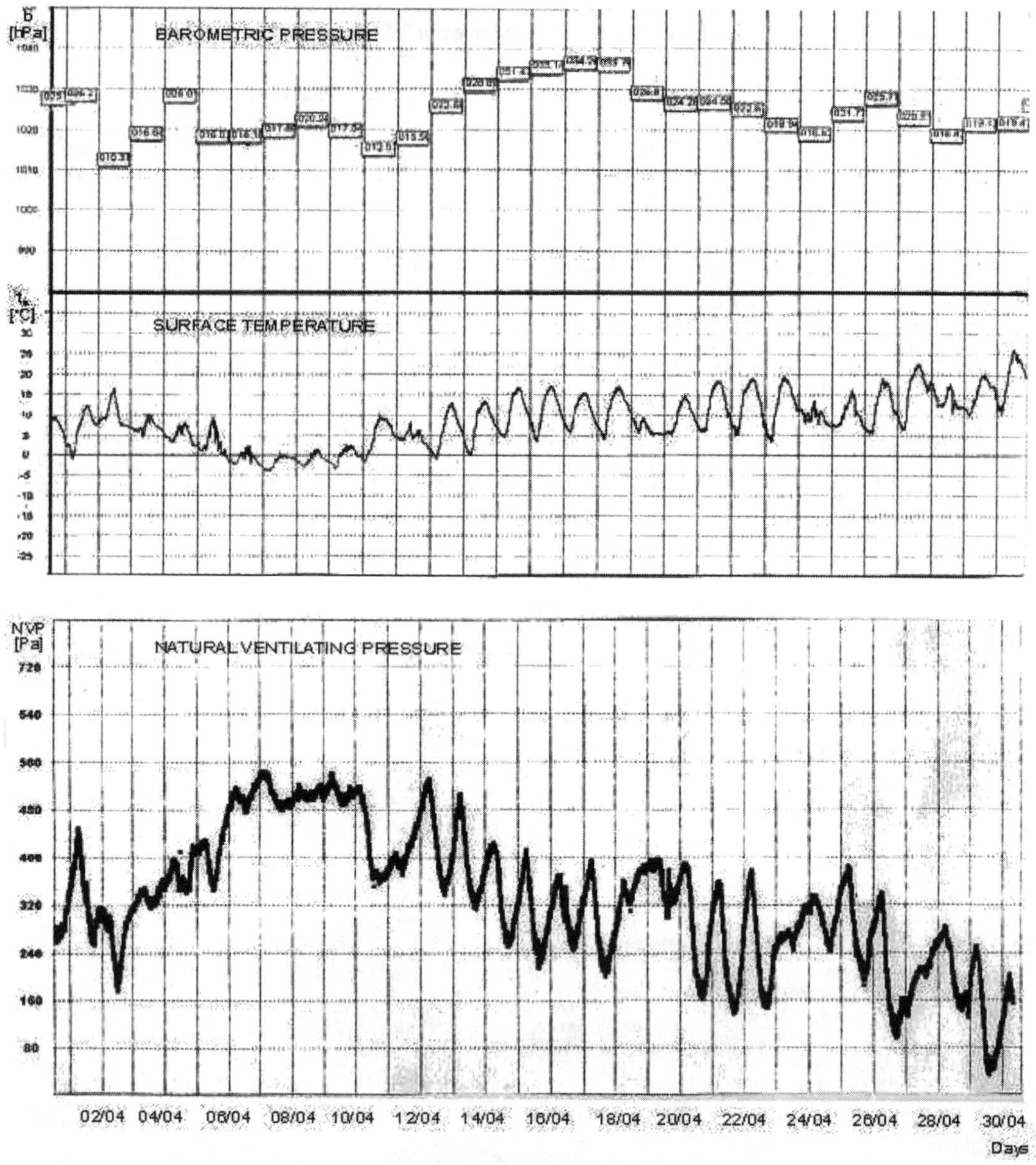

Figure 8. Surface temperature, natural ventilating pressure in a month-period.

Then for barometric pressure $b = P_1$ and dry air, the authors can write in section a_1 that:

$$\rho_1 = \frac{0{,}00348.0{,}5.\left[b + \left(P_2 - \Delta p_{1-2}\right)\right]}{0{,}5.\left(273{,}1 + t_1 + 273{,}1 + t_2\right)} \text{ [kgm}^{-3)}\text{]} \qquad (07)$$

where for equation (04), (05) and (07) t_1 and t_2 = temperatures in points No 1 (surface) and No 2; b = barometric pressure; P_2 = pressure in point No 2; Δp_{1-2} = pressure difference between points No 1 and No 2.

Further the calculation for densities ρ_1 to ρ_{13} in sections a_1 to a_{13} will be led in accordance with equations, see Figure 5.

Therefore, formulas (01), (05) and (06) give the NVP mine action. It means that the amount of NVP depends on continual measurement of temperatures t_1 to t_{13}, barometric pressure b, MF airflow Q_V and pressure difference Δp_{13} from suction part of the main fan.

The control screen shows all monitored data including fan operating point (FOP), characteristic curves of MF and activated characteristic of resistance of colliery with NVP, see Figure 6 and Figure 7. In other words the FOP is determined by MF airflow ($Q_V = 145\,\mathrm{m^3s^{-1}}$) and pressure difference ($\Delta p_{13} = 2138$ Pa), both are measured continually.

The system of operating point (SOP) is identified by equation:

$$Q_V = \sqrt{\frac{1}{R}\left[\Delta p_{13} - (\pm NVP)\right]} \quad [\mathrm{m^3s^{-1}}] \qquad (08)$$

where, R = resistance of colliery ($Q_V = 145\,\mathrm{m^3s^{-1}}$ and $\Delta p_{SOP} = 2138 + 646 = 2784$ Pa).

In the expression of the equation (08) it is apparent that the "V" apex of the activated parabola is – for constituent effect – situated in the minus part of vertical different pressure coordinate and vice versa.

The safety regulations in the Czech Republic make the MF measuring obligatory. The final effect of the measurement is the output in three forms of fan characteristic curves (Kopáček & Taufer 2001a), as follows:

- fan total pressure dependence on the airflow $\Delta p_{fv} = f(QV)$;
- fan static pressure dependence on the airflow $\Delta p_{13} = f(QV)$; fan motor power input dependence on the airflow $P_{fmi} = f(QV)$.

Now the dependence $\Delta p_{13} = f(Q_V)$ becomes the base on control screen MF and NVP ventilation, see Figure 6 and Figure 7.

The set of the Darkov Mine MF regulation curves take the form of equation:

$$\Delta p_{13} = A_0 + A_1 Q_V + A_2 (Q_V)^2 \quad [\mathrm{Pa}] \qquad (09)$$

where A_0, A_1 and A_2 = multinomial coefficients of the second rate. It is important to say that the set of equation (09) is valid for standard density $\rho_0 = 1{,}2\,\mathrm{kgm^{-3}}$. On the contrary, the dry air density in the point No 13 is as follows:

$$\rho_{13} = \frac{0{,}00348.(b - \Delta p_{13})}{273{,}1 + t_{13}} \quad [\mathrm{kgm^{-3}}] \qquad (10)$$

Therefore, it is useful to modify the MF equation (09) into a new form (Kopáček et al 2001b),

$$\Delta p_{13} = \frac{\rho_{13}}{\rho_0}\left[A_0 + A_1 Q_V + A_2 (Q_V)^2\right] \quad [\mathrm{Pa}] \qquad (11)$$

It means that on the basis of monitoring ρ_{13} by equation (10) the MF curves (09) become the so-called "floating", in accordance with the resistance of colliery by equation (08).

4 CONCLUSIONS

The authors have presented one of the methods of continually determined natural ventilating pressure and main fan running. It is important to say that operating results are very hopeful for economic return of investments for future. At present it is possible to assign saved energy of main plant ventilation. There are simple and reliable operations led from control room of Ventilation Manager, see Figure 7. So the Ventilation Manager from his PC gets the time course of surface temperatures, NVP and data of three months backwards, see Figure 8.

ACKNOWLEDGEMENT

This paper has been written as a part of the solution of the Grant project No. 105/03/0589 with a financial support from GA of the Czech Republic.

REFERENCES

Kopáček, F. & Taufer, A. 2001a: One of Methods of Determining the Ventilating Pressure. In *Proceeding of the 7th International Mine Ventilation Congress, Cracow* pg. 29–33.

Kopáček, F. 2001b: Zpracování výpočetního algoritmu pro kontinuální snímání a vyhodnocování přirozeného tlakového spádu na závodě Darkov 2. In *Závěrečná zpráva č. 3/2001, Trolex Ostrava-Radvanice, 19 s.*

Hartman, H.L., Mutmansky, J.M., Ramani, R.V. & Wang, Y.J. 1997: *Mine Ventilation and Air Conditioning. (3rd ed.).* John Wiley & Sons, Inc., New York, pg. 730.

Mc Pherson, M.J. 1993: *Subsurface Ventilation and Environmental Engineering.* Chapman & Hill, London, pg. 905.

Mine Ventilation, Ganguli & Bandopadhyay (eds)
© 2004 C.D. Taylor, ISBN 90 5809 633 5

Use of a test box to measure response times for machine-mounted monitors

C.D. Taylor, J.E. Chilton & J.A. Zimmer
NIOSH/Pittsburgh Research Lab

ABSTRACT: A technique for use in underground coal mines measures the response times of machine-mounted methane monitors. This technique uses a calibration cup to deliver gas to the sensor head to measure response times. For comparison, in the current study, response times were measured by placing methane sensor heads in a specially designed test box where they were exposed to an atmosphere containing a constant concentration of methane. Response time readings measured in the test box were shorter than the readings obtained using the calibration cup. The response times measured in the test box were comparable for sensors from three different manufacturers. The effects of dust cap and flame arrestor design on response times were also evaluated using the test box.

1 INTRODUCTION

During coal mining, methane concentrations are usually highest near the face where the continuous mining machine is extracting coal. To ensure the safety of workers, methanometers are mounted on the mining machine to continuously monitor methane levels. Each type of machine-mounted methane monitor must be tested and approved by MSHA (30 CFR 27.22 through 27.40) before it can be used in a mine. The monitors are designed to alert face workers when methane concentrations reach 1 pct (30 CFR 75.342, (b) (1)). The monitors must also warn workers and cut off electrical power to the machine when concentrations reach 2 pct (30 CFR 75.342 (c) (1)).

Clearly, the methanometer can provide protection only if it accurately records the methane readings. Accurate readings require the instrument to be calibrated and require that it respond quickly to changes in concentration. Calibration procedures are specified by the equipment manufacturer. The mine operator is required to check the methanometer calibration at least once every 31 days (30 CFR 75.342 (a) (4)).

Currently there are no criteria for the measurement of methanometer response times or response time limits. The response time for a methanometer is defined as the time interval for the methanometer output to change from a steady state or zero reading in pure air to a given percent value of the final steady state reading upon application of a calibrated gas. Since the sensor output approaches the final response limit at a steady state, the response time is usually defined as 90 pct of the final output value for the known gas concentration, although lower percentages can also be used to define response time.

To measure response times the instrument sensor head is first exposed to zero gas and then, instantaneously, exposed to the presence of a constant concentration of a calibration gas. The instrument response reading is then measured as a function of time from initial exposure to the gas. Techniques that have been described for making response time measurements with gas sampling instruments are not practical for use in an underground mine Environment (Hughlett 1982, Foster-Miller 1985).

During an earlier NIOSH study to measure response times (Taylor 2002), calibration gas was applied directly to the sensor head through the manufacturer-supplied calibration cup at a stated flow rate. Use of the calibration cup allowed the delivery of a known concentration of methane to the sensor head with minimum interference from the surrounding atmosphere and provided a means for measuring response times on equipment underground. One problem with this technique is that the installation of the calibration cup over the dust cap alters the normal airflow around and through the sensor head. Therefore, a technique for measuring response time that does not require use of the calibration cup was desirable for comparison.

This report describes a laboratory technique for measuring response time in a test box that does not require the use of a calibration cup. In the test box, response times were measured by exposing methane sensor heads supplied by three different manufacturers to a constant methane concentration. Response time measurements obtained in the test box and with the calibration cups are compared.

2 TEST PROCEDURE

2.1 *Methanometers*

Methanometers obtained from three different manufacturers, which were also used during an earlier study, are identified as "Monitor A," "Monitor C," and "Monitor G." Each of the monitors is approved by MSHA (CFR 30, part 27) for use on mining equipment and included a visual readout unit and power supply provided by the manufacturer. Three identical sensor heads were obtained from each manufacturer.

Each manufacturer's sensor head design was different but the basic components included a dust cap, a porous flame arrester that covers an internal chamber, and a sensor element. The sensor head for each type of monitor used a catalytic heat of combustion sensor to measure methane. Figure 1 shows one sensor head obtained from each of the three manufacturers.

2.2 *Test box*

A 14- × 14- × 14- inch plywood box was built to contain a methane-air mixture (see Figure 2). A bracket mounted near the center of the box held a single sensor head during testing. The box had a removable top that allowed the placement and removal of the particular sensor head being tested. A foam seal on the top helped to reduce leakage of gas from the box. Methane was injected and air samples withdrawn through plastic hoses that were inserted through holes on one side of the box. Electrical cables to power the sensor heads and transmit data also passed through holes in the opposite side of the box. Two 4-inch fans attached to the bottom interior of the box were used to mix the injected methane gas with box air.

2.3 *Gas delivery and mixing*

The three sensor heads were individually exposed to gas-air mixtures in the test box. Methane (99 pct) was injected into the box using a plastic syringe Super Syringe Hamilton Co Reno NV that had markings to indicate volumes from 0 to 1500 cc. The gas was transferred from a cylinder to the syringe (see Figure 3). The methane flowed through the plastic tubing into the water-filled container for approximately 30 seconds prior to the syringe being filled. The end of the syringe was capped until the gas was injected into the box. Approximately 4 seconds were required to manually depress the plunger and transfer the gas from the syringe to the box.

Figure 1. Sensor heads for monitors A, C and G.

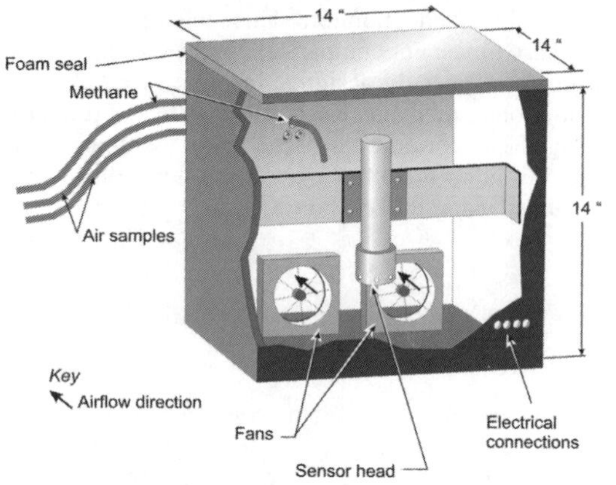

Figure 2. Test box.

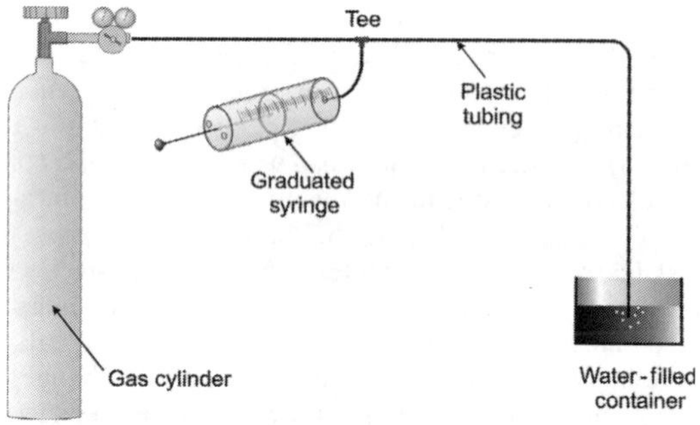

Figure 3. Apparatus for filling syringe.

The quantities of methane gas injected into the box were varied (300 to 1200) and the average concentrations in the box measured with a Horiba Model PIR-2000 infrared gas analyzer. Figure 4 shows the relationship between the quantity of methane gas injected into the box and the final average concentration. For most tests, 1000 cc of gas were injected into the box resulting in concentrations between 2.0 and 2.5 pct methane.

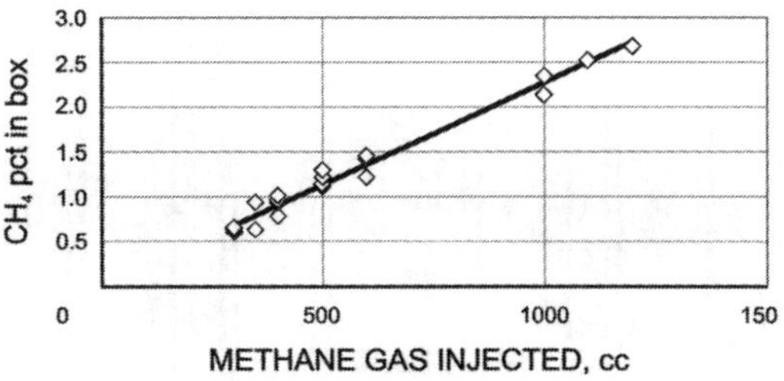

Figure 4. Relationship between gas injected and concentration in box.

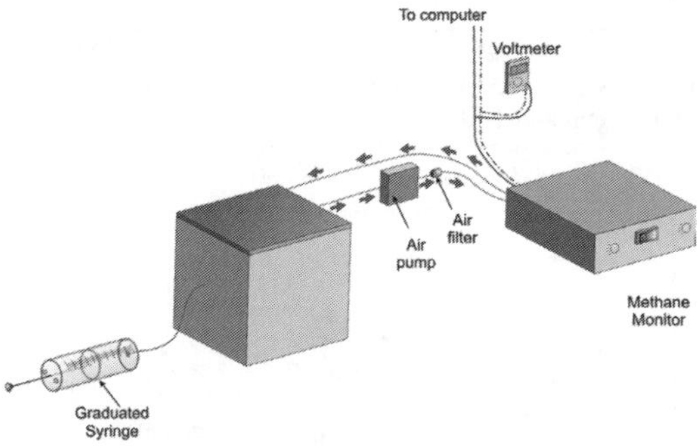

Figure 5. Test setup with Horiba monitor.

The Horiba Model PIR-2000 infrared gas analyzer, which has a very fast response rate for methane gas, was also used to measure how quickly concentration increased in the box following gas injection and how long that concentration was maintained. Air samples from the box were transferred to the Horiba analyzer gas cell using a MSA Flowlite air pump (see Figure 5). The pump was operated at 1.5 lpm and has inlet and outlet ports that allow the direct transfer of gas between the box and Horiba. The tubing between the box and Horiba was as short as practical to minimize transport time. The output signal from the Horiba, which is proportional to the methane concentration, was recorded by a computer-based data acquisition system.

Figure 6 shows the response time measurement obtained with the Horiba monitor for gas injected into the test box at time zero. Concentrations reached 90 pct of their maximum value in less than 9 seconds and 99 pct of final value in less than 10 seconds. The 4-second delay between gas injection and and any signal response was probably due to the time required for the gas to travel from the box to the Horiba sensor. Assuming this delay in response was due to gas travel time, concentrations in the Horiba monitor reached 99 pct of their final level less than 6 seconds after the gas was injected. Methane levels in the box decreased about 2.5 pct per minute after reaching their maximum level due to leakage from the test box.

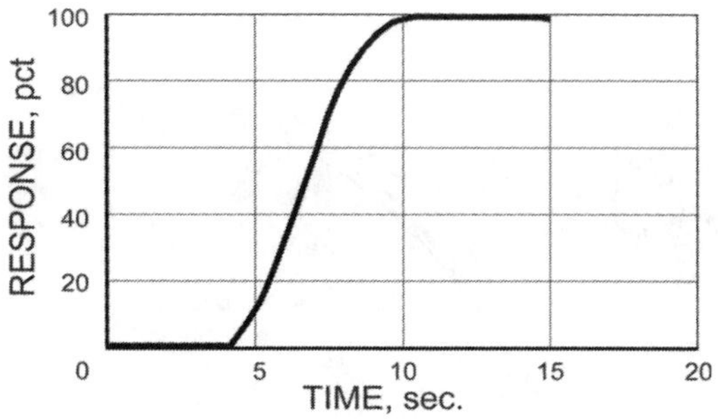

Figure 6. Response curve measured with Horiba monitor.

2.4 *Data acquisition*

Before each series of tests with a sensor head, instrument calibration was checked using 1 pct calibration gas applied by the manufacturer-supplied calibration cups. To measure response time the sensor head was placed in the bracket, the lid installed, and the fans started. After confirming a zero reading on the monitor readout, the gas was injected. A large display stopwatch was started when the gas was injected. Two observers independently recorded the methane concentrations at 5-second intervals. Data were collected for 2 to 3 minutes after injection of the gas and each test was repeated one time. The response time for each sensor head was based on the average of concentrations measured by both observers during the two tests. Readings for the three sensors from each manufacturer were averaged to determine response times.

3 RESULTS

3.1 *Response time readings*

The response curves (see Figure 7) for each type of methanometer monitor were drawn by plotting the elapsed time versus the response percentages. The response percentages were calculated by dividing the average methane concentration for each 5-second sampling interval by the maximum methane concentration measured for that test. The response curves shown in Figure 7 are based on the average responses for all tests with the three identical sensor heads. The 90, 80, and 40 pct response times, which were determined from the response curves, are given in Table 1. Table 1 also includes the response time Measurements made earlier using the calibration cup. For these tests 2.5 pct calibration gas was used.

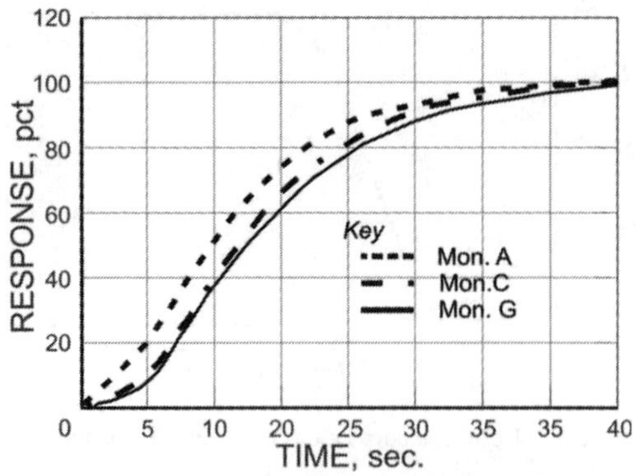

Figure 7. Response curves for three monitors.

Table 1. Response time, calibration cup, and box.

	Monitor A		Monitor C		Monitor G	
Resp. time (%)	Cal. cup (sec)	Box (sec)	Cal. cup (sec)	Box (sec)	Cal. cup (sec)	Box (sec)
90	29	23	40	25	36	29
80	23	17	30	20	28	21
40	13	8	14	10	14	10

The 80 and 40 pct response times correspond to 2 and 1 pct methane readings when using 2.5 pct gas.

4 DISCUSSION

4.1 *Calibration cup*

For each type of monitor tested, response times were shorter when measured in the box. Part of the differences in response times can be attributed to how fast the gas passes through the sensor head. Flow rates recommended by each manufacturer were used for the calibration cup tests. Gas must travel through the dust cap and flame arrestor to reach the catalytic head of combustion sensor whether the measurements are made with the calibration cup or in the box. However, the airflow paths are different depending on whether a calibration cup is or is not installed over the dust cap.

Figure 8 shows a cross-sectional view of each sensor head with the calibration cup installed. The arrows show the likely path the calibration gas takes as it enters the sensor head from the calibration cup, travels through the sensor head to the sensor elements, and finally exits the sensor head. Installing the calibration cup on the dust cap forces the air to enter the holes in the top of the cap and exit through holes in the side.

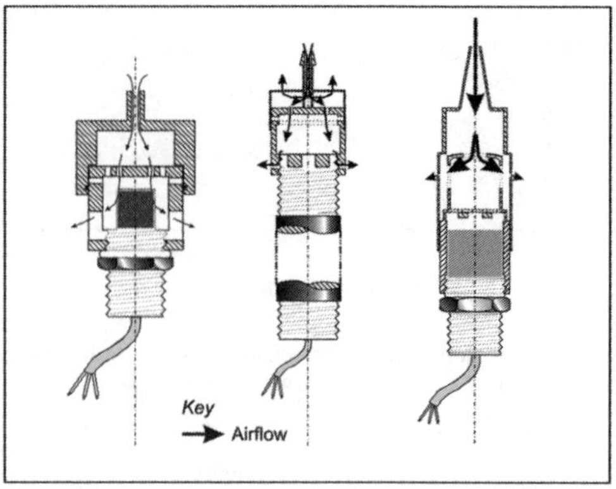

Figure 8. Sensor heads (monitors A, C, and G) with calibration cup attached.

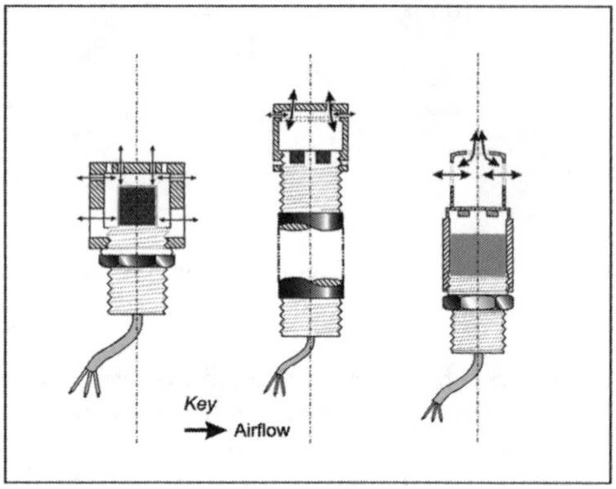

Figure 9. Sensor heads (monitors A, C, and G) with calibration cup removed.

Figure 9 shows the same sensor heads with the calibration cups removed from the dust caps. Without the cup, air can enter or exit through holes in the sides and top of the dust cap. The improved exchange of air provided by calibration cup removal was probably a factor in the reduced response times.

4.2 *Dust cap*

Response time measurements were made in the box, as described above, except that the dust caps were removed. Dust caps must be installed on the sensor head when operating the machine-mounted methanometers underground. Comparative measurements were made without the dust cap to estimate how much of the response time is due to gas flow through the cap. The 90, 80, and 40 pct response times for each monitor, with and without the dust caps in place, are

Table 2. Response time – cap on and cap off.

	Monitor A		Monitor C		Monitor G	
Resp. time (%)	Cap on (sec)	Cap off (sec)	Cap on (sec)	Cap off (sec)	Cap on (sec)	Cap off (sec)
90	23	12	25	4	29	14
80	17	9	20	3	21	10
40	8	4	10	2	10	4

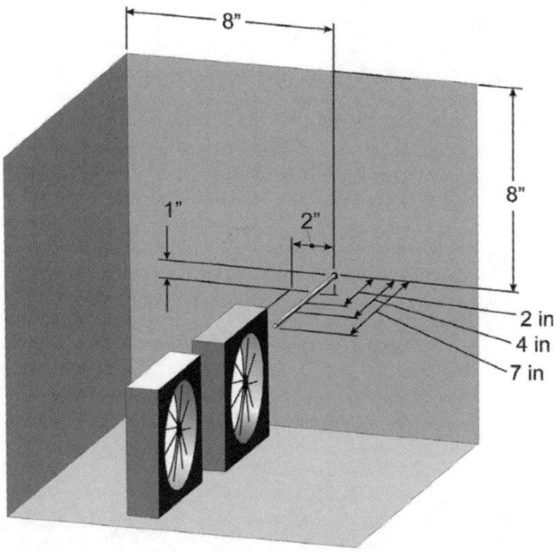

Figure 10. Flow measurement in test box.

given in Table 2. For all response times a large part of the total response time can be attributed to the time required for the gas to flow through the dust cap.

4.3 *Ambient air velocity*

The amount of air moving in and out of the sensor head is affected by the ambient air velocity and orientation of the sensor head relative to the flow. Flow conditions in the box were measured at three locations (see Figure 10) using a hot wire anemometer. Flows measured in the vertical direction varied from 35 to 45 fpm and from 65 to 150 fpm measured in the horizontal direction. These were within the range of conditions that would be encountered near the head of an operating mining machine.

4.4 *Flame arrestor*

The 90 pct response times for the three monitors varied from 4 to 14 seconds when the dust cap was removed. The differences in time are likely due to how quickly the gas traveled through the flame arrestor and reached the sensor elements. The time for the instrument to electronically record voltage changes due to the heating of the sensing element would be similar for each instrument.

The flame arrestors used with monitors A, C, and G are made either of porous metal or a fine mesh screen. The sensor elements are located close to the flame arrestors. Monitor A has a cylindrically shaped porous metal filter that surrounds the sensor element. In Monitor C, two small round porous metal plugs are located in a plate above the sensor elements. In Monitor G, a screen is located above the sensor element.

5 CONCLUSIONS

Response time readings were made by placing methane monitors inside a 14-inch square fabricated test box and exposing them to an atmosphere containing methane. These readings were compared to response times measured earlier by applying methane directly through the calibration cup to the sensor head. Use of the calibration cup was designed for underground use where the technique must be simple and require minimal equipment. The use of the calibration cup is a satisfactory method for making response time readings underground. Obtaining this data periodically will also help to indicate if instrument performance has deteriorated and whether there is need for repairing or replacing the instrumentation.

Use of the test box provided a way to expose the sensor heads to gas without using the calibration cup, and also provided conditions that closely simulate underground airflow conditions on a mining machine where a methane monitor is mounted. The response time readings made in the box were approximately equivalent for all three monitors and were faster than the response measurements made with the calibration cups.

The effects of dust cap and flame arrester design on response time were also examined. It was found that shortening response time is one way to improve instrument performance, and response time is affected by the design of the dust cap and the flame arrestor. Shortening the flow path through the dust cap and decreasing the porosity of the flame arrestor might shorten response time. However, it must be kept in mind that the dust cap helps prevent water spray and dust from contacting the sensor element and the flame arrestor prevents the propagation of any flame outside the sensor head. Therefore, no changes should be made to the dust cap or flame arrestor that would lessen the protection of the sensor element.

REFERENCES

1. Emmett Hughlett R. & George H. Schnakenberg, Jr., Statistical Approach to the analysis of Bias and Precision in Reusable Gas-Detecting Devices, BuMines Rpt of Investigation 8715, 1982, pp. 27.
2. Improved Diffuser and Sprayfan Systems for Ventilation of Coal Mine Working Faces, Foster-Miller, Inc., BuMines OFR 18–86, Jul. 1985, pp. 190–210.
3. Taylor, C.D., Chilton, J.E. & Mal, T. 2002, Evaluating Performance Characteristics of Machine-Mounted Methane Monitors by Measuring Response Time, Proc. 9th US Mine Vent Symp, pp. 315–321.

Mine Ventilation, Ganguli & Bandopadhyay (eds)

The early detection of spontaneous combustion and the evaluation of the effectiveness of fire extinguishing countermeasures based on CO time series data

Y. Tominaga
Hokkaido University, Sapporo, Japan

S. Bandopadhyay
University of Alaska Fairbanks, Fairbanks, USA

ABSTRACT: To detect the occurrence of an unusual state in a ventilation network, the extremums on the wave shape depicting time series data of a moving average concentration of CO were specified. As minimum value and maximum value on the graph appear in turn, the increase from minimum to maximum can be assumed to be due to mining activity. The occurrence frequency of the local peak value showed linear correlation when plotted on a semi-logarithmic scale with a correlation coefficient value of 0.99. This suggests that the occurrence probability of a local peak value (LPV) follows an exponential distribution. The characteristic value of the distribution is estimated from the inclination of a LPV – ln(1 − P(LPV)) diagram where P(LPV) is the occurrence probability of random variable from 0 to LPV. It is believed that the LPV can be used to monitor and detect abnormal condition occurrence even if the concentration of CO is in the three-sigma range of the trend curve. Moreover, it is believed that the effectiveness of a fire extinguishing countermeasure can be determined by using time series data from different censor positions. To evaluate this approach, an experimental formula curve composed of both power series and trigonometric functions of elapsing time was estimated by using measured carbon monoxide data from fire extinguishing situations.

1 INTRODUCTION

The signs of spontaneous combustion were indicated in the measuring chart of carbon monoxide in the Taiheiyo coalmine, located in Hokkaido, Japan. Concentration of carbon monoxide exceeded the three-sigma (standard deviation) range estimated by the difference between the measured value and the tendency curve. As the starting point of an unusual increasing concentration of CO is often located within the three-sigma range, it is essential to detect the occurrence of spontaneous combustion as soon as possible.

To have early detection of the occurrence of an unusual state, the extremums of the wave shape given by a moving average concentration data of CO was determined under normal conditions. As minimum value and maximum value on the graph appear in turn, the increase from minimum to maximum can be assumed to be due to mining activity. The occurrence frequency of a local peak value, the increase from a minimum to maximum CO concentration showed a linear correlation when plotted semi-logarithmically and a correlation coefficient value of 0.99. This relation means that the occurrence probability of the local peak value (LPV) follows an exponential distribution, characterized by the parameter lambda. The value of lambda is estimated from the inclination of an LPV – ln(1−P(LPV)) diagram where P(LPV) is the occurrence probability of a random variable from 0 to LPV.

It is concluded that the amount of CO increase from the minimum normal level can be used to monitor and detect abnormal condition occurrence, even if the concentration of CO is in the three-sigma range of the trend curve.

If the occurrence of spontaneous combustion is detected, many kinds of countermeasure can be taken. It is important to know whether a given countermeasure is effective. To evaluate the countermeasure, a concentration curve composed of both the power series and trigonometric functions of elapsing time were estimated using measured carbon monoxide data. The number of term in the polynomial for the concentration curve was increased until the value of the correlation ratio between the curve and moving average concentration data exceeded 0.9. This curve was then available to estimate the effectiveness of countermeasures using measured carbon monoxide concentration for two weeks.

2 CONCENTRATION DATA COLLECTION FOR CARBON MONOXIDE

Figure 1 shows a schematic network of the mine. In the figure, the bold line represents a mine working with a face length of 250 m, and a mining height of 2 m. The three black triangles in Figure 1, are the three CO sensor locations, A, B and C in the air flow direction.

Figure 2 shows a time series data of the carbon monoxide concentrations (in ppm) collected over a period of three-weeks at a sensor location A.

3 TREND CURVE OF CARBON MONOXIDE CONCENTRATION

Since the time series data of carbon monoxide can be impacted by uncontrollable events such as methane explosion, air quantity changes due primarily to the mining activity, short circuiting of the air due to opening and closing of ventilation doors, etc. a moving average on the time series data was used in this study.

A time interval of 30 seconds was used in collecting concentration values. Time intervals of a half-hour were used for calculating the moving average for the estimated trend curve.

A power-series trend curve of elapsed time introduced by Tominaga and Bandopadhyay (2002a) is shown in Equation 1.

$$C(t) = \sum_{k=0}^{n} a_k \cdot t^k \tag{1}$$

where, C, t and a_k ($k = 0, 1, 2 \ldots..$) are concentration in ppm, elapsed time (day/day: for non-dimensional procedure) and coefficients, respectively.

The middle curve depicted in figure 3 is the trend curve given by equation 1. The difference (ΔC) between the moving average and the trend curve showed a linear correlation when plotted on a normal probability paper and a correlation coefficient value of 0.99.

The upper and the lower curves show the three-sigma value. In a normal distribution, 99.7 percent of all data are included within the three-sigma range.

From the figure, it can be concluded that the trend curve can be used as a control chart of the ventilation control in a mine. At any point in time, if the three-sigma value exceeded by the corresponding moving average values in the next stage, then the CO level would indicate abnormal ventilation conditions.

4 THE LOCAL PEAK VALUE IN A CHART

When spontaneous combustion occurs, it is necessary to detect it as soon as possible. The extremums on the wave shape of time series data given by the moving

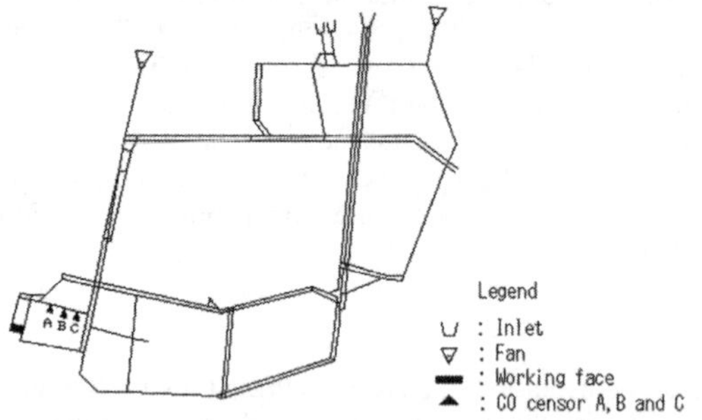

Figure 1. A schematic of mine development network in Taiheiyo coal mine.

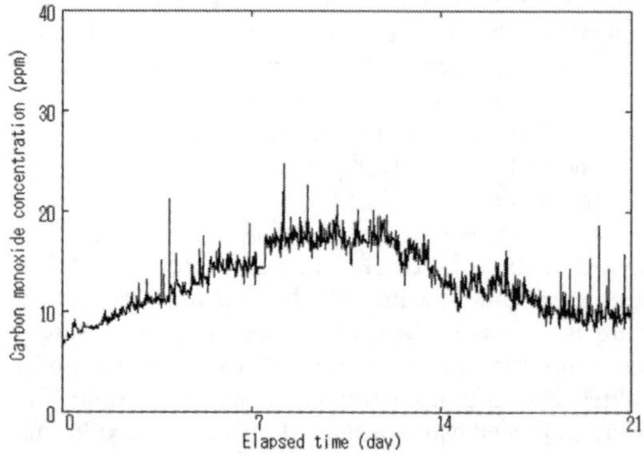

Figure 2. Carbon monoxide concentrations in Taiheiyo coal mine.

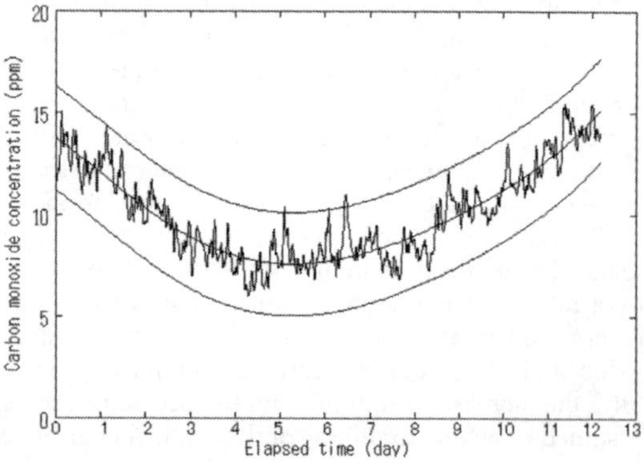

Figure 3. A power-series trend curve of elapsed time and the three-sigma range.

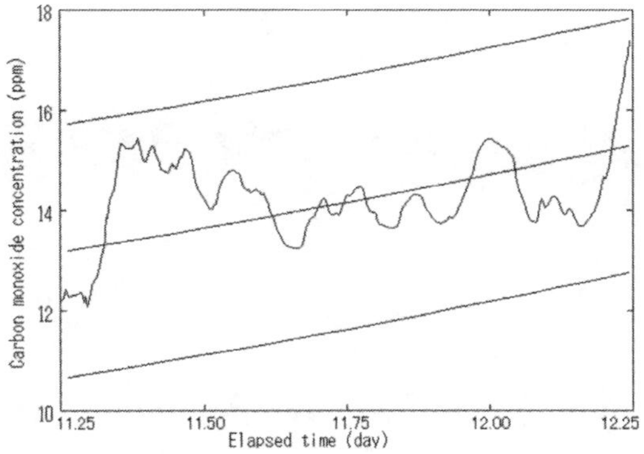

Figure 4. A typical example for local peaks in the moving average of carbon monoxide concentrations.

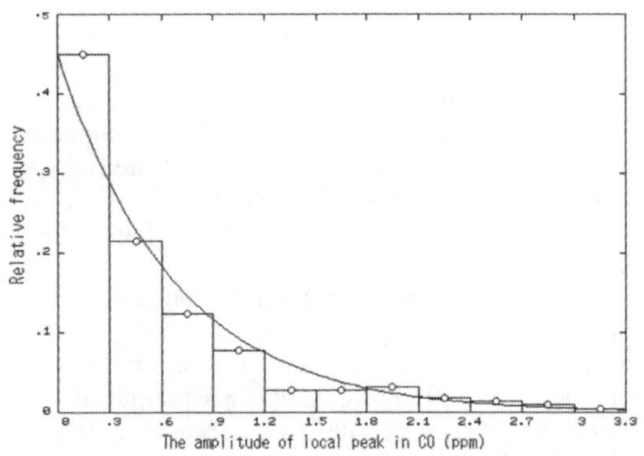

Figure 5. Histogram for the amplitude of local peak value.

average concentration of CO were specified in this study.

The expansion figure of the time axis for a 24-hour period from 6 o'clock on the11th day indicated on figure 3 is shown in figure 4. As before, minimum value and maximum value on the graph appear in turn, with the increase from minimum to maximum assumed to be due to mining activity.

The occurrence frequency of the local peak value was counted in the time series data shown in figure 3.

Figure 5 shows a histogram for the amplitude of local peak value. As the LPV has a continuous value x, the probability of occurrence x is estimated by a probability density function. Assuming that the occurrence of x follows an exponential distribution, the probability of occurrence between 0 and x is given in Equation 2.

$$\begin{aligned} P(x) &= \int_0^x \lambda \cdot \exp(-\lambda \cdot x)dx \\ &= 1 - \exp(-\lambda \cdot x) \end{aligned} \tag{2}$$

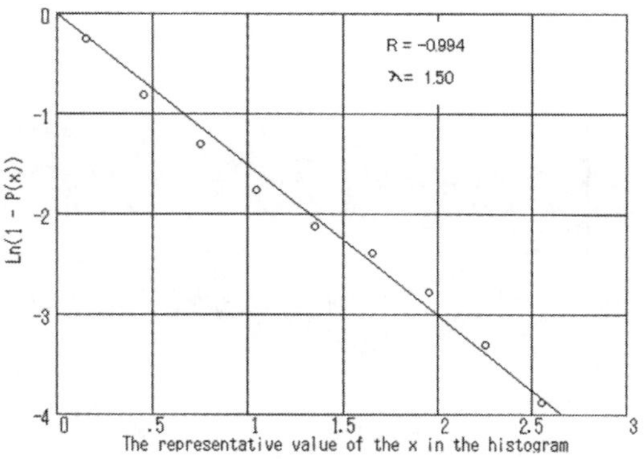

Figure 6. Histogram for the amplitude of local peak value.

Since the probability of occurrence from x to infinity is $1 - P(x)$, the following equation is derived from Equation 2.

$$\ln(1 - P(x)) = -\lambda \cdot x \tag{3}$$

The value of $\ln(1 - P(x))$ at the mid-point of each class range of histogram given in figure 5 are shown in figure 6.

The straight line in figure 6 is fitted to estimate λ from equation 3.

Figure 6 shows a linear correlation when plotted on a semi-logarithmic scale with a correlation coefficient value of 0.994. This relation means that the occurrence probability of the LPV follows an exponential distribution characterized by a one-parameter λ. The value of λ is estimated as 1.50 from the inclination of a straight line in figure 6.

Because the curve line in Figure 5 almost conforms to the distribution of the histogram, the occurrence probability of the LPV is considered to be a good fit to an exponential distribution.

5 AN APPLICATION EXAMPLE

Unusual points haven't been recognized on the control chart shown in figures 3 and 4. However, CO concentration shown in figure 7 exceeded the range of 3σ of the trend curve.

Figure 7 shows the unusual occurrence and an indicator for assessment of the occurrence probability of the LPV. Though there are not many differences between the 3σ range and the measured value, it is strongly believed that the last local peak of the figure 7 is an unusual point.

Figure 7 shows that abnormal situations may be able to be detected by comparing the occurrence of LPVs, even within the three-sigma range of control that is judged normal.

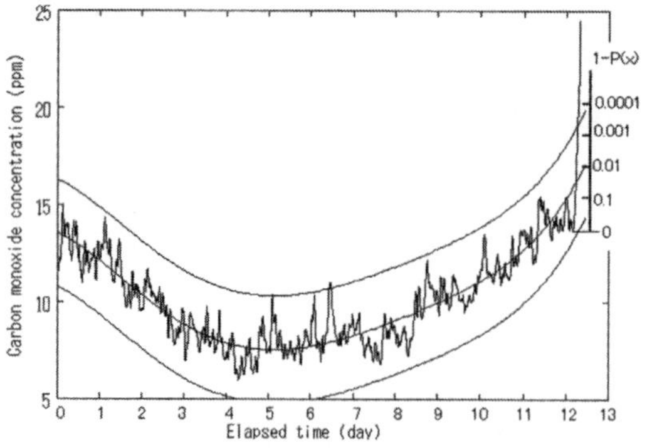

Figure 7. An example of the unusual value of local peaks in the moving average of CO and the occurrence probability.

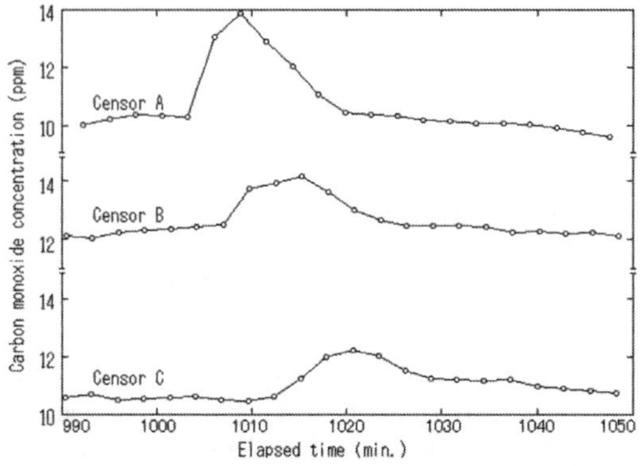

Figure 8. A typical example of time series data measured at the three CO sensors positions located at A, B and C in the air flow direction.

Because the LPV is not effective at the upper range of a control chart, it can be concluded that the combination of LPV-occurrence probability and a control chart with three-sigma can be used to monitor abnormal situations effectively at an early stage.

6 EVALUATION OF THE COUNTERMEASURE

The injection of water to roofs supported by self-advancing roof supports, pumping water into bore holes for gas-drainage in a gob, and the closing of intake/outlet drift are being carried out at the Taiheiyo Coal Mine. These procedures are used as the temperature inside the roof/gob and the concentration of the carbon monoxide dictate.

The reliability of the measurement value must be obtained when the control chart is used for evaluating the effectiveness of the countermeasure used.

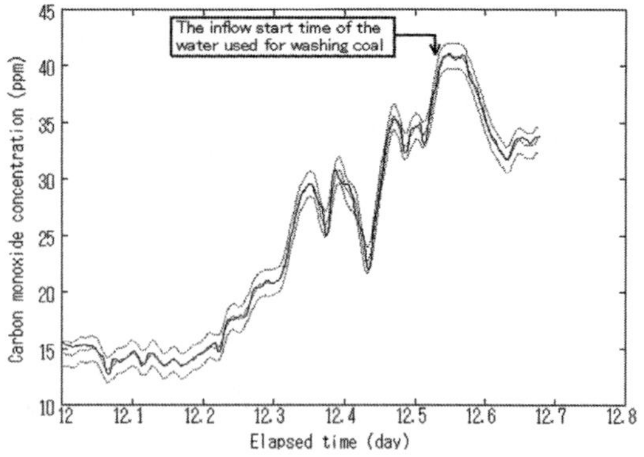

Figure 9. A typical example of time series data using moving average over a fifteen-minute interval.

Tominaga and Bandopadhyay (2002b) reported that there was correlation between the value of a sensor and its relative position along the air current.

Figure 8 shows a typical example of time series data measured at the three CO sensors located at positions A, B and C in figure 1. The reliability of the sensor can be confirmed from the relative values of these sensors.

When time series data is used for evaluation of the effectiveness of a countermeasure, it is necessary that the moving average be found over a smaller time interval. Figure 9 shows a typical example of time series data using moving average over a fifteen-minute interval.

The middle curve depicted in figure 9 is the experimental formula shown in equation 4 for fitting the time series data.

$$C(t) = \sum_{i=0}^{l} a_i \cdot t^i + \sum_{j=1}^{m} b_j \cdot \cos(j \cdot \pi \cdot t) + \sum_{k=1}^{n} c_k \cdot \sin(k \cdot \pi \cdot t) \quad (4)$$

where, C, t, a_i, b_j and c_k are concentrations in ppm, elapsed time (day/day: for non-dimensional procedure) and coefficients in ppm, respectively,

The maximum values of l, m, and n in equation 4 were determined to obtain a correlation coefficient value of 0.999.

The upper and the lower curves in figure 9 show the range of three-sigma calculated difference between the time series data and the experimental formula using an algorithm reported by Tominaga and Bandopadhyay (2002a). Thus, the algorithm predicts that effectiveness of a countermeasure can be judged from the curve of the experimental formula shown in figure 9.

7 CONCLUSIONS

It was explained that by using the time series data of CO that a trend curve and local peak value were effective in the monitoring of the mine ventilation environment.

As soon as it is possible, a countermeasure at an early stage of operation is necessary, and it is possible to use data on the continuous CO measurement to effectively judge the effectiveness of the countermeasures in case of spontaneous combustion. When the reliability of the measuring data was secured, it was shown that an experimental formula fitted from the time series data was available for the evaluation of effect on execution of the countermeasure.

ACKNOWLEDGEMENT

The authors would like to thank the Taiheiyo Coal Mine for permission to use the carbon monoxide concentration data used in this paper.

REFERENCES

Y. Tominaga and S. Bandopadhyay, 2002(a), Ventilation Safety Monitoring System Using a Trend Curve of Moving Average of Carbon Monoxide Concentration, APCOM 2002-30th Application of Computer and Operations Research in the Mineral Industry, 669–676.

Y. Tominaga and S. Bandopadhyay, 2002(b), Monitoring of spontaneous combustion based on time series data, North American/9th US mine ventilation symposium, Kingston, Canada, 335–339.

Mine fires and explosions

Mine Ventilation, Ganguli & Bandopadhyay (eds)
© 2004 Taylor & Francis Group, London, ISBN 90 5809 633 5

Use of an inertisation unit during a coal mine fire recovery operation

John E. Urosek
Chief, Ventilation Division, Mine Safety and Health Administration

Richard T. Stoltz & Dennis A. Beiter
Supervisory Mining Engineer, Mine Safety and Health Administration

George N. Aul & Thomas A. Morley
Mining Engineer, Mine Safety and Health Administration

ABSTRACT: On February 13, 2003, a fire occurred in a large underground coal mine located in West Virginia. Firefighting activities were unsuccessful. The mine was sealed at the surface openings on February 14, with the intent of inerting the mine atmosphere naturally. However, the mine operator proposed a novel approach for controlling the mine atmosphere. The Mine Safety and Health Administration worked closely with the mine operator on a plan to recover the mine. Queensland Mine Rescue Service Limited was consulted and contracted to provide technology, equipment and personnel to operate an Inertisation Unit. The primary component of the unit was a jet engine. The engine was capable of producing large quantities of inert gases and forcing the gases into the mine. The unit was installed on the surface at the slope and the injection of inert gases was started on April 4, 2003. After the mine atmosphere was successfully stabilized to a non-explosive condition, re-ventilation of the remainder of the mine was initiated. Mine rescue teams entered the mine and began exploration toward the fire area. The progress of the exploration work was hampered by roof falls, water, and excessive heat in the vicinity of the fire. The fire area could not be adequately accessed and a decision was made to permanently seal this area from the rest of the mine. Operation of the Inertisation Unit was discontinued after the fire area was partially isolated and a safe and stable atmosphere could be maintained. Eventually, construction of explosion resistant seals was completed around the fire area and rehabilitation of the remainder of the mine began. This paper will discuss the recovery operation with a focus on the use of the Inertisation Unit.

1 INTRODUCTION

A fire occurred within several crosscuts of the slope bottom of a large underground coal mine located in West Virginia. Direct firefighting activities were abandoned when elevated methane and carbon monoxide concentrations were detected near the fire. Subsequently, the mine was sealed at the surface openings. Water was pumped into the mine through the slope to cool the fire. Sample tubes were installed through the seals at each of the mine openings to monitor the atmosphere in the sealed mine. Figure 1 depicts the general layout of the mine and the locations of the shafts and slope openings. Early in the operation, a Command Center, which included members of mine management, the federal and state mine safety and health organizations, and labor representatives, was established. Formal procedures were followed to approve, direct and oversee the work activities during the operation.

Sealing of the mine was expected to result in consumption of the oxygen by the fire in the affected area, resulting in an inert atmosphere that could eventually extinguish the fire. Liberation of methane in a sealed mine can also assist the inerting process, however, methane can also create hazardous conditions. A mine fire is only considered extinguished after the fire area has cooled sufficiently to prevent re-ignition upon re-ventilation of the affected area. Generally, this process requires maintaining an inert atmosphere to prevent continued combustion for a lengthy period of time, often several months or more, thus allowing the fire area to cool sufficiently.

During this operation, monitoring of the mine atmosphere indicated the oxygen concentration in and near the fire area remained elevated. Further, much of the atmosphere in the remainder of the mine was entering into explosive methane–air mixtures and concerns arose about the potential for ignition sources other

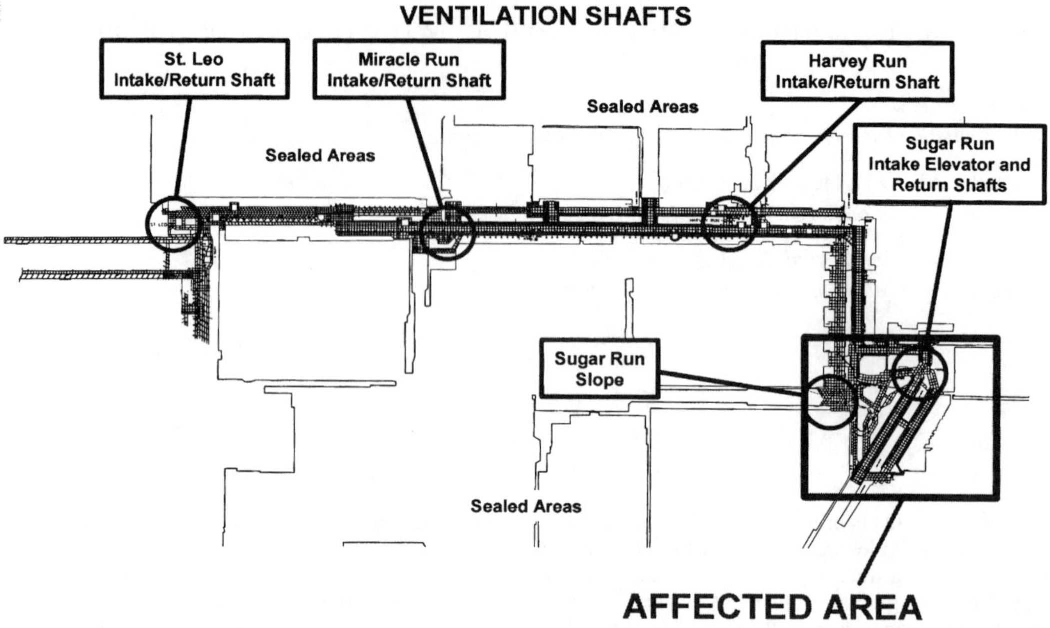

Figure 1. General mine layout.

than the fire to cause an underground explosion. Natural ventilation pressures continued to draw air into the mine through the shaft and slope seals adjacent to the fire area and move it throughout the mine. Attempts to sufficiently reduce the leakage through the seals had only limited success. Alternative approaches to reduce the oxygen concentrations in the fire area and decrease the amount of time the mine needed to be sealed were considered.

Mine management made a decision to use an Inertisation Unit (Unit) to force large volumes of mostly inert gases throughout the entire mine. The Unit was owned and operated by Queensland Mine Rescue Service Limited, located in Queensland, Australia. The Unit consisted of a jet engine with modified exhaust that was capable of continuously producing large volumes of inert gases and water vapor. The gas mixture of the exhaust consisted primarily of nitrogen and carbon dioxide. Oxygen concentrations of 2 percent or less were anticipated. This technology was initially developed in Poland in the early 1970s for fighting mine fires. The procedure was later adopted and used in Australia. Prior to this incident, the Unit had not been used for mine recovery in the United States.

The conditions and circumstances surrounding previous applications of the Unit were significantly different than the proposed undertaking. Some officials were initially optimistic that the Unit could be used to extinguish the fire. However, usage of the Unit resulted in stabilization and control of the atmosphere in the fire area that enabled the fire to be addressed by other means.

2 INITIAL OPERATIONS AND SAMPLING

Air that entered the mine near the fire area permitted continued combustion and created difficulties with monitoring the mine atmosphere at the surface openings closest to the fire. Additional information was desired to determine the extent and more accurately evaluate the state of the fire. Vertical boreholes were drilled into the mine at locations surrounding the last known extent of the fire. The boreholes were used for injecting water, inserting a borehole camera and temperature probe into the mine to determine the extent of the fire, and sampling the underground atmosphere in close proximity to the fire area. Figure 2 shows a detailed layout of the slope bottom area, the location of the initial fire area, and the locations of the six vertical boreholes.

Road and site construction for drilling the boreholes was started on February 14. Survey traverses were conducted to locate the borehole sites on the surface. The boreholes were drilled 29.2 cm (11.5 inches) in diameter with a 20.3 cm (8 inch) diameter casing installed to a depth near the mine roof. The depths of the boreholes varied from about 250 m (820 feet) to 274 m (900 feet).

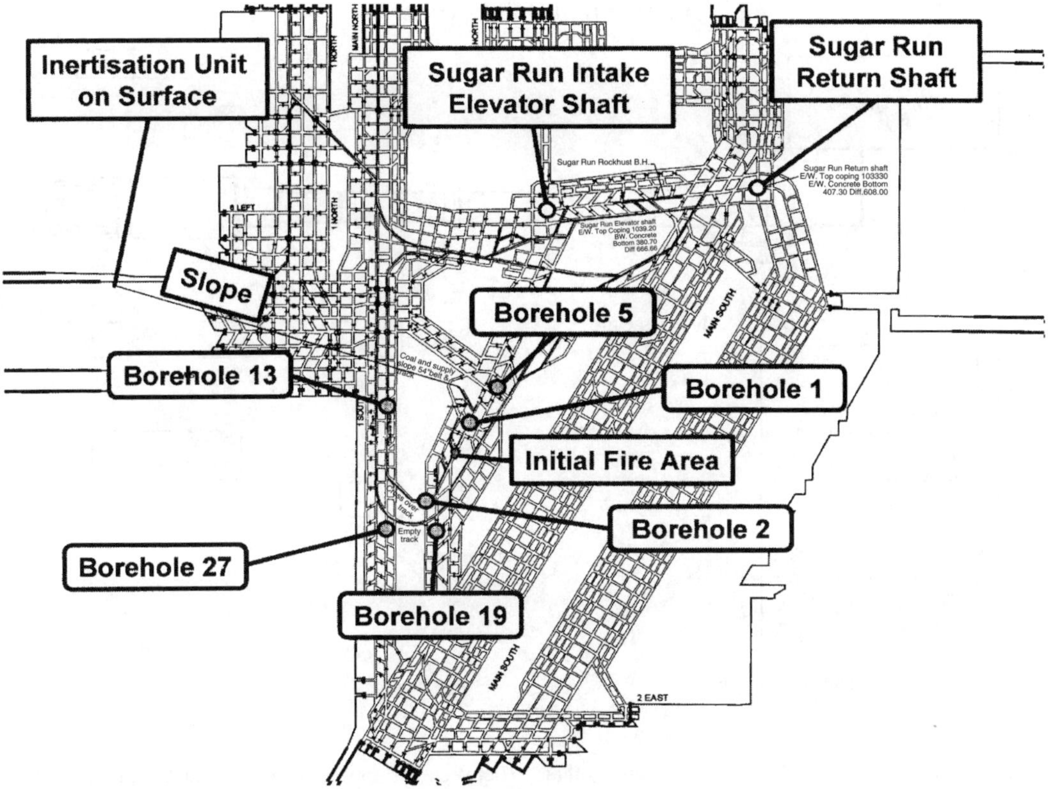

Figure 2. Affected area and borehole locations.

Grouting of the casings minimized the migration of strata gases into the sampling location. All of the boreholes were completed by March 13.

A mine atmosphere sampling and monitoring program was established to evaluate the atmosphere in the mine and the state of the fire. Samples were collected at regular intervals and analyzed by gas chromatography. Continuous monitoring of critical locations was conducted using infrared and electro-chemical instruments. Engineers evaluated the air sample analyses and provided interpretations of the results to Command Center personnel. The information was critical for controlling and directing work activities on the surface near the mine openings and underground. Information gained from the results of the mine air samples was also used to evaluate ventilation patterns in the mine.

A sampling line was inserted into each cased borehole. All sampling lines were extended into mine entries. Each sampling line consisted of a 2.5 cm (1 inch) diameter metal pipe that was lowered into the borehole to the mine floor. The bottom of the pipe was plugged. Holes were drilled through the lower 1.2 m to 1.5 m (4 feet to 5 feet) of the pipe to obtain a representative sample of the atmosphere in the mine opening at the bottom of each borehole. Vacuum sampling pumps were used to draw samples to a central location for collection. The sampling lines were designed to ensure that air samples collected were representative of the mine atmosphere and not contaminated with gases liberated from overlying strata in the ungrouted bottom portion of the boreholes. As the recovery operation progressed, the need to more accurately determine the mine atmosphere near the fire area required modifications to the sampling system. Roof falls and differing mine opening heights at certain sampling locations necessitated the sample lines be customized to address the specific conditions. Some sample lines had to be withdrawn and re-inserted in a trial and error method to ensure representative samples were obtained. Accumulated strata gases had to be vented at some boreholes while preventing air from entering the mine.

3 DESCRIPTION AND INSTALLATION OF INERTISATION UNIT

The components of the Unit included a modified jet engine, after-burner, cooling duct, diffusive cooler and

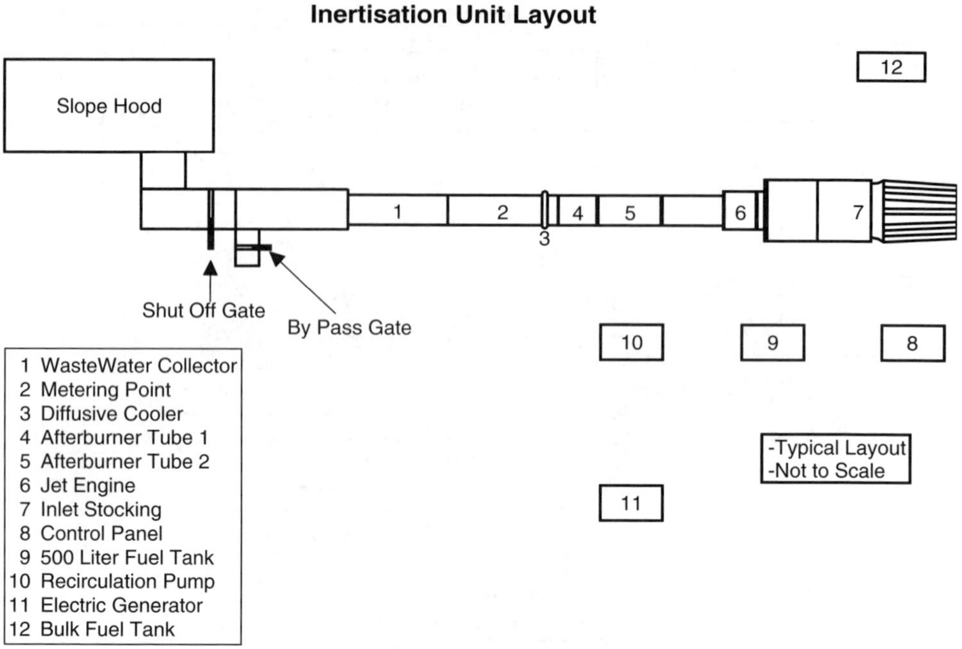

Figure 3. Inertisation unit components.

waste water collection duct. Figure 3 is a drawing that shows the various components of the Unit and the ducting fabricated to allow injection of the Unit exhaust into the mine at the slope opening. A fuel–air mixture was burned in the jet engine, with additional combustion occurring in the after-burner. Most of the oxygen was consumed in the combustion process, resulting in an exhaust composed primarily of nitrogen, carbon dioxide and water vapor.

The location of the fire, near the slope bottom, allowed for use of the slope as a conduit through which the exhaust could effectively be directed into the fire area. Preparation for installation of the Unit was conducted according to a detailed plan. The slope coping was constructed of concrete. A plywood seal had been constructed across the slope mouth when the mine was sealed. A pre-existing opening in the concrete at the top of the slope originally designed to facilitate conveyor belt change-out had been covered by a steel panel. A section of duct work was designed to replace the panel and connect the Unit to the slope. The ductwork contained a by-pass gate to enable the Unit output to be vented to the atmosphere during engine startup. To determine the existence of a safe work environment at the slope opening and minimize the potential for changes to the mine atmosphere in the fire area during installation of the ductwork, the following parameters were evaluated and monitored: pressure differentials between the mine and surface at the slope opening, barometric pressure, and the mine atmosphere.

On March 15, when all preparations were completed and readings were favorable, the command center gave approval to replace the steel panel with the fabricated ductwork. A temporary brattice was placed over the breached opening during the exchange which lasted approximately 30 minutes. Exposure of workers during the installation process was minimized through strict control. No work was performed directly in front of the slope opening.

A suitable pad, declined toward the slope at a 3 to 5 degree angle from horizontal, was also required to permit operation of the Unit. The pad was constructed with a retaining wall of pre-cast concrete blocks filled with gravel to obtain the necessary grade. A level pad was also established for the bulk fuel tanks. Components of the Unit began arriving at the mine site on April 1. Installation was completed on April 4 and the Unit was connected to the mine ductwork.

4 INERTISATION UNIT OPERATION

The Unit was operated by Queensland Mine Rescue Service. The engine weighed approximately 5 tons. A 14-man crew maintained continuous operation of the Unit. A control panel continuously monitored

Figure 4. Inertisation unit and control panel.

Table 1. Operational parameters of the inertisation unit.

Parameter	Rated		Actual
Engine speed	7,200 rpm	11,000 rpm	8,000 rpm
Flow	13.95 m^3/sec	33.25 m^3/sec	14.3 m^3/sec
Fuel consumption	17 liters/min	32.5 liters/min	370–480 gph
Cooling water	300 liters/min	300 liters/min	
Exhaust temperature	85°C	85°C	85°C
Exhaust gas constituents			(typical ranges)
– Oxygen	0 to 0.5%	0.5 to 2.0%	2.5 to 7%
– Carbon dioxide	13 to 16%	13 to 16%	8 to 13.5%
– Nitrogen	80 to 85%	80 to 85%	79 to 83%
– Carbon monoxide	3 ppm	3 ppm	100 to 7,000 ppm

temperature, pressure, and engine speed. Adjustments in engine operation were made to control the composition of the Unit exhaust. The engine was operated at speeds that did not develop significant thrust. Table 1 summarizes the operational parameters of the Unit. The rated information was provided by Queensland Mine Rescue Service.

The engine was operated at a speed of approximately 8,000 rpm. The after-burner was cooled by a re-circulating water cooling system. The exhaust was cooled to approximately 85°C (185°F) before entering the mine. The Unit could produce approximately 14.2 m^3/s (30,000 cfm) of exhaust gases and water vapor. The gas composition of the exhaust varied, depending on the operation parameters. Exhaust gases were monitored from a sample line located in the slope and analyzed by gas chromatography. Oxygen content in the exhaust was also monitored with a sensor in the Unit.

Typically, gas composition of the exhaust gas was 5 percent oxygen, 82 percent nitrogen, 1700 ppm carbon monoxide, and 11.5 percent carbon dioxide. The anticipated lower oxygen concentrations were not attainable because back pressure on the engine affected the operation. The back pressure resulted from the buoyancy of the hot exhaust gases in the slope and the restrictions in the mine caused by roof falls, accumulation of materials, and ventilation controls near the slope bottom. The back pressure reached values in excess of 1.7 kPa (7 inches water gauge).

5 INERTISATION OF THE FIRE AREA

Initial startup of the Unit was hindered by brief setbacks. The Unit was started at 7:45 p.m. on April 4, 2003, with the Unit's exhaust vented to the atmosphere. After 15 minutes of operation, the Unit was shut down because of overheating. Work continued on the Unit until 11:45 p.m., when the Unit was restarted. The by-pass gate was closed and the inert gases were directed into the slope. However, pressure from the initial introduction of the exhaust caused damage to the plywood slope seal, resulting in substantial loss of exhaust. After 15 minutes, the Unit was again shut off to repair the slope seal. Leaks in the ductwork attached to the slope were sealed with polyurethane foam and sand bags.

Figure 5. Exhaust leakage at slope.

The Unit was restarted at approximately 4:00 a.m. on April 5, and the exhaust again directed into the mine. By approximately 5:00 a.m., gases from the exhaust were detected at Borehole 1. Within eight hours, the effect on the other sampling boreholes was also evident. The exhaust gases arrived at Borehole 2 at approximately 5:25 a.m.

To facilitate movement of the exhaust gases throughout the entire mine, the seal on the St. Leo Return Shaft, located at the opposite end of the mine, was breached. After two days of continued Unit operation, samples indicated the mine atmosphere at the St. Leo Return Shaft had been adequately controlled. However, the atmosphere at several other shafts continued to be influenced by barometric changes and natural ventilation pressures and did not show evidence that the exhaust gases had reached those locations.

A decision was eventually made to reseal St. Leo Return Shaft to direct the exhaust to the other openings. The shaft was again sealed at approximately 4:15 p.m. on April 8. Within hours, the desired effect was evident at all openings. Oxygen concentrations at all openings had stabilized below 12 percent by the morning of April 9. Although the oxygen concentration of the exhaust remained too high to halt the combustion process in the fire area, the atmosphere in the fire area stabilized.

On April 10, the seal on the elevator shaft at Sugar Run portal was opened to begin installing a submersible pump to remove water from the bottom of the shaft. The shaft outgassed high levels of carbon monoxide, requiring the work to be performed with self-contained breathing apparatus. On April 12, the elevator shaft was dewatered and work began the following day to return the elevator to a serviceable condition.

Plans and preparations were also made to remotely start the mine fans at the St. Leo, Miracle Run and Harvey Run Return Shafts. On April 13, personnel began removing foam from the intake shaft seals. The fans were started sequentially to better control the gases at the remaining shafts. At 4:04 p.m., the St. Leo Return fan was started. The Miracle Run Return fan was started several minutes later. By 10:45 p.m., the Harvey Run Return fan was started and areas of the mine outside the fire area were re-ventilated. The Harvey Run Return fan was shut down on the morning of April 14 because its operation caused air to enter the fire area and adversely affect the stability of the atmosphere in that critical area.

6 MINE EXPLORATION AND RECOVERY

The mine atmosphere within the fire area remained stable with the exhaust from the Unit. Preparations were made to re-enter the mine and begin exploration of the fire area. On April 14, mine rescue teams entered the mine via the Sugar Run elevator shaft.

Mine rescue teams encountered poor roof conditions, roof falls, water, and temperatures in excess of 43°C (110°F) as they advanced toward the fire area. Exploration was slowed as ventilation changes were needed to continue advancement and confine the exhaust gases in the fire area. The temperature and water vapor from the exhaust created difficult working conditions near the slope bottom. The heat and humidity near the slope bottom prevented mine rescue teams from traveling in those areas. Access to the fire area from the side nearest the slope was blocked by roof falls and extreme heat. Efforts were made to explore the opposite side of the fire area. After several days of arduous work, exploration advanced to a location beyond Borehole 13. The rescue team encountered evidence that the fire had propagated in that direction from its known location on the day the fire occurred. Attempts were made to cool hot spots directly with water sprays. The roof falls and heat resulted in temporarily abandoning further advancement into the fire area from that direction.

Assessment of the conditions underground concluded the fire was not extinguished by the exhaust of the Unit and rehabilitation of the entire fire area was not feasible. The focus of the operation shifted toward sealing the fire area from the remainder of the mine and developing a new track turnout near the bottom of the slope. Ventilation changes that had been made by the mine rescue teams had reduced the size of, and partially isolated, the affected area. A reduced quantity of inert gas was now needed to maintain a stable atmosphere in the fire area.

A decision was made to replace the high quantity Unit exhaust with a lower quantity nitrogen injection through selected boreholes to maintain a stable atmosphere. The use of nitrogen injection would eliminate the heat and humidity associated with the Unit exhaust. These changes in conditions were expected to improve the working environment for the mine rescue teams and improve the possibility of advancing exploration into the fire area beyond the slope bottom.

During the afternoon of April 18, the Unit was shut down and nitrogen injection began at Boreholes 1 and 2. The buoyancy of the hot gas mixture and retained heat in the slope sustained an outgassing condition at the slope seal. The by-pass gate in the slope ductwork was partially opened to permit controlled venting of the hot gases in the slope. Eventually, the flow was increased to permit cooling of the slope and allow the fan induced ventilation pressures to reverse the flow.

The work environment improved and mine rescue teams advanced slowly beyond the slope bottom but still could not access the fire. Another attempt was also made to explore into the fire area from the opposite side. Mine rescue teams were able to advance exploration closer to the fire area. Evidence of continuing combustion, including flames, was encountered at roof falls. Eventually, the decision was made to abandon exploration work and to seal the fire area, including the track turnout near the bottom of the slope, from the remaining portion of the mine. Permanent bulkheads and explosion resistant ventilation seals were constructed to completely isolate the affected fire area from the remainder of the mine. The sealing process was completed on May 24, 2003. Water was pumped into the sealed area and impounded behind the bulkheads, flooding much of the affected area. Rehabilitation of the rest of the mine continued simultaneous with the construction of a new track turnout near the slope bottom. Coal production resumed in early October, 2003.

7 CONCLUSIONS

The Unit provided an effective means to stabilize the atmosphere in the fire area. Sampling results also indicated the Unit provided a sufficient quantity of inert gases to assist in eliminating explosive methane–air mixtures in remote portions of the large underground coal mine. Determination of the extent of the fire area was made possible and, ultimately, the fire area was permanently isolated from the remainder of the mine. Without use of the Inertisation Unit, recovery of the mine would not have been possible in such a short period of time. However, oxygen concentrations of the exhaust remained too high to halt combustion in the fire area and extinguish the fire. During the exploration and the sealing of the fire area, elevated temperatures and humidity from the exhaust of the Inertisation Unit presented some adverse conditions that hampered operations. The elevated temperatures and humidity may also have contributed to some of the numerous roof falls found in the area.

Mine Ventilation, Ganguli & Bandopadhyay (eds)
© 2004 Taylor & Francis Group, London, ISBN 90 5809 633 5

Fundamental study on locating spontaneous combustion of coal by self-potential method

Li Bingrui, K. Uchino & M. Inoue
Department of earth resources engineering, Kyushu University, Japan

ABSTRACT: It is already known that a corresponding distribution of self-potentials is generated in a porous medium like an aggregation of broken rocks if a temperature distribution exists in it. This phenomenon is known as thermoelectric coupling. Based on this theory, the authors propounded a method to locate the heat source of spontaneous combustion of coal by the self-potential method, and performed model experiments. It was also attempted to locate the heat source of the actual spontaneous combustion in a coal waste pile. The results of both the model experiments and the actual survey show that there is a center of the electrical potential distribution above the heat source, particularly clearly in the actual survey in the coal waste pile, the potential of the center being about 140 mV lower than that in the peripheries. Also these results showed that the electrical potential distribution corresponds to the position of the heat source. Therefore, it is possible in principle to locate the heat source of spontaneous combustion of coal by the self-potential method.

1 INTRODUCTION

The early detection of spontaneous combustion in coal mines is possible by means of the microanalysis of gases and temperature measurement. However, the fires have already spread out in many cases when they are detected. Furthermore, the most of spontaneous combustion of coal occur where they are not visible directly with naked eyes. Therefore, locating the heat source is indispensable not only to prevent expansion of fire, but also to extinguish it promptly. Mining engineers are so greatly concerned about the detection of the heat source of spontaneous combustion, that there have been various proposed detection methods such as those by the measurement of temperature distribution in gob by temperature sensors like thermocouple and by the scan of surface temperature of roadways by the means of infrared detector. Currently, the heat source of spontaneous combustion in coal mine is actually located by synthetical methods, which combines the indirect observation and experience.

If a temperature gradient exists in a porous medium like an aggregation of broken rocks a corresponding gradient of self-potentials will appear across it. This phenomenon is known as thermoelectric coupling, and may be caused by differential thermal diffusion of ions in the pore fluid and of electrons and donor ions in the rock matrix. The ratio of the generated voltage to the temperature difference is called the thermoelectric coupling coefficient. If these self-potential variations generated by heat are of measurable magnitude, it will be possible in principle to locate the heat source by measuring the self-potentials distribution. Based on this theory, the authors proposed an idea of locating spontaneous combustion of coal by self-potential method. In order to verify the possibility of this idea, the trend of self-potentials in heating process was observed by the model experiments. Furthermore, based on the results of the model experiments, it was also attempted to locate the heat source of the actual spontaneous combustion in a coal waste pile. This paper describes the result of these experiments.

2 MODEL EXPERIMENT

2.1 *Experiment apparatus*

The schematic diagram of the experimental apparatus is shown in Fig. 1. The upper surface of the experimental container of 100 cm × 100 cm × 80 cm is open to the air and the container was filled with sands. In the center of 30 cm level from the base of the container, a heater is embedded. This heater is a square board of 5 cm × 5 cm × 0.6 cm and its resistance is about 36 ohm. A thermocouple observes the temperature of the heater, and a transformer controls the power of heater.

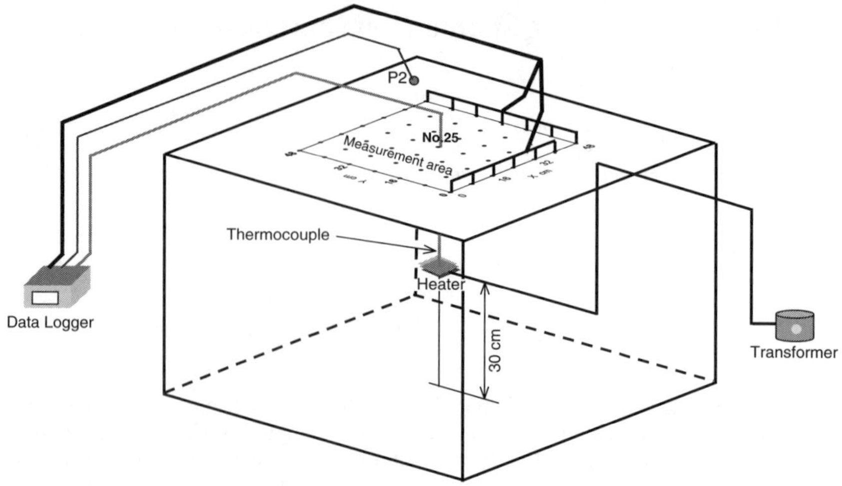

Figure 1. Experimental apparatus.

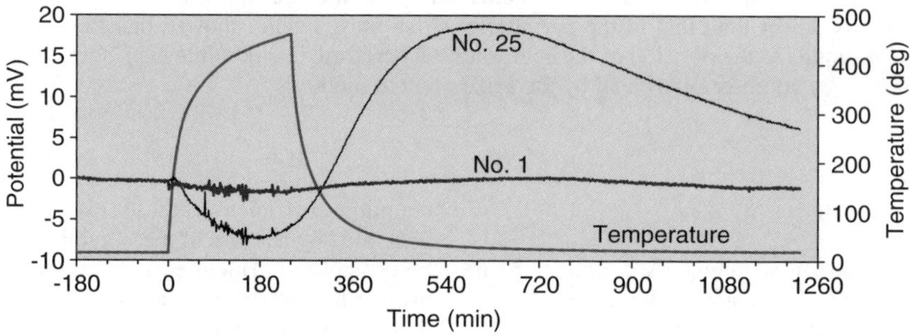

Figure 2. Potential-time data and temperature-time data.

As shown in Fig. 1, in the center part of the upper surface of the model, a square area of 48 cm × 48 cm was selected as the measurement area. In this area 49 electrodes are arranged at an interval of 8 cm. In this figure, No. 25 electrode is located right above the center of the heater, its coordinates being (24, 24), and P2 is a reference electrode. A data logger records experimental data at the time-interval of one-minute.

2.2 *Results and discussion*

In order to substantiate that the potential distributions that are generated by the heat source and the potential variations are of measurable magnitude, the development of self-potential in the heating process was observed in the condition that the power of heater is 150 W and the heating time 240 minutes.

The measurement of potential variations was begun about 180 minutes before the start of heating in order to assure the uniformity of the initial potential distribution. At the start of heating the import voltage of the heater was set at 73.5 V (the heater power is about 150 W) and this voltage was maintained through the heating process. About 240 minutes after the start of heating, the electricity supply was shut off. If time at the heating start is set at T = 0, the temperature-time data at the heater center and the potential-time data at No. 1 and No. 25 electrodes are shown in Fig. 2. No. 1 electrode is that on the boundary of the measurement area (see Fig. 1, coordinates are (0, 48)).

As shown in Fig. 2, the potential variations at No. 1 and No. 25 are not observed before the heating start. After the heating started, a negative variation of the potential at No. 25 electrode was seen. At about 180 minutes, however, this potential began to rise. After this, the potential rises rapidly as the heating continues. When time reached the predetermined heating time of 240 minutes the power supply of heater was shut off, but it still kept on rising. At about T = 600 minutes, it reached the peak of 18.7 mV and then came down slowly. The temperatures of the heater and its vicinity start to decrease after the power supply was shut off.

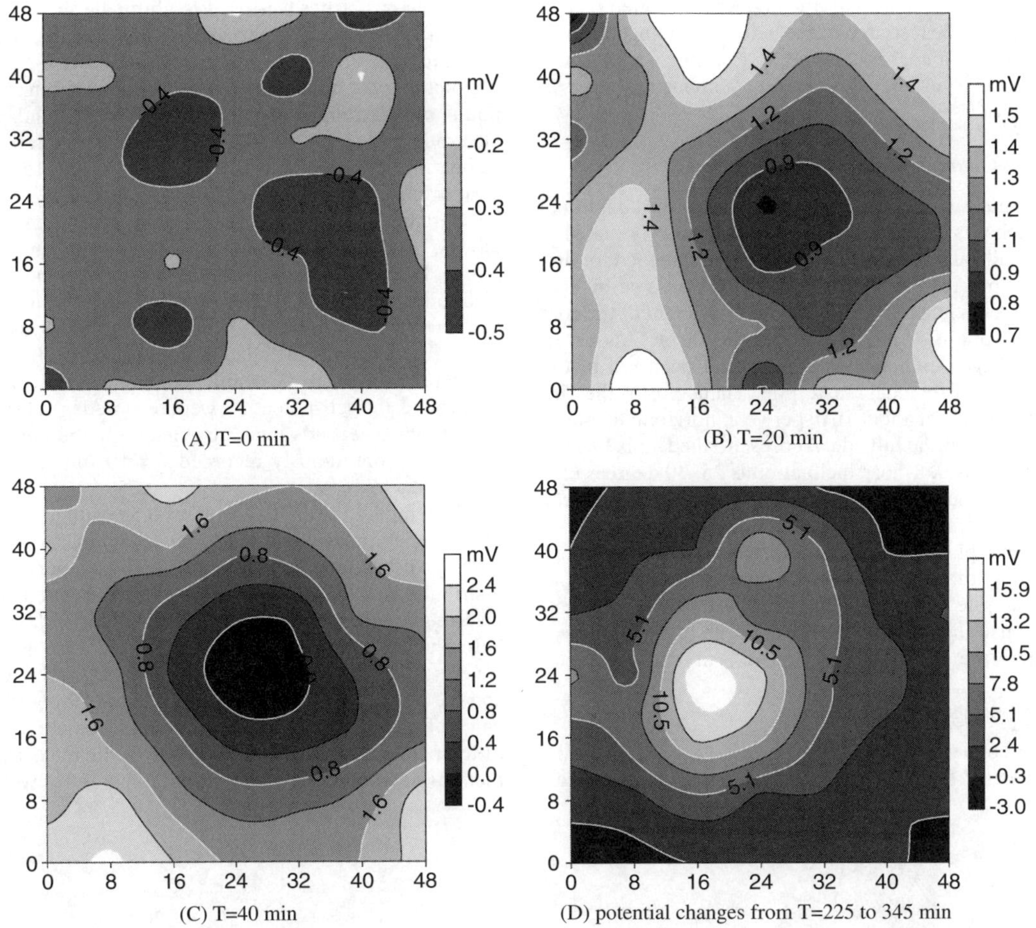

Figure 3. Potential distribution.

However the potential at No. 25 electrode did not decrease. Namely, these two curves do not coincide in their trends. This is because much thermal energy is still accumulated around the heater and the influences of heat convection and conduction in the medium are still persisting. On the other hand, the potential variation at No. 1 electrode did not change clearly because it is in the furthest from the heater. According to these observations, it can be concluded that the potential variations around the heat source are caused by heat and its magnitude is of measurable order.

In order to examine the potential distribution throughout the measurement area, the contour maps of the potential distribution are drawn as shown in Fig. 3 (Note: each map has its own scales to express potential variations clearly). Fig. 3(A) is a potential distribution map at T = 0 minute (right before the heating start). Although there are small potential variations in the map, there is no any strong abnormal zone. Fig. 3(B) and (C) are potential distribution maps at T = 20 and T = 40 minutes respectively. When T = 20 minutes, a clear potential distribution around No. 25 was observed. Since then, this potential distribution expands gradually, it spread throughout the measurement area at about T = 40 minutes. As shown in Fig. 2, the rate of potential variation was the largest from T = 285 to T = 405 minutes. The quantities of change in this 120 minutes were shown in Fig. 3(D). The potential variations around No. 25 are always the largest in all the heating process. In other words, the center of potential distribution is observed above the heat source.

3 ACTUAL SURVEY IN COAL WASTE PILE

It was shown clearly that the self-potential variations generated by heat are of measurable magnitude by the model experiments. This method was then attempted to locate the heat source of the actual spontaneous combustion in a coal waste pile, to examine whether it is applicable to the spontaneous combustion in coal mines.

3.1 *Outline of coal waste pile*

The measurement site is a waste pile of the coal mine, which was closed about 40 years ago, and is situated in a small valley. The top of the pile was flattened for disaster prevention, and its surface was covered by clay with a thickness of about 30 cm. The height of the coal waste pile is 36 m and the area 56,700 m^2. About ten years ago, spontaneous combustion occurred in the hillside of the coal waste pile. Thenceforth, the fire front moves at about 10 m per year, horizontally along the slope of the hillside. At present, the fire is located at the hillside whose inclination is 25–30 degrees (see Fig. 4). There are many small cracks and two big cracks in the combustion area, and fumes are observed in several places. The lengths of big cracks are more than 15 m. The spontaneous combustion of the coal waste pile not only pollutes the nearby environment, but also the polluted air flows to the residential area. Therefore, the control of the fire is urgently needed.

3.2 *Measurement method*

The copper-copper sulfate nonpolarizable electrodes were used for the in-situ measurement of self-potential, and a low pass filter was used to eliminate the influence of noises such as the telluric current, etc. The measurement area is a quadrangle of 70 m × 50 m surrounding the high temperature zone, and 48 survey points were arranged at interval of 10 m. A standard electrode was set in the place where the temperature is not influenced by the heat source, about 150 m away from the survey point (70, 50) (see Fig. 4). The measurement area was divided into several small blocks, each of which contains 9 (3 × 3) survey points. The self-potentials were measured by 9 mobile electrodes in the order of the blocks. Considering the condition that the measurement is not simultaneous, a fixed electrode was installed at point (20, 30). In the measurement, when 9 mobile electrodes contacted the ground, the self-potentials were recorded by a Data Logger at 30 seconds intervals. Generally the potentials were continuously recorded for 10 minutes or more in each block.

The temperatures in the depth of 0.5 m and the surface temperatures were measured to compare them with the distribution of self-potential. The temperature distribution in the depth of 0.5 m was measured by inserting the thermocouple into a guide pipe installed beforehand. The thermocouples of K type were used, covering the measurement range of 0–500 degree C, and the precision was one degree C. The X-Y coordinates of the survey points and the measurement area of temperature are the same as those of the measurement of self-potential. The surface temperatures were

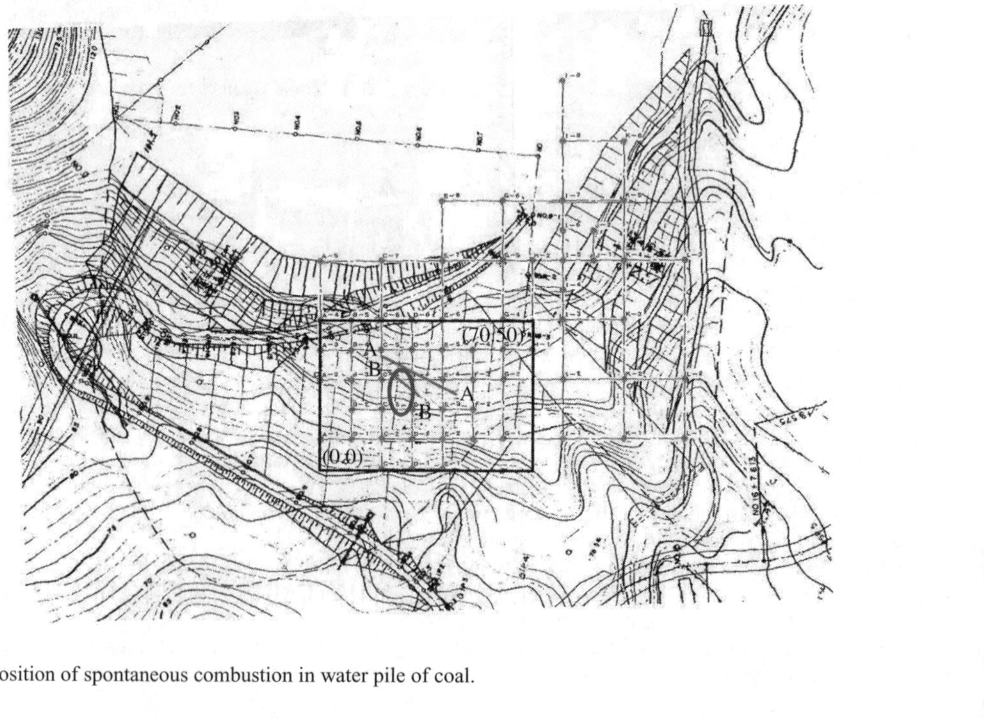

Figure 4. Position of spontaneous combustion in water pile of coal.

measured by a thermo-camera. The targeted area was a square of about 1 m^2. The average temperature in the square was defined as the temperature of the survey point. The temperature was not measured in a part covered by the dense grasses. Therefore, the measurement area of the surface temperature is 60 m × 60 m.

3.3 *Measurement results and discussion*

Measurement was finished in one day. Fig. 5 shows a part of the obtained data in the neighbouring blocks A and B. In this figure, 0–19 minutes are the measurement time in block A, 20–31 minutes are the time taken to move the electrodes from A to B block, and 32–48 minutes are the measurement time in block B. Curve 0 is the potential of the fixed electrode and curve 1 and 2 are the potentials of two mobile electrodes, respectively. Although the potential of each electrode has fluctuations, their modes are similar. Therefore, in obtaining the self-potential distribution in the same block, the potential at any time can be used. However, the influence of measurement time on the self-potential must be considered to calculate the potential distribution in the whole measurement area. The time base correction uses the data of the fixed electrode.

The potential of the survey point (70, 0) is the highest in the measurement area, therefore, all the measured values are corrected, making this point as a reference position. Moreover, the survey point (50, 30) was removed, because the potential of this point is far higher than that of its periphery.

After the data were corrected as mentioned above, the potential distribution in the measurement area is shown in Fig. 6. There is a negative center of the potential distribution in the measurement area, and the potential is about 140 mV lower than that in its peripheries. Fig. 7 shows the temperature distribution in the depth of 0.5 m and there is also a high temperature center in the measurement area. The temperature in the center is about 240 degree C, and is about 230 degree C higher than those in the boundaries. By comparison between Fig. 6 and Fig. 7, it is found that the potential distribution was a little different form that of temperature at the lower level, but the center of the potential distribution is in agreement with that of the temperature. Fig. 8 shows the surface temperature distribution, which is macroscopically similar to the potential distribution too.

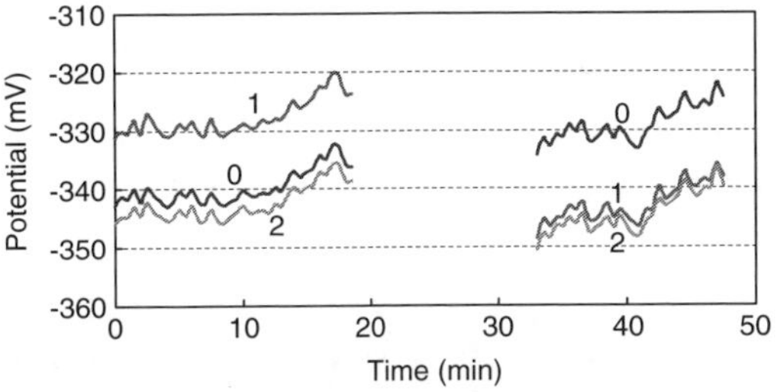

Figure 5. Potential-time data.

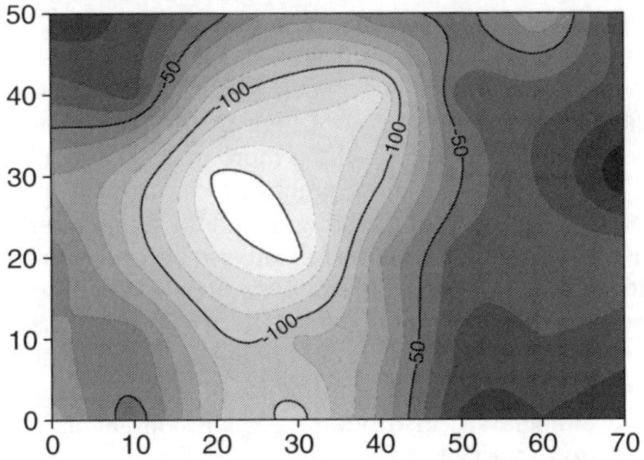

Figure 6. Potential distribution.

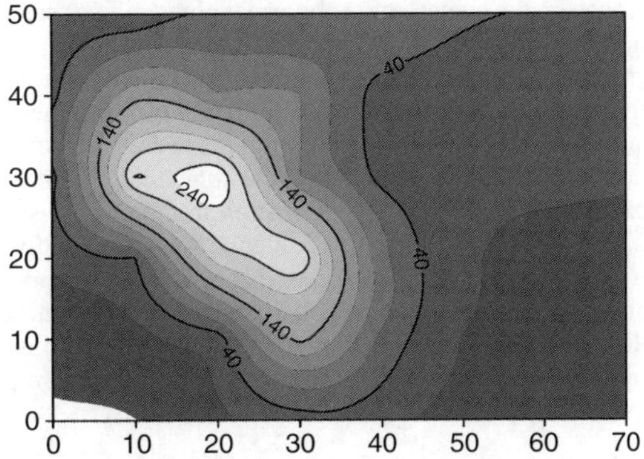

Figure 7. Temperature distribution in the depth of 0.5 m.

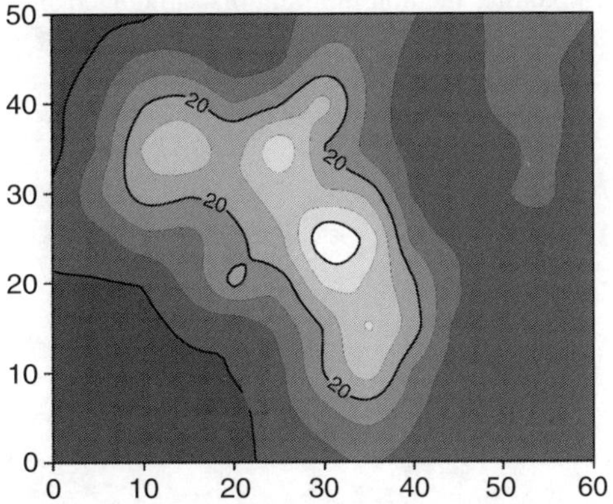

Figure 8. Surface temperature distribution.

The center of potential distribution is almost in agreement with the zone of the high temperature. But the patterns of equipotential-lines are different from those of the isotherm. This is considered to be the influence of the cracks. As shown in Fig. 4, there are two big cracks in the measurement area. One (A-A) of them extends from the point (20, 38) to (45, 25), and other (B-B) extends from (22, 33) to (35, 22). The width of the cracks is between 5 and 20 cm. The surface temperatures are very high in the parts where fumes are observed, reaching more than 100 degrees C in some zones. These cracks are thought to have become passages of the hot gas, and the temperatures are raised and the patterns of isotherm become similar to the direction of the cracks. By comparison of Fig. 7 and Fig. 8, it is found that the pattern of the isotherms are similar, however, the zones of the high temperature are apart by 3~5 m. The influence of the cracks upon the temperature distribution becomes larger as the position approaches the surface.

It is known that there is a negative correlation between the self-potential and altitude. This phenomenon is produced by groundwater flows downwards. However, the center of the abnormal potential was seen not on the summit of the coal waste pile but on its hillside. It is thought that the influence of the altitude on the potential distribution is small in the measurement area. According to investigation of the scene, there are not the materials that generate the abnormal potential except the spontaneous combustion in the coal waste pile. Moreover, the negative center of the potential distribution is in agreement with the zone of high temperature. Therefore, it can be concluded that the potential distribution was generated by the spontaneous combustion, and is dependent on the position of the heat source.

4 SUMMARY

As basic research of the heat source detection by the self-potential method, the authors conducted model experiments and in-situ measurement in a coal waste pile which been having spontaneous combustion. The model experiments and the actual surveys have many common features:

(a) The self-potential distributions generated by the heat are of measurable magnitude. Especially in the coal waste pile, the potential at the center of the heat source is about 140 mV lower than that in the peripheries.
(b) The self-potential distribution is similar to that of the temperature, and there is a center of the self-potential distribution right above the heat source.

The results of model experiments and actual surveys clearly show that the electric potential are generated by heat and the potential distribution is dependent on the position of the heat source. Therefore, it is possible in principle to locate the heat source by the self-potential method.

REFERENCES

Bingrui Li, K. Uchino and M. Inoue. 2001. Possibility of locating heat source of spontaneous combustion of coal by the electric potential. Proc. of the 7th international mine ventilation congress, 657–662, Cracow, Poland.

David V. Fitterman and Robert F. Corwin. 1982. Inversion of self-potential data form the Cerro Prieto geothermal field, Mexico. Geophysics 47(6), 938–945.

Robert F. Corwin and Donald B. Hoover. 1979. The self-potential method in geothermal exploration. Geophysics 44(2), 226–245.

Ishido T. and T. Kikuchi. 1987. The self-potential method. Geothermal energy 12(4), 425–443.

Ishido T. and H. Mizutani. 1981. Experimental and theoretical basis of electrokinetic phenomena in rock-water system and its applications to geophysics. Journal of Geophysical Research 86(B3), 1763–1775.

Eve S. Sprunt, Tony B. Mercer and Nizar F. Djabbarah. 1994. Stream potentila from multiphase flow. Geophysics 59(5), 707–711.

Mine Ventilation, Ganguli & Bandopadhyay (eds)
© 2004 Taylor & Francis Group, London, ISBN 90 5809 633 5

The use of blasting fumes to inertise underground coal mine atmospheres

D. Chalmers
School of Mining Engineering, The University of New South Wales, Sydney, Australia

ABSTRACT: This paper investigates the utilization of the blasting fumes in rendering the atmosphere inert so that explosive mixtures cannot form. And discusses the monitoring undertaken to confirm the conditions during and post blasting.

Collieries in the Bulli Seam in NSW, Australia, have areas of coal that will not drain below outburst threshold limits. As these coals will not desorb their gas, the mines have been successfully grunching in coal using P5 permitted explosives. If these coals were to desorb, then a "Permit to Mine" would be issued and there would be no need to grunch. Grunching is the only approved method in these circumstances and the mine would close if this development could not continue.

In some collieries the gas regime within these coals has been predominantly carbon dioxide, incapable of forming an explosive atmosphere, however this has changed so in as much as the gas will pass close to the nose limit for this gas mixture.

Until now P5 explosive has been available and this has been able to be used. However P5 explosive is limited in supply, has a short shelf life, and is Nitroglycerine based. This has prompted the investigation and use of P1 explosive (Powergel 3000®) that is usually reserved for stone work within the seam. Questions have been raised as to the suitability of P1 explosives for this application prompting this investigation.

Considerable knowledge is available regarding the explosibility of methane. This work includes tests conducted by Coward, United States Bureau of Mines Research and Hughes and Raybould. These reports were generally but not exclusively utilize nitrogen as the depression gas. This is useful in determining atmospheres after mine fires or in sealed areas.

Work done by Paul Mackenzie-Wood utilizes the USBM method for determination of explosive atmospheres. This method is slightly conservative compared with Coward's determination. The USBM method adds Carbon dioxide to air and then adds Methane to this mixture. This gives a conservative estimate of the ratio CH_4/CO_2 as the CO_2 and the O_2 are diluted by the methane.

The explosibility of Illawarra Bottom Gas can be determined in normal atmospheres as the depression gas in this situation is Carbon dioxide. Such a determination utilizing the work Hughes and Raybould, can be set up in a spreadsheet to determine conditions under which grunching can continue.

1 INTRODUCTION

Underground coal mine development methods favor the use of continuous mining machines to cut coal, allowing rapid development rates. One major hazard that must be controlled to make mining safe is that of outbursts. One major cause of outbursts in the Bulli seam is the high concentration of a seam gas, commonly called *Illawarra Bottom Gas*. This gas consists of a mixture of carbon dioxide and methane. Outburst risks in the Bulli seam are also noticeably increased when in close proximity to geological structures. These geological structures provide massive surface areas that enable the storage of considerable gas reservoir and usually decrease the permeability of the coal between the structure and the development roadway. In-situ gas pressures present in such coal seams have the potential to burst hundreds or even thousands of tonnes of coal over mine personnel in a violent manner. Outburst events also involve the release of these large volumes of gas.

Some coal types, such as those which have been mined at a Southern NSW Colliery, do not readily release their gas to atmosphere or to a gas drainage system.

When this occurs, the gas content can be higher that the threshold limit and a permit to mine is declined. Alternate mining strategies have to be sought if development is to continue. One mining method that allows the extraction of coal in outburst prone conditions is grunching (shooting coal from the solid). Grunching has been successfully used at this colliery, with over a kilometre of development being driven without incident.

There is potential to create explosive atmospheres posing a serious risk when explosives are being detonated. This risk has been managed by using explosives that have passed a series of tests called the Buxton tests. The guidelines that have been followed until now is that only explosives that have passed the rigorous P5 tests can be used in this situation.

Apart from this application, P5 explosives are rarely used, are nitro-glycerine based and have a limited shelf life.

Supply of P5 explosive is patchy at best and the lead time and shipping difficulties make the day to day utilization problematic. Delays in supply can exceed three months.

Collieries, caught by non draining coal and no P5 explosive have sought exemption to allow for the use of a higher powered P1 explosive that is readily available.

Risk assessments were conducted to mitigate this risk. One mine was given the go ahead on the basis that the Illawarra Bottom gas mixture was incapable of forming an explosive mixture. As mining progressed the percentage of CO_2 decreased and development work was halted until the explosive risk could be evaluated.

A second risk assessment was undertaken and it was suggested that blasting fumes would be of such quantities that they could be a significant factor in controlling the explosion risk.

This paper attempts to outline the methods used to determine the explosibility of liberated gases and the data gathered that can be used to assess the gases released in shotfiring, with the aim of confirming that no explosive mixtures were being generated.

Figure 1. Deteriorating P5 explosive 2002.

1.1 *Short term solution*

Mackenzie-Wood utilizing work done by Hughes and Raybould, Coward, US Bureau of Mines, and indicated that a ratio of 77/23 Carbon dioxide/methane ratio is the limiting factor when mixed with air. This is represented by a Coward diagram, Figure 2 where the depressant gas is Carbon dioxide rather than nitrogen.

Seam gas with methane content below this level would not be capable of passing through the explosive triangle. A slight safety margin of 80% CO_2 was suggested and P1 emulsion could be used in place of the P5 explosive.

1.2 *The effect of blasting fumes*

The use of explosives to combat oil well fires is well known. The violence of the blast and the inert atmosphere that is generated will extinguish the blaze. It was suggested that if sufficient fumes were liberated into the atmosphere and mixed with the emitting seam gas then there might be a case to lower this ratio.

Table 1 lists the gaseous quantities produced from P1 emulsion explosive.

If this is released into the atmosphere as blasting continues then it will also dilute any methane that is present.

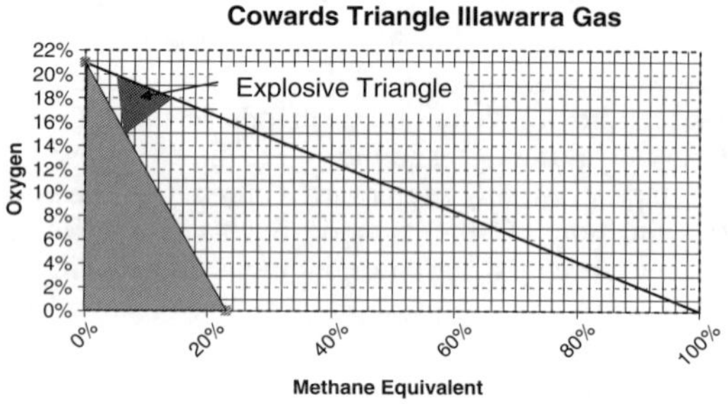

Figure 2. Explosibility of Illawarra Bottom Gas.

Table 1. Gas liberated from 20 kg P1 emulsion (Orica, 2002).

Gas	Quantity/20 kg
CO	900 litres
CO_2	1.81 m^3
NO	0.0004960 millilitres
N_2	4.04 m^3
H_2O	11.92 m^3
CH_4	0.06 m^3
Total	17.82 m^3

Figure 3 shows the effects of adding these gases to the explosibility diagram. The gas dilution line is the line that the gas mixture containing 25% blasting fumes would take on dilution with air. It can be seen that it does not pass into the explosive triangle.

The percentage of methane can be increased and still be incapable of forming an explosive mixture. By lowering the Carbon dioxide/methane ratio to 71/29 with 25% blasting fumes mixed with it will bring the gas dilution line to the nose point of the explosive triangle.

By varying the quantity of blast fumes present in the atmosphere the relationship between the amount of CO_2 and blast fumes that will render the atmosphere inert is determined. And conditions for blasting in coal where seam gases rich in carbon dioxide can be developed. Figure 4 shows this relationship. Mixtures below the line are capable of forming an explosive atmosphere.

1.3 *Desorption of seam gas*

The next question to be answered is how much gas will be liberated when the blast is fired?

Firstly the reason for grunching is that this coal will not release its gas, so it would be unreasonable to suggest that the total gas content will liberate during firing.

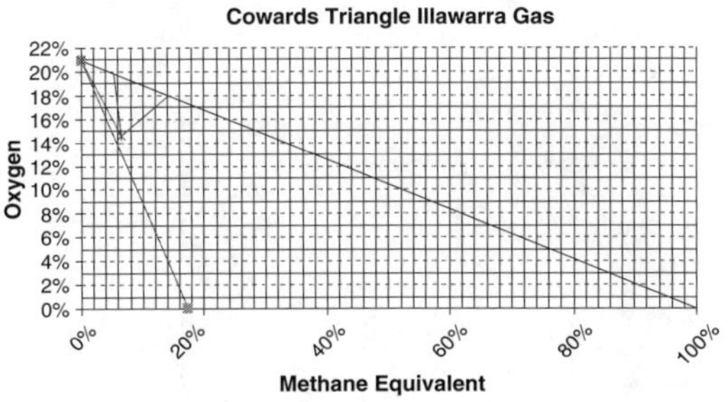

Figure 3. 25% blasting fumes mixed with 75% Illawarra Bottom Gas 77/23 CO_2/CH_4.

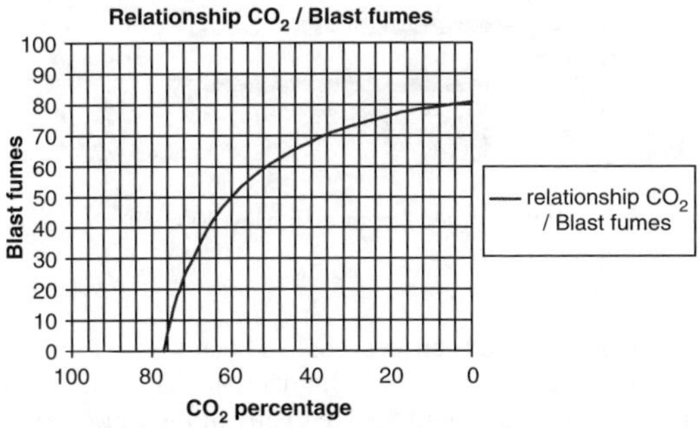

Figure 4. Relationship between carbon dioxide in the seam gas and blast fumes.

Secondly the duration of the blast is restricted to less than 150 ms by law; therefore the time that must be taken into consideration for gas liberation is also very short.

Thirdly, desorption is a function of pore pressure and this will be momentarily increased by the detonation pressure of the explosive. This would inhibit desorption.

Lastly, the major byproduct of the explosive detonation is water vapor. This would be injected into some of the connected pore spaces in the coal. It would condense and inhibit gas flow.

Many authors quote an empirical relationship, describing the rate of desorption from coal, initially proposed by Airey, 1968.

$$Qt = Q\infty \left(1 - \exp(-(t / to)^{n}\right) \ m^3/t$$

Q_t = gas desorbed in time t
t = desorption time
Q_∞ = gas desorbed at $t = \infty$
t_o = time for 63% gas to desorb
n = 0.33 bituminous coal
n = 0.50 for anthracite

Times t and t_o may be expressed in seconds, minutes or days. The time taken for 63% of gas to desorb will increase with increased fragmentation with typical values of 5 to 20 days for cores and large fragments to around 33 mins for particles less than 0.3 mm. Values for t_o are best obtained from slow desorption rate tests, but, for the purposes of this analysis, can be obtained from initial lost gas tests.

Predicting the amount of gas that might be liberated in the first second from blasting, for a coal with a gas content of $13 m^3/t$ and t_o = 5days. The total volume of gas liberated in the first second of the blast can be predicted as 178 litres/tonne. With a typical round, this would amount to a ratio of 75% blast fumes to seam gas.

At t_o = 20 days the gas liberated in the first second would be 113 litres/tonne. Again with a typical round, this would amount to a ratio of 83% blast fumes to seam gas.

1.4 *Field trial*

It was considered that blasting could continue at seam gas mixtures as low as 65% CO_2. This was based on the following:

- Firstly having determined that a carbon dioxide/methane ratio to 71/29 with 25% blasting fumes mixed with it will bring the gas dilution line to the nose point of the explosive triangle.
- The likely gas release in the 1st second would be 178 litres/tonne and therefore the 75% blasting fumes would be present.

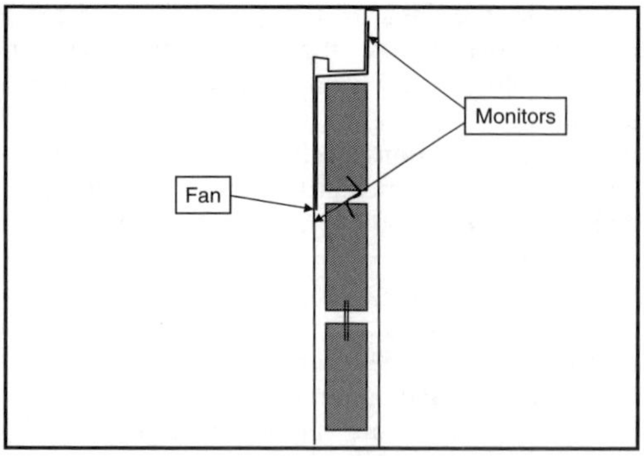

Figure 5. Fan and monitor locations.

- The face area would be heavily stone dusted.
- All personnel would be in a safe area if an ignition did take place.
- Monitoring of the face area and the return to confirm the quantities of gas liberated from the blast.

Additionally, samples were taken from the broken coal were tested for gas content.

Monitoring utilized Draeger multiwarn gas detectors. One monitor was placed 20 m from the face and one was placed in the exhaust stream of the auxiliary fan. Both multiwarns were able to log the data and this was analyzed using Draeger's proprietary software GasVision.

1.5 *Results*

Typical results from the multiwarn are shown in Figure 6. This shows a slight rise in methane and carbon dioxide when the fan is switched off. This is followed by a sharp rise as the short goes off. Then the fan is restarted leading to the drop off.

The return trace (Figure 7) shows a similar peak generated from the shot being fired. The earlier peak in figure 7 shows a gas event that has emanated from the face and has been drawn into the ventilation ducting. The noticeable difference between the two is the absence of carbon monoxide in the first peak.

Initial gas content testing of the broken coal after firing showed higher gas content than that indicated prior to grunching. The purpose of this test was to indicate the amount of gas that would have been released during shotfiring. These tests were discontinued, as they are costly and were not able to yield useful results.

Peak gas readings measured on the multiwarn are listed in Table 2.

These readings ranged from 0.21% to 0.67% CH_4, 0.77%–1.46% CO_2 and 192-in excess of 500 ppm CO. In all these peaks there was only a slight depression of the O_2 reading.

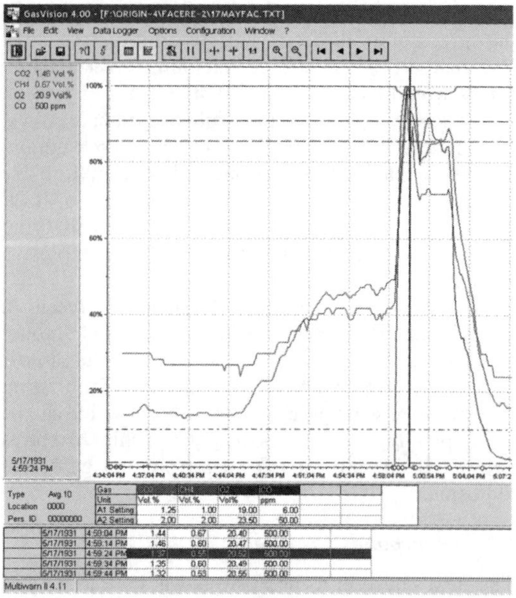

Figure 6. Multiwarn trace for shot 17 May 2003.

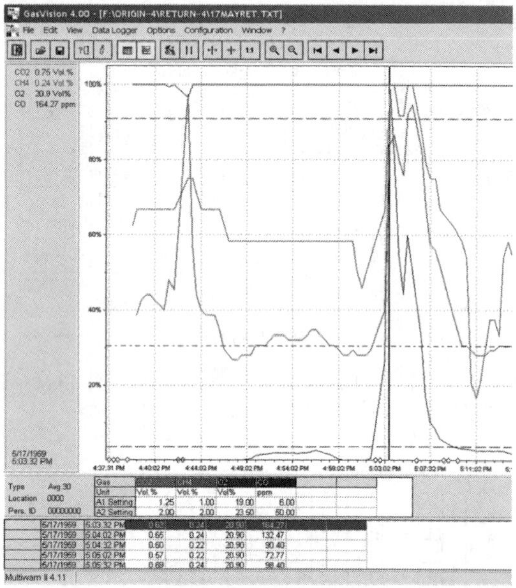

Figure 7. Return trace for shot 17 May 2003.

1.6 *Analysis*

Gas content of the broken coal indicated in the initial tests that the gas content was higher than that of the solid gas content. This could be due to any of the following:

- Samples selected from coal that has a lower ash content and thus a higher gas content.

Table 2. Peak gas readings at the face during grunching.

	CO_2 (%)	CH_4 (%)	CO (ppm)
Blast 1	1.46	0.67	500
Blast 2	1	0.21	301
Blast 3	0.77	0.36	420.4
Blast 4	0.83	0.33	198
Blast 5	0.86	0.49	189.6
Blast 6	1.31	0.24	191.9
Blast 7	0.84	0.34	368

- Increased pore pressure has forced additional gas into coal.
- Micro-fractures generated by explosives have increased the rate of gas release during the test leading to a higher than expected lost gas determination.
- Water vapor and fine dust have blocked pathways that free gas would have otherwise flowed.

Unfortunately, there was no opportunity to continue these tests due to tight budgetary constraints and production pressures. It would have been useful to have analyzed the mechanisms involved.

The amount of gas released from each shot was then determined from the area under each peak. These indicated that the gas release was less than 10% of the measured gas content of the in situ coal.

The high oxygen content, low methane and carbon dioxide values measured at the face also indicate that there were no incidences of secondary ignition at the face.

Gas readings also indicate that the amount of gas liberated is consistent with Airley's empirical predictions.

2 CONCLUSIONS

The reliance on the Buxton tests as a means of ensuring the safety of using explosives in underground mines, coupled with the limited use due to modern mining methods has provided an environment where the knowledge base is dwindling.

The underground coal industry is fast approaching the point where there will be insufficient knowledge about the behavior of explosives in this context that may bring about the banning of their use.

This paper seeks to address the concerns that the industry in New South Wales and Queensland coal mining operators have about the safety of explosive use.

Information is available that will predict the amount of gas that can be effectively liberated from broken coal over the time that the blast is conducted.

Seam gas that is liberated can be inertised by the gases liberated from the explosive. However it must be strongly stated that this inertisation can only occur if the seam gas and blast products are sufficiently intermixed. This study concedes that highly turbulent atmospheres can be generated at the face resulting in several mixtures being present.

BIBLIOGRAPHY

Gray, I, 1995. The use of drilling in outburst detection and alleviation, *Management and Control of High Gas Emissions and Outbursts in Underground Coal Mines 1995,* pp115–122 (National Organizing Committee of the Symposium: Wollongong)

Hargraves, A, 1995. Some properties of coals containing gas and some related incidents, *Management and Control of High Gas Emissions and Outbursts in Underground Coal Mines 1995,* pp335–348 (National Organizing Committee of the Symposium: Wollongong)

Hargraves, A, 1980. A review of instantaneous outburst data, *The Occurrence, Prediction, and Control of Outbursts in Coal Mines 1980,* pp1–18 (The Australian Institute of Mining and Metallurgy – Southern Queensland Branch)

Hargraves, A, 1982. Background to seam gas drainage in Australia, *Seam Gas Drainage with particular reference to the Working Seam,* pp21–35 (The Australasian Institute of Mining and Metallurgy – Illawarra Branch)

Lama, R, 1980. Adsorption and Desorption Technique in Predicting Outbursts of Gas and Coal, *The Occurrence, Prediction, and Control of Outbursts in Coal Mines 1980,* pp1–18 (The Australasian Institute of Mining and Metallurgy – Southern Queensland Branch)

Lama, R, 1995. Safe gas content threshold value for safety against outbursts in the mining of the Bulli seam, *Management and Control of High Gas Emissions and Outbursts in Underground Coal Mines 1995,* pp175–190 (National Organizing Committee of the Symposium: Wollongong)

Lama, R, 1995. Effect of stress, gas pressure and vacuum on permeability of Bulli coal samples, *Management and Control of High Gas Emissions and Outbursts in Underground Coal Mines 1995,* pp293–302 (National Organizing Committee of the Symposium: Wollongong)

Mackenzie-Wood, P, 1998. Explosions, *Emergency Preparedness and Mines Rescue,* (The Mines Rescue Service of New South Wales)

Moreby, R, 2001. Mine Gases and Gas Laws, *Graduate Diploma in Mine Ventilation,* (The University of New South Wales School of Mining Engineering)

Moreby, R, 2001. Gas Reservoir Characteristics, *Graduate Diploma in Mine Ventilation,* (The University of New South Wales School of Mining Engineering)

Strang, J and Mackenzie-Wood, P, 1990. *A Manual On Mines Rescue and Gas Detection,* (Colorado School of Mines)

Tonegato, S, 1998. Gas Outbursts, *Emergency Preparedness and Mines Rescue,* pp348–388 (Mines Rescue Board NSW)

Wynne, P and Case, R, 1995. Outburst remote control mining at Tahmoor Colliery, *Management and Control of High Gas Emissions and Outbursts in Underground Coal Mines 1995,* pp503–508 (National Organizing Committee of the Symposium: Wollongong)

Mine Ventilation, Ganguli & Bandopadhyay (eds)
© 2004 Taylor & Francis Group, London, ISBN 90 5809 633 5

Controlling fire and smoke movements in road tunnels by means of ventilation: state of the art

E. Jacques
Université Catholique de Louvain, Louvain-la-Neuve, Belgium

ABSTRACT: In a first part the paper aims at giving an overview and state of the art where tests and modeling are concerned. It shows that these simulations make by far the most important source of scientific and technical information on fire behavior and smoke movements. They provide essential knowledge for designing and checking ventilation systems and dimensioning their smoke exhaust capacity or ability to stop back-layering. Stability of smoke stratification and conditions for maintaining this stratification are important criteria for organizing escape routes. Knowledge of temperature distribution and spot temperatures would guide on thermal resistance values of structures and equipment.

For many years this study material has been used mainly in view of getting standardized reference values or curves versus time of heat release rates, temperature evolution, smoke production, critical velocity, etc., the kind of information that is usually put into rules, guidelines and recommendations for new tunnel design. Complying with these rules was generally considered to provide sufficient safety for tunnel users.

But these last years, mainly since the Mont Blanc tunnel fire, public authorities and managers had to experience that safety is not only a matter of guidelines and rules, but should include function-based requirements. Fires differ in terms of quality and quantity and never are alike to the ones that have been modeled. Therefore, it must be true that safety may be enhanced by active fire response management and on-line ventilation control, started soon after detection of the fire. This more developed approach, obviously, would imply additional equipment and therefore should be studied before commissioning in the tunnel.

The paper discusses how far we stand with ventilation control, examines the problems that are still to be solved and the implications active ventilation management got on operating the tunnel.

1 INTRODUCTION

During the last decades fires in tunnels have been the object of many studies and research projects. The first full scale tests have been carried out already in the sixties. Later on, huge efforts have been put into developing models, i.e. small-scale models but most of all numerical ones (1D, zone models, 2 & 3D).

Nonetheless, the least three years we assisted at huge catastrophic fires with tens of fatalities in the Mont Blanc tunnel (between France and Italy), the Gotthard tunnel (Switzerland), the funicular of Kitzensteinhorn (Austria), Tauern tunnel (Austria), metro of Daegu (South Korea).

These catastrophic events raise several questions: What did go wrong? Where do we stand? What is the missing piece? Can we do better in the future?

Post-mortem analyses of these disastrous events did, in a few cases, point into the direction of deficiencies of the ventilation system and operational errors. However, the common denominator that explains most of the fatalities is the human behavior of tunnel users. Findings after the event and organized tests show that too many people are most likely not aware of the upcoming danger. Most of them stay in their cars for whatever good reason. By the time they realize that they are really endangered, it is mostly too late to escape. The human behavior is certainly the key factor to master, but technicians do not have the slightest hold on it. All they can do is creating a tunnel environment as safe as possible.

If nevertheless a disastrous event has to happen it is the task of tunnel managers and operators to control the only (efficient) tool they have at their disposal, i.e. the ventilation and exhaust system. Not only by dimensioning it correctly, but also by operating it adequately they would provide the best chances for users to escape and for rescuing services to access.

All interventions, whether they deal with actions on the ventilation, with instructions to the rescuing services or with switching on signing that indicates the most appropriate escape route, they all need a thorough knowledge of the actual fire situation and the to be expected smoke movements.

2 OUTLINE OF THE PAPER

The paper aims at giving a general survey of the needs and the difficulties encountered with tunnel ventilation control in case of fire.

The paper should provide elements to decide what kind of ventilation control system would be most suitable and draw the attention to particular aspects and difficulties that should not be overlooked.

It is not the aim of this paper to provide cut-and-dried solutions. It only describes briefly the modeling techniques that are used for tunnel ventilation design and operation.

3 OBJECTIVE OF FIRE AND SMOKE CONTROL

The objective of controlling fire and smoke is rather trivial by definition: all constitutional elements, whether they address design, construction, operation, training and maintenance aspects should contribute to ensure a safe haven for all tunnel users.

If a fire hazard has to occur, then the ventilation should be operated in a manner that a part of the traffic room and the escape routes are kept smoke-free, so allowing tunnel users to rescue themselves or at least be rescued.

4 TUNNEL AND VENTILATION DESIGN

Tunnel designers who have to deal with fire hazards know about the different threats that tunnel users may face: toxicity of dangerous gas components, lack of visibility, radiation, heat and high temperatures, explosion and possibility of objects falling down. All these fire risks must, of course, be taken into consideration during the design stage of the tunnel i.e. engineering of the tunnel structure, fire resistance of structural elements and equipment, design of the ventilation system and its appropriate capacity, communication infrastructure, exits and egress routes, signing towards shelters, easy access to phones, etc. No doubt, the safety level in the tunnel is higher if these facilities are available and adequately designed.

Infact, during many years safety was thought to be solely a matter of rules, guidelines and recommendations for the engineering work. Theoretical and empirical studies, risk analyses, choice of heat or smoke release rates, simulations (Kumar et al. 1988), (Rhodes 2000) and extensive full scale fire tests (Augin et al. 2003) would provide the subject matter for formulating these rules and recommendations. Complying with them was then usually considered as a kind of guarantee that a reasonable safety level had been achieved. And up to a certain point this approach is right and, therefore, should be continued in the future.

A good physical insight and understanding of fire dynamics is certainly a great help, if not a must for planning fire safety in tunnel design. Flashover and post-flashover need to be studied particularly from the point of view of fire resistance of structures and fire fighting equipment and safety devices. Also good understanding of the growth phase is important since escape safety is directly time-related. This phase of fire development involves the complicated interaction of several phenomena such as the combustion of the material itself, the convection of energy, the diffusion mechanisms of the gaseous products and the transmission of energy by radiation. This phenomenon is so extremely complex that for decades a lot of engineers and scientists worked on different and partial approaches of the problem. Using all kind of models and experimental tools, they tried to pinpoint the most decisive parameters that control the fire and predict the smoke production and its behavior inside a tunnel. Depending on the purpose they were supposed to serve, some fire behaviors have been highlighted, mostly at the expense of others. This is why study methods cover a variety of models and experimental investigation techniques.

5 MODELS AND EXPERIMENTS

Considering the enormous complexity of the fire and smoke dynamics, huge amounts of work have been produced and a lot of experimental testing has been carried out, and certainly will go on.

Three fundamental investigation techniques may be recognized.

5.1 *Full scale tests*

A lot of information has been retrieved from full size testing. This category includes large research programs, the objective of which is mainly to investigate in a systematic way the relationship between fire sizes and ventilation, the heat release rates versus time, the smoke production and its behavior, and the temperature fields along walls and ceiling. These findings enable one to define design fire scenarios, which ultimately will determine the necessary ventilation capacity to be installed.

An inventory of the many research programs can be found in (Piarc-Aipcr 1999). The most impressive one is still the Memorial Tunnel Fire Ventilation Test Program (1995). Ninety-eight tests of controlled fires of various sizes up to 100 MW were generated with several ventilation configurations. They included both natural and mechanical ventilation, the latter one being produced by a longitudinal, transverse or semi-transverse system. This last configuration has been tested with and without single point extraction.

These and other research programs provided interesting qualitative observations. However, this experimental technique cannot be made use of on a regular basis because each experiment requires important financial means.

To this category also belongs all the individual tests with calibrated fires such as fuel pools (Pérard et al. 1994) or occasionally the burning of a small car. They are usually performed in order to check the fire detection system and the ventilation response in terms of its extraction capacity. Since the heat release rate does almost never exceed 5 MW and the instrumentation is often limited, their input to general knowledge is rather poor.

5.2 *Small-scale models*

Small-scale models use often cold fumes or gas mixtures with tracers to simulate the behavior of smoke, and therefore, their application domain is rather limited. For instance, they aim to explore particular geometric characteristics of exhaust systems and the optimal location of extraction points. They are also used to check the operating efficiency of particular exhaust system configurations (Lacroix et al. 1994).

Of course, they are only representative insofar that a strong link between the model and the full-scale situation can be demonstrated, i.e. the experiment validity is relative to the used similarity law.

5.3 *Computer models*

With the development of the computers and their calculating power, numerical modeling represents today the most used tool to investigate the fire development and the aerodynamic smoke behavior in tunnels. These models, which are a deterministic approach of the fire development, give further insight on the influence of many parameters, provided that they have been calibrated by full-scale tests.

The models can be put in three main classes according to the use it is made of:

1. empirical models, usually 1D models, some of them including transient flow behavior: they assume an homogeneous mix of air and smoke in the cross-section and therefore they cannot take the layering phenomenon into account;
2. zone models: they split the overall fire environment into several zones, each zone being governed by a sub-model that acts as a node and that exchanges mass, energy and momentum with other nodes, and,
3. field models: they deal with the full description of the differential variables governing the process and are usually referred to as CFD (Computational Fluid Dynamics) models.

Apart from a large number of case studies, numerical simulation work reveals commonly particular aspects that are not always known or picked up in real fire tests with pools.

Specific for tunnel developed 1D software are MFIRE, SES, VENDIS-FS and probably a few more that are less known.

Zone models are occasionally used, e.g. when dealing with stratification. Since it is not always an easy task to choose a set of zones that represents a real fire and to set together these zones such in such a way that the interactions among zones are adequately accounted for, the predicted results could be doubtful.

Several general CFD solvers such as FLUENT, CFX, PHOENIX, CFD2000 are more and more applied to provide the user with local values of flow velocity, temperature and smoke concentration. They are able to take specific tunnel geometry's into account. With the latest dynamic meshing techniques they can also simulate car movements, so accounting for the piston effect.

According to experimented CFD users, all the solvers provide solutions within an acceptable error margin. But the same code handled by different experimenters may result into quite a variety of solutions. This fact demonstrates the necessity of skilful users with large experimental background. The CFD user must also be very cautious with extrapolating the results, even with interpolating them: fluid dynamics are non-linear and therefore certain minor changes may result into a significantly different solution.

Apart from the amount of time needed for the calculation, the major restriction for using these models extensively are the preliminary validations (den Boer et al. 2003) and the necessity of skillfulness, otherwise the results may be quite misleading. Further, development of these models must be encouraged because one day they may probably be used in a dynamic version, i.e. as a real-time tool. Then they would gain their highest interest as may appear from the discussions in the next paragraph.

6 OPERATING VENTILATION

It is commonly accepted that safety can still be enhanced by adequately operating the ventilation. Even if this affirmation would not be true, the contrary is certainly true: danger may be increased by an unconsidered or untimely ventilation intervention. One unfortunate example is the disastrous fire in the subway of BAKU (Hedefalk 1998) that caused the death of 289 passengers and severely injured another 265 people because: "*the smoke, originally mainly moving towards ULDUS, suddenly turned and filled the entire tunnel ... Somebody at the tunnel centre had changed the directional mode of the air intake fans ...*"

But before tackling the ventilation operating aspect a few observations are worth being made:

- Fires in tunnels have been observed to exhibit substantially different heat release rates than fires involving similar combustibles in the open air, but these differences have not been entirely investigated. Unfortunate experience shows that the consequences of fires in tunnels may be far more damaging than in the open (Drysdale 1999).
- The fire usually starts as a smoldering one. Then, from a certain point on it develops very fast and goes up to temperatures that may largely exceed 1000°C. The phenomenon of flash over is observed in many cases.
- Where ventilation interventions are concerned, the operator may be faced with different and most of the time conflicting requirements. On the one hand, the fire should be confined and possibly quenched by reducing as much as possible the admission of air to the fire site. On the other hand, the atmosphere must be kept breathable for people trapped in this area. The air volumes should also be sufficient for preventing temperature from raising too rapidly. The solution to this conflicting problem must be, therefore, a compromising airflow rate that might be expressed in terms of a certain bulk air velocity upstream of the fire. Usually, the ventilation capacity has been dimensioned to achieve this velocity. But, at the time the fire starts, there is usually a significant longitudinal flow. It is then not sure that the air velocity would have the so-called "optimal" value. In some cases where the mechanical ventilation is switched off, the actual flow could be in the reversed direction due to a strong influence of the natural ventilation. To be aware of this situation, one needs to measure the air velocity, which is not a very accurate value to obtain. And even if it were, there is still a real-time aspect that deals mainly with the way in which the fans and possible regulators or dampers has to be operated.

A well-trained operator who understands how the ventilation works and who got a good overall picture of the fire situation and the traffic is certainly able to control the ventilation in a most adequate manner. This statement, that might also be an overstatement, is implicitly based on a few assumptions:

1. "... *well trained operator* ...": although it seems obvious, the everyday reality is sometimes different. Some operators are known to have had only a one-shot training by way of an introduction into their new job. When later on they are faced with a severe fire hazard they could lack the necessary confidence in their capability to handle efficiently the situation. Therefore, training sessions should be the rule and occur on a regular basis.
2. "... *who understands how the ventilation works* ...": if the tunnel is a tube with a fan system that acts as a single fan unit, changing stepwise its duty point in order to obtain the expected airflow velocity is not very difficult. But if the tunnel is equipped with several independent fan units the problem is much more complex. This configuration might be the case in a semi-transverse system with more than one ventilation section and to which a multiple point exhaust system is added. Operating the fans in a balanced way together with opening certain dampers in view of realizing, for instance, the expected airflow velocity underneath the smoke layer, needs a full understanding of the underlying ventilation network and the way it behaves. The problem would be all the more tricky if the operator has to take account of the ventilation situation that existed before the fire occurred.
3. "... *who got a good overall picture of the fire situation and the traffic* ...": if, for instance, a certain number of CCTV cameras are available, the operator may be reasonably well informed on the growing fire and the traffic situation. But, in a longitudinal ventilation system he still has to decide whether he should put on full fan power in view of pushing the smoke out of the tunnel or, on the contrary, lower the air velocity to small values in order to maintain the initial smoke stratification.

This brief analysis clearly shows that the decision making process of the operator could sometimes be very difficult. It would certainly be a great help to him if he could briefly simulate the action on the ventilation that he is planning to take. Here the classical models cannot help him. Even the fastest ones take far too much time. Predicted scenarios calculated by elaborated models could, and most of the time would, provide operating guidelines. Unfortunately, the models produce these scenarios on basis of hypothetical fire situations and suppose that all mechanical equipment works perfectly. This is not always the case in a real tunnel environment where equipment may fail. It would be unreasonable to blame the operator for errors, as ineffective operation is often due to the complexity of ventilation systems and the rigidity of procedures. In this case the operator has to take over full control and would certainly need assistance of an interactive model that would simulate and predict the consequences of his planned intervention.

Of course, with nowadays technology no accurate numerical model could run sufficiently fast. However, it must be possible to build a model that could be seen as a patchwork of a series of recorded and previously calculated local scenarios. Any change of the ventilation configuration that needs to be tested would force the computer to select the closest scenarios and put them together in a way that the overall simulation shows approximately the response to the planned intervention.

Up to now the most advanced ventilation control system in case of fire is probably the newly developed one that is installed in the Mont Blanc tunnel. The control system runs a software that sequences automatically the different operations to be carried out. The chosen sequence depends on certain input data such as the location of the fire and the airflow velocities in the tunnel. Once the fire has been detected, the control unit starts up the exhaust fans, acts on a certain number of smoke extracting points by opening the remote controlled dampers, and activates certain banks of jet fans, so to confine the smoke within a certain section of the tunnel. All these predefined actions have been calculated by means of CFD codes. But, the ventilation control software seems not being able to come up with an alternative solution if, for instance, part of the mechanical equipment would not properly function. Extensive numerical modeling has only been used to dimension the ventilation capacity and to define the best operation rules for controlling the fire situation. The entire approach is based more on the phenomenon description obtained from a 3D flow pattern simulation than on its quantification. Subsequent full size fire tests in the Mont Blanc tunnel before its putting in operation confirmed the adequacy of this approach and showed that the developed control system met the expected safety targets (Houseaux et al. 2003).

But, whether the control system would work adequately in the next real situation can only be hoped for. Indeed, part of the equipment could be unavailable for whatever reason or fail to respond in the expected manner. Then the operator has to take over the entire control command and run the sequence of actions the best he can. For ventilation systems that are as sophisticated as the Mont Blanc system, there would be no guarantee, whatsoever, that he would succeed.

The idea of an interactive model had already been discussed by (Day et al. 2000). He pointed out that this kind of interactive model would also be useful as a simulator for training the operators and members of emergency services to learn about the dynamics of smoke flows (Crausaz et al. 2003).

Today, the most advanced control software in case of more or less complex ventilation systems is based on running predefined procedures. There may be several and different ones according to the section of the tunnel where the fire is supposed to take place. The Cointe tunnel in Liège, Belgium, with a length of about 1.6 km is one such example (Jacques et al. 1998).

One step further in controlling the fire and smoke movements is answering the question whether these procedures should be started up manually by the operator, or could be triggered automatically once the fire has been detected and confirmed.

It is common knowledge that the first few minutes are the most crucial ones. If the ventilation system is rather complex, it is most likely true that for most operators it would take time to activate the right fans at the right moment with the right flow rate, particularly where fresh air injecting fans are concerned. If, moreover, dampers need to be operated, one must agree that, considering the limitations of the actual technology, predefined procedures are still the best solution for fast and targeted intervention. Then the overall intervention would be restricted to a mere pressing of a "start/stop" button, which could also be achieved by the remote control unit probably in a faster and certainly more reliable way. In this latter case the operator must still have the possibility to stop the automatic procedure if something had to go wrong, for example, a false alarm, the fire site being erroneously located, etc.

The answer to this question is more an administrative and economical one than a technical one. If the control room is permanently manned by more than one (well trained) operator, the manager could easily be convinced in favor of a button to start up the fire procedure. On the contrary, if the control room is undermanned or the operators are not particularly well experienced, valuable time could be lost due to hesitation, unawareness, etc.

It is, nevertheless, the responsibility of the tunnel operator to activate the smoke extraction the as soon as possible. This responsibility cannot be passed on to the fire brigade who are not on the scene during the first minutes.

This latter argument is thus certainly in favor of an automatically started up fire procedure.

7 FIRE DETECTION EQUIPMENT

The efficiency of a fire and smoke movement control system is, of course, dependent on the capability of a fast and reliable detection of the fire. It is well known that actions during the first minutes have the greatest impact.

However, it is beyond the frame of this paper to describe the different techniques that are used for detecting the fire as fast as possible. A comparative study of this kind of equipment has been presented by Haack (in prep.) at the recent PIARC XXIInd World Road Congress in Durban.

8 CONTROL IN VENTILATION SYSTEMS

The way of controlling fire and smoke movements dependent on the ventilation system of the tunnel.

Obviously, longitudinal systems can only rely on longitudinal control of the smoke movements whilst transverse and semi-transverse systems only can rely on smoke extraction without efficient means of longitudinal control. Since the ultimate objective is always to keep a smoke-free area for as long as possible to allow trapped people to egress and rescue teams to access, a few general thumb-rules apply.

In one-directional tubes with a longitudinal ventilation system the only feasible way to evacuate smoke is pushing it through the tunnel to the exit portal. Experiments show that the air velocity must be at least 3 m/s in a tunnel with no slope (Wu et al. 2000). This way of proceeding keeps free of smoke that section where the vehicles are held up by the fire. The assumption is that all vehicles downstream of the fire site would have left the tunnel or are about to do so. But, it must not be overlooked that in some cases, for instance, due to a second accident, an engine breakdown or any incident, the downstream section might not be empty. This situation is not unusual in urban tunnels where frequently congestion occurs. Then it is advisable to lower the ventilation in order to obtain an airflow velocity that would not exceed 1.5 m/s. By proceeding in this manner the natural stratification of the smoke close to the ceiling could be maintained for several minutes, allowing the users to reach the shelters or escape exits.

This procedure certainly applies in a tunnel with two-way traffic where stopped vehicles and thus people, could be both sides of the fire. More can be learned in (Piarc-Aipcr 1999).

In tunnels with a transverse or semi-transverse system with point extraction, the rule is often that the initial natural stratification of the smoke at the ceiling, which occurs in the early stage of the fire development, should not be disturbed by an unconsidered intervention of the mechanical ventilation. From Japanese full scale tests it appears that the longitudinal resulting airflow velocity must certainly stay below 2 m/s. Higher velocities would indeed cause a faster cooling of the smoke layer that results into a stronger vertical mixing over the entire cross-section of smoke with fresh air.

In cut-and-cover tunnels with twin tubes special attention must to be paid to an adequate separation in terms of tightness between both the tubes. If from a civil engineering point of view complete tightness cannot be achieved, the operational procedure should include the creation of an appropriate overpressure in the other clear tube. This may be obtained by bringing the fresh air supply in the clear tube to full capacity and if possible, in reversed direction in order to avoid any effect of re-entrainment of smoke coming out from the exit portal into the entrance portal of the clear tube.

9 CONCLUSIONS

- Tunnel safety is not only a matter of meeting passive safety rules and recommendations where infrastructure and ventilation capacity are concerned, but must also include the entire set of ventilation operating rules for various fire and emergency scenarios.
- If a manual intervention is the option, then the most efficient procedure would still be to let the operator chose between in advance prepared sequences of the operations.
- After an automatic start up of the control procedure, it is imperative that any subsequent manual action or procedure remains compatible with the ventilation already in operation.
- Numerical and experimental models are useful to provide the designer with quantitative observations. They enable him to develop adequate operating procedures for controlling fire and smoke movements.
- If the ventilation system is a little bit complex, there may be a need for a discriminatory control depending on the section of a tunnel where the fire is located. Further model developments must, therefore, include the possibility of an interactive real-time simulation. This kind of model would then also be suitable for training the operators and members of emergency services to learn about the dynamics of smoke flows.

REFERENCES

Augin, G., Forestier, B., Giovannelli, G. & Casalé, E., 2003. Numerical simulations of a full-scale fire test in a road tunnel with an extended low-Reynold-number turbulence model, *Aerodynamics & Ventilation of Vehicle Tunnels Vol I*: 259–273. Bedford: BHR Group Conferences Series, Publication No.43.

Casalé, E. & Biollay, H., 2001. A fully controlled ventilation response in the case of a tunnel fire – the confinement velocity, *Safety in Road and Rail tunnels*: 477–485. University of Dundee & ITC Ltd.

Crausaz, B., Wheatherill, A. & Guigas, X., 2003. A real time 3D ventilation simulator for tunnel, *Aerodynamics & Ventilation of Vehicle Tunnels Vol I*: 275–286. Bedford: BHR Group Conferences Series, Publication No.43.

Day, J.R., Spoerri, M. & Turi, A., 2000. A 'footprint' technique for rapidly assessing the effectiveness of emergency ventilation systems, *Aerodynamics & Ventilation of Vehicle Tunnels*: 671–677. Bedford: BHR Group Conferences Series No.43.

den Boer, D.J., Mertens, J.J., van Oerle, N.J. & Wijnia, J.F., 2003. Validation of CFD calculations of full scale medium sized fires in a two lane road tunnel, *Aerodynamics & Ventilation of Vehicle Tunnels Vol II*: 657–667.

Drysdale, D., 1999. *An Introduction to Fire Dynamics*: 451. John Wiley & Sons Ltd, Chichester.

Haack, A. Fire specific equipment, in preparation.

Hedefalk, J., Wahlstrom, B., et al., 1998. Lessons from the BAKU subway fire, *Safety in Road & Rail Tunnels, 3rd International Conference*: 15–28. University of Dundee & ITC Ltd.

Houseaux, B., Biollay, H. & Casalé, E., 2003. Fires in Tunnels: Integration of numerical simulation in the automatic ventilation responses – Lessons drawn from the full-scale fire tests in the Mont Blanc tunnel, *Aerodynamics & Ventilation of Vehicle Tunnels Vol I*: 103–115. Bedford: BHR Group Conferences Series, Publication No.43.

Jacques, E. & Seynhaeve, J.M., 1998. Fire in the Cointe tunnel: A case study, *Proceedings of Eurofire '98*, Brussels: CD-ROM.

Kumar, S. & Cox, G., 1988. Radiant heat and surface roughness effects in the numerical modelling of tunnel fires, 6th Int. *Symposium on Aerodynamics & Ventilation of Vehicle Tunnels*: 515–526. Bedford: BHRA.

Lacroix, D., Chassé, P. & Muller, T., 1994. Small scale study of smoke trap door systems, *Aerodynamics & Ventilation of Vehicle Tunnels*: 409–438. Bedford: BHR Group, Conferences Series, Publication No.12.

Memorial Tunnel Fire Ventilation Test Program, 1995. Interactive CD-ROM & Comprehensive Test Report, Massachusetts Highway Department, US Federal Highway Administration.

Pérard, M. & Brousse, B., 1994. Full size tests before opening two French tunnels, *Aerodynamics & Ventilation of Vehicle Tunnels*: 383–408. Bedford: BHR Group, Conferences Series, Publication No.12.

Piarc-Aipcr, 1999. *Fire and smoke control in road tunnels*, Committee on road tunnels (C5): report.

Rhodes, N., 2000. Modelling of tunnel fires – verification and validation, *Aerodynamics & Ventilation of Vehicle Tunnels*: 49–62. Bedford: BHR Group Conferences Series, Publication No.43.

Wu, Y., Abu Bakar, M.Z., Jagger, S., Bettis, R. & Allen, J., 2000. Control of smoke flow in tunnel fires using longitudinal ventilation systems – a study of the critical velocity, *Aerodynamics & Ventilation of Vehicle Tunnels*: 639–658. Bedford: BHR Group Conferences Series, Publication No.43.

Dust generation, transport and control

Mine Ventilation, Ganguli & Bandopadhyay (eds)
© 2004 G.J.Chekan, ISBN 90 5809 633 5

Evaluating ventilating air movement in underground limestone mines by monitoring respirable dust generated from production shots

G.J. Chekan, J.F. Colinet, & Roy Grau III
National Institute of Occupational Safety and Health, Pittsburgh Research Center, Pittsburgh, PA, USA

ABSTRACT: Underground limestone mines typically have large entries, ranging from 9.1 to 18.3 m (30 to 60 ft) wide and 4.9 to 12.1 m (16 to 40 ft) high, which may result in low-velocity airflow throughout the mine. Air velocities less than 0.13 m/s (25 fpm) are common, and airflow in the entry can be stratified or readily affected by seasonal patterns of natural ventilation. In addition, limestone mines may have minimal or no mechanical ventilation and often do not have extensive stopping lines, constructed to control air movement throughout the mine. Consequently, it can be difficult to measure air velocities and define airflow patterns by using conventional airflow measuring equipment such as anemometers and smoke tubes. This research describes a different approach for evaluating air movement and flow patterns throughout the mine by measuring respirable dust levels generated from production shots as the dust moves through the mine air circuit. Production shots generate a considerable volume of dust in a very short time and serve as a distinctive point source of dust that can be monitored. Typically, faces are shot on an off-shift with no mining activity, so dust levels from the shots are not influenced by dust generation from loading, hauling, and crushing operations. The respirable dust that becomes airborne after the shot can remain entrained in the air even at very low velocities. This dust moves with the general airflow patterns in the mine. In this study, light-scattering instruments, which log dust concentrations in real-time, were used to monitor the dust movement throughout two limestone mines. Studies were conducted using these instruments to determine if they may provide a viable means to assess overall mine ventilation patterns. Dust samplers were positioned at selected locations throughout each mine to record respirable dust movement after production shots. Dust concentration results indicate that this method can be used to quickly assess mine ventilation patterns. This sampling also identified the length of time that was needed for the ventilation in each mine to clear the dust from the mines after the shots. A summary of the sampling instrumentation, sampling technique, and results of the studies will be discussed.

1 INTRODUCTION

The National Institute for Occupational Safety and Health (NIOSH) at the Pittsburgh Research Laboratory is currently involved in various research projects related to worker health and safety in underground metal/nonmetal mines. A primary area of research involves health issues in underground stone mines, a growing segment of the aggregates industry. Methods for reducing worker exposure to noise, silica dust, and diesel particulate matter (DPM) are being addressed through various research programs. One engineering control being addressed is the improvement in mine air quality by developing mine-wide ventilation techniques as a means to reduce worker exposure to silica dust and DPM.

Various research studies have defined some of the primary ventilation considerations for improving airflow in large-opening mines. These studies show that two key parameters for improving airflow are use of booster fans, such as axial vane or jet fans to improve localized ventilation, and more recently the application of high-volume propeller fans, and the construction of stoppings to direct and control airflow on a larger mine-wide scale (Grau et al, 2002a; Grau et al, 2002b; Head, 2001; Kissel and Volkwein, 2002; Timko and Thimons, 1987). Typically, underground stone operations are drift mines developed after the quarry reserves have been exhausted. Room-and-pillar mining methods are utilized, with pillars of either square or rectangular dimensions ranging from 10.6 to 18.3 m (35 to 60 ft). The entries are considered large mine openings with entry widths ranging from 9.1 to 18.3 m (30 to 60 ft) and entry heights on development ranging from 4.9 to 12.1 m (16 to 40 ft). After benching, entries can be over 18.3 m (60 ft) high. Due to these large openings, ventilation fan pressure is very low, even if significant quantities of air move through the mine. Ventilation pressures of less than 24.9 Pa (0.1 in w.g.) are common whether airflow is induced by fans, natural, or a combination of both (Grau et al, 2002a). Depending on the extent of the workings, air velocities <0.13 m/s (25 fpm) are common, or in some idle areas of the mine, virtually nonexistent. In addition, airflow in the entries can be stratified, or the direction of airflow can be readily affected by the movement of mine equipment.

Instruments for measuring air velocity in large-opening mines include smoke tubes, low flow anemometers (for airflow >0.26 m/s), or more recently, ultrasonic anemometers. These methods work well for measuring and evaluating localized air movement in main air courses or at working faces, but are limited in use when attempting to assess the overall mine ventilation patterns. Sulfur hexafluoride (SF_6) tracer gas is a commonly used method for assessing mine ventilation systems on a larger scale. Studies show (Thimons et al, 1974; Timko and Thimons, 1982) that it is an effective technique for evaluating the air velocity and airflow patterns for all types of mines. However, this method has two primary limitations. First, depending on the scope of the ventilation study, it can be very time consuming. To conduct a thorough study may require many individuals, positioned at different locations in the mine, collecting hand samples with vacuum tubes at regular intervals over an extended period of time. Second, depending on the number of samples to be analyzed, turn around time for results can be slow and costly.

As a result, NIOSH is investigating a technique to assess airflow patterns in large-opening limestone mines by monitoring the travel path of respirable dust generated by production shots. Production shots are a regular part of the mining cycle and can vary in number and location depending on the mining plan and the grade of stone required to meet production schedules. Typically, faces are shot during an off-shift with no mining activity, so dust levels are not influenced by other sources of dust generated by loading, haulage, and crushing operations or diesel particulate. Production shots, like most rock breakage mechanisms, generate a considerable amount of dust in various size ranges. A majority of the dust is considered oversized and settles out of the air stream. However, the respirable portion, dust particles <10 microns, once airborne can remain entrained in the air even at very low velocities (Baron et al, 1993; Welby et al, 1988; Xu and Bhaskar, 1995). As a result, the respirable dust generated by production shots offers a distinctive point source in that this entrained "dust cloud" can be measured and tracked with real-time dust-measuring instrumentation as it is transported through the general mine air circuit.

The advantages of this production shot monitoring technique are twofold: (1) the technique requires little manpower as the monitors are positioned at various locations in the mine, before the faces are shot, to record the "dust cloud"; (2) the real-time monitors have a logging feature so that dust concentration data can be downloaded into database software and analyzed in a short time frame.

This paper describes the instrument setup and method for monitoring production shot dust, and details how the results are analyzed to determine the "dust cloud" velocity. Studies were conducted at two limestone mines using this technique to determine if it may be a viable means to assess overall mine airflow patterns. The results proved encouraging as an effective method to quickly evaluate air movement in large-opening mines.

2 SAMPLING INSTRUMENTS AND SURVEY TECHNIQUE

The method employed in these surveys requires a real-time aerosol monitor, which uses light scattering to determine dust concentrations. These concentrations are recorded on an internal logger along with time. Although these light-scattering instruments offer only a relative measure of concentrations, they provide a continuous record of dust levels so that concentrations can be evaluated over any time interval during the sampling period. There are several types of real-time aerosol monitors available, and these investigations used the Thermo-MIE personal DataRam (pDR). In these studies, the instrument was operated in the active mode to monitor respirable dust. In this active mode, dust enters the pDR after it is classified using a 10-mm Dorr-Oliver cyclone combined with a pump operated at flow rate of 1.7 lpm resulting in a D_{50} cut-point of 4.0 microns. This set-up is shown in figure 1.

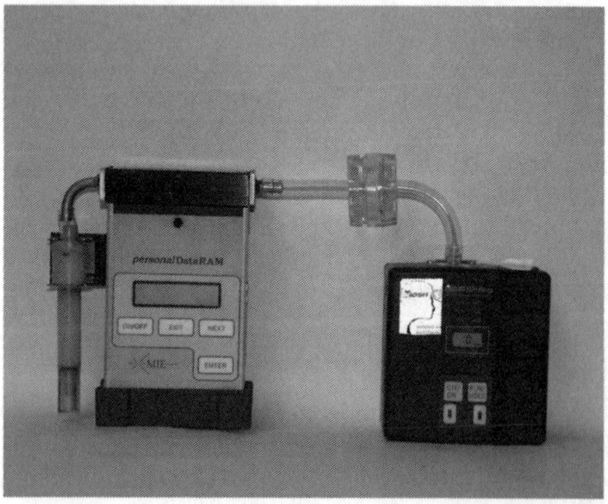

Figure 1. Thermo-Mie personal DataRam (pDR) set-up in the active mode with 10-mm Door-Oliver cyclone and pump operated at 1.7 lpm.

The pDR operates on one 9-volt battery, providing the unit with power for up to 18 hrs, and can store 13,000 data points (concentration values). The data can be downloaded from the unit and analyzed with software provided by the manufacturer or in other types of database software such as Excel.

The objectives of the studies were twofold: (1) to record the travel path and time arrival to determine the average velocity of the "dust cloud" as a means to evaluate the mine's airflow patterns; and (2) to determine the retention time of the dust or the length of time needed to clear the dust from the mine. The sampling strategy was to set up sampling stations in key locations in the mine's air course and begin sampling before the faces were shot. Sampling locations were based on suspected airflow patterns in the mine and the potential of dust from the production shots in the working developments to pass that particular location. In most cases, instruments were positioned at least 61 m (200 ft) from the shot locations to protect the pDRs from the initial blast pressure and to give the dust time to settle to the normal mine air velocity since blast pressures can provide an initial "velocity assist" to the dust cloud. All instruments were positioned in the same manner, on the rib approximately 1.8 m (6 ft) above the floor.

In both studies, the mines shot the faces on the evening shift between 4:00 p.m. and 12:00 a.m. Sampling for the surveys began about 2 hours before the end of the day shift (around 2:00 pm) and continued for approximately 18 hrs until the beginning of the next day shift at 8:00 am. This long sampling period provided sufficient data so as to differentiate between normal background particulate levels generated by mining activity and the dust generated by the shots. In all studies, the pDRs were set to log a concentration every 10 seconds, providing approximately 6,500 concentration values for analysis. In most surveys, 1 to 4 faces were shot at the same time or within minutes of each other. This provided an ample volume of dust whose movement could be monitored.

3 CHARACTERISTICS OF DUST GENERATED BY SHOTS AND METHOD FOR DETERMINING DUST CLUD VELOCITY

Production shots generate a visible "dust cloud" that contains dust in various size ranges. As the dust cloud travels from the source, the oversize dust begins to settle from the airstream. The respirable dust, <10 microns, remains entrained but begins to break up and dilute becoming visibly less distinguishable as it travels further from the source. Figure 2 shows a characteristic curve when the pDR dust concentrations from a shot are graphed in relation to time. This example is typical of the "dust cloud" signature of most shots. The definitions for the nine terms in figure 2 are as follows:

1. *Baseline Concentration* – This is the average concentration recorded on the pDR prior to the time of the shots. It represents the background particulate levels under normal operating conditions.
2. *Shot Time* – The time when the shot took place after sampling began, obtained from mine records.
3. *Arrival Time* – This is the length of time from the shot time to when the dust was first detected on the pDR. It is the leading edge of the dust cloud.
4. *Peak Concentration* – This is the highest concentration recorded on the pDR as the dust cloud passes the station.
5. *Peak Time* – This is the length of time from the shot time to the time of the peak concentration.
6. *Average Concentration* – This is the average concentration recorded on the pDR during the duration of the dust cloud. It is the area under the curve and can be compared to the average baseline concentration as a means to evaluate dilution of the dust cloud.
7. *End Time* – This is the length of time from the shot time to the tail end of the dust cloud. The dust cloud is no longer detected as concentrations return back to baseline values.
8. *Duration of Dust Cloud* – This is the length of time it takes the cloud to pass a particular station or the difference between the arrival time and end time.
9. *Return to Baseline* – This represents the time when the pDR concentrations return back to baseline values.

The shape of the curve depicted in figure 2 will change depending upon the distance between the shot location and the pDR sampling location and the mine air velocity. If the pDR is close to the shot location (in most cases <220 m), the curve is characterized by a well-defined, high peak and short duration of the dust cloud. As the distance between the shot and pDR locations increases and the air velocity decreases, the peak concentration decreases. However, the duration of the dust cloud increases as it begins to dilute and lengthen relative to the lower airflow.

The method used to estimate the velocity of the dust cloud is based on calculating an average velocity from three different time points on the curve as shown in the example in figure 2. The three points are the "arrival time," "peak time," and "end time" of the dust cloud. The time is estimated for each of these events and a velocity calculated for each time based on the distance of the pDR location from the shot location. The velocity is then the average of these three velocities. This method was chosen for several reasons: (1) it defines both the beginning and end point of the dust

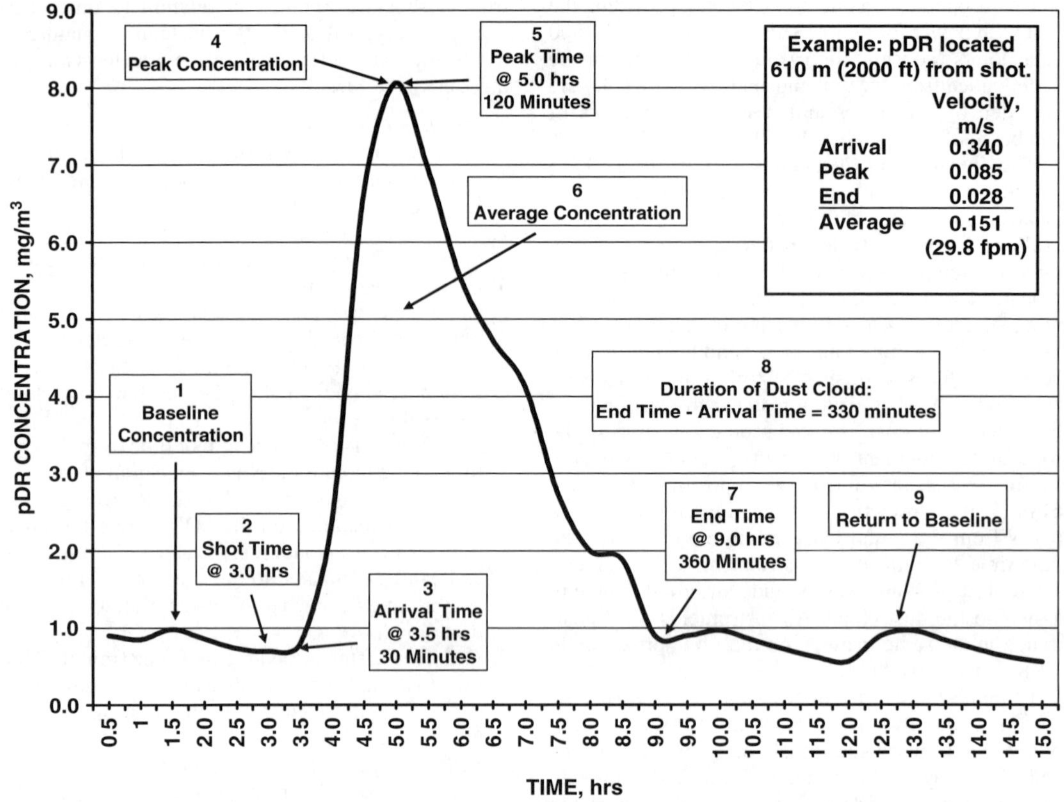

Figure 2. Characteristic curve when the pDR concentrations are graphed in relation to time, with example showing the estimated velocity of the "dust cloud" using the three time points.

cloud as the dust completely clears the station; (2) it takes into consideration the increasing time duration or length of the cloud as it moves further from the source, which gives equal weight to the assessment of the dust cloud velocity as it passes each station; (3) it gives the most conservative estimate of velocity, as compared to using only the arrival time or the peak time.

The optimum results are achieved when one or several shots are in the same location of the mine at the same time. This creates a dust cloud from one general location and gives an easily distinguishable signature on the pDRs. In contrast, when faces are shot at distant locations from each other, or at different times, multiple signatures are logged on the pDR and can be difficult to distinguish from each another. The more pDRs that are in use, the better the general airflow pattern in the mine can be characterized. If mine personnel have a knowledge of the general flow patterns or direction of airflow in the mine, this information proves useful in positioning the pDRs. Otherwise, smoke tube tests at different areas of interest in the mine can be used to determine where airflow is consistent to position the pDRs.

4 CASE STUDIES

The main mine fan used in case 1-study A was a Pope 1.8 m (6 ft) dia., 185.5 kW (250 hp), 164.9 m^3/s (350,000 cfm), axial vane fan used in the exhaust mode. In addition, two booster fans were used: a Joy 1.53m (5 ft) diameter axial vane fan rated at 74.2 kW (100 hp) and 70.7 m^3/s (150,000 cfm); a Buffalo 1.37 m (4.5 ft) diameter axial vane fan rated at 92.7 kW (125 hp) and 42.4 m^3/s (90,000 cfm). All of the fans were free-standing, not bulkheaded to separate intake air from return air. In case 1-study B, the Pope main mine fan was replaced with two Hartzell Series 10S, 3.66 m (12 ft) dia., high-volume, low-pressure propeller fans each rated at 212 m^3/s (450,000 cfm). These two fans were bulkheaded at a portal and exhausted to the outside air.

An important difference between study A and study B is the increase of ventilating air into the mine and the decrease in total power requirements for the fans. With the addition of the propeller fans, the total airflow increased from 141.3 m^3/s (300,000 cfm) to 353.4 m^3/s (750,000 cfm), while the combined total power requirement of the fans decreased from 352.5 kW (475 hp) to 315.4 kW (425 hp).

The mine for the case 2 study utilized the same Hartzell Series 10S propeller fans as in case 1. The two fans were bulkheaded and installed in the exhaust mode. In addition, the mine utilized one free-standing booster fan, a Spendrup Model 1000-50-26H axial vane fan rated at 37.1 kW (50 hp) and 21.2 m^3/s (45,000 cfm). The Hartzell Series 10S propeller fans are shown in figure 3.

Figure 3. Hartzell Series 10S propeller fans used in both case studies.

In both studies, the ventilation air was coursed, and intake entries were separated from return entries using various forms of brattice type stoppings. The brattice type ranged from inexpensive tarp material to heavier mine brattice material. This brattice was sealed as tightly as possible to the ribs, roof, and floor to minimize air leakage. In some instances, air leakage from stopping damage due to either face blast pressures or fatigue by fan pressures was observed.

4.1 *Case 1*

4.1.1 *Background*

Nine sampling locations in the mine were selected as shown in figure 4. Site locations were based on suspected airflow patterns in the mine and the potential of dust from the production shots in the working developments to pass that particular location and allow for the recording of the dust cloud arrival. Table 1 identifies the reasoning behind placing the pDRs at each sampling location.

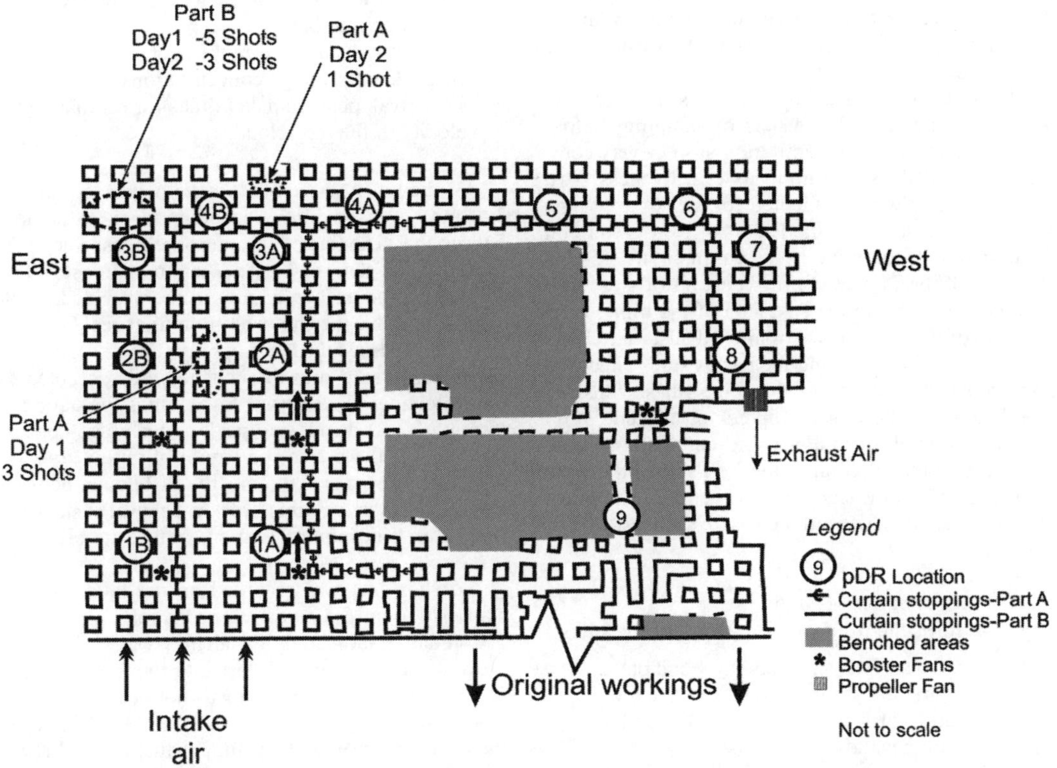

Figure 4. Mine workings for case study 1 showing location of pDR samplers, curtain stoppings, and fans.

Table 1. Case study 1 – pDR location and description for production shot dust surveys.

Station	Purpose
1	Intake air – monitor intake dust levels and dust that might roll back into the fresh air circuit from face shots on east side of mine.
2	General air circuit – 2 stations were located on east side of mine on the inby side of the curtain stoppings to
3	monitor dust migration from face shots on east side of mine.
4	General air circuit – 3 stations were located on the south end of the mine on the inby side of the curtain
5	stoppings. They monitored dust migration from face shots on the eastern side of the mine and were used to
6	assess the effectiveness of the air circuit being planned.
7	General air circuit – 2 stations were located on the western side of the mine on the inby side of the curtain
8	stoppings. They monitored dust migration from face shots on the eastern side of the mine and were used to assess the effectiveness of the air circuit being planned.
9	Return air – monitor dust that might migrate past the location of the main exhaust fan and the propeller fans and thus recirculate into the intake air.

This study was initiated to assess the current ventilation system, particularly on the east side of the mine. In this study, the mine was planning ventilation changes by constructing approximately 30 to 40 brattice type stoppings and installing two-low pressure propeller fans, each rated at 212 m^3/s (450,000 cfm), to better ventilate the working faces in the mine. It was assumed that the stoppings and fans would increase the volume of air to the faces and establish a directional flow of air from the eastern to the western side of the mine.

The strategy for the dust survey is a commonly used A-B study which consists of sampling before and then after the new ventilation system was completely installed to determine its effectiveness. Two sampling shifts were conducted for the A part of the study and two shifts for the B part. All shots were on the east side of the mine as shown in figure 4. Also shown in figure 4 are: (1) the location of two axial vane booster fans on the east side of the mine, their blowing direction, and the main exhaust fan (Pope 1.8 m dia. used only during the part A study) near the portal on the west side of the mine (these fans are mobile and can be moved depending on ventilation patterns required for mining); (2) the location of the low-pressure propeller fan at the west side of the mine (used in part B of the study) and; (3) curtain stoppings used to direct airflow from the east to the west side of the mine.

4.1.2 *Results of part A study*

Figure 4 shows the following for part A of the study:

1. The locations of the shots on day 1 and day 2.
2. The stopping line on the east side of the mine.
3. The location of stations 1A, 2A, 3A, and 4A for this part of the study.
4. The location of the last stopping by station 4A.
5. The two booster fans on the east side of the mine.

For part A of the study, the propeller fans were not yet installed. Dust from the shots exited the mine at the portal at the Pope 1.8 m (6 ft) dia. main exhaust fan between stations 8 and 9.

Table 2 summarizes the results of the dust surveys for the two days of sampling. Shown in the table are:

1. The number of shots for each sampling day.
2. The station number.
3. The station distance from the shot.
4. The peak and average concentrations.
5. The arrival, peak, and end times for estimating the velocity of the dust cloud.

As shown in the table, the velocities at stations 3A and 4A were higher in comparison to stations 5 through 9. The line of curtains on the east side of the mine up to station 4A and the two booster fans located on the east side curtain line provided ample airflow to move the shot dust from this area. Once the dust cloud passed the last stopping (figure 4), the dust began dispersing into the benched area as noted by the following: (1) decreasing velocities for stations 5 through 9; (2) decreasing peak and average concentrations; (3) the increasing length of time it took the dust cloud to pass a station (the difference between end time and arrival time); (4) dust arrival at station 9, the return side control station, before its arrival at station 8.

No shot dust was recorded on the intake side at stations 1A and 2A on day 1 and stations 1A, 2A, and 3A on day 2, indicating that airflow was sufficient to prevent the migration of dust into the intake air and indicating that recirculation was not occurring. Also, since dust was recorded at station 9, it indicates the dust was passing the main exhaust fan and it was not effectively moving the shot dust out the nearby portal.

Table 2. Part A of case study 1: estimated velocity of dust cloud before fans and curtains were installed.

Station	Approximate distance from Shot, m (ft)	Peak conc., mg/m³	Average conc., mg/m³	Arrival time, min	Peak time, min	End time, min	Estimated average velocity, m/s (fpm)
	Day 1 – 3 shots						
3A	45.7 (150)	45.6	11.3	1	2.8	17.5	0.36 (70.5)
4A	121.9 (400)	17.8	5.3	3.8	14.0	81.3	0.23 (45.9)
5	228.6 (750)	2.5	1.3	17.2	31.7	153.2	0.12 (24.1)
6	304.8 (1000)	2.9	1.4	28.8	60.0	196.5	0.095 (18.8)
7	350.5 (1150)	2.2	1.1	38.5	61.8	264.8	0.089 (17.6)
8	411.5 (1350)	1.7	1.0	48.2	96.7	255.7	0.081 (15.8)
9	426.7 (1400)	2.2	1.0	40.8	74.8	244.3	0.11 (21.0)
2A	No dust recorded from shot						
1A	No dust recorded from shot						
	Day 2 – 1 shot						
4A	24.7 (150)	11.8	4.1	1.2	7.5	23.8	0.26 (51.6)
5	152.4 (500)	1.2	0.7	15.2	74.5	175.3	0.072 (14.2)
6	228.6 (750)	1.2	0.7	32.3	60.8	175.0	0.068 (13.3)
7	274.3 (900)	1.1	0.6	38.0	49.3	177.5	0.080 (15.7)
8	335.2 (1100)	1.7	0.7	39.5	57.0	187.0	0.090 (17.7)
9	350.5 (1150)	1.6	0.7	31.7	49.2	222.2	0.11 (21.6)
3A	No dust recorded from shot						
2A	No dust recorded from shot						
1A	No dust recorded from shot						

4.1.3 *Results of part B study*

Figure 4 shows the following for Part B of the study:

1. The locations of the shots on day 1 and day 2.
2. The new stopping line on the east side of the mine that was moved as the mine developed more rooms on the east side of the mine.
3. The location of stations 1B, 2B, 3B, and 4B that were moved for this part of the study. Stations 5 through 9 remained in the same locations.
4. The completion of all stopping around the perimeter of the active faces which completely sealed off the benched area.
5. The relocation of the two booster fans on the east side of the mine.
6. The two Hartzell Series 10S propeller fans which were now installed and operational and replaced the Pope fan.

Table 3 summarizes the results of the dust surveys for the two days of sampling. Shown in the table are:

1. The number of shots for each sampling day.
2. The station number.
3 The station distance from the shot.
4 The peak and average concentrations.
5. The arrival, peak, and end times for estimating the velocity as in the example in figure 2.

As shown in the table, the velocities are much higher and more consistent after the installation of the two propeller fans and the remaining curtains on the south and west side of the mine. The shots on both days were located in the same area of the mine and dust was recorded at stations 4B through 8. The improvement in airflow with the completion of the ventilation system is evident and noted by several points. First, dust from the shots was not recorded at station 9 (return side), indicating that the curtain line was adequately sealing off the benched area and all the dust was now exiting the mine at the propeller fans by station 8, preventing recirculation. Second, the peak and average concentrations are much higher and the duration of the dust cloud (the difference between the arrival and end time) is much shorter in comparison to part A of the study. This demonstrates that the dust cloud was not dispersing but rather staying together and exiting the mine more effectively. Third, no dust was recorded at station 1B through 3B, indicating that airflow on the east side of the mine was sufficient to prevent rollback of the dust from the shots into the intake air.

Also evident is the reduction in the retention time or the length of time it takes the dust cloud to clear the mine. The "end time" at station 8 in tables 2 and 3 provides a means to assess the retention time. Based on the average end time for both sampling days, the retention time for dust shots generated on the east side of the mine in Part A was approximately 3.7 hrs and for Part B was 2 hrs.

Table 3. Part B of case study 1: estimated velocity of dust cloud after fans and curtains were installed.

Station	Approximate distance from Shot, m (ft)	Peak conc., mg/m^3	Average conc., mg/m^3	Arrival time, min	Peak time, min	End time, min	Estimated average velocity, m/s (fpm)
	Day 1 – 5 Shots						
4B	91.4 (300)	61.6	17.6	1.7	4.3	69.8	0.43 (84.5)
5	234.8 (800)	17.6	5.7	5.7	12.3	71.7	0.37 (72.4)
6	320.1 (1050)	14.5	4.6	9.3	15.1	111.0	0.32 (63.7)
7	365.8 (1200)	14.9	3.9	10.2	15.8	118.0	0.35 (68.0)
8-Fan	426.7 (1400)	9.7	4.3	12.2	22.3	121.7	0.32 (62.6)
9	No dust recorded from shots						
3B	No dust recorded from shots						
2B	No dust recorded from shots						
1B	No dust recorded from shots						
	Day 2 – 3 Shots						
4B	91.4 (300)	44.5	6.3	2.2	10.0	61.3	0.29 (57.8)
5	234.8 (800)	15.9	5.5	7.3	16.0	73.8	0.29 (56.6)
6	320.1 (1050)	16.3	4.9	10.7	21.7	88.5	0.26 (52.9)
7	365.8 (1200)	13.1	4.9	11.5	23.0	92.5	0.29 (56.5)
8-Fan	426.7 (1400)	12.3	4.5	13.8	35.8	119.2	0.26 (50.7)
9	No dust recorded from shots						
3B	No dust recorded from shots						
2B	No dust recorded from shots						
1B	No dust recorded from shots						

4.2 *Case 2*

4.2.1 *Background*

In this case study, the mine was recently purchased by new owners. Figure 5 shows the mine's development, which was very extensive and originally not well-planned for most of the life of the mine. The new management has begun a more systematic approach to mining, as shown by the projections for future developments. The mine's original ventilation plan relied solely on natural ventilation without the assistance of booster fans. Airflow to the faces especially in the northern and southern areas of the mine was minimal, as air velocity and direction were difficult to characterize with anemometer or smoke tube measurements. Dust from shots tended to migrate through the mine with little directional flow.

These circumstances required some extensive rehabilitation work which included major upgrades in the ventilation system to improve airflow at the faces. As shown in figure 5, this ventilation system included: 1) two Hartzell Series 10S propeller fans that established a directional airflow from the southern portals to the fan; 2) the Spendrup Model 1000-50-26H axial vane booster fan to provide localized ventilation to the working faces; 3) the installation of approximately 50 curtain stoppings (the stopping line is shown in figure 5) on the southern and eastern side of the mine to provide improved airflow to the active faces.

Due to the large extent of the workings and to provide sufficient ventilating air to the new faces, the site selection for the two propeller fans was different from the typical application in that they were located inside the mine rather than at a portal. On the exhaust side of the fan, to the western side of the mine, was a large abandoned benched area. The plan was to dump the exhaust air into this area where the dust would dilute and eventually exit the mine at portals to the west.

In this case, the study was conducted after the new ventilation system had been completed. Figure 5 also shows the 10 sampling stations selected for the study and location of the shots for the two days of sampling. The location of the stations was based on smoke tube measurements placing the pDRs near the faces and spacing them to cover areas of interest where airflow was consistent. Table 4 identifies the purpose of the pDRs at each sampling location.

4.2.2 *Results of study: Day 1 and Day 2*

Table 5 summarizes the results of the dust surveys for the two days of sampling. Shown in the table are:

1. The number of shots for each day.
2. The station number.
3. The station distance from the shot.
4. The peak and average concentrations.
5. The arrival, peak, and end times for estimating the velocity of the dust cloud.

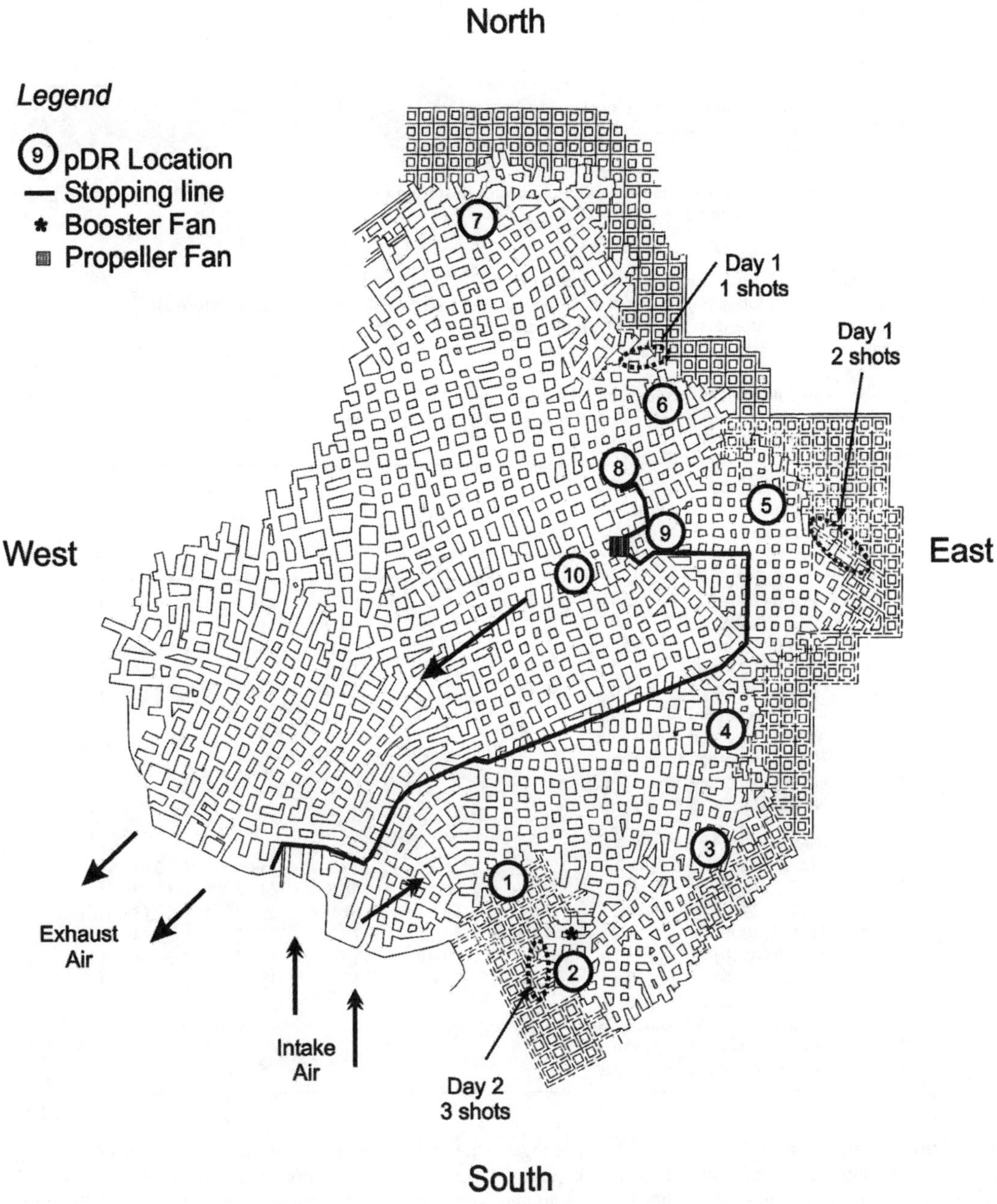

Figure 5. Mine workings for case study 2, showing location of pDR samplers, curtain line, and fans.

There were 3 shots on day 1 of sampling as shown in figure 5. This study was designed to monitor the movement of dust from nearby faces located east of the fan and determine if the ventilation was sufficient to prevent the dust from migrating to other areas of the mine. As shown in figure 5, shot 1 was located near station 6 and shots 2 and 3 were located near station 5. This distance and time spacing of shots (shot 1 was 4 minutes before shots 2 and 3) provided two different signatures on the pDRs. The pDR logs showed that dust from the shot 1 migrated past station 6, to station 8, and then to station 9 at the fan. Dust from shots 2 and 3 migrated

Table 4. Case study 2: pDR locations and descriptions for production shot dust surveys.

Station	Purpose
1	Intake air – monitor fresh air dust levels.
2 3 4 5 6 7	General air circuit – 6 stations were used to monitor dust migration from shots in the southern, eastern, and northern faces of the mine.
8	Control station – this station monitored dust that might migrate past the fan.
9	Main fan – monitor dust levels from shots on the intake side of fan before entering return air.
10	Return air – monitor dust levels in the return on the exhaust side of fan.

Table 5. Estimated velocity of dust cloud for case study 2.

	Day 1 – 3 Shots						
Station	Approximate distance from shot, m (ft)	Peak conc., mg/m^3	Average conc., mg/m^3	Arrival time, min	Peak time, min	End time, min	Estimated average velocity, m/s (fpm)
6	61.0 (200)	7.9	1.4	1.2	6.8	152.8	0.34 (67.3)
8	260.0 (850)	3.2	1.2	12.8	25.0	211.7	0.18 (34.7)
5	198.0 (650)	22.2	3.4	4.5	8.0	66.2	0.31 (60.4)
9-Fan	365.8 (1200)	13.2	3.5	10.3	18.0	74.2	0.34 (66.3)
10	579.1 (1900)	4.0	1.8	13.3	27.0	198.5	0.38 (74.1)
1	No dust recorded from shots						
2	No dust recorded from shots						
3	No dust recorded from shots						
4	No dust recorded from shots						
7	No dust recorded from shots						
	Day 2 – 3 shots						
2	122.0 (400)	18.8	5.0	5.2	24.3	284.0	0.16 (31.8)
3	487.7 (1600)	1.9	1.1	30.5	58.8	237.2	0.15 (28.8)
4	731.5 (2400)	1.2	0.8	36.8	61.0	292.0	0.19 (37.6)
5	1188.7 (3900)	1.2	0.6	55.2	75.7	325.7	0.23 (44.7)
9-Fan	1371.6 (4500)	1.1	0.7	60.0	85.5	328.7	0.24 (47.2)
10	1585.0 (5200)	0.7	0.4	72.8	119.3	354.7	0.23 (43.2)
1	No dust recorded from shots						
6	No dust recorded from shots						
7	No dust recorded from shots						
8	No dust recorded from shots						

past station 5 then to station 9 at the fan. The pDR log signatures at station 9 (fan) showed only one peak value indicating that the dust from the different shot locations merged and reached the fan at approximately the same time. This is reflected in table 5.

Since dust was recorded at station 8, some dust was passing the fans. However, the lower velocity of 0.18 m/s (34.7 fpm), lower peak value, and longer duration of the dust cloud in comparison to the other stations all indicate that most of the dust from shot 1 made a more direct path to the fans. Station 10, located approximately 152 m (550 ft), from the fan on the exhaust side, showed that velocities remained high but dilution was occurring, as evidenced by the lower peak and average concentrations and the longer duration of the dust cloud. No dust was recorded at stations 1, 2, 3, 4, or 7, showing that the established ventilation was sufficient to keep the dust from migrating to the northern and southern areas of the mine. Consistent velocities of the dust cloud were recorded at the 3 stations located in the main circuit averaging 0.33 m/s (65 fpm).

There were 3 shots on day 2 of sampling as shown in figure 5. This study was designed to evaluate the airflow pattern from distant face locations in the southern section of the mine. These shots were approximately 1432 m (4700 ft) from the main fans as all three faces were shot at the same time and in the same general area. Table 5 shows the pDR results, which are notably different in comparison to day 1, but do show a distinguishable airflow pattern towards the fans even from this distance.

As expected, station 2, located 122 m (400 ft) from the shots, shows the highest peak and average concentrations (18.8 and 5.0 mg/m^3, respectively), but the duration of the dust cloud was long (278 minutes) indicating that the nearby booster fan was moving air to these headings but not effectively enough to remove the dust. The leading edge of the dust cloud then moved in sequence to stations 3, 4, and 5, located in the general air circuit, and then to station 9 at the main fans. The peak and average values at station 2 were high in comparison to the remaining stations. Stations 3, 4, 5, and 9 all show a similar pattern of long duration and low average and peak concentrations of the dust cloud. Even at these low values, the signature of the dust cloud was evident at all the stations. These low values may indicate that a majority of the dust took a more direct path to the main fans; but since the arrival times at each successive station are in sequence, the data more likely suggest that the dust cloud was diluting in the southeastern area of the mine and gradually migrated to the main fans.

The calculated velocities at each station were very consistent, increasing slightly as the dust cloud approached the fan. The velocity on the exhaust side of the fan (station 10) was also consistent with these velocities. No dust was recorded at station 1, the intake air, indicating that no dust was migrating back towards the intake portals. Also, no dust was recorded at stations 6, 7, or 8, showing that the dust was not migrating past the fans to the northern areas of the mine.

The study showed that the new ventilation system provided an improved directional airflow in the overall mine air circuit. Day 1 of sampling gave higher velocities than day 2, showing a relation of airflow with proximity to the fans. The average velocity of the dust cloud from the two days of sampling was 0.25 m/s (48.7 fpm). The retention time or the length of time it takes the dust cloud to clear the fans can be estimated by using the "end time" at station 9. This gives dust retention times on day 1 and day 2 of 1.2 hrs and 5.5 hrs, respectively.

5 DISCUSSION

The method described in this paper offers an alternative to evaluating airflow patterns for large-opening mines with low air velocities. The primary feature of this technique is that it yields results that can be used to evaluate the overall mine airflow patterns. This technique requires little manpower as the pDRs are positioned at various locations in the mine, before the faces are shot, to record the travel path of the "dust cloud." The dust concentration data can then be downloaded into database software and analyzed in a short time frame. In both of these case studies, the two primary features of the ventilation system was the use of low-pressure propeller fans to increase the volume of air and the construction of brattice stopping lines to direct airflow. The results of the dust survey showed that the planned ventilation systems worked well and provided a directional flow of air which did not exist before the systems were installed. In addition, the retention time of the dust, or the length of time for the dust to travel from the shot location to the main fans, was substantially reduced.

This method provides an alternative approach to the way light-scattering instruments are commonly used. Usually, the time logging capabilities of the instrument are used to determine dust levels in the evaluation of different dust control technologies. In these studies, the instrument is used to assess the movement of air. In contrast to more traditional and widely accepted methods such as anemometers, this method does not directly measure air velocity, but rather the movement of dust as it is transported by the air. Therefore, it must be emphasized that this method may not yield mine air velocities that are comparable to anemometer measurements. Anemometer readings are a point measurement taken at center entry or as a traverse across the entry, and thus may give higher or even lower velocities depending on where the measurements are conducted. As an example, anemometer measurements located at stations near the propeller fans (case 1: station 8, case 2: station 9) gave velocities in excess of 1.3 m/s (250 fpm) – 4 times higher than those estimated by this method. However, this method can provide a realistic indication of how airborne contaminants are being transported throughout the mine.

The application of this method is still experimental and is not meant to replace traditional methods of assessing mine ventilation. Rather, this method can be used as another tool for the ventilation engineer. By monitoring the movement of dust, this method can determine if the ventilating air is moving in the planned direction, to areas in the mine where ventilation is not required, or if recirculation is occurring. This information proved useful to the operators at both mines. It indicated that modifications to the mine-wide ventilation systems were resulting in improved velocities and directional air movement.

REFERENCES

Baron, P.A., and Willeke, K., 1993. "Aerosol Fundamentals," "*Aerosol Measurement – Principles Techniques and*

Applications". Weilke K., and Baron P.A., eds, Van Nostrand Reinhold, pp. 8–22.
Grau, R.H., Robertson, S.B., Mucho, T.P., Garcia, F., and Smith, A.C., 2002a. "NIOSH Ventilation Research Addressing Diesel Emissions and other Air Quality Issues in Nonmetal Mines," SME Pre-print 02-187, SME Annual Meeting, Phoenix, Arizona, February 25–27, 2002, 7 pp.
Grau, R.H., Mucho, T.P., Robertson, S.B., Smith, A.C., and Garcia, F., 2002b. "Practical Techniques to Improve the Air Quality in Underground Stone Mines," Proceedings of the North American/Ninth US Ventilation Symposium, Kingston, Ontario, June 8–12, 2002, pp. 123–129.
Head, H.J., 2001. "Proper Ventilation for Underground Stone Mines," *Aggregates Manager,* January, 2001, pp. 20–22.
Kissell, F.N. and Volkwein, J.C, 2002. "Improving Ventilation in Underground Stone Mines," *Aggregates Manager,* April, 2002, pp. 20–25.
Timko, R.J., and Thimons, E.D., 1987. "Damage Resistant Brattice Stoppings in Mines With Large Entries," *Engineering Mining Journal*, Vol. 188, No. 5, pp. 34–36.
Timko, R.J., and Thimons, E.D., 1982. "Sulfur Hexafluoride as a Mine Ventilation Research Tool – Recent Field Applications," US Bureau of Mines Report of Investigations 8735, 15 pp.
Thimons, E.D., Bielicki, R.J., and Kissell, F.N., 1974. "Using Sulfur Hexafluoride as a Gaseous Tracer to Study Ventilation Systems in Mines," US Bureau of Mines Report of Investigations 7916, 22 pp.
Welby, C.G., Cheng, L., Divers, E.F., 1988. "Deposition of Respirable Dust in an Airway," Respirable Dust in the Mineral Industries: Health Effects, Characterization and Control, The Pennsylvania State University, University Park, PA. 1988, pp. 298–304.
Xu, L., and Bhaskar, R., 1995. "A Simple Model for Turbulent Deposition of Mine Dust," Proceedings of the 7th US Mine Ventilation Symposium, Lexington, Ky., June 5–7, 1995, pp. 337–343.

Mine Ventilation, Ganguli & Bandopadhyay (eds)
© 2004 Taylor & Francis Group, London, ISBN 90 5809 633 5

Identifying improvements for a modified wet scrubber through simplified modelling

Yoginder P. Chugh & Krishna K. Thatavarthy
Department of Mining and Mineral Resources Engineering, Southern Illinois University, Carbondale, Illinois

Ronald Carty
Illinois Clean Coal Institute, Carterville, Illinois

ABSTRACT: Since the enactment of the 2 mg/m^3 respirable dust standard in the 1969 Federal Mine Safety and Health Act, extensive research has led to considerable progress in coal mine dust control. Perhaps the most significant achievement in the dust control field was the development of the continuous miner wet scrubber, which has become a standard throughout the industry. However, as production demands continue to increase, mines are struggling to comply with regulatory standards. In an effort to assist the mine operators, Southern Illinois University at Carbondale has collaborated with Joy Mining Machinery and the State of Illinois to develop a dust control research laboratory. The laboratory's centerpiece is a continuous miner wet scrubber for testing scrubber efficiency under various conditions. The laboratory environment allows some control of the many variables involved. The goal of the studies is to improve the design and efficiency of the scrubber to minimize repsirable dust exposure.

Toward the above goal, a simplified wet scrubber model was developed to identify interactions between airflow, water sprays and filters. This was done to understand the relative assessment of different variables: (1) airflow through scrubber, (2) water spray pressure and volume, (3) inclination of filter/s, (4) geometry of water sprays, (5) filter design, and (6) relative values of air pressure within the scrubber and water spray pressure. Mechanics principles were utilized to understand interaction between airflow, water spray pressure, and water volume. These principles were utilized to modify an existing wet scrubber to: (1) reduce dust concentrations from about 3.6 mg/m^3 to about 2.8 mg/m^3, (2) reduce water requirement by about 40%, (3) reduce spray water pressure to about 30 psi, and (4) increase airflow through the scrubber.

1 INTRODUCTION

Since the 2 mg/m^3 dust standard was established in 1977, dust control research has focused on assisting mines with compliance (Chugh et al. 2002). Currently, most coal mines rely primarily on two methods for controlling dust. The first method utilizes large volumes of outside air to ventilate and dilute the dust to acceptable concentrations. The ventilation process alone can be ineffective, or prohibitively expensive, as a means of reducing the level of respirable dust to the standard level. The second method involves water used in various ways, such as water spray systems. Numerous studies have measured the levels of dust reduction achievable from water sprays. Studies have also been done on the effect of using surfactants in water spray systems in order to achieve greater dust suppression (Kim & Tien, 1994). Dust collectors have been increasingly used in the mining industry over the last several decades as both health concerns about dust and social consciousness have risen. For underground mining, the most prolific success story has been the use of wet scrubbers that utilize both the principles of ventilation and water sprays. These wet scrubbers are now widely used on continuous miners and are highly successful in reducing the amount of dust that is released at the working face in coal mining operations.

Over the last 20 years, there has been significant push to improve productivity in coal mines so that it can compete economically. Significant progress has been made toward increased productivity and production rates 2–3 times larger than in 1970's are being consistently achieved. However, with increased productivity, concentration of dust has increased and in MSHA reports, dust concentration standard of 2 mg/m^3

is not being achieved by all mines. There is a significant need for research and development to reduce dust concentration below 2 mg/m^3. This study is directed toward achieving that goal through improved wet scrubbers on continuous mining machines.

This first goal of this study was to create a simplified model of the wet scrubber to understand interactions between airflow, water sprays and filters. The second goal focused on modifying the wet scrubber system in several phases to improve its efficiency and obtain reduced dust concentrations.

2 SIMPLIFIED MODELING OF A WET SCRUBBER

2.1 *Static pressure drop across the length of the scrubber*

Consider a simplified wet scrubber model with a vertical filter as shown in the Figure 1. The static and velocity pressure heads at the inlet are $P_{s\text{-}in}$, $P_{v\text{-}in}$ and at the outlet are $P_{s\text{-}out}$, and $P_{v\text{-}out}$, respectively. The graph of the total pressure versus length of the scrubber is shown in Figure 2.

The following observations can be made from Figure 2.

Case 1: Scrubber with no vertical filter

- Velocity pressure (P_v) should be the same throughout if the area of cross-section is the same.
- Static pressure (P_s) increases linearly from the inlet to the outlet.
- P_s should be small.

Case 2: Scrubber with vertical filter but no water spray.

- P_v is variable around the filter zone.
- P_s drops across the filter.
- P_s drop slope may decrease due to the reduced quantity of airflow (Q) through filter.

Case 3: Scrubber with vertical filter and water spray

- P_v is highly variable along the length because water spray affects air velocity.
- P_s loss slope may decrease further due to the further reduction in the quantity of airflow (Q) through scrubber.

The effect of increase in spray pressure on the pressure drop across the filter is shown in Figure 3. The results indicate that pressure drops increase with the increase in spray pressure for different spray volumes of the spray system. This is because smaller areas of the filter are available for airflow when water pressures are high.

2.2 *Understanding airflow through the scrubber*

The two stresses that act on the water spray are the radial stress (σ_{rr}) and tangential stress ($\sigma_{\theta\theta}$). Away from the spray nozzle, both σ_{rr} and $\sigma_{\theta\theta}$ decrease. The contours of these two stresses based on a spray angle are shown in Figure 4.

The following observations can be made from the above modeling.

- $\sigma_{\theta\theta}$ along the outer surfaces of the conical spray is zero while the σ_{rr} values are variable along the length with high values near the spray nozzle.
- Since the scrubber duct is rectangular, spray pressure (σ_{rr}) values along the sides should be lower than along the top and bottom surfaces.
- When the scrubber is turned on, it has a certain pressure $P_{s\text{-}in}$, $P_{v\text{-}in}$ that vary along the scrubber length up to the filter because of the relatively rigid conical water spray.

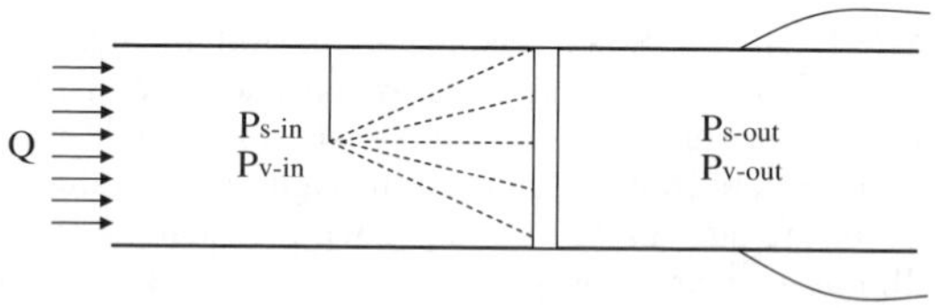

Figure 1. Simplified wet scrubber model with a vertical filter.

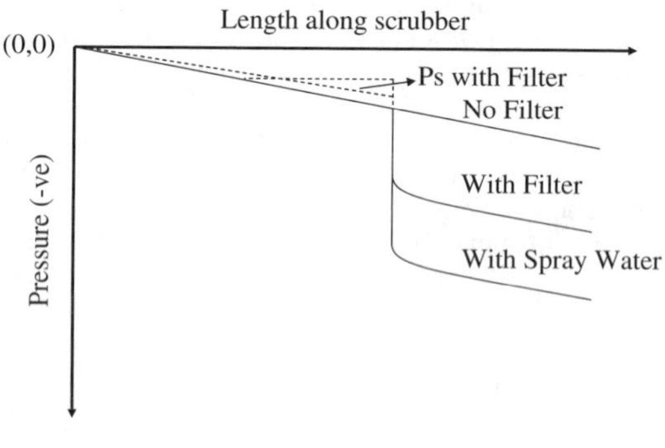

Figure 2. Total pressure vs. length for a wet scrubber.

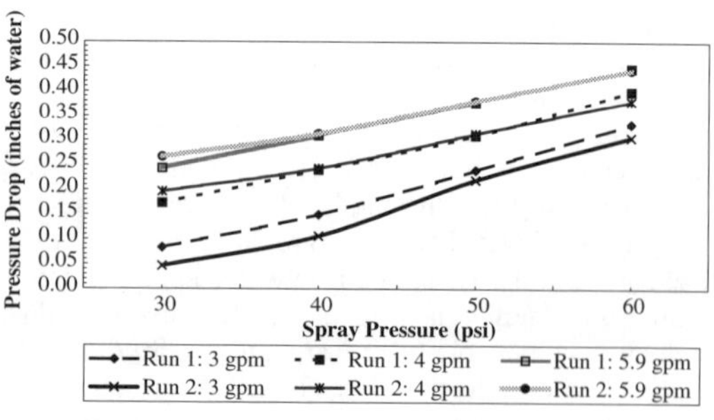

Figure 3. Pressure drop vs. spray pressure.

- Due to interaction between air pressure (A_p), and water pressure, conical geometry of the spray is deformed. The deformations are higher along the sides and smaller at the top and bottom because of higher $\sigma_{\theta\theta}$ values. The deformed geometry may be shown in Figure 5 depending upon relative values of the spray pressure, air pressure and resistance offered by the filter (number of filter layers).

2.3 *Effect of air pressure on available area for airflow*

The effect of air pressure on the available area for airflow through the scrubber is shown in Figure 6.

- With the increasing air pressure or decreasing spray water pressure, area for the airflow increases.

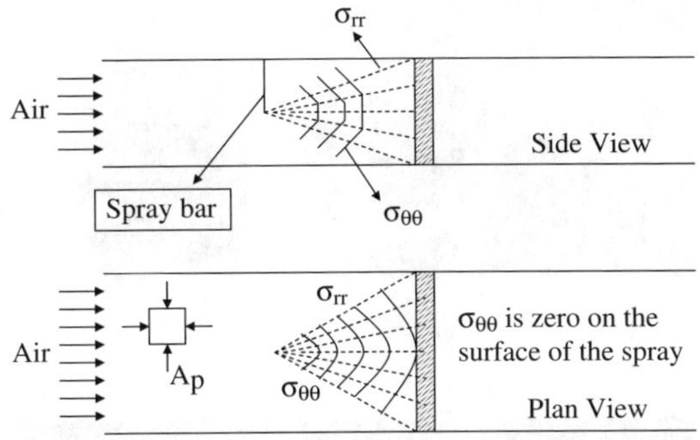

Figure 4. Conceptual contours of σ_{rr} and $\sigma_{\theta\theta}$ (not to scale).

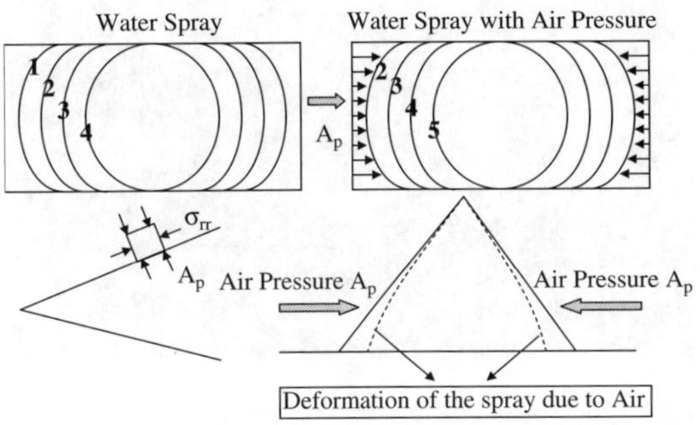

Figure 5. Deformation of spray due to air pressure.

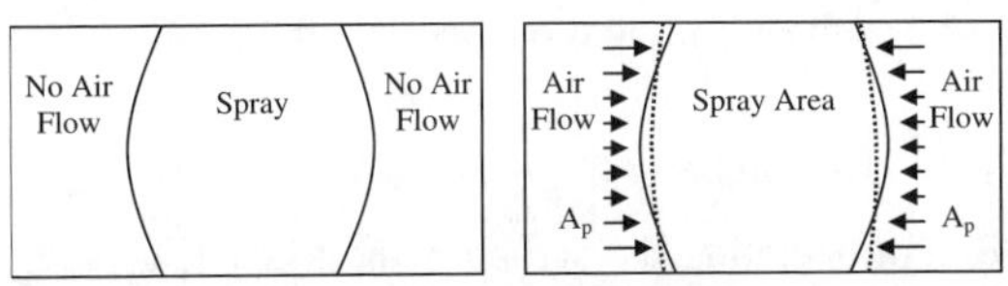

Figure 6. Effect of air pressure on available area for airflow.

- With reduced spray water volume, back pressure decreases and area for airflow increases.

2.4 *Consequences of modeling observations*

The following are the consequences of the simplified modeling:

- Dust laden air is passing mostly along the sides of the filter and small amount from the top and bottom when using solid cone spray.
- Most of the water spray with or without surfactant may not be mixing with the dust-laden air.
- If the above two hypothesis are correct, then dust control efficiency should be a function of number of filter layers (residence time), spray pressure (amount of water mixed in the air) and surfactant concentration in the order listed. Saha (2002) documented this in his work.
- Spray water fills the voids of the filter over the spray area. Depending upon spray volume, number of filter layers and spray pressure, there may be back pressure acting on the airflow (equal to the spray water pressure).
- If the air pressure is not high enough, the quantity of airflow will be significantly reduced and dust control efficiency significantly reduced.
- Spray water volume, spray pressure and filter resistance must be matched to provide maximum airflow with minimum pressure drop.
- Spraying water at the filter increases $P_{v\text{-}out}$ because of higher velocity. Therefore, power requirements are increased.

2.5 *Desired goals for a modified wet scrubber design*

The following are the three main design features for a modified wet scrubber design:

- Maximize airflow through the scrubber by allowing largest area of the filter for airflow and air-water mixing.
- Minimize the pressure drop across the filter to maximize the quantity of airflow.
- Minimize spray volume and surfactant requirements to reduce costs.

3 SIU-JOY DUST CONTROL LABORATORY

3.1 *Description*

The SIU dust control laboratory is shown in Figure 7. The laboratory consists of the scrubber system, a controllable dust feeder, and a sampling room modeled after a mine entry or tunnel. A detailed description of the primary equipment used in the scrubber simulation laboratory follows.

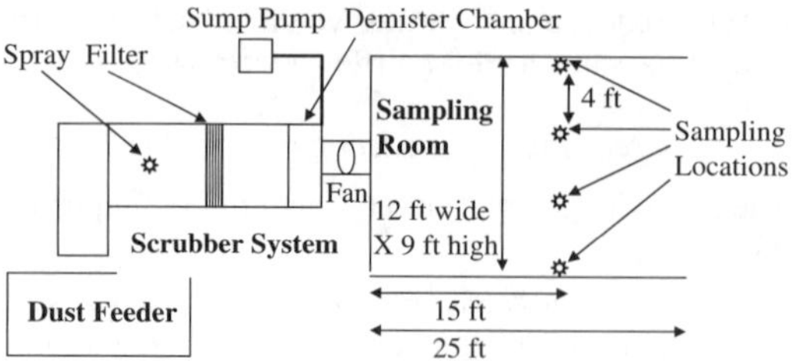

Figure 7. Layout of the dust control laboratory (not to scale).

Figure 8. Continuous miner scrubber system.

3.1.1 *Flooded-bed scrubber system*

An 8000 cubic foot per minute (cfm) flooded-bed scrubber system manufactured by Joy Mining Machinery for a model 12 CM continuous miner was set up at the Illinois Coal Development Park in Carterville, Illinois. The scrubber was equipped with two 60-gallon drums and a two-stage (1/2 HP + 3/4 HP) electric pump system to supply water to the scrubber system at a maximum rate of 10 gallons per minute and a maximum pressure of 100 psi. An appropriate amount of surfactant to achieve the desired concentration was mixed with water in the two tanks as they were filled. A 3-HP gasoline-powered sump pump removed the water and "scrubbed" dust from the system at the bottom of the demister chamber. The demister chamber is a rectangular box located on the right hand side of Figure 8 adjacent to the round fan housing.

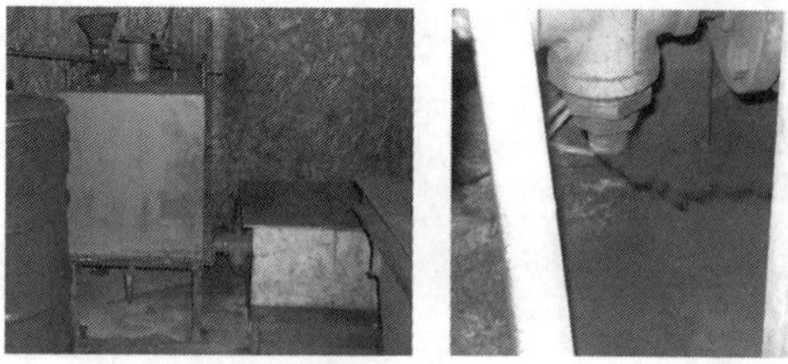

Figure 9. Dust feeder system.

3.1.2 *Dust feeder*

Raw coal from a mine was pulverized with a compressed air crusher. The resultant "dust" was fed into a small hopper, shown in Figure 9, which fed dust through a two-inch screw conveyor at the bottom. The screw conveyor was driven by a 1/8 HP motor and a Boston Gear Ratiotrol for speed control from 0 to 100% or from a minimum feed rate of 0 grams per minute. To achieve a consistent feed rate, the dust feeder was run at 20% feed rate or higher. All testing here was done at 20% feed rate and the input dust concentration was 48 mg/m^3.

Figure 10. Sampling room and stations.

3.1.3 *Dust sampling room and locations*

The sampling room was modeled after a typical mine entry as shown in Figure 10. In the Illinois Basin, mine openings average 6 to 7 feet in height and 18 to 20 feet in width. The laboratory is not part of a closed ventilation system. The scrubber is housed in a small building with large sliding doors that are opened when the scrubber is operating. The capacity of the scrubber system in use in the laboratory is about 8000 cfm. This volume of intake air is pulled from the surrounding air, which is relatively motionless. Windy conditions do not affect the laboratory scrubber but can affect the laboratory results if the wind is from the north as the sampling room opens to the north.

3.2 *Description of scrubber testing procedure*

Before operating the scrubber, the MSA Escort LC Dust Sampling Pumps were calibrated at 2 liters/min using the Wet test meter. Then, the pumps along with the Dorr-Oliver cyclone assembly with the MSA dust filter cassette fitted on them were placed in the sampling locations. The two 60 gallon barrels on top of the scrubber system were filled with water and surfactants were added as necessary to achieve appropriate concentrations. The dust feeder system was loaded with enough dust to last for one hour.

First, the compressor for the dust feeder system was turned on and the moisture in the pipeline was bled through a relief valve. Next, the sump pump and the scrubber fan motor were started. The water spray pumps were turned on and the appropriate spray pressure was established by means of relief valves in the system. Then the dust feeder system was started and set at the 20% fixed rate. A small stream of compressed air was introduced at the screw conveyor discharge point to maintain a continuous flow of dust. Once everything was running, the dust sampling pumps were turned on to initiate the one-hour test.

At the end of one hour, the pumps were shut off and the dust filter cassettes were taken out of the Dorr-Oliver cyclone assembly. The outer surface of the cassette was wiped clean with a rag cloth. A dust data card was supplied with each filter cassette. The card contained an identification number (cassette number) and the initial weight, in grams, of the filter capsule inside the filter cassette. The sample date, time, and length were filled in on the dust data card. The sample was then shipped to MSHA for analysis.

4 MODIFIED DESIGN FOR THE SCRUBBER

4.1 *Identifications of concepts for scrubber modification*

The following are some concepts for scrubber modifications to achieve the desired goals discussed above:

- Minimize filter resistance by reducing the number of filter layers.
- Minimize spray pressure to be more or less equal to or slightly less than air pressure in the scrubber.
- Change the positions of sprays to avoid spraying water directly on the filter.
- Spray water as a small particle mist with uniform pressure over the entire area rather than non-uniform pressure with high-pressure gradients.

4.2 *Modifications*

The following are the modifications made to the wet scrubber upon identification of the important concepts for improved dust control:

- The filter resistance was reduced to 5 and 10 layer filter from 20 to 30 layer filter screens used in earlier studies (Saha, 2002).
- The spray pressure was significantly reduced from 80 psi to 40 psi, thereby reducing spray volume and surfactant requirements.
- A pre-wet water only spray system was introduced in front of the scrubber unit to knock off the coarser particles and then treat the fine respirable dust particles with the surfactant spray.
- To get uniform pressure distribution, multi-nozzle sprays were designed instead of the single-nozzle sprays used in earlier studies (Saha, 2002). These sprays were designed for both vertical and horizontal type and also to accommodate full cone and hollow cone spray nozzles.
- To increase the residence time a filter was introduced after the pre-wet water only spray to effectively trap the coarser particles.

This study was conducted to modify the wet scrubber system in different phases to improve the efficiency and obtain reduced dust concentrations.

4.3 *Phase 1 studies – reduced spray pressure*

Phase 1 scrubber studies involved the study of reduced spray pressure on dust concentration. The other variables held constant were a 20-layer filter, 0.005% surfactant concentration and a single full-cone nozzle rated at 40 psi, 5.9 gpm, and 120° spray angle. The spray pressure was decreased from 80 psi to 30 psi. Results of Phase 1 studies are shown in Table 1. The results indicate that dust concentration at the lower spray pressure (30 psi) do not change significantly when compared to the higher spray pressure. The main advantages of lowering the spray pressure are (1) reduced spray volume that helps in lower water requirements, (2) lower surfactant costs, and (3) increased air flow through the scrubber.

4.4 *Phase 2 studies – two spray system*

The scrubber was modified to incorporate a two-spray system consisting of a pre-wet water-only spray

Table 1. Effect of spray pressure on dust concentration.

Surfactant spray (gpm)	Filter layers (#)	Spray pressure (psi)	Surfactant concentration (%)	Dust concentration (mg/m^3)	Standard deviation (mg/m^3)	Number of samples
5.5	20	30	0.005	3.576	0.988	8
7.5	20	80	0.005	3.503	0.388	4

system and a surfactant spray system. It was hypothesized that a two-spray system should be considered for a wet scrubber to improve its efficiency. The first water-only spray would wet the coarser dust, and the second water-surfactant spray would be used to wet the very fine dust non-wettable by water alone. Figure 11 shows the model of the two-spray system designed for the scrubber unit.

Figure 12 shows the pre-wet water only spray system on the scrubber unit. Five 1-gpm full-cone sprays were fitted to the ductwork at equal distances to obtain good interaction between the water spray and the coarser particles.

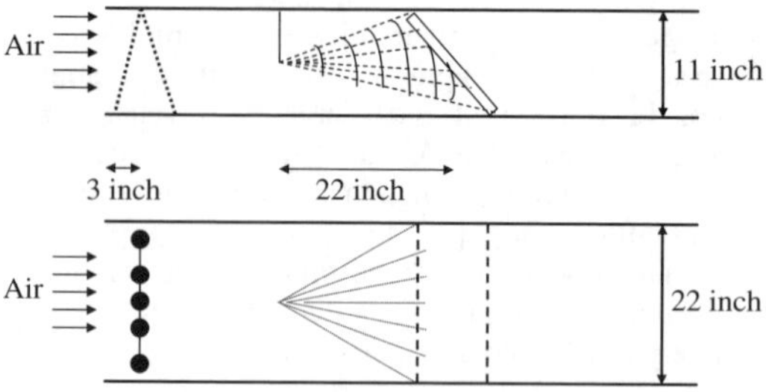

Figure 11. Two spray system.

Phase 2 involved scrubber tests on the modified two-spray scrubber system. The other variables that were held constant were a 20-layer screen filter, 0.005% surfactant concentration, and 40 psi spray pressure (120° spray angle). The spray volume was varied over two levels (2- and 3-gpm) for both the sprays.

Results of Phase 2 tests are shown in Table 2. The results indicate that using a pre-wet water-only spray system reduces dust concentrations only slightly. This is reasonable because the tests were done using only fine dust. However, in a commercial unit operating in the field, a pre-wet water spray should help significantly since a large weight percentage of dust is very coarse. The results also suggest that the spatial distribution of sprays increased the effectiveness of the spray system for dust control.

4.5 *Phase 3 studies – modification of the surfactant spray system*

To improve the interaction between the surfactant water and the fine dust-laden air in the scrubber, the surfactant spray was modified from a single nozzle to a three-nozzle spray system. The modified surfactant spray system is shown in Figure 13.

Figure 12. Pre-wet water only spray system.

Figure 13. Surfactant spray modification – Phase 3.

Table 2. Effect of two-spray system on dust concentration.

Case	Water spray (gpm)	Surfactant spray (gpm)	Filter layers (#)	Spray pressure (psi)	Surfactant concentration (%)	Dust concentration (mg/m^3)
I	0	2	20	40	0.005	3.916
	0	2	20	40	0.005	3.697
II	5	2	20	40	0.005	3.651
	5	2	20	40	0.005	3.036
III	0	3	20	40	0.005	2.335
IV	4	3	20	40	0.005	2.156

The modifications to the surfactant spray system included:

- Move the spray system closer to the filter for better coverage of the filter area as shown in Figure 14.
- Reduce the quantity of water.
- Increase the number of sprays with smaller spray angle.
- Operate sprays at 20, 30 and 40 psi spray pressures.
- Use of full-cone and hollow-cone spray nozzles.

Phase 3 scrubber tests examined only the three-nozzle surfactant spray system. Therefore, the pre-wet water-only spray system was not used. Results of the Phase 3 scrubber studies are shown in Table 3.

The results show that with a 20-layer filter, the three-nozzle surfactant spray system reduces dust concentrations to 2.8 mg/m^3 as compared to 3.57 mg/m^3 for the single nozzle surfactant spray system used in Phase I scrubber studies. The benefits of the three-nozzle surfactant spray system are (1) larger airflow through the scrubber, (2) approximately 40% reduction in water requirements, (3) reduced load on the demister, and (4) approximately 40% reduction in surfactant requirements.

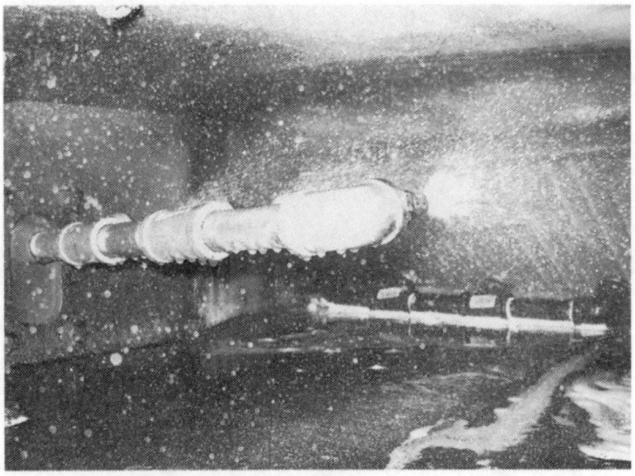

Figure 14. Filter coverage area of surfactant spray.

5 CONCLUSIONS

A simplified wet scrubber model was developed to identify interactions between airflow, water sprays and filters. The significance of different variables were identified from the simplified wet scrubber modeling and then applied to modify the design of the wet scrubber. It was concluded that spray water volume, spray pressure and filter resistance must be matched to provide maximum airflow with minimum pressure drop.

The guiding principles from the simplified model were utilized to modify an existing wet scrubber in different phases to obtained reduced dust concentrations. The highlights of the modified wet scrubber are reduced spray pressure, reduced water and surfactant requirements and the three-nozzle surfactant spray system. The results from the modified wet scrubber: (1) reduced dust concentrations from about 3.6 mg/m^3 to about 2.8 mg/m^3, (2) reduced water requirement by about 40%, (3) reduced spray water pressure to about 30 psi, and (4) increased airflow through the scrubber.

Based on the above conclusions, further modifications may be made to the wet scrubber. A vertical multi-nozzle surfactant spray system can help in increasing the interaction between the dust-laden air and surfactant. A filter screen can be introduced after the pre-wet water only spray system to knock the coarser particles. Hollow-cone spray nozzles may be used instead of the full-cone spray nozzles.

ACKNOWLEDGEMENT

This work was performed by the authors at the Dust Control Laboratory, Southern Illinois University, Carbondale as a part of a more general project to develop advanced mining technologies to reduce underground production costs. The project was funded by the Illinois Clean Coal Institute (ICCI). The authors wish to thank ICCI for their support.

Table 3. Dust concentration results for the three-nozzle surfactant spray system.

Filter layers (#)	Spray pressure (psi)	Surfactant concentration (%)	Spray volume (gpm)	Dust concentration (mg/m^3)	Sample standard deviation
20	40	0.005	3.9	2.78	0.011
10	40	0.005	3.9	5.86	0.014
20	30	0.005	3.3	5.06	0.011
10	30	0.005	3.3	7.04	0.057
20	20	0.005	2.7	4.39	0.017
10	20	0.005	2.7	6.06	0.047

REFERENCES

Campbell, J.A.L., Moynihan, D.J., Roper, W.D., and Wilis, E.C., 1983, "Dust Control System and Method of Operation", U.S. Patent 4,380,353.

Cheng, L., 1973, "Collection of Airborne Dust by Water Sprays", *Industrial Engineering Chemistry Proceedings*, Vol. 12, No 3, p. 221–225.

Chugh, Y.P., Hirschi, J., Saha, A., and Mohanty, M., 2002, "Evaluating the Use of Surfactants to Enhance Dust Control Efficiency of Wet Scrubbers for Illinois Coal Seams", *U.S. Mine Ventilation Symposium*, 2002, pp.

Colinet, J.F., 1999, "Respirable dust control technology utilized in underground coal mines in the United States", *Proceedings of the 2nd International Underground Coal Conference*, p. 171–179.

Colinet, J.F., and Listak, J.M., 1999, "Reducing Worker Exposure to Respirable Dust at Longwall Mining Operations", *Proceedings of the International Conference on Control of Dust Hazards in the Mining Industry*, Wisla, Poland, Oct, Chapter XI.

Foster Miller, Inc., 1986, "Development of Optimal Spray Systems for Dust Control in Underground Mines", Contract HO199070, *U.S. Bureau of Mines*, 146 pp.

Goodman, G.V.R., and Listak, J.M., 1998, "Variations in Dust Levels with Continuous Miner Position", *Soc. for Mining, Metallurgy, and Exploration*, preprint no. 98–130.

Grayson, R.L., 1998, "Setting Priorities for Mine Safety and Health Research", preprint no. 98–107, *SME Annual Meeting*, Orlando, FL, 7 pp.

Hartman, H.L., Mutmansky, J.M., Ramani, R.V., and Wang, Y.J., 1997, "*Mine Ventilation and Air Conditioning*", Third edition, 730 pp.

Jayaraman, N.I., Jankowski, R.A., and Stritzel, D., 1987, "Improving Health, Safety, and Productivity Through the Use of Machine Mounted Scrubbers", *Proc. Annual Meeting Illinois Mining Institute*, Marion, IL, Oct., pp. 62–75.

Jayaraman, N.I., Jankowski, R.A., and Whitehead, K.L., 1992, "Optimizing Continuous Miner Scrubbers for Dust Control in High Coal Seams", *Proc. Of New Technology in Mine Health and Safety Symposium, SME Annual Meeting*, Phoenix, AZ, Chapter 20, pp. 193–205.

Kim, J., and Tien, J.C., 1994, "The effect of Added Base on Coal Wetting Ability of Nonionic Surfactant Solution Used for Dust Control", *Mining Engineering*, Dec., pp. 151–154.

Niewiadomski, G.E., 1983, "Improving Dust Control Technology for U.S. Coal Mines", *Proc. Control of Respirable Coal Mine Dust*, Beckley, WV, p. 41–72.

Niewiadomski, G.E., and Jankowski, R.A., 1993, "Longwall Dust Trends and Developments in Longwall Dust Controls", *Proc. 6th U.S. Mine Vent. Symposium*, SME, Littleton, CO, pp. 551–557.

Saha, A., 2001, "An Assessment of the Effectiveness of Surfactants on Dust Control using a Continuous Miner Scrubber", M.S. Thesis, Department of Mining and Mineral Resources Engineering, Southern Illinois University, Carbondale, pp. 78.

Mine Ventilation, Ganguli & Bandopadhyay (eds)
© 2004 G.V.R. Goodman, ISBN 90 5809 633 5

A comparison of a directional spray system and a flooded-bed scrubber for controlling respirable dust exposures and face gas concentrations

Gerrit V.R. Goodman, Douglas E. Pollock & Timothy W. Beck
National Institute for Occupational Safety and Health, Pittsburgh Research Laboratory, Pittsburgh, PA, USA

ABSTRACT: A continuous mining machine was operated with either a directional spray system (sprayfan) or a flooded-bed dust scrubber. A series of laboratory tests examined the impacts of water pressure and curtain setback distance on respirable dust concentrations and tracer gas levels. Face ventilation was provided by an exhaust curtain. Several combinations of external and underboom sprays also were evaluated with the scrubber. Dust concentrations were measured at locations representing the mining machine operator and in the return airway, while tracer gas levels were measured around the cutting head.

When testing with the sprayfan-equipped mining machine, changes in water pressure and setback distance significantly affected miner operator dust levels. Return dust levels varied only slightly. With the dust scrubber, changes in curtain setback impacted miner operator dust concentrations. Use of an external spray manifold significantly increased both operator and return dust levels, apparently by pushing dust beyond reach of the scrubber inlets.

Tracer gas levels around the cutting head were affected by changes in sprayfan water pressure. Variations in curtain setback distance affected gas levels on the off-curtain side of the cutting head. With use of the dust scrubber, cutting head gas levels were affected mostly by changes in water pressure, setback distance, and use of the external and underboom sprays.

The scrubber was more effective than the sprayfan for controlling respirable dust levels. The sprayfan effectively controlled gas on the off-curtain side of the cutting head, although the scrubber plus the external sprays also provided good control of gas levels in this area. This same spray configuration also adequately controlled gas on the curtain side of the cutting head.

1 INTRODUCTION

The United States Federal Coal Mine Health and Safety Act of 1969 requires that the concentration of respirable dust in the mine environment not exceed 2.0 mg/m^3 (Code of Federal Regulations, 1999). This limit is measured gravimetrically as an 8-hour time-weighted average concentration of the respirable dust. These regulations also require that methane levels in the face area be maintained below 1 percent. Methane readings obtained with a hand-held methane monitor are usually taken as close to the face as practical, although they cannot be taken closer than 0.3 m (1 ft) from the roof, face, ribs, and floor (Code of Federal Regulations, 1999). Machine-mounted methane monitors near the cutting drum continuously monitor methane levels on the mining machine. Whenever methane readings exceed 1 percent, mining must stop until concentrations fall below this level.

Many factors affect the ability of a dust scrubber or sprayfan to control respirable dust exposures and methane gas concentrations during coal mining. These include face ventilation airflow quantity, water spray pressure, and ventilation curtain setback distance (distance from the cutting face to the mouth of the ventilation curtain or tubing). Past work showed that increased curtain airflow reduces both respirable dust and face gas concentrations (Colinet et al., 1991; Divers et al., 1982). Increased water pressure creates turbulence that can improve the dilution of face gas (Foster-Miller, Inc., 1985). However, excessive turbulence creates dust rollback that can push the dust cloud toward the machine operator (Jankowski et al., 1987). Shorter setback distances direct more airflow to the face and this can improve control of dust and gas.

Flooded-bed dust scrubbers are installed on many continuous mining machines and are responsible, in part, for decreased occupational exposures to respirable dusts (O'Green, 1990). Past work examined the impact of scrubber inlet and discharge configurations, scrubber quantity, and face airflow on the performance of the dust scrubber (Jayaraman et al.,

1992a, b). Past testing showed that this device also reduced face methane levels for a range of curtain setback distances and scrubber and intake air quantities (Taylor et al., 1996, 2001).

Although scrubbers are found on most continuous mining machines, some machines do not employ these devices, for reasons ranging from added maintenance to concerns regarding potential noise exposure. Instead, these machines often utilize a directional spray design (sprayfan), which is a water-powered system originally designed for controlling methane gas levels. This system contains several spray manifolds placed on the mining machine to direct fresh intake air to the cutting face, sweep contaminated air across the face, and direct this airflow into the return airway (Foster-Miller, Inc., 1985).

This paper compares respirable dust exposures and tracer gas levels measured around a continuous mining machine equipped with either a sprayfan or a flooded-bed dust scrubber. Dust levels were measured at typical machine operator locations and in the return airway, while tracer gas concentrations were measured at conventional machine methanometer locations. These dust and gas concentrations were measured for various water spray pressures and setback distances.

2 TEST DESIGN AND METHODS

This work was conducted in a surface test gallery at the Pittsburgh Research Laboratory of the National Institute for Occupational Safety and Health (NIOSH). The test facility simulated a cut 5.5 m (18 ft) wide, 2.0 m (6.4 ft) high, and up to 15.2 m (50 ft) deep. A full-scale mockup of a continuous mining machine used for this testing featured a 0.9-m (3-ft) diameter cutting drum rotating at 50 rpm. The machine was positioned at the end of the sump or box cut for testing. A coal slab measuring 2.4 m (8 ft) by 6.1 m (20 ft) remained to the right of the machine (Figure 1).

The mining machine was equipped with either a sprayfan (directional sprays) or a flooded-bed dust scrubber. The sprayfan was installed according to established specifications and featured thirty-four hollow cone spray nozzles positioned above and below the cutting boom and along the sides of the mining machine frame (Foster-Miller, Inc., 1985). The flooded-bed dust scrubber, rated at 3.3 m^3/s (7000 cfm), used a 30-layer stainless steel filter screen. When testing with the scrubber, the mining machine used a standard water spray system consisting of twenty-four hollow cone sprays positioned above, below, and along the sides of the cutting head.

In practice, the sprayfan is used only with exhaust curtain face ventilation while the dust scrubber can be used with either a blowing or exhausting curtain arrangement. To reduce the number of tests, all

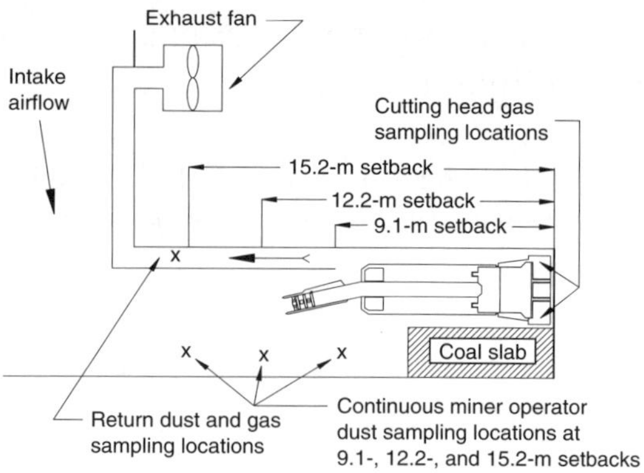

Figure 1. Test gallery setup.

evaluations used exhaust curtain ventilation which was created by drawing air via a main gallery fan through a brattice curtain positioned on the left side of the test gallery (Figure 1). Airflow was measured at the mouth of the exhaust curtain using a vane anemometer.

Past work showed that high methane gas levels could occur at the face when using exhaust curtain ventilation – evidence that fresh airflow was not reaching the face but instead was flowing directly behind the curtain (Luxner, 1969). This testing indicated that the corner of the face on the right or off-curtain side of the entry often showed the highest gas levels.

Preliminary testing of the dust scrubber with the standard water spray system showed relatively high gas levels in the off-curtain side corner. To clear this area of gas, external spray manifolds were placed on the right side of the machine frame 2.4 m (8 ft) outby the hinge point of the cutting boom and on top of the cutting head at the hinge point (Figure 2). Each manifold consisted of two hollow cone sprays oriented to "flush out" this corner of the face. These external manifolds were plumbed separately from the standard spray system to evaluate the effect of these added sprays on controlling respirable dust and face gas.

Two additional spray manifolds were placed under the cutting boom, each manifold consisting of three hollow cone sprays oriented toward the cutting drum. These underboom manifolds also were plumbed separately to assess their effects on dust and gas levels.

Coal dust was introduced into the gallery at a rate of 25 grams/min through a series of compressed air hoses mounted on the machine cutting head. One hose discharged on the left side of the cutting drum while the other discharged on the right side of the drum. The rotating cutting head mixed the dust with the ventilation flow.

For dust sampling, constant flow pumps pulled dust-laden air through 10-mm nylon cyclone separators at

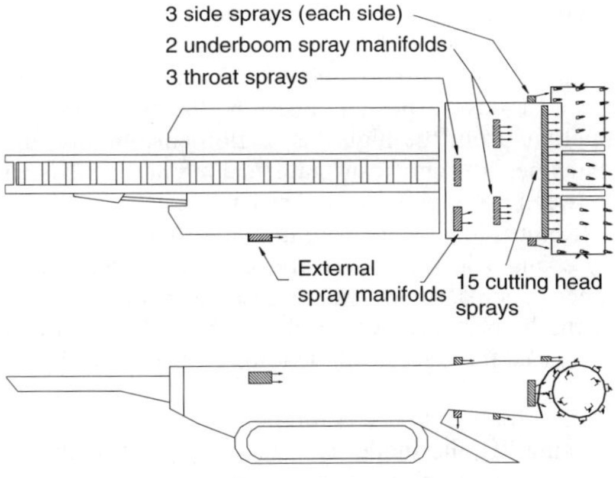

Figure 2. Water spray designs used with dust scrubber. Standard spray system consists of 15 cutter head sprays, 6 side sprays, and 3 throat sprays.

a rate of 2 liters/min (Code of Federal Regulations, 1999). The respirable mass was deposited onto pre-weighed 37-mm filters that were subsequently post-weighed to determine dust levels. A pair of gravimetric samplers placed on the off-curtain (right) side of the entry, opposite the mouth of the exhaust curtain, represented the typical location of the remote mining machine operator. This location moved with curtain setback distance. Two gravimetric samplers placed in the return monitored dust levels in that location (Figure 1).

Testing also evaluated the impact of face airflow quantity, water spray pressure, and setback distance on the removal and dilution of hazardous gas concentrations. Due to safety concerns, methane gas could not be used in this testing. Instead, sulfur hexafluoride (SF_6) was used as a surrogate or tracer gas. This gas is tasteless, odorless, not found naturally in the atmosphere, and is nontoxic in the amounts used in this testing. It is also detectable in the part per million (ppm) concentration level. This substance was introduced at the cutting drum where its rotation mixed the gas with the face airflow.

A mixture of SF_6 in air (>99% pure) was introduced at the cutting drum at 6 ml/s and measured real-time with a multipoint doser and sampler (California Analytical Instruments, Orange, CA). Gas was sampled at locations representing typical methanometer locations on a continuous mining machine – on the left and right sides of the cutting head approximately 0.6 m outby the cutting drum (Figure 1). For each test, gas was introduced into the gallery in a series of four evenly spaced pulses, each pulse lasting twelve minutes.

Tracer gas sampling ran concurrently with dust sampling. Disposable filters, placed at the ends of the gas sampling tubes, prevented dust from entering the gas monitoring equipment.

Table 1. Water flow rates (lpm) measured for various spray system designs used with dust scrubber.

	Line pressure	
Spray system design	621 kPa	1173 kPa
Standard sprays	76.6	98.5
Standard + externals	89.4	120.9
Standard + underbooms	94.4	127.7
Standard + externals + underbooms	106.5	145.9

When testing the sprayfan, a single exhaust curtain quantity of 2.8 m^3/sec (6000 cfm) was used. Curtain setback distances were 9.1 m (30 ft) or 15.2 m (50 ft), while water spray pressure values were 483 kPa (70 psi) or 1310 kPa (190 psi), measured in the line feeding the spray manifold. Measured water flow rates at low and high pressures were 76.9 lpm (20.3 gpm) and 125.1 lpm (33.0 gpm), respectively. The testing evaluated all four combinations of curtain setback distance and water spray pressure at the single curtain quantity. Two replicates of each test were made to obtain a measure of the error associated with each combination.

When testing the dust scrubber, only a single curtain flow of 3.8 m^3/s (8000 cfm) was used. This ensured that the face ventilation quantity exceeded the rated capacity of the dust scrubber (3.3 m^3/s). Quantities less than this were not considered; as they could lead to potential recirculation of dust and gas at the face. Quantities in excess of 3.8 m^3/s were not considered due their obvious benefits to dust and gas levels (Colinet et al., 1996; Taylor et al., 1996). Curtain setback distance was either 9.1 m or 12.2 m, as physical constraints did not permit testing of the scrubber at a distance of 15.2 m. Water pressures were set at either 621 kPa (90 psi) or 1172 kPa (170 psi).

Testing with the flooded-bed scrubber evaluated the impacts of setback distance and spray pressure for the single curtain airflow quantity. These tests also determined the effects of the spray designs shown in Table 1. The associated flow rates measured at the two line pressures are given.

Examining every combination required sixteen tests. As before, two replicates of each combination were run.

The dust concentration at each sampling location was calculated using the numerical average of the concentrations for each group of samplers. The tracer gas concentration at each location was the average of the concentrations recorded for each pulse. Respirable dust and tracer gas concentrations for each series of test conditions were calculated using the average concentration for each test comprising that set of conditions.

3 ANALYSES OF SPRAYFAN AND SCRUBBER PERFORMANCE

3.1 *Effects on respirable dust exposures*

3.1.1 *Sprayfan evaluation*

Analyses of variance (ANOVA) defined the impacts of the independent variables on respirable dust levels at a significance level of 0.05. Concerns arose regarding normality of the data and equality of the variances. The balanced design used in these analyses minimized the impact of any variability within each grouping of water pressure and setback distance.

The full-factorial model examined the two main effects of water pressure and setback distance plus the single interaction of these variables. Eight degrees of freedom remained for calculating the critical F-value in Table 2.

Water spray pressure and curtain setback distance both significantly affected dust levels for the miner operator (Table 2). Increasing water spray pressure increased dust levels, likely by aggravating dust rollback levels (Figure 3). Water sprays are good air movers and under certain conditions can push the dust cloud back to the miner operator. Previous work detailed the importance of balancing airflow and water pressure to control dust levels (Volkwein et al., 1988).

Increasing curtain setback distance to 15.2 m reduced dust exposure, possibly by removing the operator from the high dust region surrounding the machine. No significant statistical interaction existed between water pressure and setback distance.

Return dust levels were not significantly affected by changes in water pressure or curtain setback distance. Increases in water pressure reduced levels at the higher curtain setback distance of 15.2 m and increased these levels at the lower setback distance.

Table 2. F-ratios derived from ANOVA analyses for sprayfan usage (gravimetric dust data).

Source of variation	Mining machine operator	Return
A: water pressure	33.04*	0.00
B: curtain setback distance	20.08*	1.52
Interaction: AB	0.22	1.22

Notes: Critical F-value: $F_{0.95,1,8} = 5.32$.
* = significant effect at 0.05 level of confidence.

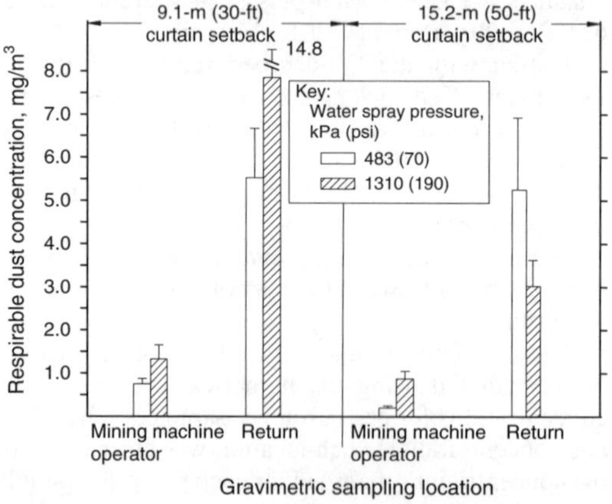

Figure 3. Gravimetric dust data for sprayfan (average plus standard deviation).

3.1.2 *Dust scrubber evaluation*

To simplify the model and its subsequent analyses, only the main effects and their two-way interactions were considered. Higher-ordered interactions were ignored. Ten degrees of freedom were used by the main effects and the interactions, leaving 37 degrees of freedom to determine the critical F-value in Table 3.

These analyses showed that setback distance significantly affected mining machine operator dust levels. The increase in dust exposures shown in Figure 4 shows that a minimal airflow of 3.8 m^3/s, though an approximate match for the scrubber capacity, could not contain the dust cloud near the cutting head. Even the larger curtain setback distance did not remove the operator from the dust cloud surrounding the machine. Dust levels generally increased with use of the external sprays, suggesting that they may have degraded the capture efficiency of the dust scrubber by blowing the dust cloud away from the scrubber inlets. The underboom sprays produced statistically insignificant variations in dust concentrations. However, their interaction with water pressure led to significant reductions, possibly through improved suppression of the underboom dust cloud.

Testing showed that changes in some test parameters significantly affected miner operator dust levels.

Table 3. F-ratios derived from ANOVA analyses for scrubber use (gravimetric dust data).

Source of variation	Mining machine operator	Return
A: water pressure	1.26	0.76
B: curtain setback distance	243.83*	2.56
C: external spray use	60.34*	124.66*
D: underboom spray use	0.92	0.02
Interaction: AB	3.39	0.24
Interaction: AC	1.08	7.43*
Interaction: AD	7.41*	13.02*
Interaction: BC	23.08*	0.68
Interaction: BD	0.00	0.15
Interaction: CD	0.07	0.00

Notes: Critical F-value: $F_{0.95,1,37} = 4.11$.
* = significant effect at 0.05 level of confidence.

However, variations in dust levels were small and, in general, dust levels were low at the miner operator location.

The external sprays significantly increased dust levels in the return, again implying that they pushed the dust cloud away from the scrubber inlets (Figure 5). Use of the underboom sprays lessened the impact of the external sprays. Dust levels also were significantly impacted by the interaction of water pressure and external spray use and by water pressure and underboom spray use.

The dust scrubber was more effective than the sprayfan in controlling miner operator and return dust levels. Use of the external or underboom sprays did not affect this trend. However, operator dust levels were low with the sprayfan running at the largest setback distance and lowest spray pressure. The sprayfan did not remove dust from the environment, as did the scrubber, but simply pushed the dust cloud around the mining machine. This led to high dust levels on the downwind side of the mining machine as evidenced by the elevated return concentrations. The higher return dust levels measured with the sprayfan suggest that any downwind workers (e.g., roof bolters) could be exposed to higher dust levels.

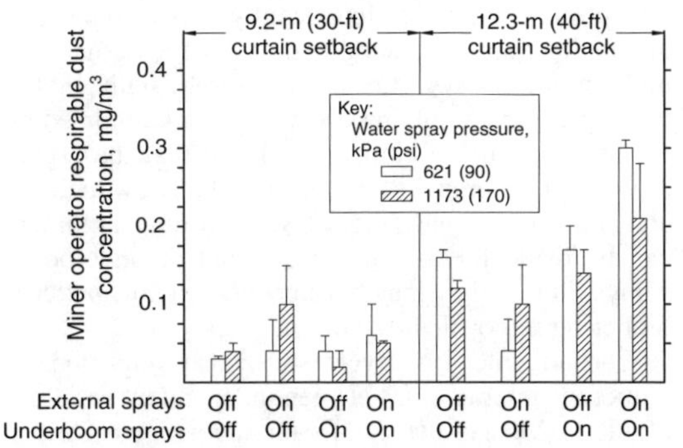

Figure 4. Miner operator dust levels with scrubber (average plus standard deviation).

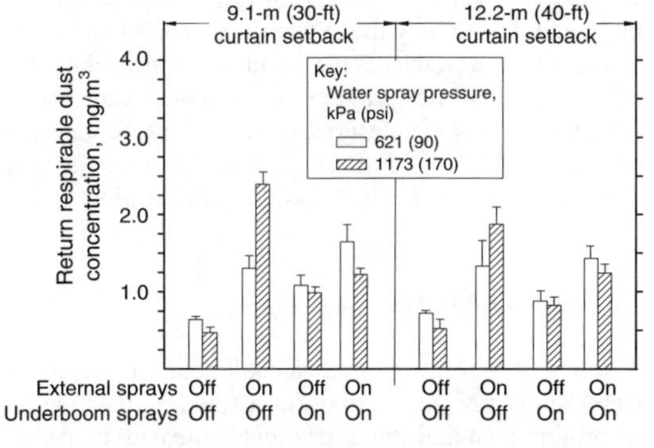

Figure 5. Return dust levels with scrubber (average plus standard deviation).

3.2 *Effects on tracer gas concentrations*

3.2.1 *Sprayfan evaluation*

Water pressure and curtain setback distance significantly affected tracer gas levels on the off-curtain side of the cutting head (Table 4). Increasing curtain setback distance adversely affected gas levels by reducing dilution airflow quantity to the face (Figure 6). Increases in water pressure resulted in slight increases in gas concentrations, suggesting that the higher pressure was hindering dilution and removal of the gas. No interactions between independent variables were significant. Gas concentrations were lowest on the off-curtain side of the cutting head, indicating that the sprayfan effectively "scoured" this region.

On the curtain side of the cutting head, higher water pressure significantly reduced tracer gas concentrations. This follows from previous work showing that higher water spray pressures create additional airflow and that this airflow can be used to dilute and remove harmful quantities of gas (Foster-Miller Inc., 1977, 1985). Slightly higher gas levels were measured at the larger curtain setback distance, suggesting that dilution airflow quantities decreased. No significant interactions were identified among the independent variables.

Table 4. F-ratios derived from ANOVA analyses for sprayfan usage (tracer gas data).

Source of variation	Off-curtain side	Curtain side
A: water pressure	8.80*	41.05*
B: curtain setback	21.06*	4.14
Interaction: AB	0.89	1.46

Notes: Critical F-value: $F_{0.95,1,8} = 5.32$.
* = significant effect at 0.05 level of confidence.

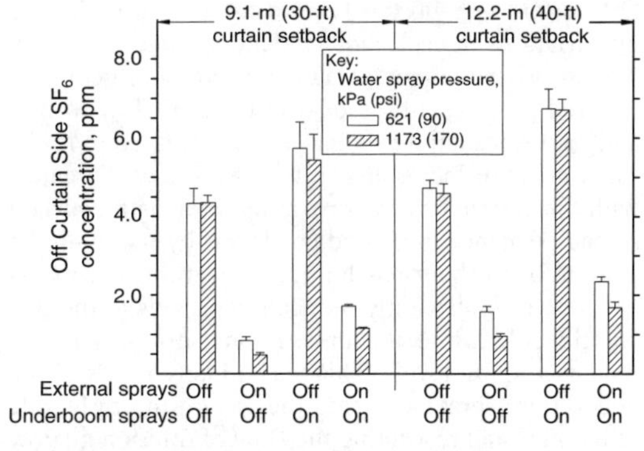

Figure 6. Tracer gas levels with sprayfan (average plus standard deviation).

Table 5. F-ratios derived from ANOVA analyses for scrubber usage (tracer gas data).

Source of variation	Off-curtain side	Curtain side
A: water pressure	13.18*	126.81*
B: curtain setback	50.81*	0.78
C: external spray use	1877.99*	86.05*
D: underboom spray use	175.37*	15.80*
Interaction: AB	0.12	15.18*
Interaction: AC	5.86*	6.24*
Interaction: AD	0.38	4.23*
Interaction: BC	0.65	14.26*
Interaction: BD	4.78*	0.08
Interaction: CD	22.80*	0.00

Notes: Critical F-value: $F_{0.95,1,37} = 4.11$.
* = significant effect at 0.05 level of confidence.

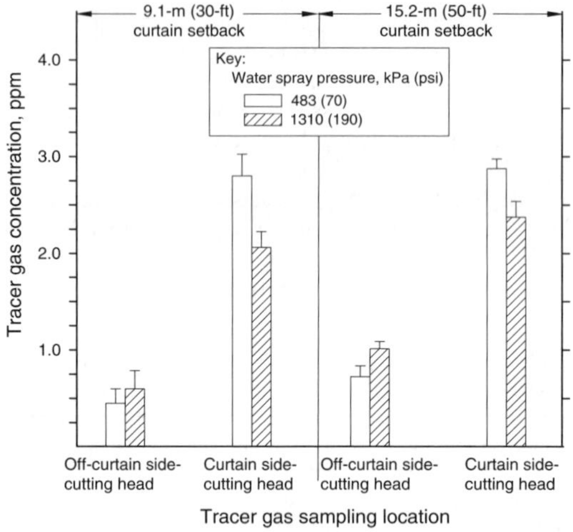

Figure 7. Off-curtain side tracer gas levels with scrubber (average plus standard deviation).

3.2.2 *Dust scrubber evaluation*

Off-curtain side gas levels were significantly affected by variations in each independent variable (Table 5). Not surprisingly, increasing water pressure generally lowered gas levels by inducing more airflow to mix and dilute accumulated gas in this area (Figure 7). The impact of increasing water pressure was greatest with the external sprays operating. Increasing curtain setback distance increased gas levels by reducing the amount of fresh air reaching this location. The external sprays significantly reduced gas levels at the off-curtain side while the underboom sprays typically increased these levels. This was a likely result of the underboom sprays trapping the gas on the right side of the face and restricting the flow of dilution airflow to this area. These sprays had a greater impact at the larger setback distance.

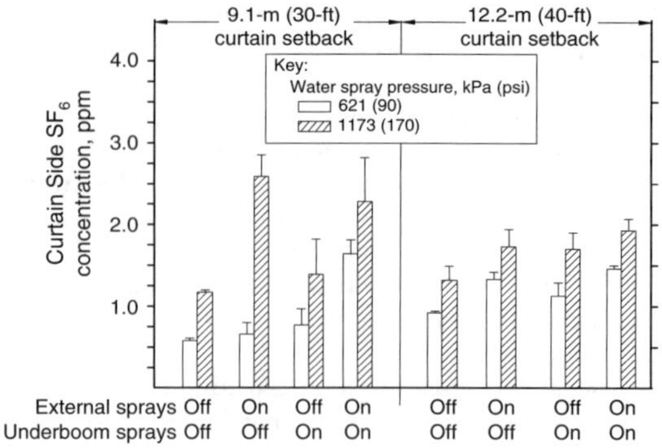

Figure 8. Curtain side tracer gas levels with scrubber (average plus standard deviation).

The significant interactions noted in Table 5 generally involved use of either the external or underboom sprays. This implies that spray placement, rather than water pressure or setback distance, impacted off-curtain side gas levels.

Gas levels measured on the curtain or left side of the cutting drum were significantly impacted by variations in water pressure and use of external and underboom sprays. Levels increased with higher water pressure, implying that the gas accumulated on the left side of the cutting head and that the higher pressure interfered with the ability of airflow to dilute and remove this gas (Figure 8). A similar argument can be made for use of the external or underboom sprays, in that they may have created airflow patterns hindering removal and dilution.

Curtain side gas levels were also impacted by numerous interactions between independent variables (Table 5). Most of the significant interactions involved water pressure, suggesting that this factor, was most important in controlling gas on the curtain side of the cutter head.

The sprayfan was more effective than the scrubber at limiting gas levels on the off-curtain side of the cutting head. Using only the external sprays and operating at the shortest setback, the scrubber approached the effectiveness of the sprayfan. At the lowest water pressure, this same spray design used with the scrubber surpassed the ability of the sprayfan for reducing gas concentrations on the curtain side of the cutting head.

4 CONCLUSIONS

A series of laboratory studies evaluated the effects of water spray pressure and curtain setback distance on respirable dust and tracer gas levels measured around a mining machine equipped with a directional spray (sprayfan) system. Subsequent testing assessed the impacts of water pressure, setback distance, and spray system design on dust and gas levels when using a flooded-bed dust scrubber. Spray system design encompassed four different combinations of standard, underboom, and external spray manifolds.

Gravimetric dust concentrations were measured at locations representing the continuous mining machine operator and in the return airway. Tracer gas levels were measured on the left and right sides of the cutting head, approximating the locations of conventional methanometers.

Changes in water pressure and setback distance significantly impacted miner operator dust levels when using the sprayfan. Increasing water spray pressure adversely affected exposures by increasing dust rollback levels. Increasing curtain setback distance reduced dust exposure by removing the operator from the high dust region surrounding the machine. Return dust levels varied slightly with changes in water pressure and setback distance, although the changes were not statistically significant.

With the scrubber, miner operator dust levels increased with setback distance. The external sprays significantly increased dust levels, implying that that these sprays may have pushed the dust cloud beyond reach of the scrubber inlets. This same mechanism also led to higher dust levels in the return airway. The underboom sprays produced insignificant changes at both sampling locations.

When using the sprayfan, water pressure affected tracer gas levels around the cutting head. An increase in water pressure produced slight increases on the off-curtain side and larger reductions on the curtain side. Increasing curtain setback distance adversely affected gas levels by reducing dilution airflow to the face.

Increasing water pressure with the dust scrubber reduced gas levels on the off-curtain side of the cutting head. The greatest impact occurred with use of external sprays. The underboom sprays increased gas in the off-curtain area by reducing airflow on this side of the cutting head. On the curtain side of the cutter head, increasing water pressure also elevated gas levels by reducing airflow to this area. A similar mechanism likely caused an increase in gas levels with use of the external or underboom sprays.

The scrubber was more effective than the sprayfan in controlling respirable dust concentrations at the miner operator and in the return. Operator dust levels did change with setback distance and water pressure, although these levels never exceeded 0.3 mg/m^3 with the scrubber. The lower return dust levels with the scrubber imply that workers may not face as great an overexposure when working downwind of the mining machine.

The sprayfan was more effective than the scrubber in reducing gas levels on the off-curtain side of the cutting head. At the shortest setback distance, the scrubber plus external sprays approached the effectiveness of the sprayfan for removing gas from this area. This same spray configuration at the lowest water pressure was more effective than the sprayfan at limiting curtain side gas levels.

REFERENCES

Code of Federal Regulations. 1999. Title 30, Mineral Resources, Parts 70 and 75, Coal Mine Safety and Health, U.S. Government Printing Office.

Colinet, J.F., McClelland, J.J., & Jankowski, R.A. 1991. Interactions and Limitations of Primary Dust Control for Continuous Miners. Report of Investigations, No. 9373. US Bureau of Mines, Pittsburgh, PA.

Colinet, J.F., & Jankowski, R.A. 1996. Dust Control Considerations for Deep-Cut Mining When Utilizing Exhaust Ventilation and a Scrubber. Report of Investigations, No. 9615, U.S. Bureau of Mines.

Divers, E.F., & Jayaraman, N.I. 1982. Coal Mine Face Ventilation Systems: New Concepts and Underground Results. In Proceedings of the First US Mine Ventilation Symp., H.L. Hartman (ed.), pp. 33–42. Littleton, CO: Society of Mining, Metallurgy, and Exploration.

Foster-Miller, Inc. 1977. Development of Optimized Diffuser and Sprayfan System for Coal Mine Face Ventilation. Contract H0230023, US Bureau of Mines, Pittsburgh, PA.

Foster-Miller, Inc. 1985. Improved Diffuser and Sprayfan Systems for Ventilation of Coal Mine Working Faces. Contract J0113010, US Bureau of Mines, Pittsburgh, PA.

Jankowski, R.A., Jayaraman, N.I., & Babbitt, C.A. 1987. Water Spray Systems for Reducing the Quartz Dust Exposure of the Continuous Miner Operator. In R.V. Ramani (ed.), Proceedings of the 3rd US Mine Ventilation Symp., University Park, PA. Littleton, CO: Society of Mining, Metallurgy, and Exploration.

Jayaraman, N.I., Colinet, J.F., & Jankowski, R.A. 1992a. Recent Basic Research on Dust Removal for Coal Mine Applications. In Proceedings of the Fifth International Mine Ventilation Congress, pp. 395–405, Johannesburg, South Africa: South African Institute of Mining and Metallurgy.

Jayaraman, N.I., Jankowski, R.A., & Whitehead, K.L. 1992b. Optimizing Continuous Miner Scrubbers for Dust Control in High Coal Seams. Proceedings, Society for Mining, Metallurgy, and Exploration Annual Meeting, Phoenix.

Luxner, J.V. 1969. Face Ventilation in Underground Bituminous Coal Mines. Report of Investigations, No. 7223. US Bureau of Mines, Pittsburgh, PA,

O'Green, J. 1990. Using Scrubber to Comply with Dust Standards in High Output Continuous Miner Sections, Preprint No. 90–102, Society for Mining, Metallurgy, and Exploration Annual Meeting, Salt Lake City, Feb.

Taylor, C.D., & Zimmer, J.A. 2001. Effects of Water Sprays and Scrubber Exhaust on Face Methane Concentrations. In S. Wasilewski (ed.), Proceedings of the 7th International Mine Ventilation Symp., Crakow, 17–22

June, 2001, Crakow: Research and Development Center for Electrical Engineering and Automation (EMAG).
Taylor, C.D., Rider, J., & Thimons, E.D. 1996. Changes in Face Methane Concentrations Using High Capacity Scrubbers with Exhausting and Blowing Ventilation, Preprint 96–167, Society for Mining, Metallurgy, and Exploration Annual Meeting, Phoenix.
Volkwein, J.C. 1988. Sprayfans Combined with Reverse Sprays Ventilate 40-foot Cuts. Technology News, No. 319. US Bureau of Mines, Pittsburgh, PA.

Mine Ventilation, Ganguli & Bandopadhyay (eds)
© 2004 R.A. Haney, ISBN 90 5809 633 5

Controlling silver dust and fumes at mine refinery

R.A. Haney & M.P. Valoski
Mine Safety and Health Administration, Pittsburgh Safety and Health Technology Center, Pittsburgh, Pennsylvania, USA

ABSTRACT: As part of the refining of gold and silver molten metal, silver dust and fumes are released into the atmosphere. The Mine Safety and Health Administration (MSHA) enforces an 8-hour, equivalent Time Weighted Average concentration limit for silver dust and fumes of 10 $\mu g/m^3$.

MSHA initiated a program to assess the controls that were being used to control silver dust and fume exposure. Refineries were visited at six mines. The layout of each refinery and the controls used varied at each refinery. At each operation, personal and area silver fume and dust samples were collected to assess worker exposures and to determine sources of fume.

Primary source of silver dust and fume exposure was the pouring of molten metal from the furnace. Secondary sources of exposure included: precipitate mixing, bar cooling, and housekeeping. Guidelines were developed addressing housekeeping, exhaust ventilation, general ventilation, administrative controls, and system monitoring. In most cases, housekeeping and general ventilation were adequate; however, the exhaust ventilation systems needed to be improved.

1 INTRODUCTION

Silver dust and fumes become airborne during the refining step of producing gold and silver. The dust and fumes becomes airborne when molten material is poured from the furnace into bars or buttons. Additionally, fumes that are not captured by the exhaust ventilation system can condense into dust and settle onto equipment in the refinery. This dust can then become entrained due to vibration or activities within the refinery.

The current exposure limit for silver fumes and dusts is found at 30 C.F.R. Section 56/57.5001. Those standards apply to numerous airborne contaminants including silver dusts and fumes. The Standards incorporate by reference the terms of the ACGIH's 1973 edition of "Threshold Limit Values for Chemicals Substances in Workroom Air." There, the exposure limit for silver fumes and dusts has been set at 0.01 mg/m^3 (10 $\mu g/m^3$) over an eight hour shift.

The one major hazard to which the silver exposure limit is directed is that of argyria. Argyria is a condition where an employee's skin, eyes, or mucous membranes turns from slate to blue-gray in color. The color may be general or localized. Additional potential health hazards associated with exposure to silver fumes and dust include diffuse pulmonary fibrosis that could be confirmed with chest radiograms. Historically, many silver producing mines have had difficulty consistently meeting this standard.

Gold and silver ore is typically mined by open pit or underground mining methods. Both consisting of drilling, blasting, and loading via front-end loaders and truck haulage. The ore contains large amounts of impurities. To concentrate the silver ore, mining companies use cyanide leaching. The low grade silver ore is finely crushed and placed onto large piles. Beneath the pile is an impervious layer. A dilute solution of cyanide trickles through the pile, dissolves the silver and gold from the ore and is trapped by the impervious layer. The cyanide solution with the silver ore is called the pregnant leachate solution. The collected pregnant solution passes through activated coconut charcoal at the mill to be concentrated and purified. Then another volume of cyanide solution passes through the charcoal to strip the silver ore from the charcoal. Finally a zinc solution precipitates out the silver ore. The liquid is pressed from the precipitate before it is dried in an oven.

The dried precipitate along with a "flux" are loaded into a furnace and heated to over 2100 degrees Fahrenheit. Once the material has been heated, the slag and metal are poured into either buttons or bars. The buttons or bars are then moved to a cooling area.

Buttons and bars typically contain 85 to 95 percent silver. Once cooled, the buttons or bars are cleaned and placed into a vault before shipping to a commercial refinery for further processing.

Silver fume and dust exposure are controlled by a number of engineering and administrative controls. Typically, these controls include:

- General refinery ventilation,
- Exhaust ventilation at the furnace,
- Material handling practices, and
- Housekeeping practices.

While the specific controls vary from refinery to refinery, there are certain consistencies among refineries. The purpose of this study was to identify and quantify those practices that gave the best results for controlling exposure to silver fumes.

2 SAMPLING AND ANALYTICAL PROCEDURES

This study included site visits to six refineries. At each operation, typically, three personal and three area silver fume samples were collected to assess worker exposures and the source of worker exposures to silver fumes. Personal samples were collected on three refiners. Area samples were collected approximately 10-feet on either side of the furnace and in the center of the refinery. Personal and area samples were collected for the full shift.

As per Metal and Nonmetal Mine Safety and Health's sampling protocol, all silver fume samples were collected using SKC Airchek® Sampler pumps (reference to specific brands of equipment is made to facilitate understanding and does not imply endorsement by the MSHA), calibrated and operated at 1.7 liters per minute (Lpm). The samples were collected on 37-millimeter diameter, MCE filters with a 5.0 micrometer pore size. The filters were mounted in two-piece Millipore filter holder. A cyclone was not used in the sample train. A control filter was utilized to verify no initial sample contamination. The samples were analyzed by Inductively Coupled Plasma (ICP), MSHA Method P-3, at the Dust Division Laboratory in Pittsburgh, Pennsylvania.

Silver fume concentrations were determined as Shift Weighted Average (SWA). The formula for calculating the SWA is:

$$\text{SWA} = \frac{\text{Weight of Contaminant (mg)}}{\text{Flow Rate (Lpm)} \times \text{Time} \times 0.001\ \text{m}^3/\text{L}}$$

As per Metal Nonmetal Safety and Health's sampling protocol, a 480-minute time is used to calculate the SWA. The PEL for silver fumes and dust, from the 1973 TLV list, is 0.01 mg/m^3.

The C/E ratio of concentration to allowable exposure (SWA/PEL) was calculated for each sample. The error factor established for silver fume determination is 1.14. When the C/E ratio exceeds 1.14, noncompliance with the silver fume standard is indicated. When the ratio is equal to or below 0.86, compliance with the standard is indicated. Ratios between 0.86 and 1.14 indicate areas of uncertainty in compliance, indicating that controls should be improved.

In addition to the personal and area samples, airflow measurements were taken with either a Pitot-static tube or a vane anemometer at exhaust hoods for the furnace. Hood and duct sizes were measured and actual air velocity and volumes were calculated.

Refinery building volume was measured and the airflow induced by wall fans was measured. Refinery air exchange rate was calculated. Other dust controls and work practices in the plant were also observed.

3 RESULTS AND DISCUSSION

In most refineries, the general mill ventilation rate exceeded 10 air changes per hour. However, in several refineries the location of exhaust fans and make-up air supply louvers caused cross drafts that over-powered the furnace exhaust ventilation system and caused silver fumes to be blown toward workers.

Several forms of administrative controls that reduce exposure to silver fumes were observed. In three of the operations, furnace controls were located more than ten feet from the furnace. In one of these three operations, a separate pouring room was established inside the refinery. The pouring room was under negative pressure and doors were closed during the pouring operation. The furnace controls were outside of the room and the pouring process was observed through a glass window. However, in the three other operations, furnace controls were close to the furnace or even inside the furnace exhaust ventilation hood.

A furnace temperature of 2200 degrees Fahrenheit was reportedly sufficient to maintain the gold and silver in a molten state for pouring. The higher temperatures produce more fumes. Furnace temperature should be monitored so that fume production is minimized.

Typically, the pouring height was two to three feet. Reducing the pouring height would reduce the amount of silver fumes emitted. Bars and buttons were generally moved with long handled carts or forklift trucks. This practice removed workers from the vicinity of molten silver bars. In one case, bars were quenched with water as soon as possible and in one case hot bars were placed under a cooling hood, to reduce the emission of fumes.

Housekeeping was generally excellent. The mills were washed down on a regular basis and floors were swept with a broom and wash several times a day.

Figure 1. Operation with exhaust ventilation.

Figure 2. Operation with housekeeping and remote furnace controls.

Additionally, in some operations, floors were waxed on a weekly basis. The primary housekeeping issue was dust that had settled on flat surfaces above the floor. This dust could become reentrained and contribute to worker exposure. In many cases these surfaces could not be washed down because of adjacent electrical fixtures. As a result these surfaces should be vacuumed on a regular basis to remove accumulations of settled dust. In addition to general housekeeping, heat-reflective clothing should be cleaned weekly. Wipe samples taken on the inside of heat-reflective clothing indicated that the clothing was a source of exposure.

The most noticeable deficiency at most of the refineries was the exhaust ventilation system for the furnace. In general, the hoods did not surround the furnace sufficiently and were not close enough to the pour spout to contain the fumes as molten material was poured. Additionally, the hood face velocity was below the ACGIH recommended 200 fpm velocity and insufficient to draw the fumes into the exhaust ventilation system. Duct transport velocities were below the ACGIH recommended 3600 fpm; however, because of the poor capture velocity, there was no observed dust settling or blockage of ducts observed.

System monitoring should include both monitoring of workers' exposure and monitoring of dust and fume controls. In most cases, workers' exposure were monitored on a regular basis. However, none of these operations monitored either the mill air exchange rate or the furnace exhaust ventilation system. Pressure gages were not installed on the dust collector or the exhaust ventilation fan. Periodic checks of ventilation systems should be made to assure that the systems are working properly.

Figure 1 shows an operation where hoods and enclosures, along with mill ventilation, are used to capture silver fumes. Additionally, furnace controls are positioned and trucks with long handles are used to keep workers away from silver fume exposure. In this operation a cooling station was provided for button cooling.

Figure 2 shows an operation which relied on mill ventilation, housekeeping, and administrative controls to reduce worker exposure. Additionally, furnace controls positioning and a forklift are used to keep workers away from silver fume exposure. In this operation a bar quenching station was provided for bar cooling.

Based on the measurements and observations made during these studies, a list of guidelines for controlling silver dust and fumes was developed. These guidelines are not meant as a list of best practices, but rather routine control measures that should be incorporated into all operations.

4 GUIDELINES

4.1 *Housekeeping*

- Wash refinery floor, daily and wash refinery, top to bottom, weekly.
- Buff (wax) refinery floor, weekly.
- Refiner's coveralls should be washed daily, and
- Wash heat-reflective clothing, weekly.
- Use a HEPA vacuum to clean up spills or dust.

4.2 *Administrative controls*

- Install furnace controls at least 10 feet from furnace discharge.
- Furnace temperature should be maintained at less than 2200 degrees Fahrenheit.
- A forklift or long-handled truck should be used to move hot bars.
- Bars or buttons should be quenched with water as-soon-as possible.

- Minimize the pouring height (distance from furnace spout to bar mold).
- Material should not be manually fed into the furnace.

4.3 *Exhaust ventilation*

- Hoods, enclosures or booths should be provided at the:
  - Furnace and furnace discharge (booth),
  - Bar or button cooling station (hood or booth),
  - Bar or button cleaner station (enclosure),
  - Grinder (enclosure), and,
  - Mixing station (enclosure).
- Hoods and booths should surround the fume source.
- Face velocities should be greater than 200 fpm (duct airflow/hood area).
- Duct transport velocity should be greater than 3600 fpm.
- Enclosures should be under negative pressure.
- Follow design guidelines in Industrial Ventilation: A Manual of Recommended Practices,
  - Metal Melting Furnaces and Mixing (Chapter 10), and
  - Exhaust System Design Procedure (Chapter 5).

4.4 *General ventilation*

- There should be at least 10 air changes per hour.
  - (Air changes/hour = 60 × Airflow/Building volume)
- Wall fans should be located behind or above the furnace to draw air away from workers.
- Makeup air inlets should be located to eliminate cross drafts.

4.5 *System monitoring*

- Refinery worker exposure should be monitored quarterly.
- Air volume monitoring stations should be established for the furnace hood, bar cooling hood and dust collector fan.
- Air volumes should be measured quarterly.
- Use smoke tubes to visualize capture and cross contamination at hoods.

REFERENCES

American Conference of Governmental Industrial Hygienists, *Industrial Ventilation: A Manual of Recommended Practices*, 26th Edition, 2002.

American Conference of Governmental Industrial Hygienists, *Documentation of the Threshold Limit Values and Biological Exposure Indices*, 2001, 6 pages (from www.acgih.org)

American Conference of Governmental Industrial Hygienists, *TLVs® Threshold Limit Values for Chemical Substances and Physical Agents in the Workroom Environment with Intended changes for 1973*, 1973, 54 pages.

Title 30 Code of Federal Regulations, Office of Federal Register, National Archives and Records Administration, July 1, 2002.

Mine Ventilation, Ganguli & Bandopadhyay (eds)
© 2004 Taylor & Francis Group, London, ISBN 90 5809 633 5

An assessment of a two-filter concept for a continuous miner wet scrubber

Yoginder P. Chugh & Krishna K. Thatavarthy
Department of Mining and Mineral Resources Engineering, Southern Illinois University, Carbondale, Illinois

Ronald Carty
Illinois Clean Coal Institute, Carterville, Illinois

ABSTRACT: The 2 mg/m^3 respirable dust standard established by the 1969 Federal Mine Safety and Health Act led to the development of the continuous miner wet scrubber, which has become an industry standard. A typical scrubber utilizes one inclined filter with horizontal water sprays. Increasing production rates and calls for lowering the dust standard to 1 mg/m^3 have intensified the compliance challenge in recent years making scrubber efficiency a critical issue. Southern Illinois University at Carbondale (SIUC) collaborated with Joy Mining Machinery and the State of Illinois to establish a dust control research laboratory focusing on increasing the wet scrubber efficiency.

This paper assesses the effectiveness of a two-filter design for a continuous miner wet scrubber used for coal dust control. In developing the design it was decided to maintain at least the same amount of air flow as with an existing single filter (~8000 cfm) scrubber. The first filter is designed to screen larger size particles and partially wet the finer dust with water-only sprays. The second filter is designed to wet fine and ultra fine dust particles with water sprays containing surfactants. Two different spray geometries, vertical and horizontal, were attempted. Number of sprays, spray water pressure, and surfactant concentration variables were evaluated using statistically designed experiments. The input dust concentration to the scrubber was in the range of 45 mg/m^3. All dust samples were collected according to MSHA guidelines including processing by MSHA's robotic weighing system. The results indicate that this two-filter design can improve scrubber efficiency and reduce dust concentrations from about 3.90 mg/m^3 for a single filter to about 1.5–2.0 mg/m^3.

1 INTRODUCTION

The issue of dust control in underground coal mines has been extensively studied by both government and industry. The dust control techniques developed either prevent the dust from becoming airborne or cleanse the air before it is inhaled. Avoidance measures minimize the exposure of miners to the airborne dust by blowing the dust away from work areas or making it unnecessary for a miner to occupy a dusty region (Wirch and Jankowski, 1996). One of several successful dust control measures in continuous miner sections is the use of a machine-mounted wet scrubber, which wets the dust-laden air, with significantly reduced dust concentration in the exiting air. Machine-mounted scrubbers have successfully controlled dust on continuous miner faces to a large degree.

In the mid '70s, research efforts led to the development of a continuous miner scrubber system (Campbell, 1983). In US underground coal mines, use of a wet scrubber mounted on a continuous miner has been the primary approach to control dust in the face area. This system applies both ventilation and water spray principles. The wet scrubber, illustrated in Figure 1, is a fan-like device which vacuums coal dust from the face atmosphere by suctioning air from behind the continuous miner's cutting head and spraying it with water. The water-laden dust in the air is collected on a screen or filter panel. The air passes through the panel where it is dewatered in the mist eliminator and discharged with significantly lower dust concentration.

Over the last 20 years, there has been a significant push to improve productivity in coal mines so that coal can compete economically in the energy market. Significant progress has been made toward increased productivity and production rates 2–3 times larger than in 1970's are being consistently achieved. However, with increased productivity, concentration of dust has increased and dust concentration standard of 2 mg/m^3 is not being achieved in larger number of MSHA

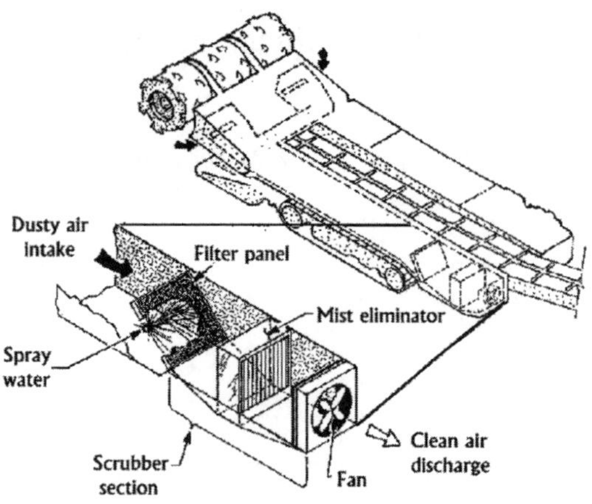

Figure 1. A Continuous miner with flooded-bed wet scrubber.

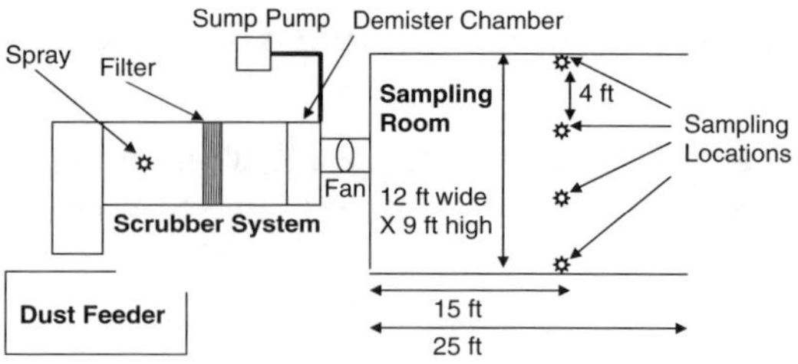

Figure 2. A layout of the dust control laboratory (not to scale).

Figure 3. Continuous miner scrubber system.

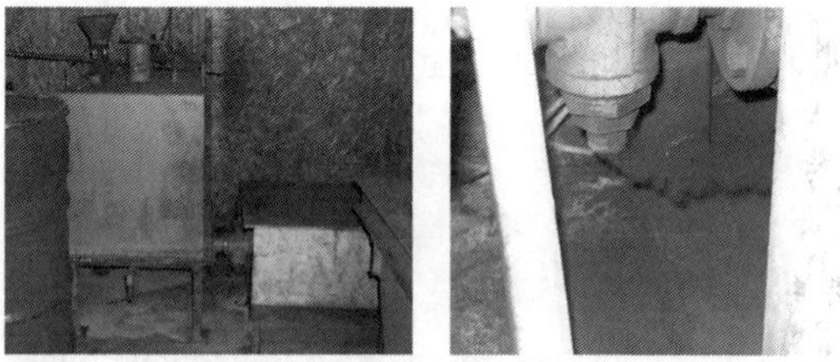

Figure 4. Dust feeder system.

collected samples. There is a significant need for research and development to reduce dust concentration below 2 mg/m^3. This research is directed toward achieving that goal through improved wet scrubbers on continuous mining machines. A successful study demonstrating significant improvements in scrubber efficiency will pave the way for healthier mine atmospheres while increasing production rates insuring a more profitable coal industry in the USA.

2 SIU-JOY DUST CONTROL LABORATORY

2.1 *Description*

Chugh (2000) developed the SIU dust control laboratory in collaboration with Joy Mining Machinery. As shown in Figure 2, the laboratory consists of a wet scrubber system, a controllable dust feeder, and a sampling room modeled after a mine entry or tunnel. A detailed description of the primary equipment used in the scrubber simulation laboratory follows.

2.1.1 *Flooded-bed scrubber system*

An 8000 cubic foot per minute (cfm) flooded-bed scrubber system manufactured by Joy Mining Machinery for a model 12 CM continuous miner was set up at the Illinois Coal Development Park in Carterville, Illinois. The scrubber was equipped with two 60-gallon drums and a two-stage (1/2 HP + 3/4 HP) electric pump system to supply water to the scrubber system at a maximum rate of 10 gallons per minute and a maximum pressure of 100 psi. An appropriate amount of surfactant to achieve the desired concentration was mixed with water in the two tanks as they were filled. A 3-HP gasoline-powered sump pump removed the water and "scrubbed" dust from the system at the bottom of the demister chamber. The demister chamber is a rectangular box located on the right hand side of Figure 3 adjacent to the round fan housing.

2.1.2 *Dust feeder*

Raw coal from a mine was pulverized with a compressed air crusher. The resultant "dust" was fed into a small hopper, shown in Figure 4, which fed dust through a two-inch screw conveyor at the bottom. The screw conveyor was driven by a 1/8 HP motor and a Boston Gear Ratiotrol for speed control from 0 to 100% or from a minimum feed rate of 0 grams per minute. To achieve a consistent feed rate, the dust feeder was run at 20% feed rate or higher. All testing here was

Figure 5. Sampling room and stations.

done at 20% feed rate and the input dust concentration was 48 mg/m^3.

2.1.3 *Dust sampling room and locations*

The sampling room was modeled after a typical mine entry as shown in Figure 5. In the Illinois Basin, mine openings average 6 to 7 feet in height and 18 to 20 feet in width. The laboratory is not part of a closed ventilation system. The scrubber is housed in a small building with large sliding doors that are opened when the scrubber is operating. The capacity of the scrubber system in use in the laboratory is about 8,000 cfm. This volume of intake air is pulled from the surrounding air, which is relatively motionless. Windy conditions do not affect the laboratory scrubber but can affect the laboratory results if the wind is from the north as the sampling room opens to the north.

3 SCRUBBER SIMULATION TESTS

3.1 *Description of scrubber testing procedure*

Before operating the scrubber, the MSA Escort LC Dust Sampling Pumps were calibrated at 2 liters/min using the Wet test meter. Then, the pumps along with the Dorr-Oliver cyclone assembly with the MSA dust filter cassette fitted on them were placed in the sampling locations. The two 60-gallon barrels on top of the scrubber system were filled with water and a surfactant was added as needed to achieve an appropriate concentration. The dust feeder was loaded with enough dust to last for a one-hour test.

First, the compressor for the dust feeder system was turned on and the moisture in the pipeline was bled through a relief valve. Next, the sump pump and the scrubber fan motor were started. The water spray pumps were turned on and the appropriate spray pressure was established by means of relief valves in the system. Then the dust feeder system was started and set at the 20% feed rate. A small stream of compressed air was introduced at the screw conveyor discharge point to maintain a continuous flow of dust. Once everything was running, the dust sampling pumps were turned on to initiate the one-hour test.

At the end of one hour, the pumps were shut off and the dust filter cassettes were taken out of the Dorr-Oliver cyclone assembly. The outer surface of the cassette was wiped clean with a rag cloth. A dust data card was supplied with each filter cassette. The card contained an identification number (cassette number) and the initial weight, in grams, of the filter capsule inside the filter cassette. The sample date, time, and length were filled in on the dust data card. The sample was then shipped to MSHA for analysis.

3.2 *Dust concentration analysis by MSHA*

The filter capsule, contained inside the cassette shell, was the only part of the cassette that was weighed. The capsule was kept free of contamination and never touched with human hands until after the weighing process was completed. To gain access to the filter capsule, the two halves of the cassette shell were pried apart using a knife or other sharp object. The capsule was then removed from the shell using forceps and placed on a weighing tray. Prior to weighing, the samples were desiccated to remove any moisture present on the samples. The samples were desiccated in a Webber Model F-3-AV vacuum chamber.

The final weight of each sample being processed was determined to the nearest 0.001 milligram using an Automated Weighing System (AWS). The AWS was composed of a Zymark Zymate System laboratory robot with a System V Controller, a sample turntable, two Mettler Model MT5 microbalances, a printer and an IBM Model AT computer with a monitor. The system was self-calibrating and was capable of automatically weighing 600 filter capsules without manual intervention.

3.3 *Data analysis*

To calculate the concentration of respirable dust in the mine simulation laboratory atmosphere, the following formula was applied:

Dust Concentration (mg/m^3) =
Weight gain (mg) * 1.38/sampling time (min) * rate of sampling (m^3/mg) (1)

where, weight gain is the number reported by MSHA (in milligrams) and sampling time is the test run period (in minutes). The 1.38 factor is a Mine Research Establishment (MRE) standard. MRE is a British group responsible for developing epidemiologic studies on the relationship between coal dust and disease. This formula is an industry standard. All the scrubber tests conducted were run for a period of 60 minutes and the sampling pump was rated for 0.002 m^3/mg.

4 EXPERIMENTAL DESIGN

4.1 *Design factors*

There are several factors, which can improve the dust control efficiency of a continuous miner scrubber. The factors selected for analysis in this study are described below.

4.1.1 *Scrubber filter layers*

The filter medium in the scrubber consisted of layers of stainless steel wire mesh secured in a flat filter panel. Filters containing 5, 10, 15, 20, 30 and 40 layers of wire mesh are commercially available. The 20-layer panel is the most common filter now being used throughout the industry. For this study, 5, 10, and 20 layer filters were employed.

4.1.2 *Spray type*

Potable water supply was available at the Illinois Coal Development Park and was used for all scrubber tests. Horizontal and vertical sprays were designed to evaluate their effectiveness for improved dust control.

4.1.3 *Spray volume*

Multiple low gpm (gallons-per-minute) spray nozzles were used instead of a single high gpm spray nozzle to improve the interaction between dust-laden air and spray water. Full-cone and hollow-cone sprays with spray volume of 0.5, 0.67, 1, 2, 3 and 4 gpm spray nozzles at 40 psi spray pressure were employed in this study.

4.1.4 *Surfactant concentration*

Surfactants were added to increase the wetting characteristics of plain water. Most tests were conducted using 0.005% concentration. A limited number of tests were conducted using 0.006% and 0.007% surfactant concentrations.

4.2 *Design methods*

In an experiment, researchers deliberately change one or more variables (or factors) in order to observe the effect of the changes on one or more response variables. The statistical design of experiments (DOE) is an efficient procedure for planning experiments so that the data obtained can be analyzed to yield valid and objective conclusions. The DOE begins with determining the objectives of an experiment and selecting the factors for the study. An experimental design involves the laying out of a detailed experimental plan in advance of doing the experiment. Well-chosen experimental designs maximize the amount of information that can be obtained for a given amount of experimental effort.

Several different types of designs are useful when one is attempting to model a response surface. D-optimal designs are one form of design provided by a computer algorithm. These types of computer-aided designs are particularly useful when classical designs do not apply. Unlike standard classical designs such as factorials and fractional factorials, D-optimal design matrices are usually not orthogonal and effect estimates are correlated (NIST/SEMATECH e-Handbook of Statistical Methods, 2002).

These types of designs are always an option regardless of the type of model the researcher wishes to fit (for example, first order, first order plus some interactions, full quadratic, cubic, etc.) or the objective specified for the experiment (for example, screening, response surface, etc.). The D-optimal design procedure generates multi-factor experiments with both quantitative and qualitative factors. The factors can have a mixed number of levels.

The reasons for using D-optimal designs instead of standard classical designs generally fall into two categories:

- Standard factorial or fractional factorial designs require too many runs for the amount of resources or time allowed for the experiment.
- The design space is constrained (the process space contains factor settings that are not feasible or are impossible to run).

For the scrubber tests, a D-optimal design experimental design was used. A test matrix of 9 tests was prepared with the assistance of the Design Expert software package developed by Stat-Ease, Inc.

Previous controlled studies at SIU (Saha, 2001) with standard one-filter in the wet scrubber achieved dust concentration values of about 3.5 mg/m^3 in the exhaust air of a typical wet scrubber. Input dust concentration to the scrubber for these studies was 48 mg/m^3. Thus, the laboratory scrubber was handling three times more respirable dust (<15 micron size) than the scrubber operating in the mine environment. The overall goal of this research is to reduce this value to about 2 mg/m^3 by utilizing two-filter design for the wet scrubber.

5 TWO-FILTER DESIGN FOR THE SCRUBBER

5.1 *Concepts*

A wettability test developed by the authors (Chugh et al., 2004), for assessing effectiveness of surfactants for dust control showed that coarser particles are easily wetted by plain water and surfactants are needed

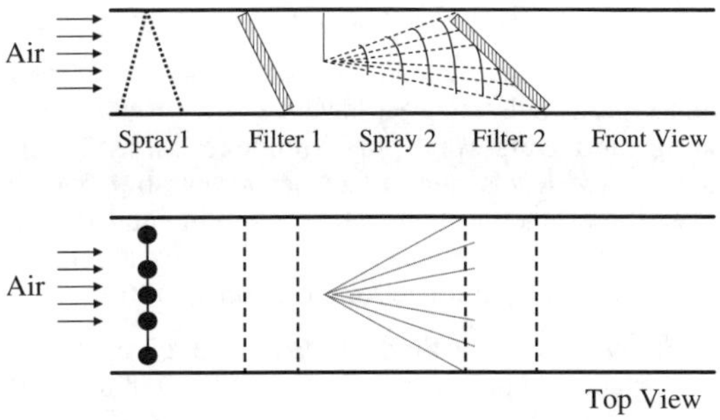

Figure 6. Two-filter design concept for a wet scrubber.

Figure 7. Actual two-filter design for the wet scrubber.

only to wet finer particles. This concept was utilized to design the wet scrubber using a two-filter system. The dust-laden air with the coarser and finer particles entering the scrubber can be wetted with a pre-wet water only spray and a filter system. This removes the coarser particles and only finer particles pass through the filter. These finer particles can then be wetted with a surfactant spray and a filter system.

5.2 *Modifications*

The scrubber was modified to incorporate a filter screen after the pre-wet water spray system to knock the coarser particles effectively and then treat the relatively fine particles that constitute the respirable dust with surfactant spray located in front of the second filter. The two-filter design for a horizontal surfactant spray system is shown in the Figures 6 and 7. The following terminology is followed here in describing the concept (Figure 6).

- Pre-wet water only spray system – Spray 1
- Filter after Spray 1 – Filter 1
- Surfactant spray system after Filter 1 – Spray 2
- Filter after Spray 2 – Filter 2

Figure 8. Spray 1, Filter 1 and Spray 2 (horizontal) in the wet scrubber.

5.3 *Experimental design*

A parametric study was conducted on coal from Mine 2 using Surfactant 2 varying three operating parameters – number of filter layers for Filter 2, type of Spray 2 and nozzles for Spray 2. The surfactants and coals used are described in the paper "Development of a wettability test for accessing surfactant effectives of surfactants for dust control". A statistical experiment consisting of eleven (11) experiments was designed using the D-Optimal method to investigate the operational effects of each parameter on the dust concentration response variable. All the operating factors were varied over two levels.

- Filter 2 – 5 layer and 10 layer filter.
- Type of Spray 2 – Horizontal spray and Vertical spray systems.
- Nozzles for Spray 2 – Full cone and Hollow cone.

Figure 8 shows Spray 1, Filter 1 and Horizontal Spray 2. The Vertical Spray 2 is shown in Figure 9. The other fixed parameters are spray volume for Spray 1 and Spray 2 at 2 and 3 gpm, respectively with spray pressure of 40-psi and surfactant concentration of 0.005%.

5.4 *Phase I experimental results*

The summarized results from the set of eleven (11) experiments and repetitions are shown in Table 1. The results show that for Spray 2, a vertical spray system is better than a horizontal spray system and use of hollow-cone nozzles give better reduction in dust concentrations than full-cone nozzles. A hollow-cone vertical spray system resulted in dust concentration of 1.9 mg/m^3 compared to 2.1 mg/m^3 with a full-cone vertical spray system, a statistically significant difference. For Filter 2, 10-layer filter gave better results than 5-layer filter but had the disadvantage of reducing airflow to 5,300 cfm. So, the conclusion was to use two 5-layer filters in a commercial wet scrubber system.

The hypothesis of the two-filter concept for a wet scrubber is valid based on the results from the Phase I experimental studies. This paved the way for Phase II experimental studies to test the validity of the two-filter concept for different coals.

Figure 9. Spray 2 (vertical) on the wet scrubber.

Table 1. Scrubber test results.

Filter 1 (#)	Filter 2 (#)	Cone type	Direction	Dust concentration (mg/m^3)
5	10	Hollow	Horizontal	2.327
5	10	Full	Horizontal	2.191
5	5	Hollow	Horizontal	2.017
5	5	Full	Horizontal	2.307
5	10	Hollow	Vertical	1.969
5	10	Full	Vertical	1.645
5	5	Hollow	Vertical	1.992
5	5	Full	Vertical	2.130

5.5 *Phase II scrubber tests for mine 1 and mine 2*

Finally, a parametric study was conducted for Mine 1 and Mine 2 coals using Surfactant 2. The variable operating parameters were spray volume for Spray 2 (varied over two levels) and surfactant concentration (varied over three levels). A statistical experimental design consisting of nine (9) tests was used to investigate the effects of each parameter on dust concentration.

- Spray volume for Spray 2 – 2 gpm and 3 gpm.
- Surfactant concentration – 0.005%, 0.006% and 0.007%.

Test results for Mine 1 and Mine 2 coals are shown in Table 2 and Table 3, respectively. The weight gain shown is the average weight gain from four filter cassettes.

Based on the above results, an empirical model for the dust concentration response variable was developed as a function of the two operating parameters. The test data were statistically analyzed using the Design Expert software package.

a) Mine 1

Several model types were evaluated and a 2FI (two-factorial) model with no transformation was found to provide the best fit for the dust concentration response variable. Table 4 gives an analysis of variance for the 2FI model. The terms A and B refer to the surfactant concentration and spray volume parameters, respectively. The term AB signifies the parameter interaction effects. Values for "Prob > F" less than 0.0500 (95% significance level) indicate model terms are significant. Surfactant concentration was found to be statistically significant for determining the response variable while spray volume was considered a categorical factor.

The developed empirical equations are given below:

- Final equation for 2-gpm spray volume:

$$\text{Dust Concentration} = 0.83 + 312 * \text{Surfactant Concentration} \qquad (2)$$

Table 2. Wet scrubber test results for Mine 1.

Run	Filter 1 (#)	Filter 2 (#)	Spray volume (gpm)	Spray pressure (psi)	Surfactant concentration (%)	Weight gain (mg)	Dust concentration (mg/m^3)
1	5	5	2	40	0.005	0.236	2.711
2	5	5	3	40	0.005	0.182	2.093
3	5	5	2	40	0.006	0.259	2.973
4	5	5	2	40	0.007	0.259	2.981
5	5	5	3	40	0.007	0.138	1.590
6	5	5	2	40	0.006	0.225	2.588
7	5	5	2	40	0.007	0.259	2.979
8	5	5	3	40	0.005	0.201	2.306
9	5	5	2	40	0.005	0.174	2.001

• Final equation for 3-gpm spray volume:
Dust Concentration = 3.7 - 303.6 * Surfactant Concentration (3)

where, dust concentration is in mg/m^3 and surfactant concentration is weight percent of water.

These equations were used to predict the response variable values, which were then compared with the actual experimental results. This comparison, shown in Figure 10, indicates that the empirical model satisfactorily fits the actual experimental data.

b) Mine 2

A linear model with no transformation was found to provide the best fit for Mine 2. Table 5 gives the analysis of variance for this model. Terms A and B have the same meaning as before. Parameter interaction effects were not present. As before, "Prob > F" values less than 0.0500 indicate significant model terms. Only surfactant concentration was found to be statistically significant.

The developed empirical equations are given below:

• Final equation for 2-gpm spray volume:
Dust Concentration = 5.42 – 409.4 * Surfactant Concentration (4)

• Final equation for 3-gpm spray volume:
Dust Concentration = 4.2 – 409.4 * Surfactant Concentration (5)

where, dust concentration is in mg/m^3 and surfactant concentration is weight percent of water.

Figure 11 provides a comparison of values predicted using these equations and actual results. It indicates that the empirical model fits the actual experimental data.

The results show that surfactant concentration is statistically significant for both Mine 1 and Mine 2 coals. The two-filter design wet scrubber with a spray volume of 2-gpm for Spray 1 and 3-gpm for Spray 2, 5-layer filters for Filter 1 and Filter 2, surfactant concentration of 0.007% and 40-psi spray pressure gave dust concentrations of 1.59 mg/m^3 for Mine 1 and 1.65 mg/m^3 for Mine 2.

Table 4. Analysis of variance for dust concentration – Mine 1.

Source	Sum of squares	DF	Mean square	F value	Prob > F
Model	1.59	3	0.53	6.43	0.0362
A	0.00	1	0.00	0.00	0.9738
B	1.13	1	1.13	13.61	0.0142
AB	0.53	1	0.53	6.44	0.0521
R – Squared			0.7941		

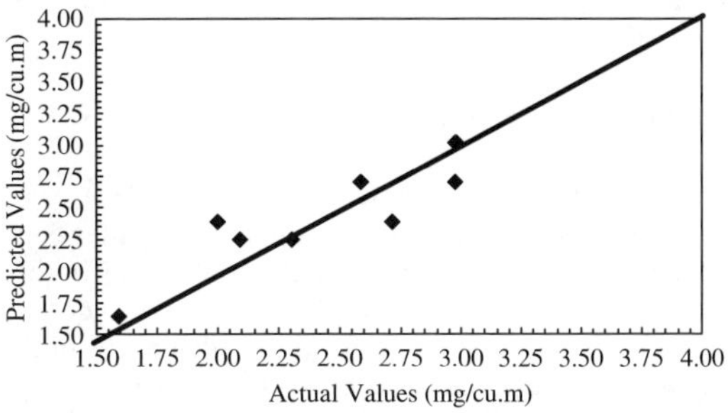

Figure 10. Predicted vs. actual values for dust concentration – Mine 1.

Table 5. Analysis of variance for dust concentration – Mine 2.

Source	Sum of squares	DF	Mean square	F value	Prob > F
Model	3.67	2	1.84	3.75	0.0877
A	1.03	1	1.03	2.11	0.1964
B	2.94	1	2.94	6.00	0.0498
R – Squared			0.5558		

Table 3. Wet scrubber test results for Mine 2.

Run	Filter 1 (#)	Filter 2 (#)	Spray volume (gpm)	Spray pressure (psi)	Surfactant concentration (%)	Weight gain (mg)	Dust concentration (mg/m^3)
1	5	5	2	40	0.007	0.165	1.895
2	5	5	3	40	0.005	0.165	1.895
3	5	5	2	40	0.007	0.303	3.487
4	5	5	2	40	0.006	0.204	2.346
5	5	5	2	40	0.005	0.373	4.290
6	5	5	3	40	0.007	0.144	1.650
7	5	5	2	40	0.006	0.215	2.473
8	5	5	2	40	0.005	0.289	3.324
9	5	5	3	40	0.005	0.167	1.915

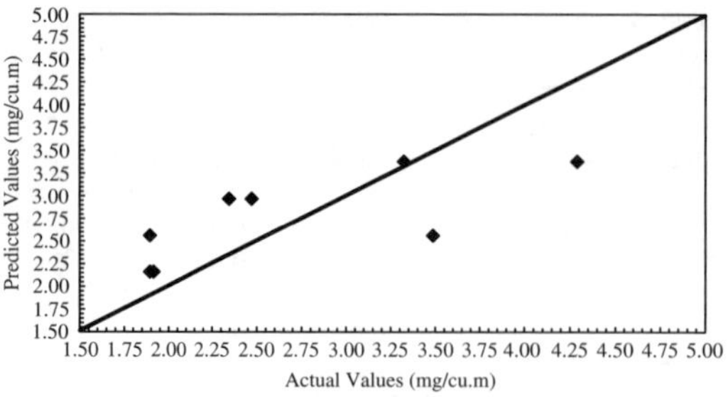

Figure 11. Predicted vs. actual values for dust concentration – Mine 2.

6 CONCLUSIONS

From the results obtained from the different phases of scrubber studies, the hypothesis that a two-filter concept for a continuous miner wet scrubber definitely improves dust control is valid. The two-filter design gives more residence time for dust-laden air to interact with the water and surfactant sprays. The wet scrubber results for a two-filter design (1.6 mg/m^3) indicate a 55.18% reduction in dust concentration compared to the original wet scrubber (3.57 mg/m^3). Multi-nozzle spray system gives more uniform spray area and with reduced spray volume, surfactant requirements and costs are further reduced.

Based on the conclusions above, the following recommendations are made for future studies:

- The distance between the two filters may be changed.
- Atomizer spray nozzles to create mist or foam between the filters may help in increasing the residence time for finer particle and surfactant interaction.

ACKNOWLEDGEMENT

This work was performed by the authors at the Dust Control Laboratory, Southern Illinois University, Carbondale as a part of a more general project to develop advanced mining technologies to reduce underground production costs. The project was funded by the Illinois Clean Coal Institute (ICCI). The authors wish to thank ICCI for their support.

REFERENCES

Box, G.E.P., and Draper, N.R., 1987, *Empirical Model Building and Response Surfaces*, John Wiley & Sons, New York, NY.

Campbell, J.A.L., Moynihan, D.J., Roper, W.D., and Wilis, E.C., 1983, "Dust Control System and Method of Operation", U.S. Patent 4,380,353.

Cheng, L., 1978, "Optimizing Water Sprays for Dust Suppression", *Engineering and Mining Journal*, Oct., p. 115–118.

Cheng, L., 1973, "Collection of Airborne Dust by Water Sprays", *Industrial and Engineering Chemistry Proceedings*, Vol. 12, No. 3, p. 221–225.

Chugh, Y.P., Hirschi, J., Saha, A., and Mohanty, M., 2002 "Evaluating the Use of Surfactants to Enhance Dust Control Efficiency of Wet Scrubbers for Illinois Coal Seams", *U.S. Mine Ventilation Symposium*, p.

Colinet, J.F., 1999, "Respirable dust control technology utilized in underground coal mines in the United States", *Proceedings of the 2nd International Underground Coal Conference*, p. 171–179.

Colinet, J.F., and Listak, J.M., 1999, "Reducing Worker Exposure to Respirable Dust at Longwall Mining Operations", *Proceedings of the International Conference on Control of Dust Hazards in the Mining Industry*, Wisla, Poland, Oct., Chapter XI.

Foster Miller, Inc., 1986, "Development of Optimal Spray Systems for Dust Control in Underground Mines", Contract HO199070, *U.S. Bureau of Mines*, 146 pp.

Goodman, G.V.R., and Listak, J.M., 1998, "Variations in Dust Levels with Continuous Miner Position", *Soc. for Mining, Metallurgy, and Exploration*, preprint no. 98-130.

Grayson, R.L., 1998, "Setting Priorities for Mine Safety and Health Research", preprint no. 98-107, *SME Annual Meeting*, Orlando, FL, 7 pp.

Hartman, H.L., Mutmansky, J.M., Ramani, R.V., and Wang, Y.J., 1997, "Mine Ventilation and Air Conditioning", John Wiley & Sons, Third edition, 730 pp.

Jayaraman, N.I., Jankowski, R.A., and Stritzel, D., 1987, "Improving Health, Safety, and Productivity Through the Use of Machine Mounted Scrubbers", *Proc. Annual Meeting Illinois Mining Institute*, Marion, IL, Oct., p. 62–75.

Jayaraman, N.I., Jankowski, R.A., and Whitehead, K.L., 1992, "Optimizing Continuous Miner Scrubbers for Dust Control in High Coal Seams", *Proc. Of New Technology in Mine Health and Safety Symposium, SME Annual Meeting*, Phoenix, AZ, Chapter 20, p. 193–205.

McPherson, M.J., 1993, "Hazardous Nature of Dusts (Chapter. 19)" and "The Aerodynamics, Sources and Control of Airborne Dust (Chapter 20)", *Subsurface Ventilation and Environmental Engineering*, Chapman & Hall, London, p. 741–811.

National Institute of Occupational Safety and Health, 1995, "Occupational Exposure to Respirable Coal Mine Dust, Criteria for a Recommended Standard", *U.S. Dept. Health and Human Services*, Center for Disease Control and Prevention, Sept., 336 pp.

Niewiadomski, G.E., 1983, "Improving Dust Control Technology for U.S. Coal Mines", *Proc. Control of Respirable Coal Mine Dust*, Beckley, WV, p. 41–72.

Niewiadomski, G.E., and Jankowski, R.A., 1993, "Longwall Dust Trends and Developments in Longwall Dust Controls", *Proc. 6th U.S. Mine Vent. Symposium*, SME, Littleton, CO, p. 551–557.

Saha, A., 2001, "An Assessment of the Effectiveness of Surfactants on Dust Control using a Continuous Miner Scrubber", M.S. Thesis, Department of Mining and

Mineral Resources Engineering, Southern Illinois University, Carbondale, 78 pp.
Sengupta, M., 1990, "Chapter 3: Dust", Mine Environmental Engineering, Vol. 1, *CRC Press Inc.*, Boca Raton, FL, p. 199–264.
Volkwein, J.C., and Thimons, E.D., 1986, "Extended Advance of Continuous Miner Successfully Ventilated With a Scrubber in the Blowing Section", preprint no. 86-308, *SME Annual Meeting*, St. Louis, Sept., 4 pp.
Wirch, S., and Jankowski. R., 1996, "Shear-Mounted Scrubbers, Are they Viable and Cost-Effective?", *U.S. Bureau of Mines*, Dept. of the Interior, Pittsburgh, PA, p. 319–325.

Strata gas and control

Mine Ventilation, Ganguli & Bandopadhyay (eds)
© 2004 Taylor & Francis Group, London, ISBN 90 5809 633 5

Migration of carbon dioxide and methane into the atmosphere in areas of closed down underground mines

T. Strumiński & B. Madeja-Strumińska
Technical University of Wrocław, Wrocław, Poland

ABSTRACT: This paper presents the results of tests concerning the reasons for the increased CH_4 and CO_2 emission into the atmosphere. Also, presented is an attempt to quantitatively evaluate the influence of the atmospheric pressure changes on the intensity of gases migrating into the subsoil zone in areas of closed down mines. The concept of a natural draught generated within the fractured rock was used during the tests. It was assumed that the gas flow in fractured rock is laminar, whereas the gas intensity involves temperature, pressure, humidity and chemical composition of gases. In the case of fractured rock and a slow increase of the water table in flooded mines, the so-called "water piston" is not of the great importance with regard to the intensity of deposit gases inflow into the subsoil atmosphere. The atmospheric pressure on the surface of closed down mines, especially its sudden drops, plays the crucial role in this process.

1 INTRODUCTION

Closing down underground mines is often combined with the presence of gas hazards in areas of an earlier mining activity. The intensity of the migration of mine gases into the atmosphere depends of the rock fracturing level, rock geological structure and its gas content. In general, the migration of gases in post-mining areas is changeable and in specific places dangerous concentrations of methane or carbon dioxide may occur (Kotarba M. & others 1999, Kulczycki Z. & Grzybek J. 1999).

Studies conducted in Poland by Kotarba (1999) in areas of closed down underground mines in the Lower Silesian Coal Basin revealed that the maximum concentration of methane and carbon dioxide in the soil reached 49.6% and 17.4%, respectively. However, the above results may include also gases occurring in buildings, cellars, garages and in different channels used for cables, water or central heating pipelines. Studies conducted in Wałbrzych Basin (Winnicki A. & others) proved that during some periods methane and especially carbon dioxide were present in many buildings. Concentration of CO_2 in some cellars and garages reached even more than 2%.

In this paper authors present the attempt of quantitative evaluation of migration of the carbon dioxide and methane into the atmosphere in the past-mining areas of the Lower Silesian Coal Basin.

2 EMISSION OF MINE GASES INTO THE ATMOSPHERE

The disturbances formed in the strata by the mining operations enhance the process of migration of gases into mine workings. In an underground operating mine released gases are carried towards the upcast shafts and than into the outer atmosphere. Mine gases, where methane or carbon dioxide dominate, may be dangerous not only for the mine workers but for the public safety as well.

The long standing mining of numerous coal beds in the Lower Silesian Coal Basin caused intensive fracturing and dewatering of the whole rock mass, what in turn impacts the migration ability of the mine gases. The fractures in the strata form a network of connections between its individual parts, with mine ventilation routes and with the ground surface. Therefore, gases released from the carbon matter may move in different directions and finally enter the outer atmosphere (Kulczycki Z. & Grzybek J. 1999).

Shafts decommissioned by filling them with ground rock material (melaphyre) become an easy migration way for gases flowing into the ground surface. It was confirmed by the measurements of CO_2 and CH_4 that concentrations in degassing (gas drainage) pipes and over the concrete plates covering the shaft tubes were even 9% and 1%, respectively.

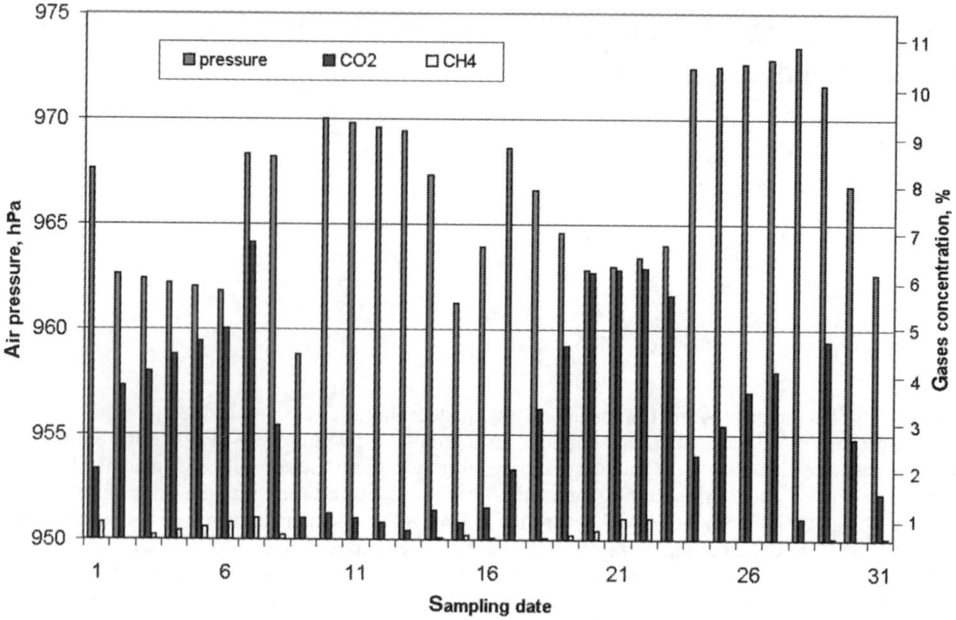

Figure 1. Time changes of barometric pressure and CO_2 and CH_4 concentrations in gas drainage pite in "Chrobry I" shaft.

High concentrations of CH_4 and CO_2 occur in zones of fissures and ravelling connected with faults as well as the edges of shallow coal mining beds and old subsurface workings (small shafts, drifts, dooks, dip-headings). High concentrations of those gases are also extended to outcrops of not-extracted coal beds. All those zones of fissures and ravelling present the complex layouts of the migration network.

According to Kotarba (1999) 28 events of methane concentration higher than 1% and 83 events of carbon dioxide concentration higher than 3% were reported between 1997 and 1998 in the subsoil layer. During the above period 28 places in the lowest floors of several buildings experienced the methane concentration reaching a maximum of 1.8% and a concentration of carbon dioxide up to 3.0%.

3 REASONS OF THE INCREASED EMISSION OF GASES INTO THE ATMOSPHERE

The presence of gas hazards in post-mining areas is often combined with the reconstruction of the water table of the carbon water horizon, deformation of the mining area surfaces and migration of the deposit gases to the land surface. In particular, the increase of methane and carbon dioxide concentration is explained by the continuous rise of the water table in the flooded mines, which causes progressive displacement of gases to the ground surface through the fissures.

The study concerning flooding of the Wałbrzych coal mines being decommissioned (Strumiński A. & Madeja Strumińska B. 2000) revealed that the water table level in those mines had increased by about 8–9 m per month, i.e. about 0.3 m per day. The water table rise in the strata of the Wałbrzych basin was very slow. We may conclude that, when the rock is heavily fractured, then the slow rise of the water table has a small impact on the intensity of gas emissions into the subsurface zone, increases only slightly the gas pressures on the level of water table, the action of the so called "water piston" does not play an important role here, and that any inflow of gases into the subsoil zone can not be attributed to the increasing level of the water table in closed underground mines.

The same study showed very strong effect of the barometric pressure on CO_2 and CH_4 concentrations in degassing pipes within decommissioned shafts. An change in atmospheric pressure has a fundamental influence on the amount of gases emitted to the atmosphere. Figure 1 illustrates the relation between atmospheric pressure and CO_2 and CH_4 concentrations in the degassing pipe located in the Chrobry I shaft in September 2000. The presented graph clearly shows that the increase of atmospheric pressure on the surface of the mine being flooded causes a decrease in concentration of CO_2 and CH_4, while the drop of this pressure gives a rise in the concentration of those gases. This regularity was observed in the case of all objects being investigated. In the author's opinion,

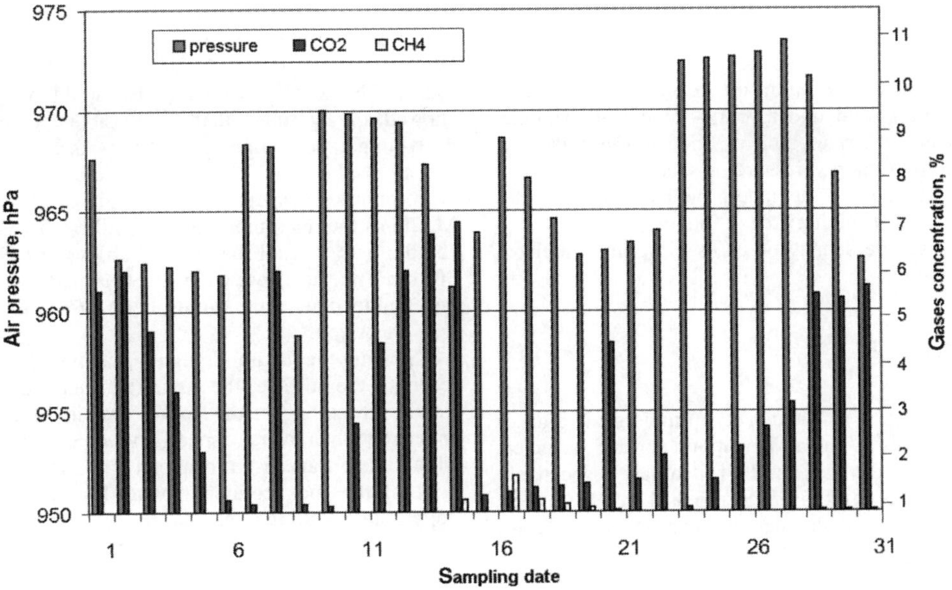

Figure 2. Time changes of barometric pressure and CO_2 and CH_4 concentrations in gas drainage pite in "Jan" shaft.

the main reason for the fluctuation in concentration of gases emitted into the subsoil zone are caused by the drops and rises in the atmospheric pressure on the surface of the closed mines. The changes in CO_2 and CH_4 concentration in an underground mine are sometimes delayed with the pressure changes on the surface by even more than one day (24 hours) (Figures 1 and 2).

4 EVALUATION OF THE INTENSITY OF GAS EMISSION INTO THE ATMOSPHERE

Usually an area of earlier mining activity has a rock mass that is strongly fractured and its separate parts are connected together and with the land surface by a network of fissures. Gases flow along those fissures, according to the produced pressure gradient or the local natural draughts (Bystroń H. 1969). The flowing substances are mine gases released from coal or other rocks or gases inflowing from the deeper parts of the Earth crust. Furthermore, the in-depth flow of the atmospheric air during a sudden growth in atmospheric pressure is accompanied by precipitation.

A study made in some underground mines (Szlązak J. 2002) showed that in fractured rock mass the gas (air) flow has laminar nature and the migration of gases into the subsurface zone on the post-mining areas takes place. The flow can be from an underground mine towards the surface or atmospheric air may enter inside the fractured strata.

It is assumed that factors such as: temperature of gases, their pressure, humidity and chemical composition has the influence on the gas flow within the rock mass. These factors determine, among others, the density of emitted gases. The work of these natural factors in a field of the gravity forces per one kilogram of the gas mass is called the natural draught (Bystroń H. 1969, Strumiński A. & Madeja Strumińska B. 2000). It turned out (Bystroń H. 1969) that the natural draught e_n (J/m^3) may be determined from the formula

$$e_{nv} = -g\int_{z}^{z_0}(\rho - \rho_s)dz \tag{1}$$

where: ρ, ρ_s are the density of gases emitted to the atmosphere and the density of the isoentropically-stratified atmosphere.

$$\rho_s = -\frac{dp_s}{gdz} \tag{2}$$

Assuming in formula (1) that ρ density is approximately equal to the average density ρ_m of gases flowing into the atmosphere and integrating the relationship (1) within the range z and z_0 while taking into consideration relation (2) we obtain

$$e_{nv} = -(p_0 - p_s) - g\rho_m(z_0 - z) \tag{3}$$

where:

p_s – pressure (static, absolute) of the dry gas solution which forms the isoentropically-stratified atmosphere within the network of fissures in the closed underground mine (for example, over the water table of flooded mine), Pa,

p_0 – atmospheric pressure on surface, Pa,

z, z_0 – ordinates of the underground and surface sites of post-mining area of interest, m,

ρ_m – average density of gases in kg/m^3 calculated from:

$$\rho_m = \frac{1}{2}(\rho_0 - \rho) \tag{4}$$

where: ρ (kg/m^3) – the density of underground gases and ρ_0 (kg/m^3) – the air density on surface are determined from ideal gas and half-ideal law, respectively (Madeja Strumińska B. & Strumiński A. 1998).

Dividing the both sides of equation (3) by ρ_m we obtain the relationship

$$e_n = -v_m(p_0 - p_s) - g(z_0 - z) \tag{5}$$

where: $v_m = 1/\rho_m$ is the average actual volume of gases emitted to the atmosphere, and $[e_n]$ is in J/kg.

After the decommissioning of shafts, by filling them, there is no access to the mine workings underground once the mine is closed. Thus, there is no possibility for making measurements of atmospheric pressure there. However, assuming that the atmosphere filling the fractured rock mass is isoentropically inter-bedded, the above pressure can be determined using the following formula

$$p_s = p_0\left[1 - \frac{\kappa - 1}{\kappa}\frac{g\rho_0}{p_0}(z - z_0)\right]^{\frac{\kappa}{\kappa - 1}} \tag{6}$$

where: $\kappa = 1{,}4$ is the isoentropic exponent.

With the assumption of the absence of fan operation and the occurrence of only natural draughts in the closed underground mine (Strumiński A. & Madeja Strumińska B. 2000), laminar inflow of gases from the fractured rock mass into the subsurface zone can be described by the formula

$$e_n = v_m R V_g \tag{7}$$

where: R (N·s/m^5) means aerodynamic drag of fractured rock mass, V_g m^3/s) – stream of gases emitted from the underground mine to the subsoil zone.

The formulas (5) and (7) show that the magnitude of the natural draught and thus the amount of gases carried up depends on differences in pressures between the mine, where gases are emitted, and the land surface.

The study of the intensity of gases emitted into the atmosphere in the area of the closed down mines in Wałbrzych Coal Basin (Kotarba M. & others 1999, Szlązak N. & Obracaj D. & Borowski M. 2002) showed, among others, that it is possible to determine the volume and composition of the gases emitted to the atmosphere by using accumulation chambers. According to (Szlązak N. & Obracaj D. & Borowski M. 2002) the maximum natural outflow of CO_2 was 150 dm^3/(m^2·h) and maximum outflow of CH_4 was 300 dm^3/(m^2·h). Those volumes of emitted gas should be combined with the atmospheric pressure during tests execution.

Knowing the natural draught e_n of the gases emitted onto the surface, the stream of their volume V_g flowing from 1 m^2 of the surface area into the atmosphere and their average specific volume v_m, it is possible to determine from formula (7) the aerodynamic drag of the rock column having 1m^2 section and height of $(z - z_0)$ in metres,

$$R = \frac{|e_n|}{v_m V_g} \tag{8}$$

While determining the aerodynamic drags in different places on the post-mining area it is possible to calculate the average drag R_m for the area being investigated. Using the equation of line filtration in the porous environment (Szlązak J. 2002) it is possible to calculate the average coefficient of permeability (k_m (m^2)) of the investigated area

$$k_m = \frac{\mu}{R_m}\frac{(z - z_0)}{A} \tag{9}$$

where:

μ – dynamic viscosity coefficient of gases, W·s/m^2,

A – area of the gases flow, m^2.

The magnitude of the natural draught e_n defined by the formula (5) depends on the pressure differences in underground and on mine surface of gas emissions. The value of draught may be positive or negative, depending on the values of pressure p_0 and p_s. Therefore, when calculating the R drag from formula (8), the absolute value of natural draught e_n is used.

Sudden changes in atmospheric pressure on the surface of a closed down underground mine (after filling the shafts and closing the central ventilation as mentioned before) are transferred to underground usually after 24 hours or even several days. If the high pressure inside the underground mine had been settled during a given period of time and a sudden drop of atmospheric pressure took place on the surface, the activation of gas emissions from the mine underground into the atmosphere starts. A sudden increase in atmospheric pressure causes the intensity of gas emissions to drop and occasionally the air may enter inside the fractured rock mass.

Trial calculations for conditions prevailing in closed down underground mines in the Wałbrzych Basin (Strumiński A. & Madeja Strumińska B. 2000) with a water table of +185 m, a surface ordinate of +443 m, an absolute surface atmospheric pressure of $p_0 = 96800$ Pa, surface air temperature –24°C and gas temperature in the subsurface zone –8°C (year average) revealed that the absolute value of the natural draught in this region was $e_{nv} = |102.5|$ J/m^3. The volume of gases emitted from the underground mine into the subsurface zone was $V_g = 720$ dm^3/(m$^2 \cdot$ h). The above calculations showed that a sudden decrease of the atmospheric pressure on the surface to $p_0 = 95880$ Pa or by 920 Pa, the intensity of gases migrated to this zone reached its highest value of 1100 dm^3/(m$^2 \cdot$ h).

5 SUMMARY

Closing down of deep underground mines can cause safety hazards due to gases released from post-mining areas, mainly CO_2 and CH_4, which migrate into the subsoil zone and further to the outer atmosphere through fracturing rock mass and along other paths.

It was discovered that the atmospheric pressure on the surface of closed down underground mines plays a crucial role in the migration of gases from the underground of that mine toward its surface. At sudden pressure drops an increased intensity of deposit gas inflows into the subsoil zone takes place. During the sudden pressure growth the intensity of deposit gas inflows is reduced which results in the much lower CH_4 and CO_2 concentrations. All those circumstances should be taken into account while considering the problem of gas hazard prevention in different buildings.

REFERENCES

Bystroń H. 1969. *Basis of potential layout of mining ventilation network (in Polish).* GIG reports, Report no 471, Katowice, Wyd. Śląsk.

Kotarba M. and others. 1999. *Gas hazards in the sub-surface zone resulting from mines closing in the Lower Silesian Coal Basin. Experience from mines decommissioning (in Polish)*, Wyd. Katowice.

Kulczycki Z., Grzybek J. *Mine gases as a risk for the public safety (in Polish).* Wyd. WUG, no 1/99.

Madeja-Strumińska B., Ŝtrumiński A. 1998. *Mine aerothermodynamics* (in Polish), Katowice, Wyd. Śląsk.

Strumiński A., Madeja-Strumińska B. 2000. *Investigations of the impact of water piston on migration and intensified emission of gases into the atmosphere on the basis of measurements of water table changes in "Julia" and "Chwalibóg" shafts and piezometers, (in Polish)*, Technical University of Wrocław, Report (not published), Wrocław.

Szlązak N., Obracaj D., Borowski M. *Gas hazard conditions on the areas of closed mines in Wałbrzych region (in Polish).* Proceedings of 2 School of Mining Aerology, Section of Mining Aerology, PAN Mining Committee, Zakopane, 07–11 October 2002.

Szlązak J. 2002. *Airflow through the tremor zone in the light of theory and experiments (in Polish).* Kraków, Wydawnictwa Naukowo-Dydaktyczne AGH. *Dissertations, Monographs no 90.*

Winnicki A. and others. *Results of measurements of gases concentration in the surface buildings in Wałbrzych and its vicinity made between January and July 2000 (in Polish).* (not published).

Mine Ventilation, Ganguli & Bandopadhyay (eds)
© 2004 Taylor & Francis Group, London, ISBN 90 5809 633 5

Model research into retentive reservoirs of methane in coal mines

J. Berger
The Mines Methane Drainage Enterprise "ZOK", Jastrzębie Zdrój, Poland

W. Dziurzyński
Strata Mechanics Research Institute of The Polish Academy of Science, Cracow, Poland

St. Nawrat & J. Roszkowski
AGH – University of Science and Technology, Cracow, Poland

ABSTRACT: The problem of using sealed coal mine headings for retentive storing of methane obtained in the process of methane drainage in coal mines is important for economic and ecological reasons, because this process enables to obtain a qualitatively and quantitatively stabilized low-methane-content fuel used in power and heat generation installations. Improvement in the degree of the utilization of methane from methane drainage influences directly the limitation of methane emission to the atmosphere.

In 2001 experimental research in the above-mentioned scope was carried out in the "Zofiówka" coalmine on three experimental objects, which were headings, sealed against explosions, of various lengths, volumes, and structures of the surrounding rocks. The technology consists in:

- establishing a methane reservoir (storage) by adapting a sealed heading,
- preparing an installation and machines forcing methane into the reservoir,
- preparing and utilizing the already operating installations and machines used in methane drainage,
- filling the reservoir cyclically with methane from methane drainage,
- removing methane cyclically from the reservoir by means of the methane drainage network,
- sending methane through a methane drainage station situated on the surface to heat and power generating installations where it is used as fuel.

The presented technology of retaining methane from methane drainage requires the use of:

- an underground gas pumping station; its current model is based on a gas compressor with rotating water-rings, known commonly as a water circulating pump,
- a methane drainage installation based on a network of underground and surface methane drainage pipelines cooperating with the surface methane drainage station.

The experimental research indicated that the sealed headings were not leak proof to various degrees, however, if they were adapted appropriately, they could be used as underground retentive reservoirs of methane from methane drainage in operating coal mines on condition that their locations and level of leakproofness guaranteed work safety and economic effectiveness. The research also confirmed the operational usability of the applied machines and technological solutions.

The development of a mathematical model, and in the next stage – methods of numerical simulation of the studied phenomena based on that model and computer software, creates a possibility to carry out preliminary evaluations of the usability of sealed headings as retentive reservoirs of methane from methane drainage and to consider optional solutions for planned and already operating reservoirs.

1 GENERAL INSTRUCTIONS

The problem of using sealed mine headings for retentive storing of methane obtained in the process of methane drainage in coal mines is important for economic and ecological reasons, because this process enables to obtain a qualitatively and quantitatively stabilized low-methane-content fuel used in power and heat generation installations. Improvement in the degree of the utilization of methane from methane drainage influences directly the limitation of methane emission to the atmosphere.

In 2001, experimental research in the above-mentioned scope was carried out in the "Zofiówka" coalmine on experimental objects which were headings sealed against explosions, of various lengths, volumes, and structures of the surrounding rocks [Berger J., 2002]. In particular experiments, the elements of the developed technology of retaining methane from methane drainage [Berger J., 2002; Nawrat S., 1999] were used. The technology consists in:

- establishing a methane reservoir (storage) by adapting a sealed heading,
- preparing an installation and machines forcing methane into the reservoir,
- preparing and utilizing the already operating installations and machines used in methane drainage,
- filling the reservoir cyclically with methane from methane drainage,
- removing methane cyclically from the reservoir by means of the methane drainage network,
- sending methane through a methane drainage station situated on the surface to heat and power generating installations where it is used as fuel.

The presented technology of retaining methane from methane drainage requires the use of:

- an underground gas pumping station; its current model is based on a gas compressor with rotating water-rings, known commonly as a water circulating pump,
- a methane drainage installation based on a network of underground and surface methane drainage pipelines cooperating with the surface methane drainage station.

The experimental research indicated that the sealed headings were not leak proof to various degrees, however, if they were adapted appropriately, they could be used as underground retentive reservoirs of methane from methane drainage in operating coal mines on condition that their locations and level of leakproofness guaranteed work safety and economic effectiveness. The research also confirmed the operational usability of the applied machines and technological solutions.

The results obtained during the research confirmed the complexity of the analyzed problem, which were the result of both the mining – geological conditions and the characteristics of the process of pumping methane into and out of the headings functioning as retentive reservoirs.

The need to create a mathematical model of describing the course of the phenomena connected with the technology of retentive storing of methane from methane drainage in sealed headings resulted first of all from the design and exploitation requirements.

The development of a mathematical model, and in the next stage – methods of numerical simulation of the studied phenomena based on that model and computer software, creates a possibility to carry out preliminary evaluations of the usability of sealed headings as retentive reservoirs of methane from methane drainage and to consider optional solutions for planned and already operating reservoirs.

The results of the experimental research enable a qualitative verification of the mathematical model, inclusive of optional calculations being an attempt to recreate the course of the processes of pumping methane into and out of the reservoirs taking place during the experiments.

2 MODEL RESEARCH – A MATHEMATICAL MODEL OF A GAS RESERVOIR

In analyzing the course of the phenomena resulting from pumping methane into the reservoir, a method of numeric simulation was used. The use of this research method requires that a mathematical model adequate to a research object be adapted. Within this scope, the most complete description can be obtained by accepting a model of the flow of the air-methane mixture in a reservoir treated as a three-dimensional movement of compressible fluid. However, a problem formulated in this way would be extremely difficult to solve because of defining unique boundary conditions, selecting the numerical method as well as further difficulties with calculations. For these reasons, simplified mathematical models are used often to describe the flow of gas mixtures in coal mine's ventilation systems. This approach made it possible to solve with practical accuracy many problems connected with the flow of air and methane in a heading network [Dziurzyński W. et al., 1987; Nawrat S., 1999] or with the development of fires [Dziurzyński W., 1998].

We assume that the methane reservoir is a sealed heading (Picture 1). The question of exploiting the methane reservoir is connected with its location and the factor resulting from the ventilation of headings. The reservoir is surrounded by rocks and is under the influence of the central ventilation fan. These two factors determine the reservoir's leakproofness, which influences the effectiveness of the exploitation of the reservoir.

Accepting the assumptions simplifying considerably a description of the phenomena taking place inside the reservoir and connected with the mixing and flow of gases allowed to move on to the consideration of a one-dimensional model of an incompressible flow of liquid, disregarding methane diffusion, the existence of the gravitational forces and, what is connected with

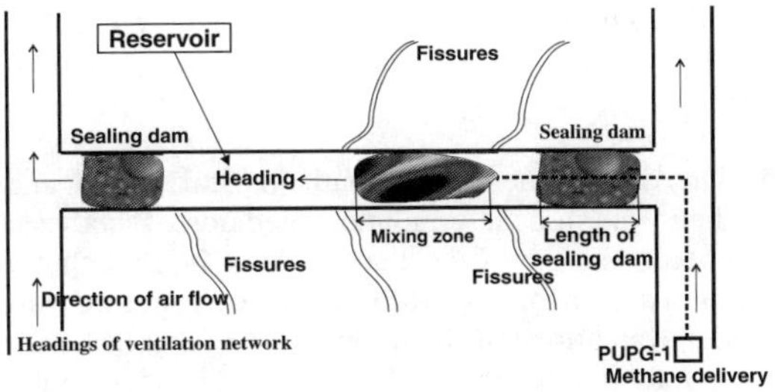

Picture 1. Schematic diagram of the reservoir and its elements.

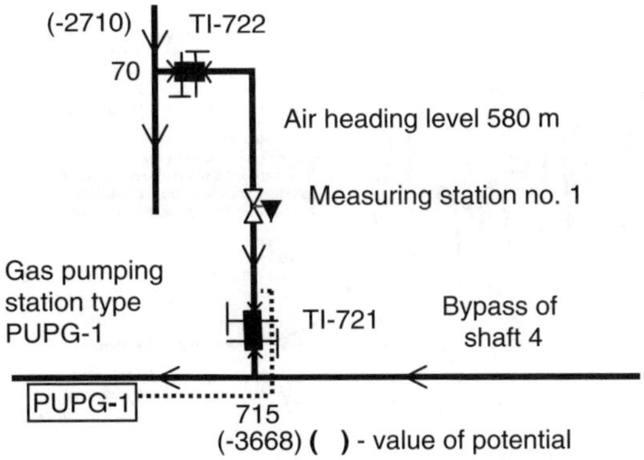

Picture 2. Spatial diagram of a section of air heading, level 580 m: reservoir, sealing dam and headings.

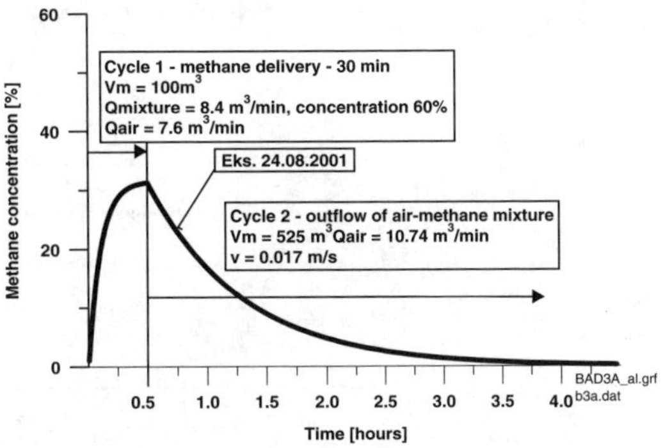

Picture 3. Changes in methane concentration in reservoir during methane delivery and outflow – experiment conducted on 24 August 2001 – a computer simulation.

this, the distribution of the mixture's density along the reservoir's height. Therefore, for further considerations, we accept a mathematical model describing the changes in methane concentration with relation to the mixture's volume, the transport of the air-methane mixture in the reservoir through the sealing dam and in the headings connected with the reservoir.

3 A COMPUTER SIMULATION OF DELIVERING METHANE TO THE RESERVOIR

Changes in the computer program METAN made on the basis of the presented mathematical model allowed to increase its usability and to attempt to recreate the course of the processes of pumping methane into and out of the reservoir taking place during the conducted experiments. In order to carry out necessary calculations we have prepared a database consisting of the following elements:

- set containing parameters characterizing the flow of the air-methane mixture through the elements of the reservoir and adjacent headings,
- data concerning the pumping of methane to the reservoir by means of the gas pumping station,
- data concerning the geometric dimensions of the reservoir, sealing dams and possible connections between the reservoir and other headings through fissures in the rock mass.

The results of the simulation of pumping methane in and out of a reservoir in the conditions of the two experiments conducted in the sealed air heading on the level of 580 m in the "Zofiówka" coal mine are presented in the form of diagrams showing changes of methane concentration in time for a measuring station localized in the middle of a 50-meter-long heading, with an average cross-section area of $10.5\,m^2$ and volume of $525\,m^3$.

The first experiment dealt with the reservoir's leakproofness and was carried out on 24 August 2001; the objective of the second experiment was the examination of the working cycles of the retentive gas storing technology, i.e. the complete cycles of delivering and removing gas.

In simulation calculations it was assumed that the volume of mixing for cycle 1 of delivering methane equaled $100\,m^3$, and for further cycles – the volume equal to that of the reservoir – $525\,m^3$. The resistance of the sealing dams was determined as well; it equaled 3000 Bd for the first and 3800 Bd for the second experiment. After taking into consideration these resistances, the determined difference of potential between the inlet and outlet of the reservoir equaled $958\,J/m^3$ ($1000\,J/m^3$ during the experiment). The volume capacity of the flow of the air-methane mixture through the reservoir oscillated between $9.62\,m^3/min$ in the first and $7.6\,m^3/min$ in the second experiment. The diagram in Picture 3 presents changes in methane concentration in the tested reservoir obtained from a computer simulation.

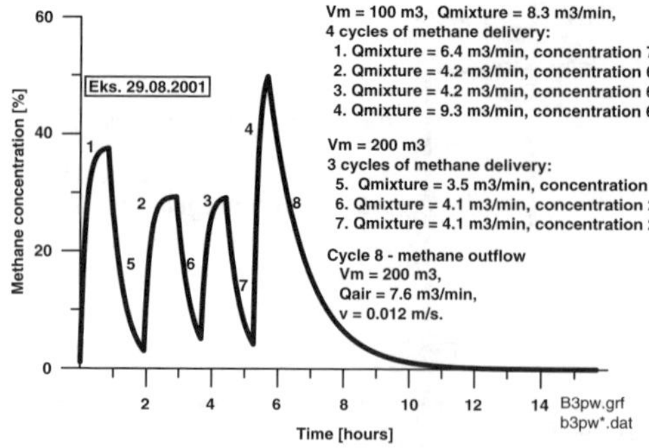

Picture 4. Changes in methane concentration in reservoir during methane delivery and removal- experiment conducted on 29 August 2001 – a computer simulation.

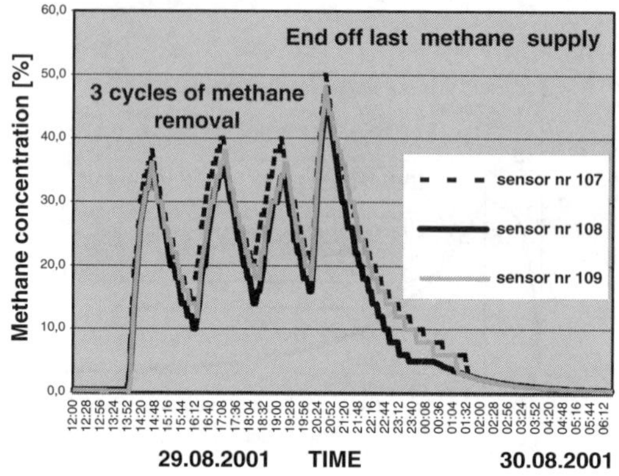

Picture 5. Changes in methane concentration in reservoir during methane delivery and removal – experiment conducted on 29 August 2001, sensor 107 under roof, sensor 108 in the middle, sensor 109 on the floor of the heading.

The following diagram (Picture 4) presents the results of a computer simulation of exploiting the reservoir consisting in four cycles of methane inflow and three cycles of methane removal.

The presented changes in methane concentration (Pictures 3 and 4) are qualitatively similar to the changes obtained during the experiment (Picture 5).

Applying the computer simulation of the exploitation of the reservoir enables also to carry out calculations whose results make it possible to evaluate the influence of selected parameters characteristic for the reservoir on the course of changes in methane concentration.

4 CONCLUSIONS

The model research and the conducted computer simulations of the operation of a retentive methane reservoir in the conditions of methane delivery and removal or outflow allowed to formulate the following conclusions:

1. The development of a mathematical model and then – numerical simulation methods of analyzed phenomena based on that model – and computer software opens opportunities for carrying out initial evaluations of the usability of sealed headings as retentive reservoirs of methane obtained in the process of methane drainage as well as for considering alternative solutions for planned and already operating reservoirs.
2. The results of simulative calculations done on the basis of data obtained from the experimental research into retentive storing of methane from methane drainage in sealed headings indicated the following:
   - a relatively good compatibility of the courses of temporal changes in methane concentration in the reservoir in relation to the experiment for the period of methane delivery, which was the result of a correct selection of the volume capacity of the mixture flowing through the reservoir, the resistance of the sealing dams and the mixing volume.
   - the deviations of the courses of temporal changes in methane concentration in the reservoir in relation to the experiment for the period of methane outflow from the reservoir.
3. Both model and experimental research confirmed that from the technical and technological point of view it is possible to guarantee safe conditions for the exploitation of endogenic reservoirs of methane obtained in the process of methane drainage in operating coalmines.

REFERENCES

Berger, J 2002. An evaluation of the possibilities of using sealed headings in coalmines for retentive storing of methane from methane drainage. *PhD Thesis, Academy of Mining and Metallurgy*, Cracow.

Budryk, W 1961. Ventilation of coal mines, part I. Ventilation of headings. Katowice. *Mining and Metallurgy Publishing House.*

Czyczuła, L 1973. A spatial and temporal distribution of methane concentration in a coalmine heading in changeable ventilation conditions. *Mining Problems Fascicles. Polish Academy of Sciences*, volume 11, fascicle 2.

Dziurzyński, W., Tracz, J., Trutwin, W 1987. On a mathematical model of the flow of air and after-breakout gases in a coal mine's ventialtion network. *Mining Archives*, volume 32, fascicle 1.

Dziurzyński, W 1998. Forecasting the process of deep mine ventilation in the conditions of underground fire, *Studies, Theses, Monographies, No.56, ISSN 0860-74-19, Institute of Mineral Resources and Energy Management, Polish Academy of Sciences*, Kraków.

Dziurzyński, W 1999. Numerical simulation of the ventilation process – computer programs. *School of Aerology – Zakopane.*

Litwiniszyn, J 1951. A problem of dynamic of flow in conduit networks. *Bulletin of Academy Polonaise Science*, vol.1, no.3.

Nawrat, S 1999. Experimental and model research on the process of methane filling of sealed headings in coalmines. *EMAG, Electrification and Automation of Coal Mining*, Katowice no. 1(11).

Pawiński, J, Roszkowski, J, Strzemiński, J 1995. Mine Ventilation. *Silesian Technical Publishing House*, Katowice.

Trutwin, W 1968. Modelling nonstationary states in ventilation networks by means of analog devices. *Scientific Papers of the Academy of Mining and Metallurgy*, no. 230, Mining, no. 19.

Trutwin, W 1972. Digital simulation of nonstationary states of the process of ventilation and regulating a mine ventilation network. *Mining Research Papers*. vol.10, book. 2.

Trutwin, W 1973. Influence of ventilation conditions on methane concentration in mine headings. *Mining Research Papers*. vol.11, book. 2.

Mine Ventilation, Ganguli & Bandopadhyay (eds)
 ISBN 90 5809 633 5

Natural heating and cooling of intake air passing through fragmented strata

J. Partyka & O. Johnson
Laurentian University, Sudbury, Ontario, Canada

ABSTRACT: In this paper a study is performed to predict psychometric properties of intake air after it passes through a mass of pit muck (gathering area), which acts as a heat exchanger. The study is done on an area which maintains an air temperature about 2°C all year round. This area contains broken rock and water that freezes in winter and melts in summer. The broken rock area may consist of one, two or three zones: the frozen zone, non-frozen zone, and ice zone at 0°C. In mid winter the upper rock layer is frozen whereas the broken rock at the bottom has a temperature few degrees above zero. In late summer all the rock has above zero temperature. The mine surface temperature changes according to the weather whereas the rock temperature at the bottom of the gathering area is assumed to be nearly constant the whole year. The Stephen Law is used to determine the location of 0°C isotherm (the border between zones). The intake air moves through a system of capillaries. During ice melting these capillaries are the way for air and water. The amount of heat exchanged between fragmented strata and ventilated air in a capillary is governed by a theory involving the Reynolds, Biot, Nusselt, and Fourier numbers. The water from melted ice wets the capillaries changing the moisture content of the air and the amount of latent heat accompanied by a condensation or an evaporation process. In the ice covered zone, where the capillaries are considered dry, air temperature changes according to the dry surface conditions. The results presented in this paper are for the geological and average weather conditions prevailing in the Sudbury region.

1 INTRODUCTION

This study is an analytical attempt to predict the degree of heating and cooling of fresh air passing through a mass of pit muck. Figures 1 and 2 illustrate an existing case and are described briefly to better understand the physical nature of the processes involving here. In Figure 1, the gathering area acts as a heat exchanger. The fresh air of lower mining areas is drawn down from surface through a large mass of broken rock lying above a series of footwall slushier drifts formerly used to mine low-grade ore using the caving method in the upper mining area. The ice field gathering area now utilizes Blocks 1, 2, 5, and 6 (Figure 2). Blocks 3 and 4 do not have enough burdens in the upper levels to be of value. Trenches below 1040 Level are too warm and not useful. The intake area measures about 244 m ≅ 183 m and the height of the broken rock averages close to 140 m. The airflows down drawpoints, across the slushier

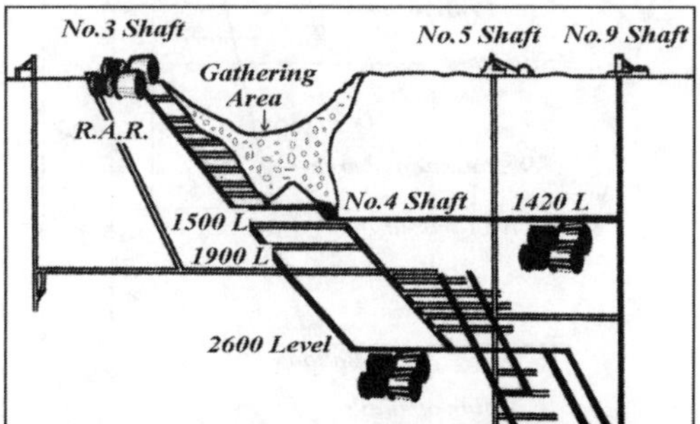

Figure 1. Primary ventilation system of a mine.

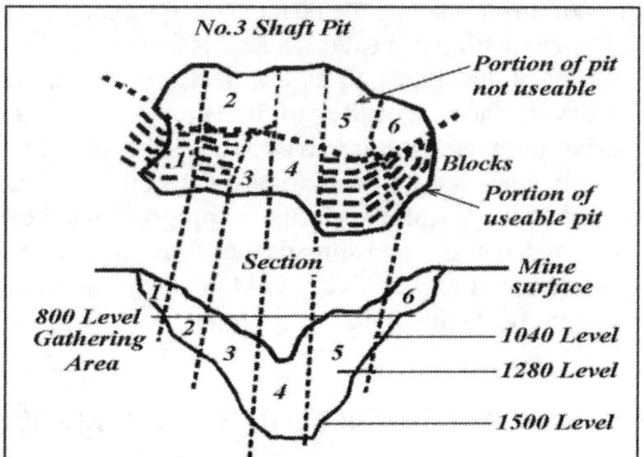

Figure 2. The ice fields gathering area.

drifts and down ore passes to 800 Level where it is transferred through large drifts to the top of the main intake airway. The broken rock acts as a heat exchanger, warming the fresh air in winter and cooling it during summer. Precipitation and water seepage result in ice formation in winter, which is then melted in the summer. Ventilation controls at the connections from slushier drifts to the ore passes are used to regulate the flow of air seasonally, with variation in the air temperatures (Golder Associates 1980).

The temperature of the fresh air at the main intake on 800 Level has averaged 3.1°C wet-bulb, with a yearly seasonal variation of plus/minus 1.4°C. During the same period, temperature at the main intake on 6600 Level has averaged 16.1°C wet-bulb, with a variation of plus/minus 0.8°C. By using this intake heat exchange area, about 730 m^3/s of cool fresh air is recently supplied to the lower mine at an almost constant temperature developing a favourable natural draft pressure of about 800 Pa (INCO 2003).

The study investigates the heat exchange between the rock which has non-stable temperature field and air for different weather conditions. The results shown in this paper are based on average monthly weather conditions prevailing in the Great Sudbury region.

2 PROBLEM DISCRIPTION

The study consists of two parts: the prediction of geothermal temperature of fragmented rock forming the heat exchanger and evaluation of the amount of heat exchanged between the air and fragmented strata at different months. The first part requires an analytical solution for determining the temperature field in broken rock for different seasons. One, two or three design zones are used here: the frozen zone, non-frozen zone, and ice zone at zero temperature. The Stephen Law that balances heat flow through the surface separating two adjacent zones is applied. The second part deals with the strata, latent and sensible heats, and the final properties of the air leaving the heat exchange area.

The evaluation of heat processes is based on a general theory developed to assess heat transfer at wet and dry surfaces. The Reynolds, Biot, Nusselt, and Fourier numbers are involved here. The air flows through a system of capillaries located in the fragmented rock. A suitable computer program has been developed, using Excel spreadsheet form that accepts small diameter airways, very cold intake air and rock temperatures below 0 EC.

3 TEMPERATURE FIELD IN BROKEN ROCK

The weather conditions have a great influence on the field temperature, especially in zones of broken rocks, which form the central part of the gathering area. In winter, part of the rock has a temperature below zero and any water present in that part of rock freezes. On the other hand, warm weather causes ice in rock capillaries to melt. These weather changes follow one another so water in the rock capillaries freezes in winter and then ice melts in spring. The freezing front formed in autumn moves from the surface down until it reaches its maximum depth, sometimes in early spring. Then, the melting front starts to move downwards and continue to move until all ice is melted, usually shortly before summer. During the remaining months of the year the whole heat exchange area has temperature above 0 EC.

3.1 *Two- zone design*

In winter the temperature field in broken rocks forms two zones: frozen and non-frozen, which are separated by a freezing front at X (Figure 3).

The following assumptions were made when developing an analytical solution for the temperature fields at that time:

1. In late summer there is no ice present in the heat exchanger.
2. The field temperature of the broken rock changes continuously but semi-steady temperature fields do exist; i.e. only average monthly temperatures are considered.
3. The rock temperature T_s at mine surface varies throughout the year and is equal to intake air temperature T_o.
4. The rock temperature T changes with time t, depth x, and the location of the melting front X.
5 In winter, the heat exchange area consists of two zones only; one filled by frozen rock and other by non-frozen rock.
6 Thermal and physical properties of the broken rock in both zones do not change in time.

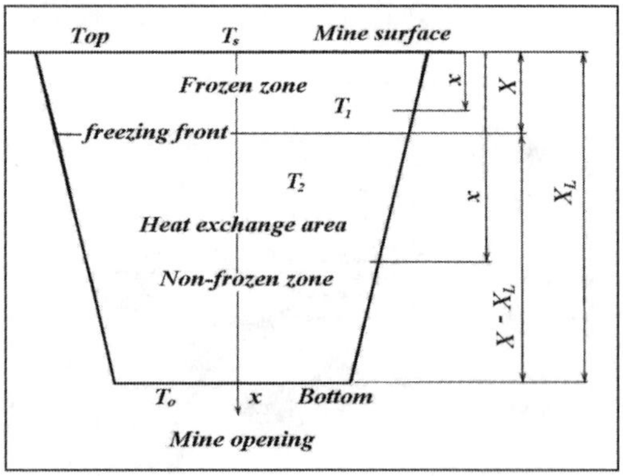

Figure 3. Design zones for winter.

The following system of equations is used here to calculate the rock temperature in both zones (Luikov 1968; Backer 1986):

$$\frac{\partial T_1}{\partial t} = \alpha_1 \cdot \frac{\partial^2 T_1}{\partial x^2}; \quad t > 0, 0 < x > X \tag{1}$$

$$\frac{\partial T_2}{\partial t} = \alpha_2 \cdot \frac{\partial^2 T_2}{\partial x^2}; \quad t > 0, X < x > X_L \tag{2}$$

where:
α_1 – thermal diffusivity of frozen zone; m^2/s
α_2 – thermal diffusivity of non-frozen zone; m^2/s
T_1 – temperature in frozen zone; EC
T_2 – temperature in non-frozen zone; EC
t – time; s
x – coordinate (independent variable); m
X – location of the melting front; m
X_L – depth of the heat exchange area; m

On the freezing front the following Stephen Law applies

$$\lambda_1 \cdot \left.\frac{\partial T_1}{\partial x}\right|_{x=X} - \lambda_2 \cdot \left.\frac{\partial T_2}{\partial x}\right|_{x=X} = L \cdot \gamma_2 \cdot w_2 \cdot \frac{dX}{dt} \tag{3}$$

where:
λ_1 – thermal conductivity of frozen zone; W/mEC
λ_2 – thermal conductivity of non-frozen zone; W/mEC
L – latent heat of fusion of ice; 335 kJ/kg
γ_2 – density of rock; kg/m^3
w_2 – moisture content in the non-frozen zone; % by weight.

All points located on the frozen front have a zero temperature, i.e. $T_1 = T_2 = 0$ when $x = X$.

The semi-steady temperature fields are described by the following equations:

$$\frac{\partial^2 T_1}{\partial x^2} = 0 \tag{4}$$

$$\frac{\partial^2 T_2}{\partial x^2} = 0 \tag{5}$$

Integrating Equations 4 and 5 we obtain

$$T_1 = C_1 \cdot x + C_2 \tag{6}$$

$$T_2 = C_3 \cdot x + C_4 \tag{7}$$

The constants of integration C_1, C_2, C_3, C_4 are found using the following boundary conditions:

1. at the top surface $T_1 = T_s, x = 0$;
2. on the melting front $T_1 = 0, x = X$;
3. at the bottom surface $T_2 = T_0, x = X_L$;
4. on the melting front $T_2 = 0, x = X$.

Thus:

$$C_1 = -\frac{T_s}{X} \tag{8}$$

$$C_2 = T_s \tag{9}$$

$$C_3 = \frac{T_0}{X_L - X} \tag{10}$$

$$C_4 = -T_0 \cdot \left[\frac{X}{X_L - X}\right] \tag{11}$$

Equations 6 and 7 can be written as

$$T_1 = T_s \cdot \left(1 - \frac{x}{X}\right) \tag{12}$$

$$T_2 = T_0 \cdot \left[\frac{x - X}{X_L - X}\right] \tag{13}$$

Applying Stephen Law we find the relation between time t and freezing front location X. From Equations 3, 12, and 13 we obtain

$$\frac{\partial T_1}{\partial x} = -\frac{T_s}{X} \tag{14}$$

$$\frac{\partial T_2}{\partial x} = \frac{T_0}{X_L - X} \tag{15}$$

$$\lambda_1 \cdot \frac{T_s}{X} - \lambda_2 \cdot \frac{T_0}{X_L - X} = L \cdot \gamma_2 \cdot w_2 \cdot \frac{dX}{dt} \tag{16}$$

Last equation can now be integrated to give (Maple V 1997)

$$t = \frac{(a \cdot X - b) \cdot X^2}{(c + d \cdot X)} \tag{17}$$

where:

$$a = \frac{1}{3} \cdot L \cdot \gamma_2 \cdot w_2; \quad b = \frac{1}{2} \cdot \gamma_2 \cdot w_2 \cdot L \cdot X_L;$$
$$c = X_L \cdot \lambda_1 \cdot T_s; \quad d = T_o \cdot \lambda_2 - T_s \cdot \lambda_1.$$

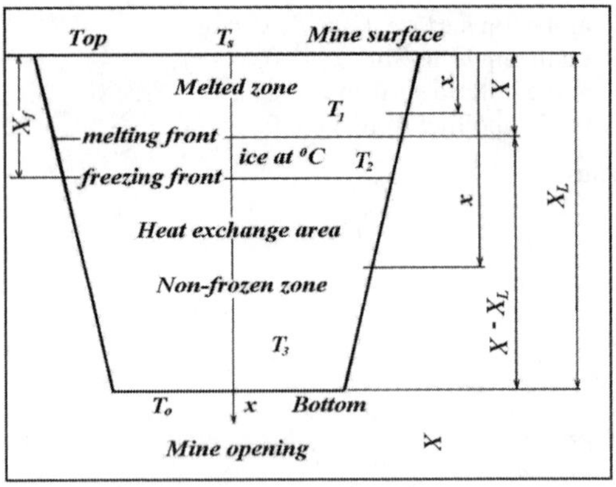

Figure 4. Design zones for spring.

3.2 *Three-zone design*

Sometime in spring melting of the frozen zone begins. When the melting front moves down a three-zone temperature field is formed. The heat exchange area then contains the melted zone, frozen zone (ice at 0°C) and non-frozen zone, as shown in Figure 4.

We assume the freezing front now stays on X_f and does not move. The melting front, however, moves from the mine surface to depth X. Stephen Law on melting front becomes

$$\lambda_1 \cdot \left.\frac{\partial T_1}{\partial x}\right|_{x=X} - \lambda_2 \cdot \left.\frac{\partial T_2}{\partial x}\right|_{x=X} = -L \cdot \gamma_1 \cdot w_1 \cdot \frac{dX}{dt} \tag{18}$$

In this case we have three equations describing the semi-stationary temperature fields

$$\frac{\partial^2 T_1}{\partial x^2} = 0 \tag{19}$$

$$\frac{\partial^2 T_2}{\partial x^2} = 0 \tag{20}$$

$$\frac{\partial^2 T_3}{\partial x^2} = 0 \tag{21}$$

After integrating it follows that

$$T_1 = C_5 \cdot x + C_6 \tag{22}$$

$$T_2 = 0 \tag{23}$$

$$T_3 = C_7 \cdot x + C_8 \tag{24}$$

where C_5, C_6, C_7, C_8 are integration constants. The boundary conditions for a melting zone are the same as those used for the two-zone design. Thus, $C_5 = C_1$, $C_6 = C_2$ and the temperature field in the melted zone is defined by Equation 12.

The freezing front does not move during forming the melting zone so it is located at a distance X_f below mine surface that represents the maximum depth of the frozen zone formed during winter. The temperature T_3 in the non-frozen zone is positive but it approaches zero on the freezing front line. At the bottom, where $x = X_L$, $T_3 = T_o$. Using these boundary conditions we obtain:

$$C_7 = \frac{T_o}{X_L - X_f} \tag{25}$$

$$C_8 = -T_o \cdot \left(\frac{X_f}{X_L - X_f} \right) \tag{26}$$

$$T_3 = T_o \cdot \left(\frac{x - X_f}{X_L - X_f} \right) \tag{27}$$

Again, applying Stephen Law on the melting front we find the relation between time t and the distance X. Since we assumed that ice zone is all time at 0 EC it yields

$$\left.\frac{\partial T_2}{\partial x}\right|_{x=X} = 0 \tag{28}$$

$$\frac{\partial T_1}{\partial x} = -\frac{T_s}{X} \tag{29}$$

$$X^2 = t \cdot \left(\frac{2 \cdot \lambda_1 \cdot T_s}{L \cdot \gamma_1 \cdot w_1} \right) \tag{30}$$

where, the depth of the melted zone X is in metres and the time t in seconds.

3.3 *One-zone design*

Sometime in summer the melting front reaches the freezing front and all ice melts. The design temperature field changes to a non-frozen zone. We have now a positive temperature field, that is

$$\frac{\partial^2 T}{\partial x^2} = 0 \tag{31}$$

From here we get

$$T = C_9 \cdot x + C_{10}. \tag{32}$$

At the top of the heat exchanger $x = 0$, $T = T_s$ whereas at the mine opening $x = X_L$, $T = T_o$. We assume, however, that a zero-temperature line will sustain at $x = X_f$ until a new freezing front is formed in the following winter. Therefore, above the zero-temperature line

$$C_9 = -\frac{T_s}{X_f} \tag{33}$$

$$C_{10} = T_s \tag{34}$$

$$T = T_s \cdot \left(1 - \frac{x}{X_f}\right) \quad \text{when} \quad 0 \leq x \geq X_f \tag{35}$$

whereas below that line

$$C_9 = \frac{T_0}{X_L - X_f} \tag{36}$$

$$C_{10} = -T_0 \cdot \left(\frac{X_f}{X_L - X_f}\right) \tag{37}$$

$$T = T_0 \cdot \left(\frac{x - X_f}{X_L - X_f}\right) \quad \text{when} \quad X_f \leq x \geq X_L. \tag{38}$$

Equations 12, 13, 27, 35 and 38 define the temperature field at any point of the broken rock in different months of the year. The melting (Equation 17) and/or freezing (Equation 30) fronts form a stratified temperature field. All the temperature zones in the heat exchange area are used for determining the temperature of air drawn down the mine through the broken rock.

4 NATURAL HEATING AND COOLING OF INTAKE AIR

The atmospheric air drawn down through broken rock is naturally heated or cooled. The amount of heat exchanged between air and the fragmented rock depends strongly on the weather conditions. Knowing the temperature field of broken rock forming the heat exchange area it is possible to determine thc temperature of intake air at any depth of the heat exchanger and any time of the year.

4.1 $T_s \leq 0°C$

As long as temperature of the intake air is below 0 EC the air is assumed to be dry. As it passes through fragmental rock it is heated sensibly. Once the air temperature becomes positive, the air also starts to pick up moisture and becomes wet air. One of the existing methods that can help to determine temperature of air moved through a gathering area is the method suggested by McPherson (1992). This method includes the size and age of an airway, airflow, barometric pressure and rock thermal properties. It predicts the heat flow applying the Reynolds, Nusselt, Biot and Fourier numbers. The same approach is used for our studies.

4.2 $T_s > 0°C$

When air temperature on mine surface is above 0 EC the intake air is wet and heat exchanged in the heat exchanger also involves latent heats from evaporation and condensation. Again, all calculations of air temperature given here are based on the general theory developed to assess heat transfer at wet surfaces (McPherson 1992). The theory is applied to air flowing through a system of capillaries located in the fragmental rock. During ice melting these capillaries are the way for air and water. The water from melted ice wets the capillaries, changes relative humidity of the air and the amount of latent heat accompanied by condensation or an evaporation process. CLIMSIM for Windows (Version 1.0) program was used here (CLIMSIM, MVS).

5 RESULTS

Calculations have been performed using monthly average temperatures of the Great Sudbury Region. The geological data, the yearly weather variations (Figure 5) and moisture content in broken rock applied in calculations are typical to the real mine conditions found in Sudbury region.

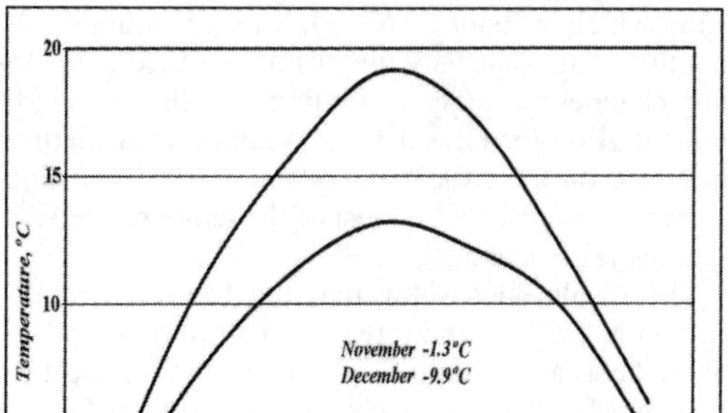

Figure 5. Average air temperature in Sudbury region.

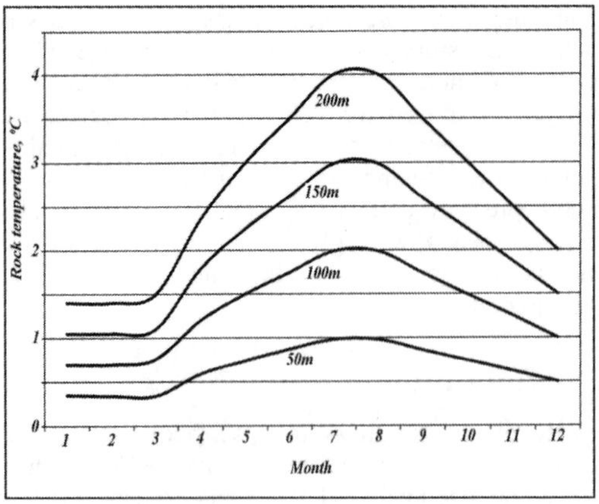

Figure 6. Rock temperature.

The following input data were used:

1. Gathering area
   $X_L = 200\,\text{m}$; $\lambda = 3.0\,\text{W/m°C}$; $L = 335\,\text{kJ/kg}$;
   $w = 0.01$; $\gamma = 2000\,\text{kg/m}^3$;
2. Intake air:
   January: $T_s = -11.6\,\text{EC}$; $T_o = 1.4\,\text{EC}$;
   June: $T_s = 15.8\,\text{EC}$; $T_o = 3.5\,\text{EC}$;
3. Capillary
   perimeter = 2.51 m; area = 0.5 m^2;
   age = 30 days; airflow = 0.1 m^3/s;
   wetness factor = 0.5;
   coefficient of friction = 0.023;
   thermal diffusivity = $1.4 \cong 10^{-6}\text{m}^2/\text{s}$;
   thermal capacity of dry air = 1005 J/kgK.

Figure 6 shows how the temperature of the fragmented rock changes with depths and throughout the year. We observe that when the depth increases the rock temperature becomes closer to the mine temperature, which is about 1.5°C to 4°C. Depending on the months of the year only the ground surface temperature changes strongly according to climate conditions. It also illustrates that the weather makes largest influence on the rock temperature in the layer adjacent to mine surface. The rest of the temperature field remains fairly stable.

The results also show that the freezing front is formed in October and it reaches its full depth of 1 m late in February (Figure 7). The melting front starts to move in March and joins the freezing front in June. A new freezing front is formed in October. The melting and freezing processes repeat on a yearly basis.

Figure 8 shows how the wet and dry temperature of the air as it reaches different depths. The intake air has the dry temperature equal to rock temperature on the mine surface. The rock temperature makes strong influence on both air temperatures, causing air to cool

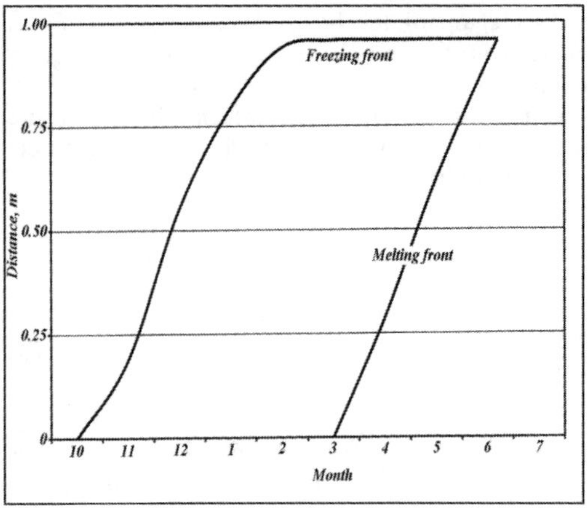

Figure 7. Freezing and melting fronts.

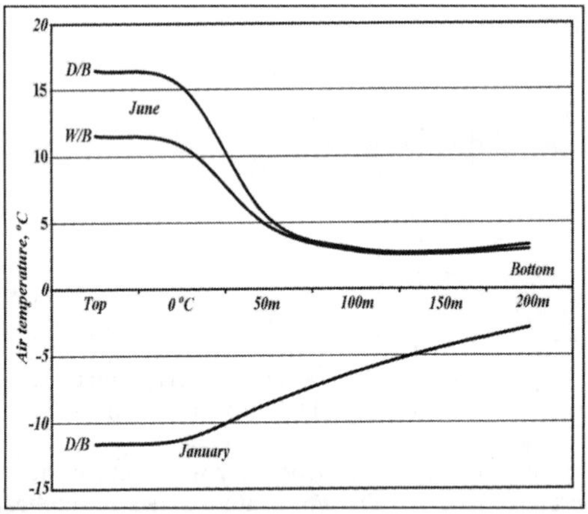

Figure 8. Air temperature.

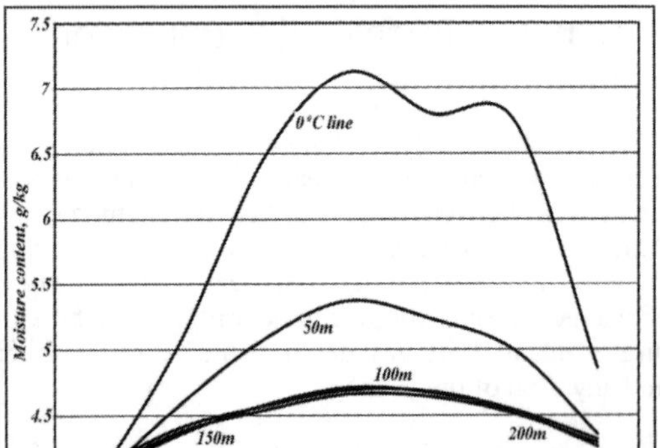

Figure 9. Moisture content.

in the summer and heat in winter. Air exiting the heat exchanger has temperatures higher than −2 EC in the coldest months and about 2 EC in the warmest months of the year.

Figure 9 illustrates how air moisture content changes throughout the year at different depths of broken rock. Moisture content varies mostly in summer and at depths less than 50 m. The greatest moisture content (about 7.12 g/kg) is found in July, August and September.

6 CONCLUSIONS

Weather conditions and the presence of water in a natural heat exchanger formed by fragmented rock are the primary factors affecting the natural process of heating and cooling intake air. In winter, temperature of part of the rock is below zero and any water present in that part freezes. The freezing front moves from the surface down to its maximum depth of 1 m sometimes in early spring. Then, the melting front starts to move downwards until all ice is melted, usually shortly before summer. During the remaining months of the year the whole heat exchange area has temperatures above 0 EC. Locations of the freezing and melting fronts are found applying the Stephen Law.

The fresh air drawn down into a mine through broken rock is naturally heated in winter and cooled in summer. The nature and amount of heat exchanged between air and the fragmented rock strongly depends on the temperature of the broken rock. However, the dry bulb temperature of air at the bottom of a 200 m deep gathering area remains reasonably stable at 0 EC, with a yearly seasonal variation of plus/minus 2°C.

To obtain more accurate results we need to use geological data and thermal properties for broken not solid rock. Also, a correct amount of moisture content is still unknown.

The presented analytical method assumes semi-steady rock and weather conditions. More advanced analysis that would apply variable boundary conditions, would result in partial differential equations that can be solved numerically or by the means of a CFM program.

REFERENCES

Backer, M., 1986, "Heat transfer – A modern approach," Plenum Press, New York.

CLIMSIM for Windows, "Users Manual and General Theory," Version 1.0, Mine Ventilation Services, Inc.

Golder Associates, 1980, "A study to determine ventilation and refrigeration requirement and costs in deep Canadian mines," Golder Associates, Consulting Geotechnical Engineers.

INCO, 2003, Private discussions.

Luikov, A.V., 1968, "Analytical heat diffusion theory," Academic Press, New York and London.

McPherson, Malcolm J., 1992, "Subsurface Ventilation and Environmental Engineering," Chapman and Hall, London, Glasgow, New York-Tokyo, Melbourne, Madras.

Mine Ventilation, Ganguli & Bandopadhyay (eds)
© 2004 Taylor & Francis Group, London, ISBN 90 5809 633 5

Coalbed methane drainage: the past, present, and future

P.C. Thakur
Consol Energy Inc., Pittsburgh, PA, USA

ABSTRACT: Coal seams and coalbed methane are syngenetic in origin. Both were formed by the biochemical decay and metamorphic transformation of plant materials over millions of years. Methane is released into coal mines as mining proceeds and it can form an explosive mixture with mine air. History of mining is replete with mine explosions and consequent loss of lives. Serious efforts in the USA to recover methane prior to mining and post-mining started about 40 years ago. Consol Energy Inc. and government agencies developed several techniques of methane drainage, e.g. in-mine horizontal drilling, vertical gob wells, hydrofracing of vertical wells, and horizontal drilling from surfaces. With the current technology, 30 to 60 percent of total mine methane emissions can be captured making the mines safer and more productive. Favorable economics have led to coalbed methane production from non-mineable coal seams as well. At present nearly 9 percent of total U.S. gas production is realized from coal formations no deeper than 1200 m. But the majority of the estimated world coalbed methane reserve, about 900 Tm^3, is in coal seams deeper than 1200 m. Efforts to produce gas from these reservoirs have had only limited success owing to extremely low permeability. The paper reviews current technology and discusses a new approach to produce gas from deeper coal formations combining horizontal drilling, hydrofacing, and carbon dioxide flooding. An additional benefit realized is the sequestering of carbon dioxide, a greenhouse gas, into methane depleted reservoirs.

INTRODUCTION

Coal seams were formed over millions of years by the biochemical decay and metamorphic transformation of plant materials. This coalification process produces large quantities of by-product gases, such as methane and carbon dioxide. Most of these gases escape to the atmosphere during the coalification process but a small fraction is retained in the coal. The amount of gas retained in coal depends on a number of factors, such as the rank of coal, the depth of burial, immediate roof and floor, geologic anomalies, and tectonic pressure and temperatures prevalent at the time. Actual gas content of coal seams vary from 0–25 m^3/t. Methane is the major component of gas in coal comprising 80 to 95 percent of the total volume.

Methane is released into coal mines from coal seams as mining proceeds. It is a colorless, odorless, combustible gas that forms an explosive mixture with mine air in the concentration range of 4.5 to 15 percent by volume. The history of coal mining in the United States of America and the rest of the world is replete with mine explosions and consequent loss of lives. The maximum concentration of methane in mine air is restricted by law to one to two percent in all major coal producing countries but methane explosions are quite common even today. To mitigate this problem, mine ventilation must be supplemented by methane drainage from coal seams prior to, during, and post-mining.

Methane drainage techniques are broadly divided into two groups depending on the mining phase, e.g. pre-mining and post-mining. Techniques used in these phases are quite different from each other but a judicious combination of methane drainage techniques can recover 30 to 60 percent of total mine methane emissions. In some cases, the recovered methane can be marketed to defray the cost of methane drainage. Favorable economic conditions have allowed coalbed methane (CBM) production from deeper, non-mineable coal seams also. At present nearly nine percent of total U.S. gas production is derived from coal seams. Mineable depths of coal generally do not exceed 1200 m and mineable world reserves of coal is only about 1 Tt but total estimated world reserves of coal range from 17 to 30 Tt (Landis and Weaver, 1993) with a methane reserve of about 900 Tm^3 (Thakur et al., 2002). The current coalbed methane production is just the tip of the iceberg. Efforts to produce CBM from deeper reservoirs have had only limited success owing to very low permeability. However, recent innovations, such as horizontal hole drilling from surface, hydrofracing of horizontal boreholes, and enhanced methane production by carbon monoxide flooding can significantly increase CBM production from deep coal seams. An additional benefit is realized by the sequestration of carbon dioxide, a greenhouse gas, in the depleted coalbed methane reservoirs. This paper reviews historical methane drainage practices, discusses current techniques of production and proposes a methodology for the extraction of CBM from deep coal seams.

1 THE EARLY DAYS (1940–1969)

1.1 *Pre-mining methane drainage*

Pre-mining methane drainage in the United States prior to 1950 was almost unknown. First references to coalbed methane drainages are found in 1953 (Spindler, 1953). In the late 1950's, Consol Energy Inc. began preliminary efforts to reduce methane emissions in development headings by drilling short (30 to 60 m) boreholes in the return entries and discharging the produced gas into mine air. This reduced methane emissions in the face area but failed to make a significant difference (Spindler and Poundstone, 1960). It was, however, concluded that methane emissions in development sections can be significantly reduced if a long horizontal borehole (300 m or longer) can be drilled ahead of the development heading. In the late 1960s, both Consol Energy Inc. and the erstwhile U.S. Bureau of Mines (USBM) began efforts to drill such long horizontal boreholes.

1.2 *Post-mining methane drainage*

Prior to 1969, there were few longwall faces working in the United States and post-mining methane drainage was unknown. However, European mines began to use the longwall method of mining in the 1940's and developed several techniques for post-mining methane drainage, e.g. packed cavity method, cross-measure boreholes, and superjacent techniques and captured 30 to 70 percent of total methane emissions in the section. Detailed descriptions of these techniques are available elsewhere (Thakur, 1981). These techniques were, however, not compatible with U.S. coal mining conditions.

2 EARLY SUCCESSES (1970–1989)

2.1 *Pre-mining methane drainage*

The most significant development in this period was the discovery of in-mine horizontal drilling techniques. Consol Energy Inc. and the USBM simultaneously discovered two different ways to drill 300 m to 600 m long boreholes in the coal seam. These boreholes drilled ahead of development headings captured about 50 percent of total methane emissions. The gas was vented to surface without mixing it with mine air, reducing the ventilation air requirement significantly. Figure 1 shows drilling scheme (Thakur and Davis, 1977).

2.2 *Post-mining methane drainage*

The discovery of vertical gob wells for post-mining methane drainage from longwall gobs was the most significant development of the 1970s. Consol Energy Inc. in collaboration with the USBM began to experiment with this technique in northern West Virginia. Simultaneously, Island Creek Coal Company (now a subsidiary of Consol Energy Inc). began to use this technique to drain methane from longwall gobs in Virginia. Construction details of gob wells are available in published literature (Mazza and Mliner, 1977; Moore and Zabetakis, 1972).

Figure 2 shows a typical layout of these gob wells on a longwall panel. Nearly 50 to 60 percent of gob

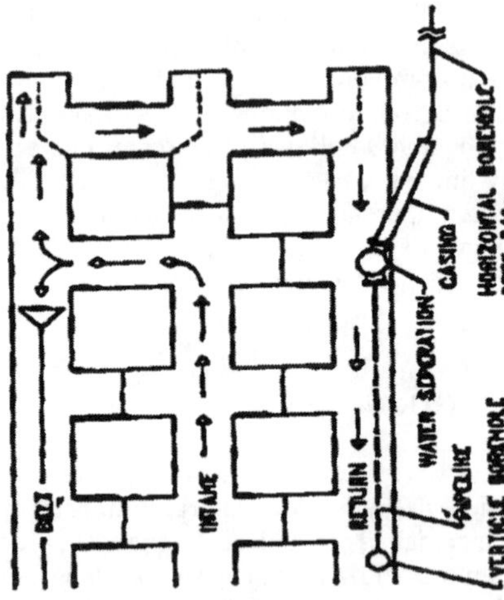

Figure 1. Horizontal drilling for coal seam degasification.

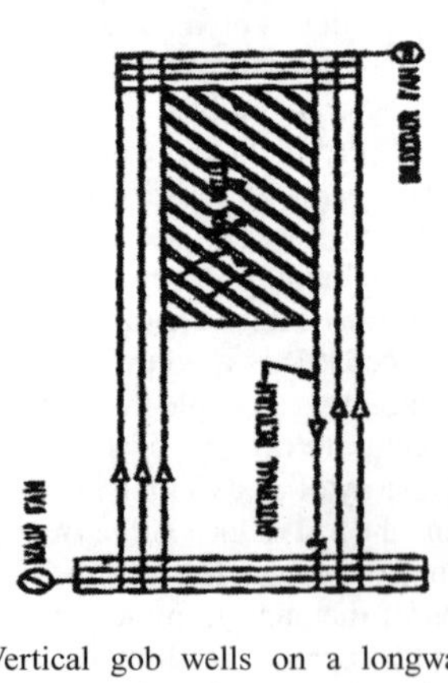

Figure 2. Vertical gob wells on a longwall panel for methane drainage.

methane emissions were captured reducing ventilation air quantities very significantly. Gob well diameters ranged from 100 mm to 228 mm and produced methane was generally vented to the atmosphere.

3 THE MIDDLE PERIOD (1984–1995)

3.1 *Pre-mining methane drainage*

The most significant discovery of this period was the successful use of vertical, hydrofraced wells to drain deep, very gassy coal seams. In-mine horizontal drilling in such seams was generally unsuccessful due to excessively high reservoir pressures (up to 5 MPa).

Hydrofracing process involves pumping sand slurry at high pressures into a formation and creating a vertical fracture that can measure several hundred meters. The technique was already widely used in the oil and gas industry but Jim Walter Resources and U.S. Steel mines in Alabama pioneered the use of this technique to degas some very gassy coal seams. In early applications the typical volume of water used was about 40 m^3 with five tons of sand. The pumping rate was 1.5–2 m^3/min. Consequently, the fractures in coal were no longer than 30 m and gas production was low at about 1 m^3/min.

In 1984, Consol Energy Inc. began to experiment with hydrofracing of vertical wells in the Pocahontas #3 seam in Virginia. In this design, the sand and water volumes were increased to 50 tons and 400 m^3 respectively and the pumping rate was also increased to 4.5 to 6 m^3/min. This resulted in much longer vertical fractures (typically 300 m long) with significantly higher gas productions. The technique became the standard primary degasification technique for very gassy coal seams removing about 50 percent of in-situ gas prior to mining. For higher rates of extraction, the longwall panels were further degassed with secondary horizontal drilling of the panel to remove an additional 20 to 30 percent of in-situ gas.

3.2 *Post-mining methane drainage*

Post-mining methane drainage techniques did not change much during this period but the application of vertical gob wells became more intense. Higher productivity needs dictated 300 m wide panels with extraction rates exceeding 30 m/day in some cases. This required bigger gob wells (up to 381 mm in diameter) at closer spacing (2.5 to 6 hectares/well). Gas production rates as high as 0.28 Mm^3/day were achieved (Thakur and Zachwieja, 2000).

During this period some experiments to introduce the European "cross-measure borehole" technique for gob degasification were also tried (Thakur et al., 1983) but the methane drainage rate was too low to cope with the fast moving U.S. longwall panels and the technique was largely abandoned.

4 THE MODERN PERIOD (1994–PRESENT)

4.1 *Methane drainage changes to commercial methane production*

During this period, several coal companies began to market coalbed methane produced from the mines as well as non-mineable coal seams. Frac wells earlier drilled to degas only the coal seam to be mined was completed in multiple coal seams to increase total gas production and also predrain the gas emission space surrounding a longwall gob. It reduced methane emissions in the post-mining phase and the number of gob wells per panel. Multiple completions in a frac well necessitated the use of nitrogen foam and sand replacing sand and water slurry as the fracing media. A frac well now could be completed in 3 to 4 coal seams in a single day. This substantially increased the productivity of frac wells. Some multiple completion wells produced in excess of 30 Mm^3 over a period of ten years.

In the late 1990s, another oil-field technique, horizontal drilling from surface, was introduced to the coalfields. In this technique, several horizontal laterals can be drilled from a vertical well to intersect different coal seams. A plan view of a single seam completion is shown in Figure 3. Typically four laterals are drilled to intersect a central well that is used for water drainage. Up to 10 Km^2 area can be drained by only five wells, minimizing the environmental footprint and concentrating production. Horizontal drilling from surface is universally applicable to shallow (depth $\leqslant$ 300 m), medium depth (300 m $<$ depth $<$ 1200 m), or deep

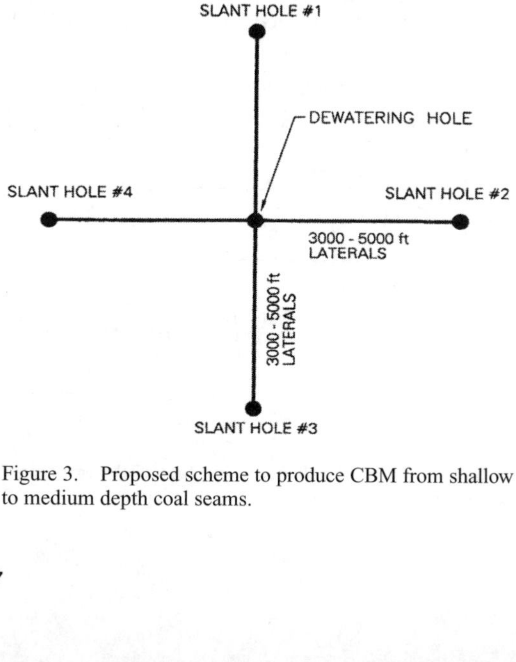

Figure 3. Proposed scheme to produce CBM from shallow to medium depth coal seams.

coal (depth ⩾ 1200 m) seams but at present, it is a quite expensive technique. The drilling pattern shown in Figure 3 can cost in excess of a million U.S. dollars. The gas production and price have to be high enough to justify such high investments. However, current higher gas prices are turning many coal reserves into gas reservoirs and the technique is likely to find wider application.

5 COALBED METHANE DRAINAGE: THE FUTURE

The world coalbed methane reserve is estimated at about 900 Tm^3 (Kuuskra, 1992; Cairns Point Publishing, 1997). The bulk of this reserve is in coal seams deeper than 1200 m. The technique described above works very well for shallow or medium depth coal seams. With increasing depth the reservoir permeability becomes exceedingly low (a few microdarcys) and efforts to produce CBM from these reservoirs have had only limited success. A successful production technique for deep coal seams must incorporate three features: (1) a permanent opening in the coal seam, (2) increased matrix permeability, and (3) gas production enhancements.

The proposed gas production technique for deep coalbeds incorporates all these features. Openings or production channels in deep coal seams are created by horizontal boreholes drilled from surface. The permeability of the coal seam is enhanced by hydrofracing, and gas production is enhanced by carbon dioxide flooding. Coal seams adsorb carbon dioxide preferentially and give out methane adsorbed on its surface. A typical scheme in shown in Figure 4. From each vertical well, three laterals (horizontal legs of the well) are drilled to 3000 m. The azimuth of these laterals should be so selected that the fractures will grow orthogonal to them, as shown, and facilitate CBM recovery. Such

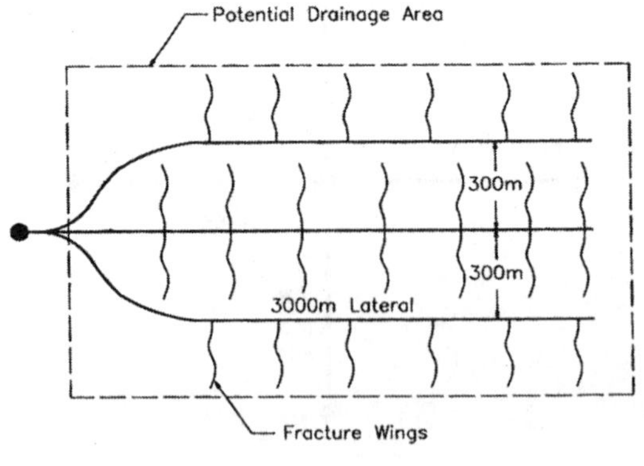

Figure 4. Proposed scheme to produce CBM from deep coal seams.

drilling can be done with both conventional rigs and coiled tubing drilling (CTD) technique (McAlister et al., 2001). The CTD technique, at present, is significantly more expensive than the conventional drilling but it is certainly the preferred way to drill because it permits real time monitoring of the drill bit and can be much faster and more accurate than conventional drilling.

Hydrofracing of horizontal wells is in the experimental stage. Some service companies have developed a special bottom hole packer assembly that permits fracing of a 3000 m lateral at 150 m interval in a single day but further evaluation is needed. Similarly, the carbon dioxide flooding technique is in the experimental stage. Laboratory experiments to displace methane from coal using nitrogen, helium, and carbon dioxide show that the latter is the most effective flooding agent. Limited field experiments with nitrogen flooding indicate a 1:1 displacement of methane (i.e., each unit volume of carbon dioxide injected displaces one unit volume of methane).

Consol Energy Inc. and the United States Department of Energy have started a joint project to demonstrate and establish working parameters of a carbon dioxide flooding program. Optimum flow rates, injection pressure, and vertical and lateral migration velocities of carbon dioxide will be determined. Gas production rates and ultimate methane recovery from the reservoir will be determined to prove economic viability of such endeavors. Since carbon dioxide is a greenhouse gas, there is some financial incentive for sequestering it in coal formations.

6 SUMMARY AND CONCLUSIONS

1. Discovery and application of coalbed methane drainage techniques have made it possible to recover 30 to 60% of methane emissions from coal mines. It has made the coal mines much safer and more productive. Produced gas is sold to defray the cost.
2. Favorable pricing of CBM has turned coalbed methane drainage into commercial coalbed methane production. Greater attention is now directed to deeper coalbeds.
3. World coalbed methane reserve is enormous at about 900 Tm^3. For comparison, the U.S. conventional gas reserve is only 25 Tm^3. With effective research and development followed by adequate investments, CBM production can become a significant source of natural gas supply.
4. It is high time to increase R&D efforts in the areas of horizontal drilling from surface, hydrofracing of horizontal wells, and CO_2 flooding for enhanced coalbed methane production. Enough knowledge is available to start the process but much more needs to be done to optimize the process.

REFERENCES

Cairn Point Publishing Inc., 1997, The International Coal Seam Gas Report

Kuuskra, Vello, A. et al., 1992, Coalbed Gas-hunt for Quality Basin Goes Abroad, Oil & Gas Journal, pp. 49–54

Landis, E.R. & Weaver, J.W., 1993, Coalbed Coal Occurrence, Hydrocarbons from Coal, AAPG Studies in Geology, No. 38, pp. 1–12

Mazza, R.L. & Mlinar, M.P., 1977, Reducing Methane in Coal Mine Gob Areas with Vertical Boreholes, Conoco Report on USBM Contract No. H0322851

McAllister, J.V. et al., 2001, Underbalanced Coiled-tubing Drilling in a Thin Gas Storage Reservoir: A Case Study, SPE 72373, SPE Eastern Regional Meeting

Moore, T.D. et al., 1972, Effects of Surface Boreholes on Longwall Gob Degasification, (Pocahontas No. 3 Seam), U.S.B.M. Report of Investigation 7657

Spindler, G.R., 1953, Degasification of Coal Mines, Coal Mine Mechanicalization, American Mining Congress

Spindler, G.R. & Poundstone, W.N., 1960, Experimental Work in the Degasification of Pittsburgh Coal Seam by Horizontal and Vertical Drilling, SME Preprint No. 60-F-106, AIME Annual Meeting, New York, NY

Thakur, P.C. & Davis, J.G., 1977, How to Plan for Methane Control in Underground Coal Mines, Mining Engineering, pp. 41–45

Thakur, P.C. et al., 1983, Methane Drainage with Cross-panel Boreholes on a Retreat Longwall Face, Preprint No. 83-398, SME-AIME Fall Meeting, Salt Lake City, Utah, October 19–21

Thakur, P.C. et al., 2000, Methane Control and Ventilation for 1000-Ft. Wide Longwall Faces, Proceedings of Longwall U.S.A., Pittsburgh, Pennsylvania

Thakur, P.C. et al., 2002, Coalbed Methane Production from Deep Coal Seams, AICHE Spring Meeting, New Orleans, LA, March 10–12, pp. 497–502

Mine Ventilation, Ganguli & Bandopadhyay (eds)
 ISBN 90 5809 633 5

Ventilation effect on radon levels in the ESF of Yucca Mountain

Romeo Jurani, YuChien Yuan, Brian Prosser
Bechtel SAIC Company, LLC, Las Vegas, Nevada, USA.

ABSTRACT: The Ventilation effect on radon levels was investigated in the main tunnel of the Exploratory Studies Facility (ESF) at the Yucca Mountain project site. Radon flux from the host rock was estimated using radon concentration data collected continuously during the first three months of 1999. Correlation between radon flux and pressure differential generated by ventilation airflow for a heavily fractured segment and a less fractured segment were compared. In the absence of adequate design control measures, radon flux and radon concentrations can reach relatively high values. Contrary to expectations, increased ventilation rates did not result in an effective decreasing of radon levels because higher exhausting airflow induced greater negative pressure differentials inside the tunnel which acted to draw increased radon flux from the fractures and fissures of the host rock into the tunnel.

1 INTRODUCTION

The Yucca Mountain site, situated on the southwestern boundary of the Nevada Test Site, is located in Nye County, Nevada, approximately 160 km northwest of Las Vegas, Nevada as shown on Figure 1. In 2002, Yucca Mountain was selected as the site for a central repository for nuclear waste after 20 years of exhaustive scientific and engineering research. The Exploratory Studies Facility (ESF), a tunnel system that consists of a 7877 m main tunnel (7.6 m diameter), a 2820 m cross-drift (5 m diameter), many test alcoves, and several smaller test niches, were built for scientific studies in laboratories up to 300 m underground. Figure 2 shows the existing layout of the ESF. The ESF contains a relatively simple ventilation system with through-flow ventilation in the main airway provided by a parallel fan installation located in the exhaust portal. The cross-drift has a dedicated auxiliary ventilation duct and fan.

Radon (^{222}Rn), produced by the decay of ^{226}Ra in underground tuffs, migrates through rock and emanates continuously from ESF surfaces into all air spaces. The emanation of radon from ESF surfaces depends on many factors, including rock properties such as density, porosity, fracture density, moisture content, and the ^{226}Ra content. Convective movements of soil gas and the resulting transport of radon within the rocks can be induced by pressure differences created by meteorological factors and operational conditions of the ESF. These factors and conditions are time-dependent

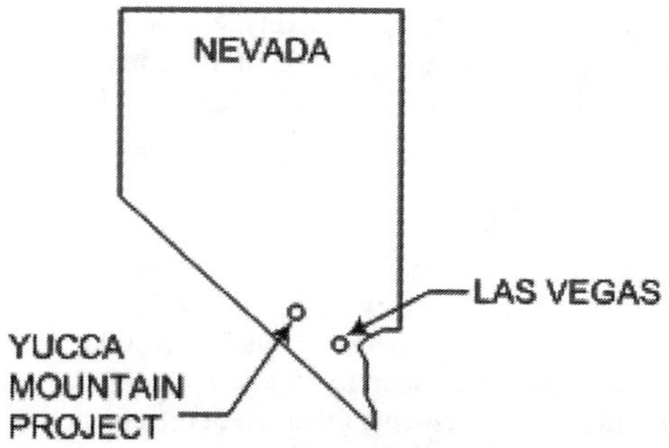

Figure 1. Location of Yucca Mountain Project.

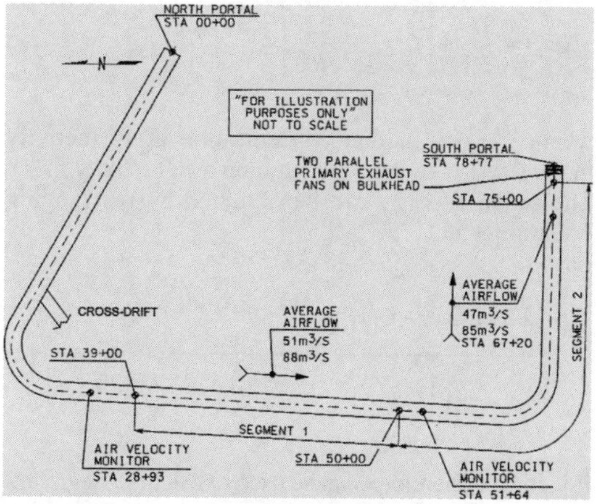

Figure 2. Layout of ESF main drift.

and essentially unpredictable. This paper presents the results of a study performed to estimate the radon flux in the ESF main tunnel and to evaluate of the effect of ventilation on the flux values. The results of this study may be used to help design an effective ventilation system for the potential repository that may be able to control the average concentration of radon and its decay products to an acceptable level.

2 METHOD/FORMULATION

The equation of continuity for radon in a tunnel can be written as:

$$\frac{\partial C}{\partial t} = -v\frac{\partial C}{\partial x} + 2\frac{F}{R} - \lambda C \qquad (1)$$

where F = average radon flux (activity $m^{-2}s^{-1}$); λ = radon decay constant (s^{-1}); R = drift radius (m); C = radon concentration (activity m^{-3}); v = ventilation flow velocity (m s^{-1}); t = radon decay half-life (s); x = incremental distance in x (m).
Equation 1 describes changes of mass concentration of radon with respect to time resulting from a ventilation flow velocity, v, and radon exhalation flux, F, from the rock surfaces.

At steady state, $(\partial c/\partial t) = 0$, and Equation 1 may be rearranged as:

$$\frac{dC}{dx} + \frac{\lambda}{v}C = \frac{2F}{vR} \qquad (2)$$

The solution of Equation 2 for the radon concentration at x_2, using a boundary condition that the radon concentration at x_1 is $C(x_1)$, is:

$$C(x_2) = \frac{2F}{R\lambda} + \left[C(x_1) - \frac{2F}{R\lambda}\right]e^{-\lambda(x_2-x_1)/v} \qquad (3)$$

where $C(x_2)$ = raddon concentration at x_2 (activity m^{-3}); $C(x_1)$ = radon concentration at x_1 (activity m^{-3}). The radon flux, F, can be obtained by rearranging Equation 3 as:

$$F = \frac{\lambda R\left[C(x_2) - C(x_1)e^{-\lambda(x_2 - x_1)/v}\right]}{2\left(1 - e^{-\lambda(x_2 - x_1)/v}\right)} \qquad (4)$$

Equation 4, therefore, can be used to calculate segment-average radon flux values using radon concentration measurements made at segment end points x_1 and x_2.

3 STUDY SITE

Yucca Mountain consists of successive layers of volcanic rocks (call tuffs), approximately 14 to 11.5 million years old, formed by eruptions of volcanic ash from ancient calderas. Most of these volcanic rocks are ash-flow tuffs of two types, welded and non-welded, that formed when hot volcanic gas and ash erupted violently and flowed quickly over the landscape. In the 7877 m main ESF tunnel, the segment between 2800 m and 5000 m from the main entrance represents the type of rock strata that may host the emplacement drifts of the potential repository. The segment between 5000 m and 7500 m, and especially between 6700 m and 7500 m, consists of highly fractured rock strata including faults and shears (Eatman et al., 1997, CRWMS M&O 1998).

3.1 *Radon monitoring data*

From 1 January 1999 through 7 April 1999, hourly radon concentrations were collected at three stations in the ESF main tunnel (near 3900 m, 5000 m, and 7500 m) using AlphaGuard (Genitron Instruments, Heerstrasse 149 D60488, Frankfurt, Germany) continuous radon pulsed ion chamber monitors. AlphaGuard also includes instruments for measuring local temperature, barometric pressure and humidity. Figure 3a presents graphically the radon concentration collected at the three stations. Figure 3b shows the barometric pressure measured at the same locations for the study period.

For this study period, the ESF ventilation system ran two fans during most of the workdays and one fan during most non-workdays. To investigate the effect of ventilation on radon concentration and flux, radon data were grouped by one-fan and two-fan operations, and average radon concentration values were calculated for each. Figure 4 shows the complimentary cumulative distribution function (CCDF) of the radon concentrations collected at the three monitoring stations. The average radon concentrations measured at these three stations (3900 m, 5000 m, and 7500 m) during the two-fan operations were 430, 670 and 1600 Bq/m^3 (becqueral per cubic meter) (12, 18 and 43 pCi/L (picocuries per liter), respectively. For one-fan operations, the corresponding average radon concentrations were 460, 660 and 1100 Bq/m^3 (13, 18 and 30 pCi/L), respectively.

3.2 *Radon flux*

Hourly average radon flux was calculated (Equation 4) for two segments of the ESF tunnel as shown in Figure 2 (Segment 1: between 3900 m to 5000 m; Segment 2: between 5000 m to 7500 m) using radon concentration measurements recorded at the end points of each segment. Ventilation rates were based on monitoring data, which include moisture, temperature and ventilation velocity measured at the ESF (Flint AL, 1999). For flux calculations, ventilation rates recorded at 6720 m were used to represent Segment 1 airflow and the averages of the ventilation rates recorded at 2893 m and 5164 m were used for Segment 2 airflow. The one fan and two fan operations during this study period correspond to airflow of about 51 m^3/s and 88 m^3/s, respectively, in Segment 1 and about 47 m^3/s and 85 m^3/s, respectively, in Segment 2 of the ESF main tunnel.

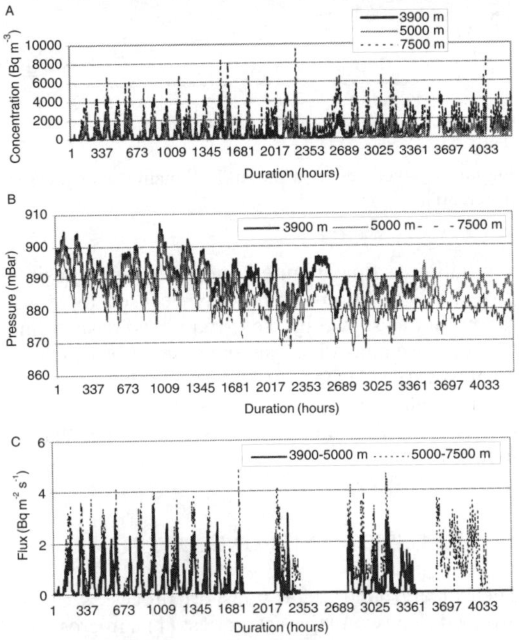

Figure 3. Measured radon concentration (a) and barometric pressure (b) and calculated radon flux (c) inside ESF main tunnel during 1 January 1999 through 7 April 1999.

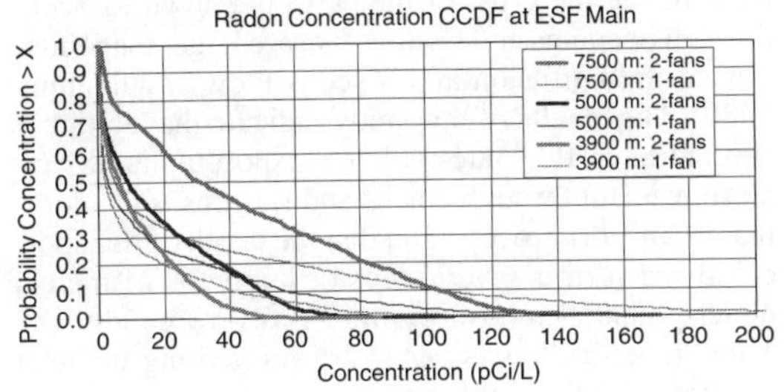

Figure 4. Complimentary cumulative distribution function of radon concentration in the ESF.

3.3 *Relative pressure distribution*

The differential pressure (static) across the fan bulkhead/airlock door (Figure 2) was monitored during single and two fan operational modes. The air velocity in the ESF airway was monitored at various stations along the ESF by United States Geological Survey (USGS) weather sensors. This velocity was used to determine the average airflow in the ESF for both one and two fan operation. The differential pressure across the fan bulkhead was measured at the beginning of each shift and during each change in ventilation. The average airflow and average differential pressure was then used to calculate the total average resistance of the existing 7877 m long ESF airway. From this ventilation data and the underground layout distances, a simple network model was developed with the VNETPC for Windows (Version 1.0a) simulation program to represent the operation of the ESF ventilation system. The average differential pressure (ambient atmosphere to subsurface) for each segment, was then determined from the network model for both one and two fan operation. Table 1 presents the pressure differential values together with the radon flux calculated for the study period.

4 RESULTS

Figure 3c presents graphically the calculated radon flux for the study period. Radon flux values were found to have an inverse relationship with barometric pressure changes (Figure 3b). For every sharp drop in barometric pressure during the measurement period, a corresponding high radon flux value was recorded. Uncertainty in the effect of weather conditions or variations in barometric pressure over time on the concentration and flux values was also assessed by constructing a cumulative probability distribution of concentration

Table 1. ESF pressure differential/radon flux values.

Segment	Fan mode	ΔP (Pa) (atm – ave rep)	Average drift Airflow (m^3/s)	Average radon flux ($pCi/m^2/s$
1	One fan	73	51.0	14
2	One fan		47.0	
1	Two fan	251	88.0	21
2	Two fan		85.0	

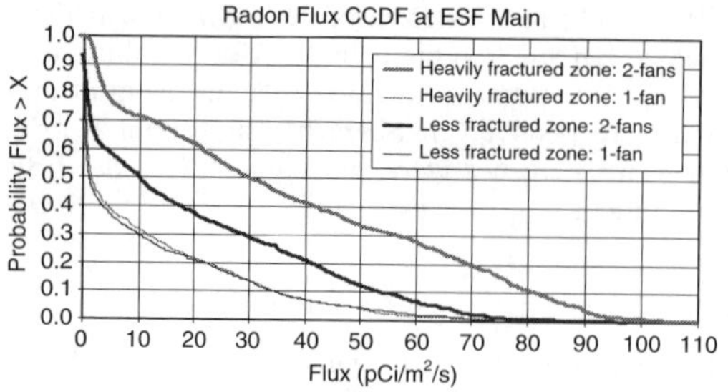

Figure 5. Complimentary cumulative distribution function of radon flux in the ESF.

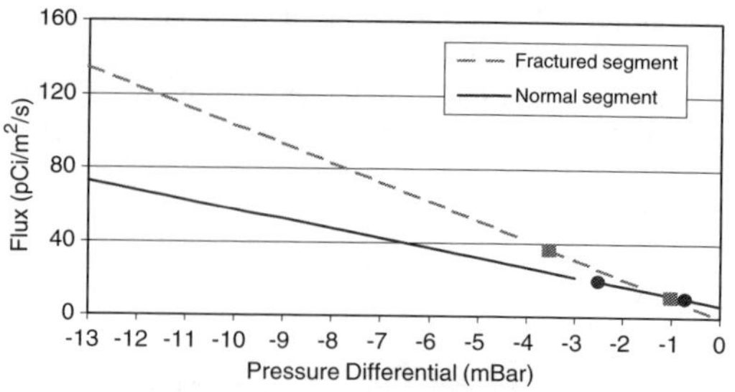

Figure 6. Average radon flux in ESF main versus pressure differential.

and flux values for the study period. Figure 5 presents the CCDF of radon flux at Segment 1 (between 3900 m to 5000 m) and Segment 2 (between 5000 m to 7500 m). The flux results indicate that under the current ESF exhausting ventilation system, higher airflows generated higher radon flux. A 70% increase in ventilation airflow resulted in a 90% increase in average radon flux in Segment 1, which represents the rock strata of the potential repository emplacement drifts. For the highly fractured Segment 2, the ventilation effect was even more significant, an 80% increase in ventilation airflow resulted in about a 240% increase in average radon flux. The higher rate of radon release apparently is related to increasing fracture intensity of the rock strata. The average radon flux for the highly fractured Segment 2 was a factor of about 90% greater than the less fractured Segment 1 when two fans in operation.

To estimate the ventilation-induced pressure effect on radon flux, pressure differentials at various locations in the ESF Main with one fan and two fans in operation were calculated (Section 3.3). Table 1 presents the pressure differential values together with the radon flux calculated for the study period. The correlation between the pressure differential (ΔP) and radon flux (F) was constructed assuming F is linearly depending on ΔP. The equations of such a correlation for the less fractured Segment 1 and heavily fractured Segment 2 are shown, respectively, as Equations 5 and 6:

$$F(\Delta P) = 6.3 - 5.1\,\Delta P, \quad (\Delta P < 0) \tag{5}$$

$$F(\Delta P) = 0.13 - 10\,\Delta P, \quad (\Delta P < 0) \tag{6}$$

where $F(\Delta P)$ = average radon flux (pCi/m²/s) at pressure differential ΔP; ΔP = pressure differential (mBar) = $P_d - P_a$; P_d = air pressure inside the repository during ventilating; P_a = ambient air pressure outside the repository.

The flux values indicate that under the ESF negative pressure (suction) system higher airflow generates higher radon flux.

Figure 6 shows graphically the correlation between radon flux and pressure differential generated by ventilation airflow. Figure 6 indicates that a positive pressure ventilation system may be used to reduce radon flux as shown by the extrapolated dashed line. It is because a positive pressure system, in which the air pressure inside the tunnel during ventilating exceeds the ambient air pressure, would serve to suppress the convective radon flux from the rocks rather than to induce it.

5 DISCUSSION AND CONCLUSION

The Underground radon transport processes can be grouped into two major categories: (1) a microscopic process in which diffusive and viscous flows are dominant and (2) a macroscopic process in which flow occurs through cracks, fissures, and underground channels (Wilkening 1980). Transport of radon-rich air from inside the crust of the earth is known to occur through openings and fissures that are larger than those treated in "Diffusion and Viscous Flow" (Wilkening 1980). The results of this study indicate that at higher ventilation airflow rates radon transport in the ESF by air movement through cracks and fissures is the dominant transport process, and that the depth of transport could reach over several tens or hundreds of meters under potential repository higher-airflow conditions. This higher-airflow is necessary for venting the heat generated by the emplaced spent nuclear fuel.

The flux results revealed that under the ESF exhausting or negative pressure ventilation system, higher-airflow generates higher a radon flux. This is because the higher exhausting airflow is created by higher fan pressures. The higher fan pressures cause a higher negative differential pressure between the atmosphere and the repository. This negative pressure acts to draw out the radon through cracks and fissures toward the tunnel. An exhausting system operating at a higher negative pressure results in a higher radon flux inside of the tunnel. This can cause problems in that while supplying a sufficient quantity of air to dilute radon inflows, the inflow of radon actually increases with the increased exhausting fan pressures. This relationship indicates that the use of a positive pressure system suppresses the convective radon flux from the host rock rather than inducing it. The ventilation system for the construction area of the repository (where the majority of personnel will be located) is currently designed with a forcing ventilation system. This will help ensure the minimization of high radon concentrations in areas where construction activities are taking place.

The continuous production of radon as a decay product in the rocks, does not allow it to be predrained like methane in a coal seam. In addition, methane in a coal mine often occurs under pressurized conditions and will be produced regardless of whether an exhausting or forcing ventilation system is utilized. The radon at Yucca Mountain once produced is retained in the rock mass or in the fractures and crevices until it decays or is drawn out by differential pressures exerted by atmospheric changes or the mechanical ventilation system.

ACKNOWLEDGEMENTS

The Yucca Mountain Project (YMP) is a U.S. Department of Energy project. Bechtel SAIC Company, LLC is the DOE's contractor for the YMP.

DISCLAIMER

This paper was prepared as an account of work sponsored by an agency of the United States Government. Neither the United States Government nor any agency thereof, nor any of their employees, nor any of their contractors, subcontractors or their employees, makes any warranty, express or implied, or assumes any legal liability or responsibility for the accuracy, completeness, or any third party's use or the results of such use of any information, apparatus, product, or process disclosed, or represents that its use would not infringe privately owned rights. Reference herein to any specific commercial product, process, or service by trade name, trademark, manufacturer, or otherwise, does not necessarily constitute or imply its endorsement, recommendation, or favoring by the United States Government or any agency thereof or its contractors or subcontractors. The views and opinions of authors expressed herein do not necessarily state or reflect those of the United States Government or any agency thereof.

REFERENCES

U.S. Department of Energy (DOE). *Yucca Mountain science and engineering report*. Washington D.C.: U.S. Department of Energy; DOE/RW-0539; 2001.

CRWMS M&O (Civilian Radioactive Waste Management System Management and Operating Contractor). *Geology of the Exploratory Studies Facility Topopah Spring Loop*. Las Vegas, Nevada: CRWMS M&O; BAB000000-01717-0200-00002 REV 01; 1998.

Eatman GLW, Singleton WL, Moyer TC, Barr DL, Albin AL, Lung RC, and Beason SC. *Geology of the South Ramp – Station 55 + 00 to 78 + 77, Exploratory Studies Facility, Yucca Mountain Project*, Yucca Mountain, Nevada. Denver, Colorado: U.S. Geological Survey; 1997.

Flint AL. *Moisture Monitoring in the ESF*, 12/98 – 7/99. (DTN: GS990908312242.011), Las Vegas, Nevada: CRWMS M&O. 1999.

Bush CA, Bunker CM, and Spengler RW. *Radioelement distribution in drill-hole USW-G1*, Yucca Mountain, Nye County, Nevada. Open file report 83–847. Denver, Colorado: U.S. Geological Survey; 1983.

Wilkening M. "Radon transport processes below the earth's surface". In: Gesell TF, Lowder WM, eds. *International symposium on natural radiation environment, 3rd*, Houston, Texas, 1978. Natural Radiation Environment III. Springfield, VA: Technical Information Center; CONF-780422 (Volume 1); 1980: 90–104.

Mine Ventilation, Ganguli & Bandopadhyay (eds)
© 2004 Taylor & Francis Group, London, ISBN 90 5809 633 5

Mitigation of methane emissions from coal mine ventilation air: an update

F. Peter Carothers
Senior Engineer, Methane Mitigation, International Resources Group, Washington DC, USA

H. Lee Schultz
Director, Environmental Programs, International Resources Group, Washington DC, USA

Clark C. Talkington
Manager, Coalbed Methane Outreach Program, US Environmental Protection Agency, Washington DC, USA

ABSTRACT: Control of methane emissions from mine ventilation systems has been a challenging goal because of the magnitude of typical airflows coupled with very low methane concentrations. The authors evaluated the technical and economic feasibility of several emerging systems that may accept up to 100 percent of the flow from a gassy underground coal mine ventilation shaft, oxidize the entrained methane, and produce marketable energy. This paper discusses flow-reversal oxidizers that provide heat for producing thermal or electric energy, lean-fueled combustion turbines, and a hybrid waste coal and methane power plant. The paper also covers research into methane concentration techniques and the use of ventilation air methane as combustion air for a variety of prime movers. When ventilation shafts emit methane concentrations approaching 1.0 percent, energy prices are high, and revenues from greenhouse gas emission reduction offsets are available, it will be possible to operate a profitable energy plant fueled with ventilation air methane.

1 INTRODUCTION

The Coalbed Methane Outreach Program (CMOP) is a part of the US Environmental Protection Agency's (USEPA) Climate Protection Partnerships Division. CMOP is a voluntary program that works with coal companies and related industries to identify technologies, markets, and means of financing the profitable recovery and use of coal mine methane (CMM), a powerful greenhouse gas (GHG), that would otherwise be vented to the atmosphere. With a global warming potential that is 21 times that of carbon dioxide[1], methane capture and use offers a particularly attractive opportunity to achieve substantial GHG emission mitigation.

CMOP assists the coal industry by profiling CMM project opportunities, conducting mine-specific technical and economic assessments, and identifying private, state, local, and federal institutions and programs that could catalyze project development. In past years CMOP has been instrumental in helping coal mines in the US and abroad to successfully implement CMM drainage and use projects, thereby markedly reducing the volumes of methane that currently enter the atmosphere. Although there remain opportunities to recover drained gas, CMOP recognizes that to achieve continued reductions in methane emissions from coal mines over the long term, it is necessary to target ventilation system emissions. Ventilation air exhaust streams at gassy underground coal mines are characterized by very large airflows, typically ranging from 47 to 470 cubic meters per second (100,000 to 1,000,000 cubic feet per minute), containing very low concentrations of methane (ranging from 0.1 to over 1.0 percent). Most VAM occurs in concentrations that are between 0.3 to 0.5 percent.

Four years ago at the 8th US Mine Ventilation Symposium, Peter Carothers of International Resources Group presented an initial overview of the then developing technologies for ventilation air oxidation (Carothers and Deo, 2000a). After four years USEPA

[1]Note that the Intergovernmental Panel on Climate Change, in its Third Assessment report *Climate Change 2001*, increased the global warming potential value for methane from 21 to 23. However, the 21 value is still used as the standard for calculating carbon dioxide equivalents in emission inventories, carbon emission reduction transactions, etc.

has witnessed encouraging progress in the pursuit of feasible and economically viable technologies that can oxidize and use the energy content of VAM. Additionally, a number of new technologies have emerged. This paper provides an update on the technological innovations occurring and presents new information on markets for ventilation air methane (VAM).

2 TECHNOLOGY OVERVIEW

USEPA has identified several types of downstream technologies for destroying or beneficially using the methane contained in ventilation air in a USEPA report published in 2000 (Carothers and Deo, 2000b). These technologies fall into the following general categories:

- Flow-reversal oxidizers
- Lean-burn combustion turbines
- Coal and VAM hybrid system
- Methane concentration
- Use as an ancillary fuel.

Each of these technology categories is described below.

2.1 *Flow-reversal oxidation*

Oxidizers use a type of recuperation that retains the heat of oxidized methane in a ceramic bed, which in turn ignites new incoming VAM. USEPA has identified two basic oxidizer types, the thermal flow-reversal reactor (TFRR) and the catalytic flow-reversal reactor (CFRR). MEGTEC Systems (De Pere, Wisconsin), has developed a TFRR known as the VOCSIDIZER® that has been adapted to oxidize VAM. Canadian Mineral and Energy Technologies (CANMET – a national laboratory in Varennes, Quebec, Canada) has developed a CFRR expressly for mine ventilation air. Both technologies employ similar principles to oxidize methane contained in mine ventilation airflows. Based on laboratory and field experience, both units can sustain operation (i.e., can maintain oxidation) with ventilation air having uniform methane concentrations down to approximately 0.1 percent. For practical field applications where methane concentrations are likely to vary over time, however, a practical average lower concentration limit at which oxidizers will function reliably is considered to be 0.15 percent.

2.1.1 *Thermal flow-reversal reactor*

Figure 1 shows a schematic of the Thermal Flow-Reversal Reactor (TFRR). The equipment consists of a bed of silica gravel or ceramic heat-exchange medium with a set of electric heating elements in the center. To start oxidation, electric heating elements preheat the middle of the bed to or above the temperature required to initiate autoignition, i.e., ⩾1000°C (1832°F). Ventilation air containing methane enters at ambient temperature and flows through the reactor in one direction, and its temperature increases until oxidation of the methane takes place near the center of the bed.

The hot products of oxidation continue through the bed, losing (transferring) heat to the far side of the bed in the process. When the far side of the bed is sufficiently hot and the near side has cooled due to the inflow of ambient-temperature ventilation air, the reactor automatically reverses the direction of ventilation airflow. The ventilation air now enters the far (hot) side of the bed, where it encounters autoignition temperatures near the center of the bed and then oxidizes. The hot gases again transfer heat to the near

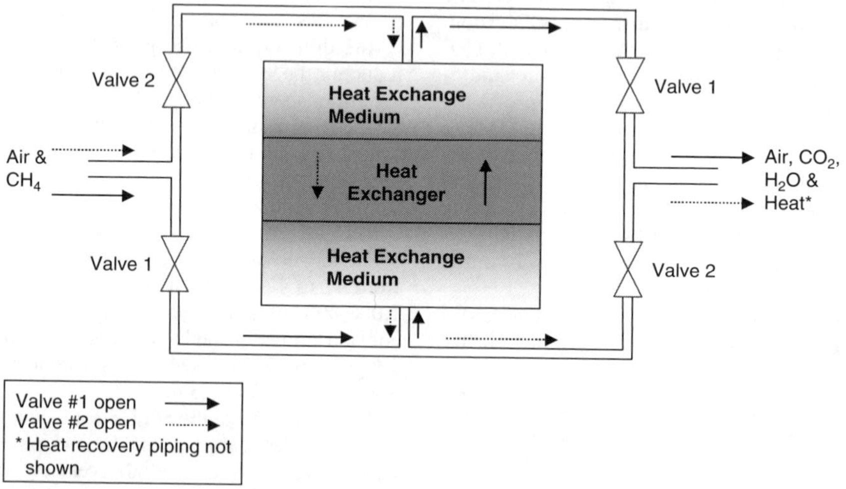

Figure 1. Thermal Flow-Reversal Reactor.

(cold) side of the bed and exit the reactor at a temperature just modestly above ambient. Then, the process again reverses.

TFRR units are effectively employed worldwide to oxidize industrial volatile organic compound (VOC) streams. Furthermore, the ability of MEGTEC's VOCSIDIZER® to oxidize VAM has been demonstrated in the field (Kallstrand and Zak, 2000–2003).

2.1.2 *Catalytic flow-reversal reactor*

Catalytic flow-reversal reactors adapt the thermal flow reversal technology described above by including a catalyst to reduce the autoignition temperature of methane by several hundred degrees Celsius, i.e., to as low as 350°C (662°F). CANMET has demonstrated this system in pilot plants and is now in the process of awarding a license to Lefebvre Freres Ltd., of Montreal, Quebec, Canada, to commercialize the design (Sapoundjiev et al., 2000–2003).

2.1.3 *Energy conversion from a flow-reversal reactor*

There are several methods of converting the heat of oxidation from a flow-reversal reactor to electric power, which is the most marketable form of energy in most locations. The two methods being studied by MEGTEC and CANMET are:

- *Use water as a working fluid.* Water is pressurized and forced through an air-to-water heat exchanger in a section of the reactor that will provide a non-destructive temperature environment, e.g., below 800°C (1472°F). The hot pressurized water is flashed to steam and the steam is used to drive a steam turbine-generator. One vendor intends to superheat the steam in a separate heat exchanger. If a market for low-quality steam or hot water is available the facility can send exhausted steam to that market. If none is available, the steam can be condensed and the water returned to the pump to repeat the process.
- *Use air as a working fluid.* Ventilation air or ambient air is pressurized in a gas turbine's compressor and sent through an air-to-air heat exchanger that is embedded in a section of the reactor that stays near 800°C (1472°F). The heated compressed air, serving as the gas turbine's working fluid, passes through the expansion section of the turbine, which powers the generator. If drained CMM such as gob gas is available, it may be used in an external combustor to raise the temperature of the working fluid to more nearly match the design temperature of the turbine inlet. The turbine exhaust may be used for cogeneration (see Figure 2), if thermal markets are available, or for heating incoming air.

Since affordable heat exchanger temperature limits are below those used in modern prime movers, efficiencies for both of the energy conversion strategies listed above will be constrained by such limits and fairly modest. The overall plant energy conversion efficiencies (with a supplemented VAM concentration of 1.0 percent) assumed for the cost estimates discussed later in this paper are expected to be in the neighborhood of 17 percent after accounting for power allocated to drive the fans that force ventilation air through the reactor.

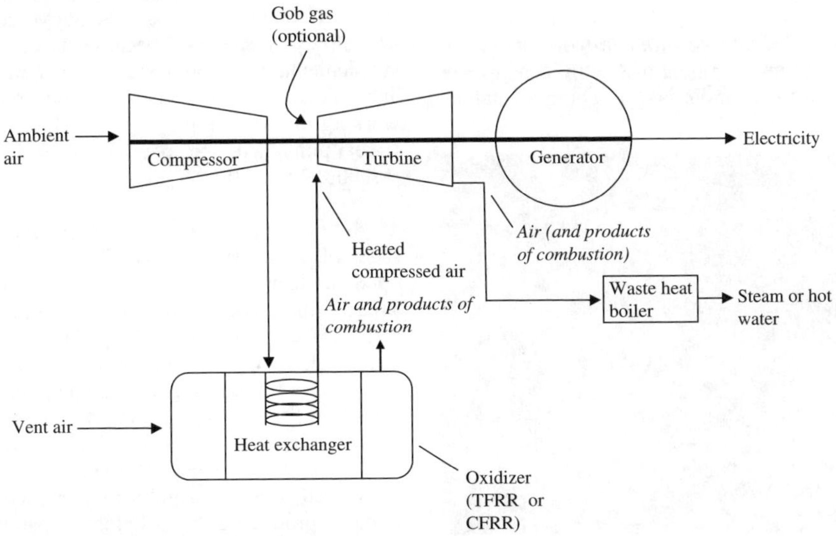

Figure 2. Schematic of cogeneration option using air as a working fluid.

2.2 *Lean-burn combustion turbines*

A number of engineering teams are striving to modify selected combustion (gas) turbine models to operate directly on VAM or on VAM that has been enhanced with more concentrated methane such as drained CMM. USEPA continually reviews the progress of research, development, and demonstration phases for several systems when and as their sponsors release information. Four of these systems are briefly described below. Two systems are based on full-sized commercial turbines, and two are modifications of "microturbines" that are currently sized in the range of 30 to 250 kW per unit.

2.2.1 *Carbureted gas turbine*

A carbureted gas turbine (CGT) is a gas turbine that employs a patented external combustor where the reaction is at a lower temperature, 1200°C (2192°F), than for a normal turbine, thus eliminating any NOx emissions. The system's fuel, a homogeneous mixture of VAM and supplemental methane, enters via the air inlet to an aspirated turbine. It is then compressed and heated in an exhaust gas recuperator to 450°C (842°F).

The CGT requires a methane/air mixture of 1.6 percent by volume, so most VAM sources, which typically contain only about 0.4 percent methane, would require enrichment. In fact, significant amounts of supplemental fuel would be required to raise the methane concentration to the 1.6 percent level.

Energy Developments Limited (EDL) of Australia is testing the CGT (see Figure 3) on ventilation air at the Appin coal mine in New South Wales, Australia. EDL is using a modified solar gas turbine model 3000 R (rated at 2.7 MW) for this demonstration (Chapman, 2001–2003).

2.2.2 *Lean-fueled turbine with catalytic combustor*

CSIRO Exploration & Mining of Australia, a government research organization, is developing a catalytic combustion gas turbine (CCGT) that can beneficially use methane in coal mine ventilation air (Su, 2001–2003). The CCGT technology being developed oxidizes VAM in conjunction with a catalyst. The turbine draws a very lean fuel/air mixture into the air intake, compresses it, and combusts it in a catalytic combustor. CSIRO's non-conventional turbine will not use combustion air for internal cooling, thus allowing the air intake to contain fuel. CSIRO hopes to operate the system on a 1.0 percent methane mixture to minimize supplemental fuel requirements. The catalyst allows the methane to ignite at a lower, more-easily achieved temperature. CSIRO continues with its research and plans for full commercialization of the VAM CCGT.

Figure 3. EDL carbureted gas turbine installation.

2.2.3 *Lean-fueled catalytic microturbine*

Two US companies, FlexEnergy and Capstone Turbine Corporation, are jointly developing a line of microturbines (starting at 30 kW) that will operate on a methane in-air mixture of 1.3 percent. FlexEnergy, using funding from US Department of Energy/National Renewable Energy Laboratory and the California Energy Commission, has successfully tested a 30 kW prototype unit in 2003 on a simulated VAM with a concentration at 1.4 percent, just above the targeted 1.3 per cent. Each unit's components fit inside a compact container that requires no field assembly. The single moving part, rotating on an air bearing, is a shaft on which is mounted the compressor and the turbine expander. Other components include: a recuperator that preheats the VAM mixture, a catalytic combustion chamber allowing low-temperature ignition, a generator, and a generator cooling section. To better serve the VAM market, FlexEnergy is investigating designs that will reduce the required VAM concentration to below 1.0 percent because large amounts of supplemental fuel would be required to raise the methane concentration to 1.3 percent. The company is also working to increase unit sizes to over 200 kW so that fewer units will be needed at each installation (Prabhu, 2001–2003).

2.2.4 *Lean-fueled recuperated microturbine*

Ingersoll-Rand (IR), a large US company, has developed a lean-fueled version of its PowerWorks Microturbine. The current prototype is rated at 70 kW, and a 250 kW version is under design. VAM is drawn into the PowerWorks microturbine and compressed. It then passes through a patented recuperator that captures heat from escaping exhaust gasses and preheats the incoming VAM. This step significantly boosts the unit's overall efficiency. The preheated, compressed VAM then enters the combustion chamber where it is ignited, producing hot, rapidly expanding gasses. These gasses flow through the turbine's first expansion section, which drives the compressor, and then through a second turbine called the "free power" turbine. Exiting gasses then pass through the hot side of the recuperator and exit the unit. IR has successfully tested its lean-fuel microturbine on a 1.0 methane concentration and the company reports that they can run on 0.86 percent methane. IR plans a field test of the 70 kW prototype as soon as they can find a mine willing to offer a test site. The 250 kW version will be ready in early 2005. The next step will be to greatly increase unit size to well over 1 MW, and that would require the cooperation of a manufacturer of conventional gas turbines to allow IR to replace the combustion section with its own design (Reinks, 2002–2003). The IR design offers the advantage of being able to operate with a smaller amount of supplemental fuel.

2.3 *Hybrid coal- and VAM-fueled gas turbine*

CSIRO and Liquatech are developing an innovative system in Australia to oxidize and generate electricity with VAM in combination with waste coal. CSIRO has constructed a 1.2 MW pilot plant that cofires waste coal and VAM in a rotary kiln, captures the heat in a high-temperature air-to-air heat exchanger, and uses the clean, hot air to power a gas turbine (see Figure 4), (Wendt et al., 2001).

Depending on site needs and economic conditions, VAM can provide from about 15 to over 80 percent (assuming a VAM mixture of 1.0 percent) of the system's fuel needs, while waste coal provides the remainder. Waste coal and ventilation air enter the rotating kiln in the same direction. The coal's heat of combustion ignites the VAM, and a large percentage of that heat is transferred to an air-to-air heat exchanger that operates at about 900°C (1652°F). Ambient air, pressurized by the gas turbine's (Allison C-18) compressor, flows through the heat exchanger's secondary loop, heats to 900°C (1652°F), and expands through the turbine's power section. Part of the compressor's output is directed to the turbine cooling path. This system is especially well suited for mines, such as those in Australia, that generate a significant percentage of high-ash content waste coal and that can market the lightweight, expanded aggregate that results from the process.

Figure 4. CSIRO-Liquatech 1.2 MW hybrid coal and gas turbine.

2.4 *Concentrators*

VOC concentrators offer another possible option for application to VAM. During the past 10 years the use of such units to raise the concentration of VOCs in industrial-process air exhaust streams that are sent to VOC oxidizers has increased. Smaller oxidizer units are now used to treat these exhaust streams which in turn has reduced capital and operating costs for the oxidizer systems. Ventilation air typically contains about 0.4 percent methane by volume. Conceivably, a concentrator might be capable of increasing the methane concentration in ventilation airflows to about 20 percent. The highly reduced gas volume with a higher concentration of methane might serve beneficially as a fuel in a gas turbine, reciprocating engine, etc. Concentrators also might prove effective in raising the methane concentration of very dilute VAM flows to levels that will support economically viable oxidation or lean-fuel gas turbine projects.

Of the various styles of concentrators employed in industrial applications, fluid bed concentrators are expected to offer the greatest chance for VAM concentration. The fluid bed concentrator consists of a series of perforated plates or trays supporting an adsorbent medium (e.g., activated carbon beads). The process exhaust stream enters from the bottom, passing upward through the adsorption trays, fluidizing the adsorbent medium to enhance capture of organic compounds. The adsorbent medium, which is now heavier because of the adsorbed organic material, falls to the bottom of the adsorber section and is fed to the desorber. The desorber increases the temperature of the medium, causing it to release the concentrated organic material into a low-volume, inert gas stream. In this continuous operation, the regenerated medium is fed back to the adsorber vessel for reuse.

Environmental C & C, Inc. (Clifton Park, New York) manufactures a fluid bed concentrator. With USEPA assistance, Environmental C & C tested that system's efficacy on simulated VAM using a series of methane-in-air mixtures. While technically successful, the tests indicated that a more efficient, faster reacting adsorbent medium will be needed to permit VAM concentration to be an economically viable system component. The company will continue to search for an adsorbent that can offer an affordable concentration element to VAM mitigation plants (Cowles, 2002).

2.5 *VAM used as an ancillary fuel*

While the primary focus of this paper is on strategies that can oxidize major fractions of global VAM emissions, a brief mention of technologies that use VAM only as an ancillary or supplemental fuel is in order. Such technologies rely on a primary fuel other than VAM and are able to accept VAM as all or part of their combustion air to replace a small fraction of the primary fuel. The largest example of ancillary VAM use occurred at the Appin Colliery in Australia, where 54 one-megawatt Caterpillar engines used mine ventilation air containing VAM as combustion air. Although the VAM portion of the project currently is inactive, the success of the project in the field proved IC engine's capability in handling the VAM stream for this specific use.

Figure 5. Appin project, New South Wales, Australia.

Similarly, the Australian utility, Powercoal, plans to install a system that uses VAM as combustion air for a large coal-fired steam power plant. In addition, the US Department of Energy funded a research project to use VAM in concentrations up to 0.5 percent as combustion air in a gas turbine manufactured by Solar. Even the CSIRO hybrid coal and VAM project described above falls in the category of ancillary VAM use when waste coal combustion is maximized and VAM use is limited to stoichiometric levels of combustion air.

3 TECHNOLOGY SUMMARY

The Table 1 below summarizes the descriptions and status of development of the systems that USEPA is tracking.

4 ENERGY MARKETS

Each of the technologies described above (with the exception of concentration and ancillary use) produces a heated working fluid that rotates a turbine that turns an electric generator. It is feasible for the oxidizers to simply generate a form of thermal energy to serve markets near the host mine. It may also be possible for the gas turbine systems to remove the power turbine and end up with only a compressed heated gas that could serve a thermal energy market.

Thermal markets might include mine ventilation air heating, coal drying, space heating, process heating,

Table 1. Technology used by USEPA.

Vendor/System	Description	Country	Development status
MEGTEC/VOCSIDIZER®	Thermal flow-reversal reactor (oxidizer)	US and Sweden	Completed small field trial; produced steam
CANMET – Lefebvre Freres Ltd./CH4MIN	Catalytic flow-reversal reactor (oxidizer)	Canada	Completed laboratory trials; no heat recovery
EDL/Carbureted gas turbine (CGT); Isentropic Systems Ltd	Lean-fueled (1.6%) Solar gas turbine with patented combustor	Australia	Undergoing full-scale trials with simulated VAM; results delayed
CSIRO/Lean-fueled turbine with catalytic combustor	Lean-fueled (1.0%) gas turbine with catalytic combustor	Australia	Research phase scheduled to be complete in 1 to 2 years
FlexEnergy/Lean-fueled catalytic microturbine	Lean-fueled Capstone microturbine (1.3%)	US	Field-tested a 30 kW unit on simulated VAM
Ingersoll-Rand/Lean-fueled recuperated microturbine	Lean-fueled (1.0%) IR power works microturbine	US	Factory-tested a 70 kW unit on simulated VAM
CSIRO/Hybrid coal and VAM fueled gas turbine	Waste coal and VAM cofired in rotary kiln; compressed air heated in heat exchanger powers a gas turbine	Australia	Factory-tested a 1.2 MW unit on waste coal and simulated VAM
Environmental C&C/ VOC concentrator	Fluid bed adsorbent concentrator	US	Bench-scale tests did not produce satisfactory results
EDL/Ancillary VAM use	VAM used as combustion air in Caterpillar 1 MW engines	Australia	Successfully operated at commercial scale

etc. USEPA, however, has decided to concentrate on projects that deliver electric power to either the mine's electric demand or to the grid (Schultz et al., 2003). There are several practical reasons for this:

- Electric power markets are everywhere, while suitable thermal markets are often remote from the host mine.
- Thermal markets are often seasonal, single shift, or otherwise non-continuous.
- Electric power usually carries a higher price per unit of energy.
- Electric power markets receive power continuously, and that is very important because a VAM facility must receive fuel continuously; the plant will be designated as "must run".

If a high-value, continuous thermal market were located near to the ventilation shaft, however, the developer should certainly evaluate serving such a market. But electric power will usually be the energy commodity of choice.

The production of electric power has the further advantage of being able to produce a constant supply of lower grade, but often useful, thermal energy by the use of cogeneration. For example, each of the gas turbines described above produce a waste heat that is in the temperature range of many needs.

Another market that may be available is the market for carbon emission reductions (see discussion below). By merely oxidizing VAM in a thermal oxidizer that has no embedded heat exchanger the project will produce emission reductions that may be sold to buyers who need offsets.

5 CARBON EMISSION REDUCTION OFFSETS

As mentioned earlier, methane has a global warming potential 21 times that of carbon dioxide. Thus, methane emission mitigation offers the possibility of generating substantial streams of carbon emission reductions (in terms of carbon dioxide equivalent, or CO_2e). Those emission reductions in turn can offer substantial monetary value through the public and private markets that exist in the U.S., and around the world. Generally, project developers realize the economic benefits either by trading the actual emission reductions through a commodity trading program, receiving tax credits or similar incentives, or receiving equity funds to invest in recovery projects in exchange for the title to the emission reductions.

According to a report from the World Bank's Prototype Carbon Fund (Lecocq and Capoor, 2002), trading in greenhouse gas emissions in 2002 more than doubled over that in 2001, rising from 29 million tonnes CO_2 equivalent to 71 million tonnes. The prices paid range from US$4.00–$6.00 per tonne. In the U.S. prices in the private markets have tended to be much lower.

Although reported individual emissions transactions have ranged from $<10{,}000$ to $>10{,}000{,}000$ tonnes, typical project-based volumes are in the 1,000,000 tonne range with unit prices varying considerably depending on a number of factors. Unit sale prices have been reported to range from roughly US$1.50 to as high as US$18.00 per tonne CO_2e. Thus, even at the lower end of the unit price range, the internal rate of return of projects mitigating substantial quantities of methane can be significantly enhanced through the sale of the carbon credits that they generate.

6 COST ANALYSIS

Some of the vendors listed above have supplied USEPA with preliminary estimates of performance and cost (both capital and operating) of their systems. The authors anticipate that more information will be available as soon as field demonstrations at commercial scale have been completed for each system. As expected, the costs for installations that oxidize VAM, recover heat, and generate electricity are higher than for conventional systems because of the very nature of VAM – dilute and moving in huge volumes. USEPA compiled available early projections of performance and costs and found the following:

- Projected cost estimates for the TFRR, CFRR, and one lean-fueled turbine fell in a reasonably narrow range.
- Lean-fuel turbines exhibit better efficiency in converting VAM to electric power.
- Thus, for a power facility of comparable size lean-fuel turbines will oxidize less VAM and earn fewer carbon credits than oxidizer systems.
- The parameter that most significantly affected performance and, consequently, project profitability, is VAM concentration; the higher the better, up to about 1.0 percent.
- Several mine sites that exhibit favorable physical and economic conditions would produce attractive power projects that yield a before-tax internal rate of return of 25 percent or better.
- Many installations with blended VAM concentrations at or near 1.0 percent could be economically viable where the project receives about $0.05 per kWh, or receives some favorable combination of electric revenue and carbon credit payments.

Using cost assumptions that combine two oxidizer system estimates, USEPA constructed a simple cost model for a project that can supply VAM supplemented, if necessary, to reach a consistent concentration of 1.0 percent. The model contains conservative economic and financial assumptions, and it assumes that the project owner would require a 25 percent before-tax IRR. With no carbon credit payments the project would have to sell its power for about $0.05 per kWh, but with a carbon credit sale price of about $2.28 per tonne of CO_2e the project could achieve its target IRR with a power price of only $0.035 per kWh. Naturally these results would change with significant variations of the many parameters upon which the project depends, but the example shows that even with current power and carbon emission reduction pricing many VAM projects can be economically viable.

REFERENCES

Carothers, P. and Deo, M. 2000a. *Mitigation of Methane Emissions from Coal Mine Ventilation Air*. In *Proc. 8th US Mine Ventilation Symposium*: 73–80. Rolla, MO.

Carothers, P. and Deo, M. 2000b. *Technical and Economic Assessment: Mitigation of Methane Emissions from Coal Mine Ventilation Air.* USEPA. EPA-430-R-001.

Chapman, T., pers. comm. 2001–2003.

Cowles, H., unpubl. 2002. *Coalbed Methane Adsorption Study.* USEPA.

Kallstrand, A. and Zak, K., MEGTEC Systems. pers. comm. 2000–2003.

Lecocq, F. and Capoor, K. 2002. State and Trends of the Carbon Market, prepared for PCF*Plus* Research.

Prabhu, E., pers. comm. 2001–2003.

Reinks, P., pers. comm. 2002–2003.

Sapoundjiev, H. et al., pers. comm. 2000–2003.

Schultz, L. et al., 2003. *Assessment of the Worldwide Market Potential for Oxidizing Coal Mine Ventilation Air Methane.* USEPA. EPA-430-R-03-002.

Su, S., pers. comm. 2001–2003.

Wendt, M. et al., 2001. *Hybrid Coal and Gas Turbine System Development for Mitigation of Greenhouse Gas at Coal Mines.* Coal Conference at Pittsburgh Energy Technology Center. Pittsburgh, PA.

Face ventilation/case studies

Mine Ventilation, Ganguli & Bandopadhyay (eds)
© 2004 Taylor & Francis Group, London, ISBN 90 5809 633 5

How scrubbers help ventilate the face during deep cut mining with a blowing curtain

A.M. Wala
Department of Mining Engineering, University of Kentucky, Lexington, KY, USA

J.D. Jacob, G. Huang, J. Brown & L. Rangubhotla
Department of Mechanical Engineering, University of Kentucky, Lexington, KY, USA

ABSTRACT: In the Department of Mining Engineering at the University of Kentucky a comprehensive validation study of current Computational Fluid Dynamics (CFD) tools against mining-related benchmark experiments has been conducted. A wide array of ventilation schemes and mining configurations has been considered. In this paper, the authors will present, compare, and discuss the experimental and numerical data collected during the part of the study dedicated to the effect of machine-mounted dust scrubber on the performance of a face ventilation system for deep cut mining. These scrubbers, whose primary purpose is dust collection, have been shown to be very useful ventilation devices in directing air into the immediate face area where ventilation is poor during the box cut mining sequence. Based on studies carried out by the former U.S. Bureau of Mines, NIOSH, MSHA and industry, the recommended ratio between the scrubber airflow $\mathbf{Q_s}$ and the ventilation airflow $\mathbf{Q_{in}}$, should be equal to one. The authors will demonstrate and discuss the flow behaviors in the face area for the different values of the $\mathbf{Q_s/Q_{in}}$ ratios.

1 BACKGROUND

During the past twenty years, since the advent of remote controlled continuous miners and deep cut (extended advance) for room and pillar mining operations (Wellman, 1981), the safety and productivity of mining operations has increased significantly. However, this high production presents environmental problems in that, per unit time, more dust is generated and higher levels of methane are liberated at the face during coal extraction. These conditions require the supply of more air into the face of the excavated entry for safe dilution. One of the ventilation techniques being used in conjunction with this modern mining method (Goodman et al., 1990) is the blowing curtain in conjunction with a machine-mounted dust scrubber. Implementation of this technique has a significant influence on the airflow patterns in the entry working area and the design of their air handling capacities.

The former U.S. Bureau of Mines (currently NIOSH), Mine Safety and Health Administration, and the coal industry have conducted significant studies to evaluate the performance of face ventilation systems through the use of full scale tests (underground/surface) or scaled physical modeling. These studies led to recommendations that make these systems more effective. However, because of the aerodynamic complexity in the face area due to the variety of ventilation arrangements and the limitations of the experimental methods, some doubts still exist when evaluating the effectiveness of such a system.

The design of a balanced ventilation system for the above-described scenarios requires considering the system in three rather than two dimensions. With Computational Fluid Dynamics (CFD) many design alternatives can be explored in a relatively short period, and detailed three-dimensional results can be viewed graphically. This graphical type of information concerning airflow, gas, and dust distribution established by CFD are readily rationalized, and could lead to confidence in the final design of a mine face ventilation system. However, to develop this confidence, extensive validation studies are required involving comparisons between physical and numerical (CFD) modeling (Hargreaves and Lownsdes, 2001).

2 INTRODUCTION

During the last three years a validation study, funded by NIOSH Grant # R01 CCR415822, has been carried out in the Department of Mining Engineering at

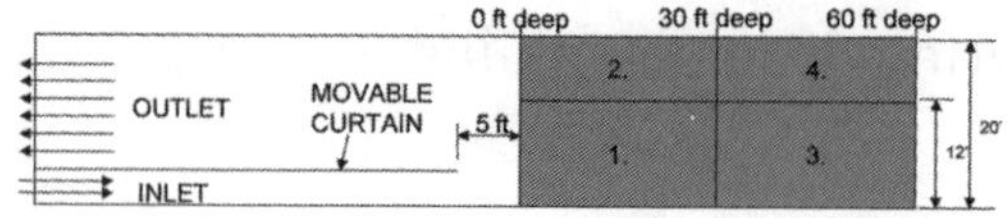

Figure 1. Test section geometry and the scenarios examined.

the University of Kentucky. The goal of this study was to prove a CFD code's ability to predict, evaluate, and design effective face ventilation systems. To achieve this goal the authors performed the following tasks:

- Designed and built a scaled physical model of selected face ventilation systems (see Wala et al., 2003),
- Performed a series of measurements using Particle Image Velocimetry (PIV) (Turner et al., 2002), to determine the velocity fields in the model,
- Developed three-dimensional numerical models of the tested systems and performed computer simulations using numerous CFD codes, and
- Compared experimental and numerical data for the purpose of CFD code validation.

One of the scenarios being tested for validation purposes was to ventilate equipment-free entry for cutting depth of approximately 18.3 m (60 ft). According to the mining plane this 18.3 m (60 ft) deep cut should be completed in a four 9.2 m (30 ft) cuts (passes), using blowing curtain ventilation system, see Figure 1. The results from the simulation study, using a scaled physical model, are shown in Figures 2, 3, 4 and 5 (Wala et al., 2000, 2002 and 2003). These figures increased an understanding of the flow behavior in cases of mining a deep cut using a blowing curtain for ventilation.

The major characteristic of the flow pattern, specifically for the box cutting sequence, is that 70 to 80% of the air leaving the end of the blowing curtain almost immediately separates from the rib and as a primary flow turns back toward the return. This separation is driven by an adverse pressure gradient which builds up as the width of the entry decreases. The rest of the curtain's airflow, approximately 20 to 30%, called the secondary flow, continues to flow toward the immediate face area in the opposite direction. As can be seen in general, the ventilation air delivered by the blowing curtain does not flow into the immediate face area.

During the validation studies the authors tested over a hundred scenarios for different combinations of such parameters as curtain setback, intake air quantity delivered by the curtain, tight-rib distance (curtain distance from the rib), and mining sequence. The data collected during these studies allows direct conclusions concerning the flow behavior.

The authors presumed that, until that time, the flow patterns for these mining scenarios had not been fully

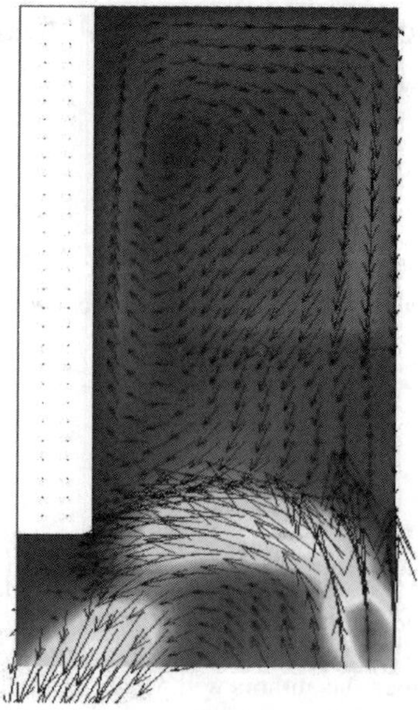

Figure 2. PIV velocity magnitude and vectors for a 30 ft deep cut with a slab.

Figure 3. PIV velocity magnitude and vectors for a 30 ft deep cut without a slab.

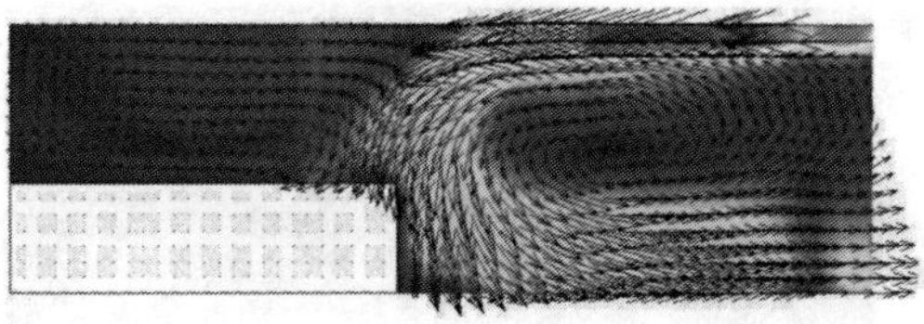

Figure 4. PIV velocity magnitude and vectors for a 60 ft deep cut with a slab.

Figure 5. PIV velocity magnitude and vectors for a 60 ft deep cut without a slab.

understood because of limited opportunity to monitor the flow patterns during full-scale experimental studies performed underground or on the surface. The only sources of information found by authors concerning the similar flow behavior as shown in Figures 2–5 were recognized and described by Luxner (1969) and Taylor et al., (1997).

To ensure that scaled model test results interpret conditions reliably in the underground face area the cooperative validation testing has been performed by the Mining Engineering Department, University of Kentucky and NIOSH, using full-scale testing gallery at the NIOSH Pittsburgh Research Center. These studies proved that a 1:15 scaled laboratory model used for the project is a valid tool and realistically simulated underground conditions for the face ventilation systems. Observation showed the flow patterns for one of the scenario tested using laboratory scaled model and at the full-scale model as very similar, including recirculation.

As mentioned before, since remote controlled continuous miners were introduced, deep cut mining was an opportunity for mine operators to increase productivity. However, the presence of airborne coal dust in the face area was a major obstruction. This coal dust can cause poor visibility around the cutting head and can expose the miners to high levels of respirable dust. A practical means of removing this dust is with a dust-collecting scrubber mounted directly on the mining machine (U.S. Bureau of Mines, 1983). When the scrubbers were added to the face ventilation system of coal mines, the existing arrangements of the face ventilation systems were altered to accommodate these scrubbers.

Many studies concerning face ventilation with scrubbers were carried out in the United States in the last twenty years. The authors would like to mention some of them, which have contributed greatly to this topic; Gillies (1982), Clouse (1982), Volkwein and Thimons (1985), Jayaraman et al., (1992), Taylor et al., (1997), Thimons et al., (1999), Kissell (2003). Most of the researchers, followed the Luxner comments concerning higher effectiveness of the blowing system versus the exhaust system, recommended the blowing curtain as the best suited ventilation system for scrubber applications, particularly because the free jet of high initial momentum leaving the end of the line curtain should penetrate into the immediate face area sweep the face and flow into the return.

This is true, according to Figures 2–5, as long as the entry width is more than 16 foot, as was discovered during this study (Wala et al., 2003). This means that only during the slab cutting scenario does the advantage of the blowing curtain prevail. For the rest of the mining scenarios the ventilation air does not like to flow to the immediate face area where it is mostly needed in the gassy mines for methane dilution.

The effects of the machine-mounted scrubber system on levels of methane and respirable dust were tested first by Ingersoll Rand Research Inc., (Clouse, 1982). These studies have shown that properly designed machine-mounted scrubber systems can allow curtain setbacks of 12.2 m (40 ft), while still controlling respirable dust and methane levels. At the same time, the machine-mounted scrubber, whose primary role was to collect respirable dust at the face (dust source location), was found to be also an air-moving device (ventilation device).

Based on the Ingersoll Rand studies, MSHA recommended a ratio between the scrubber flow $\mathbf{Q_s}$ and the intake flow $\mathbf{Q_{in}}$ of close to one. A study performed by Jayaraman, 1992, supports these finding, however, latest studies performed by Taylor et al., 1997 and by Thimons et al., 1999 show that the relations between face ventilation performance and scrubber/ventilation flow ratios are more complicated than was previously thought.

The authors, based on the results from the experimental studies using physical and numerical modeling, will demonstrate and discuss the flow behavior in the face area for the varied $\mathbf{Q_s/Q_{in}}$ ratios.

3 EXPERIMENTAL STUDIES

In order to study the effect of the scrubber on the face ventilation system, a scaled physical model of such a system was constructed in the Ventilation Laboratory at the Mining Engineering Department, University of Kentucky.

3.1 *Description of the physical model*

3.1.1 *Entry model with fans*

The analysis of face airflow patterns was undertaken using 1:15 physically scaled model. This scaled model of a typical underground room and pillar coal mine with an entry of 2.1 m (7 ft) high and 6.1 m (20 ft) wide was designed and built out of transparent Plexiglas. Scaling of the entry flow rate was performed using dimensional analysis based on the Reynolds Number, $Re = UL/\nu$, where U and L are the velocity at the brattice line mouth and width of the inflow (distance between the brattice line and entry rib), respectively. The model consists of three significant parts:

- The working section, which represents a face area, where movable Plexiglas walls can be used to replicate many different ventilation configurations such as box-cut or slab-cut arrangements with brattice setback distances varying from zero to 18.3 m (60 ft).
- The middle section, which represents an ordinary mining entry with a partition (brattice line/curtain), divides the entry into a narrow 0.6 m (2 ft) wide intake and an 5.5 m (18 ft) wide return airways.
- The intake/return section, where the ventilation system can be arranged as either blowing or exhausting by changing the fan location.

The primary airflow through the entry was drawn by a main fan in series with a smaller axial speed controllable auxiliary fan. The maximum available airflow induced through the model corresponds to the 3.0 m^3/s (6,500 CFM) for the underground conditions.

3.1.2 *Scrubber model*

The scrubber system for this modeling study did not exactly simulate the underground scrubber arrangements. Since the scrubber system could not be modeled exactly as it appears in a mine due to limited space, a simplified model of the scrubber was simulated using open ended 2 inch diameter PVC pipe. The airflow through this pipe was drawn by a suction system. The maximum airflow induced by the scrubber system corresponds to 1.4 m^3/s (3,000 CFM) for the underground conditions.

For validation purposes two different arrangements with the scrubber were tested. First, as shown in Figure 6, a cylindrical "scrubber" was simulated by the straight PVC pipe placed in the middle on the entry floor, 1.5 m (5 ft) from the face. Second, as shown in Figure 7, the same PVC pipe was placed on the top of a continuous miner model, corresponding to the Joy Model 14CM Continuous Miner, located in the face area.

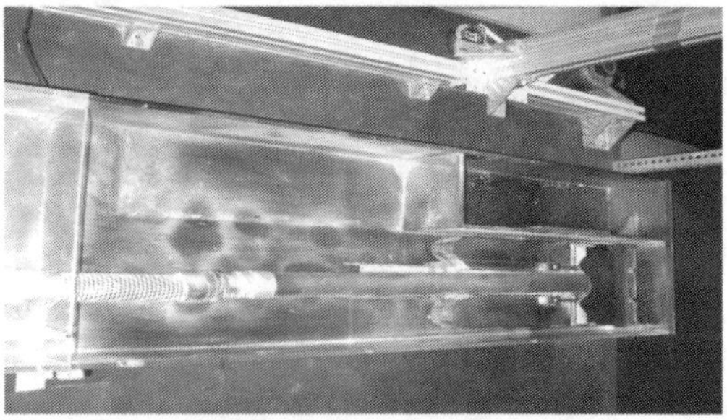

Figure 6. Simplified scrubber model in test section.

Figure 7. Machine mounted scrubber model in test section.

The model was arranged to simulate two critical ventilation scenarios as follows:

- SCENARIO #1
  The end of the box cut during the first mining pass (sequence), with a setback distance being 10.6 m (35 ft) (see Figure 1).
- SCENARIO #2
  The end of the box cut during the third mining pass (sequence), with a setback distance being 18.3 m (60 ft) (see Figure 1).

3.2 *Particle Image Velocimetry (PIV)*

PIV was used to measure instantaneous velocity fields of the time dependent flows in this scaled model. PIV is a non-intrusive, optical technique for the measurement of planar flow velocities. The technique measures the velocity of a fluid indirectly by means of the measurement of the velocity of tracer particles within the flow. A plane of the flow is illuminated twice within a short time interval, and the images of the illuminated particles are digitally recorded on two separate image frames of a CCD camera (Turner et al., 2002). Fourier analysis is used to extract the planar particle velocity information from the image pairs. The experimental set-up of the PIV measurement system consists of four sub-systems: (1) Seeding, (2) Illumination, (3) Recording, and (4) Evaluation. The integrated

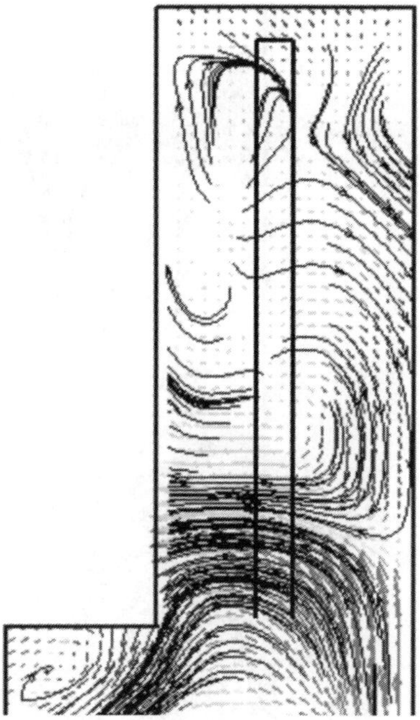

Figure 8a. Streamlines from PIV measurements with the simplified scrubber model for a 35 ft setback with a slab; $Q_s/Q_{in} = 0.14$.

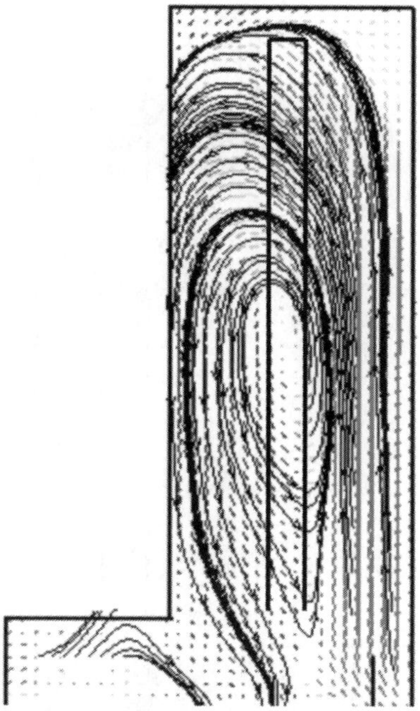

Figure 8b. Streamlines from PIV measurements with the simplified scrubber model for a 35 ft setback with a slab; $Q_s/Q_{in} = 0.34$.

PIV system used in this study is comprised of the following components: (1) laser and light sheet optics, (2) image recording system, (3) synchronizer, (4) particle generator, and (5) image analysis and display, (Turner et al., 2002).

4 RESULTS AND DISCUSSION

4.1 *Experimental data from scale modeling*

Based on the results collected during the flow distribution study in the equipment-free entry, shown in Figures 2-5, it was found that the worst ventilation conditions occur during the box cutting sequences (see Figures 2 and 4). Using the physical model of the face ventilation system, these two critical ventilation scenarios, described above, represent possible mining arrangements for a 18.3 m (60 ft) deep cut in four passes.

The performance of the face ventilation for these two extended cut mining scenarios was a goal for this validation study. The geometry of the entry examined was 2.1 m (7 ft) high and 6.1 m (20 ft) wide. The maximum intake air in volume of 3.0 m^3/s (6,500 CFM) was provided to the face area by means of a blowing curtain.

Figures 10 and 11 show the flow streamlines from PIV measurements at the horizontal middle plane of the model for SCENARIO #1 and SCENARIO #2, with and without the scrubber, respectively. As previously mentioned, the scrubber for this particular case was simulated as a simple pipe located on the floor of the empty entry. The general observation from these two tests are that scrubbers help to bring the air into the immediate face area during the box cut mining sequences, whereas it would not penetrate to the face otherwise. This makes the 18.8 m (60 ft) deep cut mining possible.

We were also interested in determining the relationship between the performance of the face ventilation system and the flow ratio between the intake and scrubber flows. Figures 8a, 8b, and 8c show the flow streamlines at the horizontal middle plane of the entry for three different flow ratios $\mathbf{Q_s/Q_{in}}$ of 0.14, 0.34, and 0.53, respectively. From a methane removal point of view for this equipment empty entry scenario, the eddies indicate poor ventilation of the face due to excessive recirculation.

Figures 9a, and 9b shows similar flow streamlines as shown in Figures 8b and 8c, except during these tests the continuous miner model was present in the face area. The measurements were taken at the horizontal plane above the miner. By comparing these figures it can be seen that equipment present in the face creates additional obstructions to the flow and decreases the effectiveness of the ventilation system.

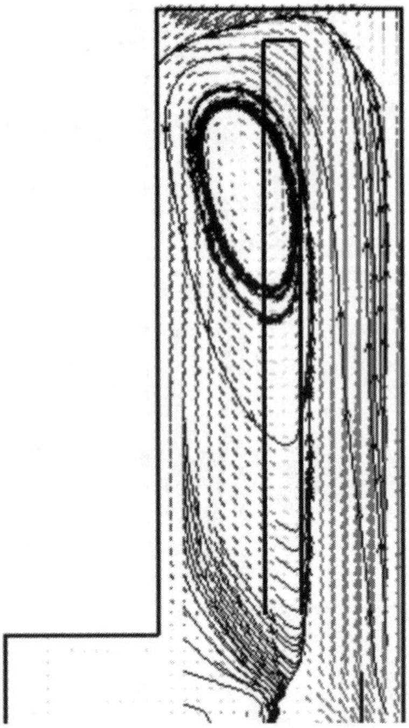

Figure 8c. Streamlines from PIV measurements with the simplified scrubber model for a 35 ft setback with a slab.; $Q_s/Q_{in} = 0.53$.

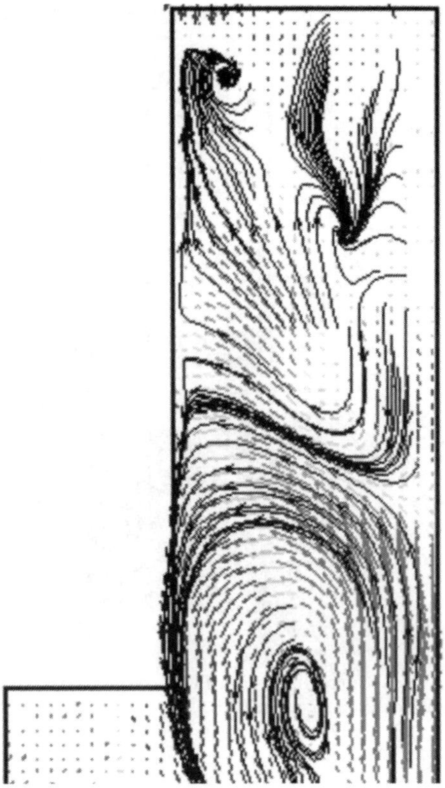

Figure 9a. Streamlines from PIV measurements with the machine mounted scrubber model for a 35 ft setback with a slab; $Q_s/Q_{in} = 0.30$.

4.2 *Numerical modeling*

The numerical simulation for these two scenarios described above was performed using CFX commercial CFD software package. Based on the intensive validation studies performed as a part of the entire project, the CFX software package along with the SST (Shear Stress Transport) turbulence model was selected for its ability to accurately predict separation due to adverse pressure gradients, as opposed to the k-ε model.

PIV results from the experimental studies of SCENARIO #1 with the cylindrical scrubber (Figure 6) and continuous miner with on-board mounted scrubber (Figure 7), served as a baseline for the numerical simulation validation. In these studies, the scrubber mass flow ratio was normalized by the mass flow at the intake flow (behind the curtain). Thus, the results presented in Figures 10 and 11 illustrate predicted flow fields for the cylindrical scrubber at the scrubber flow ratios of 0.2, 0.3, 0.4, 0.7, and 1.0 respectively. The boundary condition for the scrubber is set as a constant velocity outlet which is released outside the original geometry. These boundary conditions were set on the scaled model during the experimentations. The grid density was based on the coarse grids tested for the previous simulation exercises.

4.3 *Discussion*

In comparing the numerical (CFD) results with the experimental (PIV) results, one will see good similarity in the flow behavior in the face area. However, some flow pattern discrepancies can be noticed. The experimental results show the location of the separation point to be more sensitive to smaller values of the scrubber mass flow ratio. This flow behavior could be explained by existence of the unsteady flow along the wall (rib). However, once the scrubber mass flow ratio reaches a level near 1.0 the separation is removed and the flow carries to the immediate face area. This is notable since most scrubbers in actual mining applications are required to have a mass flow ratio equivalent to 1.0. Also, it can be seen that as the flow ratio approaches one the recirculation (eddies) in the immediate face area increased.

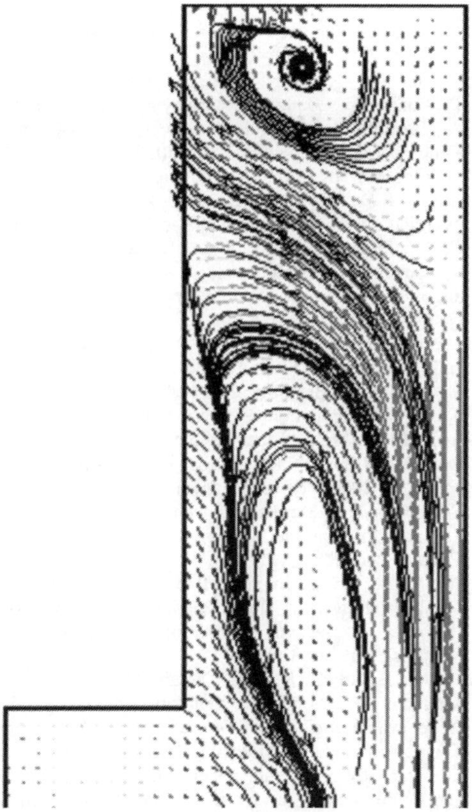

Figure 9b. Streamlines from PIV measurements with the machine mounted scrubber model for a 35 ft setback with a slab; $Q_s/Q_{in} = 0.53$.

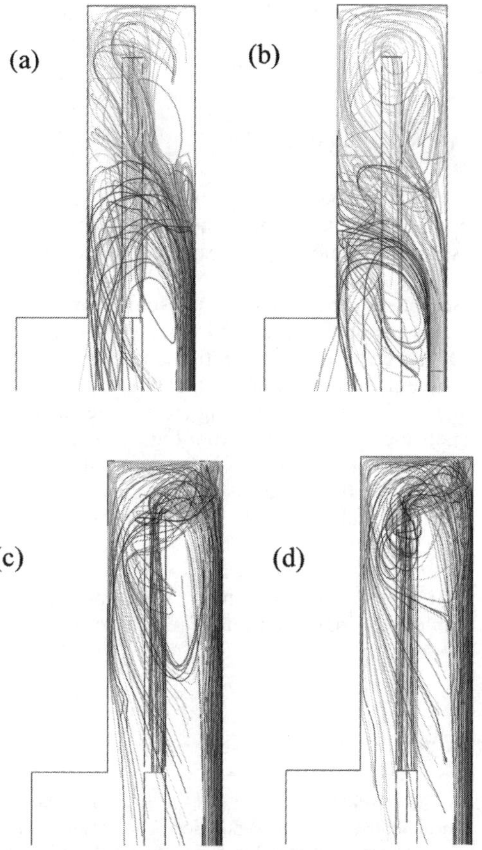

Figure 10. CFD results for a 30 ft deep cut with a simplified scrubber model for a 35 ft setback; (a) $Q_s/Q_{in} = 0.20$; (b) $Q_s/Q_{in} = 0.40$; (c) $Q_s/Q_{in} = 0.70$; (d) $Q_s/Q_{in} = 1.0$.

Figure 11 show simulated 3-D flow streamlines in the face area when a simplified virtual continuous miner with scrubber was inserted into the entry for the 9.1 m (30 ft) box cut scenario. The miner was equipped with a virtual scrubber which is incorporated into the geometry of the mine such that the scrubber's inlets are placed on each side of the miner and the exhaust is placed at its rear. Unlike the previous scrubber, this scrubber was modeled with the "fan" UDF (User Defined Function). This method of modeling allows the user to put a pressure gradient condition on a plane within the scrubber's duct essentially modeling a blower mounted to the miner. The main difference between this method and the previously used method is that the scrubber's exhaust is released within the mine as opposed to outside. This better represents the scenario of a miner with on-board scrubber and would be very difficult for an experimentalist to implement into a scaled model due to flow restrictions on fans small enough to fit these scaled geometries. Figure XX illustrates the resulting

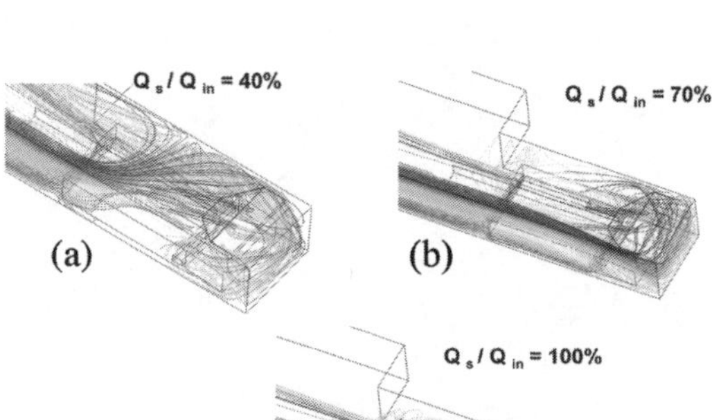

Figure 11. CFD results for a 30 ft deep cut with a machine mounted scrubber model; (a) $Q_s/Q_{in} = 0.40$; (b) $Q_s/Q_{in} = 0.70$; (c) $Q_s/Q_{in} = 1.0$.

flow over the miner without the use of a scrubber. The effect of the scrubber can clearly be seen by comparing Figures 2–5 with Figures 8–11.

5 CONCLUSIONS

- These scrubbers were shown to be very useful tools in moving air into the immediate face area where ventilation is poor, especially during the box cutting sequences.
- Scrubbers introduced as dust collection devices help to bring the air to the immediate face area and should be termed face ventilation devices, as well. Without a dust scrubber, the blowing curtain is ineffective for controlling respirable dust and methane as can be seen from Figures 2, 3, and 4.
- With scrubbers, more efficient ventilation can be obtained in the face areas without the need for miners to extend ventilation controls in by permanent roof supports.
- The flow ratio between intake and scrubber is an extremely important factor, which could be controlled for optimization of the face ventilation system.

REFERENCES

Clouse, B.E., 1982, "Assess Feasibility of Machine Mounted Scrubber System for Ventilation Coal Mine Working Face. Phase 1, Interim Report," Ingersoll Rand Research Co., New Jersey.

Gillies, A.D.S., 1982, "Improvement Coal Face Ventilation Through Use of Dust Scrubber Systems Leads to Greater Production Efficiency." *Proceedings:* First Mine Ventilation Symposium, Tuscaloosa, AL.

Goodman, G.V.R., Taylor, C.D., and Divers, E.D., 1990, "Ventilation Schemes Permit Deep Advance," *Coal*, October.

Goodman, G.V.R., Taylor, C.D., Colinet, J.F., and Thimons, E.D., 2000, "Research by NIOSH for Controlling Respirable Dust and Methane Gas on Continuous Mine Faces," *Proceedings*: Seventh International Mine Ventilation Congress, Krakow, Poland.

Thimons, E.D., Taylor, C.D., and Zimmer, A., 1999, "Ventilating the Box Cut of a Two-Pass 40-ft Extended Cut," *Proceedings*: Eight US Mine Ventilation Symposium, Kingston, Canada.

"Ventilation Schemes Permit Deep Advance," *Coal*, October, 2XXX.

Hargreaves, D.M. and Lowndes, I.S., 2001, "Measurement and Modeling Air Flow Around a Continuous Miner in a Development Heading," *Proceedings:* Seventh International Mine Ventilation Congress, Krakow, Poland.

Luxner, J.V., 1969, "Face Ventilation in Underground Bituminous Coal Mines." Bureau of Mines RI 7223.

Taylor, C.D., 2000, "Effect of Water Sprays and Scrubbers Exhaust on Face Methane Concentrations." *Proceedings*: Seventh International Mining Ventilation Congress, Crakow, Poland.

Taylor, C.D., 1997, "Impact of Unbalanced Intake and Scrubber Flows on Face Methane Concentrations," *Proceedings*: Seventh International Mine Ventilation Congress, Pittsburgh, PA.

Wala, A.M., Stoltz, J.R., and Hassan, M.I., 2000, "Scale Modeling of a Mine Face Ventilation System for CFD Code Validation" *Proceedings*: Third International Symposium on Scale Modeling, Nagoya, Japan.

Wala, A.M., Stoltz, J.R., and Jacob, J.D., 2001, "Numerical and Experimental Study of a Mine Face Ventilation System for CFD code Validation," *Proceedings:* Seventh International Mine Ventilation Congress, Krakow, Poland.

Wala, A.M., Jacob, J., Brown, J., and Huang, G., 2003. "New Approaches to Mine-Face Ventilation," *Mining Engineering*, No. 3, March.

Wellman, T.S., 1981, "Deep Cut Remote Control Mining." *Mining Congress Journal*. October.

Mine Ventilation, Ganguli & Bandopadhyay (eds)
© 2004 Taylor & Francis Group, London, ISBN 90 5809 633 5

Development of an assessment tool to minimize safe after blast re-entry time to improve the mining cycle

A.D.S. Gillies, H.W. Wu & D. Shires
The University Of Queensland, Brisbane, Australia

ABSTRACT: A study has been examining the present problems resulting from inappropriate re-entry times across Australian underground development heading mining practices. It has focused on possible solutions that enable current ventilation practices with reduced quantities of fresh air without affecting re-entry times, or shorter re-entry times without increasing fresh air requirements or a combination of both. After blast re-entry times have been identified as a potential safety problems exaggerated by advanced mining technology and complicated by the modern prevalence of 12 hour shifts. Industry surveys on blasting re-entry time indicate most mines rely on a fixed time strategy to deal with the after blast re-entry mostly based on past experiences and observation. Very few monitor blast fumes for gas composition. A series of measurements of after blast fumes in various development heading arrangements has be undertaken to improve the understanding of occurrence and magnitude of blast contaminants. A development face ventilation study has been undertaken. Rigorous analysis has produced two empirical equations that can be used with some confidence to conservatively estimate re-entry times for level development headings. It was concluded that the working space volume used in mathematical modeling might be a constant for development headings of a similar configuration and environment, regardless of the relative distance from the end of the ventilation ducting to the face. Further testing is required to confirm this hypothesis.

1 INTRODUCTION

Ventilation systems on mines primarily have to ensure adequate environmental conditions for the workforce such as sufficient oxygen, cooling, and adequate dilution of contaminants from equipment and other sources. If rock and minerals are broken with explosives, the ventilation system must also ensure the replacement of the contaminated air in a reasonable time period before re-entry of the workforce can be permitted. An appropriate re-entry time can be determined by the quantity of contaminants released, the size of the ventilation circuit, and the quantity of fresh air used, which in turn is determined by requirements during the working shift.

If a reduction in the re-entry time is required, the only option is to introduce shorter circuits, which will then have to be ventilated in parallel. This increases total air requirements and costs, and is particularly detrimental for deep level mines where the heat generated by autocompression is already a burden on the mine cooling system. To overcome this, the use of series ventilation systems or recirculation systems has often been adopted as it can reduce fresh air requirements while maintaining stope face conditions. However, since rate of removal of blast contaminants is a function of the quantity of fresh air available, the re-entry time will increase. Therefore, the requirements for satisfactory stope face conditions, low operating costs and shorter re-entry times are interlinked in such a way that they cannot be minimized simultaneously.

The extended re-entry time for blast fumes has been offset in some mines by only blasting at the end of shift. This in itself has had a productivity benefit in mines working 8 hour shifts or less when contract miners are prepared to work through their meal period. Use of 12 hour work shifts complicates this situation. A study has been undertaken to analyze the present problems resulting from inadequate re-entry times across a variety of Australian underground mining practices. It has examined possible solutions that would enable current ventilation practices with reduced quantities of fresh air without affecting re-entry times, or shorter re-entry times without increasing fresh air requirements or a combination of both.

After blast re-entry times has been identified as one of the potential safety and health problems introduced by the advanced mining technology (Fisher,

1995). An examination of Western Australian mine safety and health records in the mid 1990s indicated that at least 20 underground miners in that state suffered incidences of "fuming" resulting in loss of at least one shift over a 12 month period. An industry survey on the blasting re-entry time of 19 mines has been undertaken (Gallo, 1995) to analyze the effects of current ventilation practices such as the use of series ventilation systems on after blast re-entry time. The mines surveyed were involved with the extraction of gold, silver and base metals with annual production rates between 180,000 and 3,000,000 metric tonnes. The results show that majority of mine surveyed (75 percent) have more than four blasts per day. This indicates that the production time loss due to an after blast re-entry delay can be significant. Most of the mines (78 percent) rely on a set time strategy to deal with after blast re-entry. This strategy is mostly based on past experiences and observations (75 percent). Of the mines surveyed 83 percent did not monitor the blast fumes. More than 40 percent of mines surveyed have indicated occurrence of incidents where safety has been compromised and more than half of the mines surveyed have agreed that further study into the after blast re-entry time would be beneficial to safety and health of miners.

Modern gas measurement instrumentation with data storage allows research to be conducted to establish safe and productive blasting post detonation re-entry times. This is an area where there is a paucity of information despite the importance to health, safety and production in metalliferous mines. Many incidents and accidents occur each year due to over exposure of personnel to blast fumes and there are significant losses in production due to poor procedures for determining appropriate re-entry times after blasting (DIRWA, 2003).

Selection of re-entry times impacts on mine productivity. Over estimation of the time period reduces the time mine labor is engaged productively; underestimation will very likely lead to safety and health effects.

A development face ventilation study has been undertaken. Results from the study can offer better work productivity and a reduction of after blast incidence while minimising mine ventilation costs. Possible solutions that would enable current ventilation practices with reduced quantities of fresh air without affecting re-entry times or shorter re-entry times without increasing fresh air requirements or a combination of both were also examined. This could lead to a significant saving in mine ventilation costs.

A study has been undertaken to examine approaches to determine suitable re-entry times to improve safety and production at a 2.2 million metric tonnes per year Australian base metal mine.

2 TESTING PROCEDURES

Site tests were undertaken to monitor and record the blast fumes generated from level development headings in the test mine. Detailed testing procedures were formulated. In summary, the procedures adopted consisted of:

- Having the instrument manufacturers calibrate portable gas instruments. Gas measurement was undertaken with Draeger Multiwarn and Miniwarn instruments. These instruments can measure gas readings for a number of gases and record data for a significant period.
- Preparation and testing of instrumentation prior to traveling to site.
- Calibration of portable instrumentation with each other and with the mine gas monitoring system.
- Measurement of gas concentrations over time from development heading blasts using portable gas instrumentation.
- Measurement and recording of detailed information pertaining to each blast including airway dimensions, airflows and ventilation configurations, blast agent weights and configurations, temperatures and times.
- Downloading of data from portable instruments and the mine gas monitoring system prior to leaving site for later analysis.

3 EXPOSURE STANDARDS

Threshold limit value (TLV), time weighted average (TWA) and the short term exposure limit (STEL) are terms commonly used to describe the levels of exposure to various substances. Within this study focus is on the blast fume gases carbon monoxide (CO), nitric oxide (NO) and nitrogen dioxide (NO_2). NOHSC is a statutory body recognized in Australian that releases up to date and best practice information for the Australian industry in this area. Their current guidelines are to use a TWA of 30 ppm for CO, 25 ppm for NO and 3 ppm for NO_2.

The STEL may have some relevance for personnel exposed to blast fumes. NOHSC (2003) state that the STEL is a 15 minute exposure and should not be exceeded. The STEL should not be exceeded at anytime during a working day even if the TWA is within the exposure standard, exposures at the STEL should not be repeated more than four times per day and there should be at least 60 minutes between successive exposures at the STEL

If for example a supervisor wishes to re-enter a development heading soon after blasting to examine the face, he may do so provided that the contaminant levels are below the STEL, the duration of the visit does not exceed 15 minutes and he does not repeat this process more than four times in one day. The same could be said for a mine worker who wishes to merely position a water jet or hose to washdown the face and muckpile soon after the blast. The danger, however, is that rarely do supervisors, let alone mine workers, have access to portable gas instruments to spot measure gas readings, or rarely do robust procedures exist to prevent personnel from exposure longer than 15 minutes. There may be conditions or situations in a mine where the STEL may be used under strict guidelines for re-entry after blasting, but this should only be limited to mine supervisors and not for the general mining workforce because of the obvious dangers.

The Australian mining industry has widely moved to the adoption of 12 hour shifts with most mines adopting rosters exceeding five days straight which has implication to applicable exposure standards. The TWA exposure standard is based on 8 hour shifts and a five day working week Great care must be taken when interpreting TWA values for shifts where workers are exposed to contaminants for longer periods. To prevent over exposure the NOHSC (2003) recommend the TWA should be calculated as follows (Eq. 1)

$$TWA_{12hr} = \frac{8(24-h)TWA_{8hr}}{16h} \quad (1)$$

where h is the length of exposure, hours per day.

For a 12 hour shift the $TWA_{12\,hr}$ is half that for an 8 hour shift giving exposure limits of 15 ppm, 12.5 ppm and 1.5 ppm for CO, NO and NO_2 respectively.

In a modern Australian metalliferous mine it is unlikely that an average worker would be subject to the $TWA_{8\,hr}$ (30 ppm) for a 12 hour shift, let alone the de-rated and conservative $TWA_{12\,hr}$ of 15 ppm. This is based on the significant time in one shift an average worker may spend away from the working face in a clean environment with involvement in shift changeover, smoko/meal breaks, traveling in and out of the mine and other activities spent away from the potentially contaminated working face. This hypothesis may be simulated with the following fictitious 12 hour shift case study for a mine where blasting is carried out once per shift:

- Total of 1 hour travel time in and out of the mine, including shift changeover, with negligible exposure to CO.
- Total of 1.5 hours smoko and meal breaks, including travel to and from crib facilities, with negligible exposure to CO.
- A background CO concentration of 10 ppm, to account for diesel emissions.
- Re-entry time of 30 minutes for mucking duties into a development heading based on the blast gas decay curves provided, as examples, in Figures 1 and 5. It is noted that the operator of the mucking unit would be initially subject to CO at a concentration of about 230 ppm with this re-entry period.

The results of the simulation are provided in 15 ppm and simple calculations show that the time weighted average contaminant exposure is 9600/720 = 13.3 ppm. This is a figure that is less than the conservative $TLV_{12\,hr}$ of 15 ppm calculated previously.

Table 1 and simple calculations show that the time weighted average contaminant exposure is 9600/720 = 13.3 ppm, which is less than the conservative $TWA_{12\,hr}$ of 15 ppm as calculated previously.

The simulation demonstrates that blast re-entry times could in fact be based on gas concentrations well in excess of either the $TWA_{12\,hr}$ or $TWA_{8\,hr}$, whilst still allowing average workers to remain within the time weighted average for the 12 hour shift. The intent of this hypothesis is not for it to be used as the basis for determining re-entry times in this study, but merely to highlight the fact that the current philosophy of using the $TWA_{12\,hr}$ or the $TWA_{8\,hr}$ as the trigger for re-entry times may be overly conservative. This topic in itself could form the basis of an in-depth study in the future.

Even though the $TWA_{8\,hr}$ of 30 ppm, 25 ppm and 3 ppm for CO, NO and NO_2 may be too conservative,

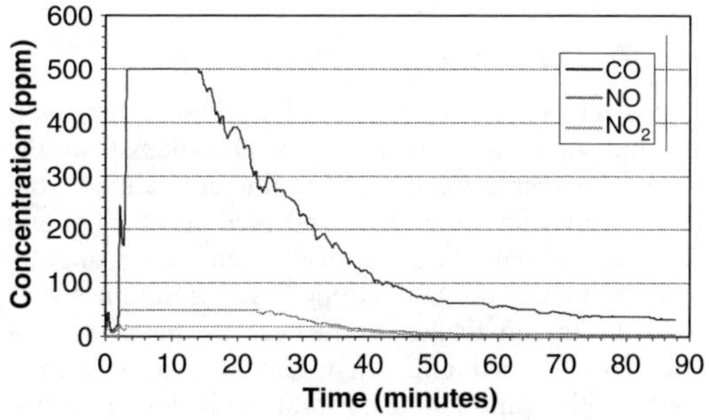

Figure 1. Blast gas decay for a fictitious case study.

Table 1. Average blast exposures for a fictitious case study.

Activity	Duration (mins)	Duration (hours)	Exposure (ppm)	Time*Exp (ppm.mins)
Travel in-out/Shift changeover	60	1	0	0
Smoko/Meals/Travel	90	1.5	0	0
Mucking out, 30–40 mins	10		170	1700
Mucking out, 40–50 mins	10		90	900
Mucking out, 50–60 mins	10		60	600
Mucking out, 60–70 mins	10	1.2	50	500
Mucking out, 70–80 mins	10		40	400
Mucking out, 80–90 mins	10		30	300
Mucking out, 90–100 mins	10		20	200
General duties	500	8.3	10	5000
	720	**12**		**9600**

they shall be adopted as the basis for determining re-entry times in this study.

4 DEVELOPMENT HEADING TESTS

During the one week testing program, 17 tests were conducted on development headings and 16 tests gave results that could be considered reliable. This paper principally examines results from seven tests (those in which there was reliable f ace ventilation delivered through ducting in the heading) and discusses subsequent analysis and interpretation.

4.1 *Typical development heading design*

Heading types varied from drag cut strippings to full face cuts. Nominal dimensions of headings were 5.2 m high × 5.0 m wide. Typically, each face was drilled out with 60, 43 mm diameter holes, with nominally seven lifter holes charged with Johnson's self stemming Easy Charge, nominally seven back holes charged with Johnson's Econotrim and the remainder of the face charged with ANFO. Four 100 mm easers completed the cut. Tunnel incline/decline grades typically varied from 1 in 20 to 1 in 50.

4.2 *Typical auxiliary ventilation system*

Standard ventilation ducting used was 900 mm in diameter and supplied in 20 m lengths. Distances from the end of the ventilation ducting to the face ranged from 15 to 53 m with an average and median of 24 m and 23 m respectively. The end of the ventilation ducting in all cases was neither "clipped" nor tied to assist in throwing the air further to the face for better sweeping. Ventilation ducting maintenance was generally good with minimal tears and satisfactory joins. Airflows ranged from 3.9 to 32 m^3/s with an average and median of 15.4 m^3/s.

4.3 *Instrument positioning*

Placement of the gas monitoring instrument was based on affording protection more so than being the most ideal location for the testing of blast gases. The instrument needed to be protected from flyrock during the blast, water during subsequent muck pile wash down and the mucking units during load out.

The instrument was typically hung from a rockbolt or low piece of steel roof mesh a considerable distance from the face and usually located in a depression in the tunnel wall. Distances from the instrument to the face ranged from 19 m to 49 m with an average of 37 m and median of 34.5 m.

In any case, instrument positioning distance from the face was not seen as critical, and any increased time for blast gases to move across the instrument at times when the additional distance was large and air velocities low was seen as relatively insignificant in the overall scheme of things. In other words, instrument positioning may have affected the calculated re-entry times by up to a minute or so as a worst case.

4.4 *Test results*

A TWA of 30 ppm for carbon monoxide (CO) determines the minimum re-entry times for each and every one of the seven tests as shown in Figure 2. Nitrogen dioxide (NO_2) with a TWA of 3 ppm is the next critical gas, whilst nitric oxide (NO) with a TWA of 25 ppm is the gas of least concern.

The results of seven tests are analyzed. These were selected as there was consistency in the operating procedure that the auxiliary fan ventilating the development heading had been switched on shortly after the blast. A mine layout for one of these tests (Test No. 4) is shown in Figure 3.

The seven tests represent near-perfect gas concentration behavior following a blast in a heading with an auxiliary fan and ducting. The plots exhibit sharp

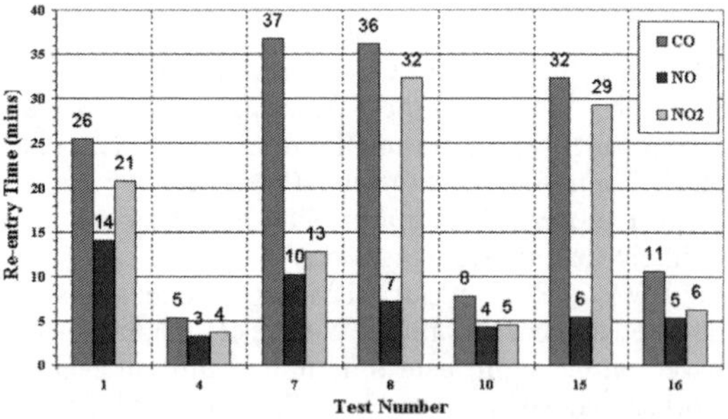

Figure 2. Re-entry times for seven development heading tests.

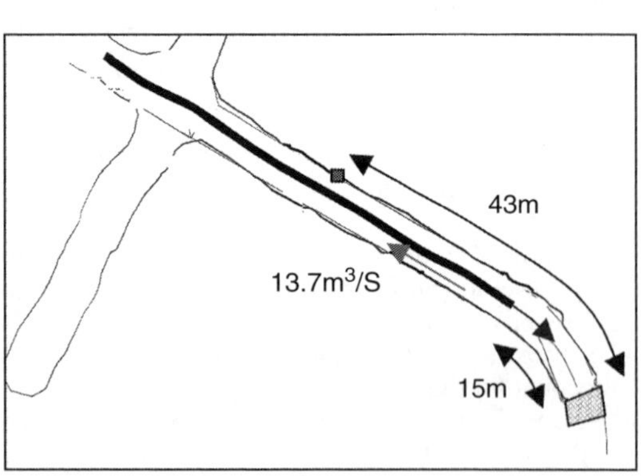

Figure 3. Example of test site layout.

rises almost immediately (~ time zero) or with minimal lag and this indicates minimal fan starting delay. Steep smooth decays represent near perfect mixing of the blast gases within the working space. An example of a plot for Test 4 is shown in Figure 4.

The gas plots reached a ceiling for a short period in many of the tests indicating that concentration was exceeding instrument limits.

4.4.1 *Calculations & results*

From the test blast gas curves, a number of important parameters can be determined (Table 2) such as:

- volume of gas liberated, V_{CO}
- peak gas concentration, x_0
- volume of actual working space, Y'
- theoretical re-entry time, Tt
- actual re-entry time, Ta

The peak carbon monoxide (CO) concentration x_0 in the working space volume can be determined by the intersection of the extrapolation of the rise and decay portions of the curves as shown for the example test in Figure 5 for Test No. 4.

The area under the extrapolated gas concentration curves can be calculated and the average CO concentration x_{AV} over the period of each test T can be determined by Eq. 2. It is noted that the area under the tail of the curve is considered to be relatively small compared to the area under the body of the curve and is therefore insignificant.

$$x_{AV} = \frac{AreaUnderCurve}{T} \quad \text{(ppm)} \qquad (2)$$

The volume of fresh air V_{AIR} supplied to the working space volume over the period of the test T can also be calculated (Eq. 3) by knowing the fresh air flow Q.

$$V_{AIR} = Q * T * 60 \quad (m^3) \qquad (3)$$

Knowing V_{AIR} and x_{AV}, the volume of CO (V_{CO}) produced from the blast is then able to be determined by re-arranging Eq. 4.

$$x_{AV} * 10^{-6} = \frac{V_{CO}}{V_{AIR}} \quad (m^3) \qquad (4)$$

The volume of the working space Y' can then be calculated using V_{CO} and x_0 as follows

$$Y' = \frac{V_{CO}}{x_0} \quad (m^3) \qquad (5)$$

Application of the theory of gas dilution (DeSouza et al, 1993) presents the equation

$$e^{\frac{-(Q_g+Q)(t-t_0)}{Y}} = \frac{(Q_g + B_g Q) - x(Q_g + Q)}{(Q_g + B_g Q) - x_0(Q_g + Q)} \qquad (6)$$

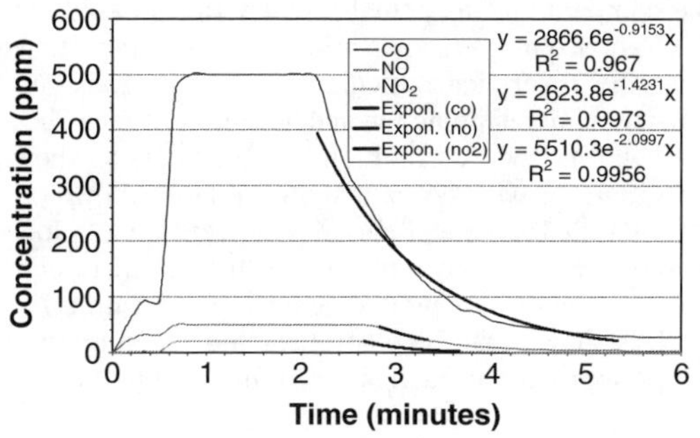

Figure 4. CO, NO and NO_2 readings versus time after blast.

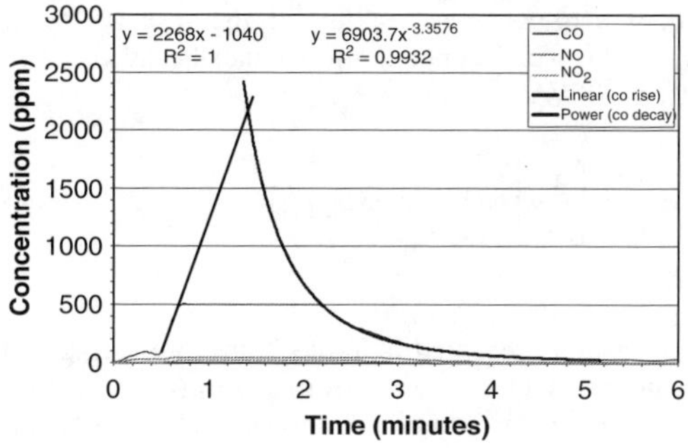

Figure 5. Extrapolated gas concentration curves.

Table 2. Test blast parameters.

Test	Area	T	x_{AV}	Q	V_{AIR}	V_{CO}	x_0	Y'	Tt	Ta
1	6500	26	250	15.4	24024	6.0	2150	2793	12.9	25.5
4	2360	6	393	13.7	4932	1.9	2150	902	4.7	5.3
7	5600	40	140	7.4	17760	2.5	1120	2220	18.1	36.8
8	6700	40	168	15.9	38160	6.4	3000	2131	10.3	36.2
10	3050	8	381	10.5	5040	1.9	2300	835	5.8	7.8
15	5800	34	171	17.2	35088	6.0	2120	2823	11.6	32.3
16	4000	11	364	20.7	13662	5.0	1900	2615	8.7	10.5

where:

Q_g	Gas contaminant inflow quantity	m^3/s
Q	Fresh air quantity	m^3/s
$t\text{-}t_0$	Time (τ)	minutes
Y	Volume of working space	m^3
B_g	Concentration of gas contaminant in intake air	fraction
x	Concentration of gas contaminant in the mixture	fraction
x_0	Concentration of gas contaminant in the initial air space	fraction

By taking natural logs of both sides of Eq. 6 and rearranging, the time τ for dilution in the unsteady state is finite and can be determined as follows (Eq. 7):

$$\tau = \frac{Y}{Q+Q_g}\ln\left[\frac{(Q_g+B_gQ)-(Q+Q_g)x_0}{(Q_g+B_gQ)-(Q+Q_g)x}\right] \quad (7)$$

If it is assumed there is no gas contaminant in the intake air ($B_g = 0$) and the gas contaminant inflow is small Q_g in comparison to the fresh air quantity (Q), Eq. 7 reduces to Eq. 8:

$$\tau = \frac{Y}{Q}\ln\left[\frac{Q_g - Qx_0}{Q_g - Qx}\right] \quad \text{(seconds)} \quad (8)$$

When the limiting gas concentration x is prescribed by mining law, then x = TWA. Assuming the gas contaminant inflow Q_g to be zero, can be reduced to Eq. 9 to give an estimation of the re-entry times, where the final concentration x can be taken as the 30 ppm TWA for CO.

$$\tau = \frac{Y'}{Q}\ln\left[\frac{x_0}{x}\right] \quad \text{(seconds)} \quad (9)$$

4.4.2 *Analysis of results*

The results presented in Table 2 are interesting. The volume of CO gas varies from 1.9 to 6.4 m^3 with an average of 4.2 m^3, but there appears to be two distinct groups; tests 4, 7 and 10 with a volume ranging from 1.9 to 2.5 m^3 and tests 1, 8, 15 and 16 ranging from 5.0 to 6.4 m^3. Obviously, some variation in the amount of fumes generated is expected with varying blast parameters as discussed by Hine et al (1985). However, most development headings at the test mine offer a fairly consistent environment with similar design and configuration; one would expect the quantity of fumes generated to be similar.

Test 10, however, was a wet face and up to half the face was charged with packaged explosives and the remainder with ANFO. This reduction in powder factor may explain the differences in the quantity of fumes generated. It is possible that varying fragmentation from tests 4 and 7 resulted in more of the blast gases becoming entrapped in the muckpile.

Hine et al (1985) suggested that up to 60 percent of the blast fumes may become entrapped in the muckpile and in the fissures of the host rock. The reduced volume of CO produced therefore has the affect of reducing the working space volume Y'.

DeSouza et al (1993) acknowledged that the gas dilution formulae represent theoretical relationships and may not represent the actual (mine environment) conditions such as gas density variations, imperfect mixing or the continuous production of contaminants from the walls or the muckpile.

The peak CO concentration in the test blast ranges from 1120 to 3000 ppm with an average of about 2100 ppm. The unusual peak concentrations of 1120 and 3000 ppm for tests 7 and 8 could simply be attributable to errors in extrapolation of the curves.

It is noted that Y' is used in lieu of Y, because the volume of the working space (Y') does not appears to correlate with the volume of tunnel beyond the end of the ventilation ducting (Y), contrary to Winn's (2002) assumption, as shown in Table 3. Table 3 also includes the distance from face to the end of ventilation ducting (L_D) and the distance from face to the gas instrument points (L_G). The volume of tunnel beyond the end of the ventilation ducting (Y) is calculated based on L_D and the cross-sectional area of the tunnel.

4.4.3 *Re-entry times*

The theoretical re-entry times (Tt) for each test can be compared to the actual re-entry times based on a TWA of 30 ppm, as shown in Table 4. It is demonstrated that the theoretical calculations for the most part grossly underestimate the actual re-entry times Ta.

There is good correlation with this data as shown in Figure 6, with an exponential relationship representing the theoretical curve as expected, and a logarithmic relationship representing the actual re-entry times when Q/Y' is plotted against re-entry time T.

It is shown that the theoretical and actual relationships approximate each other for high values of Q/Y',

Table 3. L_D and L_G and comparison between Y and Y'.

Test No.	L_D (m)	L_G (m)	Y (m^3)	Y' (m^3)	Difference between Y and Y' (%)
1	53	63	2914	2793	−4%
4	15	43	390	902	57%
7	23	34	658	2220	70%
8	32	49	968	2131	55%
10	21	33	579	835	31%
15	21	49	635	2823	78%
16	16	35	416	2615	84%

and that the difference or error between the relationships increases with a decrease in Q/Y'. The relationship between Q/Y' and the actual re-entry time T is shown in Eq. 10.

$$Q/Y' = -0.0046 \ln(Ta) + 0.0215 \quad (\text{secs}^{-1}) \qquad (10)$$

If a conservative approach is taken and the maximum Y' of 2800 m³ is adopted, which should represent the maximum for a typical level development heading, then the graph may be further simplified to give a relationship between Q and T (Figure 7).

The simplified relationship between Q and Ta becomes (Eq. 11)

$$Q = -12.924 \ln(Ta) + 60.119 \quad (\text{secs}^{-1}) \qquad (11)$$

Table 4. Re-entry times.

Test No.	Q/Y' (l/s)	Tt (mins)	Ta (mins)	Difference (%)
1	0.0055	12.9	25.5	51%
4	0.0152	4.7	5.3	88%
7	0.0033	18.1	36.8	48%
8	0.0075	10.3	36.2	28%
10	0.0126	5.8	7.8	74%
15	0.0061	11.6	32.3	36%
16	0.0079	8.7	10.5	83%

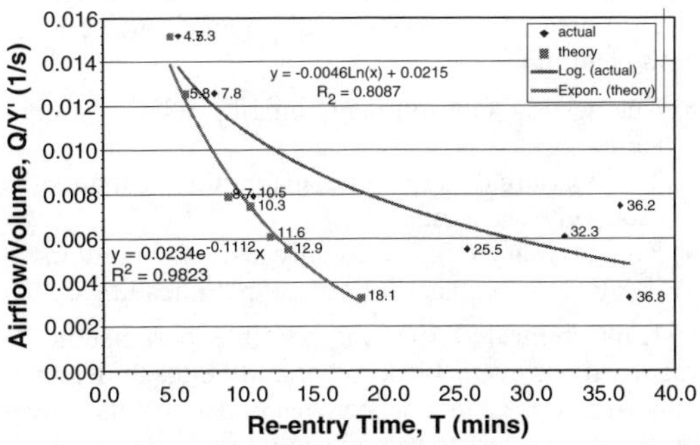

Figure 6. Re-entry time correlation (i).

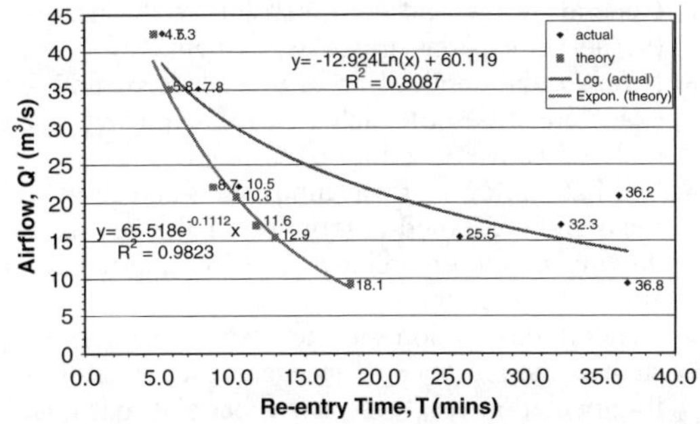

Figure 7. Re-entry times correlation (ii).

Rearrangement of Eq. 11 and taking the exponential function of each side gives an equation where the actual re-entry time (Ta) may be approximated by the following exponential equation (Eq. 12).

$$Ta = e^{-\left(\frac{Q-60.119}{12.924}\right)} \quad (\text{mins}) \qquad (12)$$

The relationship between the theoretical and actual re-entry times can be observed as shown in Figure 8. Good correlation exists, and a power function can be used to describe the relationship.

Actual re-entry times Ta can then be written as function of the theoretical re-entry times Tt as demonstrated in Eq. 13.

$$Ta = 0.5085\, Tt^{1.574} \quad (\text{mins}) \qquad (13)$$

where the Tt can be calculated in minutes using the previously simplified theory of dilution equation. Eq. 9 can therefore be re-written as:

$$Ta = 0.585\left[\left(\frac{Y'}{Q}\right)\ln\left(\frac{x_0}{x}\right) \div 60\right]^{1.574} \quad (\text{mins}) \qquad (14)$$

Once again, if a conservative approach is taken and the maximum working space volume ($Y' = 2800\,\text{m}^3$) and the maximum peak CO concentration ($x_0 = 3000$ ppm) are adopted, which represents the maximum for a typical development heading, and the TWA for CO is taken as 30 ppm then Eq. 14 may be simplified to Eq.15 to give an approximation of Ta.

$$Ta = 2383.7\, Q^{-1.574} \quad (\text{mins}) \qquad (15)$$

Two empirical equations for estimating Ta have been derived, both expressed as a function of Q. These can be used to conservatively approximate re-entry times for typical level development headings;

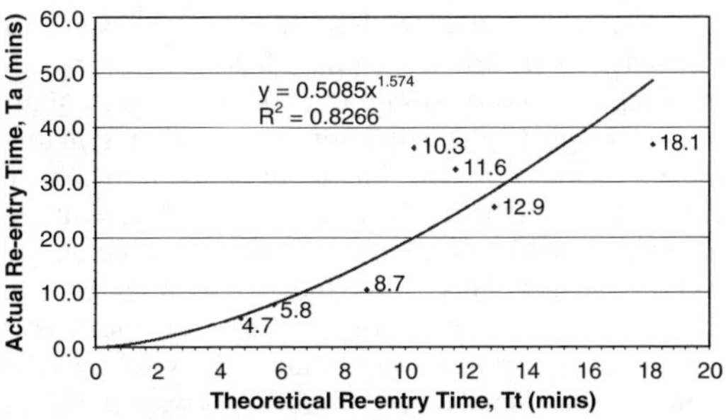

Figure 8. Relationship between Tt and Ta.

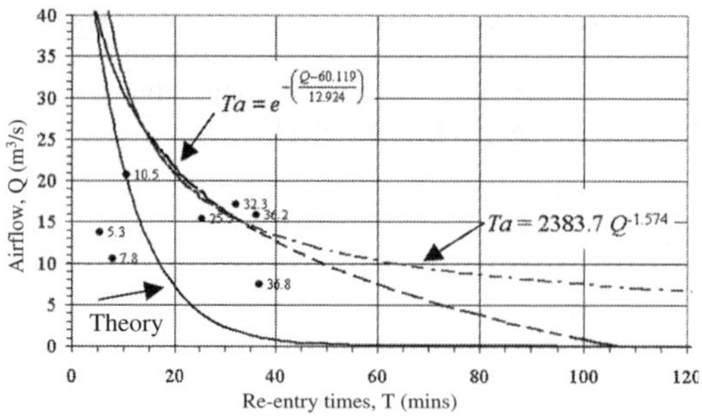

Figure 9. Correlation of re-entry equations.

Eq. 12 and Eq. 15. Figure 9 shows that reasonable correlation exists between these equations.

Additional testing data is required to firm up both equations beyond about 40 minutes.

The plot of actual re-entry times on Figure 9 demonstrates that both equations offer a reasonably conservative approach to estimation re-entry times. Of the seven tests, there are only two test results that sit marginally above the curves, whilst all others fit comfortably under.

The above analysis demonstrates that either equation would be reasonably capable of safely estimating re-entry times for typical level development headings for airflows exceeding about 8 m^3/s. Whilst these equations give a conservative approach, they should provide reasonable confidence to estimate safe re-entry times whilst maximizing productivity by ensuring headings can be re-entered as soon as gas concentrations reach safe levels.

For these equations to be effective however, good ventilation practices must be upheld in ensuring the ventilation ducting is maintained in reasonable condition and the end of the ventilation ducting kept to within 15 m from the face to maximize the sweeping effect across the face.

5 REGULATOR TESTS

The test mine currently has in place a system of remotely controlled regulators. During blasting all level regulators are opened to prevent inadvertent air blast damage to the regulator doors. All regulators remain open until such time as the blast fumes have cleared from all the levels. The doors are then returned to their pre-set levels to allow work to resume on the working levels. This in an inefficient practice and leads to longer re-entry times on the lower levels. This is because the primary ventilation airflow follows the path of least impedance leading to an oversupply of air on the upper levels. This leaves lower levels with insufficient air resulting in delays for diluting and clearing of blast fumes.

A more efficient practice may be to open the doors fully on those levels where blasting takes place whilst closing the doors (or leaving them at their normal setting) on those levels where there is no blasting. This way, blast fumes should be cleared quickly and re-entry times significantly reduced on the blasting levels with this approach.

A complication would arise when blast fumes from single blasts filter through several levels via ore passes, stopes, or rises. Through use of the "contaminants" feature of a ventilation network simulation program it should be possible to roughly predict the flow of blast gases through the mine. This would allow the Ventilation Engineer to set the regulator doors appropriately on each level to give the optimum clearing of blast fumes. Once the gas levels from the test mine's gas monitoring system indicate that the appropriate TWA has been reached the regulator doors could be returned to their pre-set levels.

6 FURTHER RESEARCH

Further research needs to be conducted at suitable mines to confirm the empirical hypotheses developed. It is suggested that detailed testing focus on one or two critical path development headings within a mine such as a priority high-speed access to a new mine area. There are a number of major advantages in doing repeated tests on a small population of critical path headings:

- The testing environment remains relatively constant.
- The heading design and configuration remains the relatively constant.
- The auxiliary fans are usually switched on without delay after the blast for critical path headings.

It is suggested that all possible parameters be recorded for every blast to try and better understand abnormalities or inconsistencies in test results if they occur. Parameters to record should include:

- Wet and dry bulb temperatures and atmospheric pressure.
- General design and configuration of the heading (tunnel dimensions, grade, type of cut etc)
- Number and depth of holes in each face,. and the exact quantities of each type of explosive used including any stemming.
- Ventilation flows (airway dimensions and air velocities) and a detailed description of the ventilation ducting condition, relative distances, orientation and end condition.
- General description of the testing environment such as the presence or absence of water, geology, fragmentation etc (these can affect blast gas quantities). Photographs may also be appropriate.
- Times of critical events (blast initiated, fan-on, instrument-on, etc).

In summary, the testing regime should focus on a discrete part of the overall project where the time schedule is paramount, the operating environment and design is consistent and where the quality and accuracy of the tests results can be assured.

7 CONCLUSIONS

The aim of this study is to examine approaches to determine suitable re-entry time to improve safety and production at mines. The literature review has highlighted a number of important aspects of previous research, namely mathematical modeling of blast fume decay (DeSouza et al, 1993), exposure standards (McPherson, 1993) and test work on blasting fumes (Hine et al, 1985) and determination of re-entry times by Winn (2002). A simulated case study has demonstrated that blast re-entry times could probably be based on concentrations well in excess of the conservative $TWA_{8\,hr}$ and $TWA_{12\,hr}$. The $TWA_{8\,hr}$ has however been adopted for determining re-entry times in this study.

Most Australian mines currently base their re-entry times on the experience of their workforce. In general they have not performed testing or attempted to apply any other means to determine safe and productive re-entry times. This approach presents potential for the overexposure of personnel and losses in productivity due to inappropriate methods for determining re-entry times.

Portable gas monitoring equipment has been used to capture the behavior of blast gases in auxiliary ventilated level development headings. An analysis of the development end test results have shown that carbon monoxide is the critical gas in terms of defining minimum re-entry times. The development end test results for auxiliary ventilated headings show that fume dilution appears to be best described by a power relationship, as opposed to the classic exponential relationship for flow-through ventilation systems such as through the regulator doors (or though a stope), and for naturally ventilated workings. This new phenomenon has not been concluded by previous research and only with further testing can this relationship be confirmed.

Rigorous analysis has produced two empirical equations that can be used with some confidence to conservatively estimate re-entry times for level development headings:

$$Ta = e^{-\left(\frac{Q-60.119}{12.924}\right)} \quad \text{(mins) or}$$

$$Ta = 2383.7\,Q^{-1.574} \quad \text{(mins)}$$

Both equations are based on a function of Q and give similar results for values of Q greater than about 8 m^3/s. It was concluded that the working space volume Y' used in mathematical modeling might be a constant for development headings of a similar configuration and environment, regardless of the relative distance from the end of the ventilation ducting to the face. Further testing is required to confirm this hypothesis.

The study results have been achieved through undertaking of the following stages.

- An extensive literature review with particular emphasis on mathematical modeling of blast fume decay, exposure standards and recent test work on blasting fumes and re-entry times.
- An analysis of appropriate exposure standards and the selection of suitable threshold limit values for the determination of re-entry times.
- Design of underground test procedures to measure post detonation fume products in level development headings at a test mine.
- Measurement of post detonation fume products, airflows and other blast parameters such as design, configuration and the operating environment in level development headings.
- An analysis of test results and mathematical models of blast fume decay to allow an empirical relationship to be established that can be used to determine suitable re-entry times.
- Recommendations on future research on this topic.

REFERENCES

Department of Industry and Resources Western Australia (DIRWA), 2003. Safety Bulletin – Re-entry After Blasting, March. At http://notesweb.dme.wa.gov.au/.

De Souza, E.M., and Katsabanis, P.D., 1993. On the Prediction of Blasting Toxic Fumes and Dilution Ventilation, *Mining Science and Technology*, 13: 223–235.

Fisher, T.H., 1995. Mining Technology Advancing Faster Than Ventilation Practices?, *Proceedings 7th US Mine Vent. Symp.,* Society of Mining Engineers, Pittsburgh.

Gallo, N., 1995. *Leakage Performance of ACME ACFAB™ Auxiliary Ventilation Ducting*, Undergraduate Thesis (unpublished), University of Queensland, Brisbane.

Haase, H.H., 1991. The Potential of Blast Contaminant Control for Reducing Re-Entry Times and Ventilation Costs, *Journal of the Mine Ventilation Society of South Africa*, 44: 86–89.

Hine, M., and Jones, I.O., 1985. The Analysis of Post Detonation Fumes in an Underground Operating Environment – A Case Study, *Proceedings 2nd U.S. Mine Vent Symp.*, Nevada-Reno 23–25 September 1985 (Ed: P Mousset-Jones) pp 689–698. Rotterdam/Boston: Balkema.

McPherson, M.J., 1993. Chapter 11 – Gases in Subsurface Openings, *Subsurface Ventilation and Environmental Engineering*, London: Chapman & Hall.

National Occupational Health and Safety Commission (NOHSC), 2003. http://www.nohsc.gov.au. June.

Winn, D., 2002. *Blast Fumes and Re-entry Times in Underground Metalliferous Mines*, Undergraduate Thesis (unpublished), University of Queensland, Brisbane.

Mine Ventilation, Ganguli & Bandopadhyay (eds)
© 2004 Taylor & Francis Group, London, ISBN 90 5809 633 5

The use of large axi-vane fans in auxiliary ventilation for multiple headings in a dieselised hot metal mine

Kuda R. Mutama
Senior Ventilation and Mechanical Engineer, Barrick Goldstrike Mines Inc., Elko, NV

ABSTRACT: This paper discusses current auxiliary ventilation practices at Barrick Underground Division in Nevada mine. Strategies being adopted in order to improve ventilation in multiple development headings in a cost effective manner using large axial flow fans is discussed. Ventilation in dieselised mines is increasingly becoming a challenge in terms of providing adequate airflow to remove contaminants and heat. Furthermore there is now legislation that mining companies are faced with that requires diesel particulate matter not to exceed 400 $\mu g/m^3$ at the present time in the USA. The old USBM guideline was to provide 125 cfm per brake horsepower for the first diesel engine in a heading and in most cases worked very well. Mechanized mining employing diesel equipment is very efficient and is not likely to be replaced in the foreseeable future. Current technologies that can replace the diesel engine including fuel cells are yet to be proven in an underground mine and are still in their infancy. It is now essential to take a hard look at auxiliary ventilation in metal mines where diesel equipment is used. Moreover at the Meikle underground complex the amount airflow required per active heading is large and is not only driven by diesel equipment but also by geothermal activity of the rock. The rock generates hot water vapors, carbon dioxide and sulfur dioxide. Cooling these headings and diluting toxic gases require a careful selection of the ventilating fan that also serve two to three headings in the same area.

1 INTRODUCTION

The need for increasing productivity and efficiency a few decades ago resulted in the employment of diesel powered vehicles in underground metal mines. The underground diesel fleet for most mining companies has grown steadily over the years from hundreds of brake horsepower (bhp) to tens of thousands bhp in fairly large gold mines where mechanization is well established.

The mining process itself has not changed whatever the degree of mechanization. Typically in a development heading the process is summarized as drilling the face, load explosives, blast, mucking and provide ground support where needed. In most mechanized mines almost all the processes in the cycle will involve some kind of diesel powered equipment to accomplish the work required. With the whole mining cycle taken into consideration, ventilation is the common denominator that enables work to take place by enabling both personnel and machine to function properly in a hospitable climate.

2 DIESEL VENTILATION

Despite so many years of the presence of diesel equipment underground there are no set ventilation standards. Several guidelines have been proposed and some of them are still in use today in many parts of the world. The USA Bureau of Mines guideline is a typical example that has been applied successfully in this country. These guidelines provided 125 cfm/bhp for the first diesel equipment and thereafter, 75% and 50% of this quantity respectively for the second and third diesel unit bhp in use in the area. Others have proposed an air quality index based on known TLVs of pollutants from the products of combustion. In all cases the desired results were achieved by providing required ventilation rates. The problem with all this is that if the ventilation supply was not monitored strictly inadequate air supply to the heading face would result. Engine manufacturers sometimes recommended a minimum airflow quantity necessary to operate the diesel equipment.

Supplying adequate ventilation to a mine is very expensive business and as much as 60% of a mine's electrical power usage can be attributed to ventilation. Therefore every consideration has to be given to come up with a very cost effective ventilation system and that is not always easy to achieve.

3 DIESEL ENGINE POLLUTION AND HEALTH RISKS

The diesel engine operates on Diesel cycle in which combustion takes place at almost constant pressure and compression ratios are typically very high. The fuel-air mixture self ignites because of the high pressures achieved. This means that diesels are more susceptible to noise generation moreover and more seriously heat, oxides of nitrogen and carbon are generated in high quantities which has to be removed by the provision of ventilation. The diesel cycle operates on a cycle efficiency of less than 40%, which means as much as 60% of the fuel burned is converted into heat.

One aspect of diesel pollution now receiving US government attention is diesel particulate matter (DPM), which includes diesel soot and solid aerosols such as ash, metallic abrasion particles, sulfates and silicates. Diesel soot particles have a solid core mainly consisting of elemental carbon, with a ranging variety of other substances attached to the surface.

It is believed that exposure to diesel particulate matter can result in adverse health effects at high enough concentrations which can include (i) sensory irritation and respiratory symptoms (ii) death caused resulting from cardiovascular, cardiopulmonary, or respiratory complications and (iii) lung cancer. Most of the diesel particulate matter is less than 1.0 micrometer in size, which makes it easier for them to be inhaled in the deeper lung system where they will accumulate with each exposure.

Although there are known adequate documented epidemiological health effects of diesel exhaust exposure underground the potential to cause harm should still be taken seriously. It is difficult to prove what threshold of each of the exhaust component will be harmful if exceeded and over what exposure times. The health statistics are difficult to prove because they are further complicated by the fact that some miners smoke cigarettes and also chew tobacco, which also cause a cumulative health risk compounding the diesel question.

The use of diesel equipment has to be addressed from a number different factors. These factors can be can summed up as follows; (i) proper maintenance of diesel engines per manufacturer recommendations, (ii) the use of low sulfur fuel and proper ventilation in workings where the equipment is in use.

4 ANSWERING THE DIESEL QUESTION

Maintenance plays a big part in reducing diesel emissions underground. All engines as received for an operating mine have to be suitable for that mine elevation and have to be adjusted for that altitude in which they operate. Maintenance instructions provided by the engine manufacturer should be followed strictly and in accordance to a number of hours of operation. Regular emission testing should be part of the routine and proper logging of what has been done should be the practice. All EPA standards pertaining to that engine should be satisfied. Improved maintenance procedures can reduce gaseous pollutants and diesel particulate matter by as much as say 65% for CO and 55% DPM (www.Deep.org). If an engine can no longer be serviced it should be replaced or rebuild.

Once an engine passes the low emissions testing program within the mine maintenance department the next step is to use low sulfur fuels. Recommended sulfur content by MSHA is now at less 0.05% by weight. If a well-maintained engine is using a good quality low sulfur content diesel fuel then the next step is to address the ventilation question for a good clean mining environment.

The mine in question should have a primary ventilation system that is able to support the amount of rock removed from the mine in terms of both ore and waste. The ventilation system should address the number of diesel equipment potentially in use underground and in particular the total brake horsepower in the mine. As an example a typical mine in the State of Nevada is a 2000–3000 tpd operation utilizing a combination of mining methods cut or drift and fill or longhole stoping with development of new headings in order to advance the mining process. Typically, if this mine utilizes diesel equipment it can have as much as 10,000 bhp. The diesel equipment in the fleet will comprise (i) muckers or LHDs of various sizes typically 2 to 8 cubic yard loaders, (ii) haulage trucks (20–30 tons), (iii) jumbo drills, (iv) bench drills and (v) roofbolters. Additionally support equipment is also present in the fleet such as (i) shotcrete machines, (ii) scissors trucks (iii) crane boom trucks, (iv) emulsion or powder trucks, (v) lube trucks, (vi) utility equipment such as underground tractors, forklifts and (vii) other miscellaneous equipment e.g. rock breakers, road headers, backhoes and jammers. All this equipment is necessary in the mining cycle to provide an efficient and cost effective operation.

The question still remains unanswered as to how much ventilation is required in the mine and in the heading or stope. Typically, in a stope or heading you will have a mucker, a haul truck and other support equipment. A six cubic yard loader and a haul truck if they are equipped with say a Detroit Diesel engine series DDEC 50 they will have a combined brake horsepower of 500. If additional support equipment is in the vicinity the horsepower can be as much as 600. On each level of the mine there is a maximum amount of airflow quantity that can be distributed without starving other areas of ventilation air. Typically, future stopes are developed from footwall drifts and sometimes before through flow ventilation is established. The challenge is now to provide enough airflow to dilute diesel emissions for the 600 bhp. In a heading typically 15 feet by 15 feet (4.6 m × 4.6 m) cross section that uses a 42″-26″-1800 RPM, 100 hp fan with 42 inch (1,067 mm) vent line and with reasonable air intake conditions of say 78/65°F (dry/wet bulb) or 25.6/18.3°C, temperatures would be satisfied by supplying a quantity of 50,000 cfm (23.5 m^3/s) to ventilate the area. Considering the presents of 600 bhp of diesel equipment this means only 83.3 cfm/bhp of fresh air is available to ventilate the area. Using the USBM guideline 100%, 75% and 50% would require a total of 61,000 cfm (28.7 m^3/s) i.e. 11,000 cfm (5.2 m^3/s) more than what the fan can provide.

The minimum required ventilation rate in a given area is only determined by practical experience in that mine i.e. (i) air intake conditions to the heading, (ii) dilution of all pollutant gases and particulate matter well below TLVs, (iii) adequate oxygen in the air stream and the secondary ventilation system available (iv) proper maintenance of the vent line. It is possible to provide enough ventilation to the area under study with as much as 30,000 cfm (14.1 m^3/s). The reason being that during the mucking cycle the haul truck is idle while the mucker does the work to load the truck. Therefore only a fraction of the diesel horsepower in the area is being applied and not all the airflow is essential for diesel ventilation. When the haul truck is full of material it will leave the area immediately and the mucker becomes idle and is now waiting for the truck to come back. Meanwhile, other support equipment in the area will not be in the heading directly when both the mucker and the truck are working because the area has only so much space available to contain a certain number of equipment.

Once a heading is removed of muck, some ground support might be necessary to avoid roof fall if the ground is poor. Typically roof bolters and shotcrete machines if they use e.g. Deutz engines, the associated brake horsepower is not is large as that of LHD or haul truck equipment and is typically 145 hp or less. Therefore, the ventilation required is not as much as for the combination of an LHD and a haul truck. However, the design criteria for any heading in any part of the mine should be based on maximum possible combination of the largest possible diesel equipment to be used in the mining cycle.

The most difficult time in mining is usually during the development phase. In most mines in Nevada

Figure 1. A typical diesel loader (or LHD) which can be 6 to 8 cubic yard bucket size with 250–350 hp engine.

breaking new rock is always hot business e.g. the Meikle Mine. Some areas can have virgin rock temperatures as high as 200°F (93.3°C). This adds heat to the airflow. The compression ignition engine or diesel cycle has a thermodynamic efficiency of about 40% and the rest is heat generated to the environment and friction in mechanical moving parts. Part of the ventilation air in a working area is to remove the heat from the area otherwise heat build up can cause very severe conditions with temperatures soaring to above 100°F (37.8°C). In dieselized mines ventilation serves several purposes and that is to provide airflow for (i) personnel to breathe, (ii) the function of the engine, (iii) remove or dilute pollutants and heat. All these factors have to be considered when designing the ventilation system.

To address the generation of pollutants by the engine good maintenance is called upon. Assuming that the engine is in its best possible state of operation, various types of soot traps are now available that can scrub almost 99% of harmful diesel particulate matter. These are installed on the exhaust side of the engine and very effective devices for clean air in the headings. Despite their effectiveness, the additional cost associated with their installation discourages mine operators to employ them on a large scale especially in those mines with a large fleet of underground diesel equipment. Although they are finding wide spread use around the world, these exhaust filters or scrubbers are not maintenance free. Here in the USA exhaust filters will make it possible to remove diesel particulate matter to well below any proposed TLV currently set at 400 $\mu g/m^3$, even if the 160 $\mu g/m^3$ becomes the final ruling in 2006. Most mines in North-Eastern Nevada are between 300 and 350 $\mu g/m^3$ DPM concentration and will have to adopt strategies going forward into the future to reduce this TLV required by legislation. Typical DPM concentration at the Meikle Mine are shown in Table 1. Only one sample in Table 1 exceeds the 400 $\mu g/m^3$ limit. This is not uncommon. Most of the measurements sampled since 2002 are in the 200 to 400 $\mu g/m^3$ range in the main exhaust drifts.

Table 1. Typical DPM concentration at Meikle mine mainly in the exhaust air stream.

Contaminant	Sample location	Total carbon ($\mu g/m^3$)	MSHA 8 hour TLV ($\mu g/m^3$)	TLV (%)
DPM	Ramp 925 to 860 level (exhaust)	303.17	400	76
DPM	Ramp from 1075 to 1125 level (exhaust)	529.68	400	132
DPM	Ramp from 1175 to 1225 Level	163.99	400	41
DPM	1750 level Intake shaft (fresh air)	54.9	400	13

Exhaust filters alone are not the only answer to the diesel question; engine technological advancements will also play a part in the future.

5 VENTILATION STRATEGIES USING LARGE AXI-VANE FANS

In most underground mines, ventilation plays a large part in removing pollutants and cooling the environment. More than 70% of the underground climatic conditions is addressed by providing adequate ventilation. Despite any future strategies employed to address the diesel question ventilation will remain the sole method to make it possible for mining to take place. It is important to look at the strategies being employed at the Meikle Mine, which includes the Griffin, and Rodeo area that makes up an underground area spread over a mile or 1.6 km. At this mine there is now over 20,000 bhp of engine diesel fleet. Diesel equipment is not the only challenge in the ventilation of headings but the presence of hot ground conditions make it harder to provide a comfortable environment. Heading drill hole temperatures in some parts of the Meikle underground complex can be as high as 300°F (149°C). When these kind of temperatures are encountered even loading explosives have to be put on hold while the holes cool off. Sometimes steam or hot saturated vapors are being discharged from the ground into the headings or open airways. Figure 2 shows such hot vapors being discharged into the ventilation air. Sites such as shown in Figure 2 are very common in the Meikle underground complex. The major challenge is to provide enough ventilation to maintain effective temperatures less than 85°F (29.4°C) so that mine personnel can work safely and efficiently. Therefore the criteria for ventilating the heading is now based on geothermal activities in the heading or stope and not so much on diesel equipment in use in that area.

The solution being applied to solve heat and diesel issues is to employ large axial flow fans of diameters 42, 45, 48 and 54 inch (1066 to 1372 mm) with 100 to 200 hp electrical motors to supply approximately 50,000 to 90,000 cfm (23.5 to 42.3 m^3/s) to multiple headings. In the Rodeo part of the mine the headings from the main footwall drifts are spaced about 50 ft (15.24 m) apart. The headings are the future stopes. The fan will be set up in the footwall drift where there is there flow through ventilation to an exhaust raise; typically about 120,000 cfm (56.4 m^3/s). The fan is then set to deliver 80,000 cfm (37.6 m^3/s) for a typical 48-inch diameter, 150 hp fan. The vent line from the fan is 42 inch in diameter made from galvanized steel. On each heading access typically there is a Tee-piece to branch the ventilation into that heading. Each heading vent line has baffles to regulate the ventilation if required. The heading with the largest horsepower

Figure 2. Hot saturated vapors or steam being discharged into drift in the Griffin zone at the Meikle Underground complex.

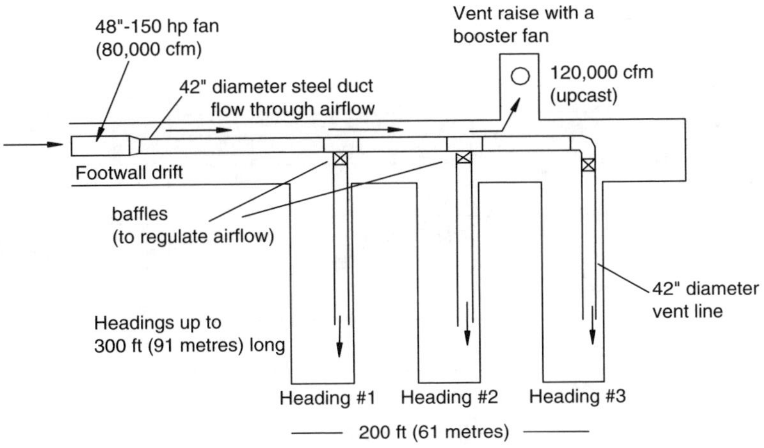

Figure 3. A schematic of a multiple heading set up at the Meikle mine.

diesel equipment e.g. a mucker will get most of the ventilation in that multi-branch auxiliary vent line. A schematic of the ventilation scheme for a multiple heading is shown in Figure 3. Each heading may receive anywhere from 20,000 to 40,000 cfm (9.4–18.8 m^3/s) from the total of 80,000 cfm (37.6 m^3/s).

The auxiliary fans can serve up to three headings sometimes up to four headings by careful regulation of the airflow. Typically drifts are 15 feet × 15 feet (4.6 m × 4.6 m) and there is a need to maintain 150 fpm (0.76 m/s) air velocity in active hot headings. This means that a minimum air flow quantity of 33,750 cfm (15.9 m^3/s) has to be maintained. This makes the air flow balancing in the common vent line very difficult especially in hot areas. In most cases the number of active headings have to be minimized to two if conditions are very hot and in that case the airflow maintained per heading is 40,000 cfm (18.8 m^3/s). In cooler areas multiple headings ventilated by one fan are easier to manage.

Figure 4 show typical 48-inch (1,219 mm) diameter fans used for multiple ventilation in headings at Meikle underground complex. Most of these fans will discharge up to 80,000 cfm (37.6 m^3/s) with a proper installation employing a steel duct 42 or 48 inches (1,066 or 1,219 mm) in diameter. Larger fans such as 54 inches (1,372 mm) in diameter are in use in areas with a larger ventilation demand. The guideline used for diesel ventilation at Meikle mine is 100 cfm/bhp (0.063 m^3/s/kW). This really works well because some of the engines have lower minimum ventilation quantity recommended by the manufacturer. Heat removal considerations become the main criteria for designing the auxiliary ventilation system.

Managing the diesel equipment in use is a strategy that works very well to reduce the required ventilation

Figure 4. 48-26-1800 RPM, 150 hp fans used for multiple heading ventilation.

in these headings. Only loaders are allowed to go in the headings to remove muck and the truck is parked in the main drift with flow through ventilation. The truck is loaded in the main drift while the loader is moving back and forth. This means that if the loader has a 250 hp engine, approximately 25,000 cfm (11.8 m^3/s) has to be supplied to the heading face to dilute diesel exhaust fumes using our 100 cfm/bhp. Additional airflow is then added for heat removal. Therefore ventilation in the heading does not need to consider the truck. Once the truck leaves the area it is being driven in fresh air or large volumes of air flow in the main ramps and drifts. In the headings during ground support or long hole stope drilling the diesel equipment employed has smaller than 150 hp engines and therefore do not need as much ventilation as loaders. The shotcreting process may generate considerable amounts of heat and will require good ventilation to

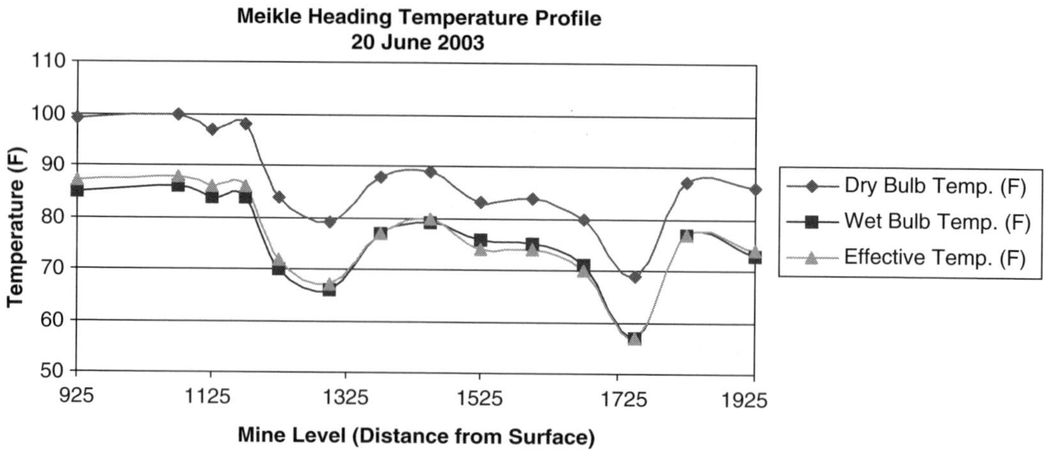

Figure 5. Typical heading temperatures at Meikle mine where diesel equipment is used.

maintain safe temperatures. The key to optimized diesel ventilation in an area is the availability of large quantities of flow through ventilation in the main footwall drifts and providing enough ventilation for the largest horsepower diesel equipment that will be used in the heading.

A typical ventilation survey showing temperature information at the Meikle mine in some of the areas with multiple headings is shown in Figure 5. In order to assess temperature conditions more appropriately, effective temperature concept is used. The dry and wet bulb temperatures together with the average airflow velocity are used to compute the effective temperature. The average minimum air speed used is 150 fpm (0.76 m/s). Table 2 shows the effective temperature chart used by miners to determine conditions in a heading where they are working. Table 2 has been very useful in headings with punishing temperature conditions in preventing heat related illnesses. Any effective temperature 81°F (27.2°C) or less is considered safe. When the effective temperatures is 82 to 89°F (27.8 to 31.7°C) caution has to be exercised by the minor working in the heading and periodic brakes have to be taken. No work is allowed at Meikle mine for effective temperatures of 90°F (32.2°C) and above unless other cooling methods are available such as ice jackets. When temperature exceeds the safe effective temperature usually as much ventilation as possible is directed in that heading to cool it down. It becomes necessary to use two fans and two vent lines in that heading to boost the ventilation quantity in order to bring the temperature down. Moreover, the effects of gases such as carbon dioxide and sulfur dioxide could also be present in large quantities and they deplete the oxygen concentration. It therefore becomes necessary to supply as much ventilation as possible using large axi-vane fans in those headings requiring large quantities of airflow.

Table 2. Effective temperature chart based on 150 feet per minute air speed, or 30,000 to 40,000 cubic feet per minute airflow in an underground heading in Meikle, Griffin and Rodeo.*

Wet bulb temperature °F
80 82 84 86 88 90 92 94
86 78 79 81 83
88 79 81 82 84 85
Dry bulb temperature °F
90 80 82 83 85 86 88
92 81 83 84 85 87 89 90
94 82 83 85 86 87 89 91 93
96 83 83 85 87 88 90 92 93
98 84 85 86 88 89 91 92 94
100 84 86 87 88 90 91 93 95
102 85 86 87 89 90 92 93 95
104 86 87 88 89 91 92 94 95
106 86 87 88 90 91 93 94 96
108 87 88 89 90 92 93 95 96
110 87 89 90 91 92 94 95 97

*Guide: 82–84 ET Caution, 84–86 ET 10 minute break every hour, 86–88 ET 15 minute break every hour, 86–88 ET 15 minute break every hour, 88–90 ET 20 minute every break every hour, 90 ET and above STOP – Administrative and engineering controls mandatory.

The key to successful auxiliary ventilation is to ensure enough flow through ventilation and a large enough fan to achieve the desired airflow. Where the drifts are large enough it is better to use large diameter rigid duct such as galvanized steel or fiberglass. These have less resistance to airflow that soft vent bag at least by 30%. More airflow can be moved in the same size vent line if rigid duct is used. When large fans are used to ventilate headings proper installation guidelines should be followed strictly. These include the following (i) using proper fan to vent line adapters, (ii) proper rigid Tee-pieces with baffles or dampers for multiple heading ventilation (iii) proper use of elbows to make gradual turns into the headings to avoid shock losses and finally (iv) taking care of leaks in the vent line when they occur. A well-designed and installed auxiliary vent system will ensure proper ventilation for the working areas. It is cheaper in the long run to have a combination of good ventilation and well-maintained diesel equipment.

6 CONCLUDING REMARKS

Large diameter fans with large horsepower motors can be used successfully to ventilate multiple headings in metal mines where large fleets of diesel equipment are in operation. These fans as used at the Meikle mine range from 42 to 54 inch (1,066 to 1,372 mm) in diameter with motors of 100 to 200 hp. The benefit of operating large fans where space is available in large headings is that more volume can be delivered to ventilate multiple headings and therefore less number of fans is required. This saves power in the long run because potentially one or two small fans are eliminated per installation by using a large fan. Successful diesel ventilation depends on the condition of the engine and a well-designed primary and auxiliary ventilation system. Before expensive programs are started on DPM each mine should have thorough knowledge of its concentration and decide on what kind of strategy to be used for that mine to be in compliance.

REFERENCES

DEEP Diesel program website
Meikle mine ventilation reports (2000–2003)
Hartman, H Mine Ventilation and Air Conditioning

Mine Ventilation, Ganguli & Bandopadhyay (eds)
© 2004 Taylor & Francis Group, London, ISBN 90 5809 633 5

Estimation of thermal environment conditions in a heading face ventilated by forcing system

Masahiro Inoue & Kenichi Uchino
Dept. Earth Resources Eng., Kyushu University

Yasuhiro Kusaba
Sumitomo Metal Mining Co., Ltd.

ABSTRACT: The strata temperature in the area of Hisikari gold mine reaches to nearly 80 degrees Celsius at only 260 m below surface. The high strata temperature causes severe thermal environmental problems particularly at the heading face in the mine. Air-cooling system using cooled water is required in the forcing system of ventilation for the heading face even though the roadway length is less than 100 m. Sometimes the heading works were to be difficult because of the thermal environment problems. Accurate estimation of the thermal environment in the roadway is inevitable for the smooth and cost effective mining operation. The authors have made a new computer program to estimate the temperature and humidity of the roadway with forcing system of ventilation considering various thermal conditions. Heat from the surrounding rock, fan, heavy machinery, air cooling coil performance, heat exchange and air leakage through the duct wall, moisture of the roadway, heading face condition etc. are considered in the program. The authors have also measured the thermal conditions in the roadway at the mine. The comparison of the estimated values and the measured values shows good agreement. The program became one of the important tools for the mining plan. The theory to estimate the thermal environment and the results of the program will be described in the paper.

1 INTRODUCTION

Several metal mines in Japan are located in the areas, which are characterized by very high strata temperatures. The highest strata temperature in Toyoha mine reaches more than 150°C (degrees Celsius) and steam outbursts have occurred in the past.

Hisikari mine also suffered from the high strata temperature. The strata temperature of the mine reaches to 65–80°C at only 260 m below surface. One of the main concern of the mine is the maintenance of appropriate thermal environment for working in underground especially at the heading face. Figure 1 shows the horizontal map of the mine. The total length of the roadways is about 80 km.

The mine is mainly divided in two parts. These parts are connected by a roadway near the center of the mine. The left part of the mine is called Yamada area and shows relatively higher strata temperature in the mine. Figure 2 shows the three-dimensional view of the mine in Yamada area. There are one inclined intake shaft (right) and one vertical intake shaft (left). Two vertical shaft (right) with same diameter are used for the main exhaust of the mine of about 100 m^3/s.

The airflow distribution in the ventilation network of the mine is often calculated to attain the most appropriate plan of mining operation. Figure 3 shows the ventilation network of the Yamada area and the example of the result of the calculation.

In addition to the analysis of airflow distribution in the network, the analysis of temperature, humidity and airflow distribution in the network is also being carried out to make reasonable mine operation plan. The analysis will be called a thermal environment analysis in the network in this paper. Figure 4 shows a result of the analysis. The numerals on the line mean the dry bulb temperature and those under the line the relative humidity.

A heading face is one of the most important place in the mine, and is prone to have severe thermal conditions if the cooling system is inappropriate. Air cooling system using cooled water is being used in the forcing system of ventilation for the heading face in the mine even though the roadway length is less than

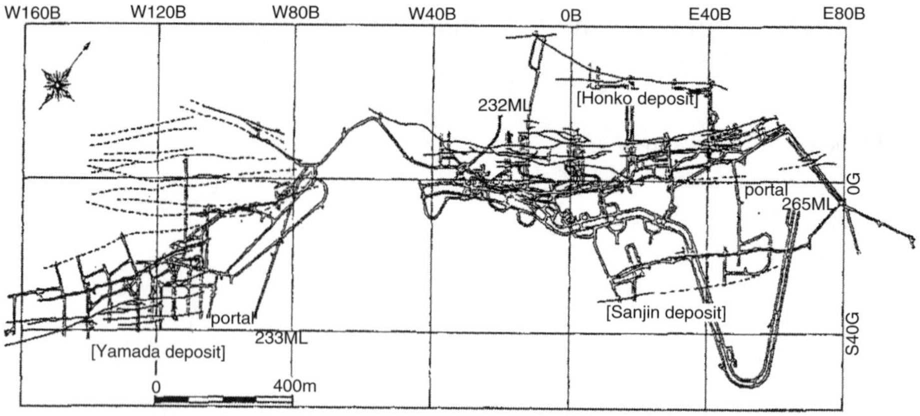

Figure 1. Horizontal map of Hisikari mine.

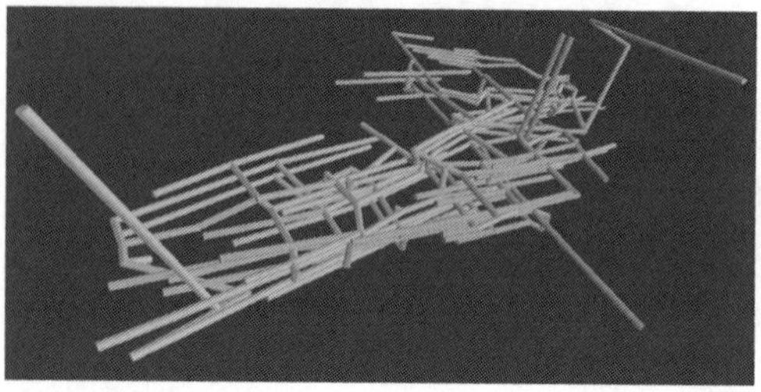

Figure 2. Three-dimensional view of the mine in Yamada area.

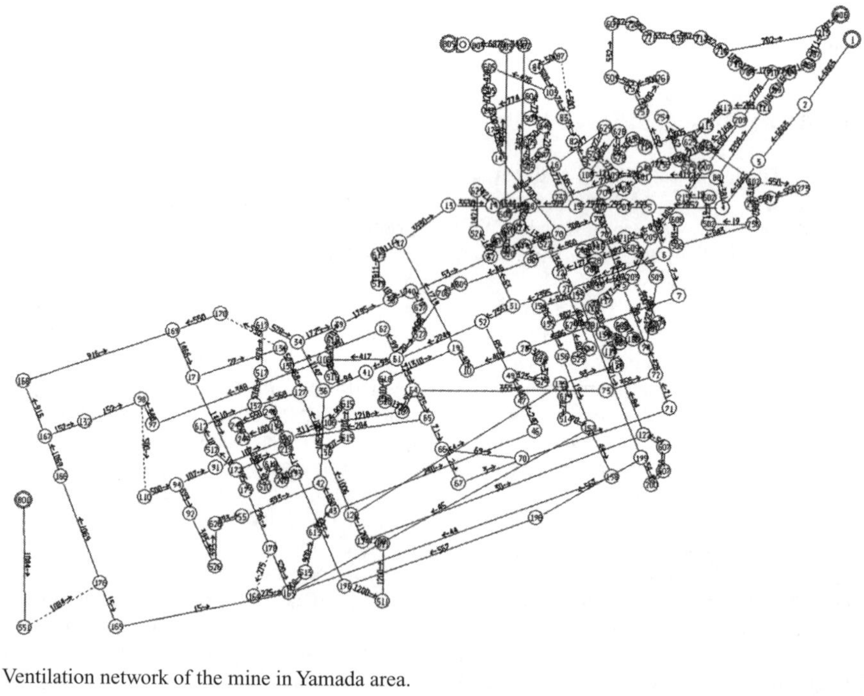

Figure 3. Ventilation network of the mine in Yamada area.

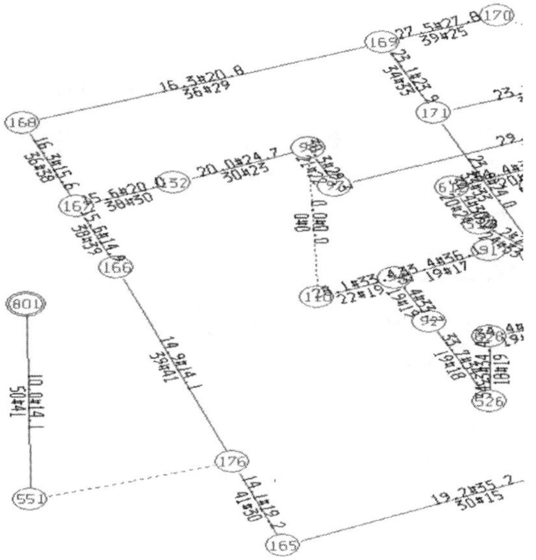

Figure 4. Example of thermal environment analysis in the network (left part of Yamada area).

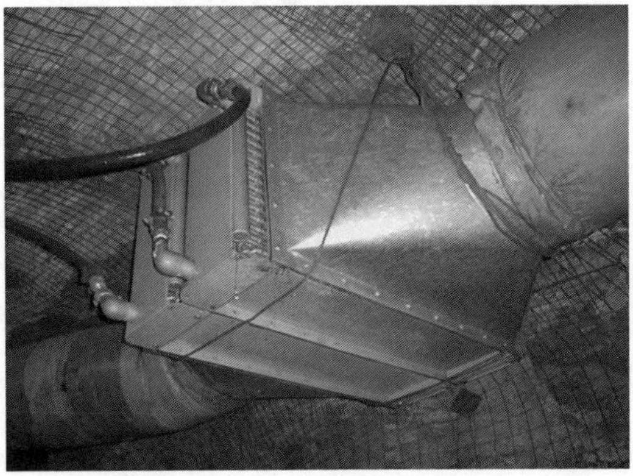

Figure 5. An air cooler used in the mine.

100 m. Figure 5 shows an air cooler employed in the mine. An accurate estimation of the thermal environment in the heading roadway is inevitable for the smooth and cost effective mine operation.

Though the thermal environment analysis in the network can predict the temperature and humidity in the whole network in the mine with relatively good accuracy, it does not predict them at the heading face accurately. There are many other factors in the heading roadway which are not generally considered in the thermal environment analysis in the network. Therefore, it was not possible to predict accurate thermal conditions at the heading face up to now.

The authors have developed a new computer program to predict the temperature and humidity in the heading roadway with forcing system of ventilation considering various thermal conditions: Heat from the surrounding rocks, fan, heavy machinery, air cooling coil performance, heat exchange and air leakage through the duct wall, moisture of the roadway, heading face condition etc. are considered in the program. The method to estimate the thermal environmental conditions and the results of the calculation by this program is described in the paper.

2 FACTORS AFFECTING ON THE THERMAL ENVIRONMENT IN THE HEADING FACE

Figure 6 shows various factors affecting the thermal environment in the heading roadway and face. Their effects are calculated in the computer program. They are as follows:

Roadway 1

Sectional area of a roadway (m^2)
Perimater of the roadway (m)
Hw: Heat transfer coefficient between the surrounding rock and air of the roadway at the wall (kcal/mhC).

Roadway 2

Elevation of several places described later in the roadway (m).
Wetness of the wall of the roadway (-)
Age of the roadway from its excavation (day)
Location of several places from air inlet of the duct as follows:

(m)0 m.....P0/Enter..1...2...3...4..P5/Outlet...Face

Air inlet of the duct is at 0 m. P0/Enter means the place of junction of two roadways, one is heading roadway and the other is one ventilated by main air stream. P5/Outlet means the outlet of the air of the duct. A part of the duct between <0 m> and <P0/Enter> is in the main ventilation stream. The rest of the duct is in the return air from the heading face.

Face

Cf: Heat transfer factor (-); Heat transfer coefficient at the face area (Hf) is larger than that at the roadway with through ventilation air by the turbulence of the air at the face. The factor is used to correct the heat transfer coefficient as follows.

Hf = Cf*Hw

Inlet

Temperature of inlet air to the duct (C).
Relative humidity of inlet air to the duct (%)

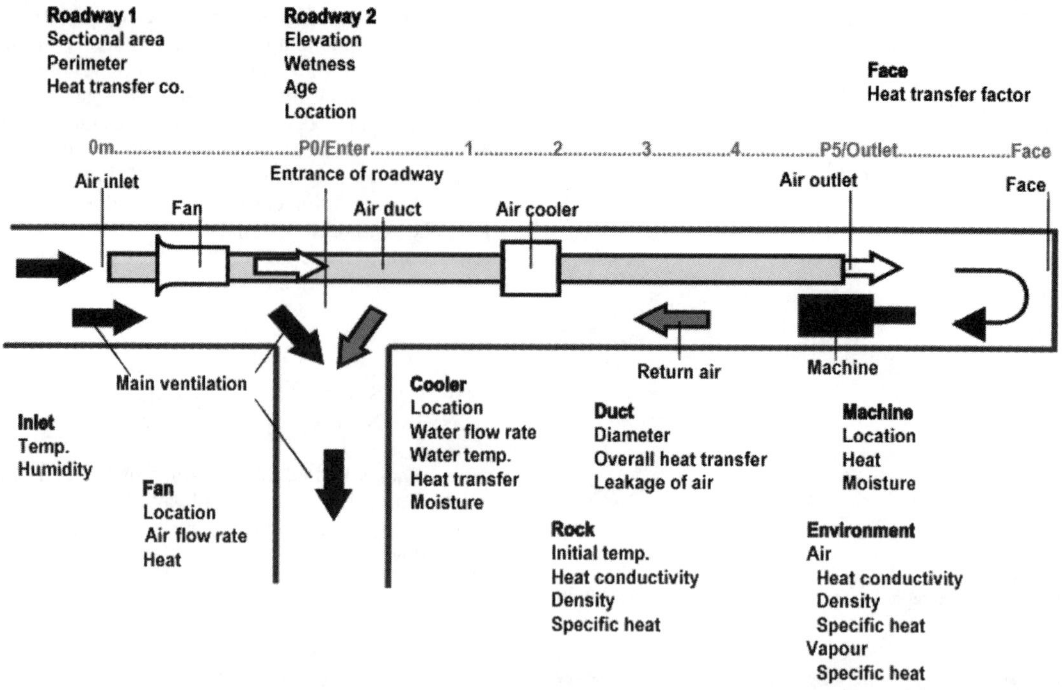

Figure 6. Factors affecting on the thermal environment in the heading face.

Fan

Location; Distance between the fan and the air inlet of the duct (m)
Air flow rate of the fan (m^3/min)
Heat generated by the fan (kcal/h)

Cooler

Location; Distance between the cooler and the air inlet of the duct (m)
Water flow rate into the air cooler (litter/min)
Water temperature into the air cooler (C).
Heat exchange rate of the air cooler (kcal/hC)
Moisture removed from the air passing through the air cooler (kg/h)

Duct

Diameter of the duct (m)
Overall heat transfer between the air in the duct and the roadway (kcal/hC)
Leakage of air per 100 m (%/100 m)

Rock

Initial temperature of surrounding rock of the roadway (C)
Heat conductivity of the rock (kcal/mhC)
Density of the surrounding rock (kg/m^3)
Specific heat of the surrounding rock (kcal/kgC)

Machine

Location; Distance between the machine and the air inlet of the duct (m)
Heat generated by the machine (kcal/h)
Moisture generated by the machine (kg/h)

* Two machines are considered in the program

Environment:

Air

Heat conductivity of air (kcal/mhC)
Density of air (kg/m^3)
Specific heat of air (kcal/kgC)

Vapour

Specific heat of vapour (kcal/kgC)

3 PROGRAM

The air temperature and absolute humidity are calculated using finite deference method in the developed computer program. The rock temperatures around the roadway are calculated analytically with some approximation to shorten the calculation time. The authors have already published a computer program(1) which also estimates the thermal environment in a heading roadway and the face using finite difference method to calculate the rock temperature. It has such an advantage as flexibility in considering boundary condition in space and time as the ordinary finite difference method has. However, it required very long time to get solutions and was thought to be impractical to make many parametric studies to investigate the thermal environment of the heading face.

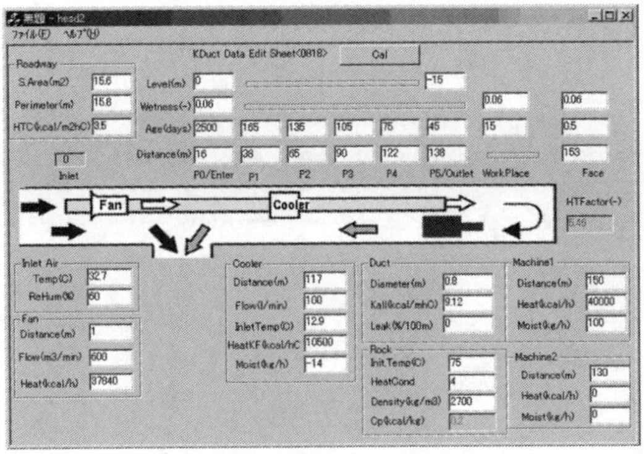

Figure 7. Data input dialog.

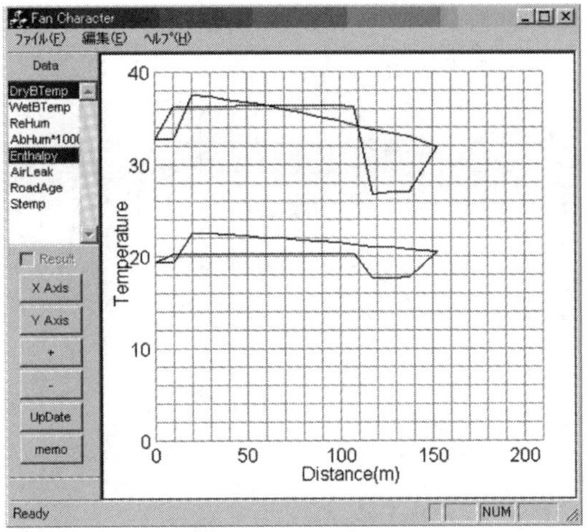

Figure 8. Display of the results.

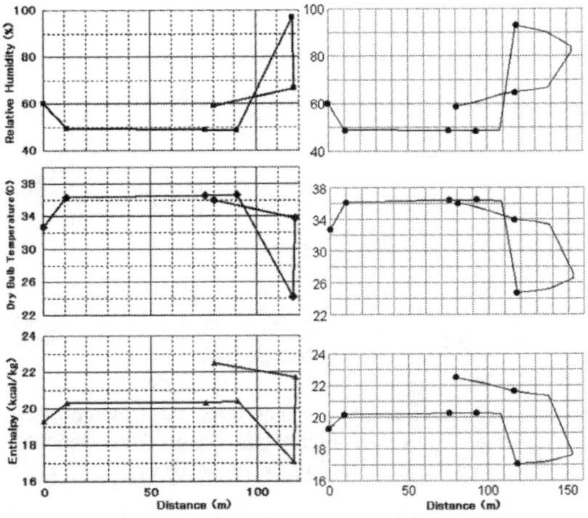

Figure 9. Comparison of measured values with calculated values by the new program.

The performance of an air cooler with the duct is one of the most critical factors affecting the thermal environment in the heading roadway. It is inappropriate in this kind of computer simulation to give the total heat transfer from the air to the cooling water. Therefore, the developed program calculate the heat transfer using several date about the air cooler.

Figure 7 shows a data input dialog of the program. Data are displayed in such a way as to understand easily what they are meaning. The calculation program will be started when the <Cal> button is clicked on the dialog. Usually, the calculation finishes in a second. The results are saved as a text format and other format suitable for the display program of the data. Figure 8 shows the example of the results of the calculation.

4 THE COMPARISON OF THE CALCULATED VALUES AND THE MEASURED VALUES

The authors have measured the thermal conditions in a heading roadway at Hisikari mine to compare the measured values with calculated values by the developed program. Figure 9 shows the comparison of dry bulb temperature, relative humidity and enthalpy of the air in duct and roadway. The comparison shows good agreement between the two. It is thought that the calculation method is sufficiently reasonable.

5 APPLICATION OF THE PROGRAM

Various effects of many factors on thermal environment in a heading roadway and the face will be discussed by the developed program. Here, the effects are investigated by the comparison of the DBT (dry bulb temperature) and WBT (wet bulb temperature) obtained in a standard condition and another condition where some factors were changed. The standard condition is as follows; inlet temperature 30 C, leakage of the duct 10% per 100 m, wetness of the wall of the roadway 0.06, cooling water temperature 12.9 C, cooling water rate 1.7 liter/s, length of roadway 153 m.

5.1 *Leakage of a duct*

The airflow at the outlet of a duct is 8.7 m^3/s (519 m^3/min) when the airflow at the inlet is 10 m^3/s (600 m^3/min) and the leakage of the duct is 10%. The airflow is decreased to 7.4 m^3/s (441 m^3/min) when the leakage of the duct is increased to 20%. Then, the DBT and WBT are decreased from [24.9, 22.2] C to [24.0, 21.9] C as the airflow rate through the air cooler decreases as shown in Figure 10. The first value in a bracket [] means DBT and latter means WBT. Therefore, the temperature is lower when the leakage is 20% where the air is coming directly from the duct.

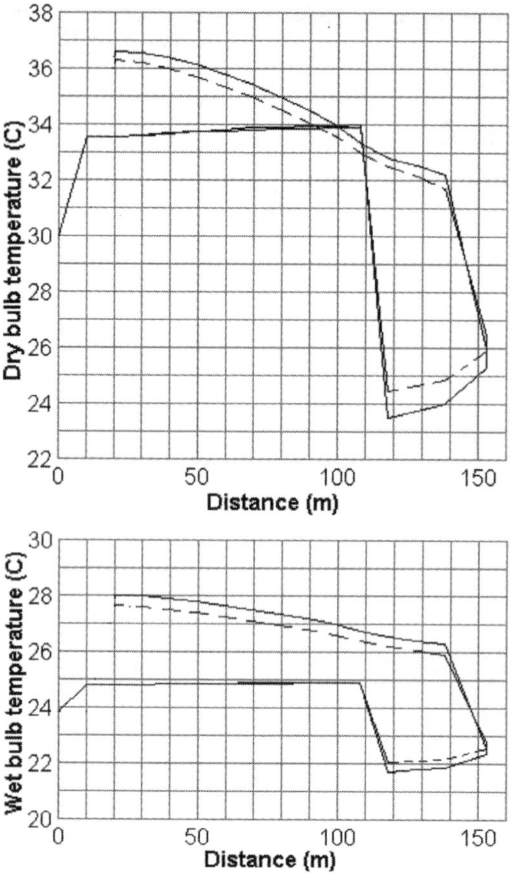

Figure 10. Temperature changes when the leakage of the duct changed from 10% to 20% per 100 m. Doted line corresponds to 10% and solid line corresponds to 20%.

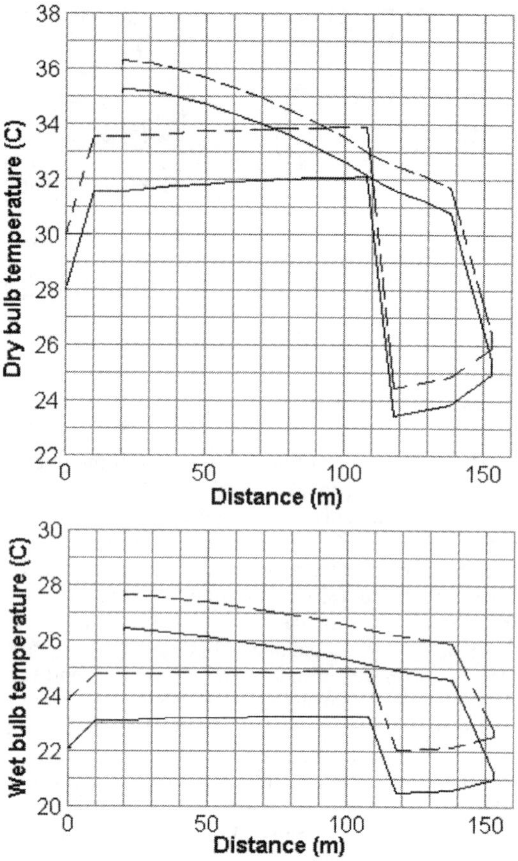

Figure 11. Temperature changes when the inlet temperature changed from 30° C to 28° C. Doted line corresponds to 30° C and solid line corresponds to 28° C.

On the other hand, the return air temperature from the face will be high. The temperature increased from [31.7, 25.9] when 10% leakage to [32.2, 26.3] when 20% leakage near the outlet of the duct. It may be possible to lower the temperature at the face by decreasing the airflow rate in the duct when the target area of cooling is limited near the face and the amount of air is enough for the face. On the contrary, the temperature will increase if the leakage will be 0.

5.2 *Decrease of inlet temperature*

The DBT and WBT decreased from [24.9, 22.2] C to [23.9, 20.6] C when the inlet temperatures of the duct are decreased from [30, 23.9] C to [28, 22.1] C as shown in Figure 11. The efficiency of the air cooler decreases as the airflow rate decreases. Then the decrease of outlet temperature is only 1.0 C despite the decrease of inlet temperature is 2.0 C. On the contrary, the decrease of outlet WBT is 1.6 C while the decrease of inlet temperature is 1.8 C. This is because that there is no addition of water in the duct.

5.3 *Wetness of the roadway*

The wall temperature of the roadway decreases when the wetness of the roadway increases. The return air temperature decreases as shown in Figure 12 when the wetness of the roadway increased from 0.06 to 0.25 (it corresponds to the condition as the floor of the roadway is covered with water). Therefore, the temperature at the outlet of the duct decreases too. On the other hand, the WBT of return air increases as the humidity increases. The thermal environment is improved a little at the face.

The heat from the surrounding rock increases when the wall temperature decreases. Then, the lowering of wall temperature may not be effective method to improve the thermal environment.

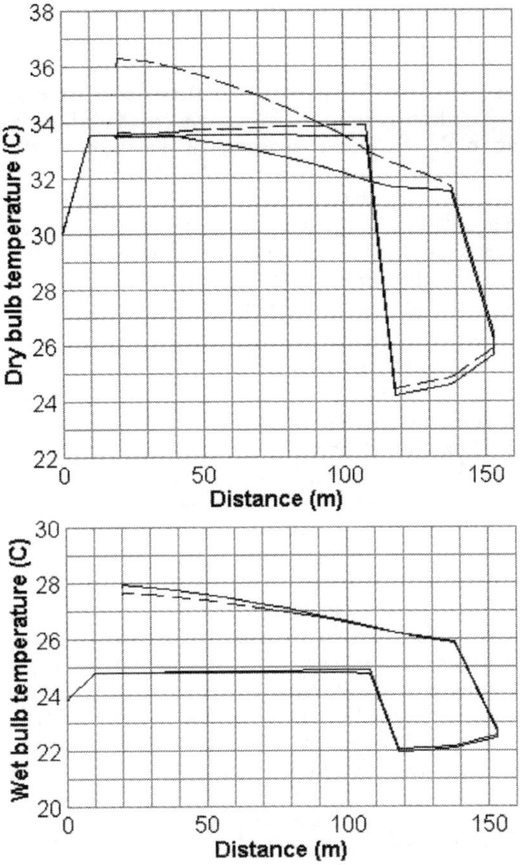

Figure 12. Temperature changes when the wetness changed from 0.06 to 0.25. Doted line corresponds to 0.06 and solid line corresponds to 0.25.

5.4 *Cooling water flow rate*

The air temperature of the outlet of the duct decreases from [24.9, 22.2] C to [23.3, 21.7] C when the cooling water rate is increased from1.7 l/s (100 l/min) to 2.5 l/s (150 l/min). The variation of cooling water temperature at the outlet of cooling coil, and the air temperature at the inlet and outlet of the air cooler with cooling water flow rate are shown in Figure 13(a). The variation of total heat transfer in the air cooler and the heat transfer per unit water flow with cooling water flow rate are shown in Figure 13(b). It is apparent that the air temperature at the outlet of the air cooler does not decrease in proportion to the cooling water flow rate from the figure. Especially, the change is small when the rate is more than 2 l/s. The heat transfer rate per unit water flow is only less than 10 kwl/s when the water flow rate is more then 3 l/s although it is almost 100 kwl/s when the water flow rate is about 1 l/s. The efficiency of utilization of cooling water is very low when the rate is more than 3 l/s. It

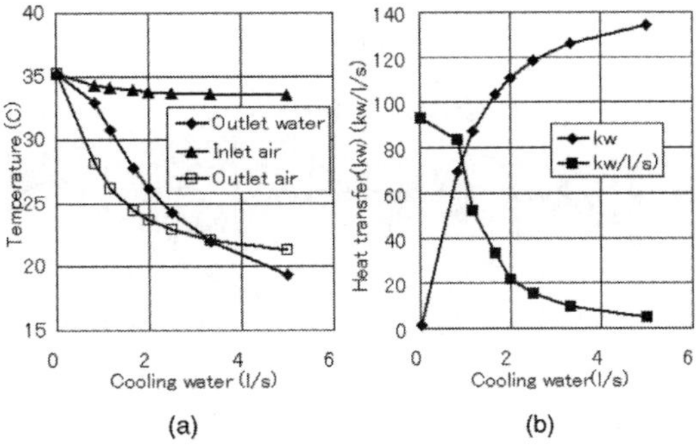

Figure 13. Performance of air cooler.

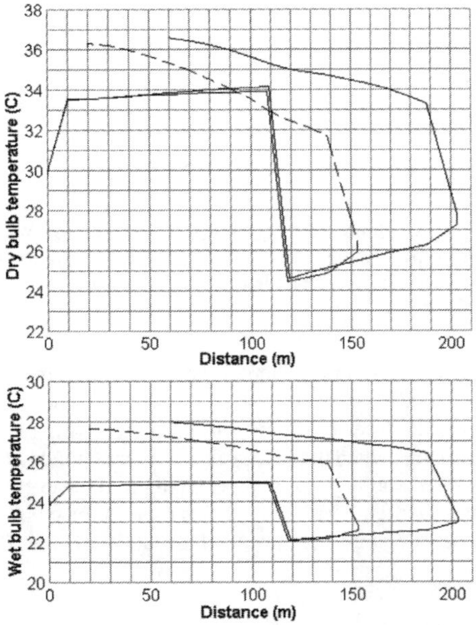

Figure 14. Temperature changes when the roadway is extended by 50 m and the air cooler remained at the same place. Doted line corresponds to 153 m and solid line corresponds to 203 m.

means that it will be in vain to use too much cooling water to decrease the air temperature unless the performance of the air cooler is not improved.

5.5 *Length of roadway*

Figure 14 shows the temperature when the roadway is extended by 50 m and the air cooler remains at the same place. The temperature at the outlet of the duct increases from [24.9, 22.2] C to [26.3, 22.6] C. The temperature of return air from the face increases considerably from [31.7, 25.9] C to [33.3, 26.4] C. The WBT does not increase much as compared to DBT.

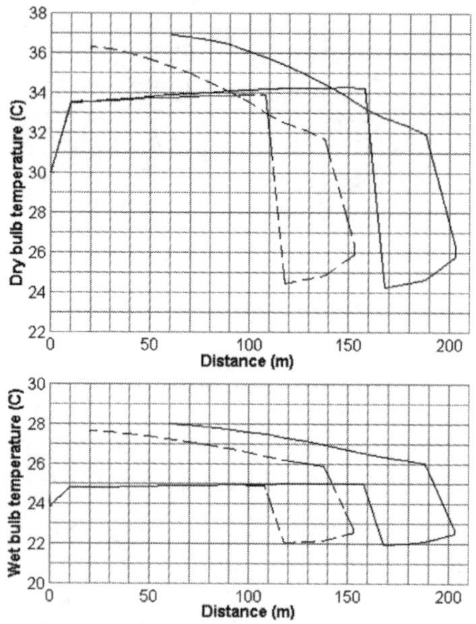

Figure 15. Temperature changes when the roadway is extended by 50 m and the air cooler moved 50 m to the face, too. Doted line corresponds to 153 m and solid line corresponds to 203 m.

Figure 15 shows the temperature when the roadway is extended by 50 m and the air cooler is also moved by 50 m to the face. The temperature of the cooling water may change when the place of air cooler changes as the length of cooling water pipe changes. However, the same water temperature 12.9 C is used in the example. The temperature of air at the outlet of the duct decreased from [24.9, 22.2] C to [24.7, 22.1] C despite the increased length of the roadway. This comes from the fact that the effect of the air cooler for unit air flow increased because the total air flow through the air cooler decreased from 8.7 m^3/s to 8.2 m^3/s by the leakage of the duct. If the same amount of air is required to the face the leakage of air should be decreased from 10% to 7.5%. The temperature will be almost same as that obtained for the standard condition. The temperature is not change so much as shown in the figure if the length of the roadway extended in this range. However, the increase of return air temperature is considerable.

6 CONCLUSION

A new computer program to predict the temperature and humidity of a heading roadway with forcing system of ventilation is developed and evaluated by comparing them with measured data in a mine. The estimated values by the program show good accordance with the measured data. The key point of the accuracy of the prediction of the thermal environment will be how to estimate the reasonable data about the mine rather than how to calculate the environment.

There still remains some inconvenience to be improved in the developed program. Almost all of them are related to the air cooler and fan performance. Nevertheless, the program has become one of the important tools for the mining plan.

REFERENCE

1. Gao, J., Uchino, K. & Inoue, M. Proceedings of the 9th US Mine Ventilation Symp. (2002), 347–353.

Mine Ventilation, Ganguli & Bandopadhyay (eds)
© 2004 Taylor & Francis Group, London, ISBN 90 5809 633 5

Auxiliary ventilation operation practices

E. De Souza
Department of Mining Engineering, Queen's University, Kingston, Canada

ABSTRACT: A series of visits to underground operations have been performed to identify installation and maintenance practices in auxiliary ventilation systems. This paper presents industry trends in duct ventilation practice and a series of guidelines are introduced for ventilation operators to follow when performing installations and inspections. This paper presents information with which mines could train their personnel on effective operational procedures to maximize the performance and guarantee compliance of their installations.

1 INTRODUCTION

Auxiliary ventilation is used to augment the primary ventilation system in supplying adequate quantities of fresh air to inadequately ventilated workplaces. Typical applications of auxiliary ventilation underground include the ventilation of development headings, and the ventilation of stopes and raises. Auxiliary ventilation systems must be designed to supply a sufficient quantity of air to provide a healthy, safe and comfortable environment in the workplace.

When designing a system one must consider government regulations which typically prescribe that 'in an underground mine, a development, exploration or production work place shall be ventilated throughout by an auxiliary ventilation system for any advance in excess of 60 m (200 ft) (two hundred feet) from a mechanical mine ventilation system; and a continuous supply of fresh air shall be provided and used to dilute and remove contaminants, in a raise, and in a sub-drift for any advance in excess of 10 m (30 ft) from a mechanical mine ventilation system, to prevent exposure of a worker to contaminants in excess of maximum allowed values (threshold limit values). The fresh air supply shall be independent of the air supplied by any drill or machine used; controlled only at the beginning of the raise or sub-drift; and operating when a blast is detonated'.

This paper presents the design aspects of auxiliary ventilation and, based on a series of visits to underground operations, introduces typical operating practices and provides guidelines to determine the most efficient ventilation system for a particular project.

2 TYPICAL AUXILIARY VENTILATION SYSTEMS

The range of auxiliary ventilation systems found in the mines ranged from 30 m (100 ft) to some 1000 m (3280 ft). Of the different categories of auxiliary systems the most common systems used are the force system, in which the intake air is led to the face through the duct, and the exhaust system, in which the air is drawn from the face through the duct.

The force system is the most commonly used method in metal mines due to its simplicity of operation and relative economical attributes. The force system supplies good quality air, at high velocity, to the face and allows a quicker clearing of the face, but sends a plug of blasting gases up the development drift, towards the main airway. Depending on the installed system, long re-entry periods after blast may be necessary. The end of the ventilation column is kept as close to the face as possible during working operations and should not exceed some 30 duct diameters away from the face (Figure 1). In general, the ventilation tubing is located at a distance between 5 and 12 m (15 and 40 ft) from the face in horizontal headings and not more than 2.5 m (8 ft) from the back in raises. Also, the tubing is installed as close as possible to the back of the drift to ensure the air sweeps downward across the working face from roof to floor (Figure 2). This method assures the airflow to pass over the workers and equipment operating in the heading, thus carrying the contaminants from the face below the worker's breathing zone. Specific design considerations are also followed in order to avoid build up of gas and dust concentrations through recirculation

Figure 1. The end of the ventilation column should be kept as close to the face as possible.

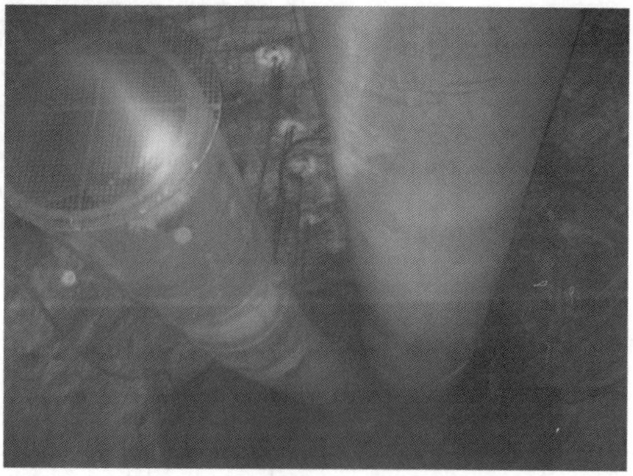

Figure 3. The fan should be positioned upstream in the main ventilation circuit to eliminate recirculation.

Figure 2. The end of the ventilation column is positioned at the centre of the drift and high at the back to provide adequate dilution of contaminants.

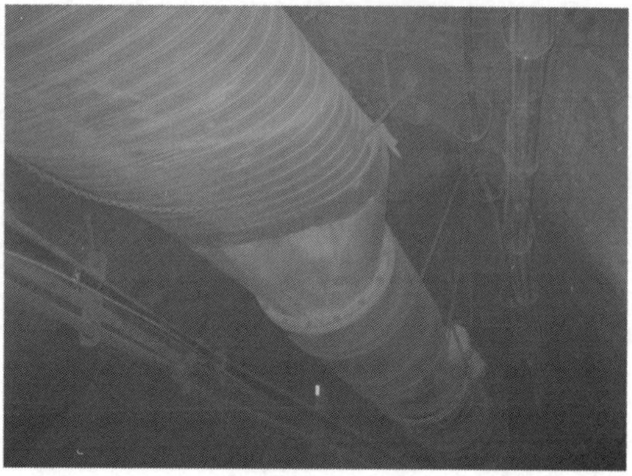

Figure 4. Circular and oval rigid duct in an auxiliary ventilation system.

of air. The fan is installed in an adequate fresh intake air base and is positioned to eliminate the possibility of recirculation. The fan is installed at a point at least 10 m (33 ft) upstream in the general mine air circuit (Figure 3). The volume of intake air available should be 2.5 times the fan capacity. In order to avoid damage when blasting, the last section of tubing at least 15 m (50 ft) is taken down to a safe distance from the blast. The last section of duct is advanced with the face until it has to be replaced; the pressure is lowest in this section of duct and minor mechanical damage due to blasting will not result in too great an air loss.

The exhaust system incorporates an exhausting fan and uses rigid ducting (Figure 4). Spiral wire reinforced flexible tubing may be used for short lengths and low pressures. In this method a single fan and column are installed to exhaust contaminated air from a point as close to the face as practical and discharge it downstream in the nearest through ventilation. The fan in this system provides a continuous, negative pressure in the tubing, and brings in fresh air through the drift, and removes it from the face and workplace through the tubing. The fan can be installed anywhere along the duct. The exhaust end of the duct is installed in an adequate exhaust air base and is positioned so there is no possibility of recirculation. The duct has to be carried to the face before drilling and mucking and the last section (15 m, 50 ft) should be removed before blasting. The advantages of this system include quick removal of smoke and gases after blasting, specially in long headings and the contaminated air from the face is removed without affecting workers in the travel way. Some disadvantages of this method include: the air enters the end of the duct at low velocity and will not sweep the face of the heading and the higher costs involved if rigid tubing is used.

The selection of the type of auxiliary ventilation system depends on the amount of air and air quality conditions required, the size and type of fan required and the size and type of duct necessary. Acceptable working conditions determine the required airflow quantities which in turn specify equipment requirements and ventilation tubing size. The main factors considered when selecting a specific auxiliary ventilation system include:

- The type of heading, i.e. drifts, ramps, raises, shafts, etc.
- The length of heading. Longer headings may require more sophisticated systems.
- The heading cross-sectional area. Larger headings require considerably higher airflows than smaller ones. This is based on a minimum air velocity requirement of 2.75 m/s per m^2 of face area (50 ft/min per ft^2 of face area).
- Equipment used to drive the heading. Headings driven by diesel powered equipment require higher air quantities than other types of equipment.
- The local temperature. In situations where high rock temperature and high humidity exist, fresh cool air must be supplied to the face via ducting in order to reduce the temperature in the working area.
- Allowed re-entry times. The time allowed to clear the heading after the blast is critical in the selection of an auxiliary ventilation system. For continuous drifting operations, where crew changes may occur at the face, any lost re-entry time is detrimental to the advance of the heading.
- Cost. The effectiveness of an auxiliary ventilation system should not be judged by cost alone. The additional cost of installing a more complicated system must be balanced against the benefits this system will provide as compared to a simpler system.

3 AUXILIARY VENTILATION DESIGN PROCEDURE

The following steps are normally followed when designing an auxiliary ventilation system:

- Step 1: Air volume calculations. This is generally based on diesel equipment operating in the heading. The following factor is considered: 100 cfm per brake horsepower (0.0633 m^3/s per kilowatt) of operating or installed diesel equipment.
- Step 2: Air leakage estimations. Leakage may be assumed to range between 20% and 30%. Calculate the flow required at the fan using: $Q_{fan} = Q_{face}/(1 - \text{leakage})$.
- Step 3: Ventilation system selection. The type of auxiliary ventilation system, the size and type of fan required and the size and type of duct necessary are decided based upon the type of heading, the length of heading, the heading cross-sectional area, the development method, allowed re-entry times, and cost considerations.
- Step 4: Airflow resistance calculations. The contribution of each component of the column to the head loss must be determined. Atkinson's equation can be used to estimate the friction head loss in the duct. The friction head loss contributed by each coupling has an equivalent length of 1.8 m (6 ft). This is multiplied by the number of couplings and added to the length of duct. Elbow losses are estimated using standard shock loss equations. If tubing larger than the fan is used, the losses (including conversion of velocity to static pressure) associated with the transition cone must be considered. Entrance and exit losses must also be included into the calculations. In all calculations appropriate airflow volumes must be used for each section of the column. The flow at the fan is different from the flow at the face and will decrease as it moves along the column.
- Step 5: Fan sizing. Fan curves are analysed to match with the system requirements of head and flow. The fan selection should be based on total pressure. The fan blade pitch and motor size are determined in this step.

A number of design factors (rules of thumb) used by mines in ventilation design calculations are listed below. It must be emphasized however, the danger of basing air requirements on ventilation factors as it may result in under designed or over designed ventilation systems.

- Air velocities to obtain gas dispersion from 0.25 to 0.30 m/s (50 to 60 ft/min).
- 0.25 to 1 m/s (50 – 200 ft/min) design face velocity.
- 0.25 m/s (50 ft/min) air velocity for headings which are not governed by diesel requirements.
- Minimum quantity of air per worker underground of 0.01 m^3/s (21.2 cfm).
- 0.014 to 0.028 m^3/s of air per square metre of face area (30–60 cfm per square foot of face area).
- 0.095 m^3/s (200 cfm) per worker at a minimum air velocity of 0.15 m/s (30 ft/min) required to dilute fumes and dust.
- 100 cfm per brake horsepower (0.0633 m^3/s per kilowatt) of operating or installed diesel equipment.
- 9290 to 13935 m^3 (100000 to 150000 ft^3) of air per imperial gallon of diesel fuel consumed.
- 0.014 m^3/s (30 cfm) of fresh air per pound of explosive used will permit re-entry in 20 minutes.

The above design steps are described in more detail in the following sections.

4 FACE AIR VOLUME REQUIREMENTS

Air quantity is governed by federal and provincial regulations. Auxiliary ventilation is required to dilute the fumes from blasting and diesel exhaust to a safe concentration required to maintain acceptable working conditions, to dilute any harmful or dangerous strata gases to safe, acceptable concentrations and to replace the oxygen used up by the workers and diesel equipment. It is also required control airborne dust and to control the temperature and humidity of the air in the working area. All these factors are considered when determining air volume requirements at the face. Some systems operate at flows as low as 1.4 m^3/s (3,000 cfm) and as high as 28 m^3/s (60,000 cfm).

Air quantity is dependent on the advance distance and dimensions. The estimation of air volume requirements is normally based on the minimum amount of airflow required at the heading during the final stages of development or at the stope face during the final stages of production; i.e. when the ventilation tubing is at its maximum length.

In the majority of cases diesel exhaust dilution requirements are used to determine the airflow volume at the face. When utilizing diesel units in development headings one must ensure the air requirements for diesel equipment are met, 0.0633 m^3/s per kilowatt (100 cfm per brake horsepower) of operating or installed diesel equipment. For example, an 8 yard scooptram with rated brake power of 179 kW (240 HP) will require 11 m^3/s (24,000 cfm) and a bolt platform with rated brake power of 63 kW (84 HP) will require 4 m^3/s (8,400 cfm).

When considering mine gases, typical regulations require that all active workings underground be ventilated by an air flow with a minimum oxygen content of 19.5% and not more than 0.5% carbon dioxide. The threshold limit values of common gases encountered in headings are: carbon monoxide 0.0025%; carbon dioxide 0.5%; nitrogen oxide 0.0025%; nitrogen dioxide 0.0003%; hydrogen sulphide 0.001%; sulphur dioxide 0.0002% and methane 1%. Such values are used to estimate the volume flow of fresh air required for appropriate dilution ventilation.

Because the fumes produced by blasting are produced only once in each drill-blast-muck cycle, and are cleared before any of the diesel equipment are employed, the volumes of fresh air required to clear the blasting fumes from the face and to dilute the diesel fumes are not added together; the air volume requirements is based on whichever amount is greater.

5 AUXILIARY FANS AND FAN REQUIREMENTS

The size of the fan is dependent on the air volume requirements, on the head losses in the tubing and air leakage in the system. Air leakage is always present in auxiliary ventilation, it is minimized when good ventilation practices are followed, and must always be incorporated during system design and fan selection. Air leakage is dependent on duct conditions, installation practices and on the static air pressure in the duct.

Under normal conditions, small fans are used in auxiliary ventilation systems. Airflow capacities range from 1.0 to 14.2 m^3/s (2,000 to 30,000 cfm), and may reach some 28 m^3/s (60,000 cfm) and static pressures normally vary between 0.25 kPa (1″ w.g.) and some 2 kPa (8″ w.g.), but may reach as much as 3.7 kPa (15″ w.g.). The electric motors running these fans normally range from 4 kW (5 HP) to 75 kW (100 HP). Larger motors, reaching 150–185 kW (200–250 HP) may be required depending on conditions. Typical fan sizes are 610, 760, 915 and 1220 mm (24, 30, 36, 42 and 48 inches).

Mines normally select fan sizes based on experience. Fan requirements for mechanized headings depend on the length of the heading and flow requirements. For example, for a short heading, 38 m (125 ft) long, requiring a flow of 14 m^3/s (30,000 cfm), a 76 cm (30″) fan with a 29 kW (40 HP) motor would be selected. A longer heading, 137 m (450 ft) long, requiring a flow of 14 m^3/s (30,000 cfm), a 91 cm (36″) fan with a 45 kW (60 HP) motor would be selected. For a flow of 21 m^3/s (45,000 cfm), a 96.5 cm (38″) fan with a 75 kW (100 HP) motor would be selected.

Typically, auxiliary axial flow fans are designed based on the final length of the drift, and the blade pitch is initially set for this requirement. Normally, instead of adjusting the blade pitch during the various phases of heading development, flow control is achieved with the use of regulators. A common method of regulating electrical fans, although improper, is by constricting the duct by means of a cord tied around it.

In long ductwork installations, booster fans are normally spaced at intervals along the auxiliary ventilation system (staged fans), rather than positioning all the system static pressure requirements (one high pressure booster or several fans in series) at the start of the column. This is done to reduce high pressures in the column thus minimizing system leakages.

6 VENTILATION TUBING AND TUBING SELECTION

The selection of the ventilation tubing is crucial in system design. In order to assure maximum air delivery to the heading, the resistance of the duct system should be minimized. The tubing resistance is governed by its size, type, length, layout and efficiency of installation and maintenance. In order to minimize resistance to flow, the tubing must be stretched tight during installation and possess good alignment and be free of unnecessary sinuosity. The type of tubing depends on the ventilation system (blowing, exhaust), the drift and workplace requirements and somewhat

Figure 5. Sufficient clearance between ducting and mobile equipment should be provided to prevent damage to column.

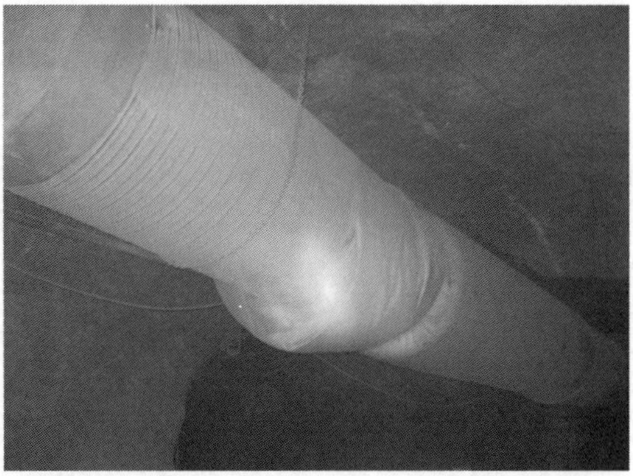

Figure 6. Flexibe and rigid tubing used in the same column.

Figure 7. Underground manufacturing of rigid tubing.

on personal preference. Duct diameters are generally selected to match the fan diameter. However, when sizing the ventilation column, the size of the drift and equipment are considered to allow for sufficient clearance between the ducting and mobile diesel equipment (Figure 5).

Three general types of tubing are commonly used to ventilate working faces: rigid, flexible and flexible wire-reinforced tubing (Figure 6). Typical tubing internal diameters range between 610 mm (24″) and 1220 mm (48″) but can go as large as 1960 mm (78″). Oval ducting is also used, and selected to have an equivalent round duct diameter. Ducting is available in various lengths, generally in 3–9 m (10–30 ft) lengths or longer.

Flexible tubing are very popular because they are lightweight, easy to handle and install and have low capital cost. They are commonly used for temporary systems and in short headings (up to 180 m, 600 ft). Their main disadvantage is their susceptibility to damage. Flexible tubing is useful in headings in which room is restricted or in headings which are not straight. Resistance characteristics are fair when hung straight and well maintained. They can only be used for blowing systems as the static pressure holds the tubing open. When abrupt changes in drift direction are encountered, the use of spiral tubing may be necessary to eliminate the possibility of crimping of the column. Flexible wire reinforced tubing are light, and easy to handle. They have poor resistance to flow unless if stretched tight during installation.

Rigid ducting are rugged, have low resistance to flow both under blowing and exhausting conditions and low maintenance costs for long, permanent systems. They are bulky, heavy, difficult to handle and install and have high capital cost. Rigid ducting can be built underground (Figure 7) thus making it competitive with flexible ducting. Although steel tubes

Figure 8. Rigid elbow and transition from oval to circular duct.

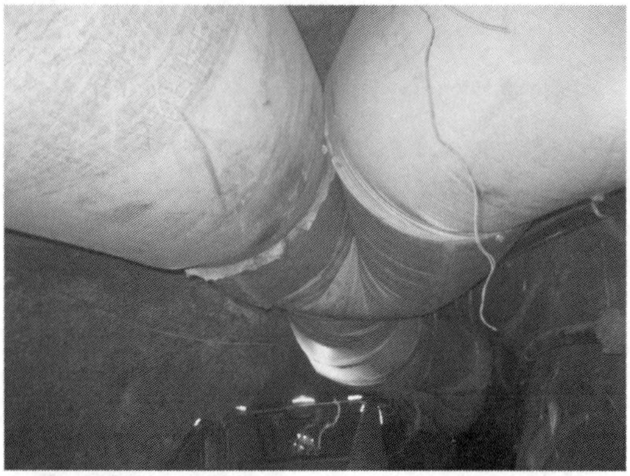

Figure 9. Y split in ventilation column.

offer extremely low leakage, they are susceptible to rust. Fiberglass tubing are smoother and have a coefficient of friction of approximately 12% less than metal tubing.

Ducting accessories include flexible and rigid elbows (Figure 8), end caps, T's, Y's (Figure 9), reducers and transition pieces.

Every duct column leaks at seams and joints. Air leakage is dependent on duct conditions, installation practices and on the static air pressure inside the duct. Air leakage can, in many cases, be very significant and must be incorporated into the ventilation balance. Leakage increases with the static pressure, which, in turn increases with the length of the column, and the volume of air flow. As a rough guide: doubling the length of column, trebles the leakage for a given volume of air flow; doubling the volume of air flow increases the leakage two and a half fold for a given

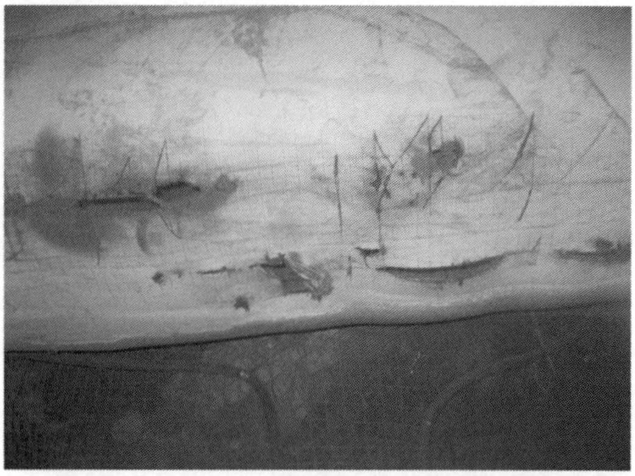

Figure 10. Severe leakage due to damage by equipment.

length of duct; approximately 0.09 m^3/s (190 cfm) of air leakage occurs per joint in the column. For these reasons, in long drifts, fans spaced at intervals along the column prove more satisfactory than positioning one (or more) high pressure fan(s) at the start of the tubing system. In general, the amount of leakage from normal ducting installations is very high. Leakage values exceeding 30–50% are not unusual due to poor maintenance practices and due to damage by blasting and equipment (Figure 10).

7 INSTALLATION PRACTICES

The proper installation and maintenance of tubing is of great importance to assure efficiency in system performance. Leakage should not exceed 10% the initial flow volume, although leakage above 50% is frequently observed due to poor installation and maintenance practices. The following installation procedures are normally followed at the mines:

- The support cable and tubing are hung and carried as straight and tight as possible with a minimum of bends (Figure 11).
- Proper fan adaptors are used to connect the duct to the fan in order to avoid leakage.
- Duct couplings, tight clamps and clips are used to assure good duct connection and avoid leakage.
- Ducts, in particular flexible ducts, are hung with hooks attached to messenger cables. Steel ducts are hung with the use of chains (Figure 6). Hooks are also directly attached to bolts, straps and screens used for strata support.
- In order to minimize damage to the column by mobile equipment, the ducting is suspended as high as possible against the back, in particular at intersections and low-lying areas (Figure 12).

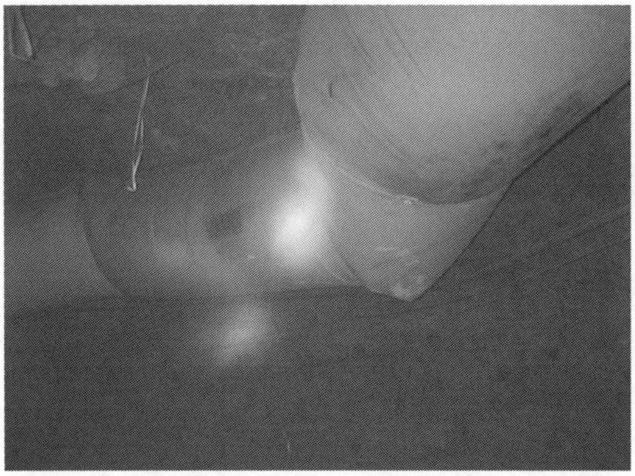

Figure 11. Crimping of column due to improper installation.

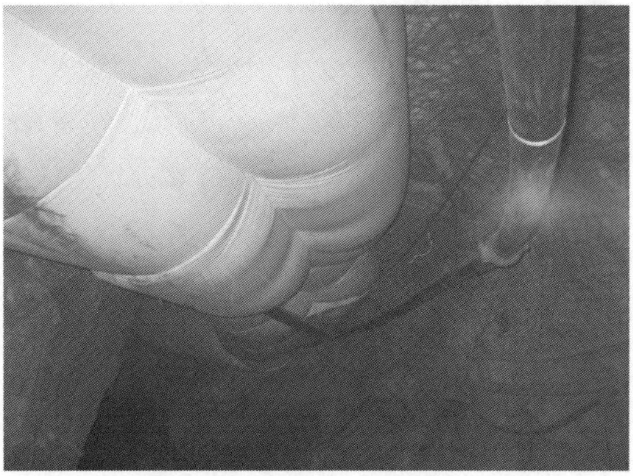

Figure 12. Special duct for narrow and low-lying drifts.

8 PRACTICAL DESIGN AND OPERATIONAL CONSIDERATIONS

The following design and operational aspects were found to be implemented in a number of mines:

- A fan should be installed in such a manner that a sufficient quantity of air reaches it at all times so as to ensure that it does not recirculate air.
- Fan installation should be conducted so as to eliminate the risk of the air which it draws being contaminated by toxic fumes and dust.
- The volume of air a fan will handle must always be less than the quantity of air passing in the airway in which the fan is installed. The capacity of the fan should not be more than one half of the quantity of air passing in the airway where the fan is situated.
- When installing fans spaced in series, at intervals along the auxiliary ventilation system, it is common to leave a gap of approximately 15 cm (6″) between the end of the column and the booster fan to prevent collapse of the column. One should however be aware of the potential for excess recirculation of contaminated air.
- When sizing the ventilation column, the size of the drift and equipment must be considered. Sufficient clearance between the ducting and mobile diesel equipment must be allowed to prevent damage to the duct. Personnel should be advised of the importance of avoiding causing damage to the column when utilizing transport and haulage equipment.
- Fans should be turned off during blasting. This step will prevent damage to the impeller blades by air blast.
- The ventilation tubing should be maintained in a condition such that air leakage is minimized. For economic operation, leakage should not exceed 10%. For design purposes, leakage may be assumed to range between 10 and 30% of the fan supply flow.
- Minor damage to tubing at different segments of the column will have a cumulative contribution to leakage. As such, even minor damage should be readily repaired during normal inspection. Simple techniques such as stitching or patching can be employed (Figure 13). The use of 'zipper tubing' may save considerable repair time.
- The last 30–45 m (100–150 ft) of tubing should be removed (or protected) prior to a blast to minimize damage.
- The end of the column should be closed prior to a blast. This will prevent or minimize the accumulation of dust and debris in the column, as such collection and build-up of material will increase the resistance to flow.
- As additional lengths of new ducting are added to the column with the advance of the heading, the last section of ducting (which is subjected to most damage) is always maintained at the end of the column.

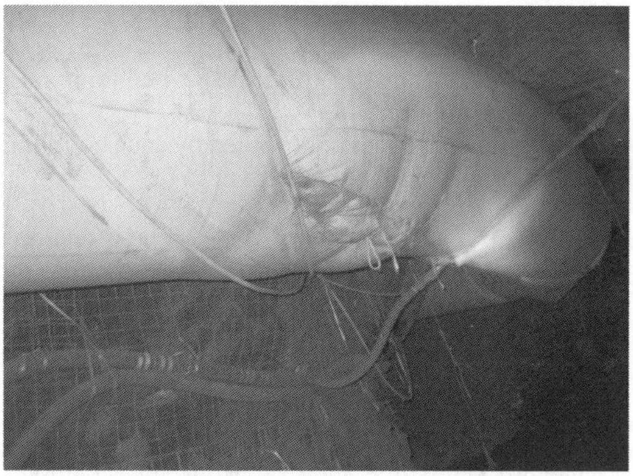

Figure 13. Stitching used to repair damage to tubing.

- Adequate air delivery to within 5 m (15 ft) of the face should be maintained.
- $0.09\,m^3/s$ (190 cfm) of air leakage per joint in a ventilation column.
- The friction head loss contributed by each coupling has an equivalent length of 1.8 m (6 ft).
- Friction factor of $0.0028\,N.s^2/m^8$ (15×10^{-10} $lb.min^2/ft^8$) for fiberglass duct, $0.0037\,N.s^2/m^8$ ($20 \times 10^{-10}\,lb.min^2/ft^8$) for canvas duct and $0.00111\,N.s^2/m^8$ ($60 \times 10^{-10}\,lb.min^2/ft^8$) for spiral-type canvas.
- One velocity pressure for discharge (exit) loss.
- Daily inspection for column damage should be performed.
- Standard ventilation surveys (with vane anemometers or pitot tubes) should be performed weekly to verify if sufficient air is reaching the face and to verify that recirculation is not developing. If the airflow is too high, the use of regulators may be required. Factors that should be included during a survey are motor amperage, fan pressure, air volumes at fan and at exit of column, gas levels, straightness of column, condition of ducting, condition of couplings, distance of last duct section from face.

9 CONCLUSIONS

This paper was presented to provide, for the ventilation operator and practising engineer, design guidelines and operational procedures which would maximize the performance of auxiliary ventilation systems. The contents of this paper are based on a series of visits to underground operations and represent common techniques practised in such operations.

REFERENCES

Bossard, F.C., LeFever, J.J., LeFever, J.B. & Stout, K.S. 1983. *A Manual of Mine Ventilation Practices*. Second edition. F.C. Bossard and Associates, Inc.

Burrows, J., Hemp, R., Holding, W. & Stroh, R.M. 1989. *Environmental Engineering in South African Mines*. Mine Ventilation Society of South Africa.

Flexible Ducting Ltd. 1957. *Tunnel Ventilation*. Maryhill: Glasgow.

Hall, C.J. 1981. *Mine Ventilation Engineering*. The American Institute of Mining, Metallurgical and Petroleum Engineers, Inc.

Hartman, H.L. (ed.), Mutmansky, J.M. & Wang, Y.J. (co-eds.) 1982. *Mine Ventilation and Air Conditioning*. Second edition. John Wiley & Sons.

Misra, G.B. 1986. *Mine Environment and Ventilation*. Oxford University Press.

National Coal Board. *The Ventilation of Headings*. Information Bulletin No. 52/74.

Notley, R.K. 1983. *Mine Ventilation Manual*. Queen's University.

Roberts, A. 1960. *Mine Ventilation*. Cleaver Hume Press Ltd.

Sandkuhler, T. et al. 1998. *Auxiliary Ventilation Systems*. Queen's University.

The Mine Ventilation Society of South Africa (1990) *Le Roux's Notes on Mine Environmental Control*. CPT Book Printers.

Thompkins, R.W. *Manual on Mine Ventilation*. Queen's University.

Mine Ventilation, Ganguli & Bandopadhyay (eds)
© 2004 R.H. Grau III, ISBN 90 5809 633 5

Raising the bar of ventilation for large-opening stone mines

R.H. Grau III, S.B. Robertson, R.B. Krog, G.J. Chekan & T.P. Mucho
National Institute for Occupational Safety and Health, Pittsburgh Research Laboratory, Pittsburgh, PA, USA

ABSTRACT: Due to the difficulties of ventilating large opening stone mines, the effective dilution of diesel particulate matter and noxious gases has often been problematic. Since stone mines have large openings that develop low air resistances, and propeller fans are designed for low pressure-high volume applications, main mine propeller fans may be a feasible option for ventilating underground stone mines. This paper describes the results of a case study where researchers from the National Institute for Occupational Safety and Health have demonstrated that significant increases in total mine airflow can be achieved and the ventilation improved in a large-opening stone mine by replacing an axial-vane mine fan with two 12-ft diameter propeller fans and adding a line of stoppings. In this study, the use of the propeller fans more than doubled the air quantity from 300,000 cfm to 750,000 cfm, while operating at 50 horsepower less than the original axial-vane fan.

1 INTRODUCTION

The National Institute for Occupational Safety and Health (NIOSH) is the primary agency responsible for conducting health and safety research pertaining to the mining industry. A portion of this research is concerned with developing methods and technologies to improve the ventilation technology of large-opening mines. Historically, the ventilation techniques used in large-opening stone mines have often been the result of trial-and-error methodology. In many cases the ventilation system consists of axial-vane fans being used as either auxiliary fans or main mine fans without the use of stoppings to direct the air. The research results presented in this report describe how the ventilation of large-opening drift mines can be dramatically improved by using large-diameter main mine propeller fans in bulkheads in combination with a stopping line. Also discussed are methods for estimating the air quantity requirements for effective dilution of diesel particulate matter (DPM), choosing appropriate fans, air coursing methods, and mine layouts.

2 CONCERNS AND LIMITS FOR WORKER EXPOSURE TO DPM

Legislation was proposed by the Mine Safety and Health Administration (MSHA) to limit DPM concentrations in underground metal/nonmetal mines. Provisions of this health standard are found in "Diesel Particulate Matter Exposure of Underground Metal and Nonmetal Mines" published in the Federal Register on January 19, 2001 and amended on February 27, 2002 (MSHA 2001). An interim DPM standard of 400 $\mu g/m^3$ of total carbon took effect on July 20, 2002 and it's proposed that the DPM limit be reduced to 160 $\mu g/m^3$ of total carbon by July 19, 2006.

To satisfy the requirements of the DPM standard, most operators will find it necessary to employ several engineering controls: increasing the air quantity entering the mine; directing air to the face areas by the use of stoppings and auxiliary fans; and decreasing DPM emissions from engines by using catalytic converters, filters, and cleaner-burning engines (Schnakenberg 2001, Grau et al. 2002a,b). In addition to the general improvements in current ventilation methods, mine planners must also consider ventilation requirements when planning future mine layouts – until now, ventilation has been overlooked in most underground stone mine planning processes (Head 2001, Grau et al. 2002a, Grau et al. 2002b, Mucho et al. 2001).

3 MINE PLANNING CONSIDERATIONS FOR EFFECTIVE VENTILATION

A large-opening stone mine ventilation system must be designed to effectively dilute all airborne contaminants by supplying sufficient ventilation air to locations where it is needed the most, i.e., working faces, worker locations, etc. In order to accomplish that goal, the ventilation plan must include a main mine fan to develop the airflow, stoppings to control the airflow, and auxiliary fans to direct air where equipment and personnel are located. Invariably, all underground stone mines use diesel equipment that emits substantial amounts of DPM and noxious gases such as NOx, SO_2, and CO. In mines that use older diesel engines, the overriding factor for determining ventilation air requirements is probably the dilution of DPM and not the noxious gases. A previous study has shown that significant ventilation air quantities are necessary to meet DPM compliance concentration levels (Grau et al. 2002a). To help determine the ventilation air quantity requirements to dilute DPM concentrations to statutory levels in the main mine air stream, NIOSH has developed an air quantity estimator (AQE) (Robertson 2001, Robertson et al. 2003). The AQE is a user-friendly computer program that comes as a stand-alone package. It will run on most personal computers and is not dependent on the installation of spreadsheet programs.

4 HISTORICAL PHYSICAL CHARACTERISTICS OF UNDERGROUND STONE MINES THAT INFLUENCE VENTILATION

Many underground large-opening stone mines have common physical characteristics that influence ventilation airflow. Multiple portal entries are common, with the portals ranging from fully open to almost completely closed due to highwall rock falls or icing during winter months. Typically, in stone drift mines, the differences in elevation between entries are small in terms of creating natural ventilation. Nevertheless, these mines do have some natural ventilation. In addition, numerous openings, truck movements, outside wind conditions, and freestanding fans provide some ventilation in the mine. Finally, the large mine void volume acts as a sizable sink of available air for diesel exhaust dilution.

Many of these mines do not utilize stoppings, and thus air moves randomly resulting in unpredictable and haphazard ventilation. It is common for freestanding auxiliary fans to move ventilation air along designated routes and to push the ventilation air to where it is needed. Generally these fan locations are determined by trial-and-error methods. The auxiliary fans, blowing in this manner, provide better ventilation than natural ventilation alone, but bulkheaded main mine fans are superior for providing consistent ventilation.

5 LARGE-OPENING MINE VENTILATION CASE STUDY

At the onset of this research effort, NIOSH engineers visited several underground stone mines to conduct ventilation surveys and observe baseline operating conditions. It was concluded from these initial surveys that the ventilation of many underground stone mines could be greatly improved by using two synchronized approaches: installing main mine propeller fans to produce the air quantity needed to sufficiently dilute the airborne contaminants to statutory levels and constructing stoppings in conjunction with auxiliary fans to direct the ventilation airflow. Propeller fans are used for a variety of circumstances; however, documentation of propeller fans being used in large-opening mines was not available. Therefore, a case study was necessary to determine whether propeller fans are truly applicable for use in underground large-opening mines. The study ran for 21 months and included monitoring the mine ventilation prior to and after the installation of a brattice stopping line and two main mine propeller fans.

The case study mine had a physical extent of approximately 0.75 square miles and entries that are about 1,125 ft^2 in cross-sectional area. However, more than one-half of the mine is benched and those entries are considerably larger, often exceeding 2,200 ft^2 in cross-sectional area. The propeller fans used for this study were selected on the basis of the parameters measured during ventilation surveys made at various stone mines (Grau et al. 2002a). Those surveys suggested that for large-entry drift stone mines, and even for some mines with shallow air shafts, propeller fans would be a good choice for improving ventilation airflow because they are designed to deliver large air quantities at pressures of less than one-inch w.g. (Grau et al. 2002a, Grau et al. 2002b).

6 VENTILATION CHARACTERISTICS PRIOR TO STUDY

Figure 1 shows the ventilation airflow patterns and fan locations before any ventilation improvements were installed. The mine exhibited noticeable air movement at various locations; but recirculation was a predominant factor. In conjunction with the fans, other factors that impacted the ventilation airflow direction included the outside air temperature, the outside wind condition, and the movement of haul trucks in the mine. During this time, one main mine fan and two auxiliary fans were all freestanding and operating without the use of stoppings. The main mine fan was a Pope[1] 250-hp, 1,750-rpm, 350,000-cfm, 72-in-diameter axial-vane fan that was located in the central portion of the mine. The two booster fans were a Joy[1] 100-hp, 1,770-rpm, 150,000-cfm, 60-in-diameter axial-vane fan and a Buffalo[1] 125-hp, 1,770-rpm, 90,000-cfm, 54-in-diameter axial-vane fan.

Ventilation airflow is needed predominantly where vehicles are most active, i.e., at the active production

[1]Product mention does not imply endorsement by NIOSH.

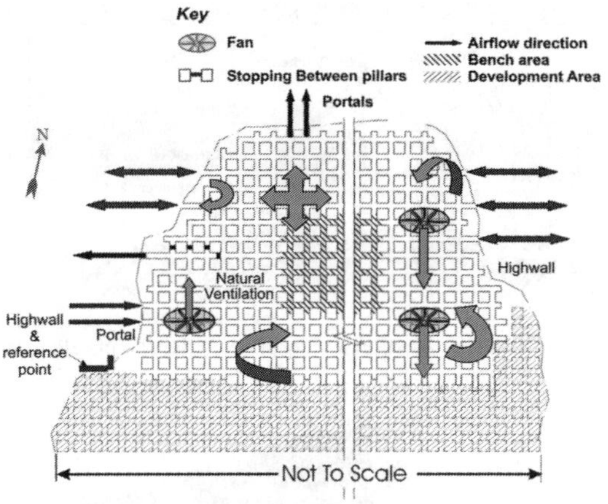

Figure 1. Ventilation airflow patterns prior to modifications.

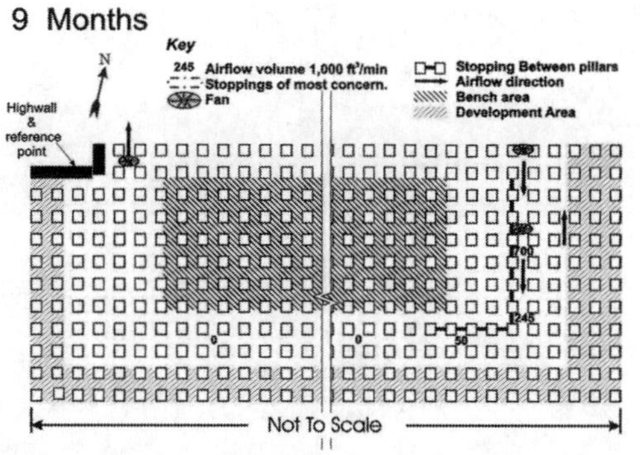

Figure 2. Airflow patterns with partial stopping lines and original fans.

Figure 3. Two propeller fans installed for this study.

faces. The active faces in this mine are located in two primary areas: in the bench area located in the central portion of the mine and at the development sections in the south and southeast portion of the mine as shown in Figure 1. From observations and discussions with various mine operators, sufficient ventilation for dilution of noxious gases is often of lesser concern in the bench areas. The large volumetric size of the entries in the bench areas sometimes provides adequate air volume for the dilution of those exhaust gases. In the study mine, as is common with many underground stone mines, the active production faces can generally utilize as much air as the most efficient ventilation system can provide, i.e., for practical purposes, they can never have too much air. Therefore, the ventilation improvements in this case study focused on the southern portion of the mine where the active production faces were located and where the air was most needed.

7 OVERHAULING THE VENTILATION SYSTEM

The mine ventilation overhaul for this case study consisted of two phases: constructing a stopping line, and installing and replacing the main mine axial-vane fan with two high-volume, low-pressure, propeller exhausting fans mounted in a bulkhead. The stopping line was oriented perpendicular to the directions of face advance and several breaks outby the working faces. This type of ventilation is typical of a perimeter mining ventilation plan (Grau et al. 2002a). Ventilation surveys were periodically taken during the construction of the stopping line to determine if the stopping line, in combination with auxiliary fans, had sufficient impact to negate the addition of a main mine propeller fan. Before completion, the stopping line which was a series of brattice curtains, initially improved the airflow, but at the end of the brattice line the air movement quickly dissipated as shown in Figure 2. Substantial air recirculation was also present at all times. It was clear that additional ventilation improvements such as completion of the stopping line and the installation of large-volume main mine fans were necessary to effectively ventilate the working faces.

Upon completion of the stopping line, two twelve-foot diameter propeller fans (Figure 3), were installed in the exhausting mode replacing the Pope[2] main mine fan. The propeller fans were Series 10 S 12-ft-diameter high-volume, low-pressure fans manufactured by Hartzell Fan Co[2] mounting 100-hp motors each. After the propeller fans were operated for several weeks, it was observed that the standard brattice stoppings would not withstand the pressure and continuous flapping created by the fans, especially at locations near the fans. Brattices were pulled from the roof and rib as shown in Figure 4, or the constant flapping of the brattices near the faces caused them to shred. Other significant forces damaging the brattices were face blast pressures and flyrock impacts.

---
[2]See Footnote 1.

Figure 4. Typical common example of brattice being pulled from rib.

The brattices used for this study were a one-piece design and were sized for the entries that were 45-ft wide and 25-ft high for a total cross-sectional area of 1,125 ft^2. Metal grommets were spaced every 3 ft on the perimeter of the brattice for added support. A base for attaching the brattice to the rib and roof was made by bolting 2-in × 6-in wood planks to the roof and rib. On the roof, the brattice was wrapped around the wood to make a secure holding for the brattice. Along the rib and roof, the brattice was attached to the wood by 1.5-in common roof nails with a fender washer to account for the grommet hole. Within a few months, many of the roof nails pulled away from the wood, suggesting that some type of a screw would be a better anchor.

8 PROPELLER FAN INSTALLATION

Fan installation techniques vary and depend upon the physical attributes at the fan site. The fans for this case study were installed in entries that were 25 ft high and 45-ft wide. The two 12-ft-diameter fans displace a combined area of 226 ft^2. Therefore, for practical purposes, a difference of about 900 ft^2 needed to be bulkheaded. The most convenient way to create the bulkhead was to first construct a ramp from waste rock across the entry to set the fan on as shown in Figure 5. This tactic effectively reduced the size of the bulkheaded area from 900 ft^2 to about 630 ft^2 (not including fan area). The ramp consisted of large boulders which were covered with smaller 2A modified gravel. The bulkhead was secured by 4-ft long split set bolts into the boulders.

Because of stresses from blast pressures, the fan bulkhead and associated framework needed to be designed carefully. Incidents have occurred where face blast pressures have blown out bulkheads or damaged fans. However, over design of bulkheads

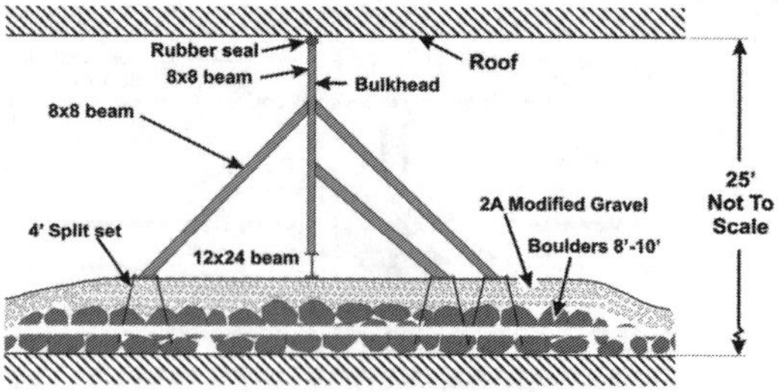

Figure 5. Cross-sectional view of propeller fan installation.

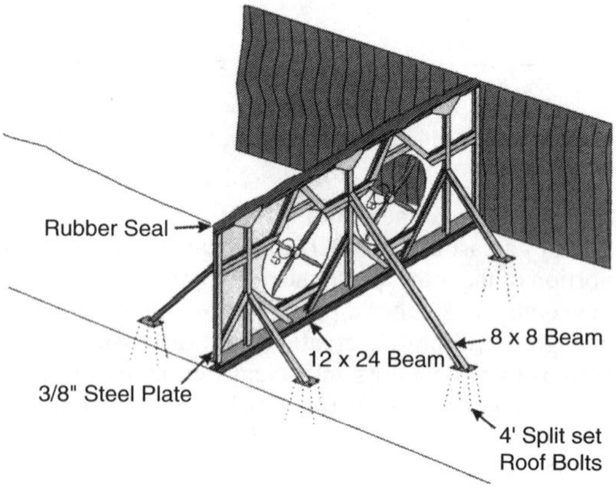

Figure 6. Plan view of propeller installation.

can be costly. The entire framework for the fan bulkhead used in this case study sits atop a 12-in × 24-in I beam. The framework consists of 8-in × 8-in beams as shown in Figure 6. The beams were oriented in a square/triangle pattern to enhance the strength of the bulkhead. The fan was bolted to the bulkhead frame through tabs available on the fan's exterior shell. Smaller sections of I-beam were positioned to increase the bearing support of the circular fans that were positioned in the square box frame. The entire frame, excluding the opening for the fan, was covered with a 3/8-in steel plate to act as a barrier to prevent short-circuiting of air. To further minimize leakage, a U-shaped rubber seal was positioned between the fan frame and the mine roof. The bulkhead framework was prefabricated in a machine shop and shipped to the mine in two sections; the bottom half with beams only, and the top half with beams and plate attached. After securing the base half, the top half was simply set on top of the base half, as shown in Figure 7, and welded in place. The final step was welding the bottom plate in place.

Mine fans are required to have a guarding grid that is less than 1 in^2. The manufacturer reports that the pressure loss on the flat fan guarding that is normally

Figure 7. Installation of prefabricated fan bulkhead.

Figure 8. Fan installation with perimeter guarding during maintenance.

used in these applications ranges from one-eighth to one-quarter of an inch of w.g., not including the resistance caused by the build up of soot deposits on the grid (Atkinson 2002). This resistance could be detrimental to the performance of a low pressure propeller fan. One alternative to flat fan guarding is area guarding. Area guarding is a fence enclosure installed several feet from the fan, and was utilized for the fan in this study. The fan area guarding consisted of a 10-ft high chain link fence with locked access doors, as shown in Figure 8. MSHA required that the guards used for this fan be at least 5-ft away from the fan. This distance not only satisfied the necessary safety requirements but also prevented any air-flow resistance.

9 MEASURED VENTILATION IMPROVEMENTS

The combination of a stopping line and propeller fans resulted in dramatic ventilation improvements at the case study mine, as shown in Table 1 and Figure 9.

Table 1. Ventilation fan criteria before and after ventilation changes.

Ventilation stage	Number of fans	Available total horsepower	Air volume; 1,000 cfm
Original ventilation	3	475	300
Improved ventilation	4	425	750
Difference	+1	−50	+450

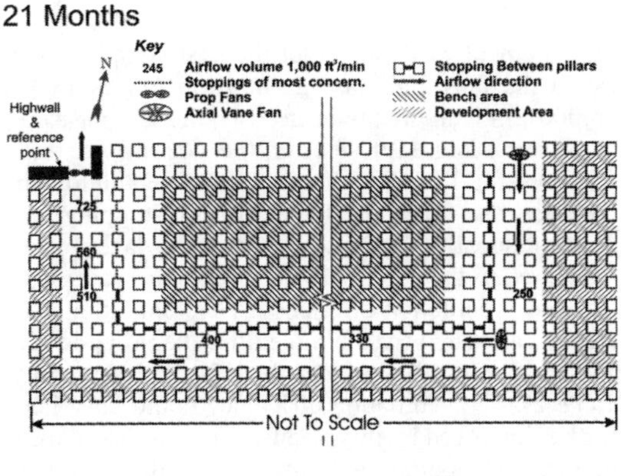

Figure 9. Airflow patterns after installation of stopping line and propeller fans.

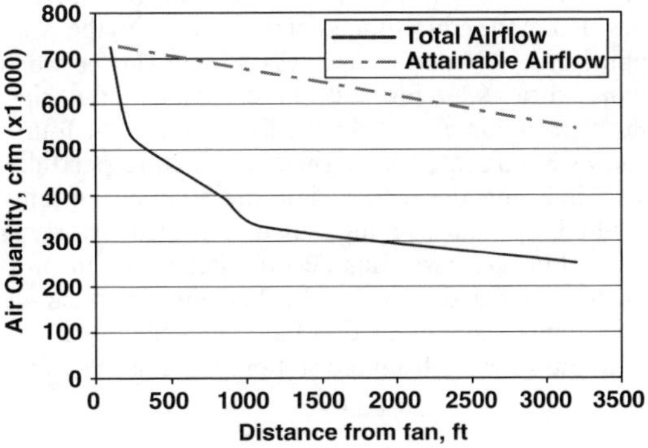

Figure 10. Airflow along stopping line.

The most important improvement was the more than doubling of the quantity of fresh airflow into the mine; from 300,000 cfm to 750,000 cfm. After the addition of the two propeller fans, the total number of fans increased from three fans to four fans, while the total rated horsepower of the fans decreased from 475–425 hp.

A graphical representation of the airflow measured along the stopping line after the propeller fans were installed is shown in Figure 10. The first airflow

Table 2. Potential improvements in air distribution by upgrading stoppings.

Scenario	Total number of brattice stoppings	Total leakage, 1,000 cfm	Available airflow at 3,200-ft, 1,000 cfm	Ventilation efficiency, pct.
1. Initial conditions	28	475	250	34
2. Improving first 8 stoppings (4 stone stoppings)	24	267	458	63
3. Improving all stoppings (14 stone stoppings)	14	179	546	75

measurement was made after the first stopping closest to the fans and was 725,000 cfm (an estimated 25,000 cfm had already been lost); the last measurement was taken 3,200-ft from the fan and was only 250,000 cfm. The 475,000 cfm loss of ventilation airflow was due to air leakage across the stopping line. The air quantity reaching the working faces, when compared to the air exhausted by the fans represents the mine ventilation efficiency. The ventilation efficiency calculated for the mine after the propeller fan and brattice installation was about 34 pct (Scenario 1, Table 2). The ventilation efficiency could have been further improved by two means; using stone pillars or permanent stoppings and/or upgrading the remaining stoppings as will be discussed latter.

One noteworthy observation from the case study was that when two parallel entries were available for ventilation airflow, about 60 pct of the total airflow traveled in the entry immediately adjacent to the stopping line. This percentage was slightly lower when more entries were present. In areas with four entries; the distribution of air from the first entry to the fourth entry was 43 pct, 29 pct, 20 pct, and 8 pct, respectfully. This indicates that when using the perimeter mining method, fresh air is being delivered near the working areas, but auxiliary fans become necessary to push the air to the active faces. This becomes increasingly important as the production faces advance more than four entries away from the stopping line as described above.

10 POTENTIAL ADDITIONAL VENTILATION IMPROVEMENTS

NIOSH is researching methods to improve the performance of both permanent and flexible stoppings (i.e. brattice). The mine in the case study utilized brattice stoppings and from observations it appears that brattice stoppings can be improved by backing with cabling or geotechnical grid and by attaching the brattice to the roof and rib in a manner to minimize high stress points. During the case study, there were 28 stoppings along the 3,200-ft brattice line perimeter (Figure 9); however, 70 pct (333,000 cfm) of the air loss occurred in the first 1,200-ft (8 stoppings) from the fan. A significant decrease in air short circuiting could have been achieved if half of the first eight brattice stoppings would have been stone stoppings or permanent stoppings of some type. If the remaining four brattice stoppings were improved to reduce losses by a conservative 25 pct, the total air loss in the first 1,200-ft of stopping line could be reduced to 125,000 cfm. The remaining 2,000-ft of stopping line had a loss of about 143,000 cfm through 20 stoppings for a total leakage of 267,000 cfm and a ventilation efficiency of 63 pct for the first potential improvement, as shown in Scenario 2, Table 2.

Figure 11. Leakage through brattice stopping.

If all stoppings were improved, i.e., if half of the remaining 20 brattice stoppings from the previous scenario were stone pillars and the remaining 10 brattice stoppings were improved to reduce leakage by 25 pct, the total air loss in the remaining 2,000-ft would be reduced to about 54,000 cfm from 143,000 cfm. This would represent a total air loss of only 179,000 cfm (75 pct efficiency) at the end of the brattice stopping line (Scenario 3, Table 2). The available air at 3,200-ft would increase to 546,000 cfm from the initial conditions of 250,000 cfm, and is shown as the attainable airflow in Figure 10. Repairing the brattice stoppings to reduce short circuiting by 25 pct is a conservative estimate since the observed stoppings had large gaps present as shown in Figure 11.

11 OTHER BENEFITS OF PROPELLER FANS

Improving ventilation is not the only benefit to using propeller fans. The two 12-ft-diameter propeller fans used in this case study are capable of producing about 750,000 cfm under loads that are less than one-quarter of an inch of w.g. with an actual 130 combined total operating horsepower. As a feasibility comparison to using traditional axial-vane fans, three 84-in diameter fans would have been required in this study with each fan producing 250,000 cfm to move a total air quantity of 750,000 cfm. Each of these axial-vane fans would require an outlet cone and a 200-hp motor that would operate at 130-hp for a total of 390-hp, as compared to the 130 hp of the two propeller fans.

The capital cost for the two propeller fans is about $25,000, as compared to about $90,000 for three axial-vane fans with outlet cones. Both sets of capital costs exclude the necessary starter motors. Annual operating costs are dependent upon the number of hours operated and the electrical unit cost. However, assuming an electrical cost of $0.08/kWh with the fans operating for 50 hrs per week for 50 weeks, the annual operating cost for the two propeller fans would be about $20,000, while the operating cost for three axial-vane fans would be about $59,000.

12 SUMMARY

The results of this case study demonstrate that dramatic improvements in ventilation airflow in underground large-opening mines can be achieved by the addition of large-diameter propeller fans. This increase in ventilation airflow should be an effective, practical, and cost-saving answer for improving the air quality in many underground large-opening mines, particularly large-opening drift mines. Not only do propeller fans produce large air quantities, they do so with relatively low initial capital cost, low operating cost, and at a higher mechanical efficiency than axial-vane fans under the typical low-pressure conditions. Additionally, a method to bulkhead propeller fans was documented and serves as an installation guide for other mine operators. The study also showed that a brattice stopping line and auxiliary fans used in conjunction with the propeller fans was an important aspect of improving the ventilation efficiency in the mine.

REFERENCES

Atkinson, G. Personnel communication (2002). Vice President of Sales, Hartzell Fan, Inc.

Grau, III, R.H., Robertson, S.B., Mucho, T.P., Garcia, F., and Smith, A.C. (2002a). NIOSH research addressing diesel emissions and other air quality issues in nonmetal mines. In 2002 Society for Mining, Metallurgy and Exploration Annual Meeting, Phoenix AZ, Feb. 26-28, 2002.

Grau, III, R.H., Mucho, T.P., Robertson, S.B., Smith A.C., and Garcia F. (2002b). Practical techniques to improve the air quality in underground stone mine, North American/Ninth U.S. Mine Ventilation Symposium, Kingston, Ontario, Canada, pp.123–129, June 8–12, 2002.

Haney, R., and Saseen, G. (1998). Estimation of diesel particulate concentrations in underground mines. Preprint 98–146, presented at the Society for Mining, Metallurgy and Exploration Annual Meeting March 9–11, Orlando, FL.

Head, R. (2001). Calculating underground mine ventilation fan requirements. Aggregates Manager, April 2001: 6(3):17–19.

Mucho, T.P., Grau III, R.H., and Robertson, S.B. (2001). Practical mine ventilation. Presentation at the Safety Seminar for Underground Stone Mines, Louisville, KY, Dec. 5.

Mine Safety and Health Administration (MSHA), U.S. Department of Labor, (2001a). MSHA Web Site, Nonpermissible diesel engines approved under Part7. http://www.msha.gov/S&HINFO/DESLREG/1909a.HTM

Mine Safety and Health Administration (MSHA), U.S. Department of Labor (2001). Final Rule 30 CFR Part 57.

Robertson, and Susan, (2001). The NIOSH Air Estimator, Underground Stone Safety Seminar, Louisville, KY, Dec. 2001.

Robertson, Susan, Grau III, R.H., and Dolgos, J. (2003). A computer software program that estimates air quantity requirements in large opening mines. North American/ Tenth U.S. Mine Ventilation Symposium, Anchorage, AL, May 16–19, 2004.

Schnakenberg, Jr., G.H. (2001). Estimate of technically feasible DPM levels for underground metal and nonmetal mines, Mining Engineering, Sept.

Mine Ventilation, Ganguli & Bandopadhyay (eds)

2500 hp variable speed mine ventilation fan installation – P&M Coal Mining Co.

T.L. Sartain & T.E. McDaniel
P&M Coal Mining Co., Berry, Alabama, USA

ABSTRACT: North River Mine's expansion into a new reserve area required additional ventilation fan capacity. The mine upgraded the existing 120-72 New Philadelphia, axial vane, exhaust fan to 2500 horsepower and installed a second (identical) fan to serve as a backup. The unique characteristics of this fan system are found in the electrical power and control system.

The 2500 horsepower motors are powered through a state of the art variable frequency drive (VFD). The VFD allows the fan performance to be controlled by adjusting the speed of the fan, in lieu of conventional blade pitch adjustments. Also, by limiting the inrush current (at startup), the VFD facilitates the use of a two mega-watt standby diesel generator, required by Alabama state law.

1 INTRODUCTION

P&M's North River Mine, located in West-Central Alabama, operates one longwall and two continuous miner sections, producing 3.5 million tons of clean coal annually. The mine was started in 1974 and currently extends 9 miles underground from the original portal to the working face. Methane is liberated from the mine at a rate of approximately 5 million cubic feet per day, via two exhausting fan shafts. The original Cedar Creek Fan primarily ventilates old works and seals, while the newer Tyro Creek Fan provides ventilation for the active areas of the mine.

2 DESIGN CONSIDERATIONS

Computer modeling indicated that future mine development would require additional fan capacity, to provide sufficient airflow volumes in the new West Area longwall panel development. As the mine expands to the west, the Tyro Creek Fan must be capable of supplying 800,000 CFM at 16″ water gage vacuum. To achieve this performance, the existing fan electric motor on the New Philadelphia 120-72 fan was upgraded from 1250 to 2500 HP.

To minimize the downtime associated with the installation of a new 2500 HP motor, a totally separate identical fan was purchased and installed. The new fan provides 100% backup capacity in the event of a mechanical failure. Additionally, the ability to alternate fans facilitates routine maintenance.

Alabama State Law requires a back-up power source, capable of providing at least 80% of normal ventilation capacity. This is usually accomplished by either a direct drive diesel engine or stand-by power generator. P&M elected to install a Caterpillar two mega-watt standby diesel-powered generator, capable of operating either fan. The generator was sized to provide 100% backup capability at full load condition, eliminating concerns of ventilation disruptions during a utility power failure.

Conventional motor starters (e.g. autotransformer, soft start, etc.) were considered when determining the most cost-effective means to start the new 2500 HP motor. Although these conventional starting methods are more economical to install, the associated inrush starting current ranges from 3–6 times full-load amperes. This would require the size of the generator to be doubled to at least four mega-watts. The VFD however, eliminates current inrush, thus minimizing the required generator size. The VFD also provides variable speed control, which was desirable. P&M selected the Allen Bradley Power Flex 7000, which was custom designed for this application. (*See Figure 1 for system layout.*)

3 VARIABLE FREQUENCY DRIVE

This VFD limits the inrush current to 100% of motor full-load amperes, allowing the power generation to be sized for operating amperes instead of starting

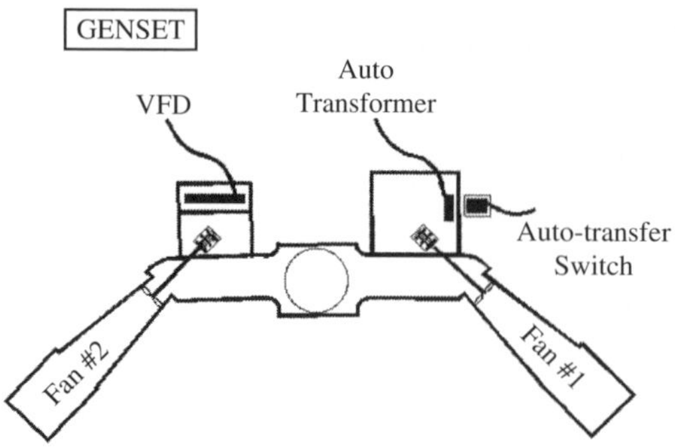

Figure 1. Fan system layout.

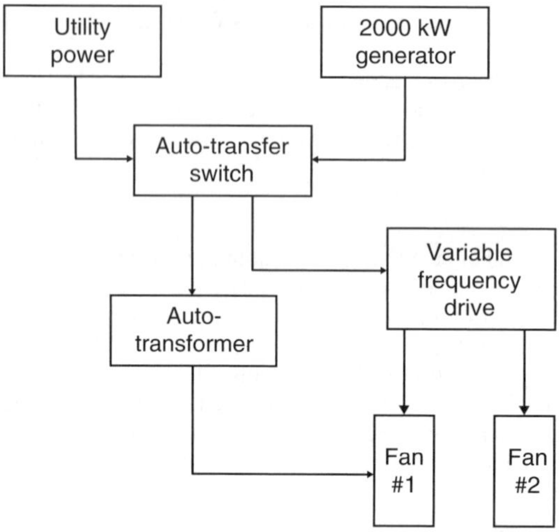

Figure 2. Basic electrical system.

Figure 3. Power Flex 7000 VFD.

Figure 4. Two mega-watt diesel-powered generator.

amperes. Figure 2 illustrates the drive's ability to start either fan on utility or generator power.

This particular drive has many beneficial characteristics such as:

- A speed operating range of 90–900 rpm.
- The power factor is increased to unity when the fan is operating under the control of the VFD.
- Speed changes can be varied in-flight, allowing ventilation adjustments to be achieved very efficiently.
- Synchronous bypass can be utilized if operating at full speed to minimize device wear. Once the motor reaches full speed at 60 Hz, bypass mode can be selected allowing the user to switch the fan directly across the line to utility power.
- During normal operation, the drive starts and stops the fan in a set ramp time.
- During power failures, the fan can coast for several minutes. The drive eliminates coast time by providing a flying start command while the blades are still rotating.
- Also, during a power failure the drive will reset and restart automatically on generator power.
- Utilizes a panel-view for its human-machine interface. It allows the user to password protect layers for extra security.

A photo of the Power Flex 7000 VFD is shown in Figure 3.

4 BACKUP POWER SOURCE

The two mega-watt Caterpillar generator was supplied as a packaged unit. (*see Figure 4*). It is totally enclosed in a sound attenuated container. The base is a 2000 gallon double wall fuel tank. An automatic transfer switch (ATS) calls the generator to start automatically when utility power is lost. After a three minute warm-up period the ATS switches generator power to the VFD. The VFD will automatically restart the fan motor and ramp up to the normal (pre-set) speed. After the fan reaches full speed, a manual transfer switch can be used to provide auxiliary power to the mine elevator, if needed.

5 REDUNDANCY

This fan installation is designed with total redundancy (i.e. no single component failure will preclude operation.) Mechanical maintenance and repairs can be performed on one fan while the other is operating, utility power is backed up with the diesel-powered generator, and an emergency starter (autotransformer) is provided to back up the VFD. The system also includes two separate communication systems to alert mine personnel in the event of a fan failure or ventilation problem.

Figure 5. No. 1 Fan noise diverter.

6 WARNING/ALARM SYSTEM

Southern Link two-way radios are used to transmit various fan warning and alarm messages. Radios are located in strategic areas on the surface and in the underground mine. Small hand-held radios are also carried by mine management and maintenance personnel. A message will broadcast if the electrical power is lost, ventilation pressure is reduced, control room temperature too high, or other problems occur with the system.

Surface and underground personnel can also receive fan alarms via the mine-wide pager phone system. Sirens are located at the nearby mine portal to alert surface personnel of a fan stoppage.

A programmable logic controller (PLC) is interfaced with the VFD to monitor the bearings of each fan and motor. This PLC is programmed to shut down the fan if excessive vibration or temperature is experienced.

Figure 6. No. 2 Fan silencer.

7 NOISE MANAGEMENT

An up-cast diverter was constructed during installation of the original fan in 1998. This structure was built at the exhaust end of the Fan #1 evase to direct the air and noise in an upward direction. The structure is lined with acoustical block on the inside and covered with lead-coated plywood on the outside. (*See Figure 5.*) A pre-fabricated metal silencer has recently been designed and installed on Fan #2. (*See Figure 6.*) A baseline sound pressure survey was conducted at the fan and at nearby residential locations to establish the reduction needed. The silencer is designed to meet a 50 dBA night time standard at the nearest residence.

As previously mentioned, the generator is totally housed in a fully sound attenuated enclosure, and has no significant impact on the environment.

8 SUMMARY

P&M Coal's recent upgrade of the Tyro Creek Fan will provide additional ventilation capacity for the West Longwall Development. The original 1250 HP motor was replaced with a 2500 HP motor. The upgrade also provided a fully redundant back-up system, by adding a second 2500 HP fan and diesel-powered generator. Both fans can be powered by either utility or the generator via the VFD starter.

The new system provides several cost saving advantages over more conventional systems such as: in-flight adjustable speed control, reduced stand-by power generation capacity and unity power factor.

The Tyro Creek Fan upgrade has been a great success, providing safety to the miners as well as flexibility for mine management.

Advances, modeling and software

Mine Ventilation, Ganguli & Bandopadhyay (eds)
© 2004 S.B. Robertson, ISBN 90 5809 633 5

A computer software program that estimates air quantity requirements in large opening stone mines

S.B. Robertson, R.H. Grau, J. Dolgos & T.P. Mucho
National Institute for Occupational Safety and Health, Pittsburgh, PA, USA

ABSTRACT: The National Institute for Occupational Safety and Health (NIOSH) has developed a computer program called the Air Quantity Estimator (AQE). The purpose of the program is to provide a starting point for estimating the air quantity needed to dilute diesel particulate matter (DPM) in an underground large opening mine, albeit taking into account various input assumptions. The AQE estimates the total air quantity needed to dilute DPM emissions that enter the main airstream of the mine to current statutory levels. It helps operators pinpoint vehicles that are high DPM contributors. In addition, it allows the user to make "what if" scenarios by varying the input parameters to achieve the most efficient and practical ventilation system. The AQE is a stand-alone software program that does not require any type of spreadsheet software. Engine data must be known or estimated with some accuracy and assumptions must be made in order to limit the variability of the program output. Program calculations and two case studies, for validation purposes, are included in this report. It should be noted that the program does not ensure adequate face ventilation and personal exposure compliance.

1 INTRODUCTION

There has been an increased recognition of worker exposure to DPM as a problem in the mining industry by various health organizations [ACGIH 2001] [NIOSH 1988] [EPA 2000]. Consequently, legislation was recently passed by the Mine Safety and Health Administration (MSHA) to limit DPM concentrations in underground metal/non-metal mines. Provisions of this DPM health standard are found in "Diesel Particulate Matter Exposure of Underground Metal and Nonmetal Mines," published in the Federal Register on January 19, 2001 (66 FR 5706) and amended on February 7, 2002 (67 FR 9180) [MSHA 2001].

Shortly after the standard was published, legal challenges ensued and an interim DPM standard of $400_{TC}\,\mu g/m^3$ went into effect on July 20, 2002, with that level to be reduced to $160_{TC}\,\mu g/m^3$ on July 19, 2006.

This paper addresses the ventilation needs of metal/nonmetal mines with large openings in particular, stone mines, although operators of other metal/nonmetal mines may find it useful. A large opening mine generally has entries at least 40–50 ft wide and 20–30 ft high. Because of the new DPM standards, operators with large opening mines will find it necessary to implement a combination of the following: increase the ventilation airflow in the mine; direct air to working areas by using stoppings or auxiliary fans; decrease DPM emissions from engines by using catalytic converters, filters, low-sulfur fuel, and cleaner burning engines; and employ administrative controls to limit DPM emissions underground. In addition, mine planners must consider ventilation needs when planning future mine layouts [Head 2001] [Grau et al. 2002a] [Grau et al. 2002b] [Mucho 2001].

A critical task for the ventilation planner is to determine the air quantity required to meet the DPM concentration limits when a particular fleet of engines is in operation. To meet this challenge, NIOSH has developed an Air Quantity Estimator (AQE) [Robertson 2001] as a mine planning tool to help underground mine operators determine the air requirements necessary to dilute DPM in the main air stream of their mine to statutory levels. The AQE is a user-friendly computer (PC) program that comes as a stand-alone package – it will run on any computer and is not dependent on the installation of spreadsheet programs, e.g., Microsoft EXCEL. The AQE also provides diesel engine performance test data from both the Environmental Protection Agency (EPA) and MSHA [EPA 2002] [MSHA 2002].

The AQE user must input specific engine data and operating parameters to run the program. The program calculates two different air quantities that will dilute DPM in the main air stream: the first to the current interim limit of $400_{TC}\,\mu g/m^3$, and the second to $160_{TC}\,\mu g/m^3$. The user can save the input parameters used in the original calculations, then take these data and theoretically add filters to selected engines, replace engines, or change other input parameters. The program will then calculate new ventilation requirements. Thus, mine planners can experiment with many scenarios until an appropriate ventilation system for the site specific conditions at their mine has been attained for the various DPM standards. This program also displays the amount of DPM each vehicle is emitting into the air stream. With this information, the ventilation planner will be able to identify vehicles that emit high DPM emissions. DPM engineering controls may be considered for these vehicles to reduce emissions [Schnakenberg 2002].

As stated earlier, the results from this program do not ensure proper working face ventilation or personal exposure compliance. If the ventilation air is not effectively directed to the working faces, high concentrations of DPM can still exist even with extremely large amounts of ventilation air in the main mine current.

2 OVERVIEW OF HISTORICAL STONE MINE VENTILATION TECHNIQUES

Recent case studies conducted by NIOSH showed that, historically, natural ventilation has been their primary means of diluting airborne contaminates in large opening stone mines [Grau et al. 2002a]. But natural ventilation, by itself, did not guarantee a consistent measurable airflow [Head, 2001]. It provided some air exchange in specific portions of a mine, but the air quantities were generally insufficient and uncontrolled, i.e., not directed to the areas where it was needed. Some mines used fans to increase ventilation airflow, but ineffectively. Commonly, the main mine fans were not installed in a bulk head making them very inefficient. In many cases, the main mine fan and auxiliary fans were improperly sized and/or not placed at the strategic locations throughout the mine. There was modest mine planning in relation with future ventilation techniques. Also, there was minimal use of stoppings to control the direction of air movement. Without consistent ventilation airflow coursing, recirculation of ventilation air and high concentrations of contaminates could potentially be found.

3 PROPER VENTILATION PLANNING IS CRITICAL

As a consequence of the DPM legislation, ventilation practices are changing in the metal/nonmetal industry, in particular, the stone industry. To effectively improve the air quality in these underground mines, bulk headed fans with stopping lines need to be incorporated into the overall mine planning process. This air coursing scenario includes an entire array of ventilation considerations and requirements. Mechanical main mine fans, auxiliary fans, stoppings, and a general ventilation concept must be integrated into mine layouts and mining sequences. Also, areas with special air requirements, such as production faces, shops, benching areas, and haulage routes need to be included in the ventilation planning process.

Criteria for proper fan selection, installation, and operation for both main mine fans and auxiliary fans must also be considered. Fan pressure and quantity characteristics should be matched to the specific operation. Utilizing stoppings as air walls helps control the mine ventilation air flow, i.e., the stoppings efficiently direct the air to where it is needed the most. The air walls also separate the intake and return airways. Stoppings can be constructed from man-made materials, by filling an opening with waste material, or by leaving areas of rock intact to direct ventilation airflow [Timko and Thimons 1987]. Furthermore, fan effectiveness is increased dramatically when used in conjunction with stoppings.

Finally, fan and stopping locations need to be an integral part of the mine layout. Stoppings will need to be built, taken down, or moved with face advances and/or changes in the direction of mining. These measures will help to ventilate the active faces while providing adequate ventilation to any special-needs area noted above. These overall ventilation concepts are discussed more fully in Grau et al. [2002a].

4 USING THE AIR QUANTITY ESTIMATOR

4.1 *Assumptions inherent with the air quantity estimator*

The AQE does not ensure adequate face ventilation, but it provides a starting point as to how much air quantity is needed to dilute DPM contaminates to statutory levels in the main airstream of the mine. There are a number of factors and assumptions that affect the accuracy of the air quantity estimate. The program assumes there are sufficient amounts and thorough mixing of ventilation air where equipment is operating. It assumes that the contaminates from all the working faces or areas where equipment is operating are eventually dumped into the main air stream of the mine and then exhausted out of the mine through one portal. Also, the program assumes that a ventilation system is in place where there is little to no air recirculation in the mine and no air leakage across the main mine fans or stoppings. Finally, it assumes that the ventilation air is free from DPM contaminates before it enters the mine. An equally important factor is that the air quantity estimates are

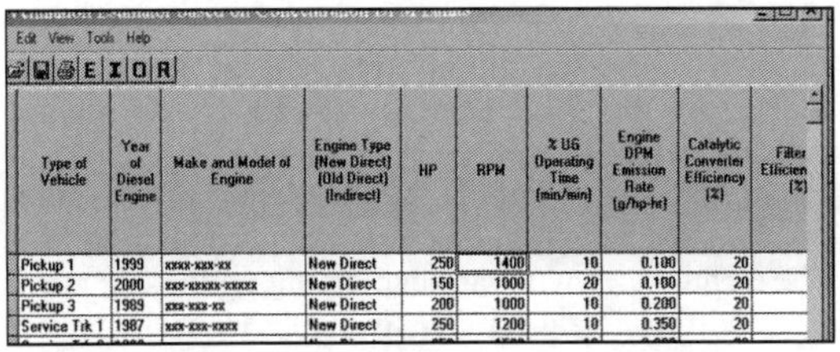

Type of Vehicle	Year of Diesel Engine	Make and Model of Engine	Engine Type (New Direct) (Old Direct) (Indirect)	HP	RPM	% UG Operating Time (min/min)	Engine DPM Emission Rate (g/hp-hr)	Catalytic Converter Efficiency (%)	Filter Efficien (%)
Pickup 1	1999	xxxx-xxx-xx	New Direct	250	1400	10	0.100	20	
Pickup 2	2000	xxx-xxxxx-xxxxx	New Direct	150	1000	20	0.100	20	
Pickup 3	1989	xxx-xxx-xx	New Direct	200	1000	10	0.200	20	
Service Trk 1	1987	xxx-xxx-xxxx	New Direct	250	1200	10	0.350	20	

Figure 1. Sample sheet of input data for the AQE.

affected by the user's presumptions for some of the input data, namely the engine DPM emission rate and the percent underground operating time.

4.2 *Getting started*

The AQE is a 32-bit computer program that runs under all Microsoft Windows operating systems of Version 95 or later. The program is distributed by NIOSH on CD-ROM and can be installed by clicking on the setup file located in the program's main directory. After installation, the program can be opened the same way as any other Microsoft Windows program, i.e., by clicking on an icon, etc. The AQE is a stand-alone program, i.e., no commercial spreadsheet program is required for its use.

When the program launches, there is an option either to open the AQE User's Guide or to continue with the program. If you choose not to open the User's Guide, the next screen is the input sheet for entering data. This screen resembles that shown in Figure 1 except that no data will be present. Data must be entered into the rows and columns of the data table (spreadsheet). Although the data can be entered by either row or column, after a data value is entered into a cell or box, the enter button on the keyboard must be pressed. The next active cell will automatically be displayed to the right of the cell where data was last entered until the end of the row is reached. Therefore, entering data by rows is the easiest method which is also the organizational principle for describing vehicles. A sample sheet of input data, shown in Figure 1, is provided in the program. There are numerous file, editing, and sorting functions in the program to assist the user in entering and handling data. In addition, there is a help menu that includes engine DPM emission data, links to the MSHA and EPA websites, and the AQE User's Guide.

5 REQUIRED INPUT DATA

Before entering input data into the program, one should first obtain all the necessary site specific performance and operating data required to run the AQE. The following input data, as shown in Figure 1, are required to be entered into the AQE program. Column headings correspond to specific input variables and are defined as follows:

Type of Vehicle: For example, Pickup 1, Water Truck, or Drill 1. This information is for the users benefit only to distinguish between vehicles/engines.
Year of Diesel Engine: The manufactured year of the engine, for example, 2002.
Make/Model of Engine: The make/model of the diesel engine, e.g., CUMMINS QSK19-C.
Engine Type: There are three choices for this item. The engine is either a New Direct, Old Direct, or Indirect Injection engine. Most engines manufactured prior to 1988 are either Old Direct Injection or Indirect Injection engines. Old Direct Injection engines lack the features found on New Direct Injection engines such as electronic controls, turbo chargers, or after-coolers. Recent Indirect Injection engines are engines have less than 100 horsepower. The Engine Type parameter will help the user determine the DPM emission rate of the engine described below [Haney 2001].
HP: This is the rated horsepower of the engine. For example, 300 hp, entered as "300."
RPM: This is the revolutions per minute of the engine at the specified horsepower. For example, 1200 rpm would be entered as "1200."
% UG Operating Time: This is the percentage of the time for which the equipment is contributing DPM into the main underground air stream of the mine during a full shift. For example, if a haul truck is operating in the mine 450 out of 600 minutes (10-hr shift), the percentage should be 75 percent (450/600), entered as "75."
Time studies can help determine these percentages.
Engine DPM Emission Rate: This is the average DPM emission rate that the engine contributes to the air stream. It is represented in grams per horsepower-hour (g/hp-hr). Because there is such little test information about the DPM emissions from diesel engines, especially for older models, the user will have to make the best estimate possible with the information available. Since there is no set method to arrive at an exact emission rate, we have provided the following steps to help guide the user in estimating an engine DPM Emission Rate:

Step 1: Check if the engine is MSHA approved under Part 7. Engines approved by MSHA will have a nameplate on the engine with an MSHA approval number on it. If the engine is an MSHA-approved engine or you believe that it is, go to step 2. If the engine is not an MSHA-approved engine, go to step 3.

Step 2: Locate the engine on the MSHA-approved engine emissions list by consulting the HELP menu under "EPA and MSHA Information." You can find the engine by looking for the MSHA Approval Number. The DPM emission rate, located in the last column, is given in g/hp-hr. There is a possibility that not all

MSHA-approved engines are on this list because the engine may have been approved after this program was developed. Therefore, if you have an approved engine, but it is not on this list, find the engine on the MSHA website at www.msha.gov/TECHSUPP/ACC/lists/07npdeng.pdf. Note that the DPM emission rate on the MSHA website is given as a Particulate Index and must be converted to g/hp-hr. To calculate the rate in g/hp-hr, multiply the Particulate Index of the engine by 0.0017, and then divide by the engine horsepower.

Step 3: The EPA has tested a number of diesel engines starting in 1998 until the present. A list is provided in the program HELP Menu of EPA-approved, nonroad engines manufactured from the years 1998 to 2003. A specific engine can be located by looking for the Make/Model, manufactured year, engine hp/rpm and then finding the corresponding DPM. The last column in the list provides the emission rate for the engine. The emissions lists for both MSHA and the EPA are based on evaluations performed on the engines during an 8-cycle test where the emission rate is based on the weighted average of all eight cycles. Therefore, there is not a direct relationship between the cycle time of the engine in an underground environment and the cycle time of the engine during the testing cycle. If information is needed on an engine that is not on the EPA list in the Help File (newer than 2003) or you would like additional information, go to step 4.

Step 4: For years not available in the help file, EPA approved engines can be found on the EPA website at www.epa.gov/otaq/certdata.htm. Make sure the DPM emission rate (Particulate Matter Rate on website) is given in g/hp-hr. If not, convert g/kW-hr to g/hp-hr by dividing by 1.34.

Step 5: If information is not available from MSHA or EPA for the particular engine of interest, check with engine manufacturers for engine DPM emission rates. If engine information is not available from the manufacturer, go to step 6.

Step 6: If information is not available for a particular engine, an estimate on the DPM emission needs to be made. The estimate is dependent on whether the engine is a New Direct, Old Direct, or Indirect Injection engine. Below is the range of DPM emission rates for each engine type [Haney et al. 2001].

NEW DIRECT	0.1 to 0.4 g/hp-hr
INDIRECT	0.3 to 0.5 g/hp-hr
OLD DIRECT	0.5 to 0.9 g/hp-hr

Estimating the DPM emission is subjective and depends on its age and how well the engine has been maintained. For example, 0.1 g/hp-hr would be an estimate for a newer, well-maintained New Direct Injection engine. A rate of 0.5 g/hp-hr would be an estimate for an Indirect Injection engine that is fairly old and not well maintained. A well maintained Old Direct Injection engine may have a DPM emission rate of 0.6 g/hp-hr.

Catalytic Converter Efficiency: If the engine has a catalytic converter, the efficiency is generally 20%. Make sure the percentage is entered as "20" on the spreadsheet.

Filter Efficiency: This is the efficiency of any filter system used in combination with a particular engine. The efficiencies will vary depending on the filter system and these data may be available from the manufacturers. If information is not available, 85% is a conservative filter efficiency estimate which should be entered as "85."

6 PROGRAM OUTPUT

As previously detailed, the AQE program output is highly dependent upon the accuracy of the input information. The program provides calculated results and displays five output values on the right side of the spread sheet, as shown in Figure 2. The first three values, "Average Vehicle DPM Emission Rate," "Sum of Average Vehicle DPM Emission Rate," and "Total HP" are general information that provides the user with a better understanding of the relationship between a particular vehicle and DPM emissions in the mine. The last two values, "Air Quantity to Meet 400_{TC} µg/m^3 DPM Standard" and "Air Quantity to Meet 160_{TC} µ/m^3 DPM Standard," are calculated air quantities that provide the mine operator with an estimate of the total air quantity needed to reduce DPM to each respective concentration level. There is an assumption in the calculations that, for these air quantity values, the total carbon is approximately 80% of the raw DPM. A description of the calculated output values are as follows:

Average Vehicle DPM Emission Rate: This is the average amount of DPM per shift that each vehicle contributes in grams per minute (g/min) into the main airstream. By examining the values in this column, a mine operator can pinpoint the vehicles that contribute higher amounts of DPM during the shift. These vehicles are leading candidates for engine replacement or DPM controls.

Sum of Average Vehicle DPM Emission Rates: This is the sum of the average DPM emission rates for all vehicles given in grams per minute (g/min).

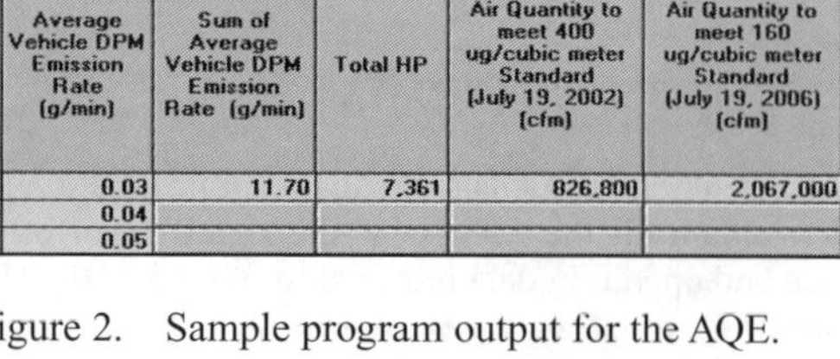

Average Vehicle DPM Emission Rate (g/min)	Sum of Average Vehicle DPM Emission Rate (g/min)	Total HP	Air Quantity to meet 400 ug/cubic meter Standard (July 19, 2002) (cfm)	Air Quantity to meet 160 ug/cubic meter Standard (July 19, 2006) (cfm)
0.03	11.70	7,361	826,800	2,067,000
0.04				
0.05				

Figure 2. Sample program output for the AQE.

Total HP: This is the sum of the rated horsepower of vehicles operated underground.
Air Quantity to Meet 400_{TC} μg/m³ DPM Standard: This is the estimated air quantity, given in cubic feet per minute (cfm), to dilute the sum of average vehicle DPM emission rates to a DPM concentration limit of 400_{TC} μg/m³.
Air Quantity to Meet 160_{TC} μg/m³ DPM Standard: This is the estimated air quantity, given in cfm, needed to dilute the sum of average vehicle DPM emission rates to a DPM concentration limit of 160_{TC} μg/m³.

7 PROGRAM CALCUL00ATIONS

Below are the input and output variable abbreviations along with the AQE program calculations:

- Horsepower = HP (hp)
- % Underground Operating Time = % UG Op (%)
- Engine DPM Emission Rate (raw DPM) = EDER (g/hp-hr)
- Catalytic Converter Efficiency = CCE (%)
- Filter Efficiency = FE (%)
- Average Vehicle DPM Emission Rate (raw DPM) = AVDER (g/min)
- Sum of Average Vehicle DPM Emission Rate (raw DPM) = SAVDER (g/min)
- Total Horsepower = Total HP (hp)
- Air Quantity to Meet 400_{TC} μg/m³ Standard = AQ 400_{TC} (cfm)
- Air Quantity to Meet 160_{TC} μg/m³ Standard = AQ 160_{TC} (cfm)

$$\textit{AVDER g/min} = (\text{hp } \textit{hp})\,[(0.01)(\%\ \text{UgOp})]\,(\text{EDER g/hp-hr})\,[0.01(100\text{-CCE})]\,[0.01(100 - \text{FE})]\,(8\ \text{hr/shift})\,(0.00208\ \text{shift/min})$$

$$\textit{SAVDER g/min} = \Sigma\text{AVDER g/min}$$

$$\textit{Total HP hp} = \Sigma\ \text{HP hp}$$

$$\textit{AQ } 400_{TC}\ \text{cfm} = 1/[(400_{TC}\ \mu\text{g/m}^3)(1^{-6}\ \text{g}/\mu\text{g})\,(1/\text{SAVDER}_{\text{raw DPM}}\ \text{min/g})\,(1.25\ \textit{raw DPM/TC})(0.0283\ \text{m}^3/\text{ft}^3)]$$

$$\textit{AQ } 160_{TC}\ \text{cfm} = 1/[(160_{TC}\ \mu\text{g/m}^3)(1^{-6}\ \text{g}/\mu\text{g})\,(1/\text{SAVDER}_{\text{raw DPM}}\ \text{min/g})\,(1.25\ \textit{raw DPM/TC})(0.0283\ \text{m}^3/\text{ft}^3)]$$

8 VALIDATION OF THE AQE

Two separate case studies were conducted by NIOSH at an underground limestone mine in southwestern PA. The objective was to verify the accuracy of the AQE by comparing the AQE estimated output results to actual air quantity measurements and DPM area sampling concentrations taken in the field.

The mine chosen for both studies utilized a perimeter mine ventilation system where all the DPM from the mine vehicles eventually flowed into the main airstream of the mine and then exhausted out of the mine through a bulkheaded fan assembly (two, 12-ft propeller fans). Stopping lines and auxiliary fans were used to direct ventilation air to the working faces. Air readings were taken in the main airstream near the sampling locations periodically throughout the shift with vane anemometers.

The primary location of the DPM area sampling was in the main airstream of the mine near the exhaust fan. Sampling was also conducted near the working areas to detect whether sufficient ventilation air was effectively being directed to the working faces. For each study, DPM sampling began at the start of the shift and generally lasted the entire shift. The sampling apparatus used for both studies were respirable dust samplers with pumps that ran at 1.7 liters/minute that were configured with a two step process to eliminate the larger particles. First, A DORR Oliver cyclone was utilized to remove the largest particles. The remaining particles are separated by a submicrometer impactor which had a cut off point of 0.9 micrometers at a flow rate of 1.7 liters/min. Particles less than 0.9 micrometers are deposited on a sampling cassette. These sample cassettes were analyzed in the lab at the Pittsburgh Research Laboratory for total carbon using the NIOSH Analytical 5040 Method and given as an eight-hour weighted average equivalent. All the pump flow rates were checked prior to, throughout, and at the end of the shift and were found to have functioned properly.

As part of collecting input data for the AQE, time studies of the haul trucks were conducted to calculate an accurate percentage of the underground operating time. The average percent of the time underground for the trucks was found to be 59%. The remaining 41% of their time was spent outside of the mine hauling stone to the crusher and stopping for maintenance.

For each case study, researchers entered the collected input data into the AQE program. An estimated ventilation air quantity was then calculated that could dilute DPM contaminates from all the vehicles contributing DPM underground down to a 400_{TC} μg/m³ concentration. This value would be compared to field study results.

8.1 *Case study 1*

The average airflow reading near the exhaust fans in the main airstream of the mine was 752,000 cfm and the DPM concentration near the same location was

Table 1. Ventilation results from 2 case studies.

Case study	Field measurements and sampling at mine outlet			AQE	
	Measure air quality (CFM)	Sampled DPM concentration ($_{TC}\mu g/m^3$)	Normalized air quality for a $400_{TC}\,\mu g/m^3$ DPM concentration (cfm)	Estimated air quality for a $400_{TC}\,\mu g/m^3$ DPM concentration (cfm)	%Difference in AQE air quality estimation
Case 1	752,000	361	678,700	794,000	+15
Case 2	760,000	285	541,000	586,000	+9

$361_{TC}\,\mu g/m^3$ as shown in Table 1. The average sampling time was 487 minutes. The AQE provided air quantities to meet a higher DPM concentration of $400_{TC}\,\mu g/m^3$, so the measured air quantity had to be normalized in order to compare the measured air quantity with the AQE estimated quantity. The normalized air quantity was 679,000 cfm. DPM sampling from locations near the working faces showed somewhat higher DPM concentrations than at the exhaust location indicating that improvements could be made directing more airflow to the working faces. The calculated result from the AQE was an estimated air quantity of 794,000 cfm to dilute DPM down to a $400_{TC}\,\mu g/m^3$ concentration. After comparison to the normalized measured air quantity, it was concluded that by using the AQE, the air quantity was overestimated by 17% (Table 1).

8.2 *Case study 2*

During this case study, the equipment underground was the same as case study 1 except for there was 1 fewer haul truck, drill, and service truck operating. In addition, a new loader was being used that had a much lower emission rate than the loader that it replaced. The average air reading near the exhaust fan in the main airstream of the mine was 760,000 cfm and the DPM concentration near the same location was $285_{TC}\,\mu g/m^3$ (Table 1). The average sampling time was 567 minutes. The normalized air quantity was 541,000 cfm. DPM sampling from locations near the working faces showed fairly low DPM concentrations indicating that the ventilation air was being effectively directed to the working faces. The calculated result from the AQE was an estimated air quantity of 586,000 cfm to dilute DPM down to a $400_{TC}\,\mu g/m^3$ concentration. Again, after comparison, it was concluded that by using the AQE, the air quantity was overestimated by 8% (Table 1).

9 SUMMARY

DPM legislation and a growing concern for miner health in underground metal/nonmetal mines has led to an increased awareness of the need to include ventilation planning in mining plans. With this in mind, the AQE is a valuable tool for estimating the total air quantity required in a mine for dilution of DPM to specified or statutory concentration levels. This estimate of required total air quantity gives the mine operator a starting point from which to plan a DPM control strategy. This user-friendly AQE program will not only provide air quantity estimates, but it will allow the user to interactively make theoretical changes to the input parameters leading to a more efficient and effective ventilation system. Results from case studies show favorable results towards the accuracy of the AQE.

REFERENCES

Grau, III, R.H., Robertson, S.B., Mucho, T.P., Garcia, F., and Smith, A.C. 2002a. NIOSH research addressing diesel emissions and other air quality issues in nonmetal mines. *Society for Mining, Metallurgy and Exploration Annual Meeting, Feb. 26–28, 2002.* Phoenix, AZ.

Grau III, R.H., Mucho, T.P., Robertson, S.B., Smith, A.C., and Garcia, F. 2002b. Practical techniques to improve the air quality in underground stone mines, *North America Ninth U.S. Mine Ventilation Sym. June 8–12, 2002.* Kingston, Ontario, Canada.

Haney, R.A., and Saseen, G.P. 2001. Estimation of Diesel Particulate Concentrations in Underground Mines, *Mining Engineering*, 52(4): 60–64.

Head, R. 2001. Calculating Underground mine ventilation fan requirements. *Aggregates Manager*, 6(3): 17–19.

Mine Safety and Health Administration (MSHA), U.S. Department of Labor, 2001, Nonpermissible diesel engines approved under Part 7. http://www.msha.gov/S&HINFO/DESLREG/909a.HTM.

Mine Safety and Health Administration (MSHA), 2001, U.S. Department of Labor, Final Rule 30 CFR Part 57. http://www.msha.gov/S&HINFO/DESLREG/1909a.HTM.

Mucho, T.P. 2001. Practical mine ventilation. *Safety Seminar for Underground Stone Mines.* Dec. 5, Louisville, KY.

National Institute for Occupational Safety and Health (NIOSH), Department of Health and Human Services, 1988. *Carcinogenic Effects of Exposure to Diesel Exhaust*. Publication 88–116: 30 pp.

Robertson, S.B. 2001. The NIOSH Mine Air Estimator. *Safety Seminar for Underground Stone Mines*. Dec. 5, 2001, Louisville, KY.

Schnakenberg, Jr., G.H. 2001. Estimate of technically feasible DPM levels for underground metal and nonmetal mines. *Mining Engineering*, Sept.: 45–51

Schnakenberg, Jr., G.H., and Bugarski, A.D. 2002. Review of Technology Available to the Underground Mining Industry for control of diesel emissions. *IC 9462:* 52 pp.

Timko, R.J., and Thimons, E.D. 1987. Damage resistant brattice stoppings in mines with large entries. *Engineering Mining Journal*, 188(5): 34–36.

U.S. Environmental Protection Agency (EPA), 2000, "Health Assessment Document for Diesel Exhaust," Report EPA/600/8-90/057E, July: 669 pp.

U.S. Environmental Protection Agency (EPA), http://www.epa.gov/otaq/certdata.htm#largeng

Mine Ventilation, Ganguli & Bandopadhyay (eds)
© 2004 Taylor & Francis Group, London, ISBN 90 5809 633 5

The application of mining industry ventilation survey techniques to a complex computational fluid dynamics analysis

Kirk H. McDaniel & Nader Shahcheraghi
Earth Tech Incorporated, Oakland, California, USA

Keith G. Wallace, Jr.
Mine Ventilation Services Incorporated, Fresno, California, USA

ABSTRACT: In order to understand the overall ventilation design for a mine or transit system, a thorough understanding of the ventilation system is required. Failure of the ventilation system can have devastating results including loss of life and property. This paper describes the emergency ventilation system in a transit station, consisting of both tunnel and station fans, in addition to a very large, open and complex architectural design. The station consisted of three tunnels entering at its lowest elevation, multiple connections between the lower and upper levels and multiple exit points from which airflows could move through the facility. Previous modeling efforts for this station had failed to find a suitable solution to the ventilation design during a station emergency such as a transit vehicle fire. This may have been influenced by the fact that the boundary conditions were established based on theoretical data as opposed to field-measured data. Using mine ventilation survey techniques, field data was measured to establish the boundary conditions for a cold flow Computational Fluid Dynamics (CFD) model. The CFD simulation results were compared to the field test measurements in order to evaluate the accuracy of the CFD model. Once validated, the model can be used to accurately model fire scenarios. This paper describes the field measurements that were performed and the resulting correlation with the CFD model. It discusses the benefits of using a traditional field measurement technique in concert with a non-traditional analytical tool such as CFD.

1 INTRODUCTION

A thorough understanding of a complex underground ventilation network is one of the most critical elements of the overall design of a safe and effective ventilation system. This is true for both the mining and transit industries where the failure of such a system can have devastating results including loss of life and property. Not only must it operate under normal conditions to provide a comfortable environment for system users, but it must also be capable of protecting those same users in the event of an emergency, such as a fire. The design of such environments must be a carefully orchestrated balance between the science of ventilation engineering and the art of applying it in a real-world situation.

Computational Fluid Dynamics (CFD) is a very powerful and useful tool in solving complex airflow problems in three dimensions and to model the distribution of heat and smoke in the event of a fire. The civil tunnel ventilation industry is at present more advanced in the use of CFD than their mining counterparts. The Subway Environment Simulation (SES) Computer Program has been the standard simulation technique since the 1970s for the analysis of underground tunnel networks associated with complex transit systems. However, it was recognized that SES did not model complex geometrical structures, such as an underground transit station very well. It was not until the development of functional, cost effective CFD software in the early 1990s that CFD analysis became a viable alternative and provided a solution to the problem of modeling complex geometries. With the advances in the computer technologies and the computing capabilities of digital computers, CFD has become an affordable, flexible and powerful method in studying complex fluid flow and heat transfer problems. Provided that the geometry of the structure can be modeled, and provided that the modeler can develop realistic assumptions and boundary conditions, CFD modeling can yield very realistic predictions of the flow of heat and smoke. However, it is important to understand that the model is only as good as these model assumptions and boundary conditions and that any model can benefit from field verification. Commercially available CFD packages have become much more "user-friendly." This has resulted in the wider use of this type of software. However, the ease of using this software can result in the modeling of systems by persons who may lack the expertise or real-world experience. Unfortunately, it is the author's opinion that CFD modeling results tend to be accepted with little question by the lay-person provided that the modeler's credentials are impressive and that the model looks good.

CFD has been used in the mining industry, although to a lesser extent than that in the civil sector. Mining applications have traditionally focused on using this tool for major fan installation analyses and to evaluate airflow patterns in dead end headings. Some of the reasons the technique is not more widely utilized in the mining industry may be related to: (1) The (perceived) high cost of the software, (2) The specialized expertise required and technical challenges associated with establishing a model, (3) The fact that no government agency mandates its use (unlike the tunnel industry where CFD is a recognized and accepted tool to analyze complex systems and is required by NFPA 130, Standard for Fixed Guideway Transit and Passenger Systems, 2003 Edition), (4) The perception that CFD is more of a academic research tool than a real-world applications tool, and (5) The fact that CFD results are subjective to user inputs and not readily verifiable.

2 TRADITIONAL APPROACH TO MODELING COMPLEX GEOMETRIES

CFD analytical techniques, even those associated with more recent software techniques, have their limitations. Like any model, they are dependent on input parameters that must be both realistic and reasonable. In the case of CFD analysis of complex geometric structures for transit ventilation systems, input parameters have traditionally been developed from two primary sources. They were either assumed (based on empirical or theoretical data) or they were derived from another model. Using SES output parameters as the "boundary conditions" for a CFD model is a common technique where only a portion of the ventilation system is to be modeled as opposed to the entire network. It is also significantly easier and significantly less expensive. Taking field measurements to collect the boundary data has not been traditionally used, possibly because of the associated expense. Attempting to take in-situ field measurement of the *complex* airflow patterns in this type of large structure can be very costly and can yield only marginal results in comparison to the level of detail obtained from a CFD analysis.

In its defense, this modeling approach works very well under most conditions and yields accurate results. However, there can be obvious problems associated with using purely theoretical data and/or the results of one model to develop another. The technique may not work so well in those cases where the CFD model is so complex that sufficient boundary conditions cannot be developed using these traditional approaches or where variances (or minor errors) in the input data can be amplified so as to produce unrealistic CFD results.

3 FIELD TESTING METHODOLOGY, PURPOSE AND RESULTS

In the case used as the basis for this paper, modeling work on an underground transit station provided the opportunity to combine these two techniques. CFD analytical techniques were to be used to determine the complex airflow patters within the facility. Traditional mine ventilation field measurement techniques were used to collect (boundary condition) data at very specific locations in order to provide sufficient input data that was both accurate and repeatable.

The field measurements were conducted using a Davis ball-bearing vane anemometer. A full entry traverse was taken across each traverse location (McPherson, 1993). At least two traverses were taken at each location (one in the opposite direction to account for variations in airflow consistency). Results of subsequent traverses were compared and correlated to within 10% in order to be considered accurate. Traverse times used were 60, 90 or 120 seconds depending on the cross sectional area of the traverse location. Cross sectional areas were measured in the field using a laser measurement device. All equipment used was calibrated to NIST traceable standards.

The station platform area is approximately 165 meters long, 20 meters wide and 11 meters high. It is accessible from one end through twin-bored tunnels. The other end is accessible through a single cut-and-cover tunnel. The platform level of the station is adjoined to the (upper) mezzanine level through a series of stairways and escalators. In addition, the mezzanine is connected to the platform through a very open architectural designed ceiling. The mezzanine is also connected to a shopping mall. The interface between the mall and the mezzanine is through multiple entrances, stairs and escalators. The very complex structure was obviously designed more for aesthetics than functionally in mind in terms of the ventilation system. While it is not a typical setting for a mine type ventilation survey, the survey techniques used for mining applications enabled the collection of data to use in validation of the CFD model. Figure 1 shows a graphical representation of the transit station that was being modeled.

During a fire emergency in the station, the distribution of ventilation flow is an important aspect of the effectiveness of the emergency ventilation system.

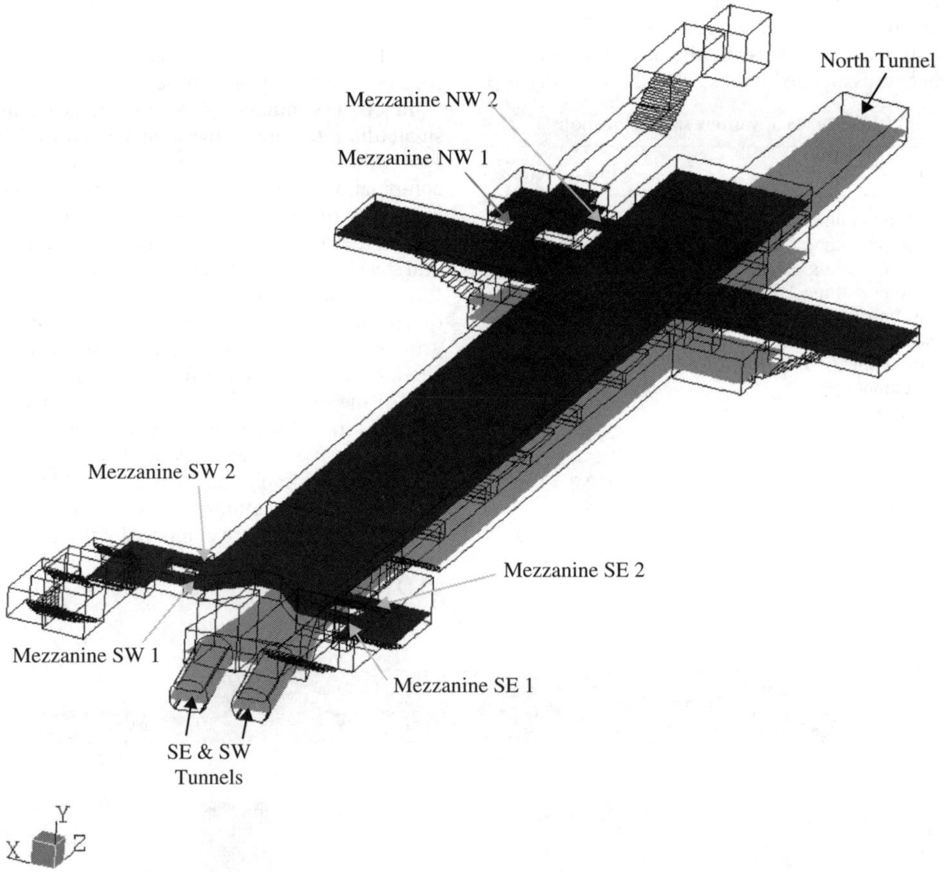

Figure 1. Station geometry and boundary locations.

The purpose of this study was to determine the complex three-dimensional flow distribution within the station after the activation of the emergency ventilation fans using a validated CFD model. The collection of field data would enable the development of a *cold-flow* model that basically simulates in field airflow distribution within the station geometry including the effects of the emergency ventilation fans but without the influences of a fire. Fire effects can easily be added to the model after confidence in the model parameters has been established. This provides a higher level of confidence in the effects of the fire on the airflow within the structure.

Airflow measurements were performed within the station at locations where the flow direction is mainly one dimensional and easy to measure and provided sufficient input data to make the CFD modeling meaningful. Figure 1 shows the station geometry and locations at which flow measurements were performed. Table 1 shows the results of the flow measurements for all the locations within the station. Comparison of the field test data in Table 1 indicates that for the entire station the overall inflow and outflow rates are balanced within 1%. These flow measurements were used to validate the three-dimensional CFD model, which was used to predict the complex flow patterns within the station during the activation of emergency ventilation system.

4 CFD MODELING

The equations that govern the three-dimensional flow in the station are the mass continuity and momentum balance equations. The control volume, tensor, form of these equations are given by CFX-TASCflow reference manual,

$$\frac{\partial}{\partial t}\left(\int_v \rho dv\right) + \int_s \rho u_j dn_j = 0$$

$$\frac{\partial}{\partial t}\left(\int_v \rho u_i dv\right) + \int_s \rho u_j u_i dn_j = -\int_s P dn_i + \int_s \mu_{eff}\left(\frac{\partial u_i}{\partial x_j} + \frac{\partial u_j}{\partial x_i}\right) dn_j + \int_v S_{u_i} dv$$

Here, t is the independent variable time, x_i indicates the three components of the Cartesian coordinate system and u_i represents the three flow velocity components associated with x_i. The fluid density is ρ, flow field pressure is P and μ_{eff} is the effective fluid viscosity. The surface area normal vector is indicated by n. The integrals are over the volume, V, or the surface, S, of the control volume.

Table 1. Field test data at various station locations.

Test location*	Flow rate(m^3/s)
North west mezzanine1 (+)	21.9
North west mezzanine 2 (+)	21.0
South west mezzanine 1 (+)	7.3
South west mezzanine 2 (+)	22.6
South east mezzanine 1 (+)	9.7
South east mezzanine 2 (+)	22.4
South west tunnel (+)	62.8
South east tunnel (+)	69.1
North tunnel (+)	262.0
Total inflow (+)	*498.7*
North fans (−)	246.4
South fans (−)	250.2
Total outflow (−)	*496.6*
Overall balance %	0.4

* (+) indicates inflow, (−) indicates outflow.

In a CFD simulation these equations are applied to small (discrete) three-dimensional control volumes that form the computational domain. In order to build the computational domain the complex three-dimensional geometric model of the station was developed. Next, the geometric model was discretized using a hexahedral mesh of approximately 195,000 nodes, which form the computational grid. Figure 2 shows various (horizontal and vertical) slices of the computational grid used in the CFD simulation. The CFX-ASCflow version 12.1 computer software was used in the solution of the governing equations. In order to validate the CFD model the flow measurements at the tunnel and fan locations were used as boundary conditions for the CFD simulation. Next, the CFD simulation results at other locations such as the mezzanine exits within the station were compared with the field measurements in order to determine the accuracy of the CFD results.

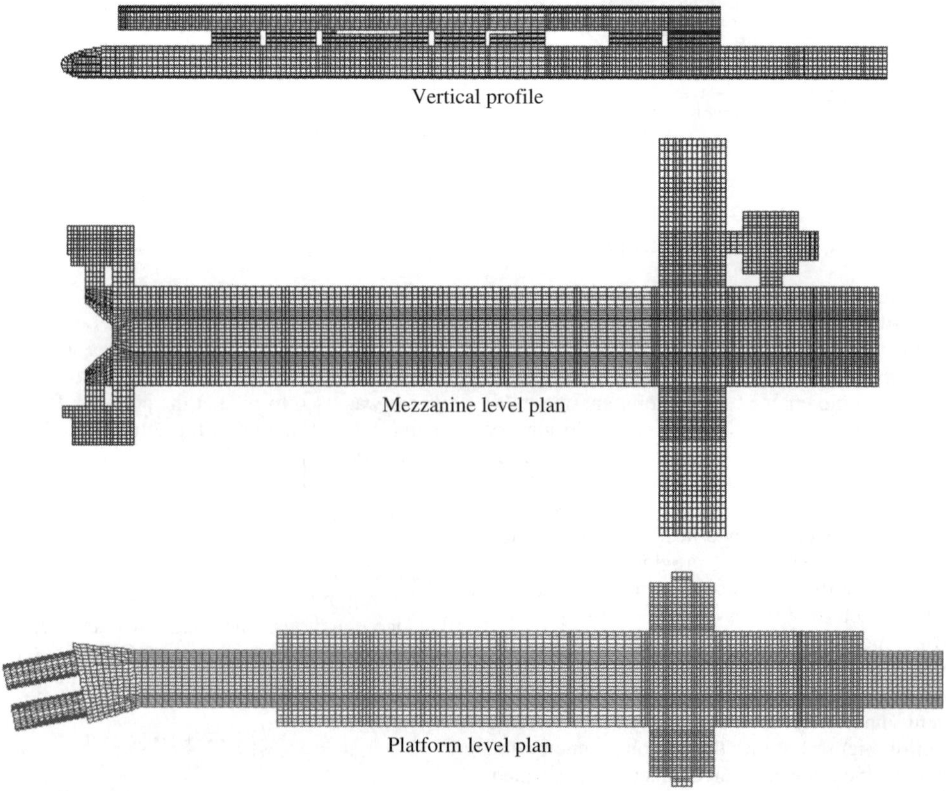

Figure 2. Computational grid slices at the platform and mezzanine plan views and station profile view.

5 CFD RESULTS AND COMPARISON WITH FIELD TEST DATA

As shown in Table 1, the field test data show excellent agreement between the ventilation fan outflow rates and the overall inflow into the station through various station exits. In order to determine the validity of the CFD model the results of the simulation were compared with the field measurements at the mezzanine exit locations. Table 2 shows the comparison between the CFD simulation results and the field measurement at same locations within the station mezzanine.

The results indicate that the CFD model predictions range from ~7% to ~10% of the field measurement data for the north mezzanine exits. In the south mezzanine area the number 2 exits are within ~7% to 14% of the test data. However, the CFD results for number 1 exit locations vary more than 15% from the field test data. The overall mass balance comparison indicates ~3% variation between the CFD and the field test results. In locations where the difference between the CFD and field test data is greater than 15% further investigation and perhaps further refinement of the CFD model is necessary. These differences can be due to inaccuracies in the geometric model or lack of grid resolution in the CFD model. In addition, the field test results have uncertainties of the order 5%. Figure 3 shows the velocity vectors in horizontal planes in the platform and mezzanine levels. The three-dimensional nature of the station flow is shown in Figures 4 and 5, which show the stream lines from the north and south mezzanine exits, respectively.

Table 2. Comparison of field test and CFD flow rate data.

North west mezz. 1	21.9	20.3	−7.1
North west mezz. 2	21.0	18.9	−9.7
South west mezz. 1	7.3	9.6	32.5
South west mezz. 2	22.6	19.4	−14.0
South east mezz. 1	9.7	15.9	63.6
South east mezz. 2	22.4	23.9	7.0
Mezzanine overall	104.8	108.1	3.2

6 CONCLUSION

The use of CFD modeling in tunnel ventilation systems has been widely used to predict the consequences of a fire. However, these models are not typically correlated with actual airflow measurements. The results described in this paper show that by using mine ventilation survey techniques an accurate and useful CFD model can be developed. By using measured information the boundary conditions necessary to perform CFD modeling are significantly enhanced. Although this technique only allows for modeling of a "cold" scenario (one without a fire), it is relatively easy to apply a fire to the model to simulate the effects this could have on the ability of passengers to escape from the transit tunnel. By correlating the "cold" model a much more accurate fire scenario is possible.

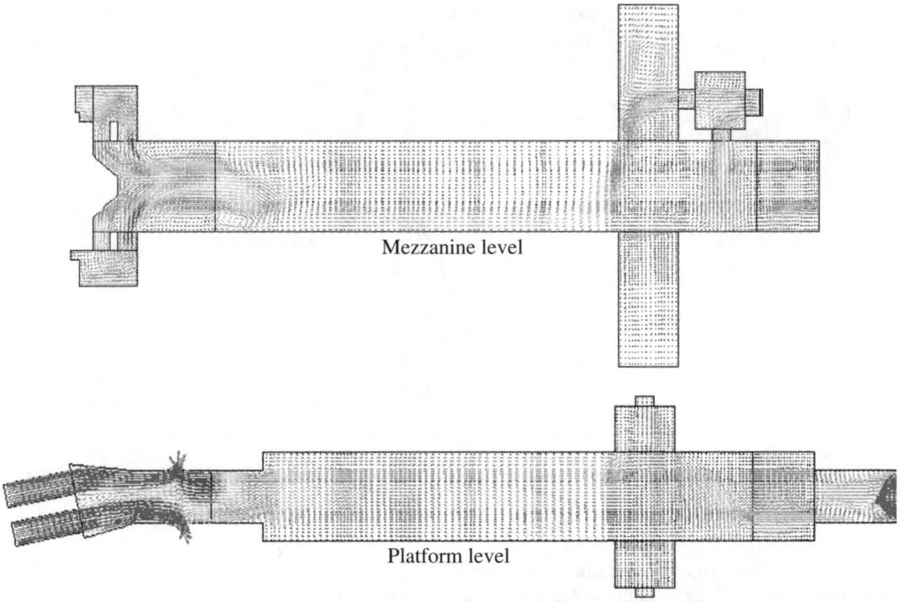

Figure 3. Velocity vectors in the platform and mezzanine levels.

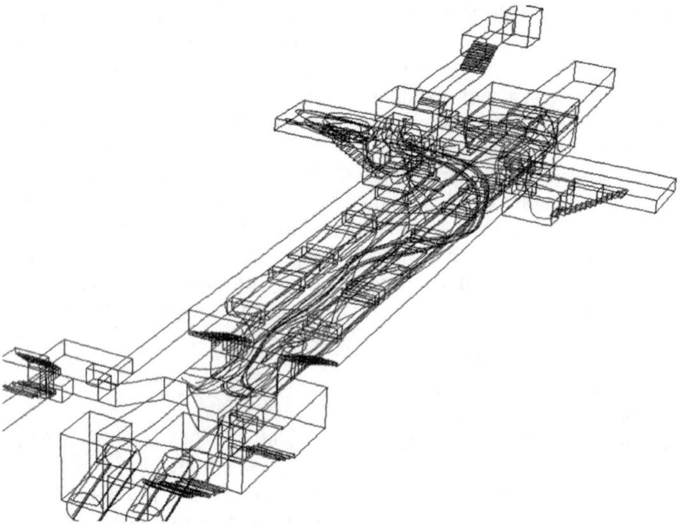

Figure 4. Streak lines from the north mezzanine exits.

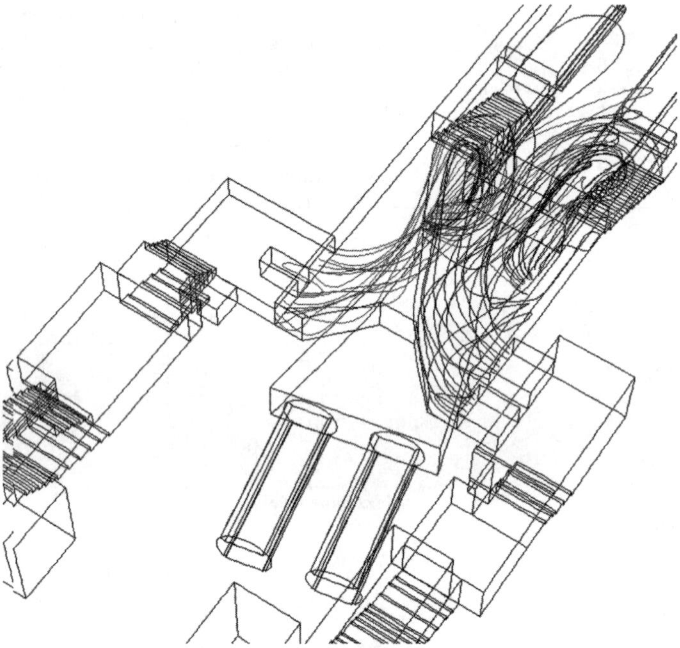

Figure 5. Streak lines from the south mezzanine exits.

REFERENCES

CFX-TASCflow Version 12.1 Reference Manual.

McPherson, M.J. 1993, Subsurface Ventilation and Environmental Engineering, London, Chapman & Hall.

Mine Ventilation, Ganguli & Bandopadhyay (eds)
© 2004 A.B. Cecala, ISBN 90 5809 633 5

Filtered recirculation – a critical component to maintaining acceptable air quality in enclosed cabs for surface mining equipment

A.B. Cecala & J.A. Zimmer
National Institute for Occupational Safety and Health, Pittsburgh, PA USA

ABSTRACT: Most types of heavy equipment used in the surface mining industry for the removal and transport of overburden and ore place the equipment operator in an enclosed cab. A review of the Mine Safety and Health Administration's dust records reveals that a substantial number of equipment operators exceed the allowable limit for silica and respirable dust while operating this equipment. Many times in the mining industry, this equipment is older and various components on the enclosed cab and the filtration system have deteriorated to levels where they are no longer effective. Another contributing factor identified during a recent field study performed by the National Institute for Occupational Safety and Health was that a significant amount of dust may be generated inside the enclosed cab by the actions of the equipment operator. When this occurs, the only method to effectively remove the dust is with a recirculation filtration system. Toward this goal, a new self-contained recirculation unit designed for the mining industry was tested on three different pieces of mining equipment at two different operations. Results of these tests, along with the critical factors for implementing an effective recirculation system for enclosed cabs, will be discussed in this paper.

1 INTRODUCTION

Many workers in the United States and throughout the world are exposed to high levels of respirable dust as they operate heavy equipment while performing many different duties and job functions in the mining, construction, and agricultural industries. Since much of the earth's surface contains silica-bearing strata, these workers are also frequently exposed to high levels of respirable silica dust. The serious health effects of workers breathing respirable silica dust and the potential for subsequently developing silicosis and other debilitating lung diseases have been known for decades. This concern has been further underscored by the 1997 ruling of the International Agency for Research on Cancer (IARC) which classified crystalline silica as a "Group 1" substance. The report stated that: "crystalline silica inhaled in the form of quartz or cristobalite from occupational sources is carcinogenic to humans." IARC (1997).

Over the past decade, most equipment operators have been placed into fully enclosed cabs while operating heavy equipment because of the increased safety and health measures provided by these enclosures. The safety component is usually designed and fabricated into the enclosed cab when the equipment is originally built, and usually includes rollover (ROPs) as well as falling object protection (FOPs). This safety component, being structural in design, does not normally change as the equipment ages. The health component of providing clean filtered air and acoustical protection is much more difficult to achieve and maintain. In fact, over the past few years, NIOSH personnel have observed a number of pieces of newer equipment that were not providing acceptable air quality in the enclosed cab. Another significant factor with this health component is that it deteriorates as the equipment ages and must be constantly maintained to provide acceptable protection to the equipment operators.

Since providing acceptable air quality in heavy equipment is such an important task, there has been a significant effort put forth by many, including NIOSH, working to further improve the technology in this area. NIOSH's efforts have been directed in two primary areas. The first has been working with the agricultural industry on the many different types of farming equipment.(Heitbrink et al. 2003; Hall et al. 1997; Hall & Heitbrink 1997; Heitbrink et al. 2000; Heitbrink & Moyer 2000) The second area has been directed towards the mining industry. When this research is completed, the information and knowledge gained from both of these research areas can then be applied to an array of different equipment used in other industries, including the construction industry.

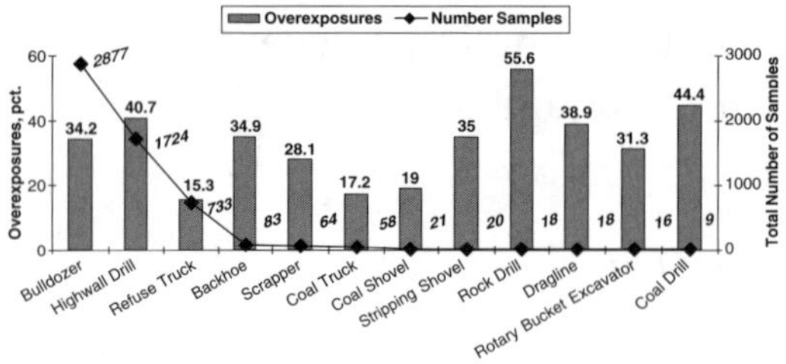

Source: Mine Safety and Health Administration: inspection respirable quartz dust mine data. Integrated Management Information System, TERA data, Denver, CO.

Figure 1. MSHA compliance overexposures in surface coal from 1993–2002.

The focus of this paper is directed towards the air quality component of enclosed cabs on surface mining equipment. When this research was initiated a few years ago, an inquiry was performed into the Mine Safety and Health Administration's (MSHA) dust compliance records to identify and quantify heavy equipment operator levels of overexposure to silica dust in surface coal operations. Figure 1 shows an updated version of this information for the 10-year period from 1993 to 2002. This graph lists various heavy equipment operator job classifications and indicates the percentage of compliance samples overexposed to silica dust and the number of samples taken during this 10-year period. When analyzing all the job classification categories for the surface coal industry, the only other workers having similar overexposure levels were: blasters/shooters at 24.2% and 236 samples, highwall drill helpers at 35.9% and 64 samples, and oilers/greasers at 20.8% and 53 samples.

Because of the high number of overexposures in surface mining equipment operators, a few years ago MSHA sponsored a number of industry seminars held throughout the United States to highlight the problems and to provide information on potential technology to improve the air quality in enclosed cabs. During these seminars, many different original manufacturers (OEMs) and companies presented their technology to improve the air quality in enclosed cabs on various types of heavy equipment. After attending the first seminar, we felt there were two noteworthy issues that emerged from the meeting. The first was the lack of focus by OEMs and other companies on the silica dust problem and the necessary filter efficiency levels that must be achieved when dealing with this dust. The second was that the vast majority of the technology developed was only directed toward improving the air quality on new equipment.

Since the majority of equipment currently used in the mining industry is older, the research performed at the NIOSH Pittsburgh Research Laboratory was mainly directed towards this older equipment. After analyzing the information in this area, we believed the most efficient and cost-effective approach was to partner our research with various manufacturers to take their new equipment technology and attempt to apply it towards older mining equipment. Over the past few years, a number of these cooperative efforts have been established and studies have been performed evaluating drills, bulldozers, haul-trucks, front-end loaders, and excavators/shovels. During these studies, we observed an array of air qualities in enclosed cabs based upon the effectiveness of the different filtration and pressurization systems. In some cases, systems were well-designed and maintained to provide acceptable air quality to the equipment operators. In many other cases, the air filtration and pressurization systems on the cabs were poorly designed and/or maintained and were providing very marginal to unacceptable air quality. A major factor in the degradation of these systems was the aging and wear of many of the components on these systems and on the integrity of the enclosed cabs. Many gaskets and seals were observed that had deteriorated to a point where they were no longer providing adequate closure or sealing. Once this occurred, it was impossible to achieve positive pressurization until they were replaced and holes and cracks in the cab were sealed.

Another serious problem identified during this research was the effects of dust generated within the enclosed cab. During one particular study, a significant increase in respirable dust was recorded inside an enclosed cab of a surface drill when a floor heater was used (Cecala et al. 2001). Testing on this cab verified that the floor heater was generating dust from the drill operator's boots grinding and stirring up materials on the floor of the cab, as well as from dust blown from the operator's clothing, Figure 2. A controlled test performed at this operation's maintenance shop indicated a 27-fold increase in respirable dust concentrations with the floor heater in use over baseline levels. In fact, this testing indicated that respirable dust levels in the cab were higher when the floor heater was used than dust levels out in the shop area.

Acceptable air quality being provided to workers in enclosed cabs is based upon many different factors, but the most critical is the cab's filtration and pressurization system. An effective filtration and pressurization system should be composed of both a recirculation and an outside (make-up) air system. As previously mentioned, there have been a number of studies directed towards the filtration component of this system, many of which have been cooperative studies with OEMs (Organiscak et al. 2000; Cecala et al. 2002; Chekan et al. 2003). A number of these studies have shown substantial air quality improvements in the enclosed cabs of older mining equipment when they were retrofitted with a new filtration and pressurization system, along with replacing gaskets and seals and plugging and sealing holes in the shell of the cab, in order to achieve positive pressurization.

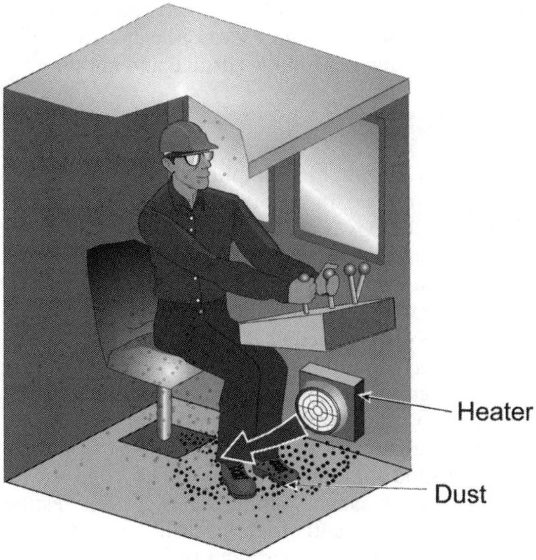

Figure 2. Artist conceptual drawing of dust being generated from floor heater in enclosed cab.

The focus of this study was directed specifically at the recirculation component. A high-quality filter is vitally important for an effective recirculation system since it is the only method of removing dust emitted or generated inside the cab. A secondary benefit from the recirculation system is that it allows the air within the cab to be conditioned to the cab operator's comfort (both heating and air conditioning) without major air changes that significantly impact the system's capability requirements. Obviously, the size of the air conditioner and heater component directly impacts the cost of a filtration/pressurization system.

A major portion of this report will discuss a partnership established with SCS/Frigette, Inc. to evaluate the use of its self-contained recirculation system for improving the air quality in enclosed cabs.[1] For this research effort, SCS/Frigette, Inc. shared 25 pct. of the cost with NIOSH in evaluating the use of its recirculation device on three different pieces of mining equipment.

The recirculation device tested here was originally developed for the automotive industry and then adapted for cabs on heavy equipment. Part of the modification was to increase the filter efficiency of this recirculation unit to HEPA quality, being 99.97 pct. effective on 0.3 micron particles. Figure 3 shows the major components of the SCS/Frigette recirculation system

[1] Mention of any company name or product does not constitute endorsement by the National Institute for Occupational Safety and Health.

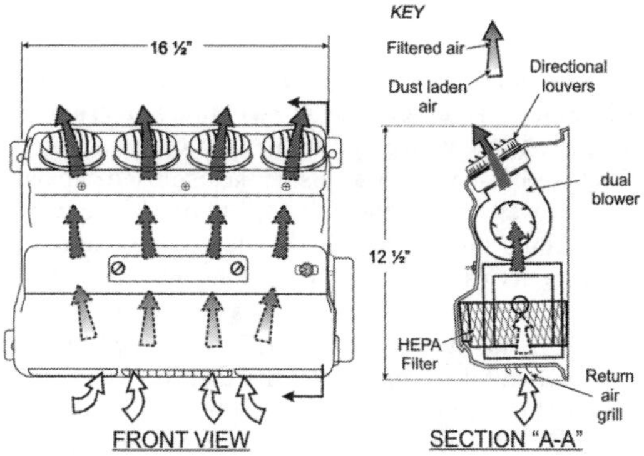

Figure 3. Design of recirculation filtration unit.

as tested in this study. Eighty to one-hundred cubic feet per minute of air from within the enclosed cab is drawn into the unit by a dual scroll blower system. This air is then drawn through a HEPA filter before being discharged back into the enclosed cab. The unit is fabricated from a high-impact ABS plastic, making it fairly light and allowing it to be easily installed inside an enclosed cab. This unit is basically self-supporting and can be mounted vertically on a wall or horizontally on the roof of the cab. The unit needs to be supplied with either 12- or 24-volt electrical power from the equipment. The unit costs approximately $850 and replacement filters (HEPA) are approximately $70.

2 FIELD TESTING OF SELF-CONTAINED RECIRCULATION SYSTEM

The objective of this research task was to quantify the air quality improvements in various enclosed cabs when implemented with a SCS/Frigette, Inc. recirculation system. A test plan was established under a cooperative agreement with SCS/Frigette, Inc. to evaluate this recirculation system on three different pieces of equipment at two different surface coal mining operations. A surface drill was tested at both operations, as well as a shovel at the second operation. The sampling strategy was to compare respirable dust levels inside the enclosed cab, with and without the recirculation system in use, to dust levels outside. The following is a brief description of the sampling setup used for this study.

2.1 *Drill testing*

For both drills evaluated, there were three main sampling areas – one inside the operator's cab and two outside locations. All in-cab sampling instruments were placed on two sampling racks located directly behind the drill operator. One sampling rack was used when the recirculation unit was operated, the other when the unit was off. Each sampling rack was composed of three gravimetric samplers and a Personal Data RAM (PDR) instantaneous dust monitor. Periodically throughout the testing, hand-held particle-counting instruments were used to quantify 0.3 and 3.0 micron size particles to indicate variations inside the enclosed cab over time, as well as to compare inside to outside levels.

It should also be noted that there were additional instruments used during this testing which will not be presented in this report due to length limitations. This additional instrumentation included cascade impactors for determining the size distribution of dust, temperature-recording devices inside and outside the enclosed cab, and pressure-recording instrumentation to monitor static pressure levels.

For the two outside dust sample locations, one was located on the drill and the other was on a tripod sampling station. All instruments used for both locations were also placed on sampling racks. The first sample location, called the "outside cab" position, was located on the handrail on the far side of the work station platform (between the drill operator and the drilling). This was the closest location to drilling that could be safely monitored, while providing the best indication of outside dust levels generated by the drill. Duplicating the inside monitoring method, sampling at this outside cab location included two identical sampling racks – one for when the recirculation unit was used, the other for when it was off. Changes inside the enclosed cab, with or without the recirculation unit would not affect the two outside sample locations; however, outside levels can vary significantly throughout the course of the workday and in an effort to most accurately evaluate the recirculation unit, the same sampling procedure needed to be duplicated outside. Additionally, in order to calculate protection factors with and without the recirculation unit, identical sample times were necessary. Each outside sampling rack also included three gravimetric samplers. Finally, one PDR device was used to provide instantaneous respirable dust concentrations continuously throughout the sampling time period.

The second outside sampling location was called the "tripod" position because it was located on a camera tripod with a modified wooden cap to hold the dust sampling equipment. The tripod was manually moved during testing by NIOSH personnel and was always placed on the windward side of the drill unit in the dust cloud. This sampling location also consisted of two racks of three gravimetric samplers, which were alternated with and without the use of the recirculation unit.

The sampling sequence varied for each piece of equipment as to when the recirculation unit was turned on and off. One-hour time segments were used during the first drill study, which appeared to be an optimum time period. However, this had to be modified during the second drill evaluation due to a number of factors, and eventually the segments were cycled based on the number of holes drilled and not time.

2.2 *Shovel testing*

For the shovel testing, a sampling strategy similar to the drill testing was incorporated. The inside cab sampling methodology was identical to that used with the drill. The sampling package was located slightly behind the shovel operator in an effort to most closely monitor the operator's respirable dust exposure. The outside sample was also located on the handrail, very close to the enclosed cab on the shovel. This sample location was also very close to the air handling system, which was used to filter and pressurize the enclosed cab.

One change from the drill testing was that there was no tripod sampling location used for the shovel due to safety reasons. Sample times for this testing also varied significantly from the drill testing because no test personnel remained with the shovel throughout the evaluation time. Periodically throughout the day, test personnel would return to the shovel and the on/off sequencing of the recirculation unit was based on these times, which varied significantly.

3 DUST TESTING AND INSTRUMENTATION

The following is a brief description of the dust sampling methods and instrumentation used in this study.

3.1 *Gravimetric dust sampling*

Airborne respirable dust concentrations using gravimetric samplers were obtained with personal sampling packages located at the various sample locations. All of the sampling locations used MSA Escort Elf sampling pumps operated at 2.0 liters/min flow rates. All dust samples were collected with the 10-mm Dorr-Oliver cyclone, which classifies the respirable portion of dust, i.e., (having aerodynamic diameters of 10 microns or less.) This respirable fraction of dust was deposited on a 37-mm MSA coal cassette filter. These filters were pre- and post-weighed to the nearest 0.001 mg on a Mettler MT5 microbalance at NIOSH's laboratory in Pittsburgh. The dust concentration of each gravimetric sampling unit was calculated based upon its own run time. After each concentration was determined, the units were averaged together to determine the average respirable dust concentration for the entire package. For every 10 gravimetric filters used during field testing, a blank cassette was set aside for calibration purposes. These blank cassettes were then used to calculate a correction factor that was applied to the final value for all field gravimetric measurements used in this evaluation.

3.2 *Instantaneous dust sampling*

Respirable dust measurements inside and outside the enclosed cab location were monitored with MIE, Inc., PDR instruments. These are real-time sampling devices that measure the light-scatter variation created when a volume of dust-laden air is drawn through an internal sensing chamber. This scatter is then correlated to a relative dust concentration and these values are stored on an internal data logger. Ten-second averages were calculated for this testing and downloaded to a laptop computer at the end of each day.

The PDR instruments were operated in active mode and were also used with a 10-mm Dorr-Oliver cyclone and MSA Escort Elf pump (2.0. liters/min), which drew an air sample into the instrument. After the respirable dust traveled through the instrument, it was deposited on a 37-mm MSA coal cassette filter cassette, identical to that of the gravimetric sampler. A new filter was used for each day of testing. A correction factor was also calculated for each PDR instrument for each day of testing and then applied to the data using a commercial spreadsheet software package.

3.3 *Particle-counter measurements*

Particle-counting measurements were taken with a Pacific Scientific Instrument, Model 227 Hand-Held Particle Counter. This instrument uses a laser-based sensor that counts two different particle sizes. The model used in this evaluation always measures 0.3 micron particles on the first channel. The second channel is capable of measuring micron particles of 0.5, 1, 3, or 5; the 3 micron size was used for this testing. This instrument uses a laser diode light source and collection optics for particle detection. As an air sample is drawn through the unit by an internal sampling pump, particles in this air sample deflect light energy from the laser onto the collection optics. The collection optic focuses the light onto a photo diode that converts these light impulses into electrical impulses. The electrical pulse height is proportional to the particle size. These pulses are counted and the magnitude is measured for particle sizing. Results are then displayed as total counts in the two specific particle sizes. These instruments were used in the manual mode setting, which recorded the particle counts after each one minute of run time once the sample mode button was activated.

4 FIELD TEST RESULTS

The self-contained SCS/Frigette, Inc. recirculation unit was tested on three different pieces of mining equipment at two mine sites. The results of all three pieces of equipment will be presented separately, followed by a general discussion of the similarities and trends observed for all three evaluations.

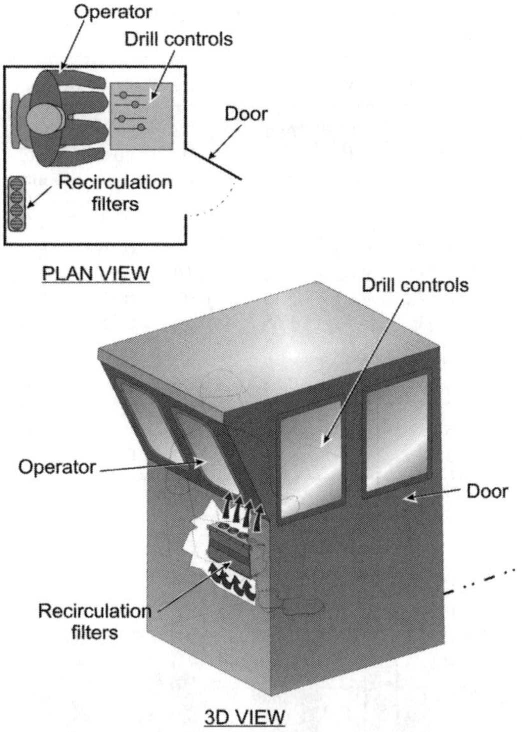

Figure 4. Location of recirculation unit in enclosed cab.

4.1 *Site #1 – drill*

Testing was performed for three consecutive days on an Ingersoll Rand DM45 drill at this mountaintop removal coal operation. For all three days, this unit was drilling into overburden material. The first two days were full days of testing (8 hours) and the third day was only a partial day, of approximately 5½ hours of testing. The recirculation unit was located mid-level on the back wall of the enclosed cab, approximately one foot below the window (Figure 4). For all three days of testing, the same testing sequence and sample times were followed for turning the recirculation unit on and off, which was on a one-hour interval.

Table 1 is an overview of the gravimetric sampling data. The values listed as "system off" refer to the measurements taken when the recirculation unit was not in use, and "system on" obviously indicates when the unit was operating. These gravimetric results indicate there was no apparent difference in the respirable dust concentrations inside the enclosed cab of this drill with and without the recirculation unit.

Protection factor is a term that can be used to evaluate and compare the air quality inside an enclosed cab. It is calculated by dividing the outside by the inside dust concentration. Obviously the higher the

Table 1. Gravimetric dust data for recirculation testing on surface drill at site 1.

	Outside tripod		Outside cab		Inside cab	
Day	Avg. dust, mg/m^3	Std error	Avg. dust, mg/m^3	Std error	Avg. dust, mg/m^3	Std error
Recirculation system off						
1	1.57	0.15	0.82	0.03	0.19	0.01
2	1.75	0.02	0.77	0.10	0.14	0.02
3	8.77	0.27	3.18	0.07	0.25	0.02
Avg.	4.03	0.15	1.59	0.07	0.25	0.02
Recirculation system on						
1	1.65	0.06	0.58	0.01	0.11	0.03
2	1.53	0.10	0.55	0.05	0.12	0.03
3	5.75	0.89	1.76	0.10	0.26	0.03
Avg.	2.98	0.35	1.76	0.10	0.26	0.03

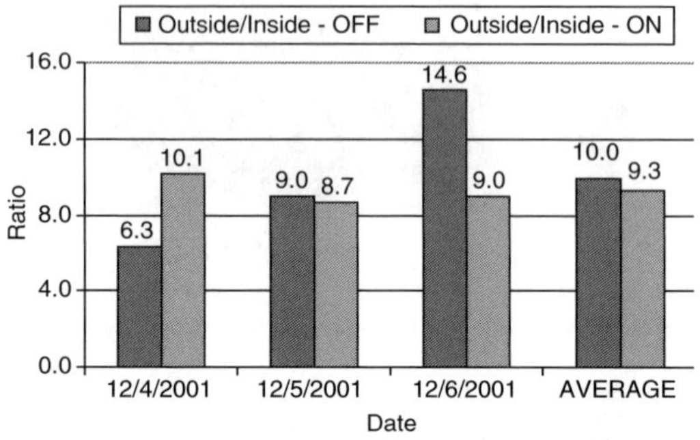

Figure 5. Protection factors in enclosed cab for drill at site #1 with and without recirculation unit.

protection factor, the better the air quality i.e., (there is less respirable dust, relative to dust levels outside the cab). The average protection factor during this drill testing was calculated to be 10 with the recirculation unit off and 9.3 with the system on. Figure 5 is an overview of the calculated protection factors from this testing with and without the recirculation unit.

A summary of the instantaneous dust measurements inside the enclosed cab on this drill can be seen in Figure 6 for all three days of testing. As previously mentioned, since the same recirculation on/off times were used for each day, the solid horizontal bars indicate time periods when the recirculation unit was off, while the dotted bars represents when the unit was operated. These instantaneous respirable dust concentrations provide very similar results to those obtained with the gravimetric dust sampling instrumentation, in which there appears to be no measurable difference in respirable dust levels inside the enclosed cab with and without the use of the recirculation unit.

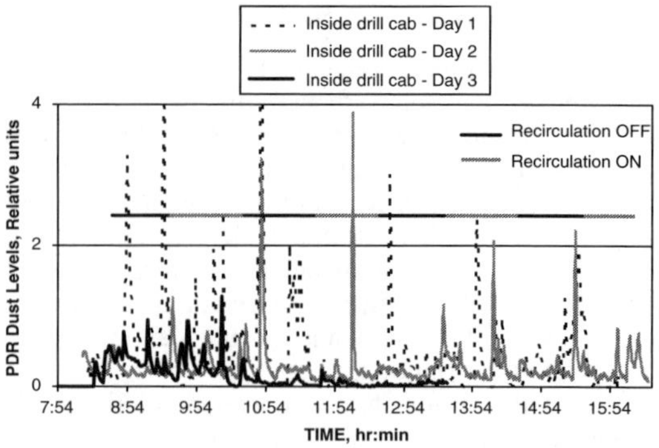

Figure 6. Respirable dust levels inside enclosed cab on drill at site #1 for three days of testing.

One noteworthy occurrence on this graph is the high spikes in respirable dust concentrations inside the enclosed cab periodically throughout the workday. These spikes occurred when the drill operator exited and then re-entered the cab to perform various functions necessary to operate the drill, such as laying out drill holes, checking and/or filling fluid levels, or other similar functions.

In addition to the gravimetric and instantaneous respirable dust measurements, intermittent readings were manually taken throughout the evaluation using hand-held particle-counting instruments. Table 2 shows an example of measurements taken on day 2 of testing and provides readings to compare particle levels with and without the recirculation unit in use. This table shows both the 0.3 and 3.0 micron size ranges and also indicates the minor differences in these particle counts with and without the recirculation unit.

4.2 *Site #2 – drill*

Testing was performed for two consecutive days on an Ingersoll Rand D45M drill at this large coal strip mining operation. For the first day, this unit was drilling 80-ft holes into the coal seam, which required two drill steels. After completing the first hole, the coupler on the drill broke and the drill was down for almost two hours. After the repair, twelve holes were drilled over the next 3.5 hours. Originally, one-hour sequences were planned but this was modified because of the amount of downtime that occurred during the repair. The test plan was then modified to cycle the recirculation unit on and off after every five drill holes. Even with this test plan, only two sampling sequences were obtained with the recirculation unit off and one sequence with the unit on.

During discussions after the first day of testing with the distributor for SCS/Frigette, Inc. and mine personnel, it was decided to place a second recirculation

Table 2. Example of particle-counting measurements taken on surface drill at site #1 for day 2 of testing.

		Time	Run #	0.3 μ	3.0 μ
System on Test #2		13:40	1	121371	4448
			2	106232	4182
		13:43	3	212959	9303
			4	135509	3840
			5	105669	3461
	Door		6	73000	1701
	opened	13:52	7	60763	1859
	Air		8	59394	1778
	turned on		9	47140	1225
			10	57878	1739
			11	69627	1452
		14:01	12	56763	1439
			Avg.	92192	3036
System off Test #2		14:37	1	150383	8287
			2	202736	6409
		14:40	3	230848	4405
			4	173337	2054
		14:43	5	158961	2345
			6	142359	1973
	Moved drill		7	133804	2084
		14:49	8	128210	2723
			9	112981	1996
		14:52	10	101393	1640
			11	96009	1683
			12	112465	1572
			Avg.	134114	2859

unit inside the enclosed cab of this drill. Obviously, this would double the air volume being cleaned by the recirculation units while also allowing a comparison of respirable dust levels inside the enclosed cab with one and two recirculation units in operation. The installation of this second recirculation unit was performed by the mine's maintenance personnel during the second day of testing and took approximately two hours to complete. After the unit was installed, the drill was also moved from operating on the coal seam to another location where it was to be used for drilling overburden. Because of both of these occurrences, data collection was also very limited for this second day of testing. For this limited testing, the unit was drilling 60-ft holes into overburden at the rate of approximately 10 minutes per hole. Because of the limited test time, three drill holes were evaluated for each of the following test sequences: both recirculation units on, one recirculation unit on, and both recirculation units off. Two sets of this test sequence were performed for this second day of testing.

Table 3 provides an overview of the gravimetric sampling data for this drill for the two partial days of testing. For the second day of testing, the "system on" refers to times when either one or both of the recirculation units were operated. It should be noted that dust levels inside the enclosed cab were very low, indicating that the primary roof-mounted filtration and pressurization system was effective. As a result, the gravimetric dust sampling results indicate no differences in respirable dust levels inside the enclosed cab with and without the recirculation unit(s). It should also be noted that for day 2, there was no measurable dust on either inside sample because of the limited test time. This very limited test time was not anticipated, nor is it desirable for this type of study, especially when using gravimetric sampling devices. However, this is a consequence of performing field testing at an active mining operation in which test plans need to be quickly modified to meet the constantly changing dynamics.

Table 3. Gravimetric dust data for recirculation testing on surface drill at site 2.

	Outside tripod		Outside cab		Inside cab	
Day	Avg. dust, mg/m³	Std error	Avg. dust, mg/m³	Std error	Avg. dust, mg/m³	Std error
Recirculation system off						
1	18.36	0.65	1.62	0.04	0.07	0.01
2	0.48	0.03	0.12	0.03	0.00	0.00
Avg.	9.42	0.34	0.87	0.04	0.04	0.01
Recirculation system on						
1	12.62	0.62	6.77	0.27	0.14	0.00
2	0.48	0.03	0.38	0.00	0.00	0.00
Avg.	6.34	0.33	3.58	0.16	0.07	0.00

The gravimetric results were used to calculate protection factors for the enclosed cab on this drill. The dust concentrations for the tripod and outside sample locations were averaged together and then divided by the inside concentration. For the first day of testing, the protection factors were determined to be 69 and 143, with and without the recirculation unit in use, respectively. This also verifies the effectiveness of the existing roof-mounted filtration and pressurization unit. No protection factors were calculated for the second day of testing since there was an insufficient weight gain on the filters from inside the enclosed cab from the limited testing.

Because of the limited testing, a more insightful analysis can be observed from the PDR instantaneous respirable dust instruments. Once again, there were two PDR instruments taking measurements on the drill, one located inside and one outside the drill cab. When analyzing the results for the first day of testing when the unit was drilling into the coal seam, the average respirable dust concentrations were 0.07 and 0.10 mg/m³ with the recirculation unit off and on, respectively, once again indicating no difference in dust levels inside the cab.

Figure 7 shows a trace of the respirable dust levels inside the enclosed cab for the second day of testing,

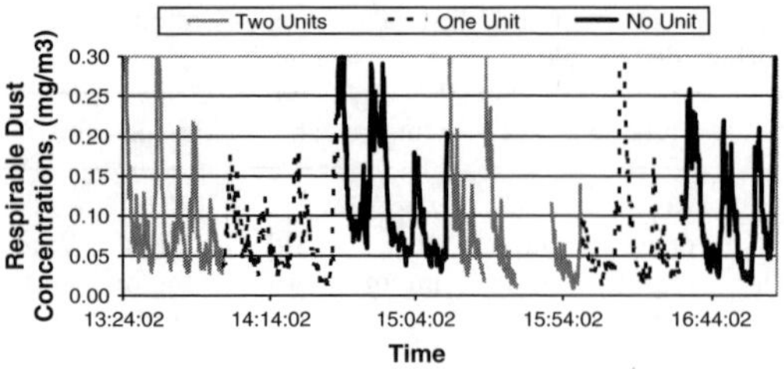

Figure 7. Respirable dust comparison for drill at site #2 for second day of testing.

which allows a visual comparison of dust levels with no recirculation units, with one recirculation unit, and with both recirculation units in use. Calculated averages of the respirable dust concentrations recorded by the PDR were determined to be 0.12 mg/m^3, 0.07 mg/m^3, and 0.09 mg/m^3, corresponding to none, one, and two recirculation unit(s), respectively. This represents a very minimal change in respirable dust levels with the recirculation unit(s) in operation.

A somewhat puzzling aspect to these results regards the comparison of the average respirable dust concentration for when one and two recirculation units were operated. Obviously one would expect that when both recirculation units were used, a lower respirable dust concentration would be obtained in the enclosed cab than when just one unit was used, but this was not the case. One possible explanation for this occurrence is the number of times the cab door was opened to allow either the drill operator or various test personnel access or egress to the cab. It is believed that each spike corresponds to when the cab door was opened and that this may have been the most influential factor for overall dust levels under each test scenario for this study. When comparing these spikes (Figure 7), only two are indicated for the second series with one recirculation unit in operation, whereas all the other series had 3 or 4 spikes or door opening episodes. This may account for the lower dust levels when one recirculation unit was operated, as well as for the significant contributor for all the variation for respirable dust levels recorded inside the enclosed cab for the three test conditions.

4.3 *Site #2 – shovel*

The 85-cubic-yard P&H 4100 shovel evaluated in this study was located in the bottom of the pit and was loading coal for both days of testing. Since there were no test personnel stationed with the shovel throughout either day of testing, the on/off times for the recirculation unit were based upon whenever test personnel were able to return to the shovel. Originally, it was hoped to sequence the recirculation on an approximate 2-hour time period, but this varied somewhat based upon the distance between the shovel and the drill and the ability to go back and forth.

When this testing was discussed with SCS/Frigette, Inc. and the mine, two recirculation systems were to be used inside the enclosed cab on this shovel. Both of these units were installed on this shovel but the shovel operator did not like the amount of "noise" produced by the fan in both units; subsequently, one unit was removed before the analysis was performed. When this unit was removed, it was re-installed in the PLC room, which is directly below and interconnected with the enclosed cab. However, because of the volume of air in the PLC room, this would have a minor impact in the air quality in the enclosed cab compartment.

NIOSH test personnel became aware that only one recirculation unit was in place in the shovel's enclosed cab when arriving for the first day of testing. The dust instrumentation was installed and started but test personnel realized that only one recirculation unit would have a minor impact on improving air quality based on the size of the enclosed cab. The cab was calculated to have a volumetric capacity of approximately 575 ft^3, which is almost 7 times greater than air volume cleaned every minute by a single recirculation unit.

The most significant contributing factor to the cab's air quality was from the original pressurization and filtration system operating on the P&H shovel. This system provided heating and cooling to the shovel operator's comfort. This system had two fan settings which delivered approximately 500 and 700 cfm of air to the enclosed cab during this testing. When considering this fact, it is even more obvious how the 85 cfm air volume cleaned by the SCS/Frigette, Inc. recirculation unit would have a minimal impact on the cab's air quality. Because of these factors, test personnel spent a significant amount of the time trying to determine the airflow and pressure parameters for the shovel rather than concentrating on the recirculation unit.

Table 4 provides an overview of the gravimetric dust measurements taken inside the enclosed cab and outside on this shovel. The outside dust concentrations were fairly consistent (0.73–0.93 mg/m^3), except for the second day of testing when the recirculation unit was tested and the average increased to 1.40 mg/m^3. Respirable dust concentrations inside the enclosed cab varied from 0.11 to 0.32 mg/m^3 over the various test segments, but the average concentration was exactly the same for when the recirculation unit was on and off. The calculated protection factors shown in this table are almost identical for both test conditions.

Once again, the recirculation unit appeared to be undersized for the enclosed cab on this shovel, especially when considering the amount of air being delivered to the cab with the equipment's existing filtration and pressurization system. In summary, there were no air quality improvements identified from the use of the recirculation unit on this shovel.

Table 4. Gravimetric dust data for recirculation testing on shovel at site #2 and protection factors.

	Gravimetric dust data			
	Outside cab		Inside cab	
Day	Avg. dust, mg/m^3	Std error	Avg. dust, mg/m^3	Std error
Recirculation system off				
1	0.95	0.02	0.32	0.00
2	0.79	0.02	0.11	0.01
Avg.	0.87	0.02	0.22	0.01
Recirculation system on				
1	0.73	0.06	0.16	0.01
2	1.40	0.04	0.27	0.00
Avg.	1.07	0.05	0.22	0.01

Protection Factors

Day	System off, mg/m^3	Recirculation unit on, mg/m^3
1	2.97	4.56
2	7.18	5.19
Avg.	5.08	4.87

5 DISCUSSION

NIOSH was interested in evaluating the SCS/Frigette, Inc. recirculation unit because it appeared to be a viable option to easily supplement a piece of surface mining equipment's existing air handling system, providing for additional air cleaning to further improve the air quality in enclosed cabs. NIOSH learned about this device when attending the first MSHA seminar on the enclosed cab problem and had also heard discussions of a number of successful applications in the field. An attractive feature with this recirculation unit is that the equipment operator can easily change the HEPA filter cartridge on a regular basis in a very minimal amount of time, and the replacement filters are relatively inexpensive. Because of this, NIOSH approached SCS/Frigette, Inc. and asked the company to consider working in a cooperative effort to perform an in-depth analysis on the effectiveness of this unit in production surface mining equipment. Subsequently, this recirculation unit was tested on three different pieces of mining equipment at two different operations. The recirculation unit installations were overseen by SCS/Frigette distributors and all units used for this testing appeared to be working properly. NIOSH's primary role was then to evaluate these recirculation units to determine their impact on improving the air quality in the enclosed cabs.

The results of testing the SCS/Frigette recirculation unit on the drill at the first field evaluation site indicated an effectiveness ranging from no detectable change, to a very minor improvement in respirable dust concentrations over the three days of testing. For this drill, it appeared that the respirable dust concentrations inside the enclosed cab were more greatly impacted by the existing filtration system located on the roof of the cab than by the implementation of this new recirculation unit. After analyzing the results of this first evaluation, it was believed that if a better correlation could be achieved between the volume of air cleaned with the recirculation unit and the volume of air within the enclosed cab, the unit could have a greater impact on lowering respirable dust concentrations inside the cab. Since this SCS/Frigette, Inc. recirculation unit is only produced in one size, the only way to increase the cleaning volume is by installing additional units.

When discussing this matter with SCS/Frigette Corp. and considering lessons learned in the first evaluation, one goal at the second field test site was to place additional recirculation units into the enclosed cabs of the equipment to be evaluated. A distributor for SCS/ Frigette Corp. was responsible for coordinating the second and final evaluation on these units. This responsibility included finding a cooperative mine willing to cooperate in this study, and second, choosing a number of pieces of equipment in which the recirculation units would be installed and found to be working properly before testing. It was decided by the distributor and the cooperating mine to test on a coal shovel and a surface drill. The P&H shovel was installed with two recirculation units, but as previously mentioned, the shovel operator disliked the amount of noise produced from both units and one unit was removed before testing was performed. Once the NIOSH test personnel arrived at the mine site, determining that only one unit was installed and calculating the air volume of the shovel cab to be 575 ft^3, it was realized that the SCS/Frigette recirculation unit would have an insignificant impact on this volume of air. An even more significant contributor to the ineffectiveness of the recirculation system was the 500 to 700 cfm of air delivered to the enclosed cab by the shovel's filtration and pressurization system. When considering these factors, it was obvious that the use of just one recirculation unit would have an insignificant impact on improving the air quality.

The enclosed cabs on both drills tested in this study were very similar and were calculated to have a cab air volume of 275 ft^3. It was believed that the low-quality filters used in the roof-mounted filtration system at the first evaluation site overshadowed the performance of the SCS/Frigette recirculation unit. The roof-mounted filtration and pressurization system on the second drill evaluated appeared to be better maintained with higher-efficiency filters, thus providing for a better analysis of the effectiveness of the recirculation unit.

After the first day of testing, once again there appeared to be very little to no impact on the drill's air quality from the recirculation unit. Because of this, it was decided to implement a second recirculation unit inside the cab on this drill for the second day of testing. Two units were supposed to have been originally installed on the drill for this analysis, and after reviewing the results from the first day of testing, it was decided to sacrifice some testing time to have a second recirculation unit installed. The first unit was located on the back wall, similar to the first study, and the second unit was located on the roof of the cab, directly above the drill operator. Because the drill needed to be moved to another location at this mine, it further reduced the amount of time available for testing. For this final analysis, the one or two recirculation units appeared to have a minor impact on the air quality within the enclosed cab for the limited testing. When examining the data more closely, it is possible that the most significant factor in variation in respirable dust levels was due to the number of times the door on the cab was opened.

In summary, the SCS/Frigette recirculation unit had a minor impact in improving the air quality in the three different pieces of surface mining equipment evaluated in this study. It appears that the main component in determining the air quality in these cabs was the existing filtration and pressurization system originally installed by the equipment manufacturer.

6 CONCLUSION

NIOSH is performing an on-going multi-faceted effort to improve the air quality to equipment operators working in enclosed cabs in the mining industry. Through this research, it has been shown there are two critical components for an effective air quality system. The first component is a well-designed filtration and pressurization system. An effective recirculation system is a critical component to this design. The second component is cab integrity in order to achieve some level of pressurization.

For an effective recirculation system, efficient filtration and flow design are the two key areas for an effective system. In the past, many OEMs were using filter materials that were not capable of removing respirable dust particles from the air. It must be remembered that the recirculation system's filtration is the only method to remove any dusts or contaminants generated inside the enclosed cab. A general rule of thumb is that the media should have a filter efficiency that exceeds 95 pct. on 0.3 micron particles. The second design feature is to consider the quantity and air flow pattern of the recirculation system inside an enclosed cab. In many cases, designs are poor because the contaminant return air and the clean filtered air are ducted within inches of each other at the ceiling. This poor design allows the opportunity for recirculated air to be continually short-circuited. Secondly, it causes the contaminated dust-laden return air to be pulled directly up over the equipment operator as it is being drawn into the ventilation system.

For this study, SCS/Frigette, Inc. took an existing recirculation system designed for the automotive industry and modified it for application in enclosed cabs of heavy equipment. A cooperative agreement was established between SCS/Frigette and NIOSH to evaluate this recirculation unit and determine its effectiveness in improving the air quality in the cabs of two surface drills and a large shovel. The results of this testing ranged from a very marginal to no difference in the air quality in these enclosed cabs. Although these recirculation units appeared to have a number of attractive features that could positively impact the air quality in enclosed cabs, it was obvious from this testing that these units were significantly overwhelmed by existing filtration and pressurization units on this equipment. It is the authors' belief that it would be more economical and effective for mining companies to focus their efforts on improving or replacing the equipment's existing filtration and pressurization system and by improving the integrity of the cab to increase the cab's pressurization.

ACKNOWLEDGEMENTS

The authors would like to thank Neil Shields of SCS/Frigette, Inc., Tom Wojcik of Triune, Inc., Dave Justice of Justice Supply Co., Roy Moore of Leslie Resources, and Kathy King-Vinot of Powder River Coal Co., for their cooperation and assistance in this work effort.

REFERENCES

Cecala, A.B., Organiscak, J.A., & Heitbrink, W.A. 2001. Dust underfoot – enclosed cab floor heaters can significantly increase operator's respirable dust exposure. Rock Products. April, V. 104, No. 4, pp. 39–44.

Cecala, A.B., Organiscak, J.A., Heitbrink, W.A., Zimmer, J.A., Fisher, T., Gresh, R.E., & Ashley II, J.D. 2002. Reducing enclosed cab drill operator's respirable dust exposure at surface coal operation with retrofitted filtration and pressurization system. SME Annual Meeting Preprint 02-105.

Chekan, G.J., Cecala, A.B., & Colinet, J.F. 2003. An evaluation of cab filtration and pressurizations systems: two case studies. *Proceedings of 2003 National Stone, Sand, & Gravel Association's Environmental, Safety & Health Forum. September 22–24, 2003, pp. 114–129.*

Hall, R.M., & Heitbrink, W.A. 1997. In-Depth Survey Report: Control Technology for Agricultural Enclosures at John Deere Manufacturing Company (NTIS publication PB-97-206288). Cincinnati, Ohio: U.S. Department of Health and Human Services, 1997.

Hall, R.M., Heitbrink, W.A., & McGlothlin, J.D. 1997. In-Depth Survey Report: control technology for agricultural environmental enclosures at Nelson Manufacturing Company in Yuba City, California (NTIS publication PB-97-206288). Cincinnati, Ohio: U.S. Department of Health and Human Services.

Heitbrink, W.A., Thimons, E.D., Organiscak, J.A., Cecala, A.B., Schmitz, M., & Adhrenholtz, E. 2000. Static pressure requirements for ventilated enclosures. *Progress and Modern Ventilation, vol. 2, Proceedings of the Ventilation 2000, 6th International Symposium on Ventilation for Contaminant Control. Helsinki, Finland*. Helsinki: Finnish Institute of Occupational Health, 2000.

Heitbrink, W.A., & Moyer, E.S. 2000. In-Depth Survey Report: control technology for environmental enclosures: an evaluation of in-use enclosures. (Report no. EPHB 223-16; NTIS publication PB2001-1016711NZ). Cincinnati, Ohio: U.S. Department of Health and Human Services.

Heitbrink, W.A., Moyer, E.S., Jensen, P.A., Watkins, D.S., & Martin Jr., S.B. 2003. Environmental agricultural tractor cab filter efficiency and field evaluation. *J of American Industrial Hygiene Association* 64: 394–400.

International Agency for Research on Cancer. 1997. Silica, some silicates, coal dust, and pararamid fibrils. Monograph 68. Evaluation of carcinogenic risks to humans. Lyon, France.

Organiscak, J.A., Cecala, A.B., Heitbrink, W.A., Thimons, E.D., Schmitz, M., & Ahrenholtz, E., 2000. Field assessment of retrofitting surface coal mine equipment cabs with air filtration systems. Virginia Tech Conference, Blacksburg, VA, August 27–30.

Mine Ventilation, Ganguli & Bandopadhyay (eds)
© 2004 Taylor & Francis Group, London, ISBN 90 5809 633 5

Development of a wettability test for assessing effectiveness of surfactants for dust control

Yoginder P. Chugh, Krishna K. Thatavarthy & A. Patwardhan
Department of Mining and Mineral Resources Engineering, Southern Illinois University, Carbondale, Illinois

Ronald Carty
Illinois Clean Coal Institute, Carterville, Illinois

ABSTRACT: Sink test has been traditionally used to assess effectiveness of a surfactant for dust control. The sink test has several limitations. The authors have developed a test Chugh, Patwardhan, Thatavarthy (CPT), 2002 that overcomes most of the disadvantages of the sink test and meets most of the requirements of a good wettability test. The authors present here results of wettability studies using the developed test and compare them with similar data using the sink test.

1 INTRODUCTION

Considerable work has been done on the issues related to respirable coal dust control in the last three decades. Studies have been done on the effect of using surfactants in water spray systems in order to achieve greater dust suppression. Mine trials have shown mixed results regarding the effectiveness of surfactant use in water sprays (Harrold, 1979; Chander, 1988; Kilau, 1990). Surfactants produced noticeable dust reduction in some mines but were ineffective in other mines (Kim, 1995). Some of these apparently inconsistent results may be explained by the selection of the right type of surfactant for the spray and on the makeup of the particles (Kim, 1994). The authors think that the selection and the amount of the surfactant used must also be appropriately selected using a physically realistic test.

The sink test was initially developed by Draves and Clarkson (1931) to determine the effect of surfactants on the wettability of cotton fibers. The "sink time" was defined as the time taken by a skein of cotton to sink after it has been immersed in the surfactant solution. Walker et al. (1952) modified this test to determine the wetting ability of various surfactants for coal dusts.

Kost et al. (1980) determined sink time as the time taken for the vast majority of the coal particles to sink through the solution. Sink times greater than 10 minutes were recorded as "greater than 600 seconds". Feldstein (1981) monitored the time taken for all the unwetted coal particles to break through the surface of the solution. Observations above two hours were cut off when necessary and a visual estimate was made for the coal remaining on the surface. During this study, the time for the last trace of coal to disappear beneath the surface of the solution was measured.

The relationship between the rate of coal dust wetting and the surfactant composition has been the object of many studies. A University of Missouri-Rolla study on respirable coal dust control using surfactants by Kim (1995) recommended the use of surfactants for dust control for both continuous miner and longwall sections. The study suggested that nonionic surfactants tend to give better coal wetting abilities due to a good balance between hydrophilic head groups and lipophilic tail groups, and are not sensitive to the mine water environment. Wetting performances with nonionic surfactant solutions were significantly improved (up to 215 percent) by adding bases such as sodium hydroxide and potassium hydroxide. Results showed that the most effective surfactants increased the ability to suppress overall coal dust levels by up to 93% when compared with plain water. More critically, the best surfactant suppressed the fine dust particles (below 3 microns) up to 125 percent more effectively than plain water. Saha (2001) also confirmed that the surfactants are indeed successful in suppressing coal dust when surfactants are used in comparison to when plain water is used.

However, studies done by Kim (1995) for selection of surfactants were based on sink tests and reported that there was a little correlation between the surfactants that were effective in the laboratory and the surfactants used for the in-mine tests. Further, it was suggested that a laboratory test be developed that can adequately simulate capturing airborne dust using surfactants. Saha (2001) also suggested that the optimum sink test concentration of surfactant is not the optimum value in wet scrubbers due to the fact that sink test is a static test whereas the wet scrubber has a dynamic mixing environment.

The goal of this study was to develop a good wettability test for selection of surfactants for different coals. The more specific objectives of this study were to develop a test that: (1) simulates the dynamic mixing condition in a continuous miner wet scrubber, and (2) gives accurate surfactant requirements for fine particle wetting.

2 SURFACTANTS AND COALS STUDIED

2.1 *Surfactants*

Four different nonionic surfactants were used in this study. The surfactants with the ingredients, solubility in water and specific gravity are shown in Table 1.

2.2 *Coals*

Three different coal samples from various companies were used in this study. The coals are: (1) Mine 1 (#6 Coal), Northern Illinois, (2) Mine 2 (#6 Coal), Southwestern Illinois, and (3) Mine 3 (#5 Coal), Central Illinois.

The coal samples were collected in 55-gallon (210 liters) barrels and transferred to the pilot-scale research facility at SIU. The barrels were marked properly and sealed airtight to minimize the effect of oxidation. Raw coal was first ground in a Gundlach hammer mill and then, this ground coal was fed into a Jetamiser compressed air pulverizer. The fine pulverized coal dust produced was stored in drums and was sealed airtight. The ground coal dust was sieved and only −200 mesh size dust was used for the laboratory wetting experiments with and without surfactants.

2.3 *Particle size analysis*

For the small-scale laboratory experiments utilized in dust control research, it is important to determine the amount of respirable dust (less than 15 microns) in a given coal dust sample. To determine the relative fine particle size fractions in the coal dust being used for the laboratory tests, particle size analysis was conducted on Mine 1, Mine 2 and Mine 3 coal samples. Raw coal was first ground in a hammer mill crusher. The ground coal was sieved to −120 mesh size and the Microtrac size analyzer was used in the ultra-fine particle analysis. Four runs were made for each of the above coals. A summary of the particle size analysis for all the three coals is shown in Table 2. The plots for the above results are shown in Figure 1.

Table 1. Properties of surfactants used in the study.

Surfactant	Ingredients	Solubility in water	Specific gravity
1	Alcohols, C9–11, ethoxylated	Complete	0.984
2	Alkyl phenol ethoxylate, Dipropylene glycol monomethyl ether and Sodium cocyl isethionate	Complete	1.219
3	Alkyl phenol, sodium stearate	Complete	>1
4	Not mentioned	Complete	1.016

Table 2. Particle size analysis of coals used in the study.

Size (micron)	Cumulative % (passing) Mine 1	Mine 2	Mine 3
125	100.0	100.0	100.0
88	94.1	92.6	93.4
62	81.1	77.0	76.8
44	66.8	52.0	60.4
31	52.1	44.7	45.6
22	38.7	31.6	32.3
16	26.9	20.9	23.8
11	16.6	13.3	17.0
7.8	10.5	7.9	13.1
5.5	6.1	4.8	9.2
3.9	2.6	2.2	5.6
2.8	1.0	0.3	3.3
1.9	0.5	0.1	1.9
1.4	0.1	0.0	0.8
0.9	0.1	0.0	0.3

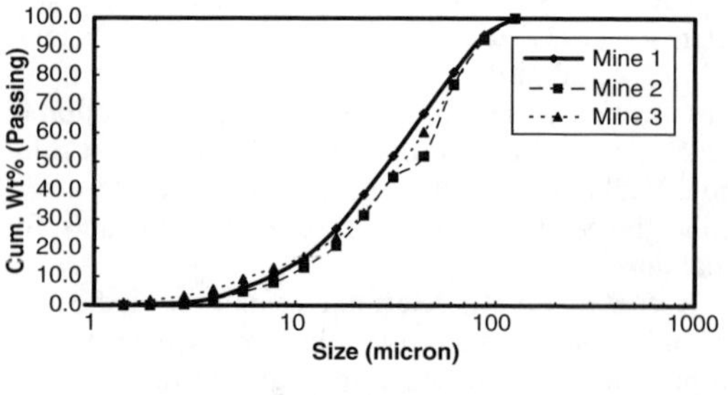

Figure 1. Particle size analysis for Mine 1, 2 and 3.

The above analysis shows that respirable dust (−400 mesh) accounts for 45 to 55 percent of the coal dust sample. This is a significant amount and it can be concluded that the results of laboratory tests performed on these coal dust samples should be applicable to the respirable coal dust control in the scrubber studies.

3 SINK TEST

3.1 *Description*

The sink tests in this study were performed as follows:

1) 200 ml solutions, with various concentrations of a surfactant, were prepared in standard laboratory beakers.
2) A 0.05-gram sample of coal dust was carefully placed onto the surface of each solution.
3) The time required for all of the coal to sink beneath the surface of the solution was measured with a stopwatch.
4) Each test was repeated three times and the mean time was used for assessing performance of a surfactant.

In the sink tests, four chemicals were tested. They were surfactant 1, surfactant 2, surfactant 3 and surfactant 4. Ten concentrations of each surfactant, ranging from 0.001% to 1.0% were prepared.

3.2 *Experimental studies*

The sink test results* for Mine 2 coal are shown in Tables 3, 4, 5 and 6 for surfactant 1, surfactant 2, surfactant 3 and surfactant 4, respectively. The sink time vs. surfactant concentration plot is shown in Figure 2.

3.3 *Limitations*

The limitations of the sink test are as follows:

- Does not predict realistic estimates of surfactant requirements under physical conditions.
- Does not account for wettability of very fine particles.

Table 3. Sink times for Mine 2 coal using surfactant 1.

	Surfactant 1					
Concentration (%)	t1 (sec)	t2 (sec)	t3 (sec)	t (avg)	Std. D.	Std. D. as % of avg
0.001	1151	700	–	926	319	34.5
0.005	930	566	–	748	257	34.4
0.010	311	366	–	339	39	11.5
0.025	33	47	49	43	9	20.3
0.050	29	15	18	21	7	35.7
0.075	13	14	16	14	2	10.7
0.100	15	15	16	15	1	3.8
0.200	14	16	14	15	1	7.9
0.500	12	13	13	13	1	4.6

Table 4. Sink times for Mine 2 coal using surfactant 2.

	Surfactant 2					
Concentration (%)	t1 (sec)	t2 (sec)	t3 (sec)	t (avg)	Std. D.	Std. D. as % of avg
0.001	960	664	–	812	209	25.8
0.005	366	585	–	476	155	32.6
0.010	40	95	107	81	36	44.3
0.025	26	31	30	29	3	9.1
0.050	18	15	16	16	2	9.4
0.075	13	20	15	16	4	22.5
0.100	15	13	15	14	1	8.1
0.200	13	12	14	13	1	7.7
0.500	12	14	8	11	3	27.0

* All the tests represent three replications around the same time.

Table 5. Sink times for Mine 2 coal using surfactant 3.

	Surfactant 3					
Concentration (%)	t1 (sec)	t2 (sec)	t3 (sec)	t (avg)	Std. D.	Std. D. as % of avg.
0.001	905	956	–	931	36	3.9
0.005	739	773	–	756	24	3.2
0.010	325	293	–	309	23	7.3
0.025	34	35	–	35	1	2.0
0.050	28	24	–	26	3	10.9
0.075	24	26	–	25	1	5.7
0.100	23	22	–	23	1	3.1
0.200	23	21	–	22	1	6.4
0.500	14	20	–	17	4	25.0

Table 6. Sink times for Mine 2 coal using surfactant 4.

	Surfactant 4					
Concentration (%)	t1 (sec)	t2 (sec)	t3 (sec)	t (avg)	Std. D.	Std. D. as % of avg.
0.001	1164	976	–	1070	133	12.4
0.005	1119	1099	–	1109	14	1.3
0.010	1092	1043	–	1068	35	3.2
0.025	829	770	–	800	42	5.2
0.050	99	116	117	111	10	9.1
0.075	38	48	51	46	7	14.9
0.100	27	28	34	30	4	12.8
0.200	26	21	26	24	3	11.9
0.500	22	17	19	19	3	13.0

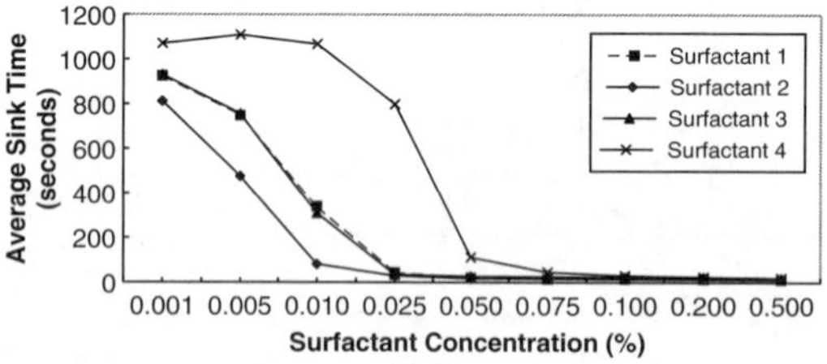

Figure 2. Average sink time vs. surfactant concentration plot for Mine 2 coal.

- Not a realistic simulation of physical practice
- Results are dependent upon how the dust is spread.
- Sink times are highly variable for low surfactant concentrations.
- Test provides only relative performance of different surfactants.

3.4 *Requirements for a good wettability test*

The sink test is laborious, time consuming and prone to errors. There remains a need for faster, cheaper, preferably automated wettability test procedure in the laboratory that can take care of the limitations of the sink test.

4 PROPOSED TEST

4.1 *Description*

The proposed wettability test is performed as follows:

1) Take 200 ml of water in a beaker at 60° F.
2) Add 2.5 gms of −200 mesh dust sample containing 40% of −400 mesh coal dust to the beaker, wait for 10 sec and then stir for 15 sec using a magnetic stirrer.
3) Let the solution sit for 15 minutes and then look at the top surface of the beaker for unwetted fine particles and film. This solution represents the base level for the test.
4) Make a 200 ml 0.001% surfactant solution in a different beaker and stir it for 10 sec.
5) Repeat steps 2–3.
6) If there are fine particles and/or film on the top surface, increase surfactant concentration in increments of 0.0002%.

7) Stir the solution for 15 sec and let the solution sit for 3–5 min.
8) Repeat step 6 until there are no fine particles and/or film on the top surface of the beaker.
9) Repeat the above procedure for different surfactants.
10) The concentration at which no fine particles and/or film are observed is an indicator of surfactant concentration for comparing different surfactants on a particular coal dust.

4.2 *Experimental studies*

Experimental studies were done utilizing different surfactants and coals. The experimental set up for the proposed test, with different surfactant solutions and the magnetic stirrer plate Thermodyne, is shown in Figure 3. The beakers with 0.001% surfactant concentrations are shown in Figure 4. The thin film seen on the top surface of the beaker is shown in Figure 5. Tests were started with a surfactant concentration of 0.001%. The concentration was increased in increments of 0.0002% until no fine coal particle film was visible.

Figure 3. Experimental setup for the proposed wettability test.

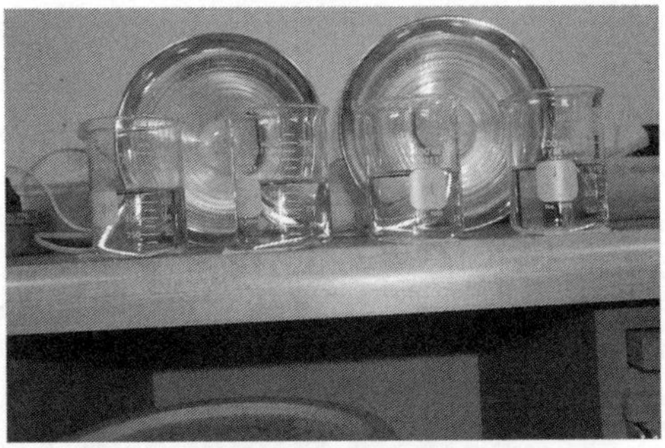

Figure 4. Beakers with different surfactant solutions.

Figure 5. Thin film seen on the top surface of the beaker.

4.3 *Results and discussion*

4.3.1 *Experimental studies using tap water*

All initial experimental studies were done as a part of small-scale laboratory tests using tap water. The tests were repeated multiple times to develop accuracy and resolution of the developed test. The proposed test was conducted on −200 mesh coal dust size using tap water for Mine 1, Mine 2 and Mine 3. The results obtained are shown in Tables 7, 8 and 9.

4.3.2 *Experimental studies using mine water*

Experimental studies were also done using mine water. The proposed wettability test was conducted on −200 mesh coal dust size using mine water for Mine 2. The results are shown in Tables 10, 11 and 12. The results indicate surfactant concentration needs to be increased 100% to get wetting results similar to tap water.

Table 7. Wettability test using tap water for Mine 1.

	Concentration (%)*					
Surfactant	Run 1	Run 2	Run 3	Avg.	Std. D.	Std. D. as % of avg.
Surfactant 1	0.0014	0.0016	0.0012	0.0014	0.00020	14.3
Surfactant 2	0.0016	0.0018	0.0016	0.0017	0.00012	6.9
Surfactant 3	0.0020	0.0022	0.0020	0.0021	0.00012	5.6
Surfactant 4	0.0024	0.0028	0.0026	0.0026	0.00020	7.7

* Runs 1, 2 and 3 were run on different days.

Table 8. Wettability test using tap water for Mine 2.

Surfactant	Concentration (%)* Run 1	Run 2	Run 3	Avg.	Std. D.	Std. D. as % of avg.
Surfactant 1	0.0014	0.0012	0.0014	0.0013	0.00012	8.7
Surfactant 2	0.0016	0.0018	0.0016	0.0017	0.00012	6.9
Surfactant 3	0.0020	0.0020	0.0022	0.0021	0.00012	5.6
Surfactant 4	0.0024	0.0026	0.0024	0.0025	0.00012	4.7

* Runs 1, 2 and 3 were run on different days.

Table 9. Wettability test using tap water for Mine 3.

Surfactant	Concentration (%)* Run 1	Run 2	Run 3	Avg.	Std. D.	Std. D. as % of avg.
Surfactant 1	0.0022	0.0024	0.0022	0.0023	0.00012	5.1
Surfactant 2	0.0022	0.0022	0.0024	0.0023	0.00012	5.1
Surfactant 3	0.0020	0.0020	0.0022	0.0021	0.00012	5.6
Surfactant 4	0.0024	0.0026	0.0024	0.0025	0.00012	4.7

* Runs 1, 2 and 3 were run on different days.

Table 10. Wettability test using mine water for Mine 1.

Surfactant	Concentration (%)* Run 1	Run 2	Run 3	Avg.	Std. D.	Std. D. as % of avg.
Surfactant 1	0.0032	0.0032	0.0034	0.0033	0.00012	3.5
Surfactant 2	0.0028	0.0030	0.0026	0.0028	0.00020	7.1
Surfactant 3	0.0040	0.0038	0.0036	0.0038	0.00020	5.3
Surfactant 4	0.0034	0.0032	0.0034	0.0033	0.00012	3.5

* Runs 1, 2 and 3 were run on different days.

Table 11. Wettability test using mine water for Mine 2.

Surfactant	Concentration (%)* Run 1	Run 2	Run 3	Avg.	Std. D.	Std. D. as % of avg.
Surfactant 1	0.0034	0.0036	0.0034	0.0035	0.00012	3.3
Surfactant 2	0.0032	0.0034	0.0032	0.0033	0.00012	3.5
Surfactant 3	0.0032	0.0036	0.0034	0.0034	0.00020	5.9
Surfactant 4	0.0038	0.0038	0.0040	0.0039	0.00012	3.0

* Runs 1, 2 and 3 were run on different days.

Table 12. Wettability test using mine water for Mine 3.

Surfactant	Concentration (%)* Run 1	Run 2	Run 3	Avg.	Std. D.	Std. D. as % of avg.
Surfactant 1	0.0036	0.0036	0.0038	0.0037	0.00012	3.1
Surfactant 2	0.0034	0.0036	0.0036	0.0035	0.00012	3.3
Surfactant 3	0.0032	0.0032	0.0034	0.0033	0.00012	3.5
Surfactant 4	0.0040	0.0042	0.0042	0.0041	0.00012	2.8

* Runs 1, 2 and 3 were run on different days.

5 CONCLUSIONS

A good wettability test for selection of surfactants for different coals was developed. The test truly represents a wettability test and also simulates the dynamic conditions in a continuous miner wet scrubber. The proposed test accounts for both coarser and finer particles. It can be observed that plain water alone is enough to wet coarser particles whereas surfactants are needed to wet the fine particles.

The proposed test gives results with low standard deviation and high accuracy in the range of 0.0002% surfactant concentration. It provides defensible data for relative performance of different surfactants and realistic estimates of surfactant requirement in the wet scrubber.

Based on the conclusions above, the following recommendations are made for future studies:

- Test other coals from different mines with different surfactants.
- X-ray diffraction studies to know the composition of the thin film remaining on the top surface of the beaker.

ACKNOWLEDGEMENT

This work was performed by the authors at the Dust Control Laboratory, Southern Illinois University, Carbondale as a part of a more general project to develop advanced mining technologies to reduce underground production costs. The project was funded by the Illinois Clean Coal Institute (ICCI). The authors wish to thank ICCI for their support.

REFERENCES

Campbell, J.A.L., Moynihan, D.J., Roper, W.D., and Wilis, E.C., 1983, "Dust Control System and Method of Operation", U.S. Patent 4,380,353.

Chander, S., Mohal, B.R., and Aplan, F.F., 1987, "Wetting Behavior of Coal in the Presence of Some Nonionic Surfactants", *Colloids and Surfaces,* Vol. 26, p. 205–216.

Chander, S., Mohal, B.R., and Aplan, F.F., 1988, "Wetting Characteristics of Particles and Their Significance in Dust Abatement", *Proc. Respirable Dust in Mineral Industries: Health Effects, Characterization and Control,* The Pennsylvania State University, University Park, PA, p. 19–23.

Colinet, J.F., and Listak, J.M., 1999, "Reducing Worker Exposure to Respirable Dust at Longwall Mining Operations", *Proceedings of the International Conference on Control of Dust Hazards in the Mining Industry*, Wisla, Poland, Oct, Chapter XI.

Draves, C.Z., and Clarkson, R.G., 1931, "A New Method for the Evaluation of Wetting Agents", *Am. Dyestuff Reporter,* Mar. 30, p. 109–116.

Feldstein, N., 1981, "Surface Chemical Technology for Improved Wetting of Coal Dust", *U.S. Bureau of Mines,* Report OFR-127-81, 47 pp.

Kim, J., 1995, "Respirable Coal Dust Control Using Surfactants – With Special Emphasis on a Liquid Spray System", Ph.D. Dissertation, 317 pp.

Kim, J., and Tien, J.C., 1994, "The effect of Added Base on Coal Wetting Ability of Nonionic Surfactant Solution Used for Dust Control", *Mining Engineering,* Dec., p. 151–154.

Kost, J.A., Shirey, G.A., and Ford, C.T., 1980, "In-Mine Tests for Wetting Agent Effectiveness", *U.S. Bureau of Mines,* Report OFR-30-82, 188 pp.

Saha, A., 2001, "An Assessment of the Effectiveness of Surfactants on Dust Control using a Continuous Miner Scrubber", M.S. Thesis, Department of Mining and Mineral Resources Engineering, Southern Illinois University, Carbondale, 78 pp.

Walker, P.L., Peterson, E.E., and Wright, C.C., 1952, "Surface Active Agent Phenomena in Dust Abatement", *Industrial and Engineering Chemistry,* Vol. 44, No. 10, Oct., p. 2389–2393.

Zeller, H.W., 1979, "Surfactants vs. Dust – Do they Work?", *Coal Age,* Jun., p. 102–105.

Mine Ventilation, Ganguli & Bandopadhyay (eds)
© 2004 S. Hardcastle, ISBN 90 5809 633 5

The benefits of replacing the diesel engine with a fuel cell in underground mining production equipment

Stephen Hardcastle & Charles Kocsis
CANMET-MMSL, Natural Resources Canada, Sudbury, Canada

Marc Bétournay
CANMET-MMSL, Natural Resources Canada, Ottawa, Canada

David Barnes
Vehicle Projects LLC, Denver, USA

ABSTRACT: Fuel cells due to their zero toxic emissions would be beneficial to both surface and sub-surface mining. Furthermore, in the underground context, they can have an additional advantage. In such mines, the need to dilute the diesel engine emissions of mobile equipment, especially for large heavy-duty production equipment, is a common ventilation design requirement. The replacement of diesel engines with fuel cells has the potential to dramatically reduce the required air volumes. This in turn, lessens a mine's energy consumption with the additional benefit of cutting greenhouse gas (GHG) emissions. However sufficient ventilation must be retained to control mineral dust conditions and blast fume clearance, plus support any remaining light-duty diesel equipment.

A ventilation-based analysis of six Canadian metal mines invariably showed that all of these mines would benefit with an operational cost reduction, from using fuel cells, of at least 20% and also a GHG emissions reduction of at least 27%. However, the magnitude of the benefits are very dependant upon the following: the relative energy usage of the primary and auxiliary ventilation systems, along with the heating system; the variable cost and GHG emissions within both electrical power supply and heating fuels; the diesel dilution requirements of different jurisdictions; heat mitigation requirements with increased depth; and, any ventilation management employed by the mine.

1 BACKGROUND

CANMET-MMSL is a member of the Fuelcell Propulsion Institute (FPI), the consortium leading the effort to replace conventional vehicle power sources with hydrogen fuel cells. The Vehicle Projects LLC contracted CANMET-MMSL to perform a cost-benefit analysis on the impact of introducing hydrogen based fuel cells, as a replacement for the diesel engine, in underground mining. This fuel cell study of mining applications was sponsored by the U.S. Department of Energy. This paper deals primarily with the ventilation and greenhouse gas (GHG) emissions analyses (Kocsis et al, 2003). Other aspects of the analysis included a comparison of mine operating costs requirements, such as vehicle maintenance and fuel delivery (Lacroix et al, 2003), and an evaluation of the regulatory and logistic aspects of using hydrogen driven equipment (Delabbio et al, 2003, Laliberté et al, 2003). In addition to the ventilation cost based benefit analysis performed in Canada, the University of Reno-Nevada carried out a similar investigation for U.S. mines (Righettini & Mousset-Jones, 2003, 2004).

2 EXTENT OF THE FUEL CELL STUDY

2.1 *Identifying the beneficiaries within the canadian mining scene*

Prior to this study, it was believed that the introduction of fuel cell powered vehicles to the underground mining industry in Canada could be extremely beneficial. Where they replace the current use of diesel-powered equipment, the benefits could include:

- A cleaner work environment for underground personnel to breath due to the elimination of diesel engine exhaust contaminants, some of which are suspected carcinogens. This would be beneficial to all mines but would be difficult to quantify due to the time delay before any changes in respiratory effects are observed in the workforce.

- Reduced ventilation operating costs, where the volume of air sent underground was dictated by the diesel exhaust dilution requirements. These costs would comprise electricity charges for the primary and auxiliary fan systems, heating fuel charges to warm the intake air in winter, and for a limited number of mines with heat issues, electricity charges for mechanical refrigeration.
- A smaller heat load addition to the work environment due to fuel cells being mechanically more efficient in their use of power. This would be beneficial when the machinery heat input is still a significant part of total heat load in mines where heat management with ventilation is the controlling design factor.
- A smaller global environmental impact from mining in terms of greenhouse gas emissions. This would result from reductions in power requirements for the ventilation systems and in both the diesel and heating fuel consumption.

These latter three more tangible benefits were the focus of the study. Within these benefits, the potential for cost savings in operating mine ventilation systems was very attractive to the industry because their ventilation costs can be significant and have generally been increasing.

However, to assess the savings it is essential to review the design criteria for ventilation systems, and how such systems are laid out with respect to mining methods. To simplify this review, it can be assumed that fuel cells, when introduced, would be on the primary production equipment, namely load-haul-dump vehicles (LHDs or scooptrams) and trucks, and even then only on the most common vehicles. The wide-scale introduction of fuel cells to the mine service vehicles is less likely due to their diversity.

The Canadian underground mining industry is heavily and extensively mechanized using mobile diesel powered or electric equipment to produce and transport ore. For the most part the Canadian industry can be subdivided into two underground mining categories: "soft" rock, typified by the extraction of sedimentary deposits such as potash, salt and coal; and "hard" rock, which encompasses the extraction of base and precious metal ores, the majority of such mines, but could also include uranium mines and the diamond mines when they start to go underground.

Within the soft-rock category, potash is the predominant mineral. In these underground mines, the production and transportation is done almost exclusively with electrical equipment and diesel equipment is only used for support purposes such as personnel transport and service functions. However, an exception to this is where diesel powered equipment is required to place mill tailings back within a mine. Consequently, due to the predominance of electrical equipment in the production process, the air volumes sent underground per tonne of mined material are relatively low when compared to hard rock mines. Therefore, at this time, fuel cell powered vehicles would probably not be introduced into potash mines due to their limited application and lack of significant ventilation benefit. In salt mines, large diesel equipment is now routinely used in the primary production and transport processes. However, the number of these mines is small, their equipment can be larger than that used in most hard-rock mines, and the volumes sent underground already result in relatively slow air velocity ventilation systems due to their large openings. Hence the introduction of fuel cells to salt mines may also be limited at this time. In Canada's only underground coal mine, although employing some diesel equipment, the ventilation rates are primarily dictated by methane dilution.

Therefore, in Canada, hard-rock operations appear to be the underground mines that could benefit most from the introduction of fuel cells, and these were the primary focus of this evaluation. However even within this sub-group uranium mines must be excluded due to their radiation protection specific requirements. This leaves the Canadian base and precious metal mining operations as the most likely beneficiaries to the introduction of fuel cells. As a consequence of this rationalization, the fuel cell study looked at six underground operations, three precious metal and three base metal.

2.2 *Canadian metal mine design criteria*

In order to find the potential savings for ventilation, it is necessary to understand how ventilation systems are designed and operated. Historically, mine ventilation systems were designed upon peak production demands. These were based upon diesel or heat criteria, and the mines were then operated at this maximum level regardless of the true demand. Furthermore, a mine practice has been to continuously ventilate every potential working level/area with air from surface, which was then distributed through auxiliary systems. This is a common scenario that still exists today, mainly due to the lack of appropriate controls. However, this is gradually changing as the industry becomes more energy conscious; it is starting to introduce controls on its fans and is exploring such possibilities as ventilation on demand (Hardcastle et al, 1999, 2004).

2.2.1 *Current practice*

Currently, diesel engine exhaust dilution or heat management determine ventilation demands, and the resultant volumes are typically sufficient to control all other pollutants. However, even with these two, the final volume required can be dependant upon the provincial jurisdiction governing the mine, or even more demanding mine imposed standards. For example, there can be as much as a 2:1 difference in diesel dilution requirements across Canada for the same sized engine (Gangal & Grenier, 2002); this would be for a certified (NRCan, 2002) "clean" according to CSA standards (CSA, 1990) as opposed to an uncertified and assumed "dirty" engine. In addition, some provinces specify a required airflow based purely upon the engine size, whereas, in other provinces, the requirement is engine-specific and based upon the quality of its emissions.

2.2.2 *Design drivers*

In Canada, the resultant required air volumes in underground metal mines, for either diesel or heat, have been increasing over the last few decades (Hardcastle & Kocsis, 2002). The drivers of this increasing trend include:

- Depth and increasing heat loads. Canada has six metal mines that are planning production 3000 m below surface.
- Depth can also increase a ventilation system's leakage potential and resistance.
- Increased mechanization. The number and size of diesel engine powered vehicles continues to increase.
- Early diesels, in the 1960s, were few, primarily production vehicles, and small, less than 75 kW.
- Today's production diesel vehicles, in Canada, include 260 kW scooptrams and 490 kW haulage trucks, with even larger units being considered.
- Increased workforce mobility. Diesel personnel carriers and service vehicles can outnumber production vehicles.
- More powerful support vehicles. Mines have been gradually increasing the size of these vehicles; early Toyota Landcruisers™ were 55 kW, today 100 kW units are more typical. Some mines even employ commercial vehicles with up to 160 kW diesel engines.
- Lower pollutant exposure limits. Depending upon the pollutant and the degree of change, a new concentration limit might be achieved with a ventilation increase.

In opposition to these trends, are the mining companies' desire to be more cost effective and such external pressures as reducing green house gas emissions. Both can benefit with reduced energy consumption. Consequently, as mine ventilation is a major consumer of energy in a mine, methods by which its demand can be reduced are constantly being sought. Fuel cell powered vehicles are one such possibility.

2.2.3 *Future practice*

Fuel cells are effectively "zero-emission" power sources, emitting only water vapour. As these emissions are harmless to humans; they have no regulated dilution requirements. However the quality of the atmosphere they operate within is still a concern. Fuel cells require oxygen to generate power, their performance can be affected by atmospheric contaminants (Bétournay et al, 2002, 2003), and heat rejection may also be an issue. However, it is doubtful that the diesel engine will be totally replaced in mines, so the diesel design criteria, although diminished, would remain. In addition, there are other historical concerns for the workforce, such as mineral dust exposure and blast fumes that will re-emerge.

3 THE BENEFIT ANALYSES

3.1 *Cost analysis case-studies*

An analysis was performed on six operating mines, which compared their current operating costs under their present or a future operating conditions using diesel powered equipment, to those with fuel cell powered production vehicles.

Under the fuel cell situation, the new flows required in the mines were derived through a combination of the following:

- $0.042\,m^3/s$ per daily tonne of material mined in shallow to medium depth mines,
- a 25% additional allowance for additional miscellaneous needs,
- regulatory requirements of the remaining service diesels,
- specified flows for such ancillary areas as crushers, loading pockets, garages etc.,
- heat removal requirements based upon tonnage modified for fuel cells, and
- a minimum air velocity of 0.5 m/s (98 fpm) in all drifts, this velocity was retained to ensure the dilution of dust and gasses, prevent stratification and the timely removal of blast fumes. This same limit was used in the U.S. study, however, that study also used a lower limit of 0.3 m/s (60 fpm) where blast fumes were not an issue (Righettini & Mousset-Jones, 2003, 2004).

Under both the diesel and fuel cell conditions, where possible, the operating costs of the primary ventilation system were established with a mine ventilation simulator.

The differing auxiliary ventilation operating costs were much harder to establish, although all mines knew the number of auxiliary fans on their property few could provide complete operational details as to which were operating, and for how long. Where this

Table 1. Mine data and a diesel engine versus fuel cell comparison of the ventilation requirements and associated costs for six Canadian underground metal mines.

Mine #	Metal mine type	Location	Tonnage depth	Design criteria	Ventilation control	Electricity charges ($/kWh)		Primary airflow m^3/s	Primary ventilation MWH	Primary ventilation $'000 (Can.)	Auxiliary ventilation MWH	Auxiliary ventilation $'000 (Can.)	Heating Source	Heating $'000 (Can.)	Cooling MWH	Cooling $'000 (Can.)	Combined $'000 (Can.)	Combined Ratios P:A:H:C%
P1	Gold	Ontario	2,300 tpd	Diesel	Shift based	($0.067/	Diesel	143	3,090	207	2,806	188	Propane	576			971	21:20:59:0
			800 m	0.063 m^3/s	95:85:65%	kW)	Fuel Cell	113	1,731	116	2,478	166		455			737	16:22:62:0
				per kW			**% Change**	**21%**	**44%**		**12%**			**21%**			**24%**	**41:10:49:0**
B1	Base	Ontario	4,000 tpd	Diesel	Constant	($0.055/	Diesel	608	61,818	3,400	27,273	1,500	Natural	0	Natural	0	4,900	69:31:0:0
			2,100 m	>0.063 m^3/s	100%	kWh)	Fuel cell	459	32,182	1,770	23,091	1,270	Strata	0	Strata	0	3,040	58:42:0:0
				per kW			**% Change**	**25%**	**48%**		**15%**						**38%**	**88:12:0:0**
				& Heat														
B1-fut	Base	Ontario	4,000 tpd	Heat	Constant	($0.055/	Diesel	652	76,364	4,200	32,727	1,800	Natural	0	Natural	0	6,000	70:30:0:0
(future			2,500 m		100%	kWh)	Fuel cell	592	58,182	3,200	29,091	1,600	Strata	0	Strata	0	4,800	67:33:0:0
deeper							**% Change**	**9%**	**24%**		**11%**						**20%**	**83:17:0:0**
scenario)																		
B2	Base	Ontario	4,000 tpd	Diesel	Constant	($0.065/	Diesel	538	42,308	2,750	33,846	2,200	Natural	350			5,300	52:41:7:0
			1,450 m	>0.063 m^3/s	100%	kWh)	Fuel cell	432	21,292	1,384	27,215	1,769	Gas	280			3,433	40:52:8:0
				per kW			**% Change**	**20%**	**50%**		**20%**			**20%**			**35%**	**73:23:4:0**
P2	Gold	Quebec	7,000 tpd	Diesel-	Constant	($0.024/	Diesel	703	100,000	2,400	38,917	934	Natural	710	14,167	340	4,384	55:21:16:8
			2,250 m	Engine	100%	kWh)	Fuel cell	605	64,333	1,544	32,542	781	Gas	610	12,167	292	3,227	48:24:19:9
				Dependant			**% Change**	**14%**	**36%**		**16%**			**14%**	**14%**		**26%**	**74:13:9:4**
				& Heat														
B3	Base	Quebec	4,300 tpd	Diesel-	Constant	($0.039/	Diesel	387	10,769	420	12,282	479	Natural	800			1,699	25:28:47:0
			900 m	Engine	100%	kWh)	Fuel cell	318	5,974	233	9,821	383	Gas	655			1,271	18:30:52:0
				Dependant			**% Change**	**18%**	**45%**		**20%**			**18%**			**25%**	**44:22:34:0**
P3	Gold	Quebec	3,700 tpd	Diesel-	Constant	($0.042/	Diesel	354	13,095	550	38,929	1,635	Natural	320			2,505	22:65:13:0
			1,000 m	Engine	100%	kWh)	Fuel cell	276	6,190	260	31,667	1,330	Gas	250			1,840	14:72:14:0
				Dependant			**% Change**	**22%**	**53%**		**19%**			**22%**			**27%**	**44:46:10:0**
							Average	**18%**	**43%**		**16%**			**19%**	**14%**		**28%**	**64:20:22:4**

Notes: The shift based control for mine P1 were as follows: ten 8-hr shifts (day & afternoon) per week, two fans at 95% of rated fan speed – for production and development, five 8-hr shifts (night) per week, two fans at 85% of rated fan speed – for production only, six 8-hr shifts (Saturday & Sunday) per week, one fan at 65% of rated speed – for maintenance.
The primary airflow and primary/auxiliary costs listed for P1 mine are the time weighted average of its three operating conditions.
The Combined Ratios P:A:H:C% refers to relative contribution of Primary Ventilation, Auxiliary Ventilation, Heating and Cooling to either the total cost or the % Change.

information was not available, the auxiliary duty was based upon the mine's measured, or estimated, average power consumption. Under the fuel cell scenario, the power reduction was based upon the mine generally downsizing to the next largest motor size in its inventory. This restriction was used so mines could retain a high-volume flush of blast fumes to avoid extended clearance times and the total replacement of all auxiliary fans.

The heating or cooling costs were based upon current expenditures for the diesel scenario, and then adjusted in direct proportion to the primary airflow change for the fuel cell scenario.

The individual and combined power consumption and their associated costs for each of the six mining operations are summarized in Table 1. This table also includes a second analysis of one base metal mine where future requirements at depth were known. The results are broken down by mine type and includes such supporting information as the regulating jurisdiction (Province), daily production, maximum depth, design criteria, ventilation control, electricity charges, heating and cooling sources.

Table 1 shows that the average reduction in the primary flow through the mines as 18%. However, it ranged from a high of 25% to a low of 9% with both of these extremes being at the B1 mine but for the present situation as opposed to a future scenario. These same reductions were replicated in the heating and cooling costs.

For the primary ventilation system, Table 1 shows the power consumption and operating cost reductions averages 43%, much larger than the flow reduction due to the cubic relationship between flow and cost. However, the reductions ranged from a high of 53% at the P3 mine, to a low of 24% for the future diesel scenario of the B1 mine.

For the auxiliary ventilation systems, Table 1 shows the power consumption and operating cost reductions averaged 16% ± 4%. These are modest compared to the primary system, as it was generally felt that mines would wish to maintain a similar flow capability in their auxiliary systems. Consequently, only the next smaller fan was used; and although it may deliver a small volume during production, as opposed to during blast clearance, it would still run at its rated power. This would be due to the auxiliary duct being throttled at discharge to keep it inflated.

Upon combining all the cost reductions, Table 1 shows the net savings averaged 28%, with the highest being 38% and the lowest being 20% with both of these extremes being at the B1 mine but respectively for present condition as opposed to a future scenario.

3.1.1 *Cost/Energy reduction variance analysis*

Although the results show some consistency in the potential benefit of fuel cells as 28 ± 7%, the table shows considerable variation in such factors as the cost of electricity, heating fuel and the relative contribution of each cost component. These warrant examination to determine whether the benefits of introducing fuel cells are unique to the technology.

3.1.1.1 The effect of depth

In the six individual mines, the relevance of depth and any need to alleviate heat was not obvious in the potential benefits of fuel cells. However, upon comparison of the present and future needs at the B1 mine, it was apparent in both the primary and total power/cost that increasing depth within a mine reduced the possible benefit. This was not surprising, as with depth, the required air volumes are more for heat mitigation than for fume control, and with increasing depth the heat from the strata becomes proportionally more significant than that from the mining machinery. This was confirmed through a climatic simulator approximation, Figure 1, which shows that with increasing depth, any primary flow savings through the adoption of fuel cells will decrease and eventually be virtually non-existent.

This would also tend to infer that fuel cells should be more beneficial at shallower depths. However, even this was not apparent as the two shallowest mines, P1 & B3, did not have the largest flow reductions and had the worst overall benefit of the current scenarios.

The diminishing cost benefit from fuel cells with depth, which generally increases the ventilation costs, is a very important consideration. This is because one of the potential drivers for developing this technology is the potential energy/cost savings and these may be limited in the more costly deep mine ventilation systems.

3.1.1.2 Primary ventilation component

In the cost benefit analysis, the relative savings were always greatest in the operating costs of the primary ventilation system. However, a mine's ranking under the primary ventilation was not reflected under total

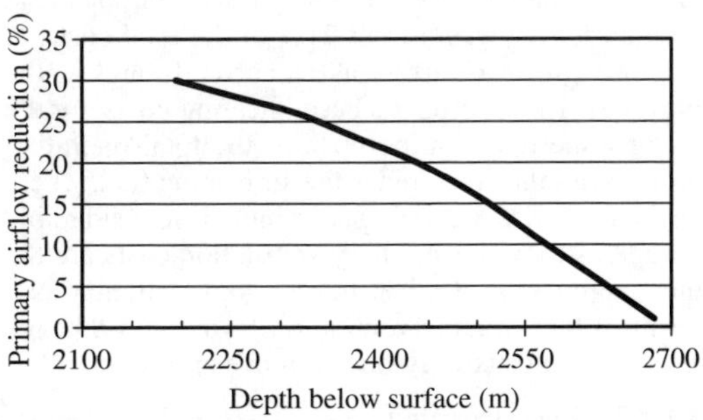

Figure 1. Diminishing fuelcell benefits with depth (approximation).

Table 2. Variance analysis of the fuel cell benefit with different electricity/heating fuel charges.

Mine #	Electricity charges/ Heating fuel		Primary ventilation MWH	Primary ventilation $′000 (Can.)	Auxiliary ventilation MWH	Auxiliary ventilation $′000 (Can.)	Heating $′000 (Can.)	Combined $′000 (Can.)	Combined Ratios P:A:H:C%	Energy price benefit
P1	($0.067/kW)	Diesel	3,090	207	2,806	188	576	971	21:20:59:0	0%
	Propane	Fuel cell	1,731	113	2,478	166	455	734	15:23:62:0	
		% Change		**45%**		**12%**	**21%**	**24%**	**40:9:51:0**	
	($0.024/kW)	Diesel	3,090	74	2,806	67	576	717	10:10:80:0	26%
	Propane	Fuel cell	1,776	43	2,478	59	455	557	8:11:81:0	
		% Change						**22%**	**20:5:75:0**	
	($0.067/kW)	Diesel	3,090	207	2,806	188	144	539	38:35:27:0	44%
	Natural Gas	Fuel cell	1,776	116	2,478	166	114	396	29:42:29:0	
		% Change						**27%**	**65:16:19:0**	
	($0.024/kW)	Diesel	3,090	74	2,806	67	144	285	26:24:50:0	71%
	Natural Gas	Fuel cell	1,776	43	2,478	59	114	216	20:27:53:0	
		% Change						**24%**	**47:12:41:0**	
B2	($0.065/kWh)	Diesel	42,308	2,750	33,846	2,200	1400	6,350	43:35:22:0	−20%
	Propane	Fuel cell	21,292	1,384	27,215	1,769	1120	4,273	32:42:26:0	
		% Change						**33%**	**66:21:13:0**	
	($0.065/kWh)	Diesel	42,308	2,750	33,846	2,200	350	5,300	52:41:7:0	0%
	Natural Gas	Fuel cell	21,292	1,384	27,215	1,769	280	3,433	40:52:8:0	
		% Change		**50%**		**20%**	**20%**	**35%**	**73:23:4:0**	
	($0.024/kWh)	Diesel	42,308	1,048	33,846	838	1400	3,286	32:26:43:0	38%
	Propane	Fuel cell	21,292	527	27,215	674	1120	2,321	23:29:48:0	
		% Change						**29%**	**54:17:29:0**	
	($0.024/kWh)	Diesel	42,308	1,048	33,846	838	350	2,236	47:37:16:0	58%
	Natural Gas	Fuel cell	21,292	527	27,215	674	280	1,481	36:45:19:0	
		% Change						**34%**	**69:22:9:0**	

cost. For example, the smallest and largest reductions of current operating conditions, the P2 & P3 mines, end up with the same combined cost benefit.

This indicates that the primary system is not the controlling factor. In Table 1, the total cost was broken down into a ratio of its contributing components, namely primary ventilation, secondary ventilation, heating and cooling. The relative savings of introducing fuel cells was similarly broken down.

This comparative cost analysis shows that the primary ventilation system with diesel engines can account for anywhere from 21% of the total costs, at P1, through to 69–70%, at B1 currently and in the future. At P1 the propane based heating costs are the major issue representing 60% of the total operating costs. At all the other mines the air heating fuel, where applicable, was natural gas, which is considerably cheaper. At P3, the auxiliary ventilation costs are the most significant at 65%; this to some extent was a result of their mining/ventilation layout and a heavy reliance upon auxiliary fans to move air around.

3.1.1.3 Energy pricing

In Table 2, two of the mines were re-analyzed using their current and the cheapest electricity rates paired with natural gas and propane based heating. To a certain degree, this would reflect the disparity between a province where the electricity price is regulated and one more subject to free market influences. For simplicity a 4:1 multiplier has been used to compare the cost of propane with natural gas, this was derived from the P1 and P3 mines, which shared very similar climates. As might be expected, the energy pricing did affect the overall cost with the order of a threefold difference between the cheapest and most expensive combinations. Somewhat surprisingly though, for the two mines evaluated, the different combinations of electricity and heating only varied the overall fuel cell benefit at each operation by a few percentage points. The maximum change in the six alternate energy combination analyses was only 6%, from 35% to 29% at mine B2.

However, one point of interest in Table 2 is that mine P1, would gain a greater cost benefit at 44% by converting to natural gas from propane, than the 23% by switching to fuel cell powered vehicles.

3.1.1.4 Ventilation control

Table 1 lists mine P1 as employing shift-based control of its ventilation. The mine employs a simple form of

Table 3. Comparitive analysis of the fuel cell benefit in a mine with/without demand based ventilation control.

Mine #	Ventilation control		Primary airflow m^3/s	Primary ventilation MWH	Primary ventilation $'000 (Can.)	Auxiliary ventilation MWH	Auxiliary ventilation $'000 (Can.)	Heating $'000 (Can.)	Combined $'000 (Can.)	Combined Ratios P:A:H:C%
P1	Constant 95%	Diesel	180	4,710	316	4,770	320	725	1361	23:24:53:0
		Fuel cell	136	2,032	136	4,200	281	548	965	14:29:57:0
		Saving	44	2,678	180	570	39	177	396	**45:10:45:0**
		% Change	**24%**		**57%**		**12%**	**24%**	**29%**	
	Shift based 95:85:65%	Diesel	143	3,090	207	2,806	188	576	971	21:19:59:0
		Fuel cell	116	1,731	116	2,478	166	467	749	15:22:62:0
		Saving	27	1,359	91	328	22	109	222	**41:10:49:0**
		% Change	**19%**		**44%**		**12%**	**19%**	**23%**	
	Constant vs Shift	Saving	37	1620	109	1964	132	149	390	**28:34:38:0**
		% Change	**21%**	**34%**		**41%**		**21%**	**29%**	

demand-based ventilation defined by the general level of activity in this mine. These control measures were introduced in recognition of the high cost to heat the mine's air with propane in the winter, and for the most part the mine was inactive across weekends and holidays, which would represent 33% of its life. The data for this mine was re-analyzed assuming continuous operation of the fans at 95% of rated speed, their maximum required operating condition, to see whether the demand based control changed the benefits of introducing fuel cells. Table 3 shows the difference that the introduction of fuel cells could make if the mine had not already introduced some form of flow management.

In Table 3, it can be seen that under a constant operation mode, the primary flow reduction increased from 19% to 24%, accordingly the primary ventilation power cost saving increased from 43% to 57%, and the net overall cost gain increased from 23% to 29%. This indicates that once a mine has already started to eliminate "unnecessary" ventilation based upon diesel criteria, the benefits of the fuel cell technology are reduced. This is because there is less to save in the non-active periods.

Table 3 also compares, for P1 mine, the operating costs under the constant delivery mode as opposed to the demand based mode. In the overall cost, it can be seen that switching to demand based ventilation or to fuel cells provides the same benefit of 29%. This indicates that mines should be considering both technologies to reduce their ventilation costs.

3.1.1.5 Diesel regulations

Table 1 listed the diesel design criteria used at each mine. In Ontario, the exhaust dilution requirement is $0.063\,m^3/s$ per rated kW (100 cfm/BHP) for all diesel equipment. In Quebec, the exhaust dilution requirements recently became exhaust quality based, whereby credit is given to the use of cleaner diesel powered equipment such as those with electronically controlled engines. Consequently, the requirement for common large diesel engines is much less, for example the P1 mine uses Detroit Diesel Series 50 electronically controlled 186 kW (250 hp) engines on its $7\,yd^3$ scooptrams; in Ontario this unit would require $11.8\,m^3/s$, in Quebec it would require $8.3\,m^3/s$ which is 30% less. To evaluate the effects of diesel regulations on the potential benefits of fuel cells, the results of the P1 mine were again re-analyzed under both regulatory scenarios.

Table 4 shows that there would be only minor benefits, 3% in total cost, upon the introduction of fuel cells if the mine's diesel design criteria was based upon engine emission quality as opposed to simple engine power. As presented, these results show that the mine would gain the same order of cost benefit if the regulation in Ontario were brought in-line with that in Quebec. This lack of benefit was surprising, considering that the Quebec mines operating under the quality based regulations showed, in Table 1, an overall cost saving of 25–27%.

3.1.2 *Limiting factors*

The preceding detailed analysis of some of the mine variances, as applied to the P1 mine, would tend to indicate that fuel cells might have limited cost benefits if the diesel regulations change. To understand this possible lack of benefit, one has to look at the detailed analysis and what has limited the airflow reductions. Furthermore, there are other possible limitations that might need to be considered.

3.1.2.1 Drift/Stope air velocity limit implications

Under a first iteration of the fuel cell benefits at the P1 mine, it was determined that the average flow could be reduced from $143\,m^3/s$ to $91\,m^3/s$ with 44% reductions on the production shifts, and a 36% average reduction. However, this resulted in air velocities

Table 4. Comparitive analysis of the fuel cell benefit in a mine as a function of diesel design criteria.

Mine #	Design Criteria		Primary airflow m^3/s	Primary ventilation MWH	Primary ventilation \$'000 (Can.)	Auxiliary ventilation MWH	Auxiliary ventilation \$'000 (Can.)	Heating \$'000 (Can.)	Combined \$'000 (Can.)	Combined Ratios P:A:H:C%
P1	Diesel	Diesel	143	3,090	207	2,806	188	576	971	21:20:59:0
	$0.063\,m^3/s$	Fuel cell	116	1,731	116	2,478	166	467	749	16:22:62:0
	per kW	Saving	27	1,358	91	328	22	109	222	**41:10:49:0**
		% Change	**19%**		**44%**		**12%**	**19%**	**23%**	
	Diesel –	Diesel	120	1,910	128	2,478	166	483	777	17:21:62:0
	engine	Fuel cell	116	1,731	116	2,478	166	467	749	16:22:62:0
	dependant	Saving	4	179	12	0	0	16	28	**42:0:58:0**
		% Change	**3%**		**9%**		**0%**	**3%**	**4%**	
	Fixed vs	Saving	23	1179	79	328	22	93	194	**41:11:48:0**
	Quality	**% Change**	**16%**		**38%**		**12%**	**16%**	**20%**	

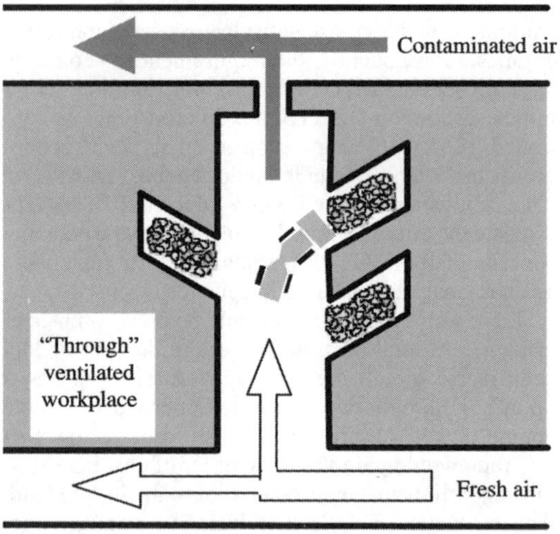

Figure 2. Schematic of "through" ventilation.

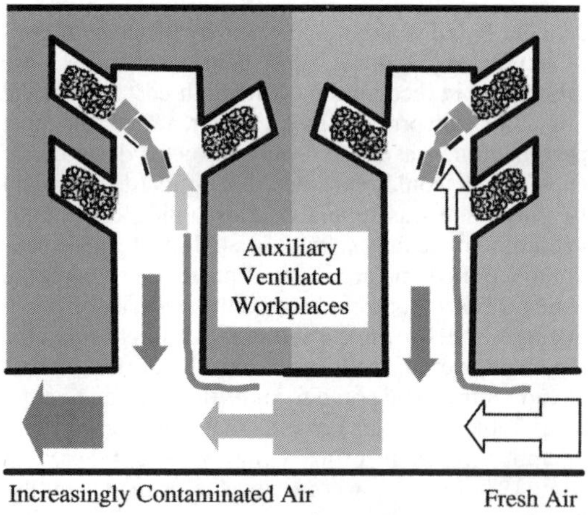

Figure 3. Schematic of "Series" ventilation.

of 0.34 m/s at most workplaces. This is lower than 0.5 m/s, which might be the practical limit for clearing blast fumes and avoiding stratification. Consequently, the reduction was limited to an average mine intake flow of $116\,m^3/s$. Under the first iteration with "clean" diesels, it was determined that the average mine flow could be reduced from $143\,m^3/s$ to $111\,m^3/s$ with 27% reductions on the production shifts, and a 22% average reduction. However, similarly to the fuel cell option, this also reduced drift velocities below 0.5 m/s and the reduction had to be limited to an average intake flow of $120\,m^3/s$.

3.1.2.2 Worker/Air relationships

The drift air velocity consideration shows that determining the benefit of fuel cells is not simple. One also has to consider mine layout and how personnel interact with contamination sources. For example, Figure 2 shows a mucking operation with a "through" ventilation layout; in such arrangements the operator is always in fresh air, and any reductions with the introduction of fuel cells should not affect dust or gas exposure because the contaminants travel directly to the exhaust. Contrary to this, as shown in Figure 3, with "series" auxiliary ventilation, any reduction in ventilation could impact upon mineral dust or gas dilution and clearance, and be detrimental to concentrations downstream.

3.1.2.3 Exhaust shaft/raise velocities

If water condenses in the vertical exhaust airways, then the exhaust air's velocity should be a consideration to avoid creating suspended water droplets and possible exhaust fan problems. This would only be an issue where the initial velocities were greater than 12 m/s and the reduction resulted in velocities greater than 7 m/s.

3.1.2.4 Heating system air velocities
In Canada, most mine heating systems are designed for a certain air velocity across the gas burner to ensure efficient combustion and mixing. This is important because low air velocities and possibly resultant air density differences can contribute to fan impeller failures (Hardcastle, 2002). Consequently, any significant reduction in flow could change their performance and should be checked. In this situation, mines with parallel intake fans may have an advantage upon switching to a single fan as reasonable velocities may be maintained across the burners.

3.1.2.5 Silica exposure
For one of the mines assessed, the average silica exposure of its personnel, with ventilation provided at diesel prescribed levels, was 0.06 mg/m^3. Consequently, if the overall ventilation were reduced the dust concentrations would probably increase. In Ontario, the exposure limit for α-quartz (the type of silica normally quantified in mine air samples), is 0.10 mg/m^3, therefore, assuming a linear relationship between airflow and dust concentration, the ventilation reductions may be limited to at most 40%. However, to avoid exposures at the limit a lower reduction in ventilation may need to be considered.

3.1.2.6 Fogging issues
Fogging within the working areas of a mine can be a serious issue. In multi-level mines, air ascending an extensive ramp system will naturally cool through decompression and mixing with colder airstreams. In certain instances this causes fogging.

With the introduction of fuel cells and reduced air volumes within mines, there is the potential for air temperatures to increase due to the larger strata component. This would permit increased evaporation of any available moisture.

A fuel cell powered vehicle would also generate more water than its equivalent diesel engine unit. If this water is not trapped it will result in more humid conditions, further exacerbated by reduced airflows. The net result could be an environment more liable for fogging.

3.2 *Environmental benefits*

At the outset of this analysis, one of the potential benefits of introducing fuel cells, was the reduction of green house gas emissions. To quantify this potential, the energy usage and reductions listed in Table 1 have been converted into equivalent carbon dioxide (CO_2) emissions. The conversions used were as follows:

- 1.90 tonnes CO_2 per 1,000,000 litres (ML) natural gas,
- 1.55 tonnes CO_2 per 1,000 litres (kL) propane,
- 2.73 tonnes CO_2 per 1 kL diesel fuel, and
- 200 tonnes CO_2 per 1 GWh of electricity usage.

These were derived, or obtained directly, from national emission analyses performed in Canada (NRCan, 1997, 1999). For electricity, it should be noted that this generation of equivalent CO_2 emissions is unique to Canada's electricity generation profile, which for the most part is from hydro-electric power stations. In other countries, where fossil fuels predominate, a value in the order of 900 tonnes CO_2 per 1 GWh of electricity would be more representative.

In this derivation of the GHG emissions and the reductions with the application of fuel cells, it has been assumed, following discussions with the mines, that diesel consumption would be reduced by approximately 75%. This reduction would be the result of replacing all the heavy-duty diesels and may not necessarily reflect the reduction of installed engine power. The results for the surveyed mines are given in Table 5; this table shows that all forms of energy used are significant emissions producers, with electricity being the prime contributor. However there are two exceptions, P1 and B3, the shallowest mines where heating is the primary contributor. If it weren't for Canada's hydro-electricity, the emission from electrical usage would invariably be the most significant.

Table 5 also shows that the potential emission reduction with fuel cells varies from 27% to 41–43%. However, similar to the costs, the reduction is dependant upon the relative magnitude of each energy source's emissions in the mine's total energy consumption and the potential reduction in that form of energy. For example, P1 at 41% shows one of the most dramatic emission reductions, but the primary contributor to this is the 75% reduction in diesel fuel usage; also the electricity component is minor as this mine already employs ventilation demand based controls. At the other extreme is the B1 mine under its future scenario with a 27% emissions reduction. At B1, the future reduction in emissions was limited by the 19% reduction in electricity, which was the major form of energy used.

4 CONCLUSIONS

This study has shown that if fuel cells were to replace the diesel engine of the primary production equipment in Canadian underground metal mines, less ventilation would be required. The degree of reduction in the six mines analyzed range from 9% for a future deep scenario through to 25%.

Upon combining all the ventilation cost components, primary and auxiliary fans, heating and cooling, the cost reductions varied from 20% for a future

Table 5. Diesel engine versus fuel cell comparison of the CO_2 equivalent emissions for six Canadian underground metal mines.

Mine #		Primary airflow m^3/s	Electricity tonnes Co_2/year	Heating fuel tonnes Co_2/year	Diesel fuel tonnes Co_2/year	Combined emissions Tonnes Co_2/year	Combined emissions Ratios E:H:D%
P1	Diesel	143	1,179	2,790	1,966	5,935	20:47:33
	Fuel cell	113	818	2,204	491	3,513	23:63:14
	% Change	**21%**	**31%**	**21%**	**75%**	**41%**	**15:24:61**
B1	Diesel	608	17,818	0	3,030	20,848	85:0:15
	Fuel cell	459	11,054	0	751	11,805	94:0:6
	% Change	**25%**	**38%**		**75%**	**43%**	**75:0:25**
B1-fut	Diesel	652	21,818	0	3,276	25,094	87:0:13
(future	Fuel cell	592	17,580	0	819	18,399	95:0:5
deeper)	**% Change**	**9%**	**19%**		**75%**	**27%**	**63:0:37**
B2	Diesel	538	15,231	2,557	1,441	19,229	79:13:8
	Fuel cell	432	9,703	2,043	360	12,106	80:17:3
	% Change	**20%**	**36%**	**20%**	**75%**	**37%**	**78:7:15**
P2	Diesel	703	30,368	3,855	5,799	40,022	76:10:14
	Fuel cell	605	21,625	3,308	1,450	26,383	82:12:6
	% Change	**14%**	**29%**	**14%**	**75%**	**34%**	**64:2:32**
B3	Diesel	387	4,645	4,343	1,441	10,429	44:42:14
	Fuel cell	318	3,179	3,563	360	7,102	45:50:5
	% Change	**18%**	**32%**	**18%**	**75%**	**32%**	**44:23:33**
P3	Diesel	354	10,402	1,737	3,063	15,202	68:12:20
	Fuel cell	276	7,572	1,359	764	9,695	78:14:8
	% Change	**22%**	**27%**	**22%**	**75%**	**36%**	**51:7:42**
	Average	**18%**	**30%**	**19%**	**75%**	**36%**	**56:9:35**

Note: The Combined Emissions E:H:D% refers to relative contribution of Electricity, Heating fuel and Diesel fuel to either the total emissions or the % Change.

deep scenario through to 38% for the same mine under its current conditions.

Similar order savings may be achieved by other means, such as changing the heating fuel, ventilation demand controls or as result of changing to quality based diesel regulations. However, this does not negate the advantages of fuel cells. These savings can be additive but the limitations imposed by such considerations as a minimum air velocity, blast clearance or dust conditions may be reached before the full combined potential benefit is reached.

One of the most dramatic benefits was the reduction in GHG emissions, which range from 27% for a future deep scenario through to 41–43% for current operations. A major component of the lowered emissions being the reduced consumption of diesel fuel.

Another advantage not considered in this study, that should be remembered, is the health benefit to underground workers through the elimination of significant amounts of diesel engine generated contaminants.

Overall, this study has shown that the introduction of fuel cells should be very beneficial to the mining industry. However, this analysis has shown various mitigating factors that may limit the reduction in airflow or overall savings, and that these factors can be mine specific.

ACKNOWLEDGEMENTS

The authors would like to express their gratitude to the mines and their management for their co-operation in this study, and for the permission to present these findings.

The authors would like to extend their gratitude for the sponsorship provided by the U.S. Department of Energy.

REFERENCES

Bétournay, M.C., Bonnell, G. and Edwardson, E., 2002, Testing a 35 W fuel cell stack in underground production conditions: Part I – Exposure to dust and diesel emissions, Natural Resources Canada, MMSL 02-024(TR).

Bétournay, M.C., Bonnell, G., Edwardson, E. and Lidkea, W., 2003, Suitability of PEM fuel cells for underground mining vehicles, Procs. 105th Canadian Institute of Mining, Metallurgy and Petroleum AGM, Montreal, on CD.

CSA, 1990, Non-rail bounded diesel powered machines for use in non-gassy underground mines, CAN/CSA-M424.2-M90, 1990, Canadian Standards Association.

Delabbio, F., Eastick, D., Graves, C., Sprott, D. and Bétournay, M.C., 2003, A risk assessment and regulatory compliance of the world's first fuel cell-powered mining equipment, Procs. 105th Canadian Institute of Mining, Metallurgy and Petroleum AGM, Montreal, on CD.

Gangal, M. and Grenier, M., 2002, An overview of regulations to control diesel emissions in Canadian mines, Proc. 1st North American/9th U.S. Mine Ventilation Symposium, Kingston, Ontario, Canada, 8–12 June, Ed. E. De Souza, Balkema, 2002, ISBN 90 5809 387 5, pp. 427–432.

Hardcastle, S.G., Gangal, M.K., Schreer, M. and Gauthier, P., 1999, Ventilation-on-demand – quantity or quality – A pilot trial at Barrick Gold's Bousquet mine, Proc. 8th U.S. Mine Ventilation Symposium, Rolla, Missouri, June 11–17, Ed. J. Tien, on CD, University of Missouri-Rolla, pp. 31–38.

Hardcastle, S.G., Gilbert, J.B., Lachance, E. and Verreault, M., 2002, Airflow considerations in the blade failure of axial fans, Proc. 1st N. American/9th U.S. Mine Ventilation Symposium, Kingston, Canada, June 8–12, Ed. E. De Souza, Balkema, ISBN 90 5809 387 5, pp. 31–38.

Hardcastle, S.G. and Kocsis, C., 2002, The ventilation challenge – A Canadian perspective on maintaining a good working environment in deep mines. Procs. ACG Deep Mining Symposium, Nov 4–9, Perth, Western Australia.

Hardcastle, S.G., and Kocsis, C., 2004, Reducing costs not capacity – A feasibility study into ventilation-on-demand, Publication pending.

Kocsis, C., Hardcastle, S.G. and Bonnell, G., 2003, Ventilation benefit analysis for Canadian mines – fuel cell loader project, Natural Resources Canada, MMSL 03-040(CR).

Laliberté, P., Desrivières, G., Lacroix, R., Bétournay, M.C. and Laflamme, M., 2003, Best methods of hydrogen transfer, Natural Resources Canada, MMSL 03-030(CR).

Lacroix, R., Laliberté, P., Kocsis, C., Bétournay, M.C. and Laflamme, M., 2003, Capital and recurring cost benefit analysis for Canadian mines – fuel cell loader project, Natural Resources Canada, MMSL 03-045(CR).

NRCan, 1997, Canada's energy outlook 1996-2020, Natural Resources Canada, Energy Forecasting Group, http://www.nrcan.gc.ca/es/ceo/toc-96E.html.

NRCan, 1999, Canada's emissions outlook: An update, Natural Resources Canada, National Climate Change Process, Analysis and Modelling Group, December 1999, http://www.nrcan.gc.ca/es/ceo/update.htm.

NRCan, 2002, List of CANMET approved diesel engines in accordance with CSA Standards M424.2-M90 & M424.1-88, Natural Resources Canada.

Righettini, G. and Mousset-Jones, P., 2003, Ventilation savings with fuel cell vehicles – A cost-benefit analysis for selected US metal/non-metal mines, University of Nevada, Contract report.

Righettini, G. and Mousset-Jones, P., 2004, Ventilation savings with fuel cell vehicles, Proc. 10th U.S. Mine Ventilation Symposium, Anchorage, USA, May 16–19, Publication pending, Balkema.

Mine Ventilation, Ganguli & Bandopadhyay (eds)
© 2004 Taylor & Francis Group, London, ISBN 90 5809 633 5

The impact of changing employment and mining practices on the mine ventilation profession in Australia

Rick Brake & Tony Nixon
Mine Ventilation Australia

ABSTRACT: The ventilation profession in Australia is at something of a crossroads. The changing nature of the mining industry nationally and globally poses numerous challenges for the profession in its role as the "repository" of ventilation knowledge and experience. However, underground mines will continue to need to be ventilated, and in fact in many respects will need to be ventilated to a higher standard than in the past. The factors and trends identified in this paper therefore provide both threats and opportunities for the profession and are changing the way ventilation services are delivered in Australia. Whilst there will remain a need for both "GP" and specialist ventilation practitioners, it is likely that the number of experienced "GP" ventilation professionals will diminish, but the number of specialist "niche" players will increase.

1 INTRODUCTION

Major changes have occurred in Australia over the past 20 years with respect to employment practices of both technical staff, such as mine ventilation practitioners, and underground workers. Some of these changes include: downsizing of the workforce, outsourcing of technical skills away from the mine site itself, FIFO (fly-in, fly-out) and "back to back" rosters, abandonment of centralised technical services at corporate headquarters, the trend away from prescriptive mining legislation (with specific underground mine ventilation standards) to more general legislation based on "duty of care", removal of legislative requirements for formal "ticketing" of ventilation staff, the difficulty of attracting high-quality school leavers into the mining industry, the development of "global" mining companies, often with head offices in overseas countries, the trend to less variety in mining methods (away from the wide range of selective, high cost, high labour methods to a narrower range of bulk, lower cost, more capital intensive methods), changes in professional indemnity insurance costs and availability, a more "blurred" demarcation between ventilation professionals and occupational hygienists, safety professionals and others, the trend to the use of mining contractors (rather than companies having their own operating workforce), the loss of "status" of the profession to other technical groups and the growth of underground civil engineering activities as major sub-surface projects in their own right. Each of these developments is having a significant effect on an overall re-shaping of the nature of mine ventilation profession in Australia. This paper lists these factors, discusses the impact they are having on the industry and the profession, and examines the likely future impact on the profession in Australia, and by implication, the potential impact on other Western countries. Note that many of these factors "overlap" or have several impacts, so that there is no easy "classification" or "ordering" of factors.

2 MINING REGULATION IN AUSTRALIA

The Commonwealth of Australia is an independent nation of just over 20 million people with a historically very strong mining industry. It has a land surface the same size as that of the continental USA and over 35 000 km of coastline. There are six states and two mainland territories in Australia, all of which have some mining activity. Mines are regulated at the State/Territory (not Federal) level, under state not federal legislation, using state and not federal inspectors.

3 CHANGING STYLE OF LEGISLATION

There are now three styles of legislation in Australia, depending on the industry and the State.

- *Conventional prescriptive legislation*: in which detailed regulations prescribe how a mining activity is to occur.
- *Duty of care legislation*: where the outcome is prescribed, but the method of achieving that outcome is left to the mining company. In some States, there is now no specific mining legislation at all, with mines being covered under the same safety legislation as other industrial workplaces.
- *Hybrid legislation*: either a mix of prescription and duty of care, or a combination of duty of care with officially approved guidelines or codes of practice, which helps companies, avoid the need to develop procedures for all activities.

There has been a variable response to the "duty of care" style of legislation. Larger mining companies tend to welcome the legislation as giving them more flexibility over their operations. The expense of developing the required operating procedures using risk assessment techniques can often be spread over multiple operations. Smaller mining companies find the loss of prescriptive regulations imposes a heavy burden on them in terms of having to develop their own standards for a great variety of procedures.

With "duty of care" legislation, ventilation professionals can no longer rely on meeting specific and detailed government regulations to demonstrate that the mine is adequately ventilated. The "burden of proof" moves from the Government to the mine owner. Ventilation professionals must therefore now be much more competent in conducting or participating in risk assessments to a recognised standard, documenting procedures, and auditing against those procedures than in the past. In addition, they must be better informed about what would constitute "good practice" within the industry, as this is often the most appropriate defence against negligence.

Another impact of this new style of legislation is that government mining inspectors are now more like technical auditors rather than traditional "inspectors". Government audits tend to be done by an audit team of inspectors, each with their own specialist area, rather than an individual "GP" inspector. This means when inspectors are on site, they are more likely to want to discuss the "big picture" ventilation issues as to how and why the company has developed the procedure that it uses to control the ventilation risks, rather than review historical records of sometimes tedious regulations.

As a further example of the trend away from industry-specific legislation, the role of the mining warden has generally been replaced by that of the State Coroner. Rather than a Warden's inquiry being held into major mining disasters, it is now more likely for this to occur via a Coronial inquiry.

4 LOSS OF STATUTORY POSITION OF VENTILATION OFFICER

In the past, each jurisdiction required each mine to have an on-site ventilation officer who had responsibility for mine ventilation. This was a statutory appointment, i.e. the ventilation officer had to be personally approved by the Government Inspectorate and issued with a statutory appointment letter. This is still the case in the coal mining operations in Queensland and NSW and the metal mining operations in Western Australia, but there are no statutory requirements for ventilation staff for other mining operations in Australia. In many mines, this has left a vacuum in terms of both the need for any ventilation staff at the mine site, and the qualifications for those staff. Frequently, the individual now responsible for ventilation at many metal mines is also responsible for one or more other areas, such as pumping and drainage, or other "services" work.

4.1 *Development of portable qualifications*

The development by the Australian Government of an Australian Qualifications Framework (AQF) covering all education and accreditation standards from high school to PhD level, and the development of Career paths and National Competency Standards, means that it is now much easier to train to a recognised standard and to transfer skills around Australia. In the past, a person with a "statutory" Mine Manager's Ticket or statutory Ventilation officer's ticket would need to sit for sometimes extensive practical and oral examinations when transferring from one state to another. The development of the AQF and the adoption of nationally accredited (and recognised) competency standards has reduced the "barriers to entry" for qualified staff in terms of moving between States. However, as most jurisdictions (except coal) do not require a ventilation officer to be appointed for each mine (even a part-time ventilation officer), nor specify the qualifications of the ventilation officer, there is no formal requirement at this time for ventilation professionals to meet these National competency standards.

5 INTERNAL COMPANY STANDARDS AND THE "QUALITY ASSURANCE" TREND

Most of the larger mining companies in Australia now have their own internal standards for mine ventilation. In the coal industry, the development of "major hazard plans" for certain key risks is required as part of the legislation. This has tended, in some instances, to remove the need for discretion or the need to develop "tailored" procedures at each mine site, reducing the need for the number and skill level of ventilation staff.

6 FLY-IN, FLY-OUT (FIFO) AND "BACK TO BACK" ROSTERS

Even mines that do have a ventilation officer may not always have access to that person. This is often the case with fly-in, fly-out (FIFO) operations. Some staff positions in FIFO operations are considered to be critical, so that when one person is "away", their colleague is on site (termed "back to back" rosters). This would include surveyors, geologists and others associated with the production process. However, ventilation staff in non-coal operations are often considered to be non-critical, so that when the ventilation officer is "away", there is no-one available to provide ventilation guidance. This has particular significance when it comes to emergency situations such as fires. Many FIFO operations in Australia have re-written their procedures so that the ventilation officer does not have a formal role in the event of an underground fire, as there is not always a ventilation officer on site. In our opinion, this is a significant flaw and a retrograde step.

7 THE REDUCED INFLUENCE OF THE MINING INDUSTRY

Twenty years ago, the resources (mining and related) companies in Australia made up over 30% of the capital value of the Australian stock market. Today, that value is under 20%. The mining industry no longer has the same influence on Australia's domestic growth and wealth. This is evidenced by the fact that the minister responsible for mining at the Federal level is one of the most junior ministry positions within the Federal cabinet.

8 PROBLEMS WITH ATTRACTING CAPABLE STUDENTS INTO THE MINING INDUSTRY

Even though Australia is large, it is highly urbanised with a heavily populated "fringe" around the coastline, with most of the remainder of the country being only sparsely populated, much like Canada. Most high school students in Australia do not want to work or live away from the major cities. In addition, the general view of the population is that the mining industry is an environmental vandal, and is a "stone age" and "low tech" industry with little "sizzle" compared to more high-flying industries such as Information Technology. This means that it is becoming increasingly difficult to attract capable high-school graduates into the mining profession. This can be seen by the fact that Universities teaching mining engineering have generally had to lower their entry requirements for students over the past 15 years. In addition, whilst there were about ten Universities teaching mining engineering 20 years ago, there are now only four and this will reduce to three within the next few years.

9 PROBLEMS WITH ATTRACTING CAPABLE STUDENTS INTO THE VENTILATION PROFESSION

Even within the mining industry itself, the ventilation "profession" is often seen as being a dreary, boring job most suited to those with poor "people" skills and who are happy with their own company, squirreling away taking gas and airflow measurements and tinkering with ventilation network models. Ventilation staffs are often seen as being one of the "nerds" of the mining profession.

10 JOB INSECURITY; EARLY DEPARTURE FROM THE WORKFORCE; "CHURNING" OF THE WORKFORCE

The long-term real prices of almost all commodities have been falling and continue to fall. The pressures on mining companies to cut costs are inexorable. The heavy cost pressures on the industry, the increased proportion of mines with short mine lives and the "rationalising" of ownership has led to a loss of employment security generally.

This, along with the rigours of fly-in–fly-out operation, means that many engineers plan to leave the industry by the time they are about 40 years old, departing from the industry at a relatively early age and taking their accumulated knowledge, skills and experience with them.

The churning (movement of individuals from one company to the next) has led to few engineers wanting to invest the time and effort into developing technical skills that may result in them being so specialised that they are among the first to be retrenched in the event of a change in corporate emphasis. They see that "having more than one string to their bow" (i.e. being a generalist rather than a specialist) makes them more redundancy-proof. Therefore there is an often an unwillingness from the individual to develop specialisation in ventilation.

In addition, it takes years of exposure to ventilation issues to develop an experienced and competent ventilation engineer. This process is not easily "nurtured" when the engineer must be regularly moving between companies due to lay-offs and takeovers. Therefore even if an engineer does want to pursue ventilation as a career, "churning" tends to selectively work against his (or her) ability to develop and then retain his skills.

11 FLATTENING OF MANAGEMENT LEVELS AND LOSS OF CAREER PATHS

Up until about 15 years ago, large mining companies in Australia typically had 8 to 12 levels of management between the workforce and the CEO. Middle management provided numerous opportunities for quite senior technical specialists and managers in very fulfilling roles. Most of these positions no longer exist. Therefore the only way for ambitious engineers to rise to more powerful positions within the industry is to go via the "operating" route, i.e. becoming a mine manager. At the very best, "ventilation" is seen as only a short stepping off point for most engineers on their way to something "bigger and better".

12 TREND TOWARDS USING VENTILATION CONSULTANTS RATHER THAN IN-HOUSE VENTILATION SPECIALISTS

This flows on from mining companies wanting to turn as large a proportion of their cost base into a variable cost, rather than a fixed cost. It is very expensive to keep a technical specialist in-house, particularly when that specialist needs to write papers and attend conferences and make site visits to other operations if he/she is to remain a "specialist". In-house technical specialists are often seen as a luxury and are among the first staff to be "downsized". Most companies consider it easier, cheaper and simpler to "buy in" specialist advice when they need it. To our knowledge, Australia now only has three "career" (long-term) ventilation engineers based at a mining operation in Australia.

13 LOSS OF VENTILATION MENTORS

The flattening of management structure and loss of technical managers has often meant a loss of the mentors that were previously responsible for nurturing, encouraging and developing individuals suited to specialist roles, such as those in ventilation. These mentors no longer exist within the mining companies. This results in frustration and also "gaps" appearing in the skills of ventilation staff, who are often left to "sink or swim" in their role.

However, the reduced number of ventilation staff within the mining companies themselves, and the loss of experienced in-house mentors, have led to a growing demand for auditing services and also for training services from consultants, in addition to normal consulting design services.

14 REDUCTION IN THE SKILL LEVEL OF IN-HOUSE VENTILATION GENERAL PRACTITIONERS

The lower emphasis on ventilation (in metal mines) has, in part, led to the relatively fast "rotation" of young mining engineers through the role of ventilation officer or engineer. This is resulting in a very wide variation in technical skills within mines in Australia. In addition, many mines assume that the ventilation knowledge that a young engineer obtains as a University undergraduate is sufficient for him to take on the role of a competent ventilation officer. This is rarely the case; in effect, the engineer is thrown into the job with little on-site or off-site support and left to "sink or swim".

15 THE TREND TOWARDS THE USE OF MINING CONTRACTORS

Many mine owners in Australia no longer have operating workforces at all. A typical example would be a mine in which the mine manager, geologists and planning engineers (perhaps) are employed by the owner, but the entire underground workforce, supervisory staff and technical support staff (ventilation, rock mechanics) are employed by a mining contractor. The fact that mining contracts are often for only two years (usually with an option for the owner to extend the contract) means that the "technical" focus of the mine becomes very short term, especially in areas such as ventilation.

16 IMPROVED TRANSPORT; COMMUNICATION TECHNOLOGY CHANGES

Almost all new mining operations in Australia in the past 15 years use fly-in, fly-out arrangements, except where there is already a nearby town. This was initially driven by changes in taxation policy in Australia, but over the past 10 years, mining companies have also found that the workforce most often has a strong preference for living in a larger community with more services, than in a remote community with relatively few services.

Concurrent with this has been the development of communication technology (fax, high-speed internet, email, video-conferencing) that allows mine plans and most other information to be transferred quickly between a mine site and an off-site. It also allows meetings to be held with staff at any location.

17 CHANGES IN CAREER PLANNING STRATEGY WITHIN MINING COMPANIES

The "cradle to grave" employment mentality of the past meant that mining companies were rather paternalistic to their workforce. Each professional employee could rely on someone in the company to be planning their future for them; looking for development roles and other opportunities for personal and professional growth. Companies would complete skills inventories and endeavour to manage employee's careers to fit with long-term company goals. Long-serving employees were considered to be an asset; they are now often considered to be a liability, part of the problem rather than part of the solution; adding to the organisation's inflexibility to change. Today, tenure of four to five years in a company is often all that the company wants. Therefore, personal and professional development is the responsibility of the individual, not the company. This results in a more ad hoc approach to corporations developing or retaining specific skills. It also results in companies being unwilling to spend funds on training courses for individuals. The responsibility for (and cost of) upgrading or retaining technical skills is often now borne by the individual not the mining company.

18 THE REDUCED INFLUENCE OF MINING UNIONS AND THE DIFFERENT PERCEPTIONS OF THE IMPORTANCE OF VENTILATION TO MANAGERS AND WORKERS

Union coverage in the private (non-government) sector in Australia has fallen from over 50% 25 years ago to about 20% today. Whilst the coal mining industry is still strongly unionised, the non-coal industry is much less so. Even in the coal industry, there is now extensive use of contractors. Unions do tend to consider ventilation more important than do mining companies. We regularly ask workers what the biggest issues are in their workplace; they regularly tell us that ventilation matters (dust, fumes, heat) are the biggest or second-biggest issue. This is so even when they do not know that we are in the "ventilation business". However, if we ask the mine managers what the biggest issues are in the workplace, ventilation issues almost never even makes it to the manager's short-list. This is especially true for metal mines. We believe this dichotomy is due to the fact that the worker has to work daily and even hourly in poor environmental conditions, which affect his enjoyment and satisfaction of the workplace. However, the manager does not believe that there are any serious health risks in the workers being exposed to unpleasant but "otherwise benign" conditions, so considers truly clean and comfortable air as being in the optional luxury category.

19 THE REDUCED INFLUENCE AND SUPPORT FOR THE PROFESSIONS AND PROFESSIONAL BODIES

There is a overall trend in Australia for a loss of public confidence and support for professions in general: lawyers, doctors, engineers, teachers, academics and others. In the past, the public held most professionals in high regard; this is no longer generally the case. This may be related to the very high proportion of school leavers who now go on to tertiary education; whereas a tertiary education was only for an elite few in the past. There is a trend to regard the "self-made" person, sometimes with little formal education, as the real success story, and for the professionals to be regarded as out to feather their own nest.

20 THE EMERGING ANTI-TECHNICAL BIAS AND CHANGES IN CORPORATE GOVERNANCE PHILOSOPHY

In the past, the directors of Australian mining companies were often technical people: engineers and geologists. Today, a majority of board members are non-mining professionals: lawyers, accountants, IT or HR professionals, environmentalists, even social activists and retired politicians or public servants. In addition, there is a strong trend in Australia away from full-time managing directors on the board towards part-time independent directors. Both these trends tend to develop a subtle bias in "upwardly mobile" young engineers towards the view that "technocrats" are not likely to "get anywhere". This is evidenced by the fact that most young engineers, if they do decide to take up post-graduate study, choose an MBA or other commercial or business degree, rather than a science or engineering master's degree or a PhD.

21 THE DEVELOPMENT OF "GLOBAL" MINING INDUSTRY AND "GLOBAL" MINING HOUSES

The chronic poor shareholder returns from most mining operations around the world has led to a dramatic increase in corporate takeovers and the growth of trans-national mining houses. This is having several effects. Two important ones are:

- The trend for major technical audits or feasibility studies to be conducted from an off-shore (non-Australian) base (e.g. corporate headquarters in London or elsewhere).

– The trend for the bigger mining houses to develop "corporate" policies in major technical areas, including ventilation, which may reduce the need for technical expertise or technical decision-making at individual sites. This is part of the "one size fits all" approach.

However, there are also benefits of globalisation. There are now more opportunities for Australian ventilation professionals, both those within companies and within consulting firms, to be engaged on projects outside Australia.

22 THE HIGH LEVELS OF ECONOMIC GROWTH IN ASIA

The Asian countries are experiencing very high levels of economic growth. In addition, the standard of living (and hence wages) and expectations of the Asian workforces is growing. The demand for all resources is very high. These factors are leading to the development of a larger mining industry in these countries, and adoption of more "western" mining practices. This is leading to opportunities for Australian mining companies and also for Australian mining consultancies in this fast-growing region.

23 CHANGES IN TECHNOLOGY

Even though mine ventilation is seen by some as a low-technology speciality, the reality is that technology is creating major changes for the profession. The practice of auxiliary ventilation is still changing significantly due to much larger fans and ducts and bigger mobile equipment (bigger airways and bigger diesel engines). This is impacting on the distances that can be mined before flowthrough ventilation is established, and the required airflows for auxiliary ventilation. Gas drainage is allowing mining operations in seams that were simply too gassy to be able to be safely mined in the past. The understanding of the nature and causes of spontaneous combustion and refrigeration/cooling (including adoption of air-conditioned cabins) is having significant effects on mine design and ventilation layouts. The trend towards diesel truck haulage via ramp to the surface (rather than conventional underground crushing and shaft hoisting) is also having considerable impact.

24 THE TREND TO LESS VARIETY IN MINING METHODS

The need to reduce operating costs and improvements in metallurgical recoveries and treatment costs means that the elaborate and selective mining methods of 20 or more years ago have often been replaced by one of only a few highly productive methods of today. Along with this reduction in mining methods and trend towards "simpler" high-productivity methods is a reduced need for ventilation design. Mass production reduces (in fact, opposes) the need for tailor-made designs, or for major design changes and can tend to stifle creativity at the macro level.

25 CHANGES IN WORKERS EXPECTATIONS REGARDING HEALTH AND SAFETY, CONCEPTS OF CORPORATE RESPONSIBILITY, PROFESSIONAL INDEMNITY INSURANCE

In the past, mineworkers often felt that "atrocious" working conditions were part of their job. Whilst the workforce has been substantially focussed on dust in the past, new issues such as heat and diesel particulates are now attracting considerable workforce concern. Good workplace conditions also helps to keep workers in the industry for longer (which helps prolong careers and investments in training), reduce turnover and improve morale.

Australia has not (yet!) developed the level of litigation that is seen in the USA. Nevertheless, there is an increasing trend towards litigation of professionals. This is resulting in very substantial increases in professional indemnity insurance, which impacts on ventilation consultants more so than in-house ventilation professionals.

It is also becoming increasingly common for government investigators of workplace accidents to look beyond the "obvious" cause of an accident to the underlying causes. Two recent cases in Australia illustrate this. In one, a welder was electrocuted and the prima facie cause was not following the approved job procedure. Further investigation revealed that the conditions on the job were excessively hot, so that the welder wanted to "just get the job done and get out", which was therefore a contributing factor to the accident. In another, a miner fell down a shaft during its sinking operation. The argument was mounted that excessive temperatures resulted in an inability to concentrate for the full shift length in what is otherwise a hazardous environment.

There is also a trend to adopt the concept of "industrial manslaughter" in new legislation, which would allow managers (or any other responsible person) whose negligence resulted in the death of an employee to be prosecuted and jailed. Technical areas such as ventilation or rock mechanics, which clearly involve judgement calls at times, and where decisions can sometimes be "counter-intuitive", are therefore now considered to have higher risk profiles.

26 CHANGING NEEDS OF CORPORATE CLIENTS IN TERMS OF FEASIBILITY WORK; TREND TOWARDS BANKABLE FEASIBILITY STUDIES

The art and practice of mining is become more complex as the body of knowledge continues to grow and as orebodies become more remote and deeper. Most corporate clients no longer have the ability to conduct a major feasibility study in-house. Most clients do not even have the ability to manage a team of independent specialist consultants themselves so as to develop a feasibility study in-house, instead preferring to get a "one-stop-shop" consulting firm to undertake or at least coordinate the entire feasibility study process.

In addition, many new projects are funded by "project financing", joint ventures or other strategies in which the technical risk of failure has to be very low to obtain bank financing. Many projects therefore require a "bankable" feasibility study (i.e. one which meets the high standards required by a bank's high risk aversion). An in-house technical team is not always seen to be sufficiently unbiased to meet these requirements.

Therefore, within the mining profession, there is a trend towards larger mining consultancies with a full complement of professional skills, complemented by a growing number of small but very specialised "niche" consulting firms. Consulting firms (and academics) are now the main repository of ventilation skills and experience in Australia.

27 CHANGES IN EMPHASIS WITHIN HEALTH AND SAFETY, ROLE OF RISK ASSESSMENT

In the past, control of ventilation hazards in the underground environment was often due to the need to control the long-term health risks, e.g. lung disease. The perception of management is that these "health" risks are now under control and will not reappear. The emphasis has now swung towards controlling safety risks, e.g. rock falls.

In our opinion, this change in emphasis is driven, at least in part, by the trend to quantify risk according to one of the many modern risk assessment tools. We believe this process, accompanied by the fact that often the participants in the risk assessment (frequently young) have little experience of the long-term health effects of poor ventilation, tends to bias the "classification" of risks towards the fast-acting acute risks (fatality due to rock fall) rather than the slow-acting chronic risks (lung disease), even if the risk profile is actually the same. This opinion is also shared by Kizil and Donoghue (2001), who argue that this trend to put higher priority on the short-term acute safety risks in the underground environment over the long-term chronic health risks is not justified. Kizil and Donoghue also believe that, if this trend is reversed, there will be a renewed interest in improving the environmental conditions in which workers spend their working life.

28 REDUCED SCOPE AND INCREASED FRAGMENTATION OF THE PROFESSION

The growth in the role of occupational hygienists and safety professionals has, in part, been at the expense of ventilation professionals. Whereas measurement of contaminants in the workplace was largely conducted by ventilation professionals in the past, this is now often conducted by occupational hygienists, with the ventilation professional being left with the diminished role of adjusting the airflow circuits to keep contaminants within acceptable levels. This is partly driven by the need for hygienists in the surface facilities of most mines, the tremendous growth in the number of chemical compounds in use in mines and the far greater versatility of hygienists in occupational hygiene matters. Hygienists tend to also cover noise, vibration, illumination, confined spaces permitting, welding fumes, etc.

In addition, there is a fragmentation of the previous very-wide scope of the ventilation profession. An example is the growth of specialist "off-shoots" to general ventilation practice, such as gas drainage, spontaneous combustion, refrigeration and heat stress.

29 LOSS OF STATUS TO OTHER TECHNICAL PROFESSIONS

This follows on from the points above. Ventilation used to be near the "top of the tree" or the "top of the food chain" in terms of status as a technical profession within mining. It has generally be supplanted by other technical areas, especially geomechanics, but also even traditionally "fuzzy" areas such as safety "professionals", environmentalists and the like.

30 THE GROWTH OF NON-MINING (CIVIL) UNDERGROUND INFRASTRUCTURE PROJECTS

A recent technical conference in Australia noted that "(Sydney) will increasingly start to look like a giant Swiss cheese". Capital expenditure on underground civil engineering projects in Australia would now significantly exceed capital expenditure on underground mining projects. The construction phase of these underground civil projects can vary in complexity as can the ventilation design of the final operating phase. To date there has not been a strong transference of ventilation design concepts and skills between underground civil ventilation practice and underground mining ventilation practice.

31 RANGE AND NUMBER OF "STAKEHOLDERS" IN VENTILATION

In the past, a ventilation professional was assessed principally on his/her technical ability. However, the range of stakeholders that he/she needs to liaise with has grown considerably. In many cases, the "people" skills of a ventilation professional are now as important as the technical skills. Some of the stakeholders in ventilation now include (variously): shiftboss/supervisor, contractors (mining, maintenance or construction), hygienists, union officials, planning engineers, purchasing staff, internal medical staff, workforce safety committees, scheduling engineers, suppliers, external medical staff, operating managers, mine rescue personnel, consultants, mines inspectors, electrical and maintenance engineers and safety officers.

This need for people skills is accentuated by the flattening of management structure, loss of mentors, loss of "status" of the profession and increasingly short-term focus of mine owners, which means decision-making timeframes are much more compressed compared to years ago.

32 CHANGES IN EGRESS AND RESCUE PHILOSOPHY

Previously, if a mine emergency occurred, it was considered normal to be sending in mine rescue teams to find and then rescue workers. Whilst this is still true to some extent, the "duty of care" now enshrined in legislation and in society itself makes it difficult to risk one person's life to possibly save another person. The trend is therefore away from "aided" rescue to "self" rescue. This is starting to have some considerable consequences on mine planning and ventilation design. Another example of this is the relatively low fitness levels of the "modern" workforce. In the past, it was normal for workers to move through a mine using ladder ways. However, many workers today would be unable to climb any distance of ladders. This impacts on egress strategy (the mine still needs to provide for safe egress even for relatively unfit workers) and ultimately on ventilation design.

33 AGE PROFILE OF VENTILATION PRACTITIONERS

As a result of many of the above factors, there is an ageing profile of the remaining ventilation practitioners in Australia. There will be a substantial exodus of ventilation experience from the industry in the next 10 to 15 years.

34 SUMMARY

The ventilation profession in Australia is at something of a crossroads. The changing nature of the mining industry nationally and globally poses numerous challenges for the profession in its role as the "repository" of ventilation knowledge and experience. However, underground mines will continue to need to be ventilated, and in fact in many respects will need to be ventilated to a higher standard than in the past. The factors and trends identified in this paper therefore provide both threats and opportunities for the profession and are changing the way ventilation services are delivered in Australia. Whilst there will remain a need for both "GP" and specialist ventilation practitioners, it is likely that the number of experienced "GP" ventilation professionals will diminish, but the number of specialist "niche" players will increase.

REFERENCE

Kizil, G.V. and Donoghue, A.M. *Coal dust exposures in the longwall mines of New South Wales: a respiratory health risk assessment using Joint Coal Board exposure data from 1985–1989.* Proc Queensland Mining Industry Health and Safety Conference, Aug 2001. pp 299–312.

Mine Ventilation, Ganguli & Bandopadhyay (eds)
© 2004 Taylor & Francis Group, London, ISBN 90 5809 633 5

Studies of air leakage in underground mines using CFD

F. Calizaya
Associate Professor University of Utah, Salt Lake City, Utah, US

I.J. Duckworth
PT Freeport Indonesia, West Papua, Indonesia

K.G. Wallace Jr.
Mine Ventilation Services, Inc. Fresno, California, US

ABSTRACT: This paper presents a summary of computational fluid dynamics (CFD) modeling studies that were conducted to investigate the leakage flow through bulkheads, doors and regulators. The models were calibrated against field data collected from ventilation controls in an operating mine. PT Freeport Indonesia's Amole Dewatering development was used for this purpose. The mine includes two long parallel drifts, one intake and one return, several crosscuts, doors and regulators. To direct the air to remote headings (over 6 km from surface portals), the mine depends on 34 concrete bulkheads, 19 doors and three fans. The models were used to predict leakage flow rates for various model geometries and boundary conditions.

1 INTRODUCTION

Effective face ventilation is essential to control mine pollutants such as diesel particulate matter, gas, and dust. Therefore, it is important to provide sufficient airflow to the last open crosscut in twin development headings. However, if flow control devices are not maintained adequately, a significant portion of the ventilation air will be short-circuited through doors and bulkheads, reducing the availability of fresh air to the workings and to dilute diesel emissions.

In this study, the leakage flow problem is investigated by means of a numerical simulator, VnetPC 2000, and a fluid flow simulator, FLUENT. These programs are used to generate models to represent the mine airways and to solve a set of mass and energy balance equations. Prior to any prediction, the models are calibrated against field data collected from an operating mine. Once the calibration is completed, each model is run for various fan pressures, and the leakage flow rates for prevailing conditions are estimated. The use of a fluid flow simulator may enhance the knowledge of leakage flow not only in twin development headings, but possibly in understanding the flow in complex systems, such as the gob behind a longwall panel.

2 AIR LEAKAGE

In underground mines, bulkheads, doors and regulators are used to separate different types of airways (typically intakes, returns and conveyors). Intakes are used to direct the fresh air to the workings, and the returns to exhaust the contaminated air. Depending upon the complexity of the mine, the ventilation system may require hundreds of these structures. The air pressure differentials created across these structures will inevitably cause some loss of air, such that a quantity of air will be short-circuited to surface without being used in the working areas. Ventilation specialists have estimated that in coal mines this quantity can be as high as 60 percent of the total quantity induced by the surface fans (Richardson, 1997). In metal mines, although often reduced by means of booster fans and the lack of connecting intake and return parallel airways, the leakage quantity has been computed on order of 30 percent (Calizaya et al., 2001, van der Bank, 1983).

In ventilation design bulkheads represent highly resistive air paths. Frequently, one path is used to simulate a group of bulkheads. When pressure (p)/quantity (Q) survey data are available, the resistance (R) of each path is calculated from Atkinson's equation ($p = RQ^2$), and sometimes, this equation is modified to reflect the leakage flows. However, there is considerable uncertainty in the application of this approach. For new designs, when no survey data are available, this resistance is typically determined based on empirical data from previous studies.

A review of the ventilation literature has shown that in the early 1960s, US Bureau of Mines personnel conducted a series of tests to investigate the leakage flow through bulkheads. The problem was studied using bulkheads built in an experimental mine. The results of these studies, published between 1977 and 1983, have shown that the resistance of bulkheads varies with a number of factors, including the porosity of the construction materials, quality of the construction techniques, and time. Experimentally, it was found that this resistance can be determined from:

$$p = RQ_L^{\,n} \tag{1}$$

where p = pressure difference across the bulkhead; Q_L = leakage quantity; R = bulkhead resistance; and n = constant for a given type of bulkhead.

Based on field measurements, and by applying Equation 1, researchers have found that the resistance for a 'good' single masonry bulkhead varies between 559 and 782,800 Ns^2/m^8 (Bruce & Koenning, 1987, Duckworth, 1995). It is this wide range of values that makes ventilation design difficult and the leakage flow estimates questionable.

3 AIR LEAKAGE AND VENTILATION PLANNING

Bulkheads, overcasts, doors, and regulators are structures used to direct the air to working areas. Depending on the mining methods and the complexity of the workings, a mine may require hundreds of these constructions. In operating mines, these structures are subject to external forces that often lead to damage. In coal mines these are represented by vertical loads of the overlying strata. Over time, these forces will create cracks in the building blocks and fractures around the periphery of the bulkhead allowing the air to pass through. In metal and non-metal mines, the destructive forces are those caused by blasting and ground movement. The damage on these constructions will depend on the intensity of blasting, and the amount of ground movement. Fractures in the rock around the structures can also contribute to the air leakage.

In ventilation planning a parameter, referred to as leakage path resistance, is used to characterize each structure. Physically this can be determined from air pressure/quantity measurements or estimated from tabulated figures. Some typical leakage path resistances are presented below:

1) From Pressure/Quantity Surveys:
   - Bulkhead around a main fan: p = 2365 Pa, Q = 7.1 m^3/s, R = 46.92 Ns^2/m^8 (a shotcreted wall with a 0.1 m diameter drainage hole.)
   - Double Airlock Doors: p = 772 Pa, Q = 2.4 m^3/s, R = 138.8 Ns^2/m^8 (each construction had two manual doors)
2) From Tables and Past Experience:
   - Concrete brick bulkheads (unfavorable conditions): R = 356 Ns^2/m^8 (Deglon et al., 1992).
   - Concrete brick bulkheads (12 m^2): R = 1100 Ns^2/m^8 (Coetzer, 1985).
   - Metal prefabricated stoppings (range): R = 1090 to 1678 Ns^2/m^8 (Duckworth et al., 1995).
   - Walls for standard overcasts: R = 1780–2220 Ns^2/m^8 (Deglon et al., 1992).

The above data shows a wide range of resistances for various leakage paths. When several doors or bulkheads are used in parallel arrangements, an equivalent resistance is determined. The resistance decreases as the number of structures increases as given by parallel theory. An approximation based on the square law is often considered sufficient for planning purposes.

4 MINE VENTILATION AND COMPUTATIONAL FLUID DYNAMICS

Mine ventilation planning is traditionally conducted using Hardy-Cross based numerical simulators. There are several computer packages available in the market including: VnetPC (US), VentSim (Australia) and VUMA (South Africa). In every case the mine openings are represented by nodes and branches collectively called a ventilation network. In this network an airway, drift or shaft, is characterized by a single parameter: the airway resistance. This resistance can be measured or calculated from Atkinson's equation. The fluid used with the simulator is air or a mixture of air and airborne contaminants. In most cases this fluid is assumed to be incompressible and of constant density. The energy source is represented by a fan curve. For a given mine geometry and a few workings with fixed flow requirements, a simulator is used to determine the airflow/pressure distribution in the network. The results are then evaluated against the flow requirements and other external factors such as minimum air velocity, maximum gas concentration, allowable fan pressure, etc. If requirements are not met, the input parameters are modified and the simulator is executed repeatedly until all the constraints are satisfied.

Although these simulators may be configured to show how the energy is spent in an airway, or what percentage of it is lost against friction or airflow obstructions (such as sudden contractions of expansions), the results represent an approximation. The energy losses through bulkheads, overcasts, doors and regulators are calculated based on the square law when in fact the flow regime is actually laminar (low Reynolds number).

Computational Fluid Dynamics is a design tool used to represent a fluid flow problem by a 2D or 3D real scale model. The technique utilizes two computer programs: GAMBIT to create the geometry of the model and FLUENT to define the fluid parameters and solve the problem. The geometry of the model is constructed to physically simulate the conduits and power sources and sinks. In a mine ventilation system, the geometry is built to represent intake and return drifts, shafts, bulkheads, doors, and regulators as conduits and fans as sources of energy. The degree of approximation to reality (actual mine conditions) is limited only by the power of the computer and the accuracy of the required results.

Once the geometry is defined, the individual components (faces or volumes) are divided into a number of live cells, the inlet and outlet conditions are defined, and the fan and wall characteristics are specified. The model is then exported as a mesh file. The file is next opened by FLUENT and the fluid properties specified. These include the fluid density, viscosity, temperature, and other thermodynamics properties. Next, the boundary conditions, i.e. wall parameters and fan characteristic curves are specified. Finally, the simulation process is started and the program is run until some pre-established conditions are achieved. Fundamentally, the process involves solving momentum and energy balance equations to obtain results such as velocity components, density, and turbulent viscosity for each cell. These are then presented as contour graphs or vector profiles to allow interpretation.

5 SAMPLE PROBLEM SIMULATION

VnetPC and FLUENT airflow simulation details are illustrated through a sample problem. Figure 1 shows an example twin-drift ventilation system used for long development headings. The drifts, 200 m long, are driven to a finished cross-sectional area of 4 × 4 m. They are connected to each other by crosscuts of the same cross-section. The energy source (fan) is located in the return drift, 30 m away from the portal. It is installed as a booster fan. Crosscut 5 is blocked by a solid bulkhead, which allows minimal leakage flow. Other design details are shown in Table 1. The problem is to determine the quantity of air circulated through the last open crosscut and the average air velocities along the intake drift (upstream and downstream of the bulkhead.)

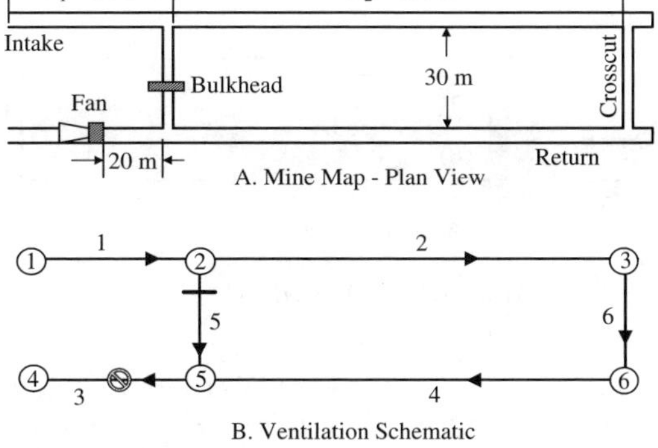

Figure 1. Twin drift ventilation system.

Table 1. Mine ventilation system details.

Airway	Area m^2	Length m	k-factor kg/m^3	Resistance Ns^2/m^8
1	16	50	0.0088	0.00172
2	16	150	0.0088	0.00516
3	16	50	0.0088	0.00172
4	16	150	0.0088	0.00516
5	Blocked	30	–	46.92
6	16	422*	0.0088	0.01451

* Includes equivalent length for two sharp bends.

Table 2. Fan characteristic data.

Quantity, Q (m^3/s)	Pressure, p (Pa)
56.7	3237
66.1	2739
75.6	1992
85.0	747

5.1 *Modeling using VnetPC*

With VnetPC, the solution to this problem is achieved iteratively. First, based on the mine geometry, a ventilation network is formulated. The network parameters are then specified on a branch-by-branch basis and finally, the fan parameters are entered (Table 2.)

In the next step, the simulator is run and its output evaluated. For this example, VnetPC produced the following results:

Fan Operating Point:
- Quantity: 89.01 m^3/s
- Pressure: 215 Pa.

Flow Distribution:
- Face quantity: 87.01 m^3/s
- Leakage flow: 2.00 m^3/s.

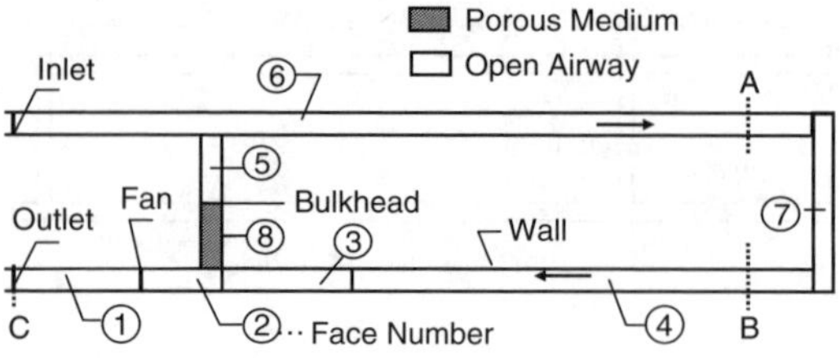

Figure 2. Basic case model specifications.

Based on the above results, 89.01 m^3/s of fresh air is taken from the surface. Of this, 87.01 m^3/s of air is available at the last open crosscut and the remainder is lost through leakage. For rectangular openings of 16 m^2 of cross sectional area, the average air velocities at two sites of the intake drift, before and after the bulkhead, are equal to 5.4 m/s and 5.6 m/s respectively.

5.2 *Modeling using CFD*

The problem is represented by a 2D airflow model. The geometry of this includes eight faces, four fixed-property edges and one porous medium (Figure 2). The Faces are used to represent the drifts and crosscuts through which the air is circulated. The edges are used to specify the fluid properties including the inlet and outlets ends, and the sites where the energy is added (fan) or absorbed (bulkhead). The porous medium is used to characterize a section of the air path where the air velocity is considerably low compared to other paths (face 8 in Figure 2.)

GAMBIT was used to create the geometry of the model. Next, the model was meshed to produce 31177 cells using Map Scheme. Finally, the fluid properties were specified and the model exported. In the next step, FLUENT was run to simulate the fluid flow problem in the geometry created by GAMBIT. Table 3 shows the fluid properties and the boundary conditions used in the simulation process. For prescribed velocity residuals (<0.01 m/s), the solution to the problem was reached after 915 iterations.

Figure 3 shows the velocity profiles in the intake and return drifts. The Intake a graph shows the velocity profile for the flow directed to the face. On the average, this yields a velocity of 5.54 m/s. The Return C graph shows the velocity profile at the model outlet. It yields an average of 6.45 m/s. The difference represents the air velocity through the porous medium (bulkhead.)

A comparison of these results with those generated by VnetPC shows that the face velocities are very close to each other, however, the velocities near the surface connections still differ by about 15%. In FLUENT, this can be adjusted by changing more than one input parameter including the porosity constant, pressure jump, porous medium thickness and wall roughness height. In this case this was corrected by changing the porous medium thickness from 0.2 to 2.0 m.

Table 3. Input parameters for CFD modeling.

Step	Description	Parameter
1	Solver	Segregate Unsteady, 1st order
2	Viscous model	$\kappa - \varepsilon$ and standard wall
3	Materials: Air	Density = 1.2 kg/m^3
4	Op. Conditions	0 Pa, Gage Pressure
5	Boundary Conditions	Inlet: Pressure inlet Outlet: Pressure outlet Booster: Fan, 215 Pa Bulkhead Conditions: – Porosity: Default – Thickness: 0.2 m – Pressure Jump C: 50 Porous Medium Fluid: – Power Law Model – Co = 10, C1 = 1 Wall Roughness: – Height = 0.6 m – Roughness C = 0.99
6	Solution Controls	Pressure: Default Initialize: inlet Time step: 0.5 s No. of time steps: 132 Iterations/step: 20

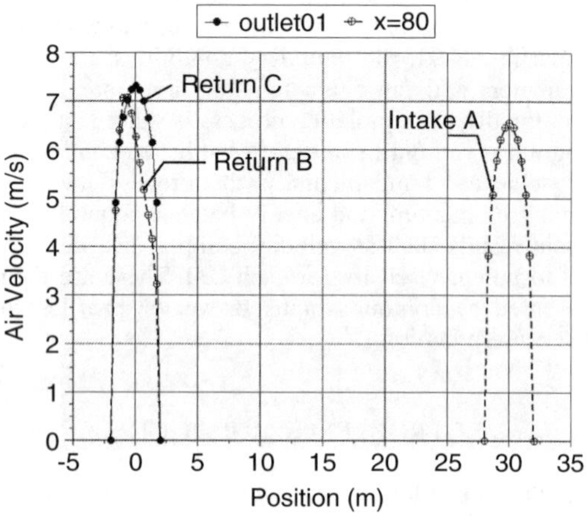

Figure 3. Velocity profiles at points A, B and C of figure 2.

6 AMOLE DEWATERING DRIFTS

P.T. Freeport Indonesia's Amole Dewatering development consists of two parallel drifts driven from the

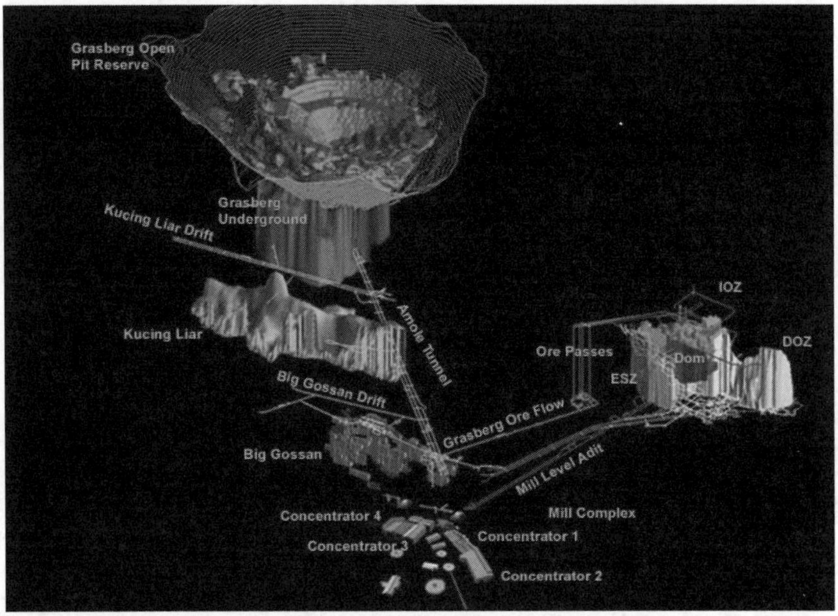

Figure 4. View of the Amole drifts in relation to the PTFI's Grasberg pit.

2930 Mill level. These drifts were developed to drain the Grasberg pit from below, to provide access for the deep drilling, and to delineate the Kucing Liar ore deposit (Figure 4). The drifts, 4.5×4.5 m in cross-section, were driven as twin headings in which one drift is used as an intake and the other as return. Every 150 m, the drifts are connected to each other by short crosscuts. To provide the headings with through flow ventilation, all the crosscuts except the last are blocked by bulkheads, doors and regulators. At cross-cut 16, the system splits into two sets: one to provide an access to the Wanagon/Kucing Liar area and another to intercept the Grasberg pit drain-wells. The drifts extend for more than six kilometers and include three water collection chambers, numerous diamond drill stations, two underground shops and one training center. Depending upon the rainfall and recharge into the ground water system, between 0.63 and 1.58 m^3/s (10,000 to 25,000 gpm) of mine water is drained through these drifts (Carlson, 1998.)

The ventilation system includes 19 doors, 34 bulkheads, three primary fans, and three floating regulators. The doors, both manual and pneumatic of 3×4 m in size, are used to separate the intakes and returns and to provide access between the drifts. Although built and installed to operate under harsh conditions, the doors allow some air leakage especially through doorframes and hinges which are not airtight. The bulkheads are built of concrete bricks as permanent constructions. Depending on their locations, these constructions include drainage holes, man-doors and/or pressure relief windows. These are also sources of air leakage. All primary fans are installed as booster fans in the return drifts, which are also used as drain-way drifts. They are all installed in concrete bulkheads and equipped with man-doors. The floating bulkheads are installed in the drain-way drifts in parallel arrangement with the main fans. Each is constructed to allow the passage of water underneath a bulkhead and minimize the flow recirculation around the fan. The main fan capacity is 139.4 m^3/s of air at a static pressure of 822 Pa. Of this quantity, 10 m^3/s is re-circulated around the floating bulkhead.

Figure 5 shows a line diagram of the main intake and return drifts near the portal area. It also shows the location of bulkheads, doors and regulators used to direct the air to the active workings (beyond crosscut 6). The drain-way drift is the only return way. Table 4 shows a summary of the airflow rates recorded in the main intake and return drifts. Based on these results, about 80% of the fresh air is supplied through an ex-conveyor drift and exploration drift, and the remainder through the Aghawagon portal. Of the total quantity (129.4 m^3/s) only 97.3 m^3/s of fresh air is directed towards the workings, with remaining (32.1 m^3/s) lost in the form of leakage through bulkheads, doors and regulators.

6.1 *Studies of air leakage using VnetPC*

The modeling was restricted to the Amole Mine Portal Area shown in Figure 5. It includes four intakes and

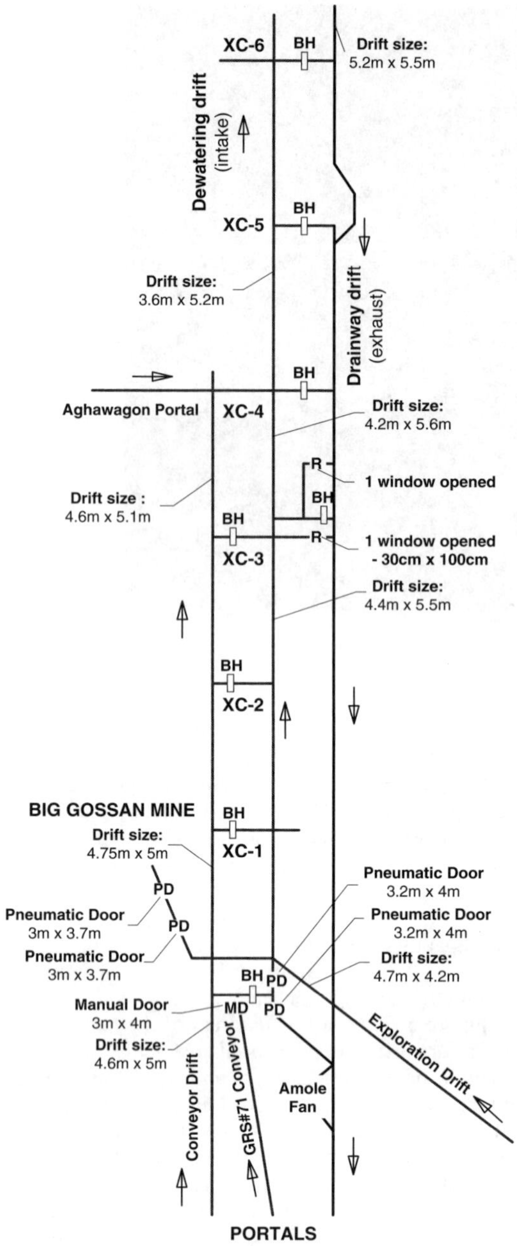

Figure 5. Amole schematic – portal area.

one return. The active area is represented by crosscut 6. For the existing conditions (fan pressure of 822 Pa), the quantity directed to the active area is equal to 97.3 m^3/s of air. This quantity is about 75% of the total quantity circulated through the mine (25% leakage). With VnetPC, a ventilation model is initially constructed, a fixed resistance is assigned to each branch, and the fan

Table 4. Airflow distribution in Amole drifts (Base case).

Airway	Quantity m^3/s	Area m^2	Velocity m/s
A. Intakes			
Conveyor Drift	51.0	18.4	2.8
Exploration Drift	53.8	19.7	2.7
Aghawagon Drift	20.3	17.2	1.2
Big Gossan Access	4.3	15.3	0.3
B. Returns			
Drain way Portal	129.4	14.7	8.8
C. Active Workings			
Near heading*	97.3	18.2	5.4

* Between cross-cuts 5 and 6.

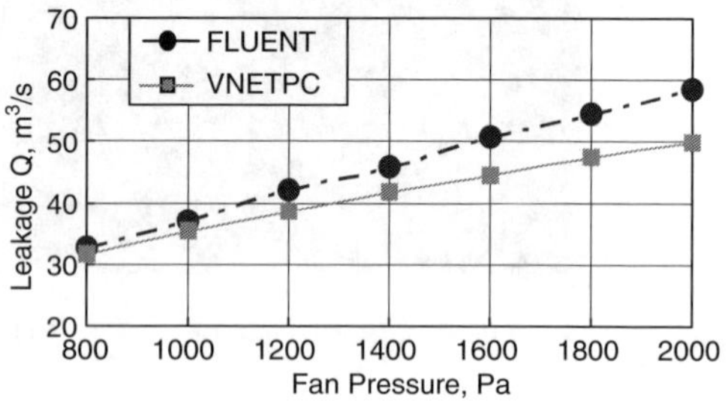

Figure 6. Fan pressure versus leakage quantity relationships.

curve is specified. Next, the model is calibrated to practically replicate the results shown in Table 4. Once the calibration process is completed, the fan pressure was varied between 800 and 2000 Pa and the leakage flow rates were estimated. The results of this study are shown in Figure 6. An evaluation of this graph shows that the leakage quantity increases almost linearly with increases in fan pressure and that the percent loss of fresh air is fairly constant (25% for this section).

6.2 *Studies of air leakage using CFD*

As described previously, the simulation studies using this approach involves the utilization of two programs: GAMBIT to generate the geometry of the model, and FLUENT to solve the model. Figure 7 shows a 2D geometry of the model developed for the Amole Mine Portal Area. This shows the faces used to represent the drifts and crosscuts. The figure also shows the location of the energy source (fan) and the airways with very low air velocities, which in this case represent the bulkheads, doors and regulators (porous zones). Prior to any prediction, the model was calibrated against the field data (Table 4). This was accomplished by modifying the fluid and wall parameters. In the case of a bulkhead, the calibration was

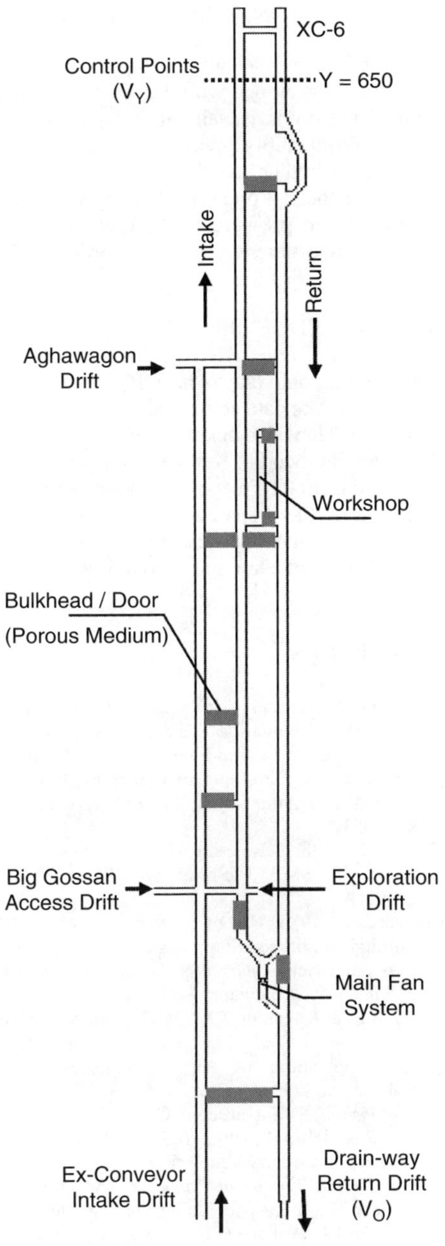

Figure 7. Amole mine geometry portal area.

performed by changing the porous medium thickness. The "correct" airway resistance was determined by changing two wall parameters: the roughness height and roughness constant. Two air velocities were used to estimate the leakage quantity: the velocity near cross-cut 6 (V_Y) and the velocity at the model outlet (V_O). The volumetric efficiency η was then calculated as the ratio of these two velocities multiplied by an area factor A_F, as follows:

$$\eta = \frac{V_Y}{V_O} * A_F \qquad (2)$$

And the leakage quantity determined by:

$$Q_L = (1 - \eta) * Q_T \qquad (3)$$

For a fan pressure of 822 Pa (base case), CFD yielded the following air velocities: $V_F = 6.3$ m/s, $V_O = 12.7$ m/s. Based on these figures, an area factor of 1.52 and a total flow rate (Q_T) of 129.4 m^3/s, the leakage quantity was estimated at 31.8 m^3/s. This quantity compares well with 32.1 m^3/s determined from field measurements.

Once the model was calibrated, the fan pressure was varied between 800 and 2000 Pa and the simulator results plotted to determine average air velocities and leakage flow rates. The results of these trials are superimposed on the VnetPC output in Figure 6.

A comparison of the two graphs shows that the VnetPC simulator produces leakage quantities lower than those determined by FLUENT, with the higher the fan pressure the larger the difference.

Figure 8 shows a velocity vector view for the base case near the model outlet. An examination of this view shows that the doors and bulkheads are not airtight instead they allow some leakage flow especially those located in the vicinity of the main fan. Another finding is that a considerable amount of air is re-circulated through the floating regulator.

7 CONCLUSIONS AND DISCUSSIONS

The leakage flow through bulkheads, doors and regulators can be a serious problem in operating mines. In coal mines, it may reduce the availability of fresh air at the workings by as much as 60%. In metal mines, although reduced by means of booster fans, this quantity can be in the order of 30%.

Every effort must be made to construct bulkheads and install doors and regulators as permanent structures to reduce leakage, especially those located near the main fans.

In ventilation planning, leakage paths are represented by airways of high resistance. This resistance varies with a number of factors including the porosity of the materials used, the design and construction techniques, and the location of each structure in relation to the main fans. In operating mines, the resistance of a "good" concrete bulkhead may vary between 47 and 2220 Ns^2/m^8. It is this range of values that makes leakage flow estimates questionable.

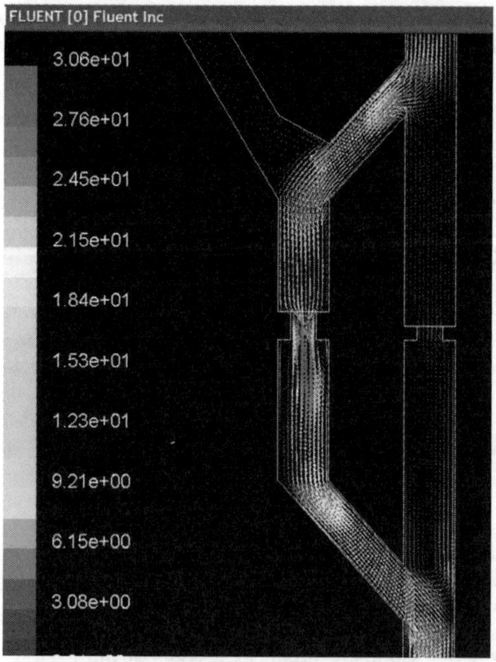

Figure 8. Velocity vectors near the main fan system colored by velocity magnitude (m/s).

Most of the simulators used in mine ventilation treat a leakage path as an airway of high resistance in which the flow rate is calculated based on Atkinson's equation (square law.) This approach can underestimate the leakage quantity.

CFDs can be used to model mine airways including air paths with very low Reynolds numbers. However, the model should be calibrated against field data. This can be accomplished by modifying the fluid and wall parameters. Once calibrated, the model can be used to predict the airflow properties such as velocity components, density and viscosity at almost any location of interest. The degree of approximation to reality (actual mine conditions) is limited only by the power of the computer and the accuracy of the required results.

The leakage flow at PT Freeport Indonesia's Amole Dewatering drifts was investigated using the FLUENT CFD program. A 2D model was built to represent a twin-drift ventilation system. The modeling results showed that leakage paths can be represented by zones of differing porosity or membranes that have known properties and that the leakage quantities can be estimated from velocity profiles. It also showed that this tool can be used to study other mine ventilation problems in more complex geometries and flow regimes such as the airflow through the gob area in a coal mine.

The CFD analyses were compared with a ventilation model developed with the VnetPC software program. A preliminary evaluation of results showed that the leakage quantity predictions using FLUENT are consistently higher than those determined by VnetPC, the higher the fan pressure the larger the difference. When the fan pressure was increased from 822 to 2000 Pa, the difference was in the order of 17%.

ACKNOWLEDGEMENTS

The authors would like to thank P.T. Freeport Indonesia for allowing the data contained within this paper to be published. They also acknowledge the cooperation of the following people: Ketut Karmawan, Dadang, and Riza Sani from PT Freeport – Indonesia for their assistance during the preparation of this paper. The assistance of the University Research Committee (URC) of the University of Utah is also gratefully acknowledged.

REFERENCES

Calizaya F., Karmawan K. & Wallace K.G., 2001. PT Freeport DOZ Mine Ventilation System. In S. Wasilewski (ed.), *Proc. of the 7th International Mine Ventilation Congress*. Research and Development Center for Electrical Engineering and Automation in Mining, Cracow, Poland, 949–953.

Carlson D., 1998. Dewatering of the Grasberg Pit, Irian Jaya, Indonesia. SME Pre-print 98–140. SME, Littleton, CO: 1–4.

Coetzer W.J., 1985. Use of Sealant Paint to Seal Leakage Through Ventilation Stoppings Constructed from Pre-cast Concrete Bricks. *Journal of the Mine Ventilation Society of South Africa*, January: 9–11.

Bruce W.E. & Koenning T.H., 1987. Computer Modeling of Underground Coal Mine Ventilation Circuits, Selection and Application of Airway Resistance Values. In J.M Mutmansky (ed.), *Proc. of the 3rd US Mine Ventilation Symposium*, SME, Littleton, CO: 519–524.

Deglon P. & Hemp R., 1992. An Evaluation of Parameters to Be Used in Colliery Ventilation. *Proc. of the 5th International Mine Ventilation Congress*. The Mine Ventilation Society of South Africa, Johannesburg: 369–376.

Duckworth I.J., Wallace G.K. & Wise R., 1995. Ventilation Planning and Design of the Skyline Mines. In A.M Wala (ed.), *Proc. of the 7th US Mine Ventilation Symposium*, SME, Littleton, CO: 9–14.

Richardson M.A., Gilbride L., Adair L. & Glasson M.W., 1997. Ventilation Planning at the Aberdeen Mine. R.V. Ramani (ed.), *Proc. of the 6th International Mine Ventilation Congress*, SME, Littleton, CO: 33–38.

Van der Bank P.J., 1983. Moving Air from Surface to the Last through Road in the Section. *Journal of the Mine Ventilation Society of South Africa*, June: 53–58.

Mine Ventilation, Ganguli & Bandopadhyay (eds)
© 2004 J.C. Edwards, ISBN 90 5809 633 5

Real-time neural network application to mine fire – nuisance emissions discrimination

J.C. Edwards, R.A. Franks, G.F. Friel, C.P. Lazzara & J.J. Opferman
National Institute for Occupational Safety and Health,
Pittsburgh Research Laboratory, Pittsburgh, USA

ABSTRACT: The National Institute for Occupational Safety and Health (NIOSH) implemented a real-time neural network system which can discriminate mine fires from nuisance diesel emissions as part of an atmospheric mine monitoring system in NIOSH's Safety Research Coal Mine. The real-time response of a neural network to fire sensor outputs was demonstrated for coal and belt combustion in the presence of diesel emissions. The fire sensors consisted of an optical path smoke sensor, a carbon monoxide (CO) sensor, and two types of metal oxide semiconductor (MOS) sensors. The real time neural network was trained with coal, wood, and belt fire experiments with and without diesel emissions background. The trained neural network successfully predicted mine fires with these combustibles in the smoldering stage prior to the onset of flames.

1 INTRODUCTION

The discrimination of mine fires from nuisance signatures such as emissions from diesel equipment, flame cutting and welding operations, and battery charging operations is an ongoing problem for early and reliable mine fire detection. Currently most in-mine fire detection systems use thermal or CO sensors. Thermal sensors are useful in the proximity of flaming combustion. CO sensors are subject to false signals from CO diesel equipment nuisance emissions and H_2 cross-interference from battery charging operations. Multiple type fire sensors and statistical methods, including neural network, have been used for mine fire detection and the elimination of nuisance emissions in non-mining applications (Ishii et al., 1994, JiJi et al., 2003). As discussed previously (Edwards et al., 2002, Friel and Edwards, 2002), a set of three sensors and a neural network was determined to predict the occurrence of a mine fire in the presence of diesel emissions. These sensors were: a carbon monoxide sensor; an optical path smoke sensor; and a MOS sensor with bimodal response to products-of-combustion (POC) from mine fires and NO_x from diesel equipment. These evaluations were based upon coal and conveyor belt combustion experiments in the presence of diesel emissions. Subsequent to that research, additional experiments included wood combustion. A new neural network (NN) was constructed based upon an optimum choice of sensors. These included the previous three sensors and a second MOS sensor. The new NN uses directly the CO sensor, optical smoke sensor, and bimodal MOS sensor output, the product of the CO sensor and optical smoke sensor outputs, and the product of the outputs from the bimodal MOS sensor, labeled 2105, and a MOS sensor, labeled 2600, which is more responsive to H_2 than CO. The utility of the program can only be successful for underground mining applications if predictions can be made accessible to the mine personnel in real-time. This required the implementation of a real-time sensor data interpretation program for the NN executable file. The objective of this research was to demonstrate how this could be accomplished.

2 SENSOR TYPES

For mine fire detection the available sensor types include gas, smoke, and thermal sensors. Although thermal sensors are widely used, their range is limited to the immediate proximity to a fire in its flaming combustion stage. A sensor commonly deployed when belt entry air is used at the mine working face is a CO sensor. An advantage of a CO sensor is its capability to quantify a fire in terms of the molecular concentration of a product gas species. A CO sensor is especially useful for early detection of a deep-seated spontaneous coal heating where smoke concentrations are low. Disadvantages are its response to emissions from diesel equipment, flame cutting, welding, and its susceptibility to interfering gases, such as H_2 from battery charging operations. Smoke sensors, which are either optical or ionization, are responsive, stable, and easy to check. The ionization smoke sensors are usually based upon the decay of an alpha emitter such as Am^{241} or a beta emitter such as Kr^{85}. A disadvantage of an ionization smoke sensor is the requirement for a license for the radioactive source dependent upon the strength of the source. Optical smoke sensors can be either a point or a path type. The point type sensor is based upon optical scattering, and the path type is based upon optical obscuration. An advantage of the path sensor is its ability to probe the cross section of an entry. A disadvantage for a path optical smoke sensor is the optical path distance, which can create a problem for its installation in a mine. Nuisance emissions, which can affect smoke sensors, include dust, water vapor, and particulate emissions from diesel equipment and flame cutting, and for the case of an optical path sensor the potential blockage by personnel and equipment. Optical smoke sensors, which are generally in the infrared optical range, are very responsive to larger smoke particulates produced by smoldering combustion, and ionization smoke sensors are generally more responsive to smaller smoke particulates produced by flaming combustion. The demarcation size is about 0.3 micrometer. Because of the optical sensor's response to larger particulates, it will be less responsive to smaller diesel particulate emissions.

Another class of sensors which respond to mine fire gases are MOS sensors. These sensors are very responsive, but not very selective in their response. Their responsiveness can be ordered in terms of their resistance change to various oxidizing and reducing gases. Methods to vary their selectivity include the addition of a catalyst, sensor element grain size selection, heating of the sensor, and filters. Another advantage of the MOS sensor is its compact size. Its cylindrical structure has a height and diameter of about 1.3 cm. A disadvantage of a MOS sensor is the high temperature, 250 to 380°C, of the sensing element. A MOS sensor, which is sensitive to NO_x, has an increase in its electrical resistance associated with adsorption of oxygen due to electron transfer from the surface element to the oxygen and the buildup of a positive space charge on the element surface. The presence of deoxidizing POC gases will remove the adsorbed oxygen and result in reduced resistance. To be useful, the MOS's sensitivity to NO_x must be sufficient to overcome its sensitivity to CO in diesel emissions, otherwise the discriminating capability of the sensor will be compromised. Generally there are gas concentration ranges over which the sensor's resistance element will respond with a power dependence upon the adsorbed target gas concentration. For an extended range of concentrations this nonlinear logarithmic dependence makes the prediction of the net effect problematic, and direct experimental verification is required.

In a refueling area where rapid flaming combustion most likely will occur, optical spectral emission sensors could be useful. In this case the distance between the source fire and sensor will be relatively short, and infrared signal attenuation by smoke will initially be minimal.

The ordinary use of fire sensors depends upon defined alarm values. Fire sensors for in-mine use have established alarm values (Code of Federal Regulations, 2001). These include a CO alert value of 5 ppm above ambient, a CO alarm value of 10 ppm above ambient, a smoke optical density of $0.022\,m^{-1}$, and a temperature of 165°F.

3 MINE FIRE DETECTION METHODS

There are three approaches for utilization of sensors for mine fire detection in the presence of nuisance emissions. One approach is to increase the sensor alarm and alert values to compensate for nuisance emissions. Without an exact knowledge of the nuisance emissions' gaseous and particulate concentrations, which will be variable in response to the operation of the emissions producing equipment under variable power loads, early and reliable fire detection will be compromised. For example, this approach would be expected to miss early smoldering stages of a mine fire. A second approach is rule making. This approach would rely upon an understanding of the generation of emissions from nuisance producing sources and combustion sources. For example, this approach could examine rates of change in a measurable emission, such as CO, so as to differentiate CO emissions from a diesel and from a mine combustible. However, the rate of change of a fluctuating time variable quantity will fluctuate considerably and can produce indistinguishable events. Another example is to rely upon the measured historical relationship between two emission components, such as NO and CO from a diesel engine. This approach can depend upon the growth rate of the fire (Edwards et al., 1999). A third approach is information processing. This approach is useful when multiple type sensors with nonlinear responses to various POC are used. In this approach a relationship between the classification of an event and the sensors' inputs can be established with a method that recognizes patterns for known data sets. An advantage is that a number of mine fire-nuisance emission experiments can be conducted to establish the functional relationships. One such information processing approach is a NN approach. Based upon these relationships classifications of unknown events similar to those for which the relationships are developed can be made. These classifications can be quantified with a probabilistic interpretation.

4 NN PROGRAM

A NN program named NeuroSolutions developed by NeuroDimension, Inc[1] was applied to fire prediction. The trained perceptron NN that was implemented was composed of thirty-eight processing elements that produced three output classifications in response to five inputs. The NN program had two-hidden-layers with 20 processing elements (PE) in the first layer and 10 PE in the second layer. A hyperbolic tangent function was used as the activation function, which operated on the linear combination of inputs to the PEs in the hidden layers. The output classification is a result of the application of a softmax function to the inputs to the PEs in the output layer. The softmax function is a probability function based upon exponential weights. Applications of the NN method require the determination of the internal weights between the input layer of sensor data and the first hidden layer, between the sequential hidden layers, and between the last hidden layer and the output layer. A backpropagation method for the errors associated with the feedforward of the input training is used to determine the weights for the PEs during the training stage. The testing stage or predicting stage uses a forward propagation method. The five inputs consisted partially of the responses of the CO sensor, the optical path smoke sensor, and the bimodal MOS sensor, all normalized by their ambient values. In addition two product functions of sensors were included. These were the product of the responses of the CO and optical smoke sensor, and the product of the bimodal MOS sensor and a MOS sensor responsive to H_2 and CO. A trained NN executable file was constructed by iterating on the training data set with known output classifications. Based upon a training data set, which consists of six in-mine combustion experiments in the presence and absence of diesel emissions, an optimum set of sensor type inputs was determined which would defect the test fires and yield the minimum number of false alarms. The combustion fuel sources were coal, coveyor, belt, and wood.

5 DATA EVALUATION

Each sensor's analog output voltage is converted into a digital signal by an accessory card for processing by the mine monitoring system. The range of the analog signal is subdivided into 255 subdivisions associated with the 8 bits of electronic information available for data representation. The subdivision represents the minimal signal variation that can be processed. For example, if the sensor's output signal range is 0 to 5 volts, then the minimum signal change, which can be processed, is 19.6 mV. As part of the mine monitoring system, an applications computer program was prepared which managed the real-time sensor data for inclusion into the NN executable file. To process the input data for the NN, the background average is computed for each sensor's response based upon a specified time increment. This time was generally selected to be about 5 minutes prior to start of the experiment. After the average values were determined for the sensors, the program proceeded to normalize the subsequent input by the average values. For all sensors exclusive of the CO sensor, this data normalization was accomplished by a division of the sensor output by the sensor's average ambient value. For the CO sensor, the ambient value was subtracted. The normalized values were found to yield more consistent input data for the NN than the actual signal values, since changes relative to ambient are important for characterization of the event. The program also developed the appropriate multiplicative combinations from the sensor data as inputs for the NN. The NN output was displayed on a monitor as columns of probability values for clear air, diesel emissions, and fire POC. When a fire was identified the monitor displayed a box with a fire alert message. This system was successfully evaluated not only with previously acquired data, but also in real-time while a mine fire detection experiment was in progress.

6 EXPERIMENTAL METHOD

Figure 1 shows a plan view of the section of the Safety Research Coal Mine (SRCM) in which the experiments were conducted. To establish a sensor atmosphere which contained diesel emissions, a diesel locomotive was positioned in B-Butt either upwind of the split formed by 10-Room, or at the entrance to 10-Room. In the latter case a tube was connected to the locomotive's exhaust port to direct the emissions directly into 10-Room. Identical fire sensor types were located at S1 and S2, which were 18 m and 148 m downwind of the fire zone. The average height and width of 10-Room were 2.0 m and 3.9 m, and the average height and width of F-Butt were 1.9 m and 4.5 m. The average air flows for the experiments were 3.0 m^3/s in 10-Room and 4.4 m^3/s in

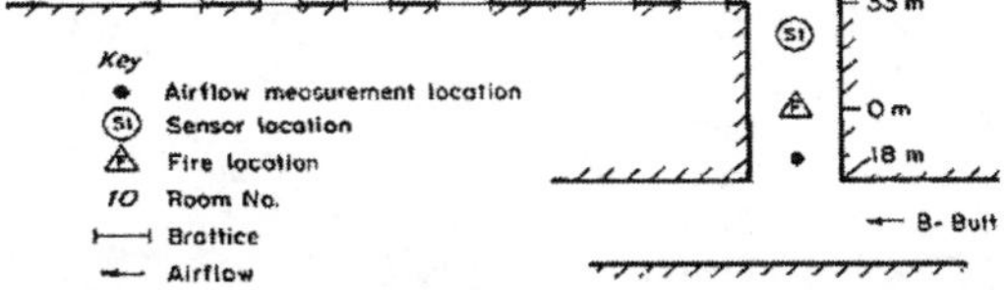

Figure 1. Plan view of the SRCM section.

[1]Mention of any company or product does not constitute endorsement by NIOSH.

F-Butt. Table 1 lists the measured airflows for the experiments. The flow ratio refers to the airflow increase in F-Butt compared to 10-Room. The `higher airflow in F-Butt was due to the leakage around brattices which block the crosscuts from parallel airways into F-Butt. The dual sensor stations with gas and smoke signature dilution between stations S1 and S2 provided a method to define two experiments for one mine fire combustion event. There was an average 47 pct increase in the ventilation between the two sensor stations.

The fires were contained in a 0.61 m square tray. In order to provide a slow heating of the solid fuels which would transition through a smoldering stage prior to a flaming combustion stage, a controllable heating source was maintained with the use of electrically powered strip heaters. For the coal fires the heater elements were embedded in approximately 14 kg of Pittsburgh Seam Coal with a diameter less than 5 cm. About 1 kg of coal fines was distributed over the coal. For the conveyor belt combustion experiments about 0.5 m square samples of belt 1.1 cm thick with a 3 kg mass were attached to a steel plate which was heated from below by the electrical strip heaters. For one combined coal and belt combustion experiment a belt sample about 0.3 m by 0.5 m was positioned on top of the coal.

Table 1. Mine fire detection experiments listed by combustible, presence or absence of diesel emissions, and airflows.

Test no. and sensor station	Combustible	Diesel emissions	Flow, m^3/s	Flow ratio
T71S2	Coal	no	4.3	1.7
T74S1	Coal	yes	2.7	
T74S2	Coal	yes	5.4	2.0
T75S1	Coal	yes	2.0	
T75S2	Coal	yes	3.9	1.9
T76S1	Coal	yes	3.3	
T76S2	Coal	yes	4.2	1.3
T77S1	Belt	yes	3.3	
T77S2	Belt	yes	5.7	1.7
T78S1	Belt	yes	1.9	
T78S2	Belt	yes	3.8	2.1
T79S1	Belt	yes	2.8	
T79S2	Belt	yes	5.4	1.9
T81S1	Wood	no	1.7	
T81S2	Wood	no	3.5	2.1
T82S1	Wood	yes	2.1	
T82S2	Wood	yes	4.8	2.3
T83S1	Wood	yes	2.9	
T83S2	Wood	yes	2.8	1.0
T84S1	Wood	yes	3.1	
T84S2	Wood	yes	4.5	1.4
T85S1	Belt	no	3.1	
T85S2	Belt	no	4.2	1.4
T87S1	Coal	yes	5.6	
T88S1	Coal/Belt	yes	4.2	
T89S1	Coal	no	3.7	

The wood combustion experiments utilized twenty eight oak sticks, 0.46 m long and approximately 1.6 cm square cross section, cut from mine support timber arranged in a two-layer crib set with an approximate total mass of 2 kg. The wood crib was placed upon five electrical strip heaters. The identification of the experiments is provided in Table 1.

7 RESULTS

The training data set for the NN consisted of experiments T71S2, T75S1, T79S1, T81S1, T82S1, and T85S1. This data set includes coal, conveyor belt, and wood fires in the presence and absence of diesel emissions. The trained neural network was used to evaluate the experiments previously conducted, and used in a real-time mode for fire prediction in experiments T87S1, T88S1, and T89S1. A conservative fire alert was marked by the probability exceeding 0.5 even though a probability of a fire near 1/3 could exceed the probabilities of the other two events.

Figure 2 shows the response of the sensors and NN probability at station S1 to a developing primary coal fire and secondary belt fire in the presence of diesel emissions as part of experiment T88S1. The power supplied to the heater elements varied from 1.6 kW to 2.9 kW over a 20 min time period. The presence of diesel emissions results in a significant response of the two MOS sensors, an increase of the CO concentration above the 5-ppm alert concentration, and an insignificant response of the optical path smoke sensor prior to heating the combustibles. Associated with the visual observation of smoke results was a rapid increase in the 2600 sensor's response, and less rapid rates of response for the CO, optical smoke, and 2105 sensors. The onset of flaming coal combustion resulted in

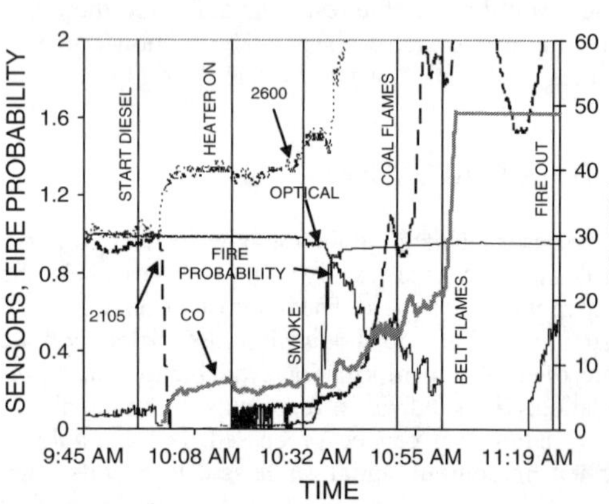

Figure 2. Sensor responses at station S1 and NN fire probability for coal and belt combustion in presence of diesel emissions.

visually obscuring smoke which saturated the optical smoke sensor's response. The identification of a fire with the NN probability curve occurred 5 min after the visual observation of smoke, and 16 min prior to coal flaming combustion. The flaming belt combustion occurred 10 min after the flaming coal combustion. When the NN identified the fire, the optical density was 0.012 m^{-1}, which is less than the 0.022 m^{-1} smoke optical density alarm value specified in 30CFR75.344 and 30CFR75.340 (Code of Federal Regulations, 2001). The maximum temperature measured at the roof above the fire was 98°F, which occurred during the belt flaming stage. This temperature is less than the thermal alarm temperature of 165°F. In this particular example, the clear advantage of the NN approach is its ability to discriminate the false CO alert value associated with the diesel emissions, and identify the incipient smoldering coal combustion stage prior to its flaming combustion, and the subsequent flaming belt combustion.

Figure 3 shows the response of the sensors at station S1 and the NN probability curve to a belt fire in the presence of diesel emissions for experiment T78S1. In this case the ventilation was adjusted to produce CO concentrations from the diesel exhaust in excess of 30 ppm prior to heating of the belt. This exceeds the 30CFR75 defined 10 ppm alarm value. Smoke emissions were observed when the temperature of the belt surface in contact with the heated plate reached approximately 560°F, and flaming combustion commenced when the temperature reached approximately 870°F. The electrical power applied to the heaters varied from 1.1 kW to 3.3 kW over a 40-min time period. The NN probability curve identified the fire 7 min after the visual observation of smoke, and 19 min prior to belt flaming combustion. At the NN identification of the fire the optical density was 0.0061 m^{-1}. A comparison of the maximum CO concentration for the experiments in figures 2 and 3 prior to heating of the fuel show measured CO values of 8 and 36 ppm respectively. In figure 2 a CO concentration of 36 ppm, which was an average of two CO sensors at the station, was not reached until flaming coal combustion occurred. This illustrates the uncertainty in increasing a fire sensor's alarm alert and alarm value. Although there will be a scaling of the CO concentration with ventilation change, this scaling provides one more parameter which would need to be known with certainty prior to identification of a fire produced CO alarm concentration.

To evaluate the effectiveness of the NN for identification of a variable quantity nuisance emissions source, the trained NN was used to identify a nuisance event, which consisted of a diesel locomotive moving along B-Butt past the air split of 10-Room. Figure 4 shows a minimal deviation of the optical smoke sensor to the diesel emissions, and the relatively sharp responses of the CO and 2105 sensor to the emissions. The NN probability curve correctly identifies the diesel emissions signature. Its response is coincident with the CO and 2105 responses. The fire probability curve shows that in no case is there a significant indication of a fire. The advantage of a probability display is the presentation in real-time of a continuous source of information, which mine personnel can use to make decisions. For example, the continuous fire probability curve can be subdivided into regimes of low, medium, and high fire alarm. The current application program provides a real-time visual display of the probability values for fire, diesel emissions, and clear air conditions.

A listing of all the mine fire experiments by combustible, presence or absence of diesel emissions and airflows is shown in Table 1. For the experiments conducted, the NN prediction of a fire is evaluated in terms of the prediction of the fire relative to the first visual observation of smoke and flames. Figure 5 shows the NN prediction of a fire based upon the fire probability exceeding 0.5 as a lag time relative to the time

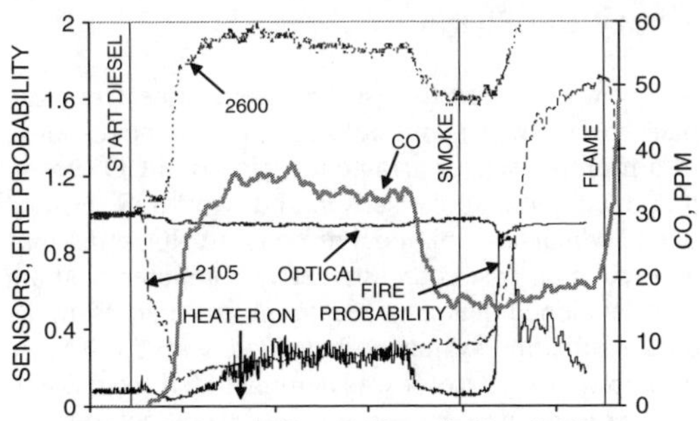

Figure 3. Sensor responses at station S1 and NN fire probability for belt fire in presence of diesel emissions.

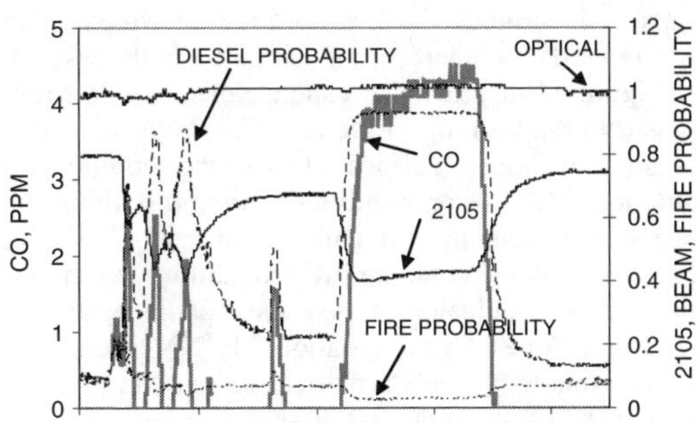

Figure 4. Sensor responses at station S1 and NN fire probabilities for diesel locomotive variable emissions.

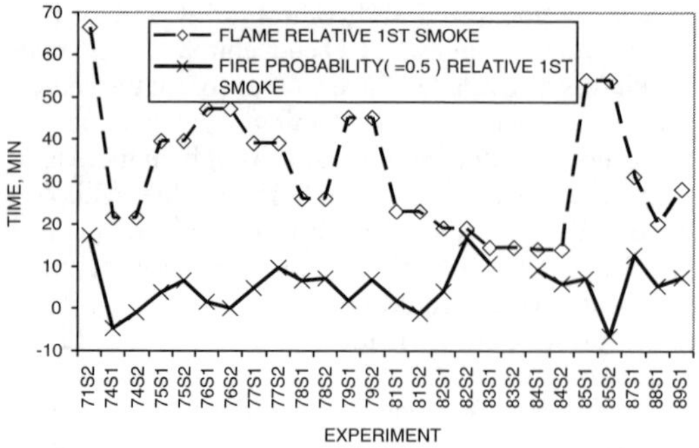

Figure 5. NN prediction of fire relative to first observation of smoke, and flaming combustion relative to smoke for each experiment.

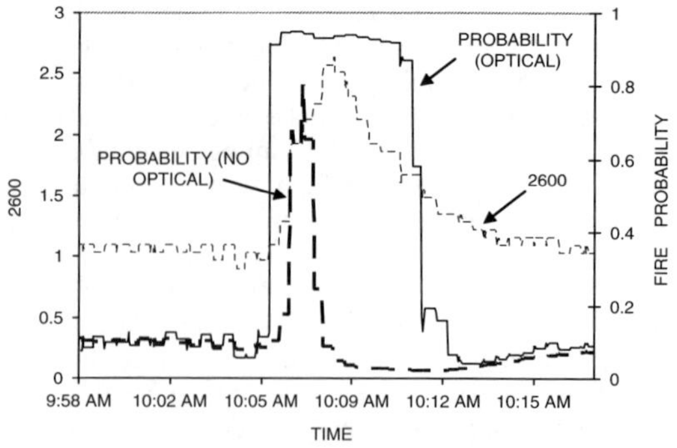

Figure 6. NN prediction of rag burn with and without optical path smoke sensor.

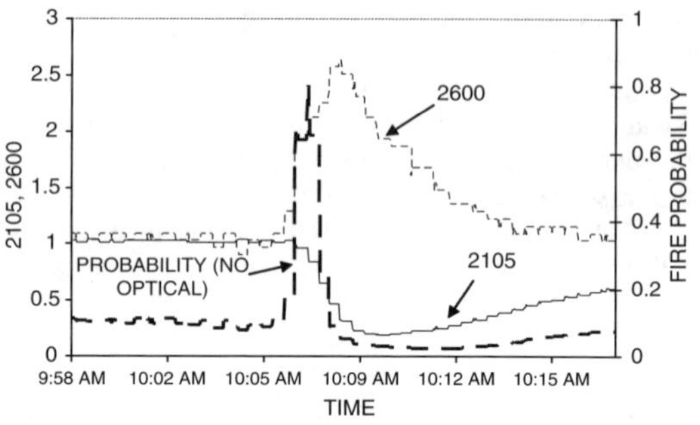

Figure 7. Comparison of NN fire prediction with response of 2105 and 2600.

of first observation of smoke. (Notation T in table 1 is omitted from experiment number on coordinate axis in Figure 5).

Also shown in the Figure is the lag time of the flaming combustion relative to first smoke. For example, for experiment T71S2 the fire probability is 17 min after the first observation of smoke, and flaming combustion occurs 67 min after smoke production. Although for experiments T74S1, T74S2, T81S2, and T85S2 the fire is identified prior to the first observation of smoke, its occurrence is after the heating of the combustible was initiated. For 25 of the 26 experiments the NN successfully predicted the occurrence of a fire. Only for one experiment, T83S2, did the NN fail to identify the fire, and that was due to a relatively small wood fire with dilution of the POC at S2. For the experiments listed in table 1, the average optical density at which the NN predicted a fire was $0.0059\,m^{-1}$ with a standard deviation (SD) of $0.0050\,m^{-1}$. This average value is more than three SD s less than the minimum smoke alarm value of $0.022\,m^{-1}$ specified in 30CFR75.

The most sensitive sensor for fire detection in the selected set of sensors is the optical path smoke sensor. This is illustrated with a simple test in which a rag was ignited upwind of sensor station S1 in the SRCM.

Figure 6 illustrates the response of the 2600 MOS sensor to the burning of the rag. The duration of this sensor's response defines clearly the combustion duration. The fire probabilities with and without the response of the optical path sensor are shown in Figure 6. It is seen that the fire probability determination with the inclusion of the optical path sensor correctly envelopes the fire duration. The NN calculation which excludes the optical path sensor results in a much shorter indication of the fire duration.

Figure 7 illustrates the reason for this result. The 2105 MOS sensor responds to the NO_x compounds generated by the burning rag. There appears to be unknown nitrogen additives in the cloth. As the NO_x concentration increases, the fire probability evaluation in the absence of the optical smoke sensor incorrectly identifies the combustion as a nuisance event.

8 CONCLUSIONS

A NN was trained to discriminate a mine fire from background diesel emissions based upon the response of a multiple sensor array consisting of a CO sensor, an optical path smoke sensor, and two MOS sensors, one of which has a bimodal response to NO_x emissions and reducing gases. An interactive computer program was developed that can accept real-time data from the data acquisition system and evaluate with the NN the probability of a fire. It was demonstrated from applications to coal, conveyor belt, and wood combustion experiments in the presence and absence of diesel emissions that a smoldering fire prior to flaming combustion could be predicted for 25 of 26 fire detection experiments conducted.

REFERENCES

Code of Federal Regulations, 30 CFR, Parts 75.340, 75.344, and 75.351, 2001. Office of the Federal Register, National Archives and Standards Administration, U.S. Government Printing Office, Washington, D.C., July 1, 2001.

Edwards, J.C., Franks, R.A., Friel, G.F., Lazzara, C.P., and J.J. Opferman. Mine Fire Detection in the Presence of Diesel Emissions. In J.C. Tien (ed), Proc. 8th U.S. Mine Ventilation Symposium, The University of Missouri-Rolla, June 11–17, 1999, (University of Missouri-Rolla Press, Rolla, MO), pp. 295–301.

Edwards, J.C., Franks, R.A., Friel, G.F., Lazzara, C.P., and J.J. Opferman. In-Mine Evaluation of discriminating Mine Fire Sensors. In E. DeSouza (ed), Proc. 1st North American & 9th U.S. Mine Ventilation Symposium, Ontario, Canada, June 2002, (A.A. Balkema, Lisse, 2002) pp. 527–532.

Friel, G.F., and J.C. Edwards. Neural Network Application to Mine-Fire Diesel-Exhaust Discrimination. In E. DeSouza (ed), Proc. 1st North American & 9th U.S. Mine Ventilation Symposium, Ontario, Canada, June 2002 (A.A. Balkema, Lisse, 2002) pp. 533–538.

Ishii, H., Ono, T., Yamauchi, Y., and S. Ohtani. Fire Detection System by Multi-Layered Neural Network with Delay Circuit, Fire Safety Science-Proc. of the Fourth International Symposium, pp. 761–772, 1994.

JiJi, R.D., Hammond, M.H., Williams, F.W., and S.L. Rose-Pehrsson. Multivariate Statistical Process Control for Continuous Monitoring of Networked Early Warning Fire Detection (EWFD) Systems, Sensors and Actuators B vol. 93, pp. 107–116, 2003.

Mine Ventilation, Ganguli & Bandopadhyay (eds)
© 2004 Taylor & Francis Group, London, ISBN 90 5809 633 5

Ventilation savings with fuel cell vehicles

G. Righettini & P. Mousset-Jones
Department of Mining Engineering, Mackay School of Mines, University of Nevada, Reno, Nevada, USA

ABSTRACT: Diesel powered vehicles have made underground mines more efficient, but with the advent of Diesel Particulate Matter (DPM) regulations, in most cases, extra ventilation and/or engine emissions control technology will be required. Using fuel cell powered vehicles will not need these measures to be taken. Nine US underground mines of varying production rates and mining methods were studied to determine possible ventilation reductions if their diesel fleets were replaced with identical fuel cell powered vehicles. Factors that might limit ventilation reductions, such as heat and the effects on other pollutants were determined for each mine. The ventilation cost savings and other benefits of fuel cell vehicles show that when they are commercially available they have the potential of being viable alternatives to diesel vehicles.

1 INTRODUCTION

Underground mines today use large numbers of diesel powered vehicles in their mining operations. Diesel powered vehicles have made underground mining more efficient and productive since they were first introduced and have changed the way mines are designed and operated. However, diesel engines require extra ventilation to control emissions of CO, NOx, and Diesel Particulate Matter (DPM), and this adds a significant cost to their operation underground. In most cases, the surface and underground fan power requirements are the single largest consumers of electricity in an underground mine.

The main benefit of fuel cell powered vehicles in underground mines is the reduced ventilation needs, but there are other benefits, which can be considerable in their own right. A fuel cell power system is more efficient than diesel power and, therefore, it outputs less heat into the air, which can be a considerable benefit in underground mines with high rock temperatures, since it reduces the need to air condition the working areas in a mine with refrigerated air. Furthermore, fuel cell power offers some advantages over conventional electric powered underground equipment using an umbilical cord power supply, which can have a high maintenance cost. A further advantage of removing diesel engines from underground mines is the significant reduction in noise levels since fuel cell/electric motor systems are much quieter. In addition, eliminating diesel engines would result in a healthier working environment. Quantifying such a benefit is difficult but it is a very definite benefit which should not be ignored.

2 METHODS OF INVESTIGATION

The information gathered for each mine usually included a computer model of the ventilation system made with VnetPC 2000, a mine ventilation simulation software program from Mine Ventilation Services, Inc. of Fresno, California. VnetPC 2000 was used to see what effect reducing the ventilation would have on various parts of the mine and in some cases to calculate the fan operating costs.

If heat appeared to be an issue then CLIMSIM for Windows (Version 1.0), a mine climate simulation program from Mine Ventilation Services, Inc. was used to analyze the contributions to the heat load from diesel and fuel cell powered equipment. This helped determine how much the ventilation air flow could be reduced without causing an increase in current underground air temperatures if fuel cell vehicles replaced diesel units. Examples of CLIMSIM model inputs are the entry dry bulb and wet bulb temperatures, barometric pressure, air flow quantity, length, cross-sectional area, and perimeter of the airway, the depth of the airway, the friction factor, wetness factor, age of the airway, virgin rock temperature, geothermal step, rock conductivity and diffusivity. CLIMSIM allows airways to be linked together to form a network.

CLIMSIM includes algorithms, which can calculate the amount of sensible, and latent heat output into the mine air by diesel equipment and conventional electric powered equipment. However, in order to be able to use CLIMSIM to estimate the workplace air temperature impact of a fuel cell powered unit, it was necessary to modify the values used in the diesel engine input data format in the following manner: CLIMSIM's input menu for diesel engines was used, but the engine power in kW's was divided by 2.8 since diesel engines put out 2.8 times as much heat as an equivalent electric motor. After dividing the engine power by 2.8 the quotient was multiplied by 1.2 to account for the heat output by the fuel cell stacks. The water vapor from a fuel cell was calculated by using a water/fuel ratio of 9.5 (Kocsis, 2003). This results in a reasonable estimate of the sensible and latent heat output of a fuel cell powered vehicle, however, more accurate values need to be determined by underground monitoring of an operating fuel cell vehicle.

3 LIMITING FACTORS

The mines were studied to see what limiting factors there might be for reducing mine ventilation. Some of the mines in the study have gases such as radon, CO_2, and H_2S to deal with, and dust control is always a requirement. When necessary, the effect on underground air temperatures was analyzed with CLIMSIM to make sure there would be no increases in workplace temperatures if the mine ventilation was reduced.

Other limiting factors used in this study included maintaining minimum air velocities of 0.3 m/sec in drifts and haulage ways where blasting did not take place, and 0.5 m/sec in production drifts where blasting takes place. These are typical recommended velocities used in mines to dilute dust and gases, and prevent stratification and ensure timely removal of blasting fumes.

Some of the mines in the study also plan their ventilation so as not to exceed a certain wet bulb temperature, wet bulb globe temperature, or effective temperature. These are identified in the Results section for the individual mines and this study used the same maximum allowable temperatures in the CLIMSIM modeling for these mines.

4 MINES IN THE STUDY

The mines in the study were chosen to represent a cross section of mining methods and scales of operation. The Mines included are shown in Table 1. Visits were made to all of these mines to gather information about their operations and ventilation systems.

5 RESULTS

5.1 *Turquoise Ridge Project*

The Turquoise Ridge Project is operated by Getchell Gold Corporation, a member of the Placer Dome Group. It is located approximately 75 km northeast of Winnemucca, Nevada. The mine was partially developed several years ago to mine a new lower zone but then it was put on standby during a period of low gold prices. At the present time the company has approved the re-opening of the mine which is currently underway, although at a lower production rate and using a different mining method than the one used in this ventilation analysis. When the mine was originally being planned a number of production

Table 1. Mines in the study.

Mine	Operator	Location	Annual production	Mining method	Product	Diesel fleet, kW
Turquoise Ridge	Getchell Gold Corp.	Nevada	1.325 million tonnes	Long hole stoping	Gold	6941
Deep Post	Newmont Mining Company	Nevada	497,000 tonnes	Drift and fill and bench stoping	Gold	8903
Meikle-Rodeo	Barrick Goldstrike Mines Inc.	Nevada	1.49 million tonnes	Bench stoping and drift and fill	Gold	13,986
Henderson	Phelps Dodge Corp.	Colorado	7.6 million tonnes	Panel caving	Molybdenum	15,828
Questa	Molycorp, Inc.	New Mexico	2.0 million tonnes	Panel caving	Molybdenum	1420
Galena	Coeur Silver Valley Inc.	Idaho	185,000 tonnes	Cut and fill	Silver-copper	680
Lucky Friday	Hecla Mining Co.	Idaho	120,000 tonnes	Cut and fill	Silver-lead-zinc	2671
Avery Island	Cargill Deicing Technologies	Louisiana	2.54 million tonnes	Room and pillar	Salt	10,947
Kerford	Kerford Limestone Co.	Nebraska	1.85 million tonnes	Room and pillar	Limestone	8645

rates, mining methods, and ventilation scenarios were considered. One of the plans considered was to mine 3630 mtpd using long hole stoping, which would have required 528.9 m^3/sec of ventilation air for the new mining areas. A new 6.1 m diameter exhaust ventilation shaft would have been required, along with two 1650 kW surface exhaust fans, and a dedicated fresh air intake drift to reduce air velocities in the haulage drifts. Figure 1 is a VnetPC schematic of this ventilation plan using diesel equipment.

Previous consultants calculated the total ventilation requirements for diesel equipment by using 0.063 m^3/sec/kW for the primary diesel loading, hauling, and backfilling equipment. A "mining allowance" of 0.047 m^3/sec/mtpd was added to account for removal of blasting fumes and dust. Finally, a miscellaneous allowance equal to 25% of the diesel and mining allowance was included.

No factors, such as heat or hazardous gasses, which might otherwise limit the reduction in ventilation, have been identified. Accordingly, the ventilation requirement for fuel cell powered vehicles was calculated using a minimum air velocity of 0.5 m/sec in the production headings. The production headings would be 3.96 m × 3.96 m and 22 headings would be sufficient to meet the production requirements using long hole stoping. Therefore, the production headings require 172.5 m^3/sec of ventilation quantity. Miscellaneous needs of 25% were added to the production heading ventilation requirements, bringing the total minimum ventilation requirement for fuel cell vehicles to 215.6 m^3/sec.

Along with the reduction in ventilation it was realized that reductions in airway sizes and main exhaust fans would also be possible. The VnetPC model of this mine was modified by reducing the exhaust shaft diameter from 6.1 m to 4.27 m. The parallel intake air way and one of the main exhaust fans were also eliminated. This still resulted in 240 m^3/sec going to the lower zone mining area, more than the minimum of 215.6 m^3/sec for fuel cell equipment.

The mine pays \$0.07/kWh for electricity and the ventilation power cost for diesel equipment would be \$2.0 million annually. The power cost for the reduced ventilation possible with fuel cell equipment would be \$0.5 million annually, a savings of \$1.5 million. The capital cost savings by reducing the exhaust shaft size, eliminating the parallel intake airway, and eliminating one of the main exhaust fans would be approximately \$3.5 million.

5.2 *Deep Post mine*

The Deep Post gold mine is owned and operated by Newmont Mining Corporation. It is on the Carlin Trend approximately 12 km north of Carlin, Nevada. Access to the mine is through a decline which is also a fresh air intake. Most of the intake air, however, is supplied by a 5.5 m diameter shaft. Figure 2 is a VnetPC schematic of the Deep Post mine.

The ventilation requirements are based on the diesel equipment needs but there is also a significant heat load from the rock and groundwater. The intake air is not refrigerated and any reduction in ventilation air flow would have to take into account the heat problem. One of the ventilation design criteria at this mine was that the Wet Bulb Globe Temperature (WBGT) should not exceed 26.7°C (Bossard, 2000).

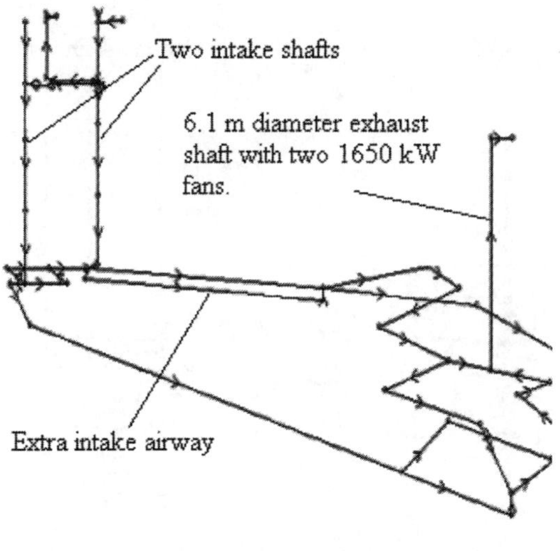

Figure 1. Turquoise Ridge project VnetPC ventilation schematic.

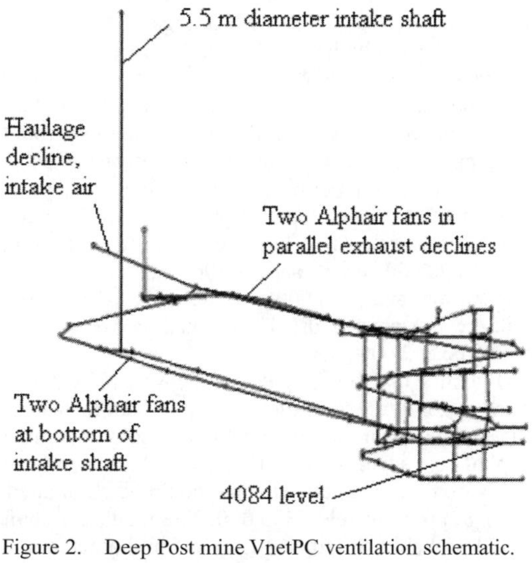

Figure 2. Deep Post mine VnetPC ventilation schematic.

The original ventilation plan called for 11.8 m^3/sec in each of the 4.11 m × 4.11 m stope headings, or a velocity of 0.7 m/sec. This is sufficient to ventilate for diesel equipment and also keep the air temperatures below 26.7°C WBGT.

A CLIMSIM model was made of the ventilation network from the main intake air shaft to a representative stope on the 4084 level to study the effect of reduced ventilation air flows on the air temperatures. The air velocity in the stope heading was set at 0.5 m/sec, equivalent to a volume of 8.45 m^3/sec, and the air flows in the upstream branches were reduced proportionately.

The heat inputs from a simulated fuel cell LHD were added to the model. The predicted WBGT in the 4084 stope heading was 23.8°C. This is lower than the predicted temperature when using diesel equipment and well below the maximum recommended WBGT of 26.7°C.

To study further the effect of reduced air flows on the predicted air temperatures, the 4084 stope air velocity in the CLIMSIM model was reduced to 0.42 m/sec, or 60% of the amount for diesel equipment. The predicted WBGT was 25.5°C, still lower than the maximum recommended level. However, when the air velocity was set at 0.3 m/sec the WBGT was 28.5°C, higher than the maximum recommended.

The ventilation reductions and cost savings calculations were based on an air velocity of 0.5 m/sec in a stope heading, equivalent to a volume of 8.45 m^3/sec. In the CLIMSIM modeling this resulted in predicted air temperatures lower than those with diesel equipment. Since the original ventilation planning for diesel equipment was based on a quantity of 11.8 m^3/sec for each of 22 stopes headings, reducing the quantity from 11.8 m^3/sec to 8.45 m^3/sec results in a total savings of 73.7 m^3/sec for all 22 headings.

The total ventilation air flow in the Deep Post mine is 396.4 m^3/sec. The ventilation reductions possible with fuel cell equipment would lower the total ventilation requirements to 322.7 m^3/sec. This amount could be supplied by the two Alphair fans in the exhaust declines, making it possible to remove the two Alphair fans at the bottom of the intake air shaft. The fan power costs calculated by VnetPC, at $0.075/kWh, are $1,090,600/year to supply 396.4 m^3/sec with four Alphair fans and $463,000/year to supply 322.7 m^3/sec with two Alphair fans. The annual power savings would be $627,000/year.

If a mine like the Deep Post mine were to be developed in the future with fuel cell equipment instead of diesel equipment then there would be some potential capital cost savings as well. The two Alphair fans at the bottom of the intake air shaft could be eliminated, saving approximately $333,000. The intake air shaft could probably be reduced from 5.5 m diameter, and possibly one of the two parallel exhaust declines could be eliminated.

5.3 *Meikle-Rodeo mine*

The Meikle and Rodeo mines are operated by Barrick Goldstrike Mines Inc. They are on the Carlin Trend in Northern Nevada, one of the most prolific gold districts in the world. Both mines are interconnected and share some of the ventilation network so are treated as one unit in this study. Total ventilation air flow for both mines is 831 m^3/sec. There are two intake shafts and three exhaust shafts (Figure 3).

The Meikle and Rodeo mines have rock temperatures as high as 60–65°C. Consequently, the intake air is cooled by a refrigeration plant at a cost of over $2 million/year. Underground air temperatures can still be as high as 37.7°C dry bulb and 30°C wet bulb. One of the ventilation planning criteria at this mine is to keep the Effective Temperature under 29.4°C. The Effective Temperature is a function of the dry bulb and wet bulb temperatures. Table 2 shows some combinations of dry and wet bulb temperatures that yield an Effective Temperature of 29.4°C.

Other ventilation planning criteria with diesel equipment are to maintain an air velocity of at least 0.66 m/sec in most headings, but increase that to 1 m/sec in hot headings. Auxiliary fans ranging from 74.6 kW to 93.3 kW deliver an average of 23.6 m^3/sec to each production heading. The auxiliary fans are not usually shut down when a heading is inactive because it takes too long to cool it down again before reactivating the heading.

A CLIMSIM model of the Meikle 1750 level, a typical example, was created that included the heat

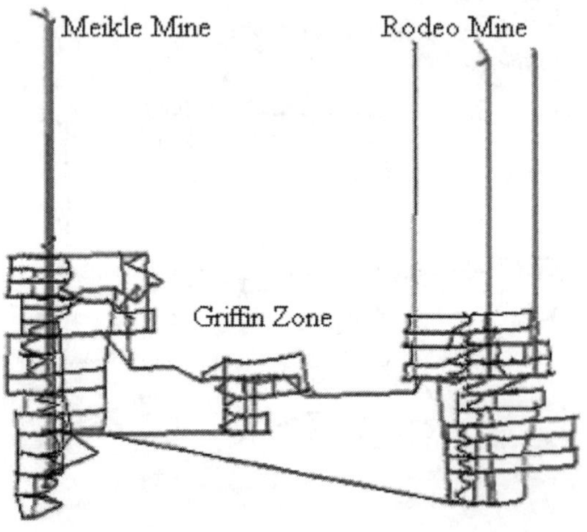

Figure 3. Meikle-Rodeo mines VnetPC ventilation schematic.

loads from the diesel equipment and accurately predicted actual air temperature measurements. The ventilation air flow in the production headings was 23.6 m^3/sec. The heat loads from the diesel equipment were then replaced with heat loads from simulated fuel cell equipment. The ventilation air flows in the production headings were reduced in the CLIMSIM model and the air flows in the upstream branches were reduced proportionately until the air temperatures in the production headings were the same as with diesel equipment. This occurred at an air flow of 20.06 m^3/sec in the production headings. The difference, 3.54 m^3/sec, is the amount that could be saved for each production heading if fuel cell equipment replaced the diesel equipment.

The Meikle and Rodeo mines have quite a large number of production headings and they are not all worked at the same rate each month. However, on the average, there are 25 production headings with ventilation air flows of at least 23.6 m^3/sec each. If 3.54 m^3/sec can be saved from each of those headings then the total mine wide savings would be 88.5 m^3/sec, or 10.6% of the total mine ventilation.

If the ventilation system at the Meikle and Rodeo mines was redesigned to yield a 10.6% reduction in ventilation quantity, the power savings would be 28.7%, based on the relationship that fan power varies as the cube of the air flow volume. The Meikle and Rodeo mines pay $0.07/kWh for electricity and the cost to run the main fans and the underground booster fans is $2,835,800/year. A 28.7% power savings would be an annual savings of $813,000.

The cost to run the cooling plant is significant at $2,130,000/year and there would be some cost savings if the plant did not have to cool as much intake air. The intake air temperatures to the cooling plant would be the same even if the air volume was less, and if it is assumed that the outlet temperatures would remain the same as they are now then the tons of refrigeration required would be directly proportional to the volume of air cooled. A 10.6% reduction in air volume cooled would save $225,780/year.

The total potential ventilation and cooling cost savings if fuel cell equipment replaced the diesel equipment would be $1,038,780/year. If the Meikle and Rodeo mines had been developed for fuel cell equipment instead of diesel equipment then there would also be the potential for some capital cost savings as well, by reducing the diameter of the exhaust shafts, using smaller fans, and reducing the capacity of the intake air cooling plant.

Table 2. Effective temperature chart, Meikle mine.

Dry bulb temperature, °C	Wet bulb temperature, °C	Effective temperature, °C
32.2	30	29.4
34.4	28.9	29.4
36.7	27.8	29.4
38.9	26.7	29.4

5.4 *Henderson mine*

The Henderson mine is a large scale panel caving molybdenum mine in Colorado, approximately 80 km west of Denver. It is operated by the Phelps Dodge Corporation. Total ventilation air flow is about 991 m^3/sec. Figure 4 is a longitudinal section and general ventilation arrangement of the mine.

The virgin rock temperatures range from 21.1°C to 48.9°C. The higher temperatures are encountered in the lower levels but they cool off after being open for a while. The workplace air temperatures range from 4.4–32.2°C dry bulb, and 1.7–32.2°C wet bulb. Heat is not a problem at this mine. In fact, the intake air has to be heated for four to five months during the winter. The main working levels are the 7755 undercut level

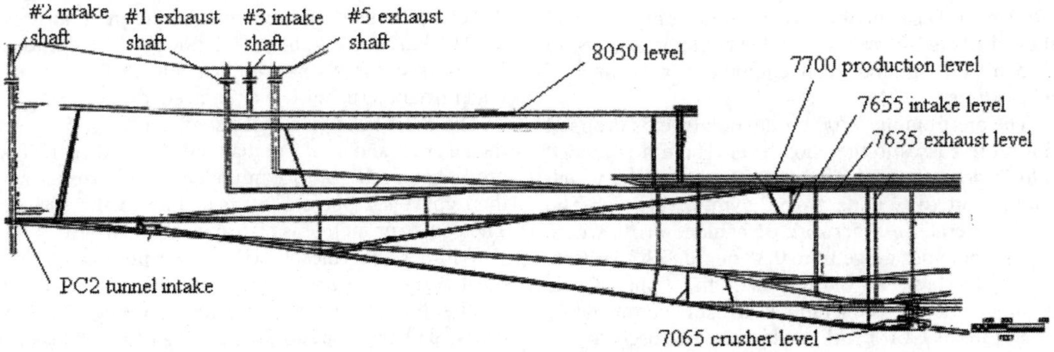

Figure 4. Henderson mine longitudinal section.

and the 7700 draw point or production level. Since heat is not a problem the ventilation design criteria for fuel cell equipment can be based on air velocities adequate enough to remove blasting fumes and to keep dust and gas below the legal limits. On the 7755 undercutting level the drifts are 14.03 m^2 in cross-sectional area and the ventilation quantities for diesel equipment average 12.98 m^3/sec in each drift, equivalent to an average air velocity of 0.92 m/sec. An air velocity of 0.5 m/sec would be adequate to remove blasting fumes and dust. Using an air velocity of 0.5 m/sec as the ventilation design criteria for fuel cell equipment results in a reduction of 45%, or 5.84 m^3/sec, in the air flow in the undercut level drifts. With nine active undercutting drifts on the 7755 level the ventilation reduction for this level could be 52.56 m^3/sec with fuel cell equipment replacing the diesel equipment.

The 7700 production level is where the caved ore is mucked by LHD's from the draw point bells. The drifts are 15.83 m^2 in cross-sectional area and the ventilation quantities with diesel equipment range from 11.7 m^3/sec to 21.3 m^3/sec at the entrance to these drifts. As the air travels along these drifts, which have active lengths up to 427 m, some of it is split off into exhaust raises. By the time the air reaches the ends of the production drifts the air quantities range from 5.76 m^3/sec to 16.61 m^3/sec, or velocities of 0.36 m/sec to 1.05 m/sec.

To determine the ventilation requirements for fuel cell equipment on the 7700 level it was assumed that the ventilation air velocity could be reduced to 0.3 m/sec at the ends of the production drifts. This should be adequate in the production drifts where no blasting takes place, except for the occasional blasting of oversize in the draw points. The new air quantities, Q, in each production drift were determined by equation 1.

$$Qnew = (0.3 \text{ m/sec} \div \text{current velocity at end of drift}) \times Qcurrent \quad (1)$$

This was done for eleven production drifts on the 7700 level. The sum of the ventilation air quantities in these drifts is 180 m^3/sec, but that could be reduced to 82.5 m^3/sec with fuel cell equipment, a savings of 97.5 m^3/sec.

The ore from the 7700 production level is dropped down ore passes to the 7065 level. There it is loaded into 70 tonne capacity Tamrock Supra trucks which haul the ore to an underground gyratory crusher. The trucks operate in a network of haulage drifts where the air velocities range from 0.38 m/sec to 1.73 m/sec. The 7065 level is warmer than the upper levels although the air temperatures are still comfortable. Therefore, a CLIMSIM model was not used to see what the effect of reduced ventilation air flows would be on the air temperatures. Instead, it was assumed that the air flows in four of the haulage drifts, where the air velocities range from 0.57 m/sec to 1.24 m/sec, could be reduced to 0.38 m/sec, since this velocity still results in comfortable air temperatures. Although the calculations are not shown, this would save approximately 41.7 m^3/sec in ventilation quantity.

The total ventilation savings for the 7755, 7700, and 7065 levels if fuel cell equipment replaced the diesel equipment would be 191.7 m^3/sec, or 19% of the current mine ventilation. The mine pays $0.04/kWh for electricity and this reduction in ventilation would save approximately $402,700/year in main fan operating costs. The intake air is also heated during four or five months during the winter at a cost of approximately $30,000 per month. If the savings in intake air heating is proportional to the ventilation reduction then this would save approximately $26,000/year. The total ventilation and air heating savings if fuel cell equipment replaced the diesel equipment would be $428,700/year.

VnetPC modeling showed that three main underground exhaust fans could be eliminated if the mine air flow was reduced by 191.7 m^3/sec. This would save approximately $500,000 in capital costs. Other possible capital cost reductions would be from reduced ventilation shaft and airway sizes.

5.5 *Questa mine*

The Questa mine is in Northern New Mexico and is operated by Molycorp, Inc. It uses panel caving to extract molybdenum ore. The mine was originally developed to use diesel equipment but current production is from conventional grizzly draw points. Future plans include re-introducing diesel LHD's in a new area of the mine known as the D ore body (Figure 5).

The diesel equipment would operate on the 7200 draw point extraction level and the 7222 undercutting level. Ventilation studies have indicated that the total mine air flow would have to be 141.6 m^3/sec when diesel equipment is used (Richardson and Gilbride, 1996). Of this amount, 71.4 m^3/sec would be for the diesel equipment requirements, based on an air flow of 0.063 m^3/sec/kW of diesel power. An additional 70.2 m^3/sec for shops, rail haulage drifts, and the underground crusher would be needed.

The mine has two service shafts, a 7.3 m diameter intake shaft and a 4.27 m diameter exhaust shaft. The ore is hauled by rail to an underground crusher and then conveyed out a 1760 m incline to the surface. The conveyor incline is also an exhaust air way.

With 112 kW diesel LHD's each production drift on the 7200 extraction level will need a minimum ventilation of 7.05 m^3/sec which, in the 10.31 m^2 drifts, will result in an air velocity of 0.68 m/sec. If fuel cell equipment replaced the diesel equipment the air velocities could be reduced to 0.3 m/sec, resulting

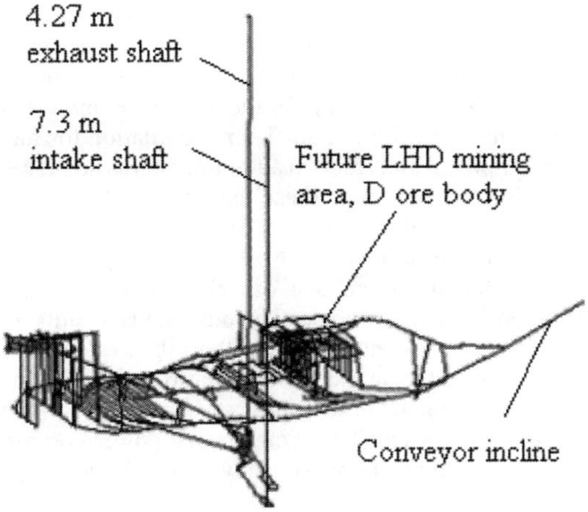

Figure 5. Questa mine VnetPC ventilation schematic.

Table 3. Questa mine VnetPC model results.

	Diesel m^3/sec	Fuel Cell m^3/sec	Reduction m^3/sec
7200 level			
drift 1	7.4	3.1	4.3
drift 2	7.3	3.1	4.2
drift 3	7.3	3.1	4.2
drift 4	7.2	3.1	4.1
drift 5	7.2	3.1	4.1
drift 6	7.3	3.1	4.2
drift 7	7.1	3.4	3.7
drift 8	7.0	3.3	3.7
7222 level	13.6	10.8	2.8
Total	71.4	36.1	35.3

in a new ventilation quantity of 3.1 m^3/sec, a savings of 3.95 m^3/sec for each production drift. The 7222 undercut level drifts would also have air velocities of 0.68 m/sec for diesel equipment. With fuel cell equipment the air flows could be reduced, but, since regular blasting will take place in these drifts, a minimum air velocity of 0.5 m/sec was used to calculate the ventilation requirements for fuel cell equipment.

Heat is not a problem so CLIMSIM modeling was not necessary. VnetPC was used to analyze the ventilation system and plan the location of main fans, booster fans, and air regulators, and calculate the fan operating costs. Table 3 summarizes the VnetPC model results comparing ventilation for diesel equipment versus fuel cell equipment on the 7200 and 7222 levels. The ventilation reduction of 35.3 m^3/sec is a 25% reduction in total mine ventilation for diesel equipment. The Questa mine pays $0.049/kWh for electricity and the VnetPC model calculated the ventilation fan power costs to be $146,300/year for diesel equipment and $81,000/year for fuel cell equipment, for a savings of $65,300/year.

The difference is not large because the Questa mine was originally designed and operated at a production rate of 16,334 mtpd, but molybdenum price declines have forced the mine to reduce production. When the mine re-introduces diesel equipment in the D ore body the production rate will be 5444 mtpd. Since the shafts and drifts were originally planned for a much greater production and also a greater air flow than is necessary to mine 5444 mtpd, the pressure loss will be relatively low. This in turn means a relatively low fan power requirement to move the necessary ventilation air flow, either for diesel or fuel cell equipment.

If a mine like the Questa mine were developed in the future with fuel cell equipment instead of diesel equipment then there would be the possibility for some capital cost reductions, such as a reduction in shaft diameters and elimination of booster fans.

5.6 *Galena mine*

The Galena mine is operated by Coeur Silver Valley Inc., and it is located in Northern Idaho's Coeur d'Alene mining district. The mine currently produces 725 mtpd of silver-copper ore from narrow, steep dipping veins. The primary mining method is conventional cut and fill with track haulage, but 40% of the total production comes from diesel powered trackless equipment in mechanized cut and fill stopes off of the 5500 level, the lowest in the mine. Future trackless mining will be used for long hole stoping on the 2400 level.

The mine is ventilated with approximately 82.45 m^3/sec of air that intakes through the Galena, #3, and Callahan vertical shafts. The air exhausts out the Coeur and Caladay shafts (Figure 6).

Of the 82.45 m^3/sec of total ventilation approximately 18.4 m^3/sec, or 22%, goes to the 5500 level mechanized cut and fill stopes. The mine is deep and hot in the lower levels, with the virgin rock temperature on the 5500 level being 47.8°C. On the 5500 level the ventilating air is cooled with spray chillers before going to the stopes.

Since heat will probably be the limiting factor for determining ventilation reductions, CLIMSIM was used to model the mechanized cut and fill stopes on the 5500 level. The CLIMSIM model included heat loads for the diesel equipment used in the mine. The largest units are 88 kW LHD's. Once the CLIMSIM model was accurately predicting actual air temperature measurements in the stopes, the heat loads from the diesel equipment were replaced with heat loads from simulated fuel cell equipment. Table 4 compares the air temperature ranges predicted by the CLIMSIM model for diesel equipment and fuel cell equipment.

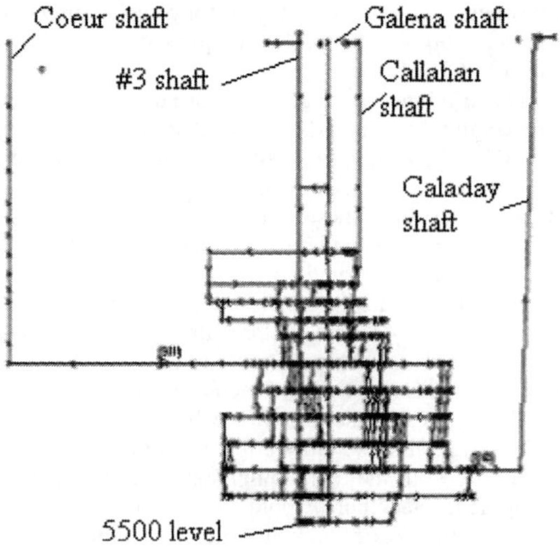

Figure 6. Galena mine VnetPC ventilation schematic.

Table 4. Range of air temperatures predicted by CLIMSIM for diesel and fuel cell equipment.

	Diesel	Fuel cell
Dry bulb, °C	32.9–34.7	28.5–31.1
Wet bulb, °C	26.6–27.6	24.5–26.5

As can be seen from Table 4, fuel cell equipment would lower the air temperatures by several degrees centigrade compared to diesel equipment. The next step in the CLIMSIM modeling was to reduce the ventilation air flows when simulating fuel cell equipment until the predicted air temperatures were the same as when using diesel equipment. The results show that the ventilation air flow to the 5500 level mechanized cut and fill stopes could be reduced from 18.4 m^3 to 15 m^3/sec.

The possible ventilation reduction is only 4% of the mine's total air flow of 82.45 m^3/sec, but this would save about 11.5% in fan power. The Galena mine pays $0.04/kWh for electricity and the main fan power costs are $221,250/year. This could be reduced to $195,750/year, a savings of $25,500/year, if fuel cell equipment replaced the diesel equipment.

The mechanized cut and fill stopes on the 5500 level use about 22% of the total mine ventilation but they produce about 40% of the ore, or 290 mtpd.

This may indicate that mechanized cut and fill stopes are more productive per volume of ventilation air than conventional stopes. If the mine production of 725 mtpd could come exclusively from mechanized cut and fill stopes using fuel cell equipment, then the total mine ventilation could be reduced even more than the 4% estimated for the 5500 level alone. The CLIMSIM modeling for the 5500 level indicated that 15 m^3/sec would be sufficient ventilation for fuel cell equipment and result in air temperatures no higher than they are now with diesel equipment. Producing 725 mtpd would require 2.5 times as many stopes as are on the 5500 level. Multiplying 15 m^3/sec by 2.5 results in 37.5 m^3/sec of ventilation. Allowing 50% additional for miscellaneous needs results in 56.25 m^3/sec of ventilation for the mine, or 68% of the current 82.45 m^3/sec. Supplying 68% of the current ventilation air flow would need about 32% of the current fan power. The savings in fan power costs in this case would be approximately $150,000/year.

5.7 *Lucky Friday mine*

The Lucky Friday mine is operated by Hecla Mining Company and is also in the Coeur d'Alene mining district of Northern Idaho. Almost all of the current production of 476 mtpd comes form mechanized cut and fill stopes in the Gold Hunter section of the mine and is hauled by truck across the 4900 level to the Silver Shaft (Figure 7).

The Lucky Friday mine introduced electric LHD's in the Gold Hunter stopes because of concerns about excessive heat from diesel engines. In the past few years, however, the mine has switched back to diesel units because of high operating costs for the electric trailing cables and take-up reels.

The total mine ventilation air flow is 107.6 m^3/sec of which 60 m^3/sec goes across the 4900 level to the Gold Hunter section of the mine where all production currently comes from. In the Gold Hunter section the air is split into a series of raises and ramps to ventilate stopes above and below the 4900 level. All exhaust air travels up an Alimak raise to the 4050 level and then across to the #2 shaft where it exhausts to the surface. Ventilation quantities adequate to dilute diesel emissions were one of the factors for planning purposes, but the main factor is the removal of heat while still keeping refrigeration to an acceptable level. One of the recommended design criteria at this mine is a maximum wet bulb temperature of 28°C at stope faces.

Because mine air temperatures would be the limiting factor for any potential ventilation reduction, CLIMSIM modeling was done for a stope that had dry and wet bulb air temperatures measured when a diesel LHD was operating in it. It was felt that this would give a good check on CLIMSIM's ability to accurately predict air temperatures. The ventilation air flow, 12.32 m^3/sec, was cooled by a spray chiller before going into the stope. Both ends of the stope were dead end headings and the vent ducts discharged

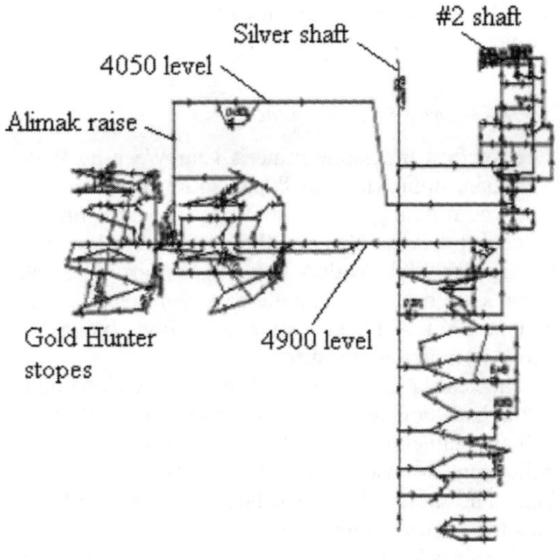

Figure 7. Lucky Friday mine VnetPC ventilation schematic.

Table 5. Comparison of air temperatures, left end of stope.

	Actual temperature with diesel LHD	CLIMSIM prediction for diesel LHD	CLIMSIM prediction for fuel cell LHD
Dry bulb, °C	29.2	29.3	28.4
Wet bulb, °C	28.3	28.2	27.5

about 16 m from the faces when air temperatures were measured in various parts of the stope. Mine Ventilation Services, Inc. had done some previous studies at this mine and determined the virgin rock temperature, geothermal step, and rock conductivity and diffusivity that were used in the CLIMSIM model (Mine Ventilation Services, Inc., 1997).

The left end of the stope was being mucked out by a 1.53 m^3 LHD with a 93 kW diesel engine so the heat load from that was added to the CLIMSIM model. The right end of the stope had already been mucked out and was idle. The CLIMSIM model accurately predicted actual air temperature measurements. The heat load from the diesel LHD was then replaced by the heat load from a simulated fuel cell LHD. Table 5 summarizes the results for the left end stope face. The air temperatures predicted for fuel cell LHD's are lower, but not appreciably, than for diesel LHD's. This is probably because the heat from the mobile equipment is not a large percentage of the total heat from the rock and water. The air flow in the CLIMSIM model was reduced until the predicted air temperatures for fuel cell LHD's were the same as actually measured with diesel LHD's operating in the stope. At air flows of 80–85% of the air flow for diesel LHD's the dry bulb temperatures were the same, but the wet bulb temperature was at the maximum recommended value, 28°C, when the air flow was 85%. Accordingly, it was assumed that the air flows for fuel cell LHD's could be 85% of those for diesel LHD's.

The mechanized cut and fill stopes use about 50 m^3/sec of the total mine air flow of 107.6 m^3/sec. If the air flow to the stopes was reduced to 85% of the amount needed for diesel equipment, but the rest of the mine air flow stayed the same, the new mine air flow would be 100.1 m^3/sec. This is 93% of the current mine air flow and supplying 93% of the air flow would require about 80% of the current fan power.

The Lucky Friday mine pays $0.035/kWh for electricity and the cost to run the ventilation fans is $315,100/year. The savings in fan power costs if fuel cell equipment replaced the diesel equipment would be about $63,000/year. Additional savings in supplying refrigerated air to the stopes would also be expected.

5.8 *Avery Island mine*

The Avery Island mine is a salt mine operated by Cargill Deicing Technology, a division of Cargill Salt. The mine is located near New Iberia, Louisiana, and has been in operation for approximately 100 years. The room and pillar method is used to mine approximately 2.5 million tonnes/year. The rooms are first developed as a top heading 8.53 m high and 27.4 m wide. After a top heading is developed a bench 16.76 m is taken, leaving a room 25.3 m high and 27.4 m wide. The pillars are the same dimensions as the rooms. After the rock salt is blasted it is loaded by Caterpillar 992 loaders into Caterpillar 775 trucks and hauled to a feed-breaker, then conveyed up a slope to a shaft loading pocket and hoisted up a 4.9 m diameter shaft (Figure 8).

Approximately 173.9 m^3/sec of ventilation air intakes down the production shaft to the 1300 level. A 932.5 kW main fan and two 149.2 kW secondary fans are on the 1300 level. The secondary fans are operated only on weekends when the main intake fan is shut down. There is approximately 59.5 m^3/sec of recirculation from the 1100 and 900 levels back into the intake shaft so that approximately 233.4 m^3/sec is handled by the main fan on the 1300 level. The air is taken from the 1300 level to the 1600 level by a 6.1 m diameter raise. The exhaust air courses up through a series of slopes to the 500 level where two shafts, the Air Shaft and the Steam Shaft, exhaust the air to the surface.

The auxiliary ventilation is handled by movable jet fans of 74.6 to 93.3 kW that handle 18.9 to 28.3 m^3/sec each. The jet fans are usually positioned near the edge

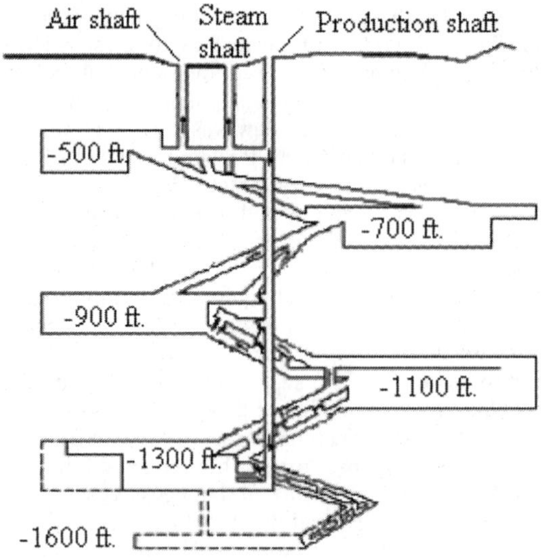

Figure 8. Avery Island mine.

of a pillar one crosscut back from the production face and blow enough air into the face to keep the air quality under control. The mine occasionally comes across pockets of CO_2 and H_2S. The virgin rock temperatures measured in drill holes range from 22.2 to 25.6°C (Lancaster, 2002).

The air velocities in the large rooms are quite low and generally are below the measuring range of standard instruments unless near an auxiliary jet fan. Air temperatures on the 1600 level range from 32.2°C dry bulb/26.9°C wet bulb at the bottom of the intake raise, to 35.6°C dry bulb/27.8°C wet bulb where the air exhausts up the conveyor slope. All of the underground trucks and loaders are equipped with air-conditioned cabs. The time allowance for clearing out fumes after a blast is 30 minutes.

Because of the already low air velocities, the time desired to clear air after a blast, and the possibility of hitting pockets of CO_2 and H_2S, it is felt that the air quantities cannot be reduced even if fuel cell vehicles replaced the diesel units.

Heat is a factor on the 1600 level so a CLIMSIM model was made to try and determine how much of the heat load is contributed by the large diesel powered loaders and trucks. The CLIMSIM analysis indicates that the diesel units are adding as much as 4°C to the air temperatures.

The mine plans to use diesel maintenance and ventilation to control DPM. Even though the current ventilation air flow cannot be decreased if fuel cell equipment replaced the diesel equipment, fuel cell equipment would make it unnecessary to increase ventilation to meet future DPM standards, air temperatures would be several degrees lower, and the air quality would improve.

5.9 *Kerford limestone mine*

The Kerford limestone mine is near Weeping Water, Nebraska, approximately 80 km southwest of Omaha. The room and pillar method is used to produce 7,260 mtpd. After blasting, the limestone is loaded by Caterpillar 988F loaders into Caterpillar 771 trucks. Rooms are mined in a single pass and are 6.7 m high and 10.67 m wide. The mine is shallow, being only about 30 m below the surface.

The ventilation air flow is 236 m^3/sec which intakes through several adit portals and exhausts through two 2.44 m diameter raises. Movable jet fans and propeller fans are used to ventilate the production headings. The mine allows two hours between shifts for blasting fumes to clear out.

The air velocities are quite low because of the large room dimensions. Heat is not a problem at this mine and the heat load from the diesel equipment is not a concern.

As with the Avery Island salt mine, it is felt that because of the already low air velocities the ventilation air flow could not be reduced even if fuel cell equipment replaced the diesel equipment. Fuel cells would improve the air quality although this mine already has good air quality. This may become a factor in the future, though, if new, stricter DPM regulations are put into effect.

6 SUMMARY

Some of the mines studied would have potentially large savings in ventilation operating costs if fuel cell powered equipment replaced the diesel equipment. Table 6 summarizes the estimated operating cost savings and, for some of the mines, capital cost savings. The amount of savings varies from mine to mine not only because of the scale of operations and mining method used, but also because some mines pay almost twice as much for electricity as others. The two room and pillar mines in the study, Avery Island, and Kerford, have very low air velocities because of the large room dimensions. Reducing the ventilation air flow is probably not possible without causing other problems, such as an unreasonable increase in the time for blasting fume removal. However, fuel cell equipment will benefit room and pillar mines in other ways, in particular, making it easier for them to continue to operate under the current and future DPM regulations. It is a difficult process to control and direct air to and from the work places in a large room and pillar underground mine. The expense incurred to do this in order to meet the DPM regulations could be quite high for some mines.

Table 6. Summary of ventilation cost savings with fuel cell equipment.

Mine	kWh of electricity saved annually	Electric power cost, $/kWh	Annual ventilation cost savings	Capital cost savings
Turquoise Ridge	21,900,000	0.07	$1,533,000	$3,445,000
Deep Post	8,368,000	0.075	$627,600	$333,000
Meikle-Rodeo	14,840,000	0.07	$1,038,800	Possible reduction in airway sizes,fans, and cooling plant.
Henderson	10,671,250	0.04	$428,700	$500,000 by eliminating three main exhaust fans and possible reductions in airway sizes.
Questa	1,346,900	0.049	$65,300	Possible reduction in airway sizes.
Galena	662,500	0.04	$25,500 with current mechanized cut and fill stopes, or $150,000 if all production was from mechanized cut and fill stopes.	
Lucky Friday	1,800,000	0.035	$63,000	

7 CONCLUSIONS

Most of the mines in this study would benefit from reduced ventilation costs if fuel cell equipment was used. There are other significant advantages to using fuel cell power in underground mines, not the least being a healthier working environment for the miners. Ultimately, there are many factors to consider when reviewing the pros and cons of diesel versus fuel cell power in underground mines, but the results presented here show that the potential benefits justify the development of a prototype fuel cell LHD so that capital and operating costs can be accurately and reliably quantified to compare with those of existing diesel powered LHD's.

ACKNOWLEDGMENTS

We would like to thank the US Department of Energy and Vehicle Projects LLC for providing the funds for this project. Also, this study would not have been possible without the cooperation and interest the mine operators had in this project. This is sincerely appreciated. We would like to thank the following people. Tom Bagan and John Porterfield, Turquoise Ridge Project, Getchell Gold Corporation, a member of the Placer Dome Group: Kuda Mutama, Anton Meyer, and Ben Lupercio, Meikle Mine, Barrick Goldstrike Mines, Inc.: John Barry, Matt Breitrick, and Ken Kins, Deep Post Mine, Newmont Mining Corporation: Lee Fronapfel, Henderson Mine, Phelps Dodge Corporation: Dave Sperling and Robert Kehrman, Questa Mine, Molycorp, Inc.: Mark Bren, Galena Mine, Coeur Silver Valley, Inc.: Terry Devoe and Douglas Bayer, Lucky Friday Mine, Hecla Mining Company: Lane Lancaster, Lisa Lopez, and Michael Bertoldi, Avery Island Mine, Cargill Deicing Technology: Frank Cranston and Bryce Pietzyk, Kerford Mine, Kerford Limestone Company: Floyd Bossard, Floyd C. Bossard and Associates, Inc.: Leo Gilbride, Agapito Associates, Inc.: Brian Prosser, formerly with Mine Ventilation Services, Inc. and now with Bechtel/SAIC: John Head, MACTEC Inc.: Charles Kocsis, CANMET: and John Marks.

REFERENCES

Bossard, Floyd. 2000. *Analysis of Ventilation and Air-Cooling Requirements for the Deep Post Underground Mine*, Floyd Bossard and Associates, Inc., Butte, MT. Private Report to Newmont Gold Company.

Kocsis, Charles, pers. comm. 2003. CANMET, Sudbury, ON.

Lancaster, Lane, pers. comm. 2002. Cargill Deicing Technology, Avery Island, LA.

Mine Ventilation Services, Inc. 1997. *Ventilation and Refrigeration Design for Hecla Mining Company's Goldhunter Deposit.* Mine Ventilation Services, Inc., Fresno, CA. Private Report to Hecla Mining Company.

Richardson, A.M., and Gilbride, L.J. 1996. *Ventilation Measurements and Analysis, Questa Mine*. Agapito and Associates, Inc., Grand Junction, CO. Private Report to Molycorp, Inc.

Mine Ventilation, Ganguli & Bandopadhyay (eds)
© 2004 Taylor & Francis Group, London, ISBN 90 5809 633 5

Case studies from application of numerical simulation software to examining the effects of fires on mine ventilation systems

H.W. Wu & A.D.S. Gillies
The University of Queensland, Brisbane, Australia

A.M. Wala
University of Kentucky, Lexington, Kentucky, USA

ABSTRACT: The structure of a comprehensive research project into mine fires study applying the "Ventgraph" mine fire simulation software, preplanning of escape scenarios and general interaction with rescue responses is outlined. Some outcomes from a project funded by the Australian Coal Association Research Program and substantial mining company site support are described. Mine site testing is essential and allows the approach to be introduced in the most creditable way. The project has assisted the Australian mining industry to attain an improved position in their understanding of mine fires and the use of modern advances to preplan actions to be taken in the advent of mine fires and possible emergency incidents. Work undertaken at individual mines is discussed as examples. The effort is built around the introduction of fire simulation computer software to the Australian mining industry and the consequent modeling of fire scenarios in selected different mine layouts. Application of the simulation software package to the changing mine layouts requires experience to achieve realistic outcomes. Most Australian mines of reasonable size currently use a ventilation network simulation program. Under the project a small subroutine has been written to transfer the input data from the existing mine ventilation network simulation program to "Ventgraph". This has been tested successfully. To understand fire simulation behavior the mine ventilation system must be understood correctly first. Some results of the project are discussed.

1 INTRODUCTION

Mine fires remain among the most serious hazards in underground mining. The threat fire presents depends upon the nature and amount of flammable material, the ventilation system arrangement, the duration of the fire, the extent of the spread of combustion products, the ignition location and, very importantly, the time of occurrence. The response to the fire by mining personnel will depend upon all of these factors.

There is a lack of knowledge about fire/heat dynamics, some unproven technology in the field of gas sensors and no general agreement on appropriate alarm response systems and measures to be taken in the event of a significant incident. There is a need to couple the detection system with the response system. A research project has been undertaken focused on the application of mine fire and ventilation software packages for contaminate tracing and fire modeling in coal mines and validation of fire modeling software against real mine incidents to reduce the effects of fire incidents and possible consequent health and safety hazards.

With the increasing complexity of technological and managerial development of mines, the effects of mine fires must be better understood. Task Group No 4 Report arising out of the Moura No 2 coalmine disaster of August 1994 in Australia made a number of recommendations including that,

"*the capability to model ventilation and the mine environment following an incident should be available at mines*".

Following these recommendations, sub-committees were formed in 1997 to further progress various findings with Sub-Committee 5 – Incident Management given certain tasks including the question,

"*that there is a need for a wider appreciation of current knowledge and improved capability of ventilation management at mines for both routine as well as emergency conditions; guidelines for modeling should include …*

- *Computer modeling of post incident mine ventilation and atmosphere to be a required element in mine safety management plans,*

- *Models interface with standard mine planning packages and be kept up-to-date, and*
- *Development of learned ventilation and fire control responses occur for different incident scenarios and locations, pre-determined for each mine and with plans prepared and personnel trained in appropriate action plans".*

The primary objective of the study has been to implement a program of research into this complex area utilizing the recently upgraded Polish mine fire simulation software, "Ventgraph". There is a need to understand the theory behind the simulation program and to allow use by those already familiar with the main existing mine ventilation analysis computer program currently popular within the Australian industry, "Ventsim", as an aid to incident and emergency management. "Ventsim" was not designed to handle fire simulation or in fact compressible flow effects in mine networks.

When a fire occurs outbye the working section, the immediate safe evacuation of miners from these areas should always be the first action during the rescue operation. Usually, the intake entries are dedicated as the primary escapeways from the working section. In many cases, the dedicated escapeways are contaminated with fire by-products from abutting entries (e.g. belt entry) due to interconnection or leakage through stoppings. It is important to keep these escapeways unobstructed and free from contamination.

It is difficult to predict the pressure imbalance and leakage created by a mine fire due to the complex interrelationships between the mine ventilation system and a mine fire situation. Depending on the rate and direction of dip of the entries (dip or rise), reversal or recirculation of the airflow could occur because of convection currents (buoyancy effect) and constrictions (throttling effect) caused by the fire. This reversal jeopardizes the functioning of the ventilation system. Stability of the ventilation system is critical for maintaining escapeways free from contamination and therefore available for travel.

Simulation software has the great advantage that underground mine fire scenarios can be analyzed and visualized. The "Ventgraph" program is described by Trutwin, Dziurzynski and Tracz (1992). The software provides a dynamic representation of the fire's progress (in real-time) and utilizes a color-graphic visualization of the spread of combustion products, oxygen and temperature throughout the ventilation system. During the simulation session the user can interact with the ventilation system (e.g. hang brattice or check curtains, breach stoppings, introduce inert gases such as those generated by a GAG unit and change fan characteristics). These changes can be simulated quickly allowing for the testing of various fire-control and suppression strategies. Validation studies on "Ventgraph" have been performed using data gathered from a real mine fire as undertaken by Wala, et al (1995).

The simulation program has recently been put in a "Windows" format for ease of use and interrelation with other software and with some ability for transferring mine ventilation planning data from programs such as "Ventsim".

The outcome of the project will be that the Australian mining industry is in an improved position in their understanding of mine fires and the use of modern advances to pre-plan actions to be taken in the advent of mine fires and the handling of possible emergency incidents. The goals of the project are built around the introduction of a modern fire simulation computer program "Ventgraph" and the consequent modeling of fire scenarios in a number of different mine layouts.

The project is undertaking simulations of the effects of common fire causes and fire progress rates. It is undertaking further validation of the simulation model through back-modeling of past fires where gas and other relevant data records are available. It is undertaking some simulation testing exercises incorporating the GAG and other inertization methods. The simulation of safe escape scenarios before or during a fire as part of emergency evacuation is being examined. Some reference is being made to work undertaken by appropriate bodies such as mines' rescue organizations during pre-planning as part of mine rescue and recovery strategies. Preparation of education (teaching) and training materials on the theory of mine fires (interactions between fire and ventilation and conversely, between ventilation and fire and so on), controlling and combating mine fires, development of evacuation management plans (escape ways) is occurring as part of a technology transfer thrust.

This paper gives an overview on the findings and results of the project to date. The main purpose of this paper is to examine the effects of fires on mine ventilation systems using numerical fire simulation software such as "Ventgraph". Various case studies based on the modeling of fire scenarios in a number of different mine layouts are discussed.

2 EFFECTS OF FIRES ON MINE VENTILATION

An open fire causes a sharp increase in the temperature of the air. The resulting expansion of the air produces a number of distinct effects. First the expansion attempts to take place in both directions along the airway. The tendency to expand against the prevailing direction produces a reduction in the airflow. Secondly, the expansion in volume increases air velocity downwind from the fire causing additional pressure loss. This is known as the choke or throttle effect. Finally, the decreased density results in the heated air becoming more buoyant and causes local effects as well as changes in the magnitudes of natural ventilating energy.

2.1 *The choke or throttling effect*

This effect results from an increase in volume of air as it passes through the fire. This increase in volume is due to gas expansion as well as the addition of combustion products such as fire gases and evaporated water. As a result the velocity of air downwind from the fire is increasing and additional pressure loss following the square law results.

The choke effect is analogous to increasing the local resistance of the airway. For the purposes of ventilation network analyses based on a standard value of air density, the raised value of this "pseudo resistance", R_t, can be estimated in terms of the air temperature as followed (McPherson, 1993).

$$R_t \propto T^2 \tag{1}$$

The value of R_t, increases with the square of the absolute temperature (T). However, it should be recalled that this somewhat artificial device is required only to represent the choke effect in an incompressible flow analysis.

Litton et al (1987) have produced an estimate of the increased resistance in terms of the carbon dioxide evolved from a fire. Because the temperature of fumes falls off rapidly downwind of the fire, the chock effect is generally limited to the vicinity of the fire. Thus, the magnitude of the flow constriction (chock or throttle) is likely to be small when compared to the ventilation pressure in the mine ventilation system.

According to the studies undertaken in USA and Japan, reduction of flow and velocity of air due to the chock effect can be in the range of 10 to 25% of the flow before the fire started. Even at this level it can create secondary hazards like accumulation of contaminants, inadequate oxygen content and smoke roll-back.

2.2 *Buoyancy (natural draft) effects*

2.2.1 *Local or roll back effect*

The most immediate effect of heat on the ventilating air stream is a very local one. The reduced density causes the mixture of hot air and products of combustion to rise and flow preferentially along the roof of the airway. The pronounced buoyancy effect causes smoke and hot gases to form a layer along the roof and, in a level or descentional airway, will back up against the direction of airflow as shown in Figure 1.

This phenomenon of roll-back creates considerable difficulties for firefighters upstream from the fire, particularly if the conflagration has become fuel-rich.

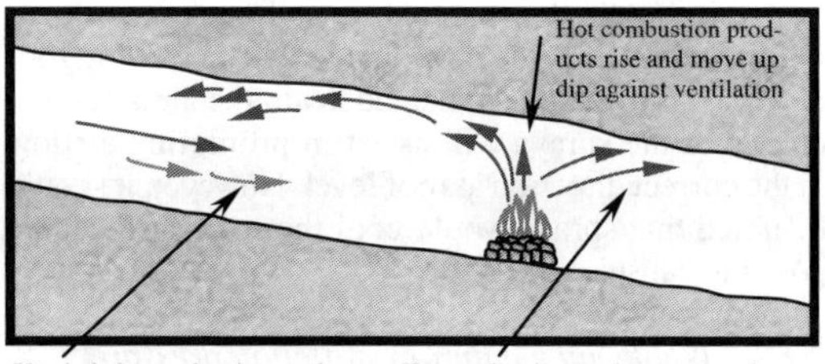

Figure 1. Buoyancy effect of fires in descentional airway (after Bofinger, 1998).

Table 1. Minimum air velocity for roll-back occurrences (after Mitchell, 1990).

Height of entry (m)	Minimum air velocity (m/s)		
	Dip 0%	Dip 10%	Dip 20%
1.2	1.00	1.20	1.50
1.8	1.25	1.45	1.80
2.4	1.40	1.68	2.10
3.0	1.60	1.88	2.35

The roll-back is visually obvious because of the smoke. However, it is likely to contain hidden but high concentrations of carbon monoxide and other toxic or explosive gases. Furthermore, the temperatures of the roll- back may initiate roof fires of any combustible material above the heads of firefighters. The most critical danger is that tidal flames or a local explosion may occur throughout the roll-back, engulfing firefighters in burning gases.

Smoke will roll-back against low airflows, except in a rising airway. How fast and far smoke rolls back depends on how much lower the air velocity is from the minimum air velocity listed in Table 1 as suggested by Mitchell (1990).

One method of reducing roll-back is to increase the airflow in the airway. This, however, will increase the rate of propagation of the fire.

A second method is to advance with hurdle cloths covering the lower 60 to 80 per cent of the airway. The increased air velocity at roof level will help to control the rollback and allow firefighters to approach closer to the fire. However, this technique may also cause the roll- back gases to mix with the air and produce an explosive mixture on the forward side of the hurdle cloth. Furthermore, the added resistance of the hurdle cloth might reduce the total airflow to the extent that a fuel-rich situation is promoted. The behavior of open fires is very sensitive to modifications to the airflow. Hence, any such changes should be made slowly, in small increments, and the effects observed carefully.

A third method of combating roll-back is to direct fog sprays towards the roof. In addition to wetting roof material and cooling smoke laden air, the air induction effects of the sprays will assist in promoting airflow in the correct direction at roof level. However, it should be noted that sprays would cool the rock surface and possible cause spalling.

2.2.2 *Whole mine natural ventilation pressure effects*

A more widespread effect of reductions in air density is the influence felt in shafts or inclined airways. The conversion of heat into mechanical energy in the ventilation system is called the buoyancy (natural draft, natural ventilating pressure or chimney) effect. Such a conversion under steady state conditions requires cyclic analysis processes that are provided by every mesh loop of the ventilation network.

The effect is most pronounced when the fire itself is in a shaft or inclined airway and promoting airflow if the ventilation is ascentional and opposing the flow in descentional airways. In addition, in the latter case flows can reverse in all parallel airways to the airway with fire. Indeed, in this case, the airflow may be reversed in the airway with fire, bringing combustion products into adjacent parallel airways and also resulting in non-steady state flow of toxic atmospheres.

Natural ventilating pressure always exist in a mine and its magnitude mostly depends on the mine's depth and difference in air density in the inclined and horizontal airways. In the case of fire, this effect is magnified due to high temperatures leading to unpredictable changes in air density and the airflow distribution.

If the air temperatures can be estimated for paths downstream of the fire then it is possible to determine the modified natural ventilating pressures. Those temperatures vary with respect to size and intensity of the fire, distance from the fire, time, leakage of cool air into the airways affected and heat transfer characteristics between the air and the surrounding strata.

At any given time, air temperatures tend to fall exponentially with respect to distance downstream from a fire. Climatic simulation models may also be employed to track the time transient behavior of air temperatures downstream from a fire, however; in that case, two matters should be checked. One is that the limits of application of the program may be exceeded for the high temperatures that are involved. Secondly, the transient heat flux between the air and strata will be much quicker than for normal climatic variations. Hence, the virgin rock temperature (VRT) in the simulation input should be replaced by a "surrounding rock temperature" (SRT), this being an estimate of the mean temperature of the immediate rock envelope around the airway before fire occurs.

Having determined air temperatures in all paths downstream from the fire, the revised natural ventilation pressures for the mine can be determined. These may then be utilized in network analysis exercises to predict the changes in flow and direction that will be caused by a fire of given output. A number of fire simulation packages have been developed to allow numerical modeling of mine fires (e.g. Greuer, 1984; Greuer, 1988; Dziurzynski et al, 1988; Deliac et al, 1985; Stefanov et al, 1984).

2.3 *Case studies of Australian longwall development panels*

One of the major goals of the study is to examine the effects of fires on mine ventilation systems. To demonstrate how the choke or throttling and buoyancy effects influence the mine ventilation system, fire scenarios were simulated for several case studies based on a typical Australian two entries longwall development panel with various panel configurations.

Panel configurations varied in the case studies covered panel lengths of 1.5 km or 3 km with panel dipping angles of plus and minus 5 degrees and 10 degrees. To generate an uniform ventilation airflow of 22.8 m^3/s through the panel at the working face, a differential pressure equal to 70 Pa was introduced across the stopping at the first cut-through between intake and return entries for the 1.5 km panel length cases studied and a differential pressure of 235 Pa was used for the 3 km panel cases studied.

Figure 2 shows typical two-entry longwall development panel ventilation systems with various panel lengths of 1.5 and 3 km. The fresh air reaches the face though one intake entry and exhausts from the face through the other, which is the belt entry and return. The fresh air intake entry is isolated from the return entry by a series of short life stoppings.

Table 2 shows a summary of the simulation results with diesel fires set in the middle of the development panel for each case study at either intake or return airways. The fire has a 5 m fire zone length, a fire intensity of 10 and a time constant 120 seconds. There were two cases under which that the face airflow almost reversed. One is observed when the diesel fire is set in the middle of the return airway at nodal points 24–25 in the 1.5 km longwall development panel mining on a 10% incline upward.

The buoyancy effect of the fire acted against the fan pressure but was not enough to overtake the fan pressure to reverse the airflow. As the airflow reduced, the oxygen supplied to the fire is decreased. This caused a significant reduction of the magnitude of the fire thus the buoyancy effect of the fire working against the fan pressure is then reduced. All this means is that now there is more air available to the fire so the fire starts to grow again (non steady state). This can be observed in the fire simulation output graphic and is illustrated in Figure 3 showing the heat production of the fire

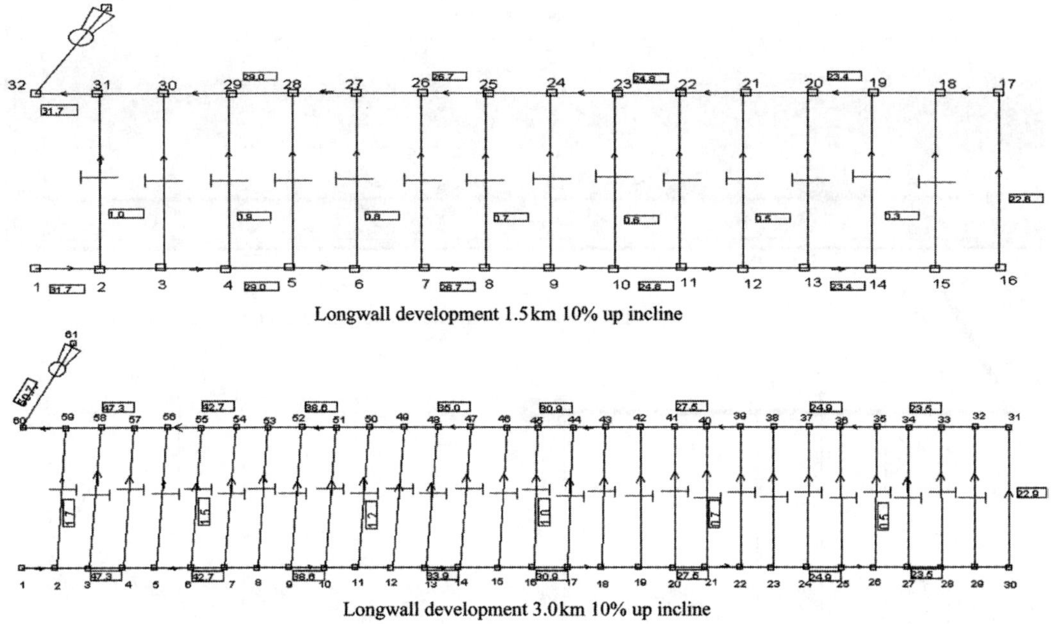

Figure 2. Two-entry longwall development panels.

Table 2. Summary of simulation results for 5 m-fire zone.

Panel length (km)	Panel inclination (%)		Face Q m^3/s Initial	Face Q m^3/s Final	Air reversal
Fire at return airways (branch 24–25 or 45–46)					
1.5	5	Up	22.8	16.1	No
1.5	5	Down	22.8	25.7	No
1.5	10	Up	22.8	1.3	Almost
1.5	10	Down	22.8	30.2	No
3.0	5	Up	22.8	21.3	No
3.0	5	Down	22.8	24.4	No
3.0	10	Up	22.8	20.1	No
3.0	10	Down	22.8	24.5	No
Fire at intake airways (branch 8–9 or 14–15)					
1.5	5	Up	22.8	26.7	No
1.5	5	Down	22.8	16.2	No
1.5	10	Up	22.8	29.3	No
1.5	10	Down	22.8	1–1.3	Almost
3.0	5	Up	22.8	24.3	No
3.0	5	Down	22.8	21.4	No
3.0	10	Up	22.8	25.7	No
3.0	10	Down	22.8	19.4	No

during simulation. It can be noted from the figure that the fire causes a sharp reduction in the airflow entering the development panel. In the case of a mine with high seam methane levels this will lead to higher methane levels in the mine air.

A similar situation was observed in the other case study when the fire was in the intake airway at nodal points 8–9 in the 1.5 km longwall development panel mining at a 10% decline down. Again, the buoyancy effect of the fire acted against the fan pressure was not enough to overtaken the fan pressure to reverse the airflow. The airflow at the faces in both cases was drastically reduced to approx. 1–1.3 m^3/s.

Simulations were re-run for these cases with a 10 m fire zone length instead of 5 m and a summary of results is shown in Table 3. Airflow reversals at the face were observed in the 10% incline case with fire in return airway and also with 10% decline with fire in the intake air.

Figure 4 shows how the smoke progresses and the amount of heat produced during different stages of the fire simulated in the 10% incline case with fire in return air. It is noted that a reduction in heat production from the fire was observed before the airflow reversal occurred. As the products of fire are forced passed or over the fire at the moment when airflow reversal occurs, the amount of oxygen available to the fire is drastically decreased.

2.4 *Modeling of fires*

One of the immediately question about fire simulation is the magnitude of the fire and how the fire can be modeled in the simulator. In the simulator, various

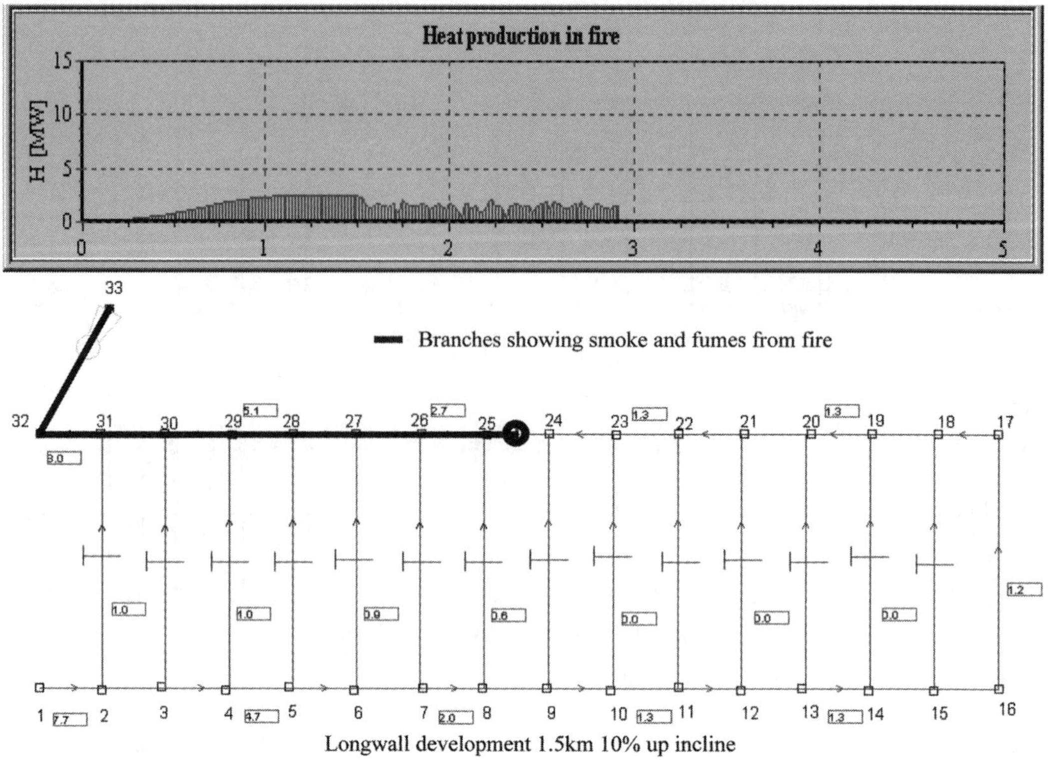

Figure 3. Smoke progressions and heat production during simulation for 5 m long fire zone in return roadway.

Table 3. Summary of simulation results for 10 m-fire zone.

			Face Q m³/s		
Panel	Panel		Initial	Final	Air reversal
Fire at return airways (branch 24–25 or 45–46)					
1.5 km	5%	Up	22.8	4.1	No
1.5 km	5%	Down	22.8	29.0	No
1.5 km	10%	Up	22.8	16.8*	Yes
1.5 km	10%	Down	22.8	33.4	No
3.0 km	5%	Up	22.8	20.9	No
3.0 km	5%	Down	22.8	24.6	No
3.0 km	10%	Up	22.8	18.4	No
3.0 km	10%	Down	22.8	26.2	No
Fire at intake airways (branch 8–9 or 14–15)					
1.5 km	5%	Up	22.8	28.1	No
1.5 km	5%	Down	22.8	3.5	No
1.5 km	10%	Up	22.8	36.6	No
1.5 km	10%	Down	22.8	1.5–4.5*	Yes
3.0 km	5%	Up	22.8	24.9	No
3.0 km	5%	Down	22.8	20.6	No
3.0 km	10%	Up	22.8	27.4	No
3.0 km	10%	Down	22.8	16.0	No

* Air flows in opposite direction to initial flow direction.

fires can to be modeled through combinations of the types of fuels, the length of fire zone (surface area) and intensity. It is necessary to investigate the effects of length of fire zone and fire intensity on the mine ventilation system. Validation exercises were undertaken to establish how location, type and intensity of fires could be established and modeled.

As for the intensity of diesel fuel fire according to Dziurzynski, 2003, some empirical work had been undertaken to establish the amount of heat produced by various types of fuels. For example, 20 liters of diesel fuel was burned in a pool of a surface of 1.72 m^2. This amount of diesel fuel burned for 17 minutes in 1.5 m/s velocity of air. With a calorific value of 34 MJ per liter for diesel fuel, a total 680 MJ of energy was produced. With a burning time of 17 minutes, this means that the heat production averaged at 0.67 MW (680 MJ/1,020 seconds).

It is understood that with a larger surface area the fire would burn faster and with a larger amount of the fuel available, the fire would burn longer. For example, if 1,000 liters of diesel is available for the same surface areas, the fire will last 850 minutes with a constant burn rate of 0.02 l/s and heat output of 0.67 MW instead of 17 minutes for 20 liters fuel with

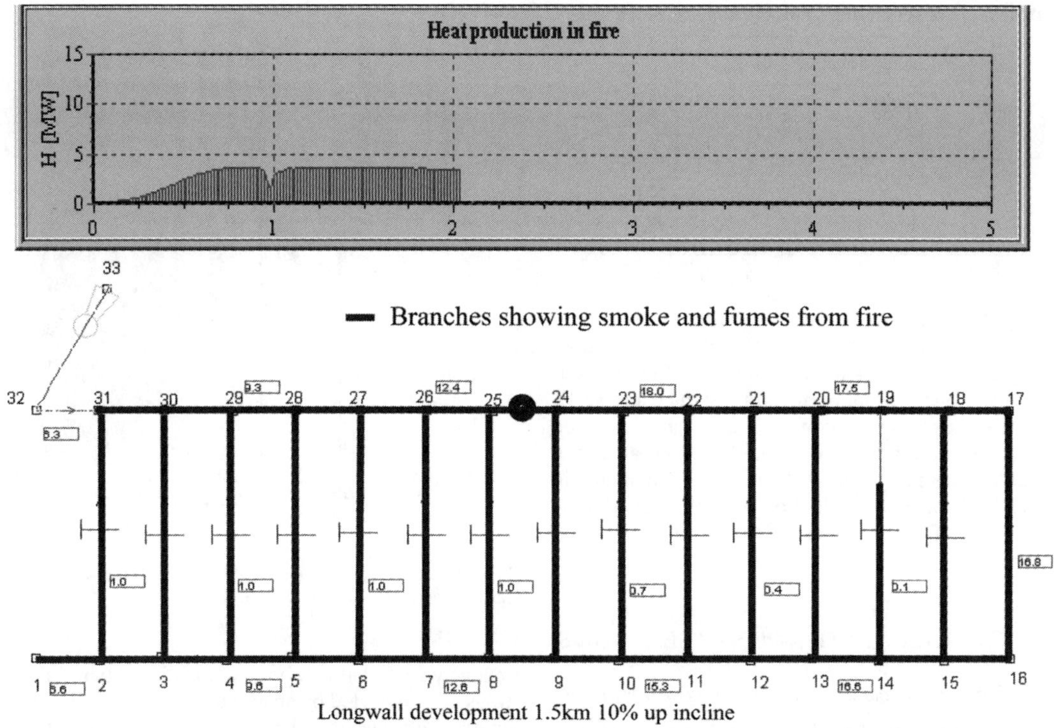

Figure 4. Smoke reversal and heat reduction observed during simulation for 10 m long fire zone in return roadway.

Table 4. Prediction of heat outputs from various fires.

Fuel (l)	Area (m^2)	Burn time (min)	Burn time (s)	Rate (l/s)	Heat output (MW)	Heat output (MW/hr)	Q m^3/s
20	1.72	17	1020	0.020	0.67	0.19	0.67
1000	1.72	850	51000	0.020	0.67	9.44	0.67
1000	10	146	8772	0.114	3.88	9.44	3.88
458	25	27	1607	0.285	9.69	4.33	9.69
818	50	24	1435	0.570	19.38	7.73	19.38
1020	100	15	877	1.140	38.76	9.63	38.76

the same burn rate and heat output. Similarly, if the fuel is spread over a larger surface area say 10 m^2 then it will burn at a fast rate providing that there is abundant oxygen supply. In this case, the fire will only last for 146 minutes but with a much faster burning rate (0.114 l/s) and a higher heat output (3.88 MW) from the fire. Table 4 shows a summary of calculated heat productions for various combinations of surface areas and amount of fuels available.

According to Yung, 2003, approximately 1 m^3/s of air is required for every 1 MW of heat produced from a fire, if the fire is not to be air/oxygen starved. Based on this, it is possible to calculate the minimum amount of air required to sustain a fire without oxygen deficiency. The minimum air quantity is also calculated and included in Table 4.

Several simulation exercises were carried out to examine the relationship between the length of fire zone and the heat production predicted by the simulator. For each of the exercises, a diesel fire was used with an intensity of nominal value 10. Three lengths of fire zone were tested namely, 25 m, 50 m and 100 m. The fires were set in a horizontal airway with about 25 m^3/s of air initially. The fires were simulated for about 160 minutes to let the fires became stable. Areas under the heat production charts were measured and

Table 5. Prediction of heat production and fuel burnt.

	Average heat output		
Length (m)	Whole (MW)	1st hr (MW)	Fuel burned (l)
25	5.26	4.32	458
50	8.86	7.73	818
100	8.97	9.64	1020

average heat outputs for the first one-hour period and the whole period were obtained. Based on heat production predicted by the simulator, it is possible to calculate the amount of fuel burned during the exercises.

The values for amount of fuel burned and averaged heat production in the fire for the first hour as shown in Table 5, compare well with the values calculated in Table 4. Therefore, it should be possible to model the magnitude of various fires by varying the length of fire zone.

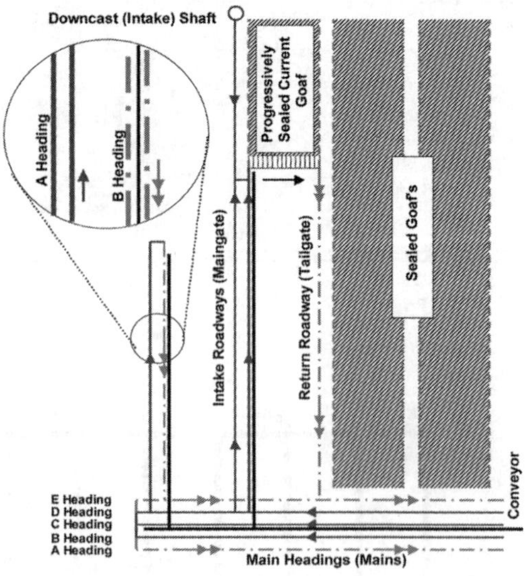

Figure 5. Typical layout of Australian longwall mining (after Mayes and Gillies, 2002).

3 LONGWALL VENTILATION CASE STUDIES

3.1 *Typical aspects of Australian longwall mining*

The typical layout of an Australian longwall mine is shown in Figure 5. In terms of ventilation nomenclature intake roadways are shown as solid, single arrow roadways whereas return roadways are shown as dashed, double arrow roadways. In this case a raisebore exists behind the current goaf and is shown as a circle with an intake roadway connecting to the longwall face roadway (Mayes and Gillies, 2002).

Australian longwalls at present normally use only two roadway maingate developments and have typically between five and seven Mains' roadways. In development, A Heading (as shown in Figure 1) is an intake roadway with B Heading the return roadway through which the panel conveyor runs.

In the Mains, B, C, and D Headings are typically intake with flanking return roadways, A and E Headings. When all longwalls are being extracted on one side of the Mains only, D and E Headings may be used as return roadways with A, B and C Headings as intake roadways. The conveyor runs in the intake headings, typically in C Heading. In Queensland this roadway is segregated from either one or both of the other intake roadways.

3.2 *Mine ventilation circuits and diagonal connections*

Ventilation systems in an operating mine generally consists of an arrangement of multiple, interconnected airway. In most of the cases, airways are interconnected in a way such that two points of different parallel airways are connected by an airway. In this case the interconnected airway can be termed a "diagonal" airway and the mine ventilation network is much more complicated than the usual parallel and/or series network (Wala, 1999).

The airflow through the diagonal connection can be unstable, and its direction depends on the resistances of the branches interconnected with it. Diagonal connections almost always exist in a mine ventilation network; however, good planning can decrease the total number of these branches. Usually it is not possible to eliminate all of them, but it is important to recognize the instability problems associated with their existence and to pay attention to their effect on the mine ventilation network.

The stability of the airflow through the diagonal connection is related to the resistance of the airways interconnected with the diagonal branch which is well known as the basic Wheatstone Bridge balance conditions for electrical circuits as shown in Figure 6.

These relationships are as follows:

a) No flow in the diagonal branch

$$\frac{R_1}{R_2} = \frac{R_3}{R_4}$$

b) Flow from A to B in the diagonal branch

$$\frac{R_1}{R_2} < \frac{R_3}{R_4}$$

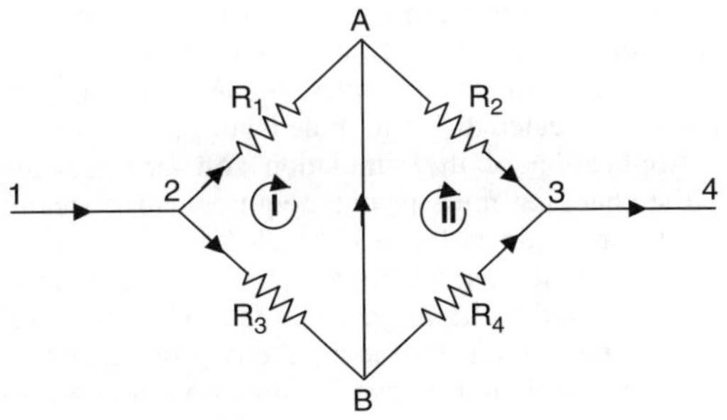

Figure 6. Ventilation network circuit – Wheatstone bridge.

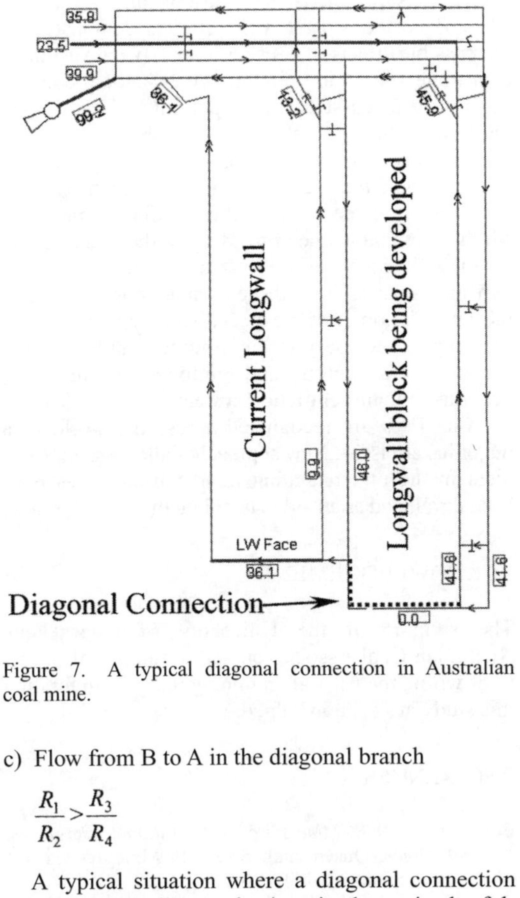

Figure 7. A typical diagonal connection in Australian coal mine.

c) Flow from B to A in the diagonal branch

$$\frac{R_1}{R_2} > \frac{R_3}{R_4}$$

A typical situation where a diagonal connection exists in Australian coal mines is shown in the following diagram. The diagonal connection is established when the next longwall block face road holes through to the existing longwall panel.

Case studies were undertaken to simulate the effects of fires on the simplified ventilation circuits before and after the next longwall face road holed through and created an airflow connection to the current longwall circuit. Figure 8 shows how the smoke progresses

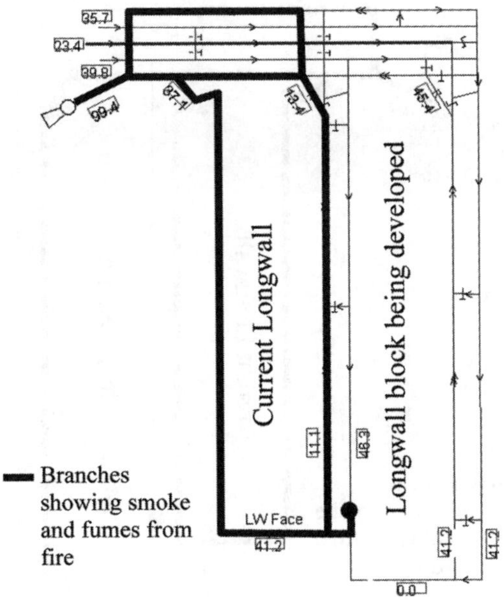

Figure 8. Smoke progression without diagonal connection.

if a diesel fire with a 30 m length of fire zone, a fire intensity of 10 and a time constant of 120 seconds is started 50 m outby of the current longwall face. In this case, the fire affects only the current longwall face.

Once the next longwall block face road holes through, an airflow ventilation connection is established. In this case, 2.2 m^3/s of air flows from next longwall block to the current longwall panel. The same fire is set at the same location for this case. It was found the fire will affect both current longwall and the next longwall development blocks, after the diagonal connection is established, as fire causes the reversal of airflow through the diagonal connection as shown in Figure 9. The fire here generates a net buoyancy effect on the airflow in the current longwall maingate. This causes an increased air pressure, which causes air to flow in reverse into the next longwall faceline.

There are two possible situations when diagonal branches can adversely affect the ventilation system (Wala, 1999).

1. When a diagonal connection already exists and a flow reversal or stagnation occurs due to the resistance changes around it (as is the above case in this example where the fire causes a net buoyancy change).
2. When a diagonal branch is added that can cause airflow redistributions in the surrounding branches.

It is important to identify the presence of existing or potential diagonal connections within the mine ventilation network as the airflow through the diagonal

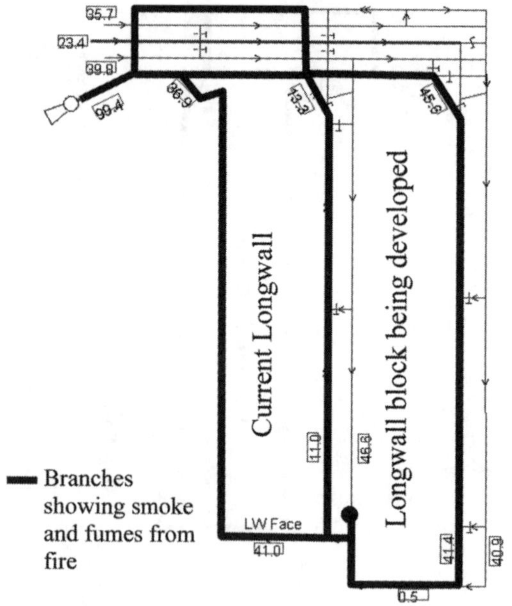

Figure 9. Smoke progression with diagonal connection.

connections could reverse or stop, due to the changes in the adjoining branches within the ventilation network. In some case, contaminated air may attempt to flow into the working areas. However, the existing diagonal branches could be useful during fire fighting or mine rescue operations, where, by alteration of the flow direction in a diagonal branch it is possible to deliver fresh air to areas where it is needed, or to redirect the contaminated air away from working areas where adverse efforts would otherwise result, if the contaminated air was present.

4 CONCLUSIONS

A comprehensive research project funded by the Australian Coal Association Research Program and substantial mining company site support into mine fires study has been described. The project involved applying the "Ventgraph" mine fire simulation software, preplanning of escape scenarios and general interaction with rescue responses. Mine site testing is essential and allows the approach to be introduced in the most creditable way. Outcomes to date from the project were discussed.

The project is aimed to assist the Australian mining industry to attain an improved position in their understanding of mine fires and the use of modern advances to preplan for mine fires and possible emergency incidents. Work undertaken so far at individual Australian coal mines is discussed as examples. The effort is built around the introduction of fire simulation computer software to the Australian mining industry and the consequent modeling of fire scenarios in selected different mine layouts.

Application of the simulation software package to the changing mine layouts requires experience to achieve realistic outcomes. Most Australian mines of size currently use a ventilation network simulation program. Under the project a small subroutine has been written to transfer the input data from the existing mine ventilation network simulation program to "Ventgraph". This has been tested successfully. To understand fire simulation behavior on the mine ventilation system, it is necessary to understood the possible effects of mine fires on various mine ventilation systems correctly first. Case studies demonstrating the possible effects of fires on some typical Australian coal mine ventilation circuits with diagonal connections were discussed. It is important to identify and understand these potential effects on the mine ventilation network as the airflow through the diagonal connections could reverse or stop due to the changes in the adjoining branches within the ventilation network. Mining companies need to identify the existing and potential diagonal connections in their ventilation system and analyze how these connections will affect their ventilation system especially in the case of fires. Training is necessary to equip mine ventilation personnel how to identify and minimize diagonal connections in their ventilation system.

Mine fires are recognized across the world as a major hazard issue. New approaches allowing improvement in the understanding their consequences have been developed as an aid in handling this complex area.

ACKNOWLEDGEMENT

The support of the University of Queensland, Australian Coal Association and a number of operations within the Australian mining industry in funding this study are acknowledged.

REFERENCES

Bofinger, C. 1998. *Mine Fires in Australian Underground Coal Mines*. Queensland: Safety In Mine Testing and Research Station (SIMTARS).

Deliac, E.P. et al. 1985. Development of Ventilation Software on Personal Computers in France and the Application to the Simulation of Mine Fires. *2nd U.S. Mine Vent. Symp.* Reno. Nevada. pp. 19–27.

Dziurzyński, W., Tracz, J. and Trutwin, W. 1988. Simulation of Mine Fires. *4th Int. Mine Vent. Cong.* (Ed, A.D.S. Gillies) Brisbane. Australia. pp. 357–363.

Dziurzyński, W. 2003. Personal communication. Director, Strata Mechanics Research Institute, Polish Academy of Sciences, Cracow, Poland.

Greuer, R.E. 1988. Computer Models of Underground Mine Ventilation and Fires. *U.S. Bureau of Mines: IC 9206 (Recent Developments in Metal and Non-Metal Mine Fire Protection)*. pp. 6–14.

Greuer, R.E. 1984. Transient-State Simulation of Ventilation Systems in Fire Conditions. *3rd Int. Mine Vent. Cong.* Harrogate. U.K. pp. 407–410.

Litton, C.D. et al. 1987. Calculating Fire-Throttling of Mine Ventilation Airflow, *U.S. Bureau of Mines Report*: *RI 9076*.

McPherson, M.J. 1993. *Subsurface Ventilation and Environment Engineering*. New York: Chap & Hall.

Mayes, T.I. and Gillies, A.D.S. 2002. Economic Modeling of Australian Longwall Ventilation, Mine Ventilation – *Proc. North American 9th US Mine Vent. Symp.*, De Souza (Ed), Netherlands: Balkema.

Mitchell, P.E. 1990. *Mine Fires – Prevention, Detection, Fighting*. Chicago: Maclen Hunter Publishing Co.

Stefanov, T.P. et al. 1984. Unsteady-state Processes during An Open Fire in a Ventilation Network. *3rd Int. Mine Vent. Cong.* Harrogate. U.K. pp. 417–420.

Trutwin, W., Dziurzyński, W. and Tracz, J. 1992. Computer Simulation of Transients in Mine Ventilation. *Proc.5th Int. Mine Vent. Cong.* (Ed. R. Hemp). pp. 193–200.

Wala, A., Dziurzyński, W., Tracz, J. and Wooton, D. 1995. Validation Study of the Mine Fire Simulation Model. *Proc. 7th US Mine Vent. Symp.* Society of Mining Engineers: (Ed. A. Wala) 1995. pp. 199–206.

Wala, A.M. 1996. Controlling Ventilation for Safe Escape from Coal Mine Fires. *Mining Engineer*. April. pp. 61–66.

Wala, A.M. 1999. Three Underground Coal Mine Explosions – Twenty Miners Killed – One Reason, *Proc. 8th US Mine Vent. Symp*. Rolla. Missouri. 11–17 June 1999. Missouri-Rolla: University of Missouri-Rolla Press.

Yung, D. 2003. Personal communication. Head of Fire Science at CSIRO's Fire Science & Technology Laboratory. North Ryde. Sydney.

Mine Ventilation, Ganguli & Bandopadhyay (eds)
© 2004 Taylor & Francis Group, London, ISBN 90 5809 633 5

A statistical approach to coal dustability

I. Göktay Ediz & Sahin Yuvka
Department of Mining Engineering, Dumlupinar University, Kütahya, Turkey

Darron W. Dixon-Hardy
Department of Mining and Mineral Engineering, University of Leeds, Leeds, UK

ABSTRACT: In this research a review was undertaken of the main parameters that affect coal dustability. These parameters were used as the input to a simple statistical model that identified which parameters were statistically significant and subsequently used to develop a multiple regression model capable of predicting coal dustability. Four sets of coal specimens were analyzed to provide data on their chemical and physical characteristics including ash content, volatile matter content, moisture content, hardness and uniaxial compressive strength. Standard coal specimens of known amount were crus hed before injection into a specially designed and manufactured dust chamber. The results of dust measurements were analyzed statistically by using a simple multiple regression technique. Among the parameters determined by chemical and physical analysis, the moisture content, hardness and uniaxial compressive strength of coal specimens were seen to be most meaningful and the mathematical relationship between coal dustability and the related coal parameters were determined.

1 INTRODUCTION

Coal is still by far, the most abundant conventional energy resource in the world. When compared with alternative energy resources, coal is seen to have several advantages, including (Central Planning Organization of Turkey, 2001):

- significant reserves and therefore a longer utilization life
- a lower price
- stability and accessibility in the supply chain

In recent years, however, the share of coal in the primary energy demand of most industrialized countries has not been proportional to the availability of the resource mainly due to environmental concerns. However, coal consumption has steadily increased since the OPEC crisis in 1973 and it is expected that it will continue in the future too. Obviously, parallel to the increase in coal utilization, new and environmentally friendly coal technologies are being developed and improved (Kural, 1994).

In underground coal mining, one of the most dangerous environmental hazards is the possibility of an explosion and the main explosive substances are methane and coal dust. In the UK it is now common practice to achieve 1.5 to 2 millions tonnes per year from a single, highly mechanized longwall coal face. This move towards greater mechanization and greater production from single coal faces can result in the generation of finer coal particles and, increased rates of methane emission from the coal face and surrounding strata and coupled with larger coal face electrical power usage have increased the possibility of a coal dust explosion in underground coal mines (Ediz et al., 1998).

Dust may cause many critical problems underground such as:

- an adverse effect on human health
- increased risk of a dust (typically coal) explosion
- exacerbates working conditions
- cause mechanical and electronic breakdown of the equipment.

Generally, there are four rules regarding dust control, which should be followed if the problem is to be tackled effectively. These are as follows (The Mine Ventilation Society of South Africa, 1989):

- keep dust production to a minimum and prevent it from contaminating the atmosphere by controlling it at its source
- dilute it as rapidly as possible

- filter it
- remove persons from high dust concentrations.

Initially, all efforts should be concentrated on the prevention of dust formation and liberation at the source. If formed, the cardinal rule is then to prevent the dust becoming airborne. This is the simplest, most effective and the cheapest method of dust control. Once the solid contaminant becomes airborne, the control task is rendered immeasurably more difficult and costly (Hartman, 1982). In order to control dust formation in the air, it is necessary to know the dust formation characteristics of solid materials from which it is formed.

2 DEFINITION AND THE FORMATION OF DUST

The term dust is applied to finely divided solid matters in the air (Sengupta, 1994). Dust which is smaller than 1 mm, can suspend in air or settle in the course of time (Güyagüler and Durucan, 1985). In mine air about 80% (by mass) of the respirable dust is smaller than 1 μm and only about 4% of it is larger than 4 μm. This is also roughly the size distribution commonly found in the lungs of a deceased pneumoconiosis patient (Cronje et al., 1997).

There are several factors affecting dust occurrence. The analysis of these factors is important to enable prediction of the amount of dust likely to result from future mining operations. These factors can be classified as the properties of the matter (as mechanical and physical properties) and the environmental conditions/processes that yield dust. It is obvious that different coals produce different amounts of dust for the same mining conditions. In other words, some coals produce dust more easily compared to others. The reason for this is the difference in their physical and mechanical properties (Güyagüler, 1982).

Environmental conditions also have certain effects on dust formation. The depth, the temperature, the strata pressure, the humidity content of the mine and how the mine is ventilated are examples of environmental parameters affecting dust formation. As the depth of the seam increases, strata pressure also increases causing a decrease in the brittleness of the rock and a similar effect may be seen because of the increase in temperature with depth i.e. as the temperature increases, the brittleness of the rock decreases. Therefore, dangerous dust occurrence can be reduced with depth. Increased temperature, on the other hand, may reduce the strength of the rock. The humidity of the mine atmosphere or the water content of the rock itself may prevent the dust becoming airborne. As the water content of the rock increases, the strength of the rocks decreases (Güyagüler, 1982) making dust liberation all the more likely.

2.1 *Dust formation characteristics of rocks*

The most important factors affecting dust formation of rocks are as follows:

- hardness
- brittleness
- impact strength
- comminution (crushing) strength

Among these factors brittleness was noted as the major parameter controlling dust formation by the researchers. The second most important factor was found to be the hardness of the rock. As the hardness of rock increases, dust formation also increases. Moreover, typical hard rocks produce dusts mainly in the range of 0–10 μm especially in drilling operations (Güyagüler, 1982).

2.2 *Dust formation characteristics of coal*

The dust formation characteristics of coal have been researched by several authors. From the results, it was generally found that carbon, volatile matter, ash and the moisture content of coal are also important factors for coal dust formation in addition to the factors stated for the rocks in section 2.1 (Güyagüler, 1982). Other important results concerning the dust formation characteristics of coal are as follows:

- An increase in volatile-combustible matter content causes an increase in dust formation for the coal. As the volatile-combustible matter content increases from 10% to 20%, dust formation doubles, as the other factors remain the same.
- There is a good correlation between dust occurrence and carbon content of coal. In other words, as the carbon content increases, so does the dust formation.
- Ash in coal may have different characteristics; therefore, it may have an effect accordingly.
- Moisture content of coal has a negative effect on dust occurrence.
- Similar to rocks, hardness increases the dust occurrence.
- Density is considered to have a positive effect on dust occurrence.
- Brittleness is regarded as the most crucial factor affecting dust occurrence.

3 DETERMINATION OF CHEMICAL AND PHYSICAL COAL PROPERTIES AND SAMPLE PREPARATION

Four different coal samples were used in the tests and these were collected from:

- Turkish Coal Enterprise Garp Lignite Company's 6th Colliery (underground).
- Turkish Coal Enterprise Garp Lignite Company's open cast mine

Table 1. Chemical analysis of the test coal samples.

Coal samples	Moisture content (%)	Ash content (%)	Volatile matter (%)	Fixed carbon (%)	Density (%)
1	10.05	11.17	19.08	69.75	1.12
2	13.55	22.10	25.07	52.83	1.23
3	35.36	24.27	41.21	34.52	1.33
4	38.11	20.80	37.91	32.01	1.28

Coals; 1 = G.L.İ. Underground Mine, 2 = G.L.İ. Surface Mine, 3 = S.L.İ. B_3 Coded Seam, 4 = S.L.İ. B_1 Coded Seam.

Table 2. Results of hardness tests by Schmidt Hammer.

Coal samples	Hardness value
1	56.3
2	53.3
3	52.4
4	54.9

Coals; 1 = G.L.İ. Underground Mine, 2 = G.L.İ. Surface Mine, 3 = S.L.İ. B_3 Coded Seam, 4 = S.L.İ. B_1 Coded Seam.

Table 3. Uniaxial compressive strength values of the samples.

Coal samples	Uniaxial compressive strength (MPa)
1	16.767
2	19.083
3	20.031
4	24.043

Coals; 1 = G.L.İ. Underground Mine, 2 = G.L.İ. Surface Mine, 3 = S.L.İ. B_3 Coded Seam, 4 = S.L.İ. B_1 Coded Seam.

- Turkish Coal Enterprise Seyitömer Lignite Company's open cast mine (B_1 coded seam)
- Turkish Coal Enterprise Seyitömer Lignite Company's open cast mine (B_3 coded seam)

A chemical analysis of the samples is shown in Table 1.

The hardness of the coal samples were determined by using a Schmidt Hammer. The results of the hardness tests are seen in Table 2.

The uniaxial compressive strength values of the coal samples were determined by Point Load Tests. The results of point load tests were corrected and converted to uniaxial compressive strength values as given in Table 3.

The samples were first comminuted in a jaw crusher before being fed into the dust chamber. To define the optimum amount of dust and size of the coal samples to be fed into the chamber, a number of tests and dust concentration measurements were made. The results of these tests are summarized in Table 4.

Table 4. Preliminary tests to define the optimum amount and the size of the samples.

Test no.	Amount of dust (g)	Dust circulation time (min)	Meas. period (min)	Delay time (min)	Dust conc. (mg/m^3)
1	40	1	5	1	1.947
2	45	1	5	1	2.837
3	45	1	5	1	2.642
4	35	1	5	0.5	1.534
5	35	1	5	1	1.309
6	40	1	5	0.5	2.536
7	40	1	5	1	2.060
8	40	1	5	2	1.092

Typical dust concentrations in mine atmospheres are in the range of 2 mg/m^3 to 5 mg/m^3. The initial tests failed to achieve these typical concentrations in the dust chamber so the inlet feed to the crusher was kept at a constant 50 g and the crusher's outlet varied between the range of 1 mm, 1.25 mm, 1.5 mm and 2 mm gap size. Tests found that the 1. 5 mm gap size was seen to produce airborne dust concentrations at about 2 to 5 mg/m^3 for 50 g of coal feed. Therefore, 50 g of coal and a jaw crusher outlet diameter of 1. 5 mm were found to be optimum.

From the tests carried out, the following parameters were found to be optimum for the subsequent experiments.

- amount of dust to be fed into the chamber: 50 g
- jaw crusher outlet diameter: 1.5 mm
- dust circulation period: 1 minute
- measurement period: 5 minutes
- delay time in measurement after circulation stops: 15 seconds.

4 THE DUST CHAMBER AND DUST MEASUREMENTS

In order to create standard dusty conditions (representing the underground environment) a specially designed and manufactured dust chamber was used and the dust concentration measurements were made under the same conditions using different coal samples for each test. Figures 1 and 2 illustrate the dust chamber and peripheral equipment. A schematic view of the dust chamber and its peripheral equipment can be seen in Figure 3 and the technical specifications are given in Table 5.

Figure 1. The dust chamber – view 1.

Figure 2. The dust chamber – view 2.

Table 5. Technical properties of the dust chamber.

Parameter	Value
Fan speed	2.2.m/sec
Diameter of air outlet (dust feeding unit)	2½″
Diameter of dust inlet (dust feeding unit)	1½″
Dimensions of dust feeding unit	Cylindrical body Ø = 20 cm, h = 15 cm conical body Ø = 20 cm, h = 15 cm
Specifications of electrical motor	230 V-50 Hz, 0.37 kW, 500 rpm
Specifications of reductor	230 V-50 Hz, 4 rpm
Dimensions of dust chamber	Cylindrical part Ø = 40 cm, h = 40 cm conical part Ø = 40 cm, h = 30 cm
Diameter of air inlet (dust chamber):	2½″
Diameter of air outlet (dust chamber):	3″
Diameter of measurement holes:	½″
Diameter of waste outlet	¾″

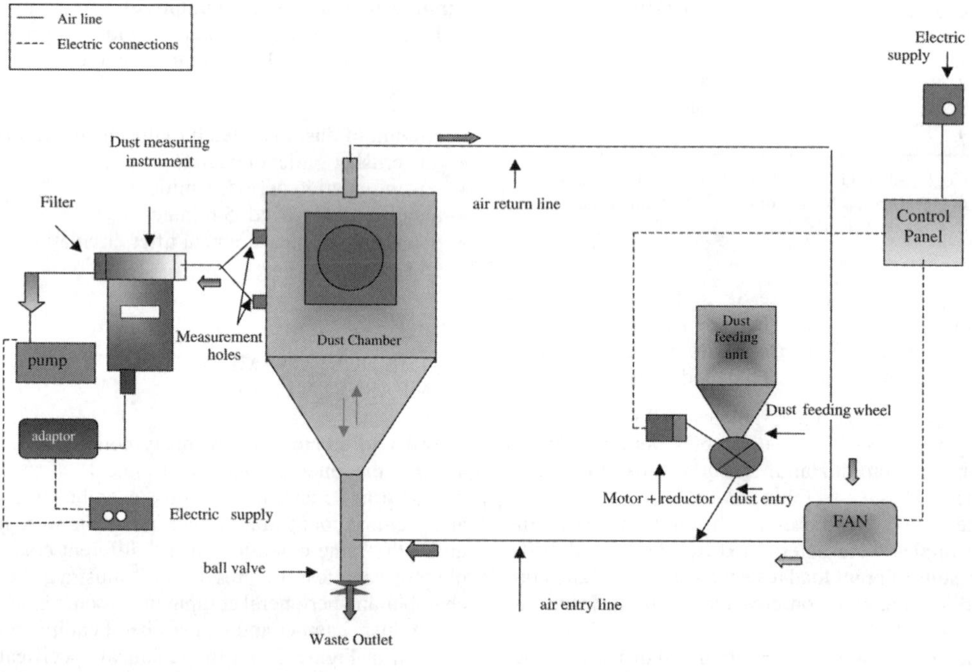

Figure 3. Schematic view of experimental set up.

The dust measuring system is comprised of 3 main parts, namely; the dust chamber, the dust feeding unit and the dust-measuring instrument. In the dust feeding unit, the coal dust sample is fed to the lower part of the ventilation pipe through which an air stream at a speed of 2.2 m/s is created by a fan. The dusty air is then circulated for 45 seconds before it stops to allow the dust particles to settle down for 15 seconds (delay time for measurements) prior to the start of the dust measurements. The fan circulating the dust through the chamber works as a closed circuit and is controlled by an electronic panel. The dust-feeding unit is controlled by a small motor and gear system connected to a 6 bladed impellor (Figure 3). This system provides a smooth feed of dust to the air stream.

The dust chamber was designed and manufactured as a cyclone to facilitate homogenous dust dispersion inside the chamber. At the bottom of the chamber, there is waste removal outlet controlled by a ball valve. There are two outlets (holes) at the outer side of the chamber to facilitate dust measurement (instrument connection). A window was also incorporated to provide visibility and cleaning of the inside of the chamber after each test.

For dust concentration measurement, an MIE personal dataRam dust-measuring instrument (PDR 1200 type) was used during the tests. A pump connected to the instrument provides a suction pressure to provide dusty airflow through the instrument from the chamber. The instantaneous dust concentrations and dust concentrations over time are measured and recorded by the instrument. The data obtained can be downloaded to a computer for analysis afterwards.

Dust concentration measurements were carried out using the parameters obtained from the preliminary test results. In other words, 50 g of coal samples for each group were comminuted using a jaw crusher with 1.5 mm jaw setting. The dust obtained was placed in the dust feeder's container and then fed into the system. The dust was then circulated for 1 minute before the resulting dust concentration was measured with the dust instrument (MIE personal dataRam 1200 model).

5 RESULTS OF EXPERIMENT AND THE STATISTICAL ANALYSIS

Three measurements were made for each test sample in order to minimize experimental errors and the results were averaged. Table 6 shows the results of the dust concentration measurements.

The results of the dust concentration measurements are given together with their physical and chemical properties and are shown in Table 7. This information was taken by the regression package for preliminary analysis to decide which parameters (listed in Table 8) were statistically significant and thus chosen for regression analysis.

The model summary of the variable analysis is given in Table 9 and illustrates an excellent correlation and

Table 6. Results of dust concentration measurements.

Coal samples	Dust concentration (mg/m^3)			
	1st test	2nd test	3rd test	Average
1	5.025	4.612	4.733	4.790
2	2.627	2.484	2.226	2.446
3	1.847	2.237	1.945	2.010
4	1.799	1.712	1.803	1.771

Coals; 1 = G.L.İ. Underground Mine, 2 = G.L.İ. Surface Mine, 3 = S.L.İ. B_3 Coded Seam, 4 = S.L.İ. B_1 Coded Seam.

Table 7. Results of dust measurements together with their physical and chemical properties.

	Coal samples			
	1	2	3	4
Dust concent (mg/m^3)	4.790	2.446	2.010	1.771
Moisture concent (%)	10.05	13.55	35.36	38.11
Ash concent (%)	11.17	22.10	24.27	20.80
Volatile matter (%)	19.08	25.07	41.21	37.91
Density (gr/cm^3)	1.12	1.23	1.33	1.28
Fixed carbon (%)	69.75	52.83	34.52	31.02
Hardness	56.3	53.3	52.4	54.9
Uniaxial strength (MPa)	16.767	19.083	20.031	24.043

Coals; 1 = G.L.İ. Underground Mine, 2 = G.L.İ. Surface Mine, 3 = S.L.İ. B_3 Coded Seam, 4 = S.L.İ. B_1 Coded Seam.

Table 8. Variables entered and excluded in the regression analysis.

Model	Variables entered	Variables removed
1	Uni-axial compressive strength Hardness Moisture content	Ash content Volatile matter cont. Density Fixed carbon cont.

Table 9. Model summary in the regression analysis.

Model	R	R square	Adjusted R square
1	1.000[a]	1.000	1.000

[a] Predictors: (Constant), Uni-axial compressive strength, Hardness.

Table 10. Results of variance analysis.

	ANOVA[b]		
Model 1	Sum of squares	Df	Mean square
Regression	5.760	3	1.920
Residual	6.39E−16	0	
Total	5.760	3	

[b]Dependent Variable: Dust Quantity.

Table 11. Coefficients obtained from the regression analysis.

	Coefficients[a]		
	Unstandardized coefficients		Standardized coefficients
Model 1	B	Std. error	beta
Constant	−17.012	0.000	
Moisture	1.6E−02	0.000	0.167
Hardness	0.489	0.000	0.624
Uniaxial compressive strength	−0.381	0.000	−0.834

[a]Dependent Variable: Dust Quantity.

the results of the analysis are given in Tables 10 and 11.

A standard regression model was applied to the multi regression analysis. This model is a method by which independent variables are evaluated as a block in one step and using the parameters from Tables 8–11 produces the regression formula:

$$y = -17.012 + 0.016X_1 + 0.498X_2 - 0.381X_3$$

where y: dust quantity (mg/m^3), X_1: moisture content, (%), X_2: hardness, X_3: compressive strength, MPa.

The regression equation can be given in terms of the variable names as below:

$$\text{Dust quantity} = -17.012 + 0.016 \times \text{moisture} + 0.498 \times \text{hardness} - 0.381 \times \text{compressive strength}$$

6 CONCLUSIONS AND SUGGESTIONS

6.1 *Results*

Dustability characteristics of coal must be known to prevent dust problems in underground mining. Dust formation is certainly related to dustability characterization of rocks and coals. Therefore, the effective parameters of the coal dustability were determined and dust formation characteristics of 4 different coal samples were investigated by a statistical method in this research. The results can be summarized as follows:

- As moisture content of the coal increases, dustability decreases.
- Hardness affects dustability. Soft coals are more dustable than hard coals since coal hardness is mainly dependent on mineral composition, strength and the bounding properties.
- As compressive strength of coal increases, coal dustability decreases.
- Coal dustability increases with increased fixed carbon content.
- The effects of ash content and density to dust formation were not determined.
- After the statistical analysis, the following dust formation equation was produced:

$$\text{Dust quantity} = -17.012 + 0.016 \times \text{moisture} + 0.498 \times \text{hardness} - 0.381 \times \text{compressive strength}$$

6.2 *Recommendations*

- In this study, four different coal samples were used for the analysis. However, it is necessary to use more coal samples to obtain not only more representative results but also to make full use of the results in mining practice.
- In order to determine the effect of mining or excavation on dust formation, experiments to model parameters such as power applied during the excavation, type of the excavation and types of equipment/machinery should be investigated and incorporated into the statistical analysis.

The simple statistical approach presented in this paper is very useful since it provides an indication of the sort of relative dustability likely when excavating a certain rock (in this case, coal) only knowing moisture, UCS and hardness of the rock. It also allows an indication of the dust concentration when the relationship between the variables change and this can be especially important when a new mine concerned. There is a need for more coal samples to be tested and where rocks are concerned whether there are other important factors affecting dust formation. Several rocks should be tested and the results should be evaluated with mining parameters such as excavation type, drilling type and advance rate etc. This is only a preliminary work and proves that there is a good correlation between coal properties and its dust producing characteristics.

REFERENCES

Baysal, F. 1979. Dust problems in working places (in Turkish), *6th Turkish mining, technical and scientific congress*: 85–88.

Cronje, G.P., Vuuren, P.J.J. & Rawlins, C.A. 1997. Respirable dust in the intake airways of a coal mine, Journal of the Mine Ventilation Society of South Africa, January: 1–14.

Central Planning Organization of Turkey (DPT), 2001. *Eighth development plan (for 5 years)* (http://e-kutup.dpt.gov.tr).

Ediz, I.G., Seyfettinoglu, M.A. & Dixon-Hardy, D.W. 1998. Suppression of coal dust using surfactants in combating coal dust explosions, *International Symposium on Environmental Issues and Mineral Production*, SWEMP'98, Ankara, Turkey: 375–379.

Güyagüler, T. 1982. Factors affecting dust occurrence (in Turkish), *Turkish 3rd coal Congress*: 409–419.

Güyagüler, T. & Durucan, Ş. 1985. Mine dusts, environmental problems and the control methods (in Turkish), Middle East Technical University seminar notes: 55–58.

Hartman, H.L. 1982. *Mine Ventilation and Air Conditioning*, second edition, John Wiley and Sons Press, New York.

Kural, O. (Editor), Uğursal, I., Tawel, A.M., Hamdullahpur, F., Özil, E. & Uyar, T.S. 1994. *Economics of Coal Utilization*, Coal: 417–419.

Sengupta, M. 1994. *Mine Environmental Engineering*, Volume 1: 199.

MIE INC, 2000. *Personal dataRAM*, instruction manual, Massachusetts, USA.

The Mine Ventilation Society of South Africa, 1989. *Environmental Engineering in South Africa Mines.*

Mine Ventilation, Ganguli & Bandopadhyay (eds)
© 2004 C. Taylor, ISBN 90 5809 633 5

Evaluating factors affecting the performance of three axis ultrasonic anemometers

Charles Taylor, Emery Chilton, Linda McWilliams & Mark Senk
NIOSH, Pittsburgh Research Laboratory, Pittsburgh, PA, USA

ABSTRACT: Three-axis ultrasonic anemometers are used to studfy airflow in the NIOSH ventilation gallery where ventilation in an underground mine is simulated. In this study, performance characteristics of this anemometer are evaluated and the effects of sensor orientation and sampler location on flow readings are examined. The study results provide guidelines for more effective use of the ultrasonic anemometers in the ventilation gallery.

1 INTRODUCTION

Monitoring airflow in underground gassy mines is essential for maintaining methane levels below the explosive concentration limit and providing a safe environment for workers. This is particularly true in areas near working faces where methane liberation rates are generally highest. Normally a handheld vane anemometer is used to measure average air velocity at the end of ventilation tubing or brattice. Smoke tubes are used to observe short distance flow patterns in face areas. However, airflow is often turbulent near the face and it is difficult to accurately measure air velocity and direction.

Three-axis ultrasonic anemometers are often used for collecting meteorological data. Two units, called "Windmasters," were purchased from Gill Instruments Ltd., Great Britain (Reference to specific products does not imply endorsement by NIOSH). Figure 1 shows one of the anemometers attached to a stand. The "Windmaster" units were used to measure airflow velocity and direction in the NIOSH Ventilation test gallery. The conditions in the gallery simulate ventilation conditions underground and the measurements are used to evaluate the effectiveness of airflow for diluting and removing methane liberated at the mining face (Taylor 2003).

In the test gallery the anemometers must frequently be moved to make measurements at multiple locations. Thus, proper calibration of the anemometers is important in order to compare flows at multiple locations or for different operating conditions.

The anemometer has a linear response and an absolute calibration that depends only on sensor spacing and transit time measurement accuracy. Ultrasonic pulses traveling back and forth between three sets of sensor probes are used to measure the air velocity vector. The instrument is robust to the surrounding environment, but physical impacts can alter the distances between the probes and affect instrument calibration.

Adjustments to an instrument to correct for changes in sensor spacing can only be made by the manufacturer. During testing in the gallery it is important to check instrument performance to ensure that readings remain constant for equal flows and that instruments exposed to the same airflow give the same flow readings.

This study describes how anemometer performance was evaluated by comparing readings from anemometers exposed to the same flows in the gallery. Factors affecting variations in readings obtained in the gallery are also discussed.

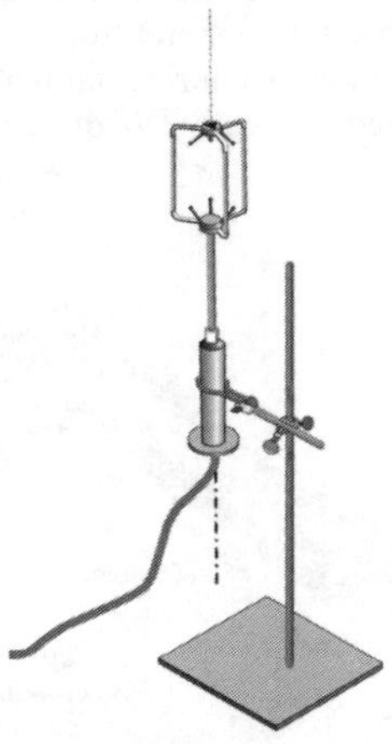

Figure 1. Ultrasonic anemometer on stand.

2 TEST PROCEDURE – COMPARING AIRFLOW IN THE VENTILATION GALLERY

2.1 *Test gallery*

Tests were conducted in the NIOSH ventilation test gallery to evaluate how operating conditions affect the flow of intake air in mines. The gallery is designed to simulate ventilation conditions near the face in an underground mine. The research focuses on ways to improve the dilution and removal of methane liberated at a working mining face.

One side of the L-shaped gallery is designed to simulate a mining entry with a 2.2-m (7 ft) high roof and ribs that are 5 m (16½ ft) apart (see Figure 2). Air enters through two windows and an exhaust fan removes air from the gallery at a rate of 5.9 m^3/s (12,500 cfm). A brattice curtain that reaches from the floor to the roof is supported by a wood frame that is constructed 0.6 m (2 ft) from the wall. The curtain directs outside air toward the face. The air quantity was increased by closing the regulator doors or decreased by opening them.

The anemometers were tested in the 2-ft wide area between the curtain and wall. The flow turbulence was less in the area because the wall and curtain were relatively straight and the cross-sectional area – 0.7 by 2.2 m (2 by 7 ft) – relatively uniform.

2.2 *Instrument orientation*

The ultrasonic anemometer measures flow in three orthogonal directions, U, V and W (see Figure 3), but airflow in the test facility moves primarily in the horizontal U and V directions. Most profiles of flow are drawn in a UV plane that is parallel to the gallery roof and floor. Using anemometer readings, airflow profiles for the gallery are drawn to show the air direction and relative velocity in the horizontal plane that is mid-way between the roof and floor.

To measure flow in a horizontal plane, the anemometer is attached vertically (W direction) to a stand. In this position, the flow velocity and direction are measured in a horizontal plane that is perpendicular to and passes through the center of the sensor head. Flow velocities in this horizontal plane are described by the resolved magnitude of the U and V vectors and calculated by the equation:

$$\text{Velocity} = \sqrt{U^2 + V^2}.$$

2.3 *Data acquisition*

Each anemometer is powered by a "Power Communication Interface box" (PCI) which collects signals from the instruments and transfers the information to a computer. To collect data from two anemometers, the two instruments are networked through separate PCI boxes as shown in Figure 4.

A software program, ANEMVENT 2003, was written and developed by NIOSH personnel for acquisition of ultrasonic anemometer data. More information about the ANEMVENT 2003 program is available in the software instruction manual (Senk 2003). ANEMVENT 2003 sequentially polls the two anemometers at intervals of 0.2 seconds. ANEMVENT 2003 then creates EXCEL spreadsheets for storing and analyzing the data. EXCEL functions are used for calculating the air flow velocity and direction.

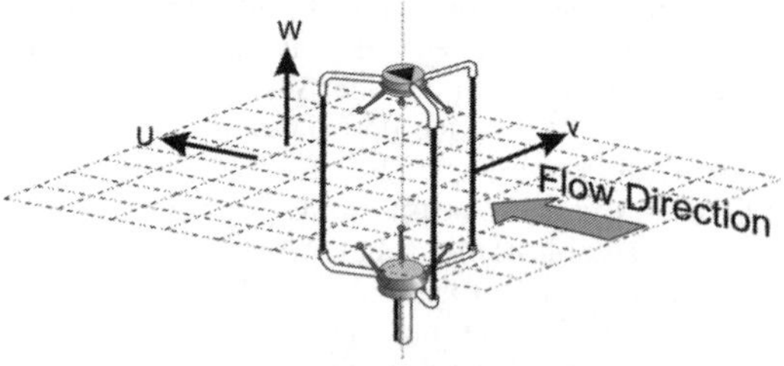

Figure 3. Airflow passing over sensor head.

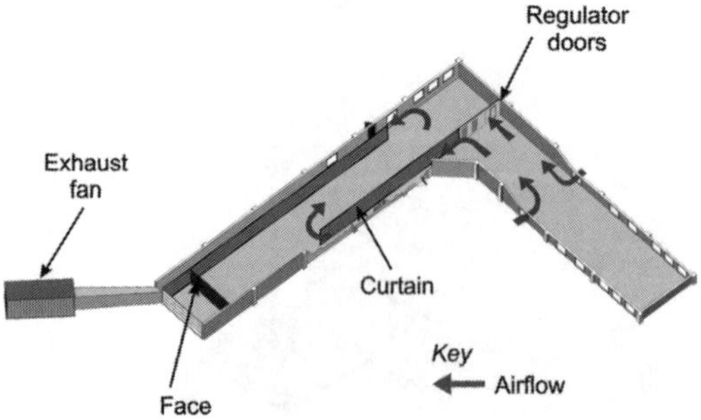

Figure 2. Ventilation test gallery.

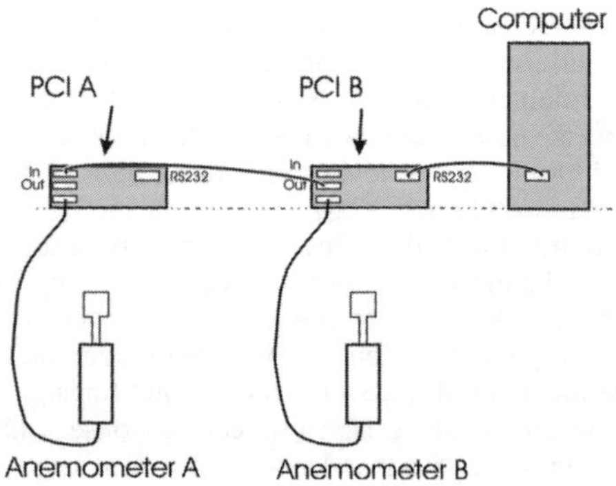

Figure 4. Networking of two anemometers.

During a test, data from each instrument was recorded once per second for a period of 3 minutes. The 3-minute sampling period (180 samples) was chosen to provide average flow magnitude and direction readings that would include the expected range of values.

2.4 *Comparing readings from the two instruments*

To compare the flow readings the two instruments (designated instruments A and B) were alternately placed at locations 3 and 6.1 m (10 or 20 ft) from the end of the curtain. The sensor heads were positioned at the centerline of the area behind the curtain i.e. 0.3 m (1 ft) from the curtain and 1.1 m (3.5 ft) from the floor (see Figure 5). To maintain the same orientation for each instrument, a reference arrow on top of the sensor head was pointed toward the face, and the anemometer body was positioned vertically. A round bubble level was used to check the vertical orientation. Flows of approximately 100, 400, and 700 fpm were directed over the two anemometers.

Table 1 presents the mean airflows and standard deviations for anemometers A and B for each of the airflow velocities.

A two-way analysis of variance (ANOVA) was conducted to compare the effect for instrument (A and B) and the effect of sampling at the two locations (20 ft and 10 ft). The results are presented in Table 6 which can be found in the Appendix. Because this was a balanced design, the analysis was robust to any deviations from normality and variance homogeneity (Keppel 1982). Due to the large sample size of 720 measurements, the level of statistical significance was set at $p < .0001$.

No main effect due to anemometer was found at any of the three airflow velocities (100, 400, or 700 fpm). There was a significant anemometer-location interaction at 100 and 400 fpm. The interaction at 100 fpm was disordinal (i.e., at 10 ft the mean for B was larger while at 20 ft the mean for A was larger); thus the significant main effect for location could not be interpreted without considering the individual anemometer. At 400 ft, the interaction was ordinal, with the means at 10 ft being consistently higher than the means at 20 ft. At 700 fpm there was only a significant main effect for sampling location (10 ft means greater than 20 ft means).

For concurrent measurements with instruments A and B, airflow velocities were 4 to 6 pct lower at the 20-ft location most of the tests. Differences in cross sectional areas at the 10- and 20 ft locations account for most of this difference. The cross sectional area at the 20-ft location was 3 percent greater than at the 10-ft location.

Flow measurements were also made at locations 3.0 and 3.7 m (10 and 12 ft) from the end of the curtain. The mean airflows and standard deviations for anemometers A and B are given for flows of 100, 400, and 700 fpm in Table 2.

A two-way ANOVA to test the effect of instrument (A and B) and the effect of sampling at the two locations (10 and 12 ft) was again performed (See Table 7 in the Appendix). No main effect was found for instrument at either the 100 or 700 fpm air flows. At 400 fpm, there were significant main effects for anemometer and location. There was a disordinal interaction at 100 fpm (instrument A having a larger mean at 10 ft and instrument B having a larger mean at 12 ft) and an ordinal interaction at 700 fpm (the mean values were consistently higher at 12 ft).

Air flow measurements at the 12-ft location were consistently higher except for the 100 fpm. The cross-sectional area measured at the 12 ft location was greater (6 pct) than at the 10 ft location. The area at the measurement location, therefore, could not account for the higher velocities.

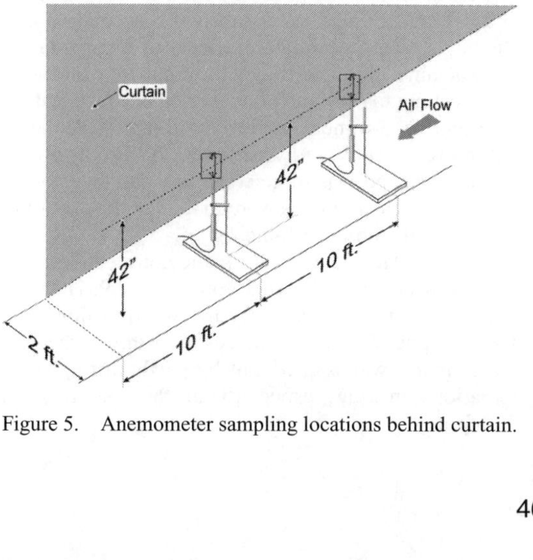

Figure 5. Anemometer sampling locations behind curtain.

Table 1. Flows measured at 10 and 20 ft (fpm).

	Anemometer A		Anemometer B	
	Mean (fpm)	Std. dev.	Mean (fpm)	Std. dev.
Air Velocity 100 fpm				
Sampling location				
10 ft	125.39	9.00	138.90	8.64
20 ft	137.87	9.56	120.22	11.23
Air Velocity 400 fpm				
Sampling location				
10 ft	415.93	11.41	405.85	12.24
20 ft	387.82	17.69	391.91	17.08
Air Velocity 700 fpm				
Sampling location				
10 ft	692.98	19.86	687.88	17.70
20 ft	658.16	26.26	656.23	25.58

Table 2. Flows measured at 10 and 12 ft (fpm).

	Anemometer A		Anemometer B	
	Mean (fpm)	Std. dev.	Mean (fpm)	Std. dev
Air Velocity 100 fpm				
Sampling location				
10 ft	96.99	9.99	82.96	7.86
12 ft	90.75	8.45	101.96	10.75
Air Velocity 400 fpm				
Sampling location				
10 ft	454.44	10.02	451.22	10.96
12 ft	482.13	12.14	476.20	10.47
Air Velocity 700 fpm				
Sampling location				
10 ft	640.34	13.61	650.09	12.71
12 ft	710.56	14.70	707.81	16.01

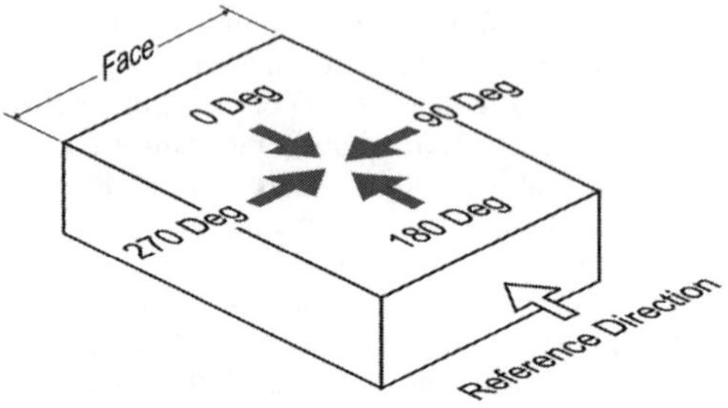

Figure 6. Varying directions of the sensor head.

The proximity of the instruments is the most likely cause for the flow differences. The two instruments were placed so that one instrument was directly upstream of the other instrument. The upstream instrument (12 ft) diverted part of the flow around the downstream instrument. This effect was seen when the instruments were 2 ft apart, but when the distance was 10 ft the upstream instrument appeared to have no effect on downstream flow readings.

2.5 *Orientation of the anemometer*

Tests were conducted to determine how rotation and tilting of the instruments affected the airflow velocity readings. To evaluate the effects of rotation, flow velocity measurements were made with the reference arrow on top of the sensor head directed toward the face, 180 degrees away from the face, 90 degrees to the right, and 90 degrees to the left of the face (see Figure 6). All measurements were made with instruments A and B located 20 and 10 ft respectively from the end of the curtain. Each instrument was oriented

Table 3. Three-minute readings (fpm).

Instrument A			
Direction, deg.	100 fpm	400 fpm	700 fpm
0	90	374	662
90	99	380	655
180	97	382	660
270	103	378	665
Instrument B			
Direction, deg.	100 fpm	400 fpm	700 fpm
0	92	384	727
90	100	393	707
180	98	392	716
270	106	396	727

vertically. Air velocities were measured for flows of approximately 100, 400, and 700 fpm. The one-second readings were recorded for 3 minutes and averaged. The 3-minute average readings for the four directions are given in Table 3.

The differences between flow measurements taken for the four rotational directions were numerically small but, in most cases statistically significant. The greatest variation occurred with the flow was approximately 100 fpm.

The results of one-way ANOVA's conducted for anemometer A found significant mean differences for rotation direction at 100 fpm [$F(3,716) = 67.48$, $p < .0001$] and 400 fpm [$F(3,716) = 8.44$, $p < .0001$]. No significant differences among the means were detected at 700 fpm. For anemometer B significant mean differences across the four degrees of rotation were shown at 100 fpm [$F(3,716) = 105.19$, $p < .0001$]; 400 fpm [$F(3,716) = 36.36$, $p < .0001$]; and 700 fpm, [$F(3,716) = 28.63$, $p < .0001$].

The differences in flow measurements could be due to the effects of sensor head orientation, or to variation in flow behind the curtain between tests. The differences in readings for the four head orientations increase as flow variation between the tests increases.

Measurements comparing readings 10 and 20 ft and 10 and 12 feet were taken 0.2 seconds apart. Readings taken to compare effects of head rotation were obtained during different 3-minute time intervals.

To estimate how much flow varied between tests, 5 consecutive 3-minute airflow readings were taken behind the curtain with instrument A. The arrow on the sensor head was directed toward the face for all readings, Measurements were made for flows of 100, 400, and 700 fpm (see Table 4).

The standard deviations for the consecutive readings shown in Table 4 varied from 1.5 to 7.9. The variation was 5 times greater for the 100 fpm flow than for the higher flows. Differences in readings when the anemometer was rotated can be partly attributed to variations in flow, especially at the low flow of 100 fpm.

Table 4. Airflows for consecutive 3-minute readings (fpm) for Instrument A.

3-minute reading	100 fpm	400 fpm	700 fpm
1	90	374	662
2	97	378	659
3	95	377	663
4	77	377	661
5	92	378	662
Std. Dev.	7.9	1.6	1.51

Table 5. Tilt angles versus velocity readings.

Tilt angle	Velocity
0	437
5	431
10	425
20	405
30	374
45	300
60	216
70	127
80	30
90	6

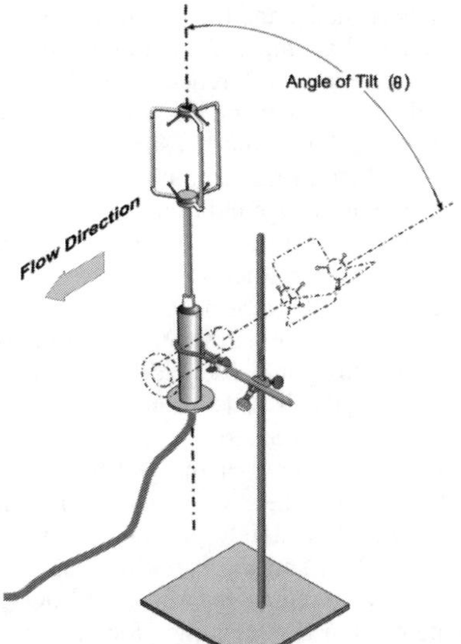

Figure 7. Anemometer tilt orientation.

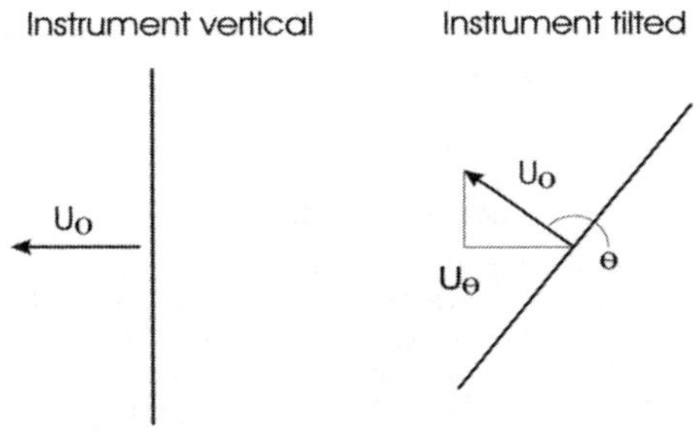

Figure 8. Effect of tilt angle (θ) on U magnitude.

To evaluate the effects of vertical orientation, measurements were made with the anemometer labeled as instrument B. It was clamped to a stand and placed in inlet air 20 ft from the end of the curtain. The reference arrow was pointed toward the face while a flow velocity of approximately 400 fpm was directed over the anemometer.

The flow velocity was first measured with the anemometer positioned vertically and then tilted, into the airflow, at angles of 5, 10, 20, 30, 45, 60, 70, 80, and 90 degrees (see Figure 7). Tilt angles were measured with an inclinometer placed on the side of the anemometer. Measurements of flow velocity were made for 3 minutes at each of the tilt angles.

Table 5 gives the average horizontal flow velocity (fpm) measured for each tilt angle. As shown, the measured velocities decreased as the tilt angle increased. The error in measurement of flow caused by tilting is approximately 1 pct at 5 degrees and 3 pct at 10 degrees. The tilt angle must be kept less than 10 degrees to maintain flow accuracy within 3 pct.

Airflow velocity in the horizontal plane is calculated from the U and V flow components. U_0 is the magnitude of the U component when the instrument is vertical (0 angle of tilt). Increasing the angle of tilt decreases the magnitude of U component. For a tilt angle of θ degrees, the measured magnitude of the U component is U_θ. The ratio U_θ / U_0 is equal to cosine θ (see Figure 8).

The graph in Figure 9 shows how the measured U vector decreased with increasing tilt angle. The magnitude of the U vector was corrected for angle of tile by dividing U_0 by the cosine of the angle of tile (θ). The values for the $U_0/\cos\theta$ remained relatively constant until the tilt angle exceeded 60 degrees, and after which the calculated values decreased rapidly.

This sudden change above 60 degrees is due to the structure of the anemometer. The three spars that hold the sensor probes in place are attached to the body of the anemometer, just below the sensor head, and to a round connector, at the top of the sensor head. At tilt angles greater than 60 degrees, the body of the anemometer and the upper spar connector interfere with flow over the sensor head. At a tilt angle of 90 degrees, flow in the U direction was essentially 0.

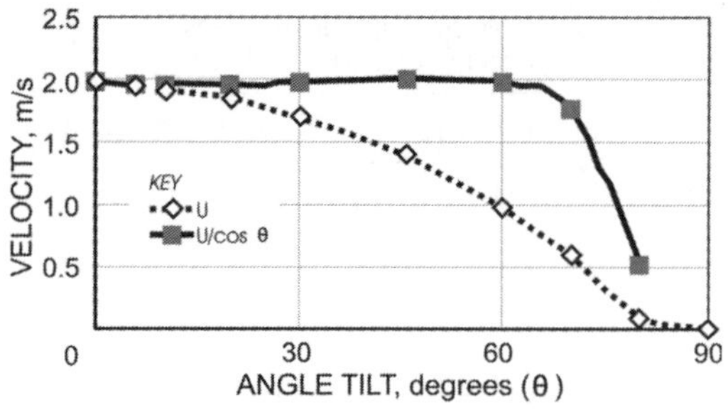

Figure 9. Change in U magnitude with tilt angle.

Table 6. ANOVA for 10 and 20 ft readings.

Source	Sum of squares	df	Mean square	F
Air Velocity 100 fpm				
Anemometer	773.07	1	773.07	8.29
Location	1730.23	1	1730.23	18.55*
Anem X Loc	43684.08	1	43684.08	468.23*
Error	66800.67	716	93.30	
Air Velocity 400 fpm				
Anemometer	1616.22	1	1616.22	7.31
Location	79608.03	1	79608.03	360.11*
Anem X Loc	9039.58	1	9039.58	40.89*
Error	158284.18	716	221.07	
Air Velocity 700 fpm				
Anemometer	2218.24	1	2218.24	4.32
Location	198819.85	1	198819.9	387.64*
Anem X Loc	452.62	1	452.62	0.88
Error	367235.69	716	512.90	

*p < .0001

Table 7. ANOVA for 10 and 12 ft readings.

Source	Sum of squares	df	Mean square	F
Air Velocity 100 fpm				
Anemometer	355.47	1	355.47	4.08
Location	7329.60	1	7329.60	84.11*
Anem X Loc	28668.83	1	28668.83	328.98*
Error	62395.03	716	87.14	
Air Velocity 400 fpm				
Anemometer	3765.91	1	3765.91	31.55*
Location	124853.72	1	124853.7	1045.93*
Anem X Loc	332.35	1	332.35	2.78
Error	85469.43	716	119.37	
Air Velocity 700 fpm				
Anemometer	2197.56	1	2197.56	10.73
Location	736606.97	1	736607.9	3595.88*
Anem X Loc	7031.22	1	7031.22	34.32*
Error	146670.93	716	204.85	

*p < .0001

3 DISCUSSION AND CONCLUSIONS

Three-axis ultrasonic anemometers are used for measuring airflow and making flow profiles in the NIOSH ventilation gallery. This study compared the performance of two anemometers in the gallery and evaluated the effect of instrument positioning and orientation on the readings.

When exposed to the same airflow and placed 10 ft from one another, the instrument readings on the anemometers were very similar. Irregularities in the cross-sectional areas of the locations where the measurements were made, and the proximity of the instruments, affected how much the readings differed.

To correct for area differences in readings it is necessary to alternate sampling locations and average the results. When making flow measurements any objects which could obstruct flow should be kept as far away as practical. For flow conditions in the gallery, the results indicate that objects 10 ft away will not interfere with flow measurements.

The rotation of the sensor head has little effect on flow velocity readings. However, any flow tilt angle does affect the flow velocity readings. As reflected in these study results, it is recommended that the tilt angle not exceed 10 degrees.

To compare airflows in the test gallery requires that the flow behind the curtain be controlled and that the flow remains relatively constant. Flow did vary behind the curtain and the variation was greatest at the lowest flow. To reduce the effects of flow variation behind the curtain flow readings for multiple instruments should be compared at higher velocities (e.g. 400 to 700 fpm).

Tests of the anemometer tilted into the airflow showed that although the U and V axis measurements were accurate, flow in the W, or vertical axis, was accurate only for airflows less than 60 degrees from the horizontal axis. The unit structure holding the six ultrasonic sensors in fixed position interferes with airflows at tilt angles greater than 60 degrees. To make accurate flow measurements when there is significant vertical flow, the anemometer should be positioned at a 90 degree tilt (horizontal).

REFERENCES

Taylor, C.D., Timko, R.J., Senk, M. & Lusin, A. Measurement of Airflow in a Simulated Underground Mine Environment using an Ultrasonic Anemometer, Annual Meeting, Preprint 03–065, 6 pp.

Senk, M. & Taylor, C.D. Anemomvent 2003 User's Guide, NIOSH Internal Document.

Keppel, G. 1982. Design and Analysis A Researcher's Handbook 2nd edition. Prentice-Hall, Inc., Englewood Cliffs, NJ.

Mine Ventilation, Ganguli & Bandopadhyay (eds)
© 2004 Taylor & Francis Group, London, ISBN 90 5809 633 5

Prediction of shock loss at splits in mine airways – a computational and experimental assessment

R.K. Jade & B.S. Sastry
Department of Mining Engineering, Indian Institute of Technology, Kharagpur, India

ABSTRACT: Investigations on combining and dividing flows in branches aim to determine the energy and pressure fluctuations, including velocity distributions. Research on this aspect of aerodynamics appears in both mining and non-mining fields. As such, the instability of flow in the turbulent flow regime, with the disturbances being caused by the division and combination of flow at a junction is a complex phenomenon. Very rarely can these disturbances be understood in a precise mathematical sense. Theoretical shortcomings for flow analysis at mine intersections, in the past, have resulted in a complete dependence on experimental studies to establish shock loss coefficients. Experimental methods gave useful but limited information regarding the design of such systems.

This paper is an attempt to present the existing concepts and to promote the computational fluid dynamics approach towards understanding these shock loss phenomena. A numerical study on loss coefficients for two-way splits is conducted and these estimations validated with experiment.

1 INTRODUCTION

1.1 *Definition*

A real fluid flow in closed conduits is accompanied with losses introduced by the viscosity of the fluid, by eddies carried by the fluid, by sudden area changes in the conduit, by secondary flow losses, and so on. For a flow in conduits encountering obstructions, change in area, direction, or division or combination a pressure head loss occurs at such locations in addition to the frictional drop. This additional loss is termed as the shock loss or the minor loss. Characteristic to an underground mine is a mesh of passageway intersections which become a huge source of pressure head losses (shock loss), in addition to the frictional drop. Therefore, shock losses would comprise a significant portion of the total head loss in mine pathways with high-density intersections.

This paper studies the airflow behavior at 90-degree square bifurcation. A coefficient which shapes the shock loss for varying flow conditions is typically found attached to the velocity head. This coefficient is termed as shock loss coefficient and defined as:

$$H_{loss} = \left(X_{1,x}\right)_x \times \left(\frac{V_x}{2g}\right) \quad (1)$$

where, $X_{1,x}$ = loss coefficient for branch x; V_x = velocity in branch x; g = gravitational constant.

Investigators have experimentally studied loss coefficients as a function of quantity ratio. Some numerical studies were also attempted to investigate the loss coefficients.

This paper includes numerical study on loss coefficient for two-way splits and the validation of the same with experimental data to ensure that the model assumptions and numerical results are accurate. To meet this objective the following tasks are performed:

- Development of two-dimensional and three-dimensional numerical models and performance of computer simulations using a CFD technique.
- Construction of a physical scaled model of intersection, and performance of a series of measurements to determine the velocity and pressure fields.
- Comparison of the simulated and measured data for the purpose of validation of present numerical

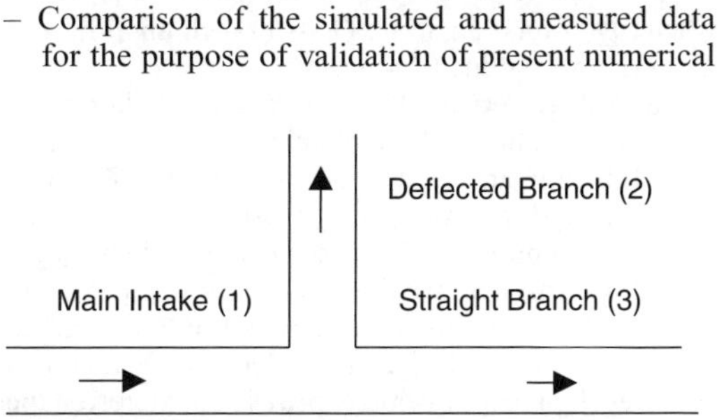

Figure 1. A two-way 90-degree square bifurcation.

results, and to compare with loss coefficients found in literature.

Branches are often encountered in mining system. Thus the flow may split from one airway at inlet to two or more airways at outlet. The flow in a bifurcation is complicated. At least two zones of boundary layer separation characterize it; the first is located downstream of the leading edge of the branch exit, and the second lies opposite to the branch in the straight through leg. Upstream, the flow anticipates these two separation zones with complex distributions of momentum associated with the wall turbulent boundary layer. In the main duct, the flow enters fully developed, but near the branch the flow in the upper part of the main duct (close to the branch exit side-wall) accelerated as a portion of the mass flow is directed into the branch and a deceleration occurs on the opposite wall as a consequence of bifurcation. A centerline pressure recovery is observed in the branching region, and then quickly a pressure drop resumes as the flow returns to fully developed state. As for the flow that enters the branching duct, it initially forms a significant flow reversal region. A recovery in pressure occurs just past the re-circulation zone, after which the fully developed flow and pressure variation characteristics return.

1.2 *Past studies*

Hartman, in 1960, proposed empirical relationships for both combination and division of flow (1982). Based on the quantity ratio the loss coefficients can be calculated.

For straight branch:

$$X = 0.5 \times \left(\frac{Q}{Q_b} - 1 \right)^2 \tag{2}$$

For deflected branch:

$$X = \left(\frac{(0.5 \times Q)}{Q_b} \right)^2 + \left(\frac{Q}{Q_b} - 1 \right)^2 + X_b \tag{3}$$

where, Q = total air flow; Q_b = branch air flow and X_b = shock loss due to bend.

Edwards & Perlee (1977) computed the shock loss at the intersections of the ventilation ducts. A computer code for laminar, steady-state incompressible, two-dimensional flow developed by Gosman was modified and used. Contrary to flow being in turbulent range, a solution with laminar flow equations was obtained. Flow in the Reynolds number range of $8 \times 10E^4$ to $2 \times 10E^5$ was tested. Use of coarser grid resulted in a slow development of velocity profile and deferred the point of separation in the deflected branch. A loss of numerical convergence is observed when 35% or more flow entered the deflected branch. Numerical solution is obtained in a small range for 65% to 89% of flow through straight duct. In this study, only straight branch loss coefficients are predicted. Over the limited range the loss coefficients for straight branch agreed well with those of Hartman's values.

Liu & Wang (1987) studied the effects of shock loss in mine ventilation using a two-dimensional finite element model – the flow being steady, incompressible, and turbulent. In this study, interestingly, an absence of shock loss was observed at the split. For the deflected branch, at a flow rate of above 30% of the total flow, the numerical solutions failed to converge. A tremendous recirculative flow is generated in the deflected branch domain resulting in convergence problem. Airway configuration in this study was a 2 m-diameter duct with 4 m length on the upstream and downstream side.

The turbulence of flow on the upstream and downstream of the bifurcation was studied by Sierra-Espinosa et al. (2000). He observed that a length of two times diameter is inadequate for the flow to develop fully downstream section.

Work on intersections by Korst et al. (1950), Vazsonyi (1980), and ASHRAE (1981) is available in the fluid mechanics literature. A set of empirical equations and tables provide for determination of loss coefficients from these works.

Literature summary: The work of Edwards & Perlee (1977) is confined to shock loss determination for straight branch only. For a slender range of 65% to 89% of total flow in the straight duct the loss coefficients were determined. Lack of numerical convergence in other flow possibilities, the use of laminar equations in numerical prediction, absence of frictional component towards loss prediction speak for the main limitations in his study.

A decade after Edwards worked with splits, Liu & Wang in 1987, modeled a steady, incompressible, turbulent flow at bifurcation using finite element method. Insufficient downstream lengths, distressed numerical convergence, lack of recirculation regions entailed to insignificant contribution being made towards final objective.

The loss coefficients obtained from various literature studies highlight a marked scatter in the estimates.

2 TREATMENT OF FLOW UNDER STUDY

The flow under study is correctly classified based on following essential considerations.

2.1 *Incompressibility*

Benedict and Carlucci (1965) tested the assumption of compressibility in the determination of pressure loss for flow in straight duct. This study revealed the degree of error in the pressure determination while comparing the compressibility and incompressibility treatments. For flows with Mach number above 0.2 the errors become significant. A similar conclusion is made by McCormack & Crane (1973). An error of 1% is observed in the evaluation of static pressure at M = 0.2. This rises to 2.25% at M = 0.3 and 28% at M = 1. Benedict et al. (1966) examined similar possibility in the determination of loss coefficients for abrupt area changes and made similar conclusions. Thus the Mach number of the flow of the gas provides a good guide to differentiate the given flow condition as incompressible. The flow in a mine never reaches a Mach number of 0.2, which decides for the mine airflow treatment to be of incompressible nature.

2.2 *Viscosity*

Inviscid condition adds to the idealization of the flow. An inviscid flow completely excludes the wall effect. The viscosity in a flow accounts for shear stresses at the wall boundary layer. Viscosity must thus be considered in obtaining the drag force.

2.3 *Turbulence*

An estimate of the character of the flow is the Reynolds number. The character is a function of the fluid properties, geometry, and the approximate magnitude of the velocity field. Reynolds number measures the relative strengths of the inertial and viscous forces. Thus, at the Reynolds number in 10^4 and above, the flow exhibits turbulence. Hence such condition needs to be modeled with the turbulent flow equations. In a numerical analysis, if turbulence is needed but not activated, the numerical analysis will diverge.

3 COMPUTATIONAL APPROACH

In the CFD finite element method, the Navier-Stokes system of partial differential equations are reduced to a finite system of algebraic equations and then solved using matrix solution techniques. Finite-element discretization equations are formed in an element-by-element fashion, independent of the global mesh. Each element in the flow field represents a separate sub-domain over which the N-S equations are integrated. For each element, these equations express the integrated form of the discretization equations in terms of nodal value of the field variables. An assembly of individual element matrix equations is solved through iterative solution techniques. Finally, a solution to the field variables is obtained at the nodal locations over the flow domain.

Computational Fluid Dynamics offers a comprehensive tool for analyzing two-dimensional and three-dimensional fluid flow. The two-equation turbulence model in ANSYS FLOTRAN enables to account for the effect of the turbulent velocity fluctuations on the mean flow. General considerations in numerical simulations are pointed as under.

3.1 *Meshing requirements*

The flow domain is meshed with elements. The adequacy of this mesh near the walls can be tested as mentioned below. Applying the turbulence model near walls involves the "Log Law of the Wall," the relationship between a non-dimensionalized distance from the wall Y^+, and the log of a non-dimensionalized velocity in a region near the wall. This law calculates a near wall viscosity that is consistent with experimentally determined velocity profiles near the wall. Y^+ below 1 indicated crowded mesh and greater than 5000 points to inadequate mesh spacing near the wall. A Y^+ value near 1000 approves of an adequate mesh.

3.2 *Turbulence models*

The Standard k-epsilon Model provides a realistic picture of the flow. It is useful for the analysis of turbulent flow in pipes. However, it over predicts the amount of turbulence in situations of flow separation and recirculation. The Re-Normalized Group Turbulence Model (RNG) is effective where the geometry has a strong dominance. The RNG model produces significantly more realistic and reliable results in regions of large strain (e.g. a converging duct) or where there is significant separation or recirculation (e.g. a duct that goes through a 180-degree change of direction). The Standard k-epsilon model encounters difficulties in handling flows with stagnation points.

3.3 *Flow boundary conditions*

Each boundary of the problem domain requires treatment. A combination of degrees of freedom (VX, VY, VZ and PRES) is specified at the boundary.

Specified flow: All velocity components are specified at the boundary.

Specified pressure: Typically, a relative pressure (usually zero) as an inlet/outlet boundary condition is applied.

If the flow has not fully developed, the numerical analysis is forced to adjust it across the last row of elements to satisfy the boundary condition. Occasionally, this adjustment may cause a mass imbalance. In order to avoid this, a development length needs to be added to the inlet and exit. The pressure drop in this "chimney" can be subtracted to obtain the desired pressure drop.

Stationary wall: This is also referred to as a no-slip condition. All the velocity components on the walls are set to zero.

4 MODELLING OF SPLITS

4.1 *Scaled physical model and flow description*

A physical scaled model of mine intersection was designed and built out of transparent Plexiglas. Keeping the available space and fan capacity limitations in mind the following model geometry of the mine-conforming to the laws of similitude- was investigated. The model consists of Masonite ducts prior to and following the Plexiglas intersection for the flow to become uniform. The intake section carries an axial forcing fan. The initial flow makes its way from fan through a honeycomb flow straightener and develops both laterally and vertically across the section on its way into the split. The split is a three-segment 90-degree square construction with a straight upstream segment length of 6.3 m and downstream segments of 3.65 m length each. The pipe wall allows for hydraulically smooth flow. The duct is of rectangular cross-section with a width of 20 cm and a height of 15 cm. The average velocity at upstream station is 6.3 m/s with a Reynolds number of 72000.

4.2 *Computational model*

The flow through the split is numerically solved with the ANSYS/FLOATRAN 6.1 CFD package. Two-dimensional (radial symmetry) and three-dimensional numerical models representing flow at mine intersections are generated. These models posses the same equivalent diameter as that of physical models so as to allow for their direct comparison of results. The flow is assumed to be steady, and the fluid to be Newtonian, incompressible, and of constant viscosity. The governing equations are solved using SIMPLE algorithm for a computational grid of 124600 elements.

Computational boundary conditions: Simulations started from fully developed velocity profiles (U_{out}, V_{out}) at two branch outlets of the split.

To simulate the most realistic velocity distribution possible at the duct outlet, the velocity distribution was determined using the 1/7 Power Law,

$$U = U_{max}(1-\frac{r}{R})^{1/7} \qquad (4)$$

where:

U = Air velocity at distance r (m) from the air duct centerline (m/s); U_{max} = Maximum air velocity (m/s); r = Lateral Distance from the air duct center line (m); R = Hydraulic Radius of the air duct (m).

A zero pressure (P_{in}) boundary condition is applied to inlet face. A no-slip zero velocity boundary condition prevails on walls of the split.

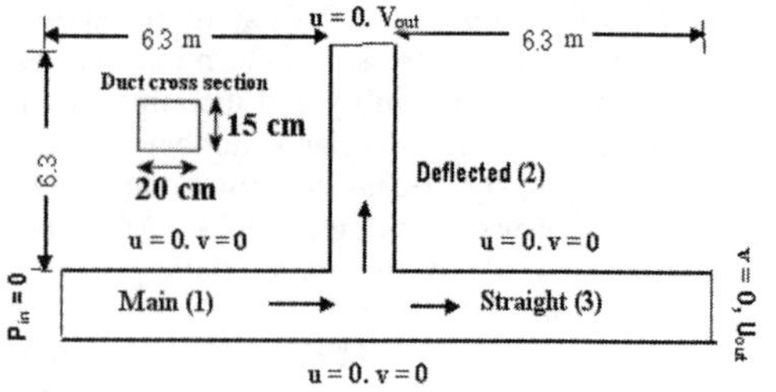

Figure 2. Boundary conditions and the extent of domain for flow at split.

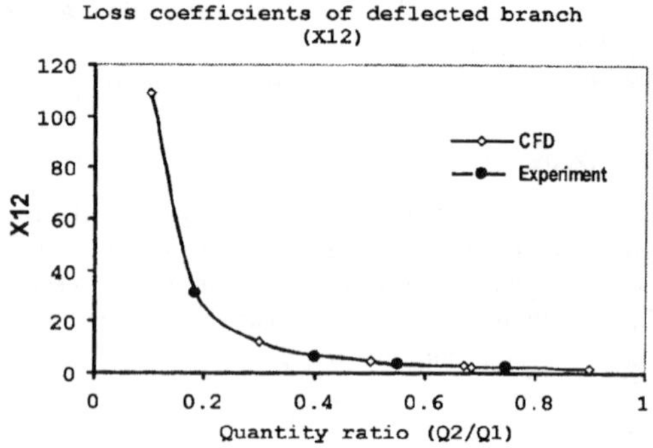

Figure 3. Shock loss coefficients for deflected branch.

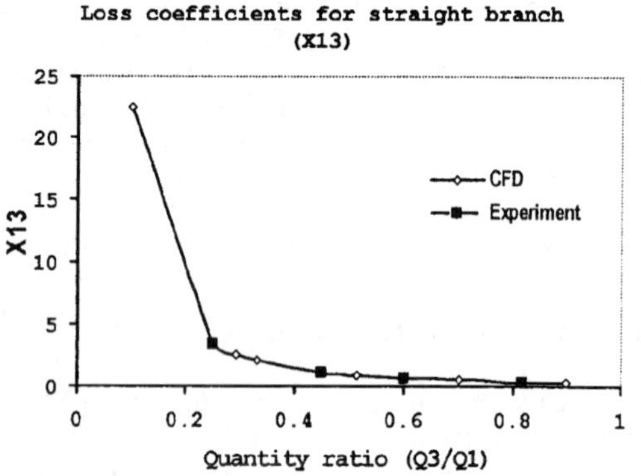

Figure 4. Shock loss coefficients for straight branch.

5 RESULTS

Loss coefficients for quantity ratio varying from 0.1 to 0.9 for both straight and deflected branch were determined both experimentally and numerically.

Figures 3 and 4 show the experimental and computationally obtained loss coefficient. Figures 5 and 6 show a comparison of loss coefficients of deflected branch (*X12*) and straight branch (*X13*) with those of Hartman, Vazsonyi, Korst and ASHRAE. The values of *X12* and *X13* obtained from various studies exhibit

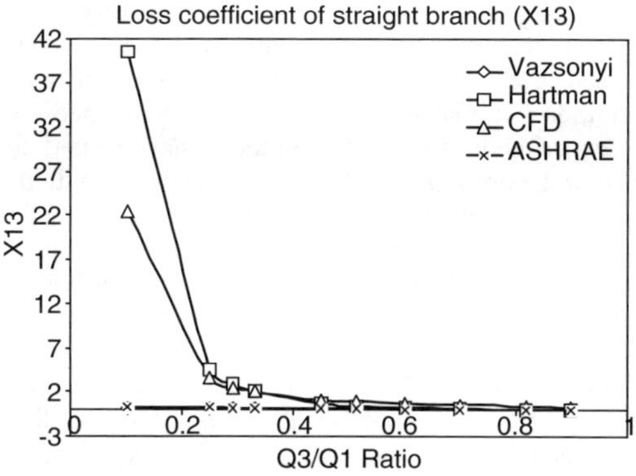

Figure 5. Comparison of shock loss coefficients for straight branch.

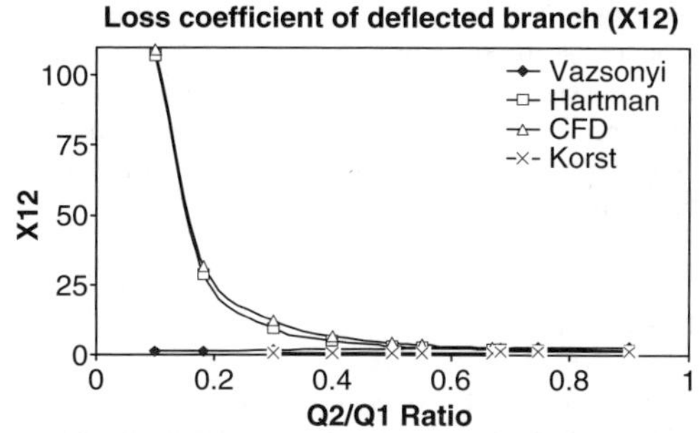

Figure 6. Comparison of shock loss coefficients for deflected branch.

a wide scatter. Only Hartman's values come close to the numerically estimated values. At the two ends of flow conditions *X12* from CFD falls within 10% of the Hartman's empirical estimate. However in remaining range of flow the *X12* from CFD is around 30% or more than empirical values.

Results from the computational solutions were compared with those obtained from physical model (Figs 3, 4). The loss coefficients are decreasing with increasing quantity ratio. The loss coefficients for the straight branch showed a 5% maximum deviation. The deflected branch loss coefficients showed a nice concurrence with experimental results. For the situation analyzed, the two-dimensional and three-dimensional computational analysis shows no difference in the results.

A comparison of computational solutions with empirical estimates of ASHRAE and Korst yields little confidence (Figs 5, 6). Observed is a minimum difference of 100% for loss coefficients of deflected branch and 20% for those of straight branch.

A comparison of computational solutions with empirical estimates of Vazsonyi showed 27% minimum difference in the deflected branch loss coefficients.

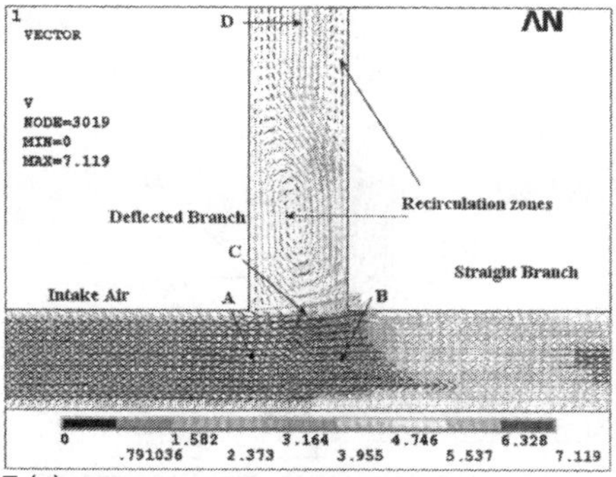

7 (a)

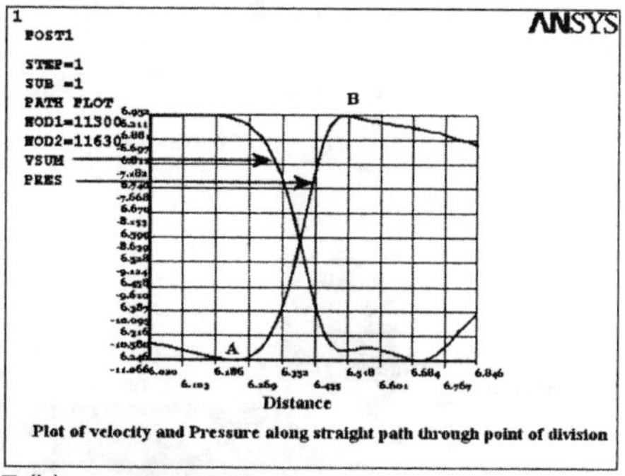

7 (b)

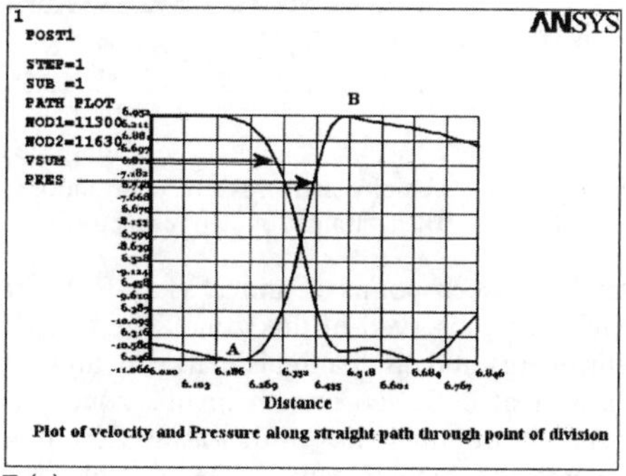

7 (c)

Figure 7 (a, b, c). Velocity and pressure distributions at a split for flows of 10% in deflected branch.

For flows of 70 % or more in straight branch the corresponding loss coefficients spoke in agreement. But, for flows less than 70% in straight branch the loss coefficients differed by at least 20%.

CFD analysis shows that, in a split, there are three recirculation regions. Figures 7a, b and c show the velocity and pressure distribution at a split for flows of 10% and 90% through deflected and straight branch respectively. For small flow in the deflected branch (Fig. 7a) there are two recirculation zones or bubbles in the deflected branch; they are due to very high negative pressure drop developed at the branch

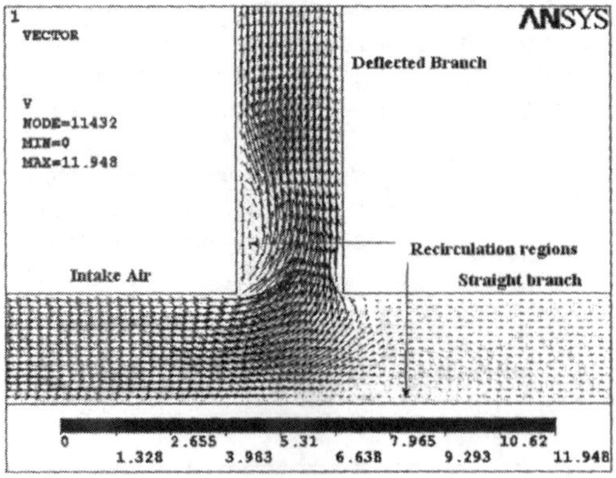

8 (a)

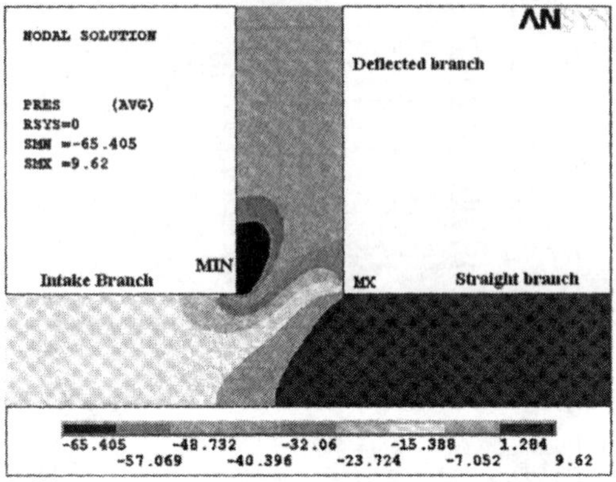

8 (b)

Figure 8 (a, b). Velocity and pressure distributions at a split for flows of 10% in straight branch respectively.

entrance between points *C* and *D* (Fig. 7b). Strong turbulence is observed in this zone. The turbulence kinetic energy and dissipation is markedly high on the outer wall of deflected branch in this zone. In this case the eye of bubble lies farther away from point A and nearer to branch centre. The straight branch shows no recirculation zone, especially with the high static pressure driving the flow (Fig. 7c). Figures 8a and 8b show the velocity and pressure distribution at a split for flows of 90% and 10% through deflected and straight branch respectively. When the flow is large in deflected branch (Fig. 8a), there is only one recirculation zone in deflected branch; this lies closer to point A and nearer to wall. A second zone occurs on the outer wall of straight branch. A maximum static pressure is observed at the stagnation point (Fig. 8b).

6 CONCLUSIONS

The numerical study on loss coefficients for two-way splits and the estimation of the losses with experimental data is conducted. Results suggest the validity of the CFD model assumptions and numerical results as performed in comparison to the experiments. Under the conditions simulated in the studies, the errors between the experiments and the CFD models have turned out to be of the order of 5%. Also, the comparison with the empirical loss coefficients as suggested by ASHRAE and other researchers has yielded little confidence in the estimations proposed in these empirical schemes. Literature suggests an absence of weightage to wall friction. Considerations for change in the aspect ratio of duct are also lacking in the studies. Experimental methods give useful information regarding the flow behavior at such geometries. However, experimental methods face with high costs, long time, and limitations of equipment for conducting detailed experiments. The CFD technique becomes advantageous in analyzing all such possible changes in the governing parameters. The computational solution obtained for splits under the assumptions of flow being viscous, incompressible and turbulent proves to be reliable as compared with that of Edwards & Perlee and Liu & Wang. The RNG turbulence model looks competent to analyze the shock loss at intersections.

REFERENCES

ASHRAE Handbook, Fundamentals Volume, Chapter 33, American Society of Heating, Refrigerating, and Air-Conditioning Engineers, Atlanta, Ga. 1981.

Benedict, R.P., Carlucci, N.A., Swetz, S.D., 1966, Flow losses in abrupt enlargements and contractions, *Transactions of ASME, Journal of Engineering for Power*, p. 73–81.

Benedict, R.P., Carlucci, N.A., 1965, Flow with losses, *Transactions of ASME, Journal of Engineering for Power*, p. 37–49.

Edwards, J.C., Perlee, H.E., 1977, Shock loss calculations across junctions and splits, *USBM, RI 8227*.

Hartman, H.L., 1982, *Mine Ventilation and Air Conditioning*, John Wiley and Sons, Inc., p. 718.

Korst, H.H., Buckley, N.A., Konzo, S., Roose, R., 1950, Fitting loss for extended plenum forced air systems, *ASHVE Trans*, vol. 56, p. 295.

Liu, C.W., Wang, Y.J., 1987, Analysis of ventilation shock losses by finite element method, *International Journal of Mining and Geological Engineering*, 5, p. 377–383.

McCormack, P.D., Crane, L., 1973, *Physical Fluid Dynamics*, Academic Press, New York.

Sierra-Espinosa, F.Z., Bates, C.J., Doherty, T.O., 2000, Turbulent flow in a 90 degree junction, Part 1: Decay of fluctuations upstream the flow bifurcation, *Computers and Fluids*, Vol. 29, p. 197–213.

Vazsonyi, S.: In Benedict, R.P. 1980, *Fundamentals of Pipe flow*, John Wiley and Sons, p. 206–207.

Case studies

Mine Ventilation, Ganguli & Bandopadhyay (eds)
© 2004 Taylor & Francis Group, London, ISBN 90 5809 633 5

Ventilation improvement at Mimosa's Mine 3 in the Sabinas Coal Basin, Coahuila, Mexico

Jose Ruben Ponce Aguirre
Minerales Monclova, Palau, Coahuila, Mexico.

ABSTRACT: Minerales Monclova, S.A. de C.V. (MIMOSA) operates 5 underground coal mines in the Los Olmos coal seam of the Sabinas Basin in the North of Mexico in a range of depth from 120 to 300 meters and methane concentration from 4 to 12 m^3/ton.

The Mine 3 started its operations in 1980, this mine was operating since its beginning in an area with an average methane concentration of 4 to 5 m^3/ton in the coal seam and a depth from 120 to 150 meters. In 2000 the mine was scheduled to change its operations to an area with 200 m depth and a methane concentration of 8.5 m^3/ton in the coal seam.

Due to the high resistance of the mine, a ventilation project was designed in order to maintain the safety of the operations and the success of the mining plan.

Besides this, a methane drainage program with short holes was started.

1 INTRODUCTION

The geological history of México shows that there has been three events that were suitable for the development and formation of coal beds. The first event happened from the Upper Triassic to Middle Jurassic Epochs (200–170 Ma); the second event of favorable geologic conditions took place at the end of the Late Cretaceous Epoch during the Maestrichtian age (75–65 Ma); and the last third event happened during the Eocene Epoch, Lutetian–Bartonian Age (50–40 Ma).

The Mexican coals that have been more explored and developed, because of their economic potential, are the Maestrichtian coals in the Coahuila State. Most of the coals in the Sabinas and Monclova sub-basins are metallurgical, whereas the coals from the Fuentes – Río Escondido basins are steam coals (long flame type); the latter ones are being used by the Mexican government company (Comision Federal de Electricidad, CFE) to generate electricity in their plants of Nava, Coahuila.

Total annual production of steam and metallurgical coal of the State of Coahuila, (Figure 1) is approximately 15.4 million metric tons. Six million is metallurgical coal and 9.4 is steam coal used to generate electricity.

MIMOSA's annual run of mine coal production is an average of 5 million metric tons. and the company coal reserves in the Sabinas sub-basin are 400 million of metric tons.

MIMOSA is presently operating the underground mines 2, 3, 5 and 6 and one open pit in the Sabinas sub-basin (Mine 4 finished it's operations in July 2003), in this basin the coal seam is located in the Los Olmos formation.

This formation contains two distinct coal seams with a rock layer between them (0.10 to 0.30 m thick), the combined thickness of these two seams is about 2.5 m.

The coal is medium to high volatile in rank and is used to supply steel making operations in the city of Monclova which is 100 miles far from the MIMOSA's mines.

Coal mining in the Sabinas sub-basin initiated in 1889, with surface and drift mining near outcrops. Mining on MIMOSA's concession in the Sabinas sub-basin initiated in 1950. The mining operations were owned by the Mexican government until 1991 when the company was purchased by Grupo Acerero del Norte (GAN).

Each mine is equipped with a Joy 4LS10 longwall system and the development of the longwall panels is performed with Alpine AM-50 roadheaders with the dual entry system.

Due to the different location of the mines in the coal-basin, the operation depth of the 4 underground

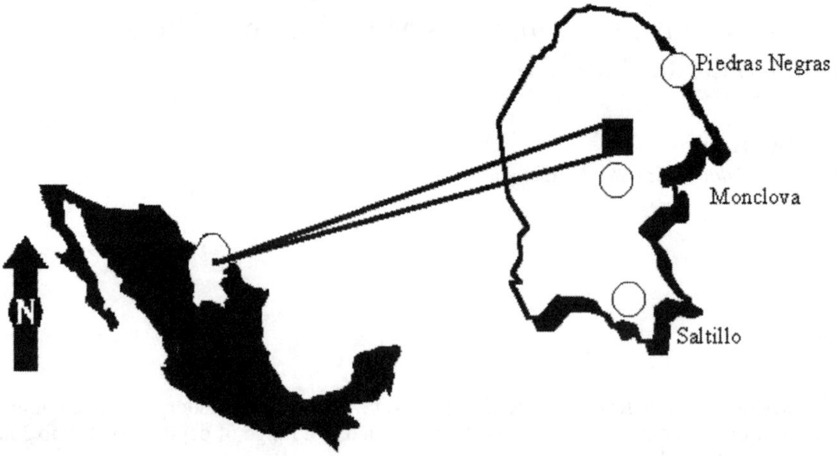

Figure 1. Location of the MIMOSA's mines in Coahuila, Mexico.

MIMOSA's mines goes from 170 to 330 m and the methane generation is bigger as the depth increases.

The total methane liberation in the ventilation systems for the last three years of the five underground mines is an average 400,000 m^3 of methane per day.

MIMOSA implemented methane drainage systems in Mines 2, 4, 5 and 6 since 1992 to 1999 under a horizontal directional drilling contract with Resource Enterprises (REI) from the United States, under this contract REI drilled almost 90,000 ft of in-seam boreholes with an average length of 2,000 ft per hole.

After 1999 MIMOSA has been doing it's own degasification drilling with small pneumatic machines and short in-seam holes with an average depth of 600 ft per hole.

This paper is focused in the activities performed at the MIMOSA's Mine 3, in order to change succesfully the operations of the mine from an old area where the mine was working since its beginning in 1980 to 2000 in a range of depth from 120 to 150 m and a coal seam methane concentration of 4 to 5 m^3/ton to a new area with a depth of 200 m and a methane concentration of 8.5 m^3/ton in the coal seam.

2 VENTILATION BACKGROUND

2.1 *Main ventilation system*

MIMOSA's mine 3 operates with the longwall system. The layout of the mine is shown in Figure 2. The access from the surface to the interior of the mine is provided with two parallel slopes with an inclination of 28% and a length of 600 m supported with steel arches 4 m wide by 3.5 m high.

Both slopes are the fresh air intake to the mine and also one of them is the access for men and supplies and the other is used to bring the coal production of the mine to the surface using the main conveyor belt.

The mine uses the exhausting ventilation system with the main fan on the surface connected to the mine through a circular shaped vertical shaft of 4.5 m diameter and a depth of 170 m.

The main fan is equipped with devices for checking continuously the fan pressure, bearing vibrations and temperature, amperimeter and methane measurement in the discharge of the air.

The main fan was used since the beginning of the mine until July of 2001 a Jeffrey 8HUA-96 with a 900 H.P. and 1200 r.p.m electric motor. This mine was the one with the highest ventilation resistance of the MIMOSA's group because all the entries excepting the bottom of the slopes were developed with a rectangular shape and supported with steel beams and wood timbers, with this system the average size of the entries was 4.0 by 1.7 m for a final cross section close to 7.0 m^2.

All the electric equipment used underground is permissible. Also there are methane monitors installed at the main returns, at the development faces and the tailgate in the longwall, this monitors shut down of the electric power if the methane concentration in the environment is more than 1.0% where there are electric equipments and 1.5% or more in the main returns.

All the monitors are connected to a computer system on the surface which is recording all the measurements.

There is a ventilation technician in each mine in charge of checking every week the airflow measurements in the mine and also ventilation supervisors in each shift, including weekends.

2.2 *Airflows and methane generation*

In 1999 and 2000 when the mine was working with a longwall 200 m wide by 800 to 1000 m long and the

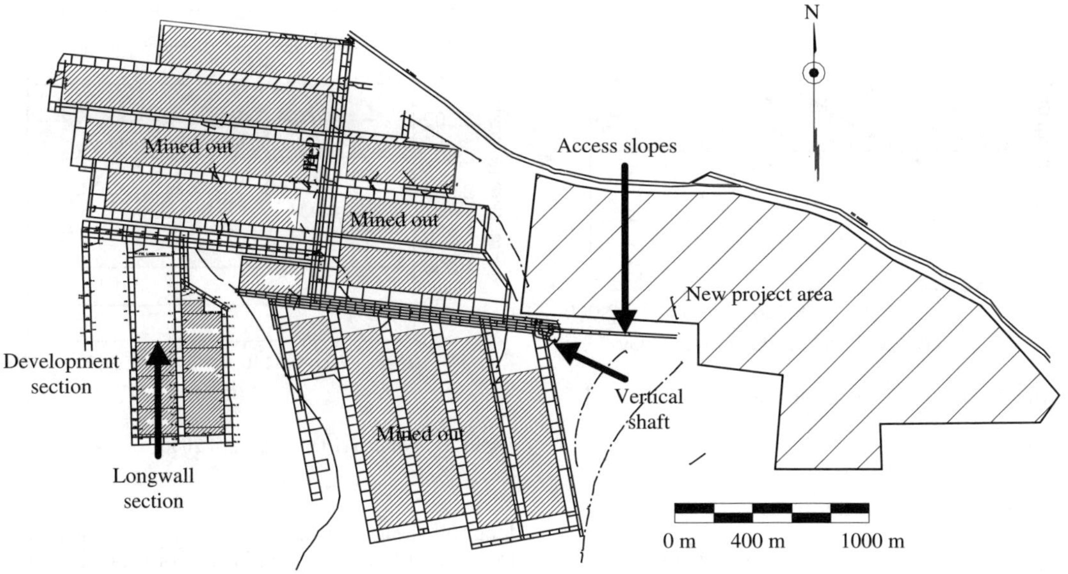

Figure 2. Mine layout in April 2000.

developments in an area with a methane concentration in the coal seam of 4 to 5 m^3/ton. The total airflow in the ventilation system of the mine was 85 m^3/s, fan pressure of 2.600 Kpa with a methane generation of 0.5 m^3/s and an average daily production of 4000 metric tons. run of mine coal. This methane generation is more than 50% less than in the other MIMOSA's mines when they have been working in deeper areas with higher methane concentrations in the coal seam and using methane drainage systems.

2.3 *Longwall ventilation*

In the longwall the ventilation was provided by the wrapped around bleeder system, the gob area is isolated with concrete seals built in each crosscut left behind the longwall, a 1 m^2 opening with a metallic door is used in each seal in order to bleed methane from the gob to the ventilation system from the immediate seal behind the longwall advance, this opening is closed when the longwall gets far from the crosscut where the seal is located. We were using 10 m^3/s across the face and 15 m^3/s for methane dilution, for a total volume of 25 m^3/s.

2.4 *Developments ventilation*

In the development section with two "Alpine AM50" roadheaders driving the longwall panels with the dual entry system and crosscuts every 60 m., the air volume used was 25 m^3/s, the ventilation of the faces with auxiliary fans Joy 36-26-1770 with 75 H.P., blowing system and fiberglass ducting of 30″ diameter (0.762 m), these auxiliary fans always were operating in one of the lower blade settings to supply an air volume to the face of around 6 to 8 m^3/s and have some air in excess to avoid recirculation. These arrangements were enough for the operation of the development and longwall sections of the mine without delays because methane for the average coal production of 4,000 metric tons/day.

3 METHANE GENERATION FORECAST

3.1 *Longwall*

For the longwall methane generation prediction in order to determine the ventilation requirements for the new area, a statistic of 8 months of other longwall in MIMOSA which operated in similar conditions of depth and coal seam methane concentration was used. In this case we used the information of coal productions vs. methane generation of the 6 South longwall of MIMOSA's mine 4 (Figure 3). This longwall operated in 1991.

The methane emission rate was 650 l/s for a coal production level of 4500 tons/day.

According to this information we determined that a total air volume of 60 m^3/s would be enough, because our maximum methane concentration allowed in the longwall return is 1.5% and with this air volume we would have between 1.0 and 1.1% of methane.

3.2 *Developments*

For the developments methane generation prediction, a statistic of a development system of a 5 months period

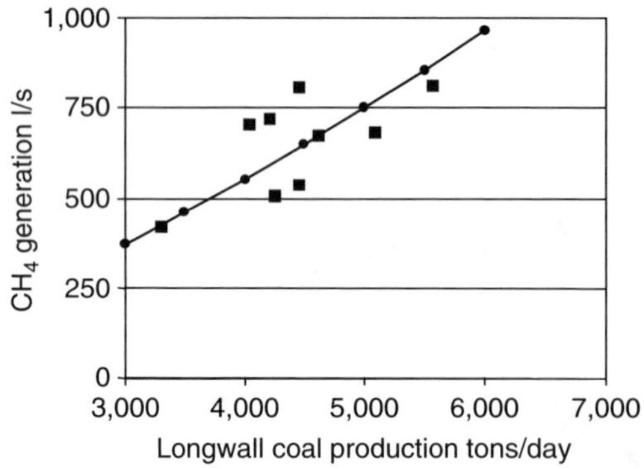

Figure 3. Longwall CH_4 generation.

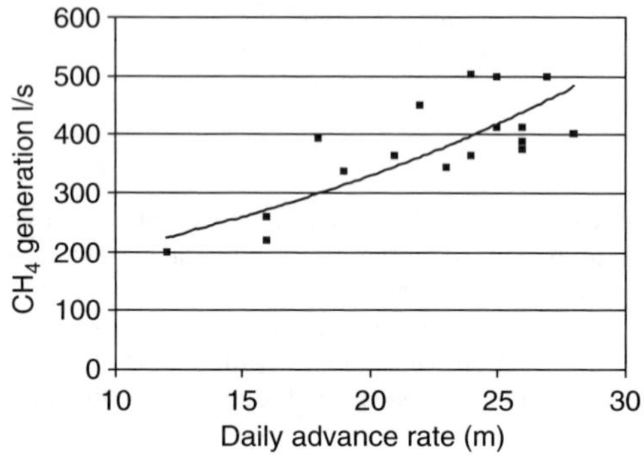

Figure 4. Developments CH_4 generation.

of other section in MIMOSA with similar conditions and methane drainage system was used.

In this case we used the information of daily advance rates vs. methane generation of the 4 north entries at MIMOSA's Mine 4 (Figure 4).

The methane emission rate was between 400 to 450 l/s for an advance rate of 25 m/day with two "Alpine AM-50" roadheaders.

With this information we determined that a total air volume of 40 m^3/s would be enough for the operation of a section with two "Alpine AM-50" roadheaders driving a two entry system for the longwall panels, because our maximum methane concentration in the developments return is 1.5% and with this air volume we would have between 1.0 and 1.12% of methane.

4 TOTAL VENTILATION REQUIREMENT

With the statistical information of methane emissions and ventilation requirements for developments and

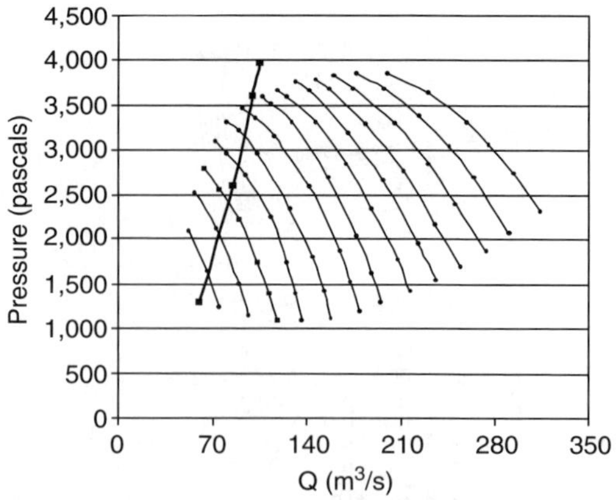

Figure 5. Mine resistance in April 2000.

longwall, the air needed for other areas, which also have to stay ventilated like pumping stations and 25 to 30% air leakage, a total future air volume requirement of 160 m^3/s for all the mine was defined.

4.1 *Fan measurement*

A survey of the main fan was conducted in April 2000, the measured airflow was 85 m^3/s with a pressure of 2.60 Kpa, the mine resistance was quite high and it is showed (Figure 5). This high resistance was because of the size and length of the main entries and returns of the old area of the mine. Concerning to the slopes a survey using the gauge and tubing method was performed, the pressure for the volume of 85 m^3/s was 115 pascals, with this value we determined no problem with the slopes for the future ventilation requirements of 160 m^3/s. The same occurred with the vertical shaft, which has a section of 15.9 m^2, an estimated pressure loss of 9 pascals in the vertical shaft was calculated for the air volume in April 2000 of 85 m^3/s.

5 VENTILATION PROJECT

The ventilation projections for the new area of the mine were reviewed using the VNETPC 2000 software, this software has been in use in MIMOSA since 1989 for the ventilation planning of all our mines and also for analyzing any major change in the ventilation systems.

- With this software in the last 14 years the correlation between the predicted results and the real numbers has been from 93 to 95% and sometimes higher.

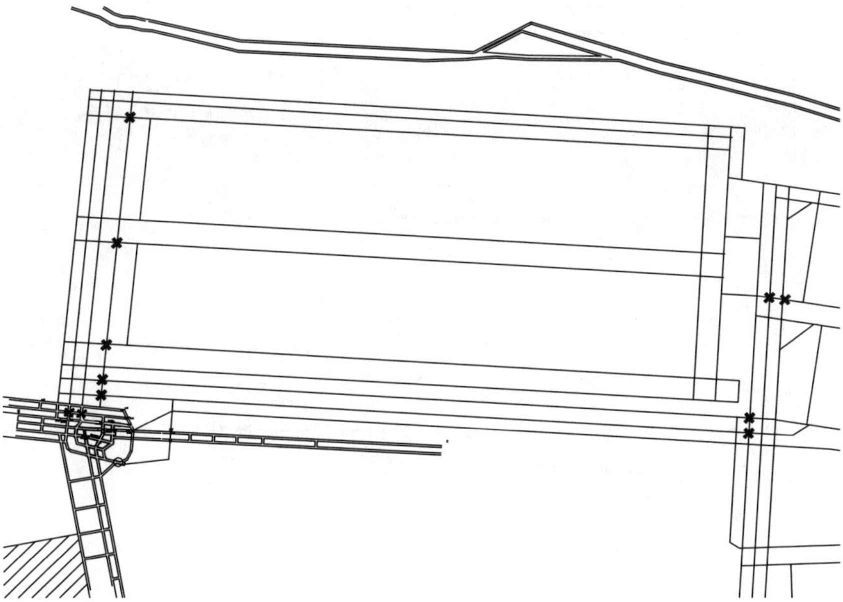

Figure 6. Mine layout for mine 3.

Table 1. Friction factors used for airways.

Airway type	Friction factor of kg/m^3
Belt entry 10 m^2	0.0228
Materials 10 m^2	0.0106
Return 10 m^2	0.0141
Belt & materials in parallel 10 m^2	0.0037
2 returns in parallel 10 m^2	0.0035

- In order to avoid the high resistance problem of the old area of the mine, a new support type and shape for the future entries in the new area was proposed, this was the roof support with steel arches of 4 m wide and 3 m high, with this support the effective size of the future entries would be increased from the average of 7.0 in the old area to 10 m^2.
- This has been a standard for the last years in mines 2 and 4 and for the gassiest and deeper MIMOSA's mines like mine 5 and 6 the same steel arches are used but with dimensions of 5 m wide by 3.5 m high for an effective size of approximately 15 m^2.
- Two new connections of the main returns to the vertical shaft were proposed also.

We already have the average friction factors for the MIMOSA's entries supported with this arches, this been obtained from ventilation surveys has performed in all the mines using the gauge and tubing method also (Table 1).

For the slopes we used the resistance value calculated with the ventilation survey.

The information of the main fan supplied to the fan data bank of the software is obtained from the manufacturer's fan curves and then we applied the air density correction, because our average air density is about 1.12 kg/m^3 vs the standard of 1.2 kg/m^3. Several situations of the mining plan were analyzed, for this paper we will explain the results of the most critical, with all the equipments, 1 longwall and 2 development sections working in the new project area.

The mine layout for the first 2 longwalls is shown in Figure 6.

6 VENTILATION SIMULATION

The mine ventilation projection for May 2002 was simulated, the schematic is showed in Figure 7.

- With the proposed arrangement, the fan operating point predicted was 165.99 m^3/s with a pressure of 2.79 Kpa, in the blade setting of 23 degrees for the main fan Joy M-96-50.
- The old Jeffrey main fan was changed for a new Joy fan in July 2001, this decision was made by our maintenance department, it was not possible to change the blade setting of the old fan for the ventilation requirements of the mine because of corrosion in the internal mechanism.
- The air volume predicted for the longwall was 56.7 m^3/s and for the development sections 45.0

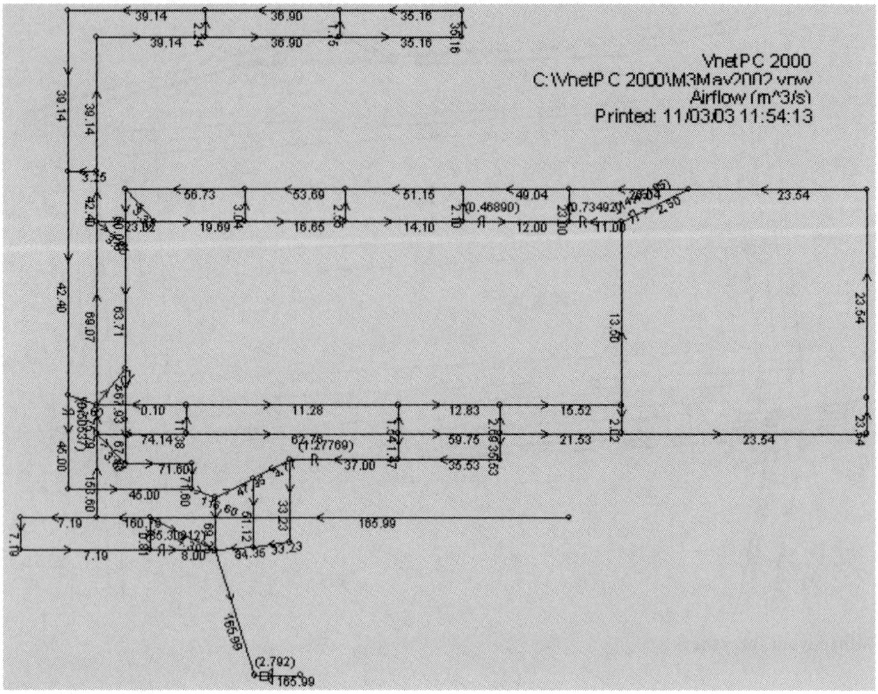

Figure 7.

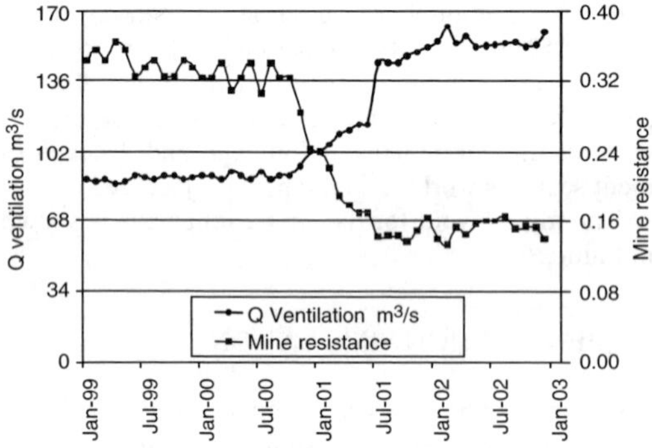

Figure 8.

and 37.0 m^3/s with 25% leakage, all these results were according to our requirements.

6.1 *Correlation between projections and the real results*

- The first longwall in the new area of the mine started in March 2002 with a measured total air volume of 57.8 m^3/s, the total air volume of the mine for 2002 was an average of 155 m^3/s with a fan pressure of 2.95 Kpa and the main fan blade setting was 23 degrees.
- There was a 98% correlation for the longwall ventilation projection and a 94% for the main fan operating point, the developments were in a similar situation.

The evolution of the ventilation system from 1999 to 2002 is shown in Figure 8.

As seen in Figure 8, the ventilation of the mine was increased from an average 88 m^3/s in 1999 to 160 m^3/s in December 2002 and the mine resistance fell almost 70%.

7 METHANE

The methane in the ventilation system was also increased from 0.5 m^3/s in 1999 to a maximum of 1.4 m^3/s in January 2002, in June 2002 a methane drainage program started.

This was performed with in seam horizontal holes drilled with pneumatic drilling machines, the average length of the holes was 200 m.

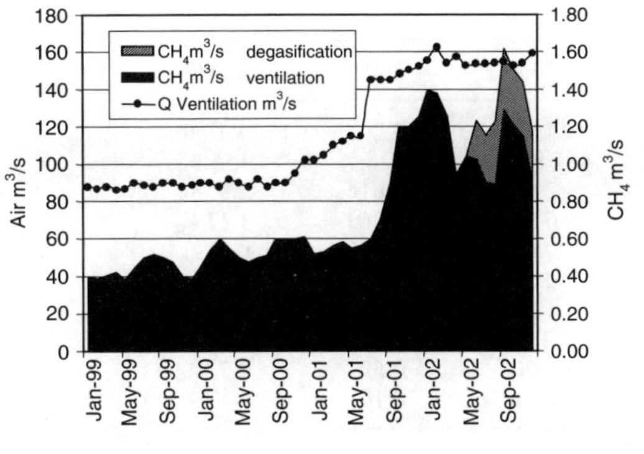

Figure 9.

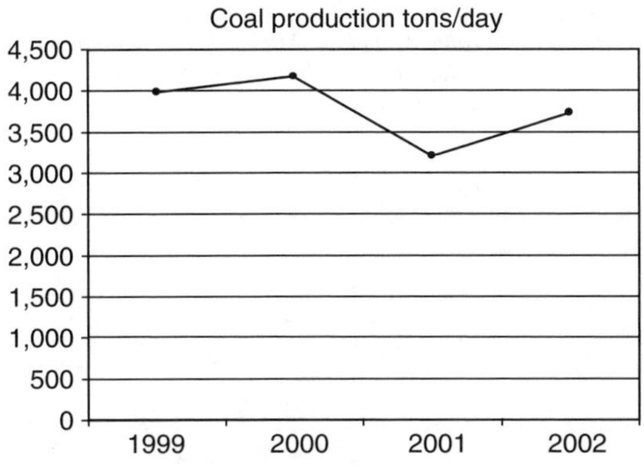

Figure 10.

The average methane volume handled by this system in 2002 was 0.280 m^3/s.

The behavior of the methane generation for the entire mine in the ventilation and degasification systems from 1999 to 2002 is shown in Figure 9.

8 COAL PRODUCTION

The average daily run of mine coal production from 1999 to 2002 was very close to 4,000 tons/day in the four year period (Figure 10).

9 METHANE DRAINAGE HOLES

The location of the horizontal methane drainage holes is shown in Figure 11.

10 CONCLUSIONS

- The ventilation project designed for MIMOSA's mine 3 between 1999 and 2000 allowed us to increase the ventilation of the mine a 75% with only 11% more fan pressure.
- The correlation between the predicted airflows with the VNETPC 2000 software and the real numbers is been more than 94%, this has been the same for the other mines of MIMOSA.
- Even that the methane generation in 2002 was three times bigger than in 1999, we have been able

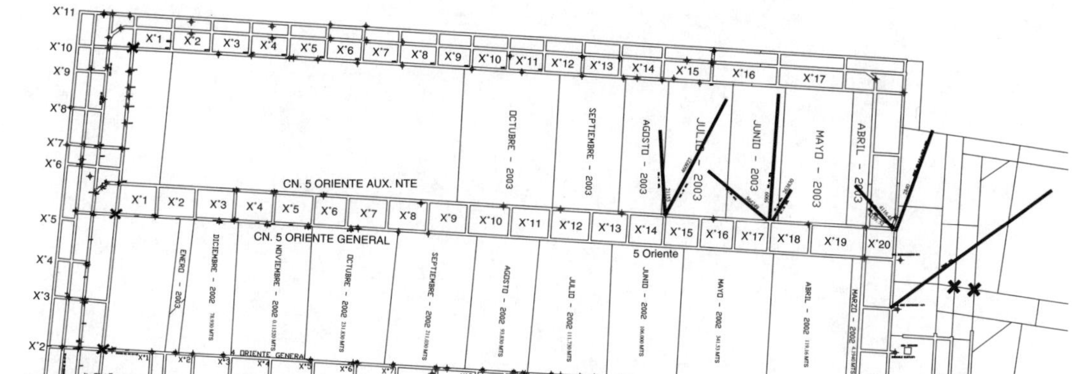

Figure 11. Horizontal methane drainage holes and mine advance in October 2003.

Table 2. Summary of results.

	1999	2000	2001	2002
Coal production tons/day	3,991	4,170	3,209	3,739
Q ventilation m^3/s	88.3	91.3	128.7	155.0
Dp Kpa	2.664	2.606	2.616	2.958
CH_4 ventilation m^3/s	0.437	0.545	0.763	1.114
CH_4 degasification m^3/s	0.000	0.000	0.000	0.282
Total CH_4 generation m^3/s	0.437	0.545	0.763	1.396
CH_4 handled by ventilation	100%	100%	100%	80%
CH_4 handled by methane drainage	0%	0%	0%	20%
% CH_4 at the main fan	0.49	0.60	0.59	0.72
Mine resistance	0.34	0.31	0.16	0.12

to acomplish and mantain the advancement and production goals of the mine.

- 80% of the methane generation of the mine is handled with the ventilation system and 20% by the methane drainage system.

A final summary is shown in Table 2.

ACKNOWLEDGMENTS

The author wish to thank MIMOSA for the support for this effort and for permission to publish this paper.

REFERENCES

Brunner Daniel, J. and Ponce Ruben 1999. *Methane Drainage at the Minerales Monclova Mines in the Sabinas coal basin, Coahuila, Mexico "Proceedings of the 8th U.S. Mine ventilation Symposium 1999."*

Mine Ventilation, Ganguli & Bandopadhyay (eds)
© 2004 Taylor & Francis Group, London, ISBN 90 5809 633 5

New techniques versus experience: a real case of mine ventilation analysis

J. Toraño, R. Rodríguez & J.M. Rivas
School of Mines, Oviedo University, Oviedo, Asturias, Spain

ABSTRACT: For mining new panels in a coal mine, changes of the mine infrastructure were neccesary. For this purpose, a new mine ventilation design with a different fan location was done. Under this conditions, it was very interesting to know if the old fan was useful or a new more powerful fan was neccessary. In this work we briefly describe how the problem was solved by two ways: the first was a complete ventilation analysis aided by computer programs while the other was only a simple reasoning based on the experience and some measures.

1 INTRODUCTION

In a coal mine in which coal was being worked by old methods it started to work by the longwall method. Under these circumstances and due to the location of the fan, the exhaust air had to exit through old galleries with small section due to the rock mass deformation which had a high resistance (Fig. 1).

A solution for this problem was to excavate a new gallery for the exhaust air and change the fan location (Fig. 2) in order to diminish the resistance. A question asked by the company was if the old fan was yet useful or if a new fan, more powerful, was neccesary. This was the most adequate moment to change the fan.

Because this mine has not a large dimensions, the problem was not very complex and the authors (all of them lecturers of the School of Mines of Oviedo University) proposed to solve it to a group of students that had to do a practical work for finishing their career.

The students, always under supervision of their lecturers, followed step by step the methodology for solving mine ventilation problems that they learned in the career which can be considered actually, more or less, a standard.

2 BRIEF DESCRIPTION OF THE SOFTWARE

In order to analyze the future mine ventilation scheme and to study if the same fan could be used, a model of the mine was elaborated using the VENPRI program, a computer program especially developed for calculating mines ventilation (Aitemin 1994).

The program follows a classical calculation scheme:

- for each branch i of the network, the frictional pressure drop Δx_i is estimated by the well known formula:

$$Q = 3.4 m^3 / kW \min Pl \tag{1}$$

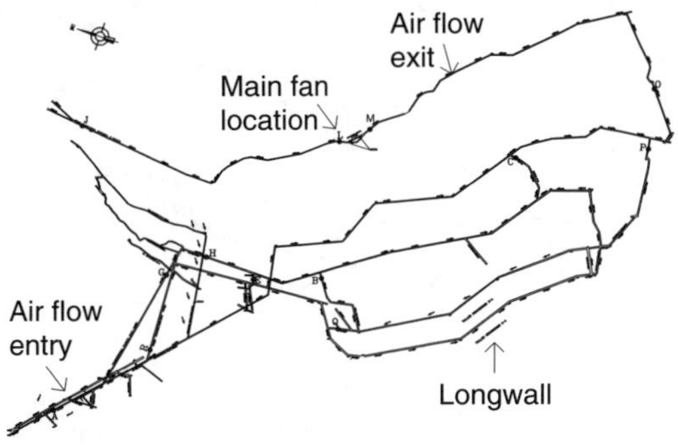

Figure 1. Old mine ventilation network.

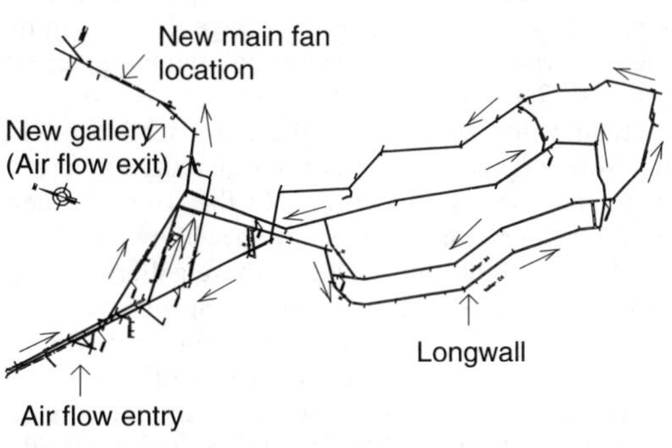

Figure 2. Future mine ventilation network.

r_i being its resistance (Ns^2/m^8) and q_i the quantity airflow (m^3/s);
– if there is a fan in the branch, the equation is:

$$\Delta x_i = r_i\, q_i^2 - f(q) \tag{2}$$

where $f(q)$ is the working point on the characteristic curve of the fan;
– for each node j of the network, the sum of the quantities of all branches is null:

$$\sum_k q_{jk} = 0 \tag{3}$$

q_{jk} being the quantity airflow that comes to the j node through k branch;
– for a closed way (mesh) from one point l to the same point l, the total pressure drop Δx_{lm} is null:

$$\sum_m \Delta x_{lm} = 0 \tag{4}$$

where Δx_{lm} is the sum of pressure drop (Pa) of the branches that form the mesh m.

Finally equations for every node and every mesh are established and the equation system is solved by Hardy-Cross method. If there is only one main fan in the mine, the total quantity and total pressure drop in the mine Q, ΔX is equal to the quantity and pressure at the fan Q, ΔH.

The program includes graphic screen for data introduction and results viewing in order to make the re-calculations easier when it is neccesary to do various simulations representative of different solutions of the problem.

3 QUANTITY AIRFLOW REQUIREMENTS

It was estimated that in the future, two longwall faces can be worked simultaneously. The mined seam was of anthracite coal type, practically without a methane content. Then, the quantity airflow requirement at the longwall was impossed by the exigence of a minimum air velocity of 0.2 m/s through the tail gateroad (Spanish Standard ASM-15 1985).

Taking into account that the cross section of the roadways was 15 m^2, the needed quantity airflow in this conditions is only 3 m^3/s. In the case of a new roadway for the next panel was been advancing the conditions are different. Due to these galleries being driven with all the section in coal, the Spanish Standard ASM-52, 1999, establish that the mininum air velocity is 0.3 m/s, i.e. a quantity airflow of 4.5 m^3/s is needed at the dead-end roadway. On the other hand, the Spanish Standard ASM-51, 1997, obliges to take the airflow for this advancing roadways from a gallery with a 30% higher quantity airflow. Then, under this conditions, the quantity airflow through the longwall gateroad required is about 6 m^3/s.

That is, the quantity airflow required varies from 3 to 6 m^3/s depending on the condicitons.

4 PREVIOUS MODELS OF THE MINE

4.1 *First model*

The first model of the mine was based on a previous mine state in which the volumetric airflow through the galleries was been measured (Fig. 3).

The model was developed following four steps:

– a ventilation network in which each branch represened a mine gallery was done;
– the characteristics introduced were length (m) and resistance (Ns^2/m^8); the latter was estimated from the results of experiences in similar mines (Luque 1988, Fig. 4)
– the main fan of the mine was represented by means of its characteristic curve provided by the manufacturer;
– the quantity airflow through all the roadways were calculated with the program (Fig. 5).

After comparing the measured and calculated volumetric airflow (Fig. 6) it was proved that the results did not fit with the experience. In those roadways in which it was difficult to determine the resistance (i.e. due to the existence of a singular element as a ventilation door or in the case of very deformed sections) the calculated quantity airflow was very different from the measured one.

Then, it was neccesary to calibrate the model following these steps:

– first, the resistance of these roadways was modified individually in order to diminish this difference between measured and calculated quantity; then a linear relationship betwen measured and

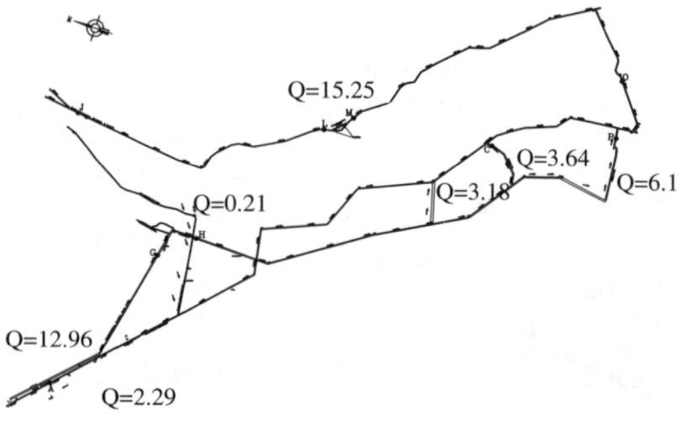

Figure 3. First model: measured volumetric airflow.

estimated airflow was obtained; the calculated airflow quantity was higher than the measured one that means that the values of resistance used in the model were less than real resistances

– for this reason the resistance of all the galleries was increased by multiplying by the factor k, regression coeficient of the linear relationship:

$$q_c^2 = k\, q_m^2 \qquad (5)$$

The new results are shown in the graph of the Figure 7.

4.2 *Second model*

As the mine infrastructure was modified, the model was re-done. Then, based on previous model, a new one corresponding to an intermediate state, in which the fan was yet in its original location but the new gallery for exhaust air was already advanced, was elaborated. The measuring of volumetric airflow, that was done by the students, involved a number of problems (as the variations in air velocity) that are interesting, but they are out of the aim of this paper. Measured quantity airflow is shown in the Figure 8.

In this new model it was neccesary to estimate the resistance of the new galleries (the resistance of the other galleries was the same as in the previous model). Due to the existence of this new part of the mine infrastructure, the previous calibration sequence was repeated. On the other hand, it was proved that some quantity losses through mined out areas exist and a branch simulating this area was needed.

In the Figures 9 and 10, the results of the second model before and after of its calibration are shown.

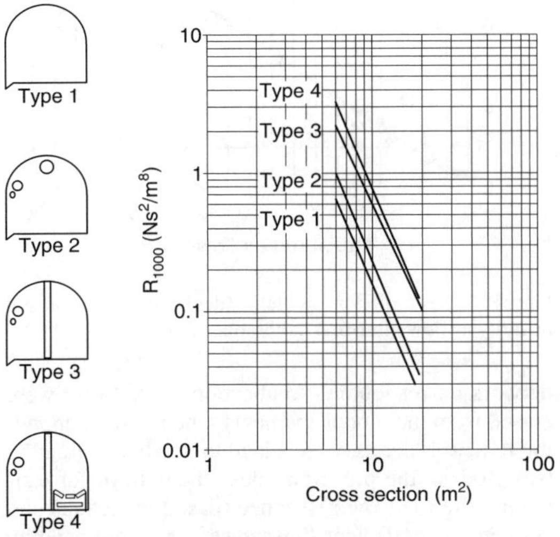

Figure 4. Roadway resistance per km (after Luque, 1988).

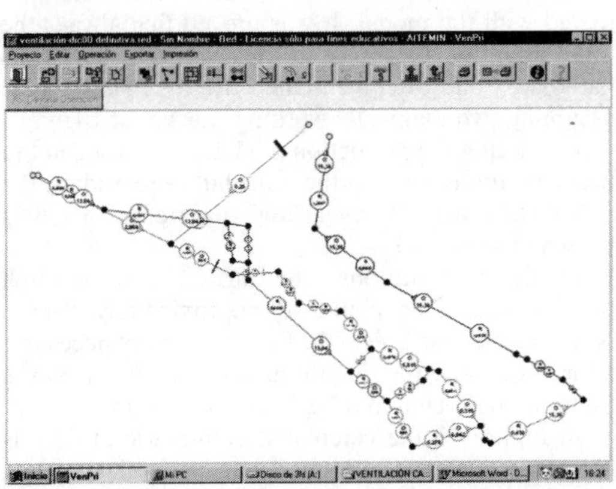

Figure 5. Results screen of the VENPRI program.

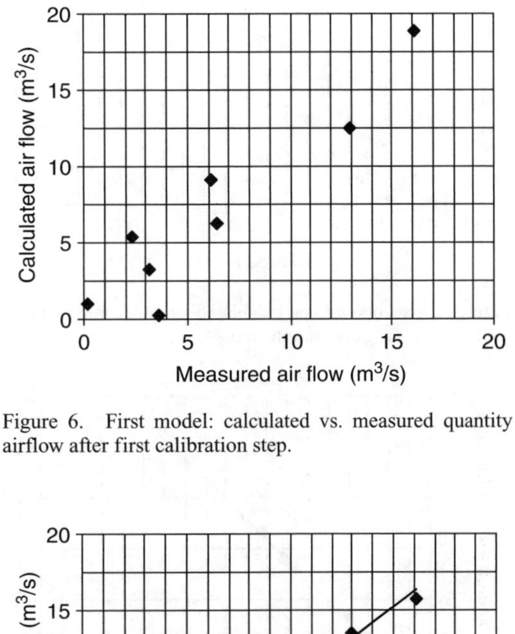

Figure 6. First model: calculated vs. measured quantity airflow after first calibration step.

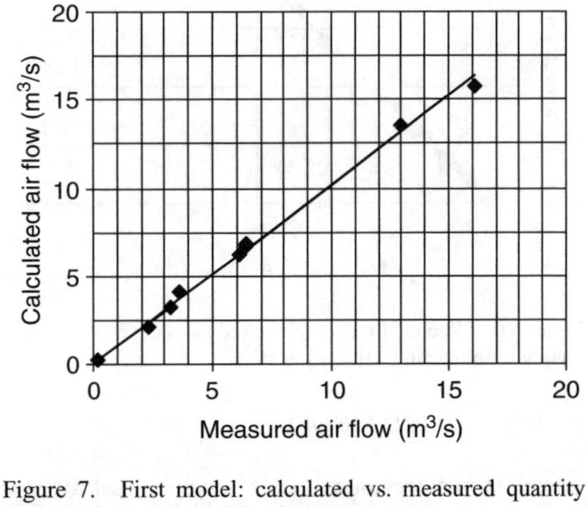

Figure 7. First model: calculated vs. measured quantity airflow after second calibration step.

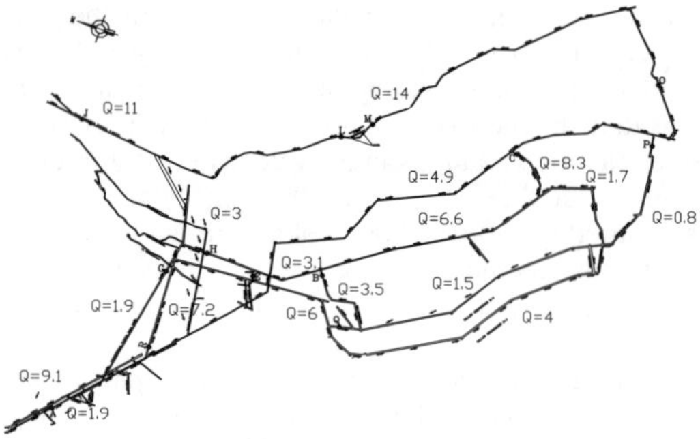

Figure 8. Second model: measured volumetric airflow.

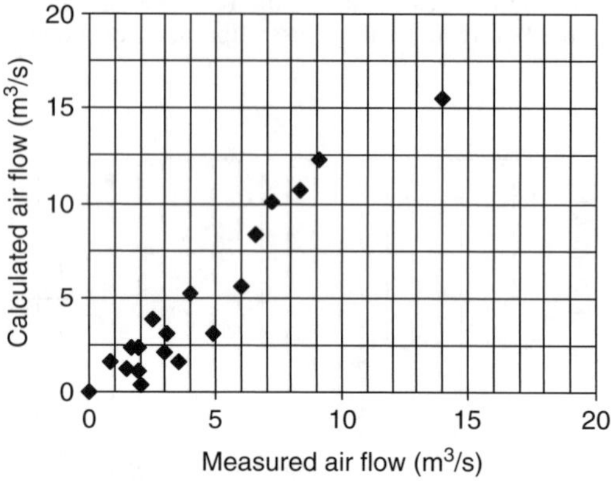

Figure 9. Second model: calculated vs. measured quantity airflow after first calibration step.

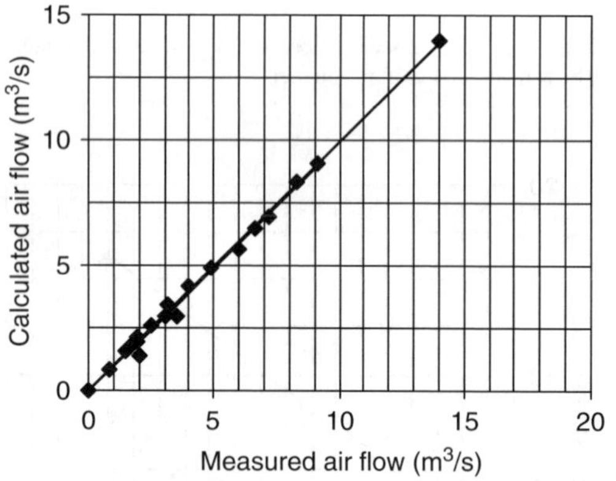

Figure 10. Second model: calculated vs. measured quantity airflow after second calibration step.

5 PREDICTION MODEL

Based on two previous models, a new model for prediction was elaborated. Branches of the ventilation network corresponding to abandoned roadways were erased from the model and new branches, corresponding to new galleries, were added to it. After calibrating two models, the procedure described above for estimating the roadway resistance (based on abaqus and applying a coefficient *k*) was useful without needing other operations.

Many different future states of the mine were simulated with the model. It was proved that always the total quantity that passed through the longwalls gateroads (and after ventilated the longwalls faces) assuming two longwalls working was about $10\,m^3/s$.

For testing the prediction model, it was used not to predict a future state of the mine but to reproduce the actual state of it. The measured volumetric airflow is shown in Figure 11.

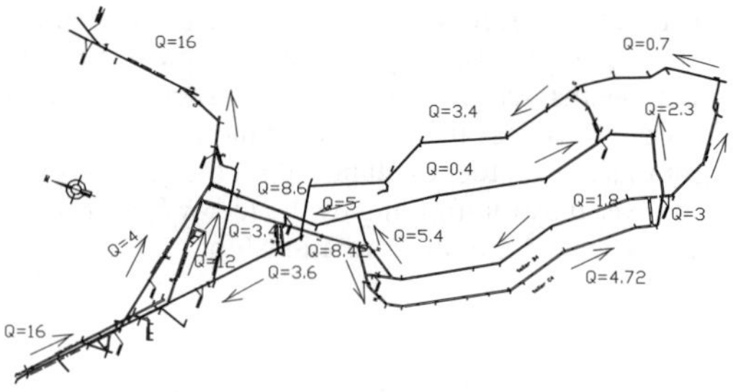

Figure 11. Actual state: measured volumetric airflow.

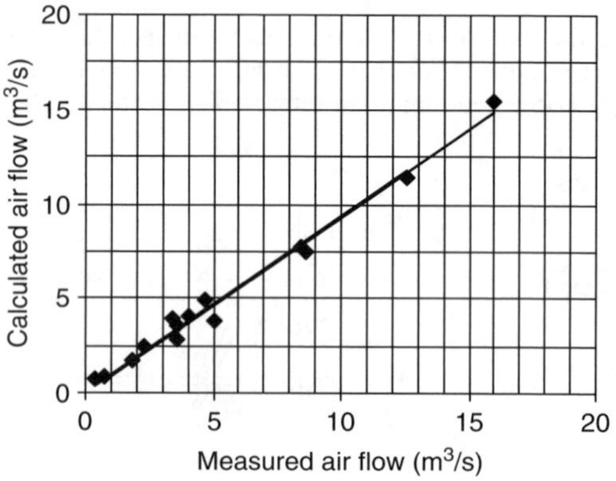

Figure 12. Prediciton model: calculated vs. measured quantity airflow after first calibration step.

In these conditions, the airflow rate through the longwall face could be approximately $8\,m^3/s$. Nevertheless, and due to the fact that it is not necessary a lot of quantity passing through the longwall, $3\,m^3/s$ of air are deflected to other parts of the mine.

In Figure 12 the calculated volumetric airflow is represented versus measured one. It is shown that, without new calibration, the model is accurate enough for calculating the ventilation in the mine.

6 CHARACTERISTIC CURVE OF THE FAN

For calculating ventilation, the program needs the characteristic curve of the fan that is always provided by the manufacturer. Nevertheless, after some years of continuous installation working and due to the wearing of fan components, the curve may change.

From this point of view it is always a good procedure to do a test in the mine in order to determine some points on the real curve of the fan. According to this, while the second model was being elaborated, the volumetric airflow and the total pressure at the fan were measured.

Improving the existence of a gallery for the personnel pass in parallel with the main fan gallery, these measures could be done for two different states (Fig. 13). One of them correspond to the normal state of the installation, it is, the doors of this gallery for personnel pass are closed. And the second measures were done with these doors partially opened, that reproduces a virtual state of the mine with less overall resistance.

In Figure 14, the two pairs of values Q, ΔH of the real characteristic curve are represented. In the same graphic, an estimation of a part of the curve is also represented. That is the curve that was used later in the models elaboration.

7 SOME REFLEXIONS ABOUT THE FAN

In the Figure 15, this real curve of the fan is represented with the two curves corresponding to the overall resistance against the airflow in the two states.

As it can be deduced from the graph, when the resistance diminish the fan forced a higher quantity without arising the extreme zone of the fan curve, i.e. that zone in which although the resistance diminish the fan is not be able to apport significantly more quantity and it becomes constant.

The mine overall resistance after the changes in its infrastructure would be less than the resistance before it, because the new galleries for the exhaust air circulation would be shorter and with a significant larger cross section. Under these conditions, and taking into the mind the real curve of the fan, it would be reasonable to think that the same fan would be able to force a quantity of about 19–20 m^3/s without problems.

Now, taking into account that the requirements for the ventilation of two longwalls were about 6 m^3/s for each one, it is simple to conclude that it was not necessary to change the fan. The solution was to distribute, in a given point of the mine, the total airflow in two parts of 10 m^3/s each one.

Then, to answer the main question, we only need to determine the actual working point of the fan and, based on the experience, in which working point the fan will work when the mine resistence diminish.

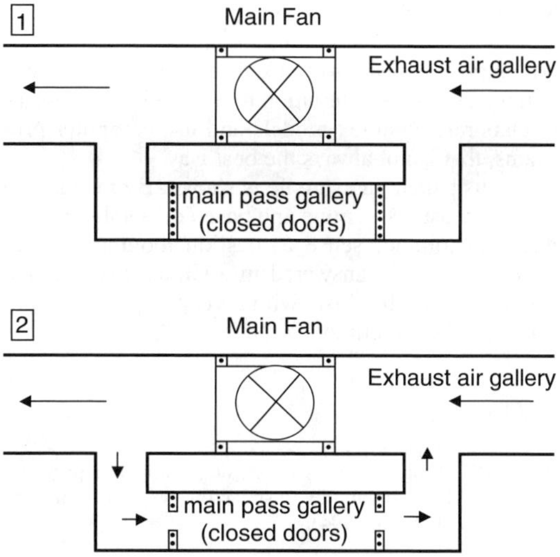

Figure 13. The two conditions in which measures at fan were done.

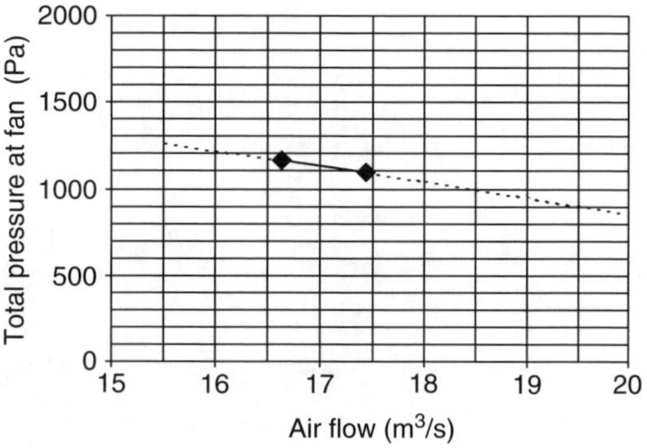

Figure 14. The two points of the real curve of the fan.

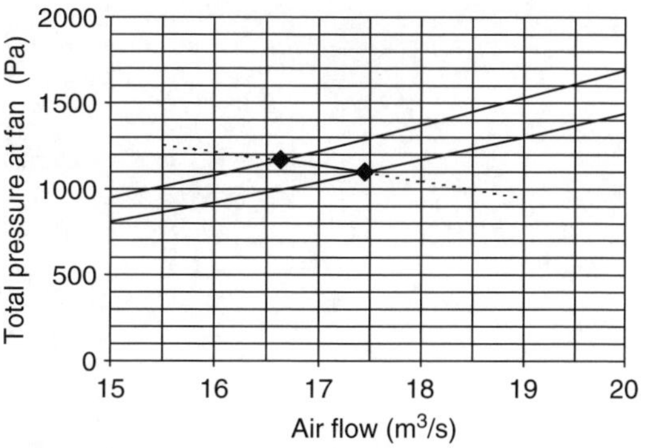

Figure 15. Real curve of the fan and curves representatives of the resistance.

8 CONCLUSIONS

Although it is usual to think that for solving a problem related to the ventilation of a mine it is necessary to elaborate complex models and use computer programs, that is not always the best way.

In this paper, an example is shown. By means of a simple measures in mine and based on a solid experience and common sense, a question about a main fan of a mine can be answered in a simple and rational way and, as in this case, with a very important economy of effort and time.

REFERENCES

AITEMIN. Programa de cálculo de ventilación principal VENPRI. Manual de instalación y uso. Aitemin. (Madrid, España, 1994).

ASM Std. 15 (05.0.01)-1985. Minas subterráneas de carbón y labores con riesgo de explosión. Ventilación y desagüe. Circulación de la corriente de aire.

ASM Std. 52 – 1999. Sistema de explotación de labores subterráneas clasificadas respecto al riesgo de grisú y otros gases inflammables.

ASM Std. 51 – 1997. Explotación de capas de carbón por el métodos de sutiraje con subniveles.

Luque Cabal, V. Manual de ventilación de Minas. Ed. PEDECA, S.C.L. Madrid, España, 1988. 732 pp.

Mine Ventilation, Ganguli & Bandopadhyay (eds)
© 2004 Taylor & Francis Group, London, ISBN 90 5809 633 5

Preliminary ventilation design for the Grasberg block cave mine

Ian J. Duckworth, Ketut Karmawan & Timothy Casten
P.T. Freeport Indonesia, West Papua, Indonesia

ABSTRACT: P.T. Freeport Indonesia operates a mining complex located in the highlands of Papua, Indonesia. This complex consists of both underground and surface operations. By approximately 2014 the open pit will have reached a depth below which it is more economic to mine from underground. Prior to this date a new underground mine will have been developed and brought into production to cave the deposit below the pit. This mine, called the Grasberg block cave, will connect to the surface via 8 km long tunnels equipped with a light rail system. These tunnels will also provide access to at least two other new mines. This paper describes pre-feasibility ventilation planning for the Grasberg block cave mine. The mine is being designed based on a nominal production rate of 115,000 tonnes/day (tpd) with a panel caving footprint of about 1 km wide by 1 km long. An option for increasing the production rate to 160,000 tpd was also considered. The present reserves for the mine exceed 780 million tons. The proposed ventilation design allows for the long production panels to be broken into five discrete ventilation zones, and ensures the economic delivery of large volumes of air to the working regions. Discussion is provided on the ventilation criteria used, the network modeling conducted, examination of shaft versus drifting options, the proposed infrastructure requirements, and recommendations for future study.

1 BACKGROUND ON PTFI

P.T. Freeport Indonesia's (PTFI) project site is located in the Sudirman Mountain range of Papua (formerly Irian Jaya). West Papua is the Eastern-most province of Indonesia, which occupies the Western half of the island of New Guinea (see Figure 1).

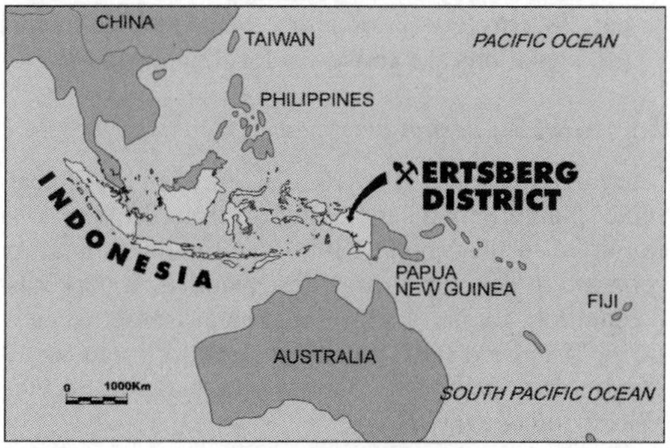

Figure 1. Location of PTFI's mining operations.

PTFI acquired and have been developing the Ertsberg district since 1967. The presently known ore reserves are located approximately 100 km North of the Southwest coast of Papua, between elevations of 2,800 m and 4,000 m above sea level. The terrain varies from mangrove swamp and jungle in the lowlands to extremely rugged mountains.

Since 1967 the Ertsberg open pit, GBT block cave, IOZ block cave and DOZ open-stope mines have all been developed and subsequently closed. Present production is from the Grasberg open pit and the DOZ block cave mines. Total ore tonnage has increased from 15,000 tonnes/day (tpd) in 1978, to 60,000 tpd in 1992, to the present tonnage of 230,000 tpd. At the time of writing this paper approximately 45,000 tpd is produced from the DOZ block cave underground mine. Present reserves consist of six copper-gold-silver orebodies and several other resources located within a 15-kilometer square area (see Table 1 & Figure 2).

The Grasberg porphyry copper-gold deposit possesses the world's largest gold deposit and is among the five largest copper deposits in an operating mine. The present reserves for the Grasberg ore body are about 1.7 billion tons.

Table 1. PTFI ore reserves.

		Recoverable reserves		
Deposit	Reserve (10^6 t)	Copper (10^9 lbs)	Gold (10^6 ozs)	Silver (10^6 ozs)
Present				
Grasberg pit	863	16.3	28.1	34.7
DOZ cave	184	3.2	3.0	14.9
Future				
Grasberg BC	782	16.3	17.6	35.8
KL cave	478	11.7	8.6	38.4
MLZ cave	50	1.3	1.3	3.9
ESZ cave	121	1.2	2.7	3.1
BG stope	33	1.7	0.8	8.7
DOM cave	44	0.8	0.3	4.0
DOM pit	27	0.7	0.2	3.8
Total	2,582	53.2	62.6	147.3

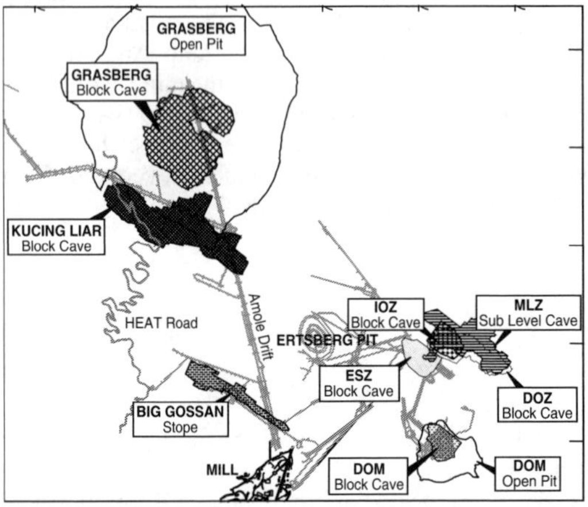

Figure 2. Plan view showing PTFI ore bodies.

2 GRASBERG BLOCK CAVE MINE

The Grasberg open pit is scheduled to cease production in 2014. Since the orebody continues at depth below the economic limits of surface mining, the reserve that is left behind will be extracted using block-caving methods. Development of the Grasberg block cave mine (GRS BC) is scheduled to commence during 2008 and will be accessed via the Ali Budiardjo (AB) Tunnel system. The mine will be a mechanized block caving operation with a nominal production rate of 115,000 tpd. The option of an increased production rate of 160,000 tpd was also considered in the planning and design.

The GRS BC will be a high-tonnage mine utilizing electric ore handling equipment on both the Extraction and Rail Haulage Levels. The tonnages will likely make it the highest production mine in the world. Caving, development, pre-production and mine support services will utilize diesel-powered mobile equipment. Due to relatively cool virgin rock and ambient air temperatures, significant cooling or heating of the air will not be required. The main access for men and materials will be via the AB Tunnels and a Service Winze.

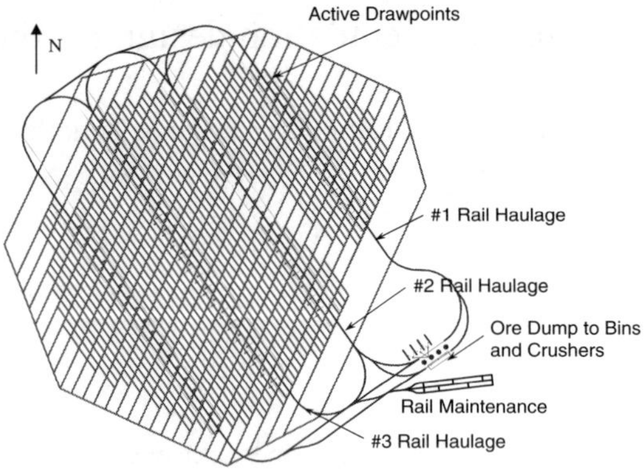

Figure 3. Plan view of 115,000 tpd extraction level with underlying rail loops.

The GRS BC operation will include the following components:

- A light rail passenger system connecting a surface station adjacent to worker living quarters with an underground terminus. The tunnel length will be approximately 8 km.
- Freight rail system providing goods delivery from surface warehouses and lay-down areas to the underground operations.
- An underground crushing system with twin conveyor drifts (2.1 m wide belts) linking the crusher bins/feeders to the surface milling facilities.
- A monolithic concrete-lined Service Winze with man cage providing access between the rail terminus and the block cave operation. The vertical length will be approximately 300 m.
- Single ramp linking the rail terminus with the orebins/conveyors and the block cave operation.
- Block caving operation consisting of 2840 m Undercut, 2820 m Extraction and 2790 m Service Levels (all elevations are above sea level datum).

2.1 *GRS BC design overview*

The mine design for the GRS BC utilized the Vulcan mine planning package. Output from Vulcan was imported to the VnetPC 2000 ventilation simulation program to develop a three dimensional network representation. Figure 4 shows a general isometric view of the proposed GRS BC. This network includes all components such as the tunnel access, ramp, Service Winze and conveyors.

Figure 3 shows a plan view of the Extraction Level for the GRS BC. For 115,000 tpd there will be

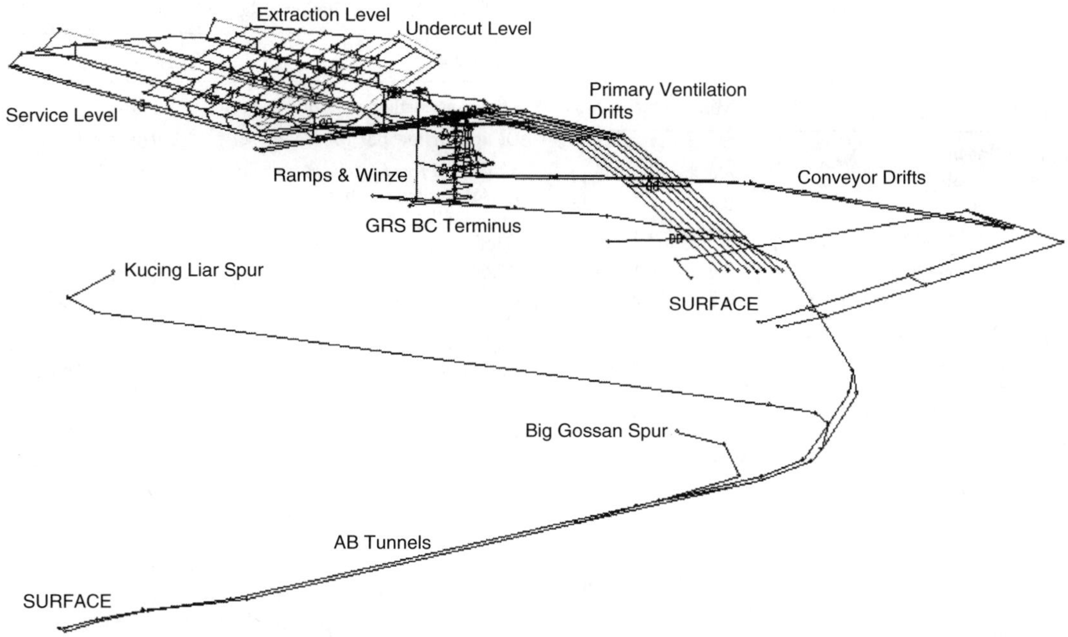

Figure 4. Schematic showing the layout of the GRS BC.

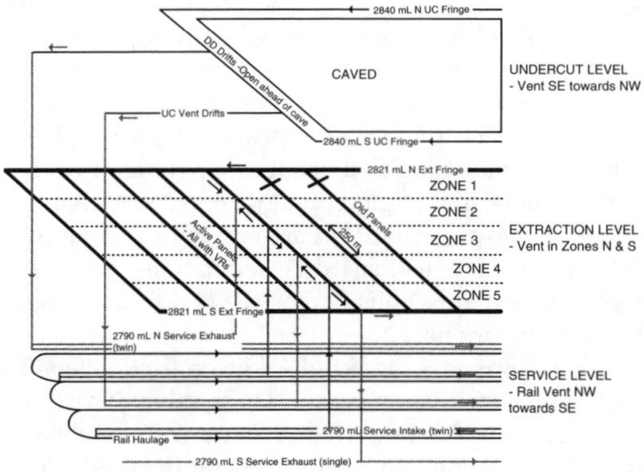

Figure 5. Sketch showing the proposed GRS BC ventilation system.

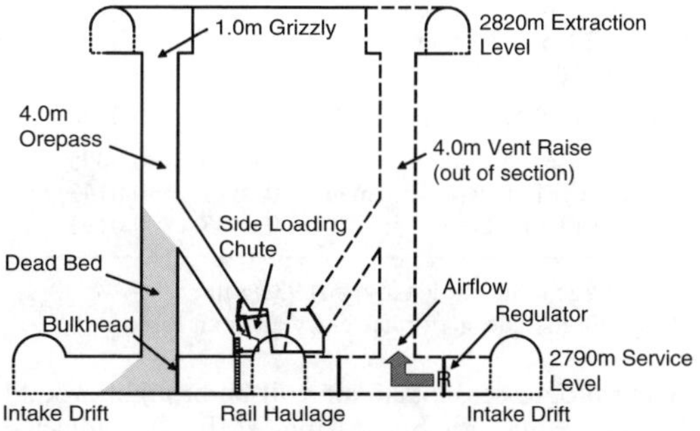

Figure 6. Cross-section showing the panel orepass and ventilation raise configuration.

approximately 870 active drawpoints that will be brought online as the cave moves from the Northeast towards the Southwest.

The ventilation and track drifts comprise the underlying Service Level. The Undercut Level overlies the Extraction with the same general footprint.

A schematic layout of the proposed ventilation system for the Extraction, Undercut and Service Levels is shown in Figure 5. The airflow is delivered and exhausted via the ventilation service drifts on the 2790 m Service Level. There will be four groups of three service drifts, with each group comprising two parallel ventilation drifts (intake or exhaust) and one-track drift. Each group of service drifts will be connected to every Extraction panel by two 4 m diameter smooth raises. One raise will serve as a chute, and the other for ventilation (see Figure 6).

These ventilation service drifts and connection raises allow the full width of the Extraction Level to be segmented into five separately ventilated zones.

The Undercut Level will be discretely ventilated. To avoid the requirement for many ventilation raises interconnecting the Undercut and Extraction Levels, two ventilation drifts will be driven across the Undercut Level ahead of caving. These drifts will be connected down to the Service Level exhaust at the West end. As

Table 2. Air velocity design criteria.

	Air velocity (m/s)		
Airway	Min	Opt	Max
Conveyor drifts			
– Homotropal	0.8	2.0	4.0
– Antitropal	0.8	1.0	2.0
Truck haulage drifts	0.8	4.1	6.1
Primary ventilation drifts	0.8	8.1	10.2
Rough-walled large raises (+4 m)	0.8	14.2	19.8
Typical ALIMAK ventilation raise	0.8	12.7	19.8
Drop ventilation raise	0.8	6.6	19.8

Note: – Optimum values are generally used for design purposes.
Velocity criteria are based on an economic assessment factoring in power and development costs.

Table 3. Atkinson friction factors.

	Friction factor (kg/m^3)	
Description	Actual*	Standard
9 m^2 to 15 m^2 – drifting	0.0102	0.0132
15 m^2 to 20 m^2 – drifting	0.0093	0.0120
20 m^2 to 30 m^2 – drifting	0.0083	0.0107
30 m^2 up – drifting	0.0074	0.0095
Conveyor drift (with 2.1 m belt)	0.0093	0.0120
Primary raise – 6.0 m ϕ+	0.0074	0.0095
Typical ALIMAK Rse – 3.0 m ϕ+	0.0111	0.0143
Drop ventilation Rse+	0.0148	0.0191

* At average mine air density = 0.93 kg/m^3.
+ Does not include shaft/raise entry and exit losses.

caving progresses, intake air will be brought around the cave using temporary fringe drifts and advance Drill Drifts (DDs). Auxiliary ventilation will be used to supply air to the active DD headings from the perimeter. The air will be exhausted via the advance Undercut ventilation drifts as shown in the sketch. The Service Level track drifts will be ventilated in the opposite direction to the Undercut, which will help balance the air velocity in the intake and exhaust service drifts.

The GRS BC area will be supplied and exhausted with air via primary intake and exhaust drifts. To facilitate the large quantity of air required, there will be four main intake drifts and four exhaust drifts as shown in Figure 4.

2.2 *Ventilation design criteria*

PTFI adopt certain ventilation criteria to assist with underground mine planning. Tables 2 and 4 present some of the criteria that have been established based on Indonesian Mining Regulations (Decree of the Minister of Mines and Energy 1995), US Mine Safety and Health Administration (MSHA), and good engineering practice. Although the design criteria do not specifically address diesel particulates, during the last year PTFI has commenced a comprehensive program to monitor and minimize diesel particulates in all underground areas. This program integrates baseline monitoring of worker exposures with reduction initiatives, including fuel changes, upgrade of equipment to tier-rated engines, and increased ventilation. Through this program PTFI are committed to gaining full compliance with MSHA regulations on diesel fuel type and emissions in existing and future underground operations.

Table 4. Other ventilation design criteria.

Criterion	Value*
Minimum airflow per 100 workers	7.1 m^3/s
Design airflow per 100 kW diesel	7.9 m^3/s
Shops and fixed facilities – exhaust	
Diesel equipment shop	40.1 m^3/s
Lube/fueling shop	28.3 m^3/s
"Hot" shop or rock breaker Stn	23.6 m^3/s
Explosives magazine	9.4 m^3/s
Common gas TLVs	
Oxygen	>19.5%
Carbon dioxide	TWA = 5,000 ppm
Carbon monoxide	TWA = 50 ppm STEL = 400 ppm
Hydrogen sulfide	TWA = 10 ppm
Nitrogen dioxide	STEL = 5 ppm

* TWA = Time weighted average for 8 hr working shift.
STEL = Short term exposure limit for 15 minutes.

Table 3 provides the list of Atkinson friction factors used for the ventilation design. These values have been obtained from detailed friction factor tests conducted in PTFI's existing underground operations, and hence there is a good level of confidence in the data.

2.2.1 *Airflow requirements*

Table 5 provides an estimate of the airflow requirements for the GRS BC operating at 115,000 tpd. This includes a comprehensive list of all mobile equipment, with actual engine power and estimated operating factors. To provide for an economic ventilation design, every attempt has been made to reduce the amount of diesel equipment operating in the mine, and a significant component of the production operation will be run remotely using automation techniques. Major contaminant producing areas of the mine such as the crusher stations and shops will be established on separate air splits direct to exhaust. The rail haulage and chute loading systems will be fully automated from a

Table 5. Estimated airflow requirements for the GRS BC – 115,000 tpd operation.

Item	No.	Operating factor(%)	Unit power (kW)	Airflow per unit (m^3/s)	Total airflow (m^3/s)
Mobile equipment					
LHD 6.0 m^3 (Dev.) – Dsl	8	70	186	14.7	82.6
Truck 40 t (Dev.) – Dsl	3	80	354	28.0	67.3
Drill jumbo (Dev.)	5	30	58	4.6	6.9
Production drills	6	30	113	9.0	16.1
Rock breaker	13	30	123	9.7	38.0
LHD 6.0 m^3 – electric	45	80	–	–	–
Truck 30 t (Dev.) – Dsl	8	70	224	17.7	99.1
Drill jumbo (sec. breaking)	16	60	43	3.4	32.3
Commando (sec. breaking)	26	60	36	2.8	44.2
Drill Jumbo (rockbolting)	2	30	42	3.3	2.0
LHD 2.7 m^3 (clean-up) – Dsl	2	50	138	10.9	10.9
Scissor lift	12	30	61	4.8	17.4
Explosives truck	4	70	123	9.7	27.3
U/G road grader	3	30	112	8.8	8.0
Shotcrete jumbo	5	30	58	4.6	6.9
Shotcrete truck	10	30	57	4.5	13.6
U/G lube trucks	6	30	115	9.1	16.4
U/G boom truck	4	75	115	9.1	27.3
U/G flatbed	7	75	115	9.1	47.7
U/G service truck	4	30	115	9.1	10.9
U/G electrical truck	6	30	115	9.1	16.4
U/G fork lift	4	15	32	2.5	1.5
U/G personnel carrier	10	30	115	9.1	27.3
U/G tractor (general)	10	30	24	1.9	5.7
U/G tractor (engineering)	2	30	24	1.9	1.1
Rail haulage locomotives	8	80	–	–	–
Subtotal mobile equipment					626.7
Fixed allowances					
Developed panel segment	45	100	–	14.2	637.2
Rail haulage drives	3	100	–	33.0	99.1
Personnel	600	100	–	0.1	42.5
Maintenance shop	4	100	–	40.1	160.5
Crusher bins and belt drifts	2	100	–	94.4	188.8
Subtotal fixed allowances					1,128.1
Total airflow required					1,755

remote control station. The nature of the Extraction Level lends itself to automation of the loader fleet although at this level of the study it has been assumed that all mobile equipment on the level will have an operator.

The total airflow estimate for the mobile equipment is about 627 m^3/s. The requirement for the active Extraction Level panel segments (electric loaders), and the Service Level electric rail loops is 736 m^3/s. Other fixed allowances include four shops, with a total airflow of 161 m^3/s exhausting directly to return ventilation drifts. There is an additional 189 m^3/s intaking the AB Tunnels which is exhausted via the GRS BC conveyors. This ventilation split requires an underground booster fan. Based on this list, the total airflow provision for the mine, excluding leakage and balancing, is about 1,755 m^3/s.

2.3 *Ventilation modeling and infrastructure sizes*

Ventilation analyses were conducted which included modeling the GRS BC during two different phases in the proposed mine life. The analyses took into account all anticipated air leakage paths (doors, bulkheads, old panels, etc.) to provide a realistic representation of the mine. The modeling incorporated the AB Tunnel system and made allowance for interconnection with the proposed Big Gossan and Kucing Liar Mines. The analyses assume primary underground exhaust fan installations close to the main portals.

The basic ventilation model was established as shown in Figure 4. The model was adapted to simulate two different scenarios:

1. Production from the first 12 panels, with a further 4 panels under development or construction. For

this case the Extraction Level South fringe has not yet been developed, so there is a temporary South fringe drift (consists of a line of future draw points) and only 4 zones per panel. (Early Mine).

2. Final (most Westerly) 4 panels are under development or construction, with 12 adjacent panels in production. The center of mining has shifted to the West, and leakage in the system has increased. (Mature Mine).

The ventilation modeling assumes that the conveyor system will exhaust air from the GRS BC (189 m^3/s) and also from the adjacent Kucing Liar Mine (additional 189 m^3/s). The ventilation modeling also makes an allowance in the AB Tunnels for the Kucing Liar and Big Gossan Mines, which will both likely be operational during the period of the GRS BC mine life.

The predicted fan requirements for these two scenarios are given in Table 6.

For the early mine the total airflow is predicted to be 1,812 m^3/s. Based on a nominal daily tonnage of 115,000 tpd this equates to 0.0157 $m^3/s/tpd$. The total installed fan motor power requirement is about 7,200 kW with a predicted annual power cost of US$2,830,000. These values assume 75% efficiency for the main fan and motor installation. For the mature mine the total airflow quantity increases to 1,888 m^3/s, or about 0.0164 $m^3/s/tpd$. This is higher than the value shown in Table 5 due to the system leakage. As a result of the efficient layout for the ventilation system, with few connections between intake and exhaust, the predicted leakage is less than 10%.

The installed fan motor requirement for the mature mine increases to 8,650 kW, with a projected annual operating cost of about US$3,410,000. Both of these operating characteristics (Early and Mature cases) will be achievable with the same fan installations.

The following drift sizes have been determined for the 115,000 tpd design (all ventilation drifts will be fully shotcreted):

- Primary Ventilation Drifts from Portals to Service Level = 6.8 × 6.8 m
- Ventilation Service Drifts below Extraction = 5.5 × 5.5 m
- Ramp and Access Drifts = 4.5 × 5.0 m
- Conveyor Drifts = 4.5 × 5.0 m
- Extraction panels = 4.5 × 4.5 m
- Undercut DDs = 3.6 × 4.2 m
- Panel Raises = Raise-bored at 4 m diameter
- Service Winze = 8.5 m diameter smooth concrete.

Figure 7 shows a graph of the relative pressure profile for the mature mine case. The relative pressure for this case is the difference between ambient (surface) and locations underground. For the profile a typical circuit has been chosen through the mine, and the trend line plotted against linear distance covered, including ventilation raises. The circuit chosen was: intake portals → primary intake drifts → service intake drifts → far Western panel region → service exhaust drifts

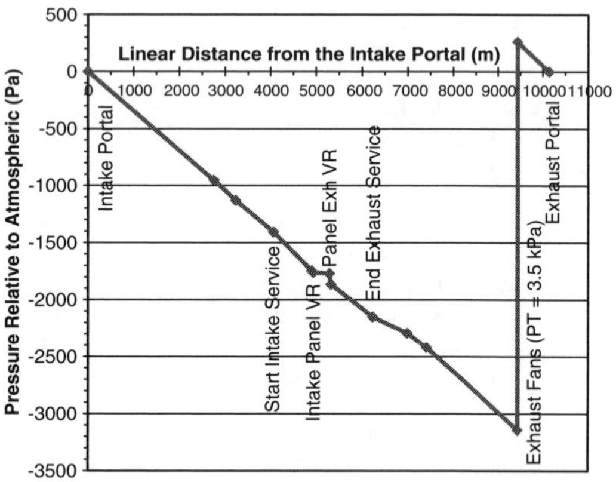

Figure 7. Relative pressure profile for the 115,000 tpd mature mine case.

Table 6. Predicted fan requirements for 115,000 tpd operation.

Scenario/location	Total pressure (Pa)*	Quantity (m^3/s)	Input power (kW)^	Operating cost (US$/yr)+
Early mine				
Main exhaust	2,989	1,623.7	6,471	$2,551,000
Boosteri installation	2,865	188.8	721	$284,000
Totals for mine	–	1,812.5	7,192	$2,835,000
Mature mine				
Main exhaust	3,487	1,699.2	7,901	$3,114,000
Booster installation	2,989	188.8	752	$297,000
Totals for mine	–	1,888.0	8,653	$3,411,000

* System pressure to be met by fan. Does not include fan losses.

^ Assumes 75% efficiency for fans and motors.

\+ Assumes 0.045 US$/kWh.

→ primary exhaust drifts → underground primary exhaust fans → exhaust portal. The trend is considered excellent, remaining generally linear and averaging about 333 Pa/1000 m. The one short steep section represents the "open-split" in the ventilation circuit where the air exhausts from the development panel region to the service exhaust system. A linear profile is indicative of balanced air velocities and frictional pressure losses, suggesting that drift and raise sizes and numbers are correct for the system.

2.4 *Adits versus shafts design comparison*

A ventilation study was conducted comparing adits versus shafts for the primary mine infrastructure. An initial shaft layout was developed in which the intake air is brought down two concrete-lined 7.5 m diameter shafts into the orebody area and back out two similar 7.5 m diameter exhaust shafts to surface. Due to the mountainous topography, the shafts would be approximately 1.0 km long. In order to place the shafts far enough away from the cave influence, primary intake and exhaust drifts would still be required, with each drift about 1,500 m long. It was assumed that two sets of four drifts will be developed at 6.8 × 6.8 m connecting the shafts to the production area.

The maximum velocities for the 7.5 m diameter shafts would be 18 m/s, which is 25% higher than the economic optimum velocities used as the current basis of mine design (see Table 2). However, it is noted that the economic velocity given in the table is for a rough shaft (2.1 m bored then slashed to size), not a fully concrete lined shaft, which will be considerably more costly to construct per meter, and have a higher optimum velocity.

The shaft study showed approximately US$69 million in capital allocated to sink the intake and exhaust shafts and develop the shorter drifts. The adit option required approximately US$46 million in capital.

The mature mine 115,000 tpd scenario was used for a fan power cost comparison. The results are provided in Table 7. These results indicate that there is a significant power savings (+US$580,000/yr) associated with drifting compared to shafts.

In addition to the economic benefits, the drifting option was considered to have certain other advantages, including:

- Shafts require specialized equipment and crews to sink. Drift development will be spread over the mine ramp-up years with multiple headings driven by PTFI development crews.
- Shafts require specialized maintenance equipment and trained crews. Repairs typically require a long time frame to accomplish.
- Historically PTFI has encountered water in long vertical shafts. In an exhaust shaft this counter-current water flow acts as a permanent drag and increases resistance. In such a long shaft it may also result in "blanketing" and cyclic loading (even potential fatigue) of the fans.
- The fan installations will be more accessible for construction, maintenance, inspections and power delivery if placed near the proposed portals. Longer-term, access to the shaft collars would be difficult being close to the final cave line and remote from other mine facilities.
- In the case of an emergency, personnel can quickly egress to a fresh air intake and if required walk completely out of the mine using that primary escapeway.
- The drifts provide large tunnel diameter access to the mine for occasional oversize pieces of equipment that could potentially disrupt the tunnel rail system.
- Current plans allow for a drill-and-blast method of developing the drifts. There is potential for a tunnel boring machine to be employed, which would result in a smoother, less resistive tunnel section and potentially reduce the number of tunnels required.

Table 7. Power cost comparison – drifting versus shafts.

Scenario/location	Total pressure (Pa)	Quantity (m^3/s)	Input power (kW)	Operating cost (US$/yr)
Shafts				
Main exhaust*	4,110	1,699.2	9,312	$3,671,000
Booster installation	3,238	188.8	815	$321,000
Totals for mine	–	1,888.0	10,127	$3,992,000
Drifts and portals				
Main exhaust	3,487	1,699.2	7,901	$3,114,000
Booster installation	2,989	188.8	752	$297,000
Totals for mine	–	1,888.0	8,653	$3,411,000

* Atk. friction factor = 0.0037 kg/m^3 (0.0048 kg/m^3 stnd) for shafts.

Table 8. Fan comparison – 115,000 tpd versus 160,000 tpd (mature mine).

Scenario/location	Total pressure (Pa)	Quantity (m^3/s)	Input power (kw)	Operating cost (US$/yr)
115 ktpd case				
Main exhaust	3,487	1,699.2	7,901	$3,114,000
Booster installation	2,989	188.8	752	$297,000
Totals for mine	–	1,888.0	8,653	$3,411,000
160 ktpd case				
Main exhaust	3,238	2,124.0	9,170	$3,615,000
Booster installation	2,615	188.8	658	$260,000
Total for mine	–	2,312.8	9,829	$3,875,000

2.5 *160,000 tpd comparison*

Ventilation analyses were also conducted assuming an expanded production rate of 160,000 tpd. The simulation incorporated an updated equipment list, and the number of panel segments in production increased from 45 to 62. This change resulted in an increase in the main fan airflow to 2,124 m^3/s, with a total mine airflow of 2,313 m^3/s (when including the GRS BC oreflow ventilation system). Certain infrastructure sizes were increased as follows:

- Primary Ventilation Drifts = 6.5 × 9.5 m (benched development)
- 2790 m Ventilation Service Drifts = 5.5 × 6.5 m.

Table 8 provides a comparison of the main fan operating requirements for both the 115,000 tpd and 160,000 tpd cases.

Although the total airflow increases, there is an actual reduction of about 0.002 $m^3/s/tpd$. The fan pressures are reduced slightly due to the increase in drift sizes.

3 RECOMMENDATIONS FOR FURTHER STUDY

The work contained in this publication supports a pre-feasibility design of the GRS BC operation. As the study progresses to feasibility level a more detailed analysis will be conducted. Certain ventilation issues have been identified for special consideration.

3.1 *Mine total airflow*

The total predicted airflow in the mine is 1888 m^3/s, which corresponds to approximately 0.016 $m^3/s/tpd$. This compares to about 0.028 $m^3/s/tpd$ for the existing PTFI DOZ Mine ventilation system. Experience with diesel panel and block cave mines suggests that 0.021 $m^3/s/tpd$ represents an average industry value. However, this figure takes no account of how well ventilated the mine is, and the value for the DOZ Mine is probably valid for rule-of-thumb design for diesel operations. The value for the GRS BC Mine is significantly lower than this, however the contaminants in the mine are expected to be much lower than those presently encountered. The GRS BC will incorporate electric panel loaders, electric oreflow handling, low emission diesel development equipment operating on underground mine fuel, and significantly reduced Undercut drilling (room-and-pillar undercutting methods). When combined it is anticipated that these changes will allow compliance with gas and diesel particulate criteria, however it will be important to continue to revisit and revise airflow requirements as the design progresses.

3.2 *Airflow control*

One of the main challenges faced by a ventilation engineer in a large mine is control of airflow. This can be particularly difficult in caving operations where there are multiple open parallel splits. On Extraction Levels the problem is compounded in longer panels, which often have more than one ventilation zone (not a simple flow-through system). PTFI have experience with operating large block cave mines and will apply the knowledge and lessons learnt from the GBT, IOZ and DOZ operations to the GRS BC. Experience suggests that automated regulation of the panel raises on the Extraction Level will be difficult due to damage from equipment and drawbell blasting activities (initial development and removal of hang-ups). Hence it is likely that remotely controlled heavy-duty regulators will be designed and installed on the Service Level accesses to the panel raises (see Figure 6).

Pneumatic door design is well proven at PTFI and heavy-duty equipment airlock doors will be installed where required. Typical high-pressure bulkheads at PTFI consist of timber frames keyed into the ribs, roof and floor, covered with two layers of wire mesh separated by a layer of ventilation brattice. Shotcrete is then sprayed onto the high-pressure side that covers onto the surrounding rock. Such a bulkhead has been found to hold up well against blasting pressure waves, and impact from equipment. Existing regulators at PTFI are typically manual window-type devices constructed from heavy-gauge steel that are fit into a shotcrete bulkhead. In all cases durability is the main priority in the design of ventilation control devices.

3.3 *Dust and diesel particulates*

Control of dust in block cave mines can be extremely challenging. Approximately 230,000 tpd of ore are presently moved through PTFI's site operations via loaders, trucks, crushers, belts and orepasses. In certain areas, such as panel drifts, belt transfers, orepass head chutes and stockpiles particular attention must be given at the design stage to ensure compliance with dust Threshold Limit Values (TLVs). This may consist of a comprehensive plan incorporating weather station data, belt moisture sensors, foggers and sprays tied to the mine PLC network. From analysis of the silica content of dust from present Grasberg ore, the dust contains significant amounts of respirable quartz (10–17%) corresponding to a TLV of about 0.70 mg/m^3. The mine design will also have to ensure compliance with MSHA's diesel particulate TLV, and as such the fuels, equipment, and ventilation will factor this in.

3.4 *Fans*

The predicted main fan operating requirement is 1699 m^3/s at 3.5 kPa fan total pressure, which corresponds to 425 m^3/s per exhaust drift. This assumes a single primary exhaust fan installation. Further work will evaluate the potential advantages of primary forcing fans and a push-pull system with both primary forcing and exhausting fans. The characteristic of 425 m^3/s per drift can be achieved with either centrifugal fans (portal structure) or underground axial fans (vane or mixed-flow). The power demand will be high, but it is anticipated that losses can be kept to a minimum through careful design of the fans, layout, and electrical supply. The underground booster fan for the conveyor drifts is sized at 189 m^3/s at 3.0 kPa fan total pressure. A similar sized fan would also be required to handle the airflow exhausting from the proposed Kucing Liar Mine crusher area.

3.5 *Air velocity*

The airflow quantity in the twin exhaust conveyor drifts is about 378 m^3/s, corresponding to an air velocity of approximately 6 m/s. This exceeds the maximum design velocity of 4 m/s for a homotropal conveyor system. During further study the GRS BC oreflow ventilation system will be examined. It may be possible to significantly reduce the airflow exhausting the belts, or even intake the belts with a dedicated exhaust to the primary mine fans.

Depending on the active segment layout, the air velocity at the East end of the service ventilation drifts may become high (exceed 10 m/s). However, when considering the entire length of these collection drifts the pressure losses are not excessive. It is planned that the two ventilation drifts in each set of three service drifts will be aerodynamically connected by equalizing cross-cuts.

3.6 *Upcasting air and service Winze design*

In the ventilation model airflow is upcasting both the lower ramp and Service Winze. Based on experience at PTFI it is likely that condensation will occur with the upcast air (Calizaya et al., 2002). In the ramp this can result in fogging and poor visibility. In the shaft this can cause fogging, water collection, and in some cases more rapid deterioration of fixed facilities. For the shaft it would be possible to downcast intake air. Further study will be devoted to the determination of shaft resistances and the dynamic effects of cage movement, and the optimization of airflow in the shaft. For the lower ramp fixed fan-heaters may be required to dry the air and maintain good visibility. Heaters have been used to good effect in existing PTFI underground ramp systems.

3.7 *AB tunnels and terminus*

Approximately 189 m^3/s is predicted to intake the lower mine section from the AB Tunnels. This air then either enters the ramp, Service Winze, or shop facilities (dedicated exhaust). This represents a significant volume of air, with a tunnel air velocity of about 5.7 m/s. During future study the airflow requirements for the tunnels will be more accurately determined, incorporating fire simulation to determine contaminant spread in the event of a tunnel vehicle fire.

3.8 *Mine fire*

Fire analysis is an essential component of an overall mine design. For the pre-feasibility design fire mitigation and control philosophies were incorporated into the mine plan. More detailed "hot-flow" fire analyses will be conducted as the design progresses. These analyses will utilize the Mine Fire package (a hybrid of the VnetPC 2003 and former US Bureau of Mines MFIRE codes), which allows full interrogation of the entire mine network. For the preliminary design PTFI has:

1. Made provision for all shops and hot-work areas to be directly ventilated to an exhaust airway and to be equipped with isolation fire doors.

2. All conveyor belts will have independent airflow splits directed to the belt portals, and will be equipped with smoke sensors and fire-suppression deluge systems.
3. The air from the AB Tunnels will be used for the lower ramps, and will exhaust via the conveyor drifts. This air will not enter any of the production areas.
4. The air from the upper ramps will be used for ventilating the conveyor drifts, and will not enter the block cave working areas.
5. Air critical velocity values (per Froude number analysis) will be sufficiently high throughout the ramps as to prevent backlayering of smoke against the direction of airflow in the event of equipment fire. This is necessary to ensure predictable smoke spread and provide a non-compromised fresh air base for fire fighting.

4 CONCLUSIONS

The GRS BC will be a very large mine that will provide many unique engineering challenges. It is anticipated that like most block cave operations, there will be issues associated with the balancing of the air, especially through the production panels. Particular attention will also need to be paid to the integration of dust and diesel particulate control strategies and systems within the ventilation design. However, it can be stated that for a pre-feasibility study, the proposed GRS BC mine layout integrates good ventilation design. The predicted pressure losses and air velocities generally agree with PTFI optimum design criteria, and peak values are within acceptable economic and operational levels. When considering operating cost, there is a definite benefit in developing the mine with portals, rather than intake and exhaust shafts, and the drifting option has been selected as the base case.

ACKNOWLEDGEMENTS

The authors would like to thank PTFI Management for allowing publication of this paper.

REFERENCES

Calizaya, F., Karmawan, K. & Wallace, K.G. 2002. Utilization of Heater Fans to Control Mine Atmospheric Fogging. In Euler De Souza (ed), *Proceedings of the 9th NA/US Mine Ventilation Symposium*. The Netherlands: A.A. Balkema Publishers.

Decree of the Minister of Mines and Energy Indonesia, 1995. Number 555.K/26/M.PE/1995, *General Mining Occupational Safety and Health*.

MSHA, *Code of Federal Regulations 30*, Part 75.

Mine Ventilation, Ganguli & Bandopadhyay (eds)
© 2004 Taylor & Francis Group, London, ISBN 90 5809 633 5

Revision of the ventilation system in Louhi mine – a case study

A.L. Martikainen & P.S. Särkkä
Helsinki University of Technology, Espoo, Finland

ABSTRACT: The Louhi limestone mine has been in operation over 50 years. Because of the mine's changing shape the ventilation system has been checked and updated several times. Due to the recent growth a revision became indispensable and was accomplished in 2002. The aim of this study was to find ways to prevent fog in the ramp, and to guarantee sufficient amounts of fresh air entering active mining areas. The methods used were field studies, measurements, calculations, and computer modeling.

The received information was combined with the measurement results to enable the creation of the computer simulation model. The model was used in testing the possible changes of the ventilation network.

Short-term as well as long-term improvement suggestions were given. For long-term operation three main alternatives were lined out. These were a new fan, a new raise as an extension to the old dead shaft, and/or changes in ventilation raise functions.

1 INTRODUCTION

The Louhi mine was opened up during the Second World War. The founder, Ruskeala Marble Ltd, had operated a limestone mine at Ruskeala in the Finnish Karelia. In 1940 the area was lost to the Soviet Union, and the company had to compensate the production.

The next best unleased limestone deposit was found to be in Eastern Central Finland at Louhi, the Kerimäki municipality, some 20 km NE from the city of Savonlinna, 300 km NE from the city of Helsinki.

The limestone layer at Louhi is steeply dipping with a thickness of some 20 to 30 meters. Mining was started as an open pit from the surface outcrop, and continued underground, first with shrinkage stoping, and from the 60'es with longitudinal sublevel stoping.

Sublevel spacing is 40 meters, and the stope sizes vary from 20000 m^3 to 40000 m^3. Loading is done with LHD loaders, and hoisting takes place with trucks via an incline to the surface.

Annual production was 212000 metric tons in the year 2002. The limestone is sold basically as Industrial and agricultural limestone to the local industry and farmers.

Nordkalk Ltd, the largest limestone producer in the Baltic Sea area, bought the mine in the year 1988. When the production with diesel equipment was advancing to new, deeper areas, the needs to intensify mine ventilation grew obvious.

The main problems to be solved were to guarantee the availability of sufficient amount of fresh air in the new areas, and to prevent the formation of fog in the ramp. The mine contracted Helsinki University of Technology, Rock Engineering, in 2002 to develop solutions in the above problem areas.

2 THE VENTILATION SYSTEM OF LOUHI MINE

2.1 *General*

Fresh air route consists of two shafts that enter the same stope between levels +90 and +130 merging fresh air flows on level +130. The air is then directed through an inclined tunnel to level +145 to enter another shaft that takes the air first to level +190 and then finally to the bottom level +230.

Exhaust air is partly taken out through the decline, and partly using exhaust shafts. The exhaust shaft take air from the bottom of the mine to level +190, then it flows through tunnels to another shaft entering level +110 and finally through an inclined tunnel from level +90 to the surface. The principles of the ventilation system are shown in Figure 1.

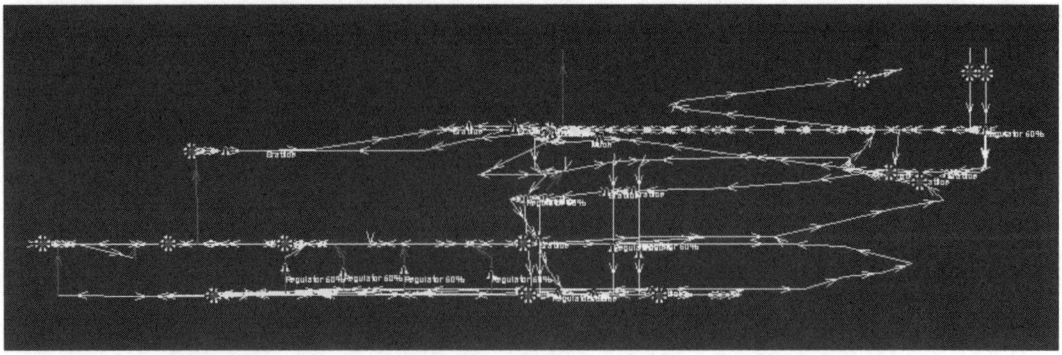

Figure 1. Principles of ventilation shown in model.

Figure 2. New main fan.

2.2 *Ventilation equipment*

Ventilation is controlled using fans, ventilation walls, and ducts. The two main fans are on the surface at the top of the fresh-air shafts. The newer of the two main fans is shown in Figure 2. Other rigidly mounted fans are used in places where air is conducted to tunnels, and in front of the uppermost part of the exhaust-air shaft. In working areas the fans and ducts are moved when needed. Ventilation walls are made of fabric and are either fixed or equipped with a drive flap.

3 MEASUREMENTS AND CALCULATIONS

3.1 *General*

Most measurements were made during the autumn of year 2002. Some were performed about a year later. Air velocities, temperatures, and airway cross-sectional areas were measured in 2002. Later these measurements were supplemented with temperature, humidity, and particle measurements. Moreover, all measurements were performed in the main airways.

The use of fresh-air raises as emergency exits helped to measure air velocity, because these raises were equipped with stairs. It was especially difficult to measure the cross-sectional areas in some of the raises and tunnels because of all the installed services like ducts, stairs and cables. Also the unevenness of tunnel walls was an error source in the cross-sectional area measurements.

3.2 *Measurements*

The first temperature and all air velocity measurements were carried out with a thermoanemometer. For cross-sectional area the length measurement pole of the mine's geologists was used. Airflows in the main airways were calculated from air velocity and cross-sectional area measurement results.

Humidity and the latter temperature measurements were made with a hygro- and thermometer, and particle measurements using an aerosolmeter. The reason for these measurements was the fog that sometimes hinders visibility in the decline.

3.3 *Operating rate analysis*

To estimate the total amount of air needed in the mine an operating rate survey of the diesel equipment was performed. The use of each machine was monitored by the operators during the shifts. The operators wrote down the actual usage times of the machines for four weeks.

The results were gathered and the average use of each machine was calculated. Furthermore, the total average usage was calculated using the diesel power and the operating rate of the machines. Finally the total amount of air needed was calculated based on this information using the formula widely accepted in Finland:

$$Q = 3.4\boldsymbol{m}^3 / \boldsymbol{kW} \min Pl \tag{1}$$

where Q is the air volume, P is the power of equipment, and l is a leakage percent, which typically varies from 30% to 50% (Seppänen, 1982). The calculations based on the operating rate analysis showed that the needed total fresh air volume needed is 71,5 m^3/s.

4 MODELING OF THE VENTILATION NETWORK

4.1 *General*

The ventilation simulation software used was Ventsim Version 3.32. The program is user-friendly, quite flexible, and has a graphical interface. Ventilation network can be handled and edited both in 2-D and 3-D. The program enables simulation of airflows, air velocities, change characteristics and increase the number of the airways, assimilate short- and long-term ventilation planning, receive information for choosing of fans and dimensioning of ventilation ducts as well as to model air contaminants like smoke, dust and gases, and their content and distribution (Haapala, 2001).

Gathered data is used as an input in creating databases. Other needed information can be found from the program either as parameters that can be chosen based on the user's needs or as constants.

The ventilation model of Louhi mine consists of eight levels, and connecting inclines, tunnels, stopes, raises and shafts. Parameters used were either from the received data or were estimated with the help of mine visits, and comments from the personnel. The chosen parameters from the program, like the excavation smoothness, which could not be checked in some of the tunnels and raises, were changed until the ventilation model correlated well enough with the measurement results. In tunnels where the cross-sectional area could not be measured an estimation was used. Shock values were chosen based on charts and pictures.

4.2 *Modeling difficulties*

The worst complications in the modeling were old tunnels and raises that were not used anymore. It was difficult to estimate the amount of air that leaks through such old openings. Some of them were well closed and not taken into account in the model, when some were noticed to leak and were therefore included in the model. Another problem was the modeling of large stopes. These were modeled as knot points so that the same amount of air that enters the stope also leaves it. Possible air re-circulation in the stope was ignored.

Some of the fans were so old, that the charts could not be found. The motor and the blade pitch were changed several times, and practically nobody could tell anything certain anymore about their function. These fans were modeled with charts of the most similar fans found or based on the information received from the fan manufacturer.

4.3 *Completing and testing of the model*

The model was adjusted until all parts of it correlated well with the measurement results and other gathered data. The completed model is shown in Figure 1.

After that changes and their effects on the model were tested. The mine personnel had some own ideas, how they would like to improve the ventilation system. The stoping order for the becoming years was also taken into account. The ventilation changes were done for the coming years, too. These improvements and planned stopes were inserted in the model one-by-one and the results of these changes were then checked and reported.

Some fascinating things were observed during the testing of the model. The most important of those was that some of the fans were actually more efficient than the manufacturer's charts showed. These fans were quite new and from a different manufacturer than the older ones. The personnel have been very pleased with these Removex fans, which was fully understandable when comparing the theoretical charts with the modeling results. Another remarkable thing was that the fans were chosen and placed so well that only one of them was restricted by another fan. So the fans were able to work with their maximum capacity without idling. No convergence problems or ill-placed ventilation walls were noticed in modeling.

The model remains to be a valuable help in testing possible changes in the ventilation system. The mine personnel are still interested in it and the model has already been updated twice after the project finished in November 2002.

5 DISCUSSION AND SUGGESTED IMPROVEMENTS

5.1 *Results and observations*

During the mine visits it was acknowledged that the ventilation was mainly well-done. The ventilation walls

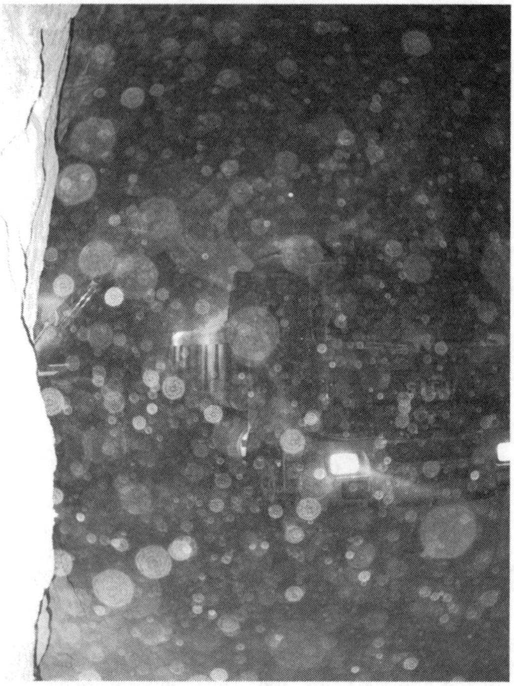

Figure 3. Loading site.

and pipes were in quite good condition, the fans were all working properly, and a lot of thought was put in the design of the system.

Humidity measurements showed that the moisture content of the air increases considerably as the air flows through the mine. High temperatures occur in Louhi mine quite seldom, since the depth of the mine is only 230 meters. Highest values in particle measurements were observed in loading sites. These were, however, less than half of the regulatory limit values given by The Finnish Ministry of Social Affairs and Health (Sosiaali- ja terveysministeriö, 2002). Figure 3 gives an impression of particle amount in a loading site. Other higher than average values were noticed in the foggy parts of the decline.

The fog problem in the decline has not yet been solved, but is still examined, because this seems to be a common problem in Finnish mines. Thus, further research for comparison purposes in other mines is at the moment a work in progress. By the end of October 2003 measurements have been performed in three Finnish mines and more research is under consideration.

5.2 *Short-term improvements*

One of the few problematic sites was the connection of the older fresh-air shaft on the level +90 to the stope to the level +130. Part of the air was leaking to the long-abandoned tunnel system of the level +90. The corrective measure suggested was to build two new ventilation walls on the level.

Another problem was air re-circulation in level +190 where part of the exhaust air from the working area was circulating back to the fresh air fan through a stope. A ventilation wall was suggested to be placed in the tunnel near the stope opening.

These above-mentioned changes were tested with the ventilation model, and they functioned as planned. The losses of fresh air were now considerably smaller and more fresh air was conducted to the bottom of the mine and to the working areas. Also the recirculation of air was minimized on level +190 and exhaust air now entered the exhaust air route.

The places for new fans were tested next. The only location where a new booster fan was considered to be beneficial was the so-called tail-end no 6, i.e. the tunnel combining the fresh-air stope from the level +130 to the fresh-air raise between the levels +145 and +190. This would increase the air velocity in fresh-air raises from level +145 to level +190 and from level +190 to level +230 and thus also diminish the total air circulation time through the mine.

On the level +145 a ventilation wall was partly open so that some of the fresh air goes in to the stope. This proved to be a good solution in decreasing the resistance of the fresh air route, because the fresh air entering the stope could not escape to the exhaust route but was actually going straight to the bottom of the mine to be used with the fresh air coming through the raises.

5.3 *Long-term improvements*

The model showed that the actual amount of fresh air entering the mine was about 55 m^3/s. This is not quite enough in relation to the equipment used today. Minimizing the fresh air losses helps temporarily, but in the long run the amount of fresh air should be increased.

Three different options for increasing the amount of fresh-air were investigated. All of these could be considered as long-time solutions. These options were

1. installing a new fan on top of the new fresh-air shaft next to the recently acquired main fan,
2. a new raise as an extension to the old dead shaft in the area where new stopes have been planned,
3. and/or changes in ventilation raise functions.

Installing a new fan parallel to the other fan would probably be the easiest and cheapest option. This possibility was taken into account already in planning of the new fresh-air shaft that is now large enough for two parallel fans.

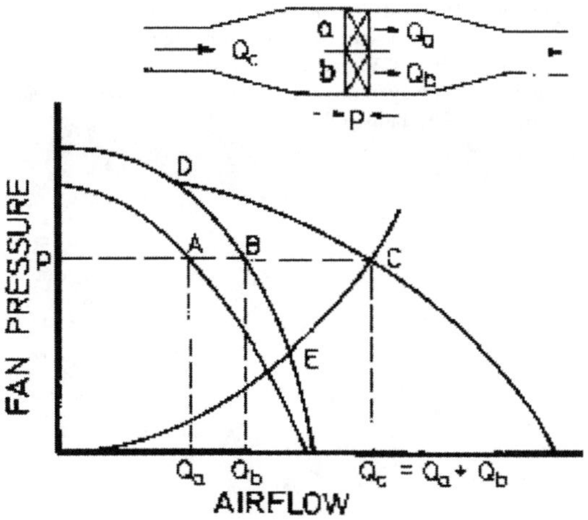

Figure 4. Characteristic curve for two fans installed parallel (McPherson, 1999).

With the fan parameters now used convergence problems might occur with a second fan. The best solution to that would be to change the blade pitches. This would result in a working combination of two fans in parallel, increasing the fresh air amount about the needed 17 m^3/s. As a result the air velocities would increase in every airway. The only problem would be that the air velocity in the decline would be quite high, changing from the earlier 1,5 m/s to about 2–2,5 m/s depending of the place. This is, however, still reasonable, and it shouldn't yet cause air circulation or pressure pumping problems by the traffic. An example curve for parallel fans is shown in Figure 4 (McPherson, 1999).

In the area to which the mine is growing, there is an old dead shaft called p75, which could be deepened and taken into use. The shaft could be used as a new fresh-air shaft. This would result in more fresh air in the mine, and the distance of fresh air from the surface to the working areas would be short. When modeled it was observed that the shaft would automatically become an exhaust shaft, so a new fan on the top of it would be needed. The shaft should reach the bottom of the mine in order to achieve all the improvements it enables.

If a fan with quite similar properties to either one of the previous main fans would be chosen, the amount of fresh air would increase about 20–35 m^3/s depending on the parameters of the fan and the properties of the shaft. This option would allow the growth of the mine without further changes in the ventilation system for a longer period than the other alternatives. Also, the exhaust gases could be well diluted even in the case of more machines. Unfortunately, the cost of this option is quite high.

The third possibility, explored was to change the shaft functions by lengthening the dead shaft discussed before, and using it as a fresh-air shaft, turning the old exhaust shaft to a fresh-air shaft, too, and changing both current fresh-air shafts to be used as exhausts. This would result in much shorter fresh-air routes to the working areas, but there would not be so much increase in the amount of fresh air. This option is quite difficult to accomplish due to many structural changes, like taking down existing ventilation walls and building of new ones. The exhaust shaft should be kept ice-free during winters because of all installed systems, thus its use as fresh-air shaft might cause troubles in wintertime.

This option could only be chosen in a case when the construction of a third fresh-air shaft would be done simultaneously to ensure sufficient fresh-air amount for further capacity increase. There seems to be no need for such large changes as the mine production has been growing quite marginally. This option should only be considered if the mine continues to be in production decades.

6 CONCLUSIONS

The ventilation system of Louhi mine is a functioning entity with a quite simple working idea. Also, ventilation equipment is mainly in good condition.

The ventilation survey performed was considered successful. A lot of observations were made and based on those as well as computer modeling both short-term and long-term improvement suggestions were found. All of the suggested short-term improvements have been implemented in Louhi mine. The long-term improvements are still under consideration, since the latest expansion plans have not yet been settled. The most useful long-term improvements at the moment are either a new fan or a new raise. Both options increase the air volume to the required level based on the calculations and modeling.

Fog problem still needs attention and more studies will be done in the future to help in solving it. This research work has already started and is proceeding well.

ACKNOWLEDGEMENTS

The authors thank Nordkalk Co. for the possibility to study the mine and to write this paper. The personnel of the Louhi mine have given their full attention to the project and we thank them. We especially like to mention Harri Taskinen, Jyrki Koljonen and Reima Kosonen, who where hosting our mine visits. We also thank Helsinki University of Technology for financial support and for providing the measurement devices.

REFERENCES

Haapala, Jari. 2001. Sipoon kaivoksen yleistuuletuksen nykytilanteen mallinnus ja parannusehdotukset. Espoo: Raportti, Kalliotekniikan laboratorio, Teknillinen korkeakoulu.

McPherson, Malcolm, J. Subsurface Ventilation and Environmental Engineering, 1999. http://www.multimedia.vt.edu/mcpherson/teaching.html: 10.11.2003.

Seppänen, Pentti. 1982. Kaivostuuletus. In Matikainen, Raimo (sen. ed.) Kaivos- ja louhintatekniikan käsikirja: 645–669. Hanko: Vuorimiesyhdistys.

Sosiaali- ja terveysministeriö, 2002. *HTP–arvot 2002*. Tampere: Kirjapaino Öhrling.

Mine Ventilation, Ganguli & Bandopadhyay (eds)
© 2004 Taylor & Francis Group, London, ISBN 90 5809 633 5

Alternatives to using flame retardant belts in underground workings: a case study at Impala Platinum

F.J. van Zyl
CSIR Mining Technology

H. Smit
Impala Platinum

ABSTRACT: Due to the high cost of fire retardant conveyor belts and their associated reduced durability, Impala Platinum investigated alternatives to using fire retardant conveyor belts in their underground operations. A solution was to use non-fire retardant conveyor belts in parallel with a conveyor belt fire suppressions system. Impala Platinum identified a commercially available system that could potentially be applied as a conveyor belt fire suppression system. This system is widely used to protect important facilities and equipment in enclosed spaces where static environments exist. In order to deploy this system underground the system had to be evaluated to ensure that it is capable of extinguishing a conveyor belt fire in a safe and effective way in the dynamic underground environment.

CSIR Miningtek's Kloppersbos test facility was approached by Impala Platinum to develop an accepted test protocol to evaluate the effectiveness of the system in suppressing belt fires in the dynamic underground environment.

This paper describes the risk assessment process followed to identify the requirements for an underground conveyor belt fire suppression system at Impala Platinum, the test protocol developed to evaluate the fire suppressions system, the test gallery preparation and testing of the system, the findings and recommendations from the tests, and the underground trial implementation of the fire suppression system.

1 INTRODUCTION

CSIR Miningtek's Kloppersbos test facility was approached by Impala Platinum Limited to quantitatively assess the effectiveness of a conveyor belt fire extinguishing system, namely the Fogtec system.

A full-scale test simulating underground conditions was used to determine if, and how effective the proposed system could control and extinguish an underground belt fire. Using the controlled environment of the test section, a quantitative assessment of the effectiveness of the Fogtec system was made.

In conjunction with the actual determination of the belt fire extinguishing ability of the Fogtec system, environmental conditions downstream from the fire were also monitored. This information was used to facilitate in the optimum placement of sensors in an airway and optimum activation levels to automatically activate the Fogtec system.

2 TEST GALLERY LAYOUT

2.1 *Scenario to be tested*

To be able to assess the effectiveness of the Fogtec system the conditions to be simulated must imitate a worst-case scenario. If the belt extinguishing system can control, and extinguish the worst-case scenario, it is reasonable to expect that the system will be able to control the majority of underground belt fire conditions.

2.2 *Worst-case scenario*

The conditions for the worst-case scenario to be tested, as decided on in a workshop, is described in Table 1.

A high-energy propane burner as used in the standard British fire retardancy test (British Coal Specification 158:1989 clause 8.5) will be used to simulate the heat energy from the seized idler.

2.3 *Test section*

The test section to be used for the worst-case scenario simulates an underground belt road with a width of 3,2 m and a height of 3,3 m.

A full-scale belt structure, excluding a drive and tail pulley assembly, as used by Impala Platinum Limited, is used in the test section. The belt structure and accompanying belt is situated in the middle of the belt road, simulating actual underground conditions. (Figures 1 and 2)

A length of seven meters conveyor is used on the structure for a test. As required by the worst-case scenario, the belt will be stationary during the test, simulating a belt stoppage. The belt will be rolled over at the front of the belt structure as if it was rolled over the drive pulley. This is necessary to simulate airflow conditions in the area between the top and bottom belt, i.e. inside the belt structure where the heat source to ignite the belt is situated. If the propane burner is exposed the general belt road velocity of 5,5 m/s, the heat transfer to the belt from the burner will be significantly affected. Another factor that will be influenced by the change in flow conditions 'inside' the belt structure is the operation of the Fogtec system. By rolling the belt over the front of the belt structure (Figures 3 and 4) this change in flow conditions is minimized, resulting in more reliable data.

Table 1. Worst-case scenario conditions.

Condition	Status
Ignition source	Seized idler
Conveyor system status	Stationary
Belt condition	Used non fire retardant belt
Ventilation conditions	Ventilation doors open, air velocity of $\pm 5{,}5$ m/s

2.4 *Fogtec system*

The Fogtec system is installed according to site-specific conditions, i.e. there is no standard installation procedure. Site conditions and intended application influence the installation of the system (Figures 5 and 6).

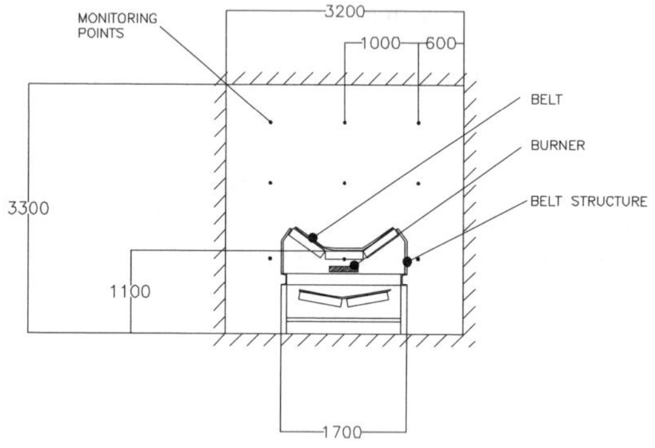

Figure 1. Schematic front view of the test section with only belt structure in place.

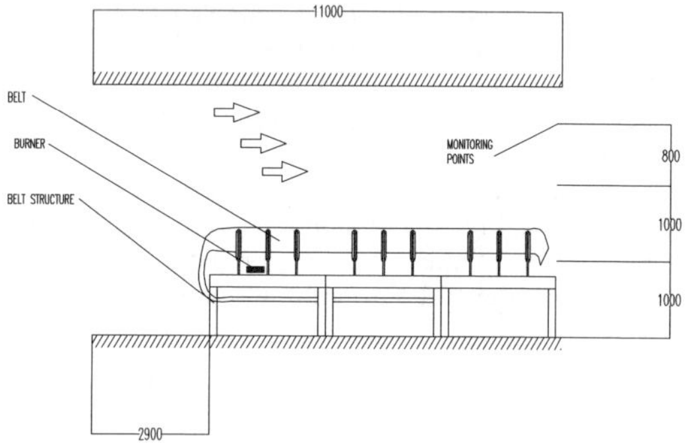

Figure 3. Left view of the test section with belt structure in place.

Figure 2. Front view of the test section without the belt installed.

Figure 4. Conveyor belt front assembly on structure.

The parameters to be measured to quantitatively assess the operation of the Fogtec system during testing are water consumption and pressure.

2.5 *Environmental conditions monitored*

In order to quantitatively assess the operation of the belt fire extinguishing system various environmental conditions are monitored.

2.5.1 *Heat for ignition*

Firstly, the heat supplied to the belt is measured. To quantify this the flame temperature and duration of exposure will be used.

2.5.2 *Carbon Monoxide (CO) levels*

At the exit from the test section CO levels will be monitored at nine different locations during testing (Figures 1 and 3). This will allow for a quantitative assessment of the levels and distribution of CO under fire conditions.

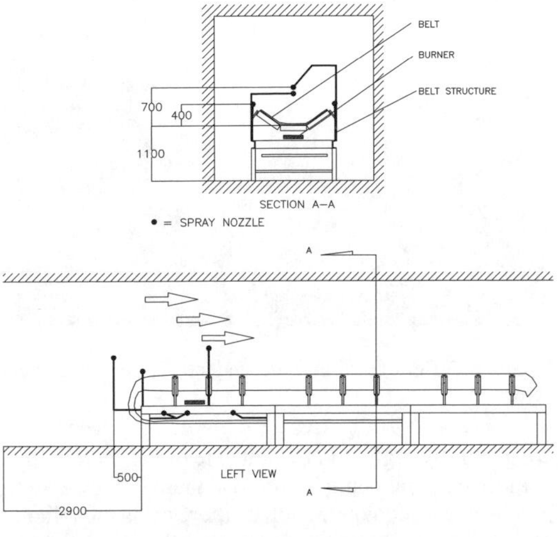

Figure 5. Fogtec installation in test section.

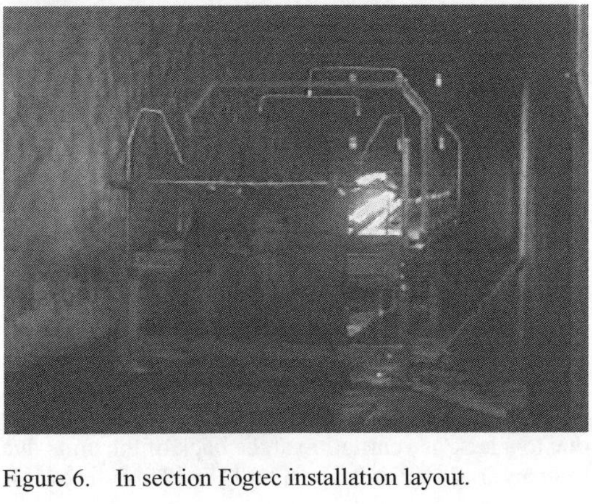
Figure 6. In section Fogtec installation layout.

2.5.3 *Ambient air temperature levels*

At the identified nine CO level monitoring locations the ambient air temperature during a test is also measured.

The time required to establish ambient air temperatures in the section will also be used to quantify operating effectiveness of the Fogtec system.

2.5.4 *Airflow conditions*

The test section air velocity is measured in the middle of the test section in front of the belt. This will be monitored manually before and after a test has been completed, and adjusted as required.

2.6 *Test procedure*

- Establish an airflow of 5,5 $\pm$ 0,5 m/s in the belt test road.
- Activate sensor bank and allow to warm up and stabilise (1 minute).
- Start the data acquisition system.
- Run system for 60 seconds to establish zero levels.
- Ignite the burner, as specified in the British belt fire retardency test procedure.
- Monitor gas and temperature levels in the control room to establish trigger levels for the Fogtec system.
- Allow the belt to start burning independently and then shut down the burner. Note the time.
- Activate the Fogtec system.
- Visually determine if the Fogtec system is containing and extinguishing the belt fire (The temperature readings can also be used as indicator if the system is functioning).
- The belt has been deemed to be extinguished if no visible glowing is present and there is no increase in temperature in the region of the belt.

3 TEST

3.1 *Belt used*

For the test a non-fire retardant belt that has been in service for two years was used.

The burner was active for 8 minutes before being switched off. At this stage the belt fire was self-sustaining. After self sustained burning of 1 minute and 30 seconds, the Fogtec system was activated.

3.2 *Test data*

From the burner temperature data it appears that the belt starts to participate in the generation of heat, i.e. starts burning, after approximately 1 minute and 30 seconds.

Table 2. Fogtec system operating conditions.

Item	Quantity	Units/No
Water supply pressure	110	Bar
Total water flow	46	L/min
DK6 Spray Nozzles	1	N/A
DK5 Spray Nozzles	7	N/A

Table 3. General test section data.

Quantity	Units	Value
Air velocity prior to test	M/s	6,0
Air velocity after test	M/s	6,0
Ambient section temperature	°C	21
Ambient CO levels	ppm	0

Table 4. Supply heat for belt fire initiation.

Action	Units	Time
Burner on	Mm:ss	01:00
Burner off	Mm:ss	09:00
Average burner temperature	°C	621
Peak burner active temperature	°C	691
Peak self sustained temperature	°C	710

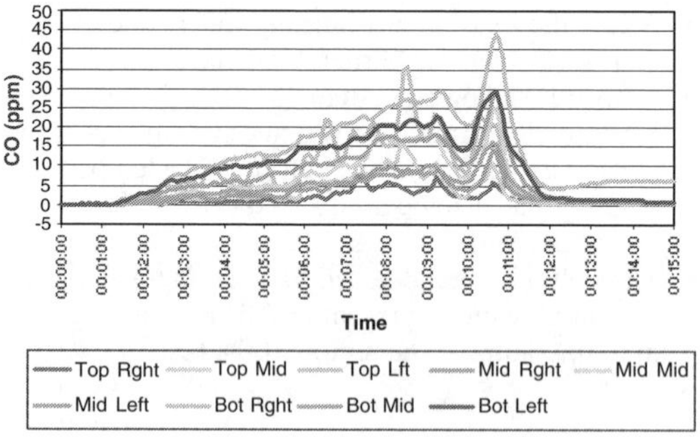

Figure 7. Test section CO levels for test 1.

In Figures 8 and 9 the burner was switched on at 01:00. After 8 minutes the burner was switched off i.e. 09:00. The belt was allowed to burn for 1 minute and 30 seconds before the Fogtec system was switched on i.e. 10:30.

3.3 *Visual observations*

The belt appears to burn more vigorously on the left side of the section (Figure 9). This is most likely due to the airflow distribution in the section being more established on the left hand side of the section. This is also evident in the CO and temperature data. The CO levels

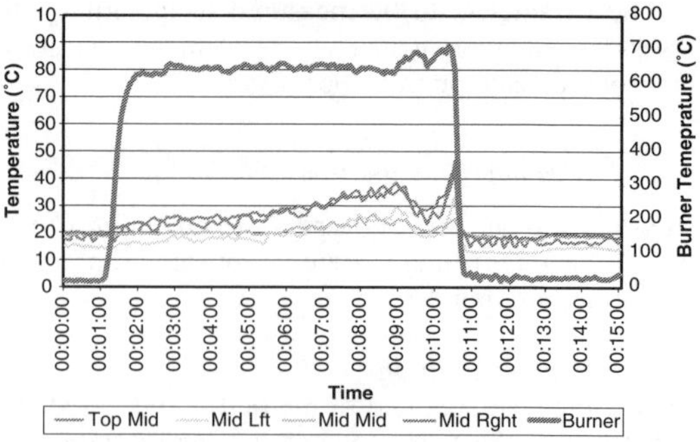

Figure 8. Test section temperatures for test 1.

Figure 9. Belt burning independently prior to system activation.

are generally higher on the left side of the section, due to more material being burnt. Also there is a lower temperature recorded on the left as more air is most likely passed through this area, increasing the 'cooling'.

The belt was allowed to nearly burn all the way through on top as to allow for maximum belt fire intensity (Figure 10).

The Fogtec system brought the top of the belt fire under control in 6 seconds from system activation, and contained the bottom of the fire to the area immediately behind the belt bend i.e. would be pulley (Figure 11). This status was maintained until the two spray heads under the belt was re-orientated. Upon re-orientation it took the Fogtec system 2 seconds per side to extinguish the under belt fire (Figure 12).

Although generally low temperatures were recorded at the sensor heads of the CO sensors (<40°C with a short exposure to 47°C), some minor structural heat damage i.e. deformation of the plastic body, was observed at the back of the units. This was most likely due to a lack of ventilation at the back of the units due to form drag, causing a heat build up. This however is

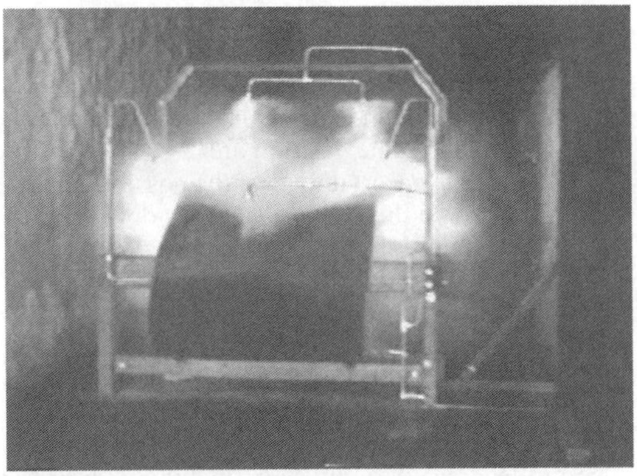

Figure 10. Two seconds after system activation.

Figure 11. Top of belt fire extinguished and belt fire contained after 6 seconds from activation.

unlikely to immediately affect the operation of the CO units.

3.4 *Comments*

It has to be remembered that the water supply (Table 5) was only to extinguish the visible flames of the belt fire. More water will be needed to cool down the belt structure and surrounding combustible material, if around, to avoid re-ignition of the in belt fire in underground conditions.

Of the 8 spray nozzles used in the Fogtec system set-up only 5 effectively participated in the control and extinguishing the belt fire, i.e. the five in front of the fire.

3.4.1 *CO level distribution*

There is steady increase in CO levels as the belt fire progress, i.e. more belt start to take part in the fire.

The levels recorded on the left side of the section tend to be higher than those on the right side. This is due to the airflow distribution in the test section causing the belt to burn more vigorously on the left.

After the burner was switched off there is a noticeable drop in CO levels as the CO from the burner is no longer present in the test section. After approximately 1 minute the belt fire becomes more vigorous, resulting in higher CO levels being recorded.

After approximately 1 minute and 30 seconds after the Fogtec system was activated, the CO levels returns to levels below 3 ppm. The response time of the sensors caused the long decay time of the CO levels in the test section.

From the evaluation of the CO levels on the left side of the test section, there appear to be no significant difference in CO levels from the top to the bottom of the section.

On the right side of the section however, where there is a lower airflow velocity, it appears that the CO levels increase from the top of the section to the bottom. Note has to be taken that after the test, the bottom right sensor has an offset of 6 ppm.

3.4.2 *Temperature distribution*

The temperature at the back the test section is generally low, i.e. <40°C. When the burner was switched off and the belt started to burn independently a sharp increase in temperature can be seen until the activation of the Fogtec system (Figure 8).

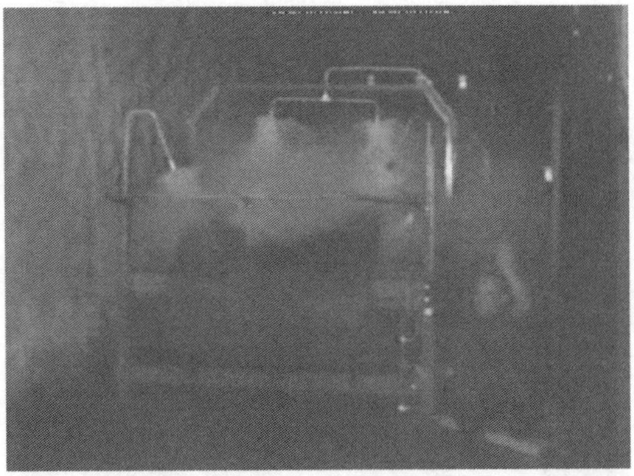

Figure 12. Belt fire extinguished after spray head adjustment (45 s).

Table 5. Fogtec operation observations.

Observation	Quantity	Units
Time delay from activation to full spray development	2–3	s
Water usage to control top belt fire	3,2	L
Water supply to complete control (including spray adjustment time)	33,1	L

Figure 13. Up wind in-effective plume dispersion.

After the activation of the Fogtec system the general body of air in the section took approximately 15 seconds to return to ambient.

The burner structure in the center of the belt took approximately 50 seconds to reach ambient temperature. This relates to a cooling rate of approximately 13°C/s.

Due to the airflow pattern in the test section the temperature on the left of the section is generally lower than the right due to the cooling effect of the increased volumetric flow rate.

4 CONCLUSIONS AND RECOMMENDATIONS

- From the non-fire retardant belt burn test it is clear that the Fogtec system can control and extinguish a significant belt fire. Although the mist developed by the Fogtec system is very susceptible to airflow conditions (velocity), it can be applied effectively to combat belt fires in extreme airflow conditions. Definite care has to be taken when designing the spray nozzle layout as this can drastically influence the operational effectiveness of the system. It is recommended that:
  - Sprays always point down stream to assist in plume integrity, which is a basic operating principle of the system to combat fires. Compare the plume in Figure 13, dispersed, to the ones formed in Figure 11, closed plumes.
  - Spray plumes must have a definite cross cover design built in i.e. more than one spray nozzle covers the same position/area.
  - If there is uncertainty over the possible airflow direction a duplicate system for down and up flow conditions needs to be considered.
- The Fogtec system is able to contain a full 2,5 m belt fire with an average burning temperature of 695°C within 6 seconds. It is estimated that the system will be able to extinguish such a fire in less than 1,5 seconds with the proper spray plume design layout.
- Due to the fine mist generation the belt fire was controlled and extinguished (including time for nozzle adjustment) using less than 35 liters of water. It is estimated that this could be done with less than 10 liters of water utilizing a proper spray nozzle design. With the low water consumption and an effective cross cover nozzle layout the system can be used without fear of 'drowning' a belt road. The low water consumption also lends the system to be tested regularly.
- The system controlled the fire in a section velocity of 5,5 m/s. Normal fire fighting procedure will be to close the ventilation doors to the section containing the fire, hence reducing the air velocity to well below 1 m/s. In the low air velocity conditions it is reasonable to expect that the Fogtec system will be even more effective, as the mist will not be moved away from the fire. This will increase cooling and oxygen displacement in the region of the fire.
- From the test it is clear that CO levels increase as a belt starts to burn. Levels increase to approximately 30 ppm prior to the Fogtec system being activated. Levels of 30 ppm approximately 4 meters from a belt fire relates to a significant belt fire (Figure 10).
- As the system utilizes high-pressure water to function, it is important that an underground system needs to have definite back-up circuits. If a broken supply line or detached nozzle should cause a drop in water pressure, the effectiveness of the system would be seriously compromised.

ACKNOWLEDGEMENTS

Kloppersbos test facility would like to acknowledge Mr. Don Bryden for the installation of the environment monitoring system and Mr. Tobie Janse van Rensburg for the construction of the belt structure. The general staff is also commended for their efficiency during testing.

Mine Ventilation, Ganguli & Bandopadhyay (eds)
© 2004 Taylor & Francis Group, London, ISBN 90 5809 633 5

A case study on dust concentration at Guleman Kef Chromium Mine

A. Dalgic* & T. Guyaguler**
**Dicle University Mining Engineering Department, Diyarbakir, Turkey*
***Middle East Technical University Mining Engineering Department, Ankara, Turkey*

ABSTRACT: The aim of this study is to develop solutions for reducing excessive dust concentration due to insufficient ventilation which causes interruptions in the mining operations of underground chromium mines at the Guleman district. Natural ventilation system is currently being applied at the chromium mines in the district. The natural ventilation pressure leads to air circulation with variable air quantity. The air quantity obtained from natural ventilation is not uniform. Morever, the air quantities circulating in the mines which are resulted from seasonal differences are insufficient.

Dust concentration exceeds the maximum allowable dust concentration limits especially during production and after blasting operations. This situation consequently halts the mining activities and results in loss of economical values.

A laboratory equipped with extensive pressure and dust concentration measuring devices is established in the vicinity of the mine in an effort to support the field studies. The natural ventilation pressure quantities and the dust concentration values are measured continuously throughout one year. Finally, optimum values for excessive dust concentration are determined and recommendations for better working environment are made.

1 INTRODUCTION

The term dust is applied to suspensions of finely solid particles in the air. Instability, coagulation and precipitation are their special characteristics. Dust includes a wide range of particle sizes from over 1 mm to under 1 μm. The size of the particles considerably influences their behavior. Dust particles are usually divided into three classes (Sengupta, 1992):

i. Particles greater than 10 μm, which settle according to the laws of gravity, with increasing velocity in still air.
ii. Particles between 0.1 to 10 μm, which in still air settle at a constant velocity, the uniform settling velocity depends on particle size and density, viscosity of the medium and acceleration due to gravity.
iii. Particles between 0.01 and 0.1 μm, which do not settle but diffuse in the air and remain in a colloidal state.

A dust cloud consisting of particle 5 μm in size is not visible to the naked eye. It cloud not be seen in the strongest beam of light underground, but would appear as motes in a bear at sunlight.

1.1 *Classification and harmfulness of dust*

Dust can be classified according to their harmful physiological effects. The following list different types of dusts in order of decreasing harm in each category:

i. Fibrogenic dusts (harmful to the respiratory system)
ii. Carcinogenic dusts (cancer – causing)
iii. Toxic dusts (poisonous to human organs)
iv. Radioactive dusts
v. Nuisance dusts (few adverse effects)

Any dust, if present in excessive quantities and inhaled for a sufficiently long time, can cause physiological damage. Factors, which determine the harmfulness of any dust, are its composition, particle size, concentration in mine air, exposure time and susceptibility of the exposed individual.

Mineralogical and chemical compositions are important. Free silica is more damaging than combined silica. The surface energy of the particles can be determining factor and solubility is a very important variable in toxic dusts.

Dust concentration in mine air is expressed either in terms of the number of particles per unit volume of air as particles per cubic centimeter (ppcc) or on the basis of weight per unit volume of air as milligram per cubic meter (mg/m^3).

In assessing the pulmonary effect of dusts the concept of respirable dust is important. At present two criteria are accepted for defining respirable dust. Both criteria approximate dust deposition in the noniliated portions of the lung. The first resulting from work performed by the U.S. Atomic Energy Commission (AEC) and the other recommended by British Medical Research Council (BMRC) are defined by the sampling efficiency curve labeled in Figure 1.

Particle sizes refer to equivalent diameter which is defined as the diameter of a spherical particle of unit density having the same falling velocity as the particle in question.

1.2 *Generation of dust*

Dust is generated and dispersed into the mine air through rock breakage, rock loading, transportation and unloading, and through the flow of ventilating air. The quantity of potentially airborne dust produced is related to the quantity of rock broken. There are two phases involved, cutting the chip and removing it from the rock face. In the cutting process the production of potentially airborne dust is inversely related to size of the rock chips (Knight, 1980).

In order to reduce the formation of such dust, the intensity of the cutting force should be applied a small area with a greater depth of cut. The design of cutting tools achieve this, is complex. The minimum specific energy consumption in rock breaking and the minimum specific dust generation are usually achieved simultaneously. The chip-removal process in drilling can generate airborne dust. Insufficient clearance or not enough flushing fluid can produce further breakage of chips, decrease cutting speed and energy loss in the drill rod with increased specific dust generation.

The handling of the broken rock has a significant effect on the dispersion of the fine dust in mine air.

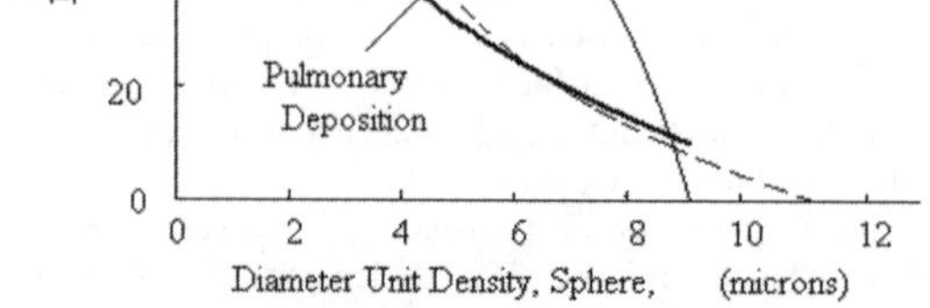

Figure 1. Pulmonary deposition curve and comparison of respirable dust criteria (ILO).

The quantity of dust dispersed in handling process is determined by the amount of fine dust created in breaking process. The extent to which the material is disturbed in handling, the energy expended to create new breakage, the kinetic energy imparted to the freshly broken rock particles by the stress relief in the material broken and ambient air velocities.

1.3 *Measurements of dust generation in underground mines*

Underground hard rock mining consists of breaking the ore into small pieces, loading, transportation and handling, drilling and ancillary operations. Measurements and laboratory analyses were performed to determine total respirable dust in these operations. Sampling stations were located in the airways both upstream and downstream of operation. The sampling station locations were chosen to give the most uniform mixture of dust and air possible.

The total amount of dust produced at each operation was determined by multiplying the dust concentration at each return station by the total airflow in the airway and subtracting the same figures obtained in the intake airway. This figure was then divided by the unit of production, tones of ore, length of holes drilled or other units to give specific values per unit of production.

Three main categories of instruments are used in underground mines to measure concentration of respirable dust in mine atmosphere. These are gravimetric dust samplers, short-term dust monitors and instrument for particle size measurements.

1.3.1 *Personal dust sampler*

The personal sampler consists of an air pump, filter assembly and a small cyclone that acts as a size selector to remove nonrespirable dust particles. In the cyclone, air enters a cylindrical chamber tangentially and passes inward through a spiral path with increasing velocity. Because of centrifugal force, large particles are removed from the airstream and collected at the bottom of the cyclone chamber. Respirable dust penetrating through the cyclone is collected on a filter paper. The filter is weighted to determine the mass of the respirable dust.

The preweighted capsulated filter is weighed after sample measurement. The weight gain due to dust is determined by extracting initial weight from final weight. Dust concentrations are expressed in milligrams of dust per cubic meter of air sampled. The calculation is as follows:

$$\text{Dust Concentration} = \frac{\text{Weight Gain}}{\text{TR}} \ \text{mg/m}^3 \qquad (1)$$

where

T = Time of sampling in s,

R = Rate of sampling in m^3/s,
Weight gain in mg.

Personal gravimetric samples have several advantages over other sampling methods. A miner can wear the sampler, permitting sampling of the dust cloud to which he is exposed. The sampler provides an estimate of the amount of respirable dust in the mine air for periods of up to 8 hours.

1.3.2 *MRE gravimetric dust sampler*

The MRE is a portable gravimetric dust sampler that continuously samples air in a working mine and separates the airborne dust into respirable and nonrespirable fractions. This separation occurs as the larger particles settle to the floor of the horizontal elutriator (size-selector) sections and are prevented from reaching the filter smaller particles having slower settling rates pass through the elutriator and are deposited on the filter. The instrument consists of an elutriator, a filter, and an electrically driven pump.

Dust-laden air enters the precision-made four-channel horizontal elutriator at an adjustable flow rate. The rear of each channel floor plate is lipped to help prevent nonrespirable dust from being accidentally dislodged from the plates onto the filter. A nose restrictor is fitted at the entrance to minimize the effects of air cross flow. A diaphragm pump, fitted with flap values of aluminum-coated polyester film, induces airflow. An adjustable crank varies the pump output, and the unit is driven through spurges at half speed by a governed electric motor. A flow meter is connected to the output side of the pump through a diaphragm-type-smoothing device.

There are many other types of dust sampling instruments. These are GCA Dust Monitor, TSI Piezobalance, Sibata Digital Dust Monitor, The Ram-1, Konimetry, Midget Impinger, Impinger-Microprojector-Coulter Counter, Thermal Precipitator, and Thermal Precipitator-Diffraction Size Frequency Analyzer.

2 FIELD STUDY ON DUST SURVEYS

2.1 *Introduction*

Eti Holding Eastern Chrome Establishment in Guleman lies 76 km to the East of Elaziğ province and is within Southeast Anatolian thrust belt, Turkey. Figure 2 shows the location of the mine area.

Eti Holding Eastern Chrome Establishment is the most important chromite ore-producing mine in Turkey. The mine area comprises 200 km^2 and has East-West elongation (Arikal et al., 1989). In this study, the Kef district of the mine, which has a 6 km^2 was chosen for field studies.

In the Guleman district about 12 million tons of chromite ore were produced by the end of 2003. The district has about 6 million tons of unmined chromium reserve. The Kef district contains 80% of the total ore deposit explored in the region.

2.2 *Mineralogy and Petrography of the Guleman Chromite*

The isotropic character of the Guleman chromites has relatively low reflectivity and considerably high Vicker hardness number, with grayish color. They have

Figure 2. Location of Eti Holding Eastern Chrome Establishment.

Table 1. Chemical composition and structural elements of the Guleman chromites.

Oxides	%	Structural elements	%
Cr_2O_3	58.06	Cr	39.72
Al_2O_3	14.56	Al	7.71
MgO	14.12	Mg	8.51
FeO^T	12.87	Fe^{-3}	2.77
SiO_2	4.02	Fe^T	7.23
Fe_2O_3	3.96	Fe^T	10.00
CaO	0.15	Ni	0.16
TiO_2	0.15	Mn	0.07

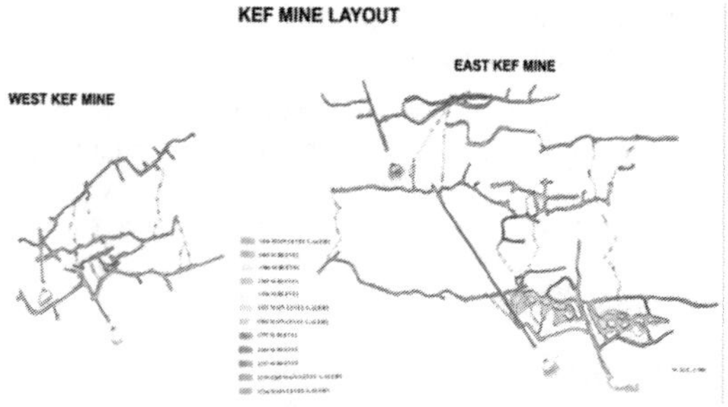

Figure 3. Kef Mine layout.

subhedral form cataclastic and pull apart textures with small surrounded silicate. Grain size ranges from few microns to over one hundred microns. By increasing deformation, cataclastic and pull apart textures developed; pull apart cracks are opened within the chromite grains and are filled by gangue minerals.

Texture of the chromite is classified as primary textures related to crystal settling and secondary textures related to magmatic deformation. The primary texture of the chromite is nodular and orbicular. It exhibits well-developed pull apart texture and lenticular cavities within the nodules, which are filled by gangue minerals. Secondary textures are marked by tendency to show layering and lineation due to thinning of chromite and gangue minerals, by showing cataclastic textures, and decreasing grain size, pull apart textures due to stretching of the mineral and especially of nodules. Chemical composition of the Guleman chromites and their structural elements are given in Table 1.

The Vickers hardness number of the chromite varies from 1225 to 1416 kg/mm^2 with an average of 1297 $\pm$ 72 kg/mm^2. The hardness increases with increasing Al_2O_3 and decreases with increasing MgO content. The reflectivity of the chromite was measured in air at 589-nm wavelength, which corresponds, to yellow light. The reflectivity ranges between 12.41 to 13.24%. The reflectivity increases with increasing Cr_2O_3 (Duzgoren).

2.3 *Dust measurements*

The presence of dust in mine air is caused by almost every mining operation. The main object of dust sampling is to obtain the amount of dust in the air in order to locate the sources of undesirably high dust concentrations, so that control or remedial measures can be introduced.

The measurements of dust concentrations were made at fixed positions assumed to be typical of all working places. Figure 3 shows 3D representation of the Guleman Kef mine. Figure 4 illustrates sample locations of the dust measurements in the mine. Dust sampling were conducted at 18 locations in the mine.

In this study, two dust sampling equipments were used to determine dust concentrations. The first one was personal dust sampler, and second one was gravimetric dust sampler.

2.3.1 *Personal dust measurements*

The measuring instrument used was the Casella AFC 123 IS type gravimetric dust sampler. It measures the dust exposure of individual men in a particular operation or location. The sampled air is drawn through a membrane filter by a small electrically driven pump.

In the sampling study Whatman's GF/A glass microfibre filters with 2.5 cm in diameter were used. The filters were weighted in the laboratory with a Sortorious 0.1-mg sense type balance before and after sampling. Therefore the difference between weights was the dust quantity in mg. For each sampling procedure the filters remained 45 minutes in a 105°C heated furnace before and after sampling.

Flow rate of the personal dust sampler was already adjusted according to individual men operation. The rate was adjusted as 2.2 l/min for drill holes equipment user, 2.1 l/min for workers and mine machine users and 1.9 l/min for technicians and engineers.

2.3.2 *Gravimetric dust measurements*

Gravimetric dust sampler Casella type 113A instrument was used for dust measurements for determining dust concentration in the mine environment. Figure 4.10 shows the gravimetric dust sampler. The gravimetric sampler was placed at a height of 1.5 m above the floor about 5 m. inby. The instrument sampled dust laden air at a flow rate of 2.5 l/min and continuously for up to 4 hours.

The filters used were Whatman's GF/A glass microfibre filters with 5.5 cm in diameter. The dust sampling process was the same as in personal dust sampling procedures. Dust sampling surveys were carried out during operation and after blasting.

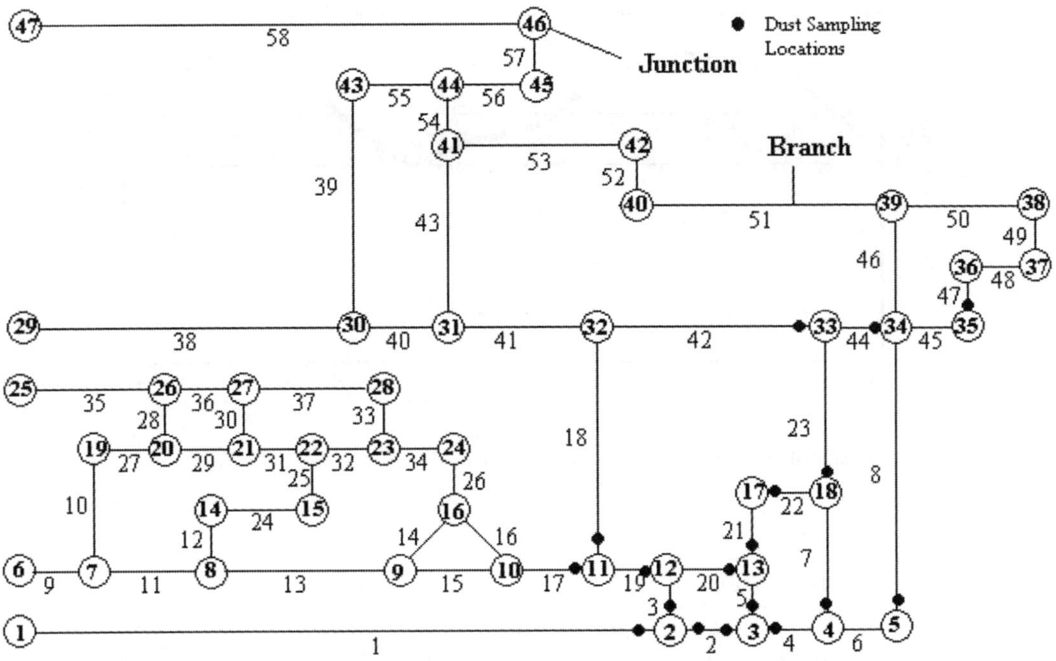

Figure 4. Dust sampling locations in the Kef Mine.

Usually after blasting, the mining operation was stopped to clear the dust generated.

3 ANALYSIS OF DUST CONCENTRATION

In this study, dust concentration measurements were carried out for two cases. These cases were dust concentration during mining operation and dust concentration after blasting. The dust samples collected from the mine were weighed and procedure for determining the dust concentration was followed (Dalgic). The dust concentration of the mine was determined by the following equation in unit of mg/m^3 during production and after blasting.

$$\textbf{Dust Concentration} = \frac{C_f - C_i}{F \times T} \quad (2)$$

where;

C_i = Corrected initial filter mass in mg,
C_f = Corrected final filter mass in mg,
F = Airflow rate in m^3/min,
T = Sampling time in min.

The dust measurements during operation in the mine were carried out for one year in the study. The dust concentration were determined monthly at working places of the 1214 level. The results have been illustrated in Table 2.

Table 2. Dust concentration during the mining operation.

Months	Dust concentration (mg/m^3)
January	38.27
February	46.42
March	48.85
April	49.91
May	45.28
June	38.34
July	35.93
August	32.82
September	39.34
October	50.51
November	50.66
December	43.61

The dust concentration is not constant and changes during in the year. It increases in the spring and fall seasons, but decreases in the winter and summer seasons as shown in Figure 5. Maximum allowable dust concentration in such mines is 5 mg/m^3 according to Turkish official law of health and safety. The dust concentration exceeds this limit in every month.

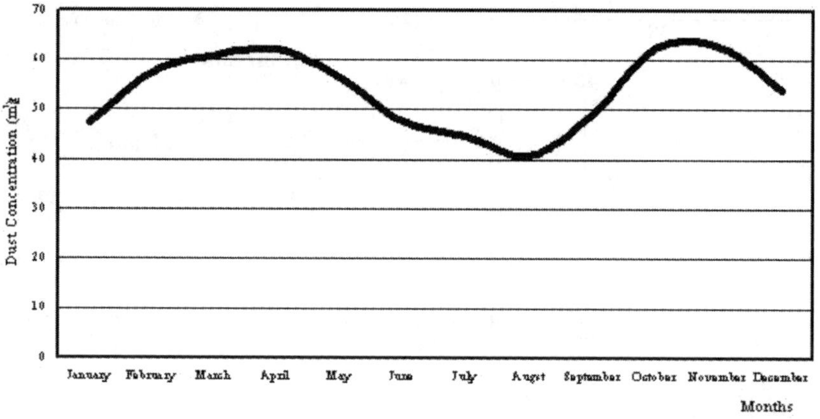

Figure 5. Dust concentration during production in the Kef Chromium Mine.

Table 3. Dust concentration after blasting.

	Dust concentration (mg/m^3)							
Months	30 (min)	60 (min)	90 (min)	120 (min)	150 (min)	180 (min)	210 (min)	240 (min)
January	25.54	19.09	14.27	9.77	6.12	3.42	1.64	0.67
February	27.57	21.45	16.23	11.10	6.95	3.93	1.89	0.78
March	28.90	22.74	17.00	11.57	7.36	4.13	2.02	0.85
April	28.97	22.87	17.53	12.07	7.68	4.30	2.19	0.96
May	26.78	20.83	15.56	10.66	6.67	3.73	1.78	0.73
June	21.76	16.90	12.46	8.41	5.34	3.01	1.53	0.57
July	19.22	14.95	11.32	7.35	4.54	2.76	1.41	0.54
August	17.95	13.78	10.02	6.27	3.82	2.14	1.09	0.41
September	24.78	19.50	14.76	10.18	6.14	3.35	1.58	0.69
October	29.47	22.93	16.90	11.41	7.13	4.13	1.98	0.81
November	29.61	23.03	16.93	11.00	7.00	4.06	1.94	0.78
December	26.02	20.25	14.92	10.07	6.14	3.32	1.55	0.65

The dust concentration measurements after blasting were also carried out during one year in the study. These measurements were for the determination of the dust concentration after blasting and observing the behavior of the dust concentration based on elapsed time. The results are illustrated in Table 3.

The dust concentration level is very high after blasting. The mining activity must be interrupted for 150 minutes to 190 minutes for providing safe dust conditions. These results graphically represented in Figure 6.

The calculation of required air quantity for dust is important in poor ventilated production places and in headings as well. In the study the main problem was excessive dust generation during production and after blasting. In order to determine the required amount of air, dust generation rate must be determined. The following equation is used to determine required air quantity;

$$Q = \frac{G}{TLV - \mathrm{B}} \quad (3)$$

where;

G = Dust production in g/min,
TLV = Threshold limit value of dust concentration in mg/m^3,
B = Dust Concentration with intake air in mg/m^3.

Dust production rate is 5.79 g/min for recovery of 180 tons/shift ore by drilling and blasting operation (Sengupta, 1992). In this study, the dust production rate was determined as 6.72 g/min. The ore production amount in the Kef mine is 600 ton/day. The threshold limit value for such mines in Turkey is

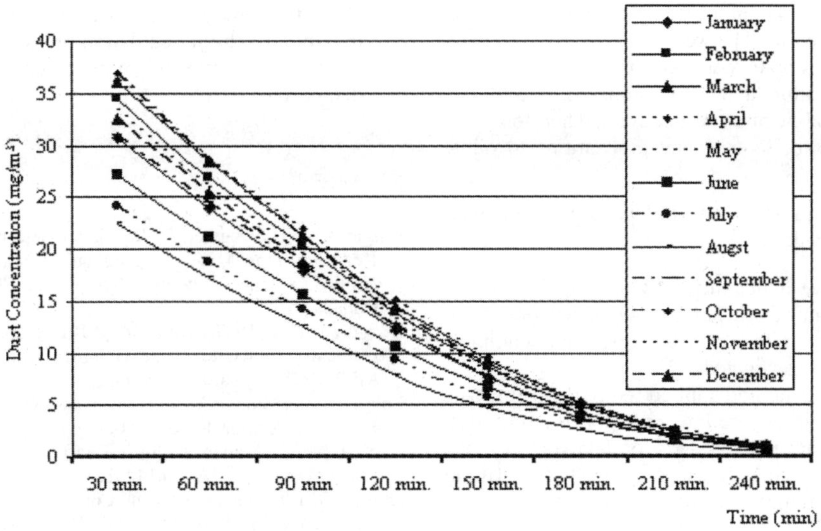

Figure 6. Dust concentration after blasting in the Kef Chromium Mine.

5 mg/m^3. Dust concentration within intake air was about 0.1 and 0.4 mg/m^3. The required air quantity was found to be 17.71 m^3/s. If a safety factor of 2 is chosen the final required air quantity is 35.42 m^3/s.

4 CONCLUSIONS

Analysis of the experimental data and results show that air quantity supply by natural means and dust concentration are at critical levels in the mine throughout the year. The major conclusions based on the field study and computer applications are summarized in the following:

i. Required air quantity of the mine is approximately 18m3/s for production rate of 600 tons ore per day. This air quantity was not achieved by natural ventilation but a few days of August and January. In order to supply the desired air quantity to the mine a fan by Korfmann Company with 1000-mm blade diameter was selected which can supply up to 38m^3/s with approximately 900-pascal pressure. Mechanical ventilation should be applied to the mine as soon as possible.
ii. There was not any air circulation in the mine due to equal density of the mine air and atmospheric air in sometime in March, April, December and November. Mining activity was completely interrupted as a result of insufficient ventilation at these days. This situation causes some economical losses for the mining company. Therefore, by mechanical ventilation this economical loss will be overcome.
iii. Natural ventilation pressure quantity of the mine has a range of between 1.58 to 321.78 Pa. in November and in August, respectively. These quantities are high enough to take into consideration for ventilation planning of a metal mine.
iv. Dust concentration level of the mine during production changes between 32.82 and 50.66 mg/m^3 in August and in November, respectively. These high dust concentration rates exceed maximum allowable dust concentration ratios throughout the year. Dust suppression techniques should be applied to the mine for decreasing dust concentration to a tolerable range. At present, the mining activity is interrupted due to excessive dust concentration for a long time, up to four hours, between two successive operation shifts.
v. Dust concentration level after blasting changes between 17.95 and 29.61 mg/m^3 in August and in November, respectively. These high dust concentrations cause interruption of the mining activity too. The maximum and minimum time for decreasing the dust concentration by natural ventilation to a tolerable level of 5 mg/m^3 after blasting is 140 and 190 minutes in August and in November, respectively. Therefore, mechanical ventilation is here also essential to the mine in order to achieve better mining operation.
vi. Three types of mine gases; namely methane, hydrogensulfur and carbonmonoxide were measured during the field study. Methane concentration was generally low and maximum occurrence was observed as 0.05% during blasting operation. Hydrogensulfur concentration was measured in a

range of 1 to 4 ppm. Carbonmonoxide concentration was measured between 1 and 8 ppm. The maximum allowable concentrations for methane, hydrogensulfur and carbonmonoxide are 2%, 50 ppm and 50 ppm, respectively. Therefore, it is not necessary to take any precautions to dilute these gases.

Based on this study the following recommendations can be made:

i. Ore reserve under the 1214 main gallery is not explored in detail. For further ventilation planning; the definite reserve and boundary conditions should be investigated.
ii. The mining method applied to ore deposit is true but pillars in ore are left for supporting purpose. These pillars contain 15% of total ore reserve. Necessary studies must be conducted on filling material strength and filling facilities.
iii. Larger natural ventilation pressure may be obtained if the 1416 gallery connected to 1627 old work places. Therefore, this situation needs some investigation and experimental work.
iv. Studies on dustability of chromite and surrounding rock may be helpful for better understanding of the dust generation and suppression subject.
v. If blasting time is selected at the time of day that maximum natural ventilation pressure exists, the interruption time of mining activity can be decreased.

REFERENCES

Dalgic, A. and Guyaguler, T. "Ventilation design of Guleman Kef Chromium Mine" Proceedings of North American Ninth U.S Mine Ventilation Symposium 2002, Ontario Canada.

Düzgören, N.S. 1987. Mineralogy, Petrography and Geochemistry of Kizildağ and Guleman Chromites. MSc. METU, Geological Engineering.

Engin, T. and Sümer, Y. 1987. Kef dağ-Kapin (Guleman, Elaziğ) yöresinin Jeolojisi. Bati Kef-Doğu Kef Krom Yataklarinin Maden Jeolojisi Raporu, Maden Etüt ve Arama Dairesi, MTA, pp.142–168.

Knight, G. 1980. Generation and Control of Mine Airborne Dust. CANMET Report 80-270E, Canada Department of Energy, Mines and Resources, Ottawa.

Sengupta, M. 1992. Mine Environmental Engineering. Vol. II, pp. 199–264.

ILO, Course on Dust Prevention in Industry.

Mine Ventilation, Ganguli & Bandopadhyay (eds)
© 2004 Taylor & Francis Group, London, ISBN 90 5809 633 5

Author index